INTRODUCTORY LINEAR ALGEBRA
WITH APPLICATIONS

Sixth Edition

INTRODUCTORY LINEAR ALGEBRA
WITH APPLICATIONS

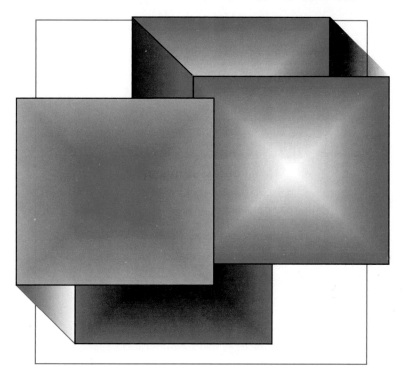

Bernard Kolman
Drexel University

With the assistance of
DAVID R. HILL
Temple University

 PRENTICE HALL, Upper Saddle River, New Jersey 07458

Library of Congress Cataloging-in-Publication Data

KOLMAN, BERNARD
  Introductory linear algebra with applications / Bernard Kolman
with the assistance of David R. Hill—6th ed.
        p.   cm.
  Includes bibliographical references and index.
  ISBN 0-13-266313-9 (alk. paper)
  1. Algebra, Linear.   I. Hill, David R. (David Ross)
II. Title.
QA184.K67   1997
512'.5—dc20   96-44292
                                                  CIP

Acquisition Editor: George Lobell
Editorial Assistant: Gale Epps
Assistant Editor: Audra J. Walsh
Editorial Director: Tim Bozik
Editor-in-Chief: Jerome Grant
AVP, Production and Manufacturing: David W. Riccardi
Production Editor: Elaine W. Wetterau
Managing Editor: Linda Mihatov Behrens
Executive Managing Editor: Kathleen Schiaparelli
Manufacturing Buyer: Alan Fischer
Manufacturing Manager: Trudy Pisciotti
Director of Marketing: John Tweeddale
Marketing Assistant: Diana Penha
Creative Director: Paula Maylahn
Art Director: Maureen Eide
Interior Design: Maureen Eide
Cover Design: Maureen Eide
Cover Illustration: Moultanville II by Frank Stella.
  Courtesy of Knoedler & Company, New York

Printed in the United States of America

10   9   8   7   6   5   4   3   2   1

ISBN 0-13-266313-9

Prentice-Hall International (UK) Limited, *London*
Prentice-Hall of Australia Pty. Limited, *Sydney*
Prentice-Hall Canada Inc., *Toronto*
Prentice-Hall Hispanoamericana, S. A., *Mexico*
Prentice-Hall of India Private Limited, *New Delhi*
Prentice-Hall of Japan, Inc., *Tokyo*
Simon & Schuster Asia Pte. Ltd., *Singapore*
Editora Prentice-Hall do Brasil, Ltda., *Rio de Janeiro*

*To the memory of Lillie
and to
Lisa and Stephen*

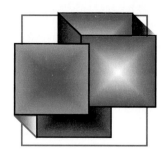

Contents

5 Eigenvalues and Eigenvectors   291

6 Linear Transformations and Matrices   327

Cumulative Review of Part I   366

Part II ■ APPLICATIONS

7 Linear Programming   371

8 Applications   417

Part III ■ NUMERICAL LINEAR ALGEBRA

9   Numerical Linear Algebra   535

Part IV ■ MATLAB FOR LINEAR ALGEBRA

10   MATLAB for Linear Algebra   573

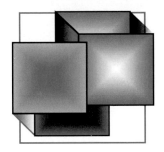

Preface

Material Covered

This book represents an introduction to linear algebra and to some of its significant applications and is designed for a course at the freshman or sophomore level. There is more than enough material for a semester or quarter course. By omitting certain sections, it is possible in a one-semester or quarter course to cover the essentials of linear algebra (including eigenvalues and eigenvectors), to show how the computer is used, and to have a little time left over for several applications. The book can also be used for a year course in linear algebra or for a second course emphasizing applications of linear algebra. A suggested pace for covering the basic material is given at the end of the Preface. The level and pace of the course can be readily changed by varying the amount of time spent on the theoretical material and on the applications. Calculus is not a prerequisite; examples and exercises using very basic calculus are included and these are labeled "Calculus Required."

The emphasis is on the computational and geometrical aspects of the subject, keeping abstraction down to a minimum. Thus we sometimes omit proofs of difficult or less rewarding theorems, while amply illustrating them with examples. The proofs that are included are presented at a level appropriate for the student. We have also concentrated our attention on the essential areas of linear algebra; the book does not attempt to cover the subject exhaustively.

What Is New in the Sixth Edition

We have been very pleased by the widespread acceptance of the first five editions of this book. Encouraged by the success of the calculus reform movement that has been going on in this country during the last few years, a start has been made on developing ways to improve the teaching of linear algebra. The *Linear Algebra Curriculum Study Group* and others have made a number of recommendations for doing this. In preparing this edition, we have considered these recommendations as well as faculty and student suggestions. Although many changes have been made in this edition, our objective has remained the same as in the earlier editions: *to develop a textbook that will help the instructor to teach and the student to learn the basic ideas of linear algebra and to see some of its applications.* To achieve this objective, the following features have been developed in this edition:

- New Chapter 3 has been added on Vectors in R^2 and R^n.

- New Appendix B has been added on Further Directions.

- New sections have been added on

  ◇ Dot product and Matrix Multiplication.
  ◇ Introduction to Linear Transformations. Early introduction to this material in the new chapter on Vectors in R^2 and R^n.
  ◇ Subspaces. Now treated separately from vector spaces.
  ◇ Orthogonal Complements.
  ◇ Inner Product Spaces (in new Appendix B).
  ◇ Composite and Invertible Linear Transformations (in new Appendix B).
  ◇ Electrical Circuits.
  ◇ QR-factorization.

- Appendix A on complex numbers has been expanded and now consists of two sections.

- More geometric material has been added.

- Many new exercises (more than 200) at all levels have been added.

- Many of these additional exercises are qualitative, rather than quantitative, and their solutions require discussion rather than just computation.

- The section on least squares has been expanded.

- More material on computational techniques has been incorporated. In particular, the QR-factorization of a matrix is now discussed.

- The Cumulative Review of Part I has been expanded.

- Brief Previews of the longer applications in Chapters 7 and 8 are presented at appropriate places in the first six chapters.

- There is now a stronger emphasis on the geometrical presentation of basic ideas, which necessitates the use of a larger number of illustrative figures.

- Pedagogically improved MATLAB M-files.

- These changes have resulted in a textbook that has more visualization, geometry, computation, and exercises whose solution calls for a verbal answer.

Exercises

The exercises in this book are grouped into three classes. The first class, **Exercises**, contains routine exercises. The second class, **Theoretical Exercises**, includes exercises that fill in gaps in some of the proofs and amplify material in the text. Some of these call for a verbal solution. In this technological age, it is especially important to be able to write with care and precision, and exercises of this type should help to sharpen that skill. These exercises can also be used to raise the level of the course and to challenge the more capable and interested student. The third class, **MATLAB Exercises**, consists of exercises developed by David R. Hill and labeled by the prefix ML (for MATLAB). These exercises are designed to be solved by MATLAB or another appropriate computer software package. Answers to all odd-numbered numerical and ML exercises appear in the back of the book. There is also a comprehensive review of Part I, the basic linear algebra material, consisting of 75 true–false questions (with answers in the back of the book). An **Answer Manual**, containing answers to all even-numbered exercises and solutions to all theoretical exercises, is available (to instructors only) at no cost from the publisher.

Presentation

We have learned from experience that at the sophomore level, abstract ideas must be introduced quite gradually and must be based on some firm foundations. Thus we begin the study of linear algebra with the treatment of matrices as mere arrays of numbers that arise naturally in the solution of systems of linear equations, a problem already familiar to the student. Much attention has been devoted from one edition to the next to refining and improving the pedagogical aspects of the exposition. The abstract ideas are carefully balanced by the considerable emphasis on the geometrical and computational aspects of the subject.

Material Covered

Part I, consisting of Chapters 1 through 6, presents the basic linear algebra material. **Chapter 1** deals with matrices and their properties; Section 1.3, new to this edition, introduces matrix multiplication by using the dot product operation for n-vectors. Methods for solving systems of linear equations are discussed in this chapter. **Chapter 2** presents the basic properties of determinants and some of their applications. **Chapter 3**, new to this edition, deals with vectors in R^2 and R^n. In this chapter we give an early introduction to linear transformations and explore some of the many geometric ideas that arise naturally. In **Chapter 4** we come to a more abstract notion, that of a vector space. Section 4.9, new to this edition, discusses orthogonal complements; it is part of the added geometric emphasis in this edition. **Chapter 5**, on eigenvalues and eigenvectors, settles the diagonalization problem for symmetric matrices. **Chapter 6** covers linear transformations and matrices. Short applications to a wide variety of areas have been included in the first six chapters. More extensive applications comprise Part II of the book. Brief Previews of these applications are given at appropriate places in the first six chapters.

Part II consists of Chapters 7 and 8. **Chapter 7** contains an introduction to linear programming, an extremely important application of linear algebra. **Chapter 8** presents several other diverse applications of linear algebra: graph theory, electrical circuits (new to this edition), Markov chains, least squares (expanded in this edition), linear economic models, differential equations, the Fibonacci sequence, quadratic forms, conic sections, quadric surfaces, and the theory of games. The applications in Chapters 1 through 8 deal with problems in mathematics, physics, biology, the social sciences, management, business, and economics.

In Part III, which consists of **Chapter 9** on numerical linear algebra, we discuss rather briefly numerical methods commonly used in linear algebra to solve linear systems of equations and to find eigenvalues and eigenvectors of matrices. Section 9.4, on QR-factorization, is new to this edition. These methods are widely used in conjunction with computers. However, the examples and exercises presented here can all be solved with handheld calculators.

Part IV, which consists of **Chapter 10**, provides a brief introduction to MATLAB (which stands for MATRIX LABORATORY), a very useful software package for linear algebra computation, described below.

Chapters 7, 8, and 9 are almost entirely independent, the one exception being Section 8.11, The Theory of Games, which requires Chapter 6 as preliminary study. The applications can either be covered after completing the entire linear algebra material in the course or they can be taken up as soon as the material required for a particular application has been developed. The chart at the end of the Preface giving the prerequisites for each of the applications and the brief Previews will be helpful in deciding which applications to cover and when to cover them. Some of the sections, in particular those in Chapters 8 and 9, can also be used as independent student projects. Classroom experience with the latter approach has met with favorable student reaction. Thus the instructor can be quite selective both in the choice of material and in the method of study of the applications.

Appendix A, on complex numbers, introduces in a brief but thorough manner complex numbers and their use in linear algebra. **Appendix B** presents two more advanced topics in linear algebra: inner product spaces and composite and invertible linear transformations.

End of Chapter Material

Every chapter contains a summary of **Key Ideas for Review**, a set of Supplementary Exercises (answers to all odd-numbered numerical exercises appear in the back of the book), and a Chapter Test (all answers appear in the back of the book).

MATLAB Software

Although the ML exercises can be solved using a number of software packages, in our judgment MATLAB is the most suitable package for this purpose. MATLAB is a versatile and powerful software package whose cornerstone is its linear algebra capabilities. MATLAB incorporates professionally developed quality computer routines for linear algebra computation. The code employed by MATLAB is written in the C language and is upgraded as new versions of MATLAB are released. MATLAB is available from The Math Works, Inc., 24 Prime Park Way, Natick, MA 01760, [(508) 653-1415], e-mail: info@mathworks.com. The Student version is available from Prentice Hall at a reasonable cost. This Student Edition of MATLAB also includes a version of Maple, thereby providing a symbolic computational capability.

Chapter 10 of this edition consists of a brief introduction to MATLAB's capabilities for solving linear algebra problems. Although programs can be written within MATLAB to implement many mathematical algorithms, *it should be noted that the reader of this book is not asked to write programs. The user is merely asked to use MATLAB (or any other comparable software package) to solve specific numerical problems.*

Approximately 18 programs, called scripts or M-files, have been developed in MATLAB and are available to users of this book through The Math Works, Inc. Complete and mail the reply card bound at the front of the book. If the card is missing, contact The Math Works's Sales Department at (508) 653-1415 [via fax at (508) 653-2997 or via e-mail at info@mathworks.com]. Indicate that you want the *Introductory Linear Algebra with Applications Toolbox*. Be sure to include your mailing address as well as noting whether you want a PC or Macintosh format diskette. These M-files are designed to transform many of MATLAB's capabilities into courseware. This is done by providing pedagogy that allows the student to interact with MATLAB, thereby letting the student think through all the steps in the solution of a problem and relegating MATLAB to act as a powerful calculator to relieve the drudgery of tedious computation. Indeed, this is the ideal role for MATLAB (or any other similar package) in a beginning linear algebra course, for in this course, more than in many others, the tedium of lengthy computations makes it almost impossible to solve a modest-size problem. Thus, by introducing pedagogy and reigning in the power of MATLAB, these M-files provide a working partnership between the student and the computer. Moreover, the introduction to a powerful tool such as MATLAB early in the student's college career opens the way for other software support in higher-level courses, especially in science and engineering.

A MATLAB Workbook

An excellent supplement to this book is *Linear Algebra Labs with MATLAB*, 2nd ed., by David R. Hill and David E. Zitarelli, Prentice Hall, 1996. This workbook provides labs and applications of the material that are designed to be solved using MATLAB. The primary goal of the workbook is to provide experiences that will aid in the understanding of the basic ideas of linear algebra. The workbook can be obtained at a reduced cost when purchased together with this book. This package can be ordered as ISBN 0-13-848409-0.

Student Solutions Manual

The *Student Solutions Manual*, prepared by Nina Edelman, Temple University, and Dennis R. Kletzing, Stetson University, contains solutions to all odd-numbered exercises, both numerical and theoretical.

Suggested Pace for Basic Material, Part I

Chapter 1	7 lectures
Chapter 2	3 lectures
Chapter 3	3 lectures (omitting Sections 3.4 to 3.6)
Chapter 4	10 lectures
Chapter 5	5 lectures
Chapter 6	4 lectures

32 lectures

Prerequisites for Applications

Chapter 7	Section 1.5
Section 8.1	Section 1.4
Section 8.2	Section 1.5
Section 8.3	Section 1.5
Section 8.4	Sections 1.5 and 1.6
Section 8.5	Chapter 1
Section 8.6	Section 5.1 and Calculus
Section 8.7	Section 5.1
Section 8.8	Section 5.2
Section 8.9	Section 8.8
Section 8.10	Section 8.8
Section 8.11	Chapter 7
Section 9.1	None
Section 9.2	Section 1.5
Section 9.3	Sections 1.6 and 9.2
Section 9.4	Section 4.8
Section 9.5	Chapter 5

Acknowledgments

We are pleased to express our thanks to the following people who thoroughly reviewed the entire manuscript in the first edition—William Arendt, the University of Missouri; and David Shedler, Virginia Commonwealth University. In the second edition—Gerald E. Bergum, South Dakota State University; James O. Brooks, Villanova University; Frank R. DeMeyer, Colorado State University; Joseph Malkevitch, York College of the City University of New York; Harry W. McLaughlin, Rensselaer Polytechnic Institute; and Lynn Arthur Steen, St. Olaf's College. In the third edition—Jerry Goldman, DePaul University; Allan Krall, The Pennsylvania State University at University Park; Stanley Lukawecki, Clemson University ; David Royster, The University of North Carolina; Sandra Welch, Stephen F. Austin State University; and Paul Zweir, Calvin College. In the fourth edition—William G. Vick, Broome Community College; Carrol G. Wells, Western Kentucky University; Andre L. Yandl, Seattle University; and Lance L. Littlejohn, Utah State University. In the fifth edition—Paul Beem, Indiana University–South Bend; John Broughton, Indiana University of Pennsylvania; Michael Gerahty, University of Iowa; Philippe Loustaunau, George Mason University; Wayne McDaniels, University of Missouri; and Larry Runyan, Shoreline Community College. In the sixth edition—Daniel D. Anderson, University of Iowa; Jürgen Gerlach, Radford University; W. L. Golik, University of Missouri at St. Louis; Charles Heuer, Concordia College; Matt Insall, University of Missouri at Rolla; Irwin Pressman, Carleton University; and James Snodgrass, Xavier University.

We also wish to thank the following people for their help with selected portions of the manuscript: Thomas I. Bartlow, Robert E. Beck, and Michael L. Levitan, all of Villanova University; Robert C. Busby, Robin Clark, the late Charles S. Duris, Herman E. Gollwitzer, Milton Schwartz, and the late John H. Staib, all of Drexel University; Avi Vardi; Seymour Lipschutz, Temple University; Oded Kariv, Technion, Israel Institute of Technology; William F. Trench, Trinity University; and Alex Stanoyevitch, the University of Hawaii; and instructors and students from many institutions in the United States and other countries, who shared with me their experiences with the book and offered helpful suggestions.

The numerous suggestions, comments, and criticisms of these people greatly improved the manuscript. To all of them goes a sincere expression of gratitude.

We thank Dennis R. Kletzing, Stetson University, who typeset the entire manuscript, the *Student Solutions Manual*, and the *Answer Manual*. He found a number of errors in the manuscript and cheerfully performed miracles under a very tight schedule. It was a pleasure working with him.

We thank John Wilton for providing some of the artwork.

We should also like to thank Michael Kaltman, for checking the answers to all the new exercises; and Nina Edelman, Temple University, and Lilian Brady, for critically reading the page proofs.

Finally, a sincere expression of thanks to Elaine Wetterau, Production Editor, who patiently and expertly guided this book from launch to publication, as she has done in the previous five editions; to George Lobell, Executive Editor; and to the entire staff of Prentice Hall for their enthusiasm, interest, and unfailing cooperation during the conception, design, production, and marketing phases of this edition.

B. K. and D. H.

PART I

INTRODUCTORY
LINEAR ALGEBRA

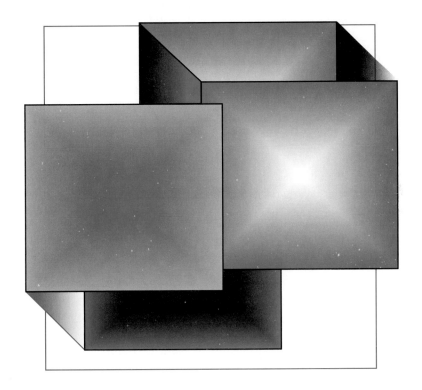

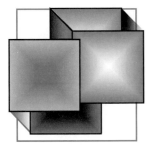

1

Linear Equations and Matrices

1.1 ▾ Linear Systems

A good many problems in the natural and social sciences as well as in engineering and the physical sciences deal with equations relating two sets of variables. An equation of the type

$$ax = b,$$

expressing the variable b in terms of the variable x and the constant a, is called a **linear equation**. The word *linear* is used here because the graph of the equation above is a straight line. Similarly, the equation

$$a_1 x_1 + a_2 x_2 + \cdots + a_n x_n = b, \tag{1}$$

expressing b in terms of the variables $x_1, x_2, \ldots, x_n$ and the known constants $a_1, a_2, \ldots, a_n$, is called a **linear equation**. In many applications we are given b and the constants $a_1, a_2, \ldots, a_n$ and must find numbers $x_1, x_2, \ldots, x_n$, called **unknowns**, satisfying (1).

A **solution** to a linear equation (1) is a sequence of n numbers $s_1, s_2, \ldots, s_n$, which has the property that (1) is satisfied when $x_1 = s_1$, $x_2 = s_2$, $\ldots$, $x_n = s_n$ are substituted in (1).

Thus $x_1 = 2$, $x_2 = 3$, and $x_3 = -4$ is a solution to the linear equation

$$6x_1 - 3x_2 + 4x_3 = -13,$$

because

$$6(2) - 3(3) + 4(-4) = -13.$$

This is not the only solution to the given linear equation, since $x_1 = 3$, $x_2 = 1$, and $x_3 = -7$ is another solution.

More generally, a **system of m linear equations in n unknowns**, or simply a **linear system**, is a set of m linear equations each in n unknowns. A linear system can be conveniently denoted by

$$
\begin{aligned}
a_{11}x_1 + a_{12}x_2 + \cdots + a_{1n}x_n &= b_1 \\
a_{21}x_1 + a_{22}x_2 + \cdots + a_{2n}x_n &= b_2 \\
\vdots \qquad \vdots \qquad\qquad \vdots \qquad \vdots & \\
a_{m1}x_1 + a_{m2}x_2 + \cdots + a_{mn}x_n &= b_m.
\end{aligned}
\tag{2}
$$

The two subscripts i and j are used as follows. The first subscript i indicates that we are dealing with the ith equation, while the second subscript j is associated with the jth variable x_j. Thus the ith equation is

$$
a_{i1}x_1 + a_{i2}x_2 + \cdots + a_{in}x_n = b_i.
$$

In (2) the a_{ij} are known constants. Given values of $b_1, b_2, \ldots, b_m$, we want to find values of $x_1, x_2, \ldots, x_n$ that will satisfy each equation in (2).

A **solution** to a linear system (2) is a sequence of n numbers $s_1, s_2, \ldots, s_n$, which has the property that each equation in (2) is satisfied when $x_1 = s_1$, $x_2 = s_2$, ..., $x_n = s_n$ are substituted in (2).

To find solutions to a linear system, we shall use a technique called the **method of elimination**. That is, we eliminate some of the unknowns by adding a multiple of one equation to another equation. Most readers have had some experience with this technique in high school algebra courses. Most likely, the reader has confined his or her earlier work with this method to linear systems in which $m = n$, that is, linear systems having as many equations as unknowns. In this course we shall broaden our outlook by dealing with systems in which we have $m = n$, $m < n$, and $m > n$. Indeed, there are numerous applications in which $m \neq n$. If we deal with two, three, or four unknowns, we shall often write them as x, y, z, and w. In this section we use the method of elimination as it was studied in high school. In Section 1.5 we shall look at this method in a much more systematic manner.

EXAMPLE 1 ■ Consider the linear system

$$
\begin{aligned}
x - 3y &= -3 \\
2x + \ y &= \ \ 8.
\end{aligned}
\tag{3}
$$

To eliminate x, we add (-2) times the first equation to the second, obtaining

$$
7y = 14,
$$

an equation having no x term. We have eliminated the unknown x. Then solving for y, we have

$$
y = 2,
$$

and substituting into the first equation of (3), we obtain

$$
x = 3.
$$

To check that $x = 3$, $y = 2$ is a solution to (3), we verify that these values of x and y satisfy *each* of the equations in the given linear system. Thus $x = 3$, $y = 2$ is a solution to (3).   ■

EXAMPLE 2 ■ Consider the linear system

$$x - 3y = -7$$
$$2x - 6y = \quad 7. \tag{4}$$

Again, we decide to eliminate x. We add (-2) times the first equation to the second one, obtaining

$$0 = 21,$$

which makes no sense. This means that (4) has no solution. We might have come to the same conclusion from observing that in (4) the left side of the second equation is twice the left side of the first equation, but the right side of the second equation is not twice the right side of the first equation.   ■

EXAMPLE 3 ■ Consider the linear system

$$x + 2y + 3z = \quad 6$$
$$2x - 3y + 2z = \quad 14$$
$$3x + \quad y - \quad z = -2. \tag{5}$$

To eliminate x, we add (-2) times the first equation to the second one and (-3) times the first equation to the third one, obtaining

$$-7y - \quad 4z = \quad 2$$
$$-5y - 10z = -20. \tag{6}$$

This is a system of two equations in the unknowns y and z. We multiply the second equation of (6) by $\left(-\frac{1}{5}\right)$, obtaining

$$-7y - 4z = 2$$
$$y + 2z = 4,$$

which we write, by interchanging equations, as

$$y + 2z = 4$$
$$-7y - 4z = 2. \tag{7}$$

We now eliminate y in (7) by adding 7 times the first equation to the second one, to obtain

$$10z = 30,$$

or

$$z = 3. \tag{8}$$

Substituting this value of z into the first equation of (7), we find $y = -2$. Substituting these values of y and z into the first equation of (5), we find $x = 1$. To check that $x = 1$, $y = -2$, $z = 3$ is a solution to (5), we verify that these values of x, y, and z satisfy *each* of the equations in (5). Thus $x = 1$, $y = -2$, $z = 3$ is a solution to the linear system (5). We might further observe that our elimination procedure has effectively produced the following linear system:

$$x + 2y + 3z = 6$$
$$y + 2z = 4 \tag{9}$$
$$z = 3,$$

obtained by using the first equations of (5) and (7) as well as (8). The importance of the procedure lies in the fact that the linear systems (5) and (9) have exactly the same solutions. System (9) has the advantage that it can be solved quite easily, giving the foregoing values for x, y, and z. ■

EXAMPLE 4 ■ Consider the linear system

$$x + 2y - 3z = -4$$
$$2x + y - 3z = 4. \tag{10}$$

Eliminating x, we add (-2) times the first equation to the second one, to obtain

$$-3y + 3z = 12. \tag{11}$$

We must now solve (11). A solution is

$$y = z - 4,$$

where z can be any real number. Then, from the first equation of (10),

$$x = -4 - 2y + 3z$$
$$= -4 - 2(z - 4) + 3z$$
$$= z + 4.$$

Thus a solution to the linear system (10) is

$$x = r + 4$$
$$y = r - 4$$
$$z = r,$$

where r is any real number. This means that the linear system (10) has infinitely many solutions. Every time we assign a value to r, we obtain another solution to (10). Thus, if $r = 1$, then

$$x = 5, \quad y = -3, \quad \text{and} \quad z = 1$$

is a solution, while if $r = -2$, then

$$x = 2, \quad y = -6, \quad \text{and} \quad z = -2$$

is another solution. ■

EXAMPLE 5 ■ Consider the linear system

$$x + 2y = 10$$
$$2x - 2y = -4 \tag{12}$$
$$3x + 5y = 26.$$

Eliminating x, we add (-2) times the first equation to the second one to obtain $-6y = -24$, or

$$y = 4.$$

Adding (-3) times the first equation to the third one, we obtain $-y = -4$, or

$$y = 4.$$

Thus we have been led to the system

$$
\begin{aligned}
x + 2y &= 10 \\
y &= \;4 \\
y &= \;4,
\end{aligned}
\tag{13}
$$

which has the same solutions as (12). Substituting $y = 4$ in the first equation of (13), we obtain $x = 2$. Hence $x = 2$, $y = 4$ is a solution to (12).   ■

EXAMPLE 6 ■ Consider the linear system

$$
\begin{aligned}
x + 2y &= \;10 \\
2x - 2y &= -4 \\
3x + 5y &= \;20.
\end{aligned}
\tag{14}
$$

To eliminate x, we add (-2) times the first equation to the second one to obtain $-6y = -24$, so

$$y = 4.$$

Adding (-3) times the first equation to the third one, we obtain $-y = -10$, or

$$y = 10.$$

Thus we have been led to the system

$$
\begin{aligned}
x + 2y &= 10 \\
y &= \;4 \\
y &= 10,
\end{aligned}
\tag{15}
$$

which has the same solutions as (14). Since (15) has no solutions, we conclude that (14) has no solutions.   ■

These examples suggest that a linear system may have one solution (a unique solution), no solution, or infinitely many solutions.

We have seen that the method of elimination consists of repeatedly performing the following operations:

1. Interchange two equations.

2. Multiply an equation by a nonzero constant.

3. Add a multiple of one equation to another.

It is not difficult to show (Exercises T.1 through T.3) that the method of elimination yields another linear system having exactly the same solutions as the given system. The new linear system can then be solved quite readily.

As you have probably already observed, the method of elimination has been described, so far, in general terms. Thus we have not indicated any rules for selecting the unknowns to be eliminated. Before providing a systematic description of the method of elimination, we introduce, in the next section, the notion of a matrix. This notion will greatly simplify our notation and will enable us to develop tools to solve many important problems.

Consider now a linear system of two equations in the unknowns x and y:

$$
\begin{aligned}
a_1 x + a_2 y &= c_1 \\
b_1 x + b_2 y &= c_2.
\end{aligned}
\tag{16}
$$

The graph of each of these equations is a straight line, which we denote by l_1 and l_2, respectively. If $x = s_1$, $y = s_2$ is a solution to the linear system (16), then the point (s_1, s_2) lies on both lines l_1 and l_2. Conversely, if the point (s_1, s_2) lies on both lines l_1 and l_2, then $x = s_1$, $y = s_2$ is a solution to the linear system (16). (See Figure 1.1.) Thus we are led geometrically to the same three possibilities mentioned above.

1. The system has a unique solution; that is, the lines l_1 and l_2 intersect at exactly one point.

2. The system has no solution; that is, the lines l_1 and l_2 do not intersect.

3. The system has infinitely many solutions; that is, the lines l_1 and l_2 coincide.

EXAMPLE 7
(*Production Planning*) ■ A manufacturer makes three different types of chemical products: A, B, and C. Each product must go through two processing machines: X and Y. The products require the following times in machines X and Y:

1. One ton of A requires 2 hours in machine X and 2 hours in machine Y.

2. One ton of B requires 3 hours in machine X and 2 hours in machine Y.

3. One ton of C requires 4 hours in machine X and 3 hours in machine Y.

Machine X is available 80 hours per week and machine Y is available 60 hours per week. Since management does not want to keep the expensive machines X

FIGURE 1.1

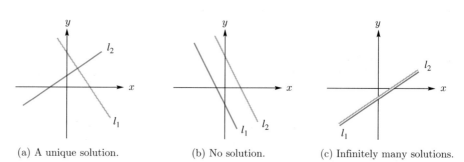

(a) A unique solution.   (b) No solution.   (c) Infinitely many solutions.

and Y idle, it would like to know how many tons of each product to make so that the machines are fully utilized. It is assumed that the manufacturer can sell as much of the products as is made.

To solve this problem, we let x_1, x_2, and x_3 denote the number of tons of products A, B, and C, respectively, to be made. The number of hours that machine X will be used is

$$2x_1 + 3x_2 + 4x_3,$$

which must equal 80. Thus we have

$$2x_1 + 3x_2 + 4x_3 = 80.$$

Similarly, the number of hours that machine Y will be used is 60, so we have

$$2x_1 + 2x_2 + 3x_3 = 60.$$

Mathematically, our problem is to find nonnegative values of x_1, x_2, and x_3 so that

$$2x_1 + 3x_2 + 4x_3 = 80$$
$$2x_1 + 2x_2 + 3x_3 = 60.$$

This linear system has infinitely many solutions. Following the method of Example 4, we see that all solutions are given by

$$x_1 = \frac{20 - x_3}{2}$$
$$x_2 = 20 - x_3$$
$$x_3 = \text{any real number such that } 0 \le x_3 \le 20,$$

since we must have $x_1 \ge 0$, $x_2 \ge 0$, and $x_3 \ge 0$. When $x_3 = 10$, we have

$$x_1 = 5, \qquad x_2 = 10, \qquad x_3 = 10$$

while

$$x_1 = \tfrac{13}{2}, \qquad x_2 = 13, \qquad x_3 = 7$$

when $x_3 = 7$. The reader should observe that one solution is just as good as the other. There is no best solution unless additional information or restrictions are given.   ■

1.1   EXERCISES

In Exercises 1 through 14, solve the given linear system by the method of elimination.

1. $x + 2y = 8$
$3x - 4y = 4.$

2. $2x - 3y + 4z = -12$
$x - 2y + z = -5$
$3x + y + 2z = 1.$

3. $3x + 2y + z = 2$
$4x + 2y + 2z = 8$
$x - y + z = 4.$

4. $x + y = 5$
$3x + 3y = 10.$

5. $2x + 4y + 6z = -12$
$2x - 3y - 4z = 15$
$3x + 4y + 5z = -8.$

6. $x + y - 2z = 5$
$2x + 3y + 4z = 2.$

7. $x + 4y - z = 12$
$3x + 8y - 2z = 4.$

8. $3x + 4y - z = 8$
$6x + 8y - 2z = 3.$

9. $x + y + 3z = 12$
$2x + 2y + 6z = 6.$

10. $x + y = 1$
$2x - y = 5$
$3x + 4y = 2.$

11. $2x + 3y = 13$
$x - 2y = 3$
$5x + 2y = 27.$

12. $x - 5y = 6$
$3x + 2y = 1$
$5x + 2y = 1.$

13. $x + 3y = -4$
$2x + 5y = -8$
$x + 3y = -5.$

14. $2x + 3y - z = 6$
$2x - y + 2z = -8$
$3x - y + z = -7.$

15. Given the linear system

$$2x - y = 5$$
$$4x - 2y = t.$$

(a) Determine a value of t so that the system has a solution.

(b) Determine a value of t so that the system has no solution.

(c) How many different values of t can be selected in part (b)?

16. Given the linear system

$$2x + 3y - z = 0$$
$$x - 4y + 5z = 0.$$

(a) Verify that $x_1 = 1$, $y_1 = -1$, $z_1 = -1$ is a solution.

(b) Verify that $x_2 = -2$, $y_2 = 2$, $z_2 = 2$ is a solution.

(c) Is $x = x_1 + x_2 = -1$, $y = y_1 + y_2 = 1$, and $z = z_1 + z_2 = 1$ a solution to the linear system?

(d) Is $3x$, $3y$, $3z$, where x, y, and z are as in part (c), a solution to the linear system?

17. Without using the method of elimination solve the linear system

$$2x + y - 2z = -5$$
$$3y + z = 7$$
$$z = 4.$$

18. Without using the method of elimination solve the linear system

$$4x = 8$$
$$-2x + 3y = -1$$
$$3x + 5y - 2z = 11.$$

19. Is there a value of r so that $x = 1$, $y = 2$, $z = r$ is a solution to the following linear system? If there is, find it.

$$2x + 3y - z = 11$$
$$x - y + 2z = -7$$
$$4x + y - 2z = 12.$$

20. Is there a value of r so that $x = r$, $y = 2$, $z = 1$ is a solution to the following linear system? If there is, find it.

$$3x - 2z = 4$$
$$x - 4y + z = -5$$
$$-2x + 3y + 2z = 9.$$

21. An oil refinery produces low-sulfur and high-sulfur fuel. Each ton of low-sulfur fuel requires 5 minutes in the blending plant and 4 minutes in the refining plant; each ton of high-sulfur fuel requires 4 minutes in the blending plant and 2 minutes in the refining plant. If the blending plant is available for 3 hours and the refining plant is available for 2 hours, how many tons of each type of fuel should be manufactured so that the plants are fully utilized?

22. A plastics manufacturer makes two types of plastic: regular and special. Each ton of regular plastic requires 2 hours in plant A and 5 hours in plant B; each ton of special plastic requires 2 hours in plant A and 3 hours in plant B. If plant A is available 8 hours per day and plant B is available 15 hours per day, how many tons of each type of plastic can be made daily so that the plants are fully utilized?

23. A dietician is preparing a meal consisting of foods A, B, and C. Each ounce of food A contains 2 units of protein, 3 units of fat, and 4 units of carbohydrate. Each ounce of food B contains 3 units of protein, 2 units of fat, and 1 unit of carbohydrate. Each ounce of food C contains 3 units of protein, 3 units of fat, and 2 units of carbohydrate. If the meal must provide exactly 25 units of protein, 24 units of fat, and 21 units of carbohydrate, how many ounces of each type of food should be used?

24. A manufacturer makes 2-minute, 6-minute, and 9-minute film developers. Each ton of 2-minute developer requires 6 minutes in plant A and 24 minutes in plant B. Each ton of 6-minute developer requires 12 minutes in plant A and 12 minutes in plant B. Each ton of 9-minute developer requires 12 minutes in plant A and 12 minutes in plant B. If plant A is available 10 hours per day and plant B is available 16 hours per day, how many tons of each type of developer can be produced so that the plants are fully utilized?

THEORETICAL EXERCISES ▩

T.1. Show that the linear system obtained by interchanging two equations in (2) has exactly the same solutions as (2).

T.2. Show that the linear system obtained by replacing an equation in (2) by a nonzero constant multiple of the equation has exactly the same solutions as (2).

T.3. Show that the linear system obtained by replacing an equation in (2) by itself plus a multiple of another equation in (2) has exactly the same solutions as (2).

T.4. Does the linear system

$$ax + by = 0$$
$$cx + dy = 0$$

always have a solution for any values of a, b, c, and d?

1.2 ▼ Matrices

If we examine the method of elimination described in Section 1.1, we make the following observation. Only the numbers in front of the unknowns $x_1, x_2, \ldots, x_n$ are being changed as we perform the steps in the method of elimination. Thus we might think of looking for a way of writing a linear system without having to carry along the unknowns. In this section we define an object that enables us to do this—that is, to write linear systems in a compact form that makes it easier to automate the elimination method on a computer in order to obtain a fast and efficient procedure for finding solutions. Their use is not, however, merely that of a convenient notation. We now develop operations on matrices and will work with matrices according to the rules they obey; this will enable us to solve systems of linear equations and to do other computational problems in a fast and efficient manner. Of course, as any good definition should do, the notion of a matrix provides not only a new way of looking at old problems, but also gives rise to a great many new questions, some of which we study in this book.

DEFINITION   An $m \times n$ **matrix** A is a rectangular array of mn real (or complex) numbers arranged in m horizontal **rows** and n vertical **columns**:

$$A = \begin{bmatrix} a_{11} & a_{12} & \cdots & a_{1n} \\ a_{21} & a_{22} & \cdots & a_{2n} \\ \vdots & \vdots & & \vdots \\ a_{m1} & a_{m2} & \cdots & a_{mn} \end{bmatrix}. \tag{1}$$

The **ith row** of A is

$$\begin{bmatrix} a_{i1} & a_{i2} & \cdots & a_{in} \end{bmatrix} \qquad (1 \le i \le m);$$

the **jth column** of A is

$$\begin{bmatrix} a_{1j} \\ a_{2j} \\ \vdots \\ a_{mj} \end{bmatrix} \qquad (1 \le j \le n).$$

We shall say that A is **m by n** (written as **$m \times n$**). If $m = n$, we say that A is a **square matrix of order n**, and that the numbers $a_{11}, a_{22}, \ldots, a_{nn}$ form the **main diagonal** of A. We refer to the number a_{ij}, which is in the ith row and jth column of A, as the **i, jth element** of A, or the **(i, j) entry** of A, and we often write (1) as

$$A = \begin{bmatrix} a_{ij} \end{bmatrix}.$$

For the sake of simplicity, we restrict our attention in this book, except for Appendix A, to matrices (plural of matrix) all of whose entries are real numbers. However, matrices with complex entries are studied and are important in applications.

EXAMPLE 1 ■ Let

$$A = \begin{bmatrix} 1 & 2 & 3 \\ -1 & 0 & 1 \end{bmatrix}, \qquad B = \begin{bmatrix} 1 & 4 \\ 2 & -3 \end{bmatrix}, \qquad C = \begin{bmatrix} 1 \\ -1 \\ 2 \end{bmatrix},$$

$$D = \begin{bmatrix} 1 & 1 & 0 \\ 2 & 0 & 1 \\ 3 & -1 & 2 \end{bmatrix}, \qquad E = \begin{bmatrix} 3 \end{bmatrix}, \qquad F = \begin{bmatrix} -1 & 0 & 2 \end{bmatrix}.$$

Then A is a 2×3 matrix with $a_{12} = 2$, $a_{13} = 3$, $a_{22} = 0$, and $a_{23} = 1$; B is a 2×2 matrix with $b_{11} = 1$, $b_{12} = 4$, $b_{21} = 2$, and $b_{22} = -3$; C is a 3×1 matrix with $c_{11} = 1$, $c_{21} = -1$, and $c_{31} = 2$; D is a 3×3 matrix; E is a 1×1 matrix; and F is a 1×3 matrix. In D, the elements $d_{11} = 1$, $d_{22} = 0$, and $d_{33} = 2$ form the main diagonal.   ■

A $1 \times n$ or an $n \times 1$ matrix is also called an **n-vector** and will be denoted by lowercase boldface letters. When n is understood, we refer to n-vectors merely as **vectors**. In Chapter 3 we discuss vectors at length.

EXAMPLE 2 ■ $\mathbf{u} = \begin{bmatrix} 1 & 2 & -1 & 0 \end{bmatrix}$ is a 4-vector and $\mathbf{v} = \begin{bmatrix} 1 \\ -1 \\ 3 \end{bmatrix}$ is a 3-vector.   ■

DEFINITION   A square matrix $A = \begin{bmatrix} a_{ij} \end{bmatrix}$ for which every term off the main diagonal is zero, that is, $a_{ij} = 0$ for $i \ne j$, is called a **diagonal matrix**.

EXAMPLE 3 ■

$$G = \begin{bmatrix} 4 & 0 \\ 0 & -2 \end{bmatrix} \quad \text{and} \quad H = \begin{bmatrix} -3 & 0 & 0 \\ 0 & -2 & 0 \\ 0 & 0 & 4 \end{bmatrix}$$

are diagonal matrices.   ■

DEFINITION   A diagonal matrix $A = \begin{bmatrix} a_{ij} \end{bmatrix}$, for which all terms on the main diagonal are equal, that is, $a_{ij} = c$ for $i = j$ and $a_{ij} = 0$ for $i \neq j$, is called a **scalar matrix**.

EXAMPLE 4 ■ The following are scalar matrices:

$$I_3 = \begin{bmatrix} 1 & 0 & 0 \\ 0 & 1 & 0 \\ 0 & 0 & 1 \end{bmatrix}, \qquad J = \begin{bmatrix} -2 & 0 \\ 0 & -2 \end{bmatrix}.$$

■

EXAMPLE 5
(*Tabular Display of Data*)

The following matrix gives the airline distances between the indicated cities (in statute miles).

	London	Madrid	New York	Tokyo
London	0	785	3469	5959
Madrid	785	0	3593	6706
New York	3469	3593	0	6757
Tokyo	5959	6706	6757	0

■

EXAMPLE 6 ■ Suppose that a manufacturer has four plants each of which makes three products.
(*Production*)   If we let a_{ij} denote the amount of product i made by plant j in 1 week, then the 3×4 matrix

	Plant 1	Plant 2	Plant 3	Plant 4
Product 1	560	360	380	0
Product 2	340	450	420	80
Product 3	280	270	210	380

gives the manufacturer's production for the week.   ■

Whenever a new object is introduced in mathematics, we must define when two such objects are equal. For example, in the set of all rational numbers, the numbers $\frac{2}{3}$ and $\frac{4}{6}$ are called equal although they are not represented in the same manner. What we have in mind is the definition that a/b equals c/d when $ad = bc$. Accordingly, we now have the following definition.

DEFINITION   Two $m \times n$ matrices $A = \begin{bmatrix} a_{ij} \end{bmatrix}$ and $B = \begin{bmatrix} b_{ij} \end{bmatrix}$ are said to be **equal** if $a_{ij} = b_{ij}$, $1 \leq i \leq m$, $1 \leq j \leq n$, that is, if corresponding elements are equal.

EXAMPLE 7 ■ The matrices

$$A = \begin{bmatrix} 1 & 2 & -1 \\ 2 & -3 & 4 \\ 0 & -4 & 5 \end{bmatrix} \quad \text{and} \quad B = \begin{bmatrix} 1 & 2 & w \\ 2 & x & 4 \\ y & -4 & z \end{bmatrix}$$

are equal if $w = -1$, $x = -3$, $y = 0$, and $z = 5$.   ■

We shall now define a number of operations that will produce new matrices out of given matrices. These operations are useful in the applications of matrices.

Matrix Addition

DEFINITION   If $A = \begin{bmatrix} a_{ij} \end{bmatrix}$ and $B = \begin{bmatrix} b_{ij} \end{bmatrix}$ are $m \times n$ matrices, then the **sum** of A and B is the $m \times n$ matrix $C = \begin{bmatrix} c_{ij} \end{bmatrix}$, defined by

$$c_{ij} = a_{ij} + b_{ij} \qquad (1 \le i \le m, \ 1 \le j \le n).$$

That is, C is obtained by adding corresponding elements of A and B.

EXAMPLE 8 ■ Let

$$A = \begin{bmatrix} 1 & -2 & 4 \\ 2 & -1 & 3 \end{bmatrix} \quad \text{and} \quad B = \begin{bmatrix} 0 & 2 & -4 \\ 1 & 3 & 1 \end{bmatrix}.$$

Then

$$A + B = \begin{bmatrix} 1+0 & -2+2 & 4+(-4) \\ 2+1 & -1+3 & 3+1 \end{bmatrix} = \begin{bmatrix} 1 & 0 & 0 \\ 3 & 2 & 4 \end{bmatrix}.$$ ■

It should be noted that the sum of the matrices A and B is defined only when A and B have the same number of rows and the same number of columns, that is, only when A and B are of the same size.

We shall now establish the convention that when $A + B$ is formed, both A and B are of the same size.

The basic properties of matrix addition are considered in the following section and are similar to those satisfied by the real numbers.

EXAMPLE 9 ■
(*Production*)

A manufacturer of a certain product makes three models, A, B, and C. Each model is partially made in factory F_1 in Taiwan and then finished in factory F_2 in the United States. The total cost of each product consists of the manufacturing cost and the shipping cost. Then the costs at each factory (in dollars) can be described by the 3×2 matrices F_1 and F_2:

	Manufacturing cost	Shipping cost	
$F_1 =$	32	40	Model A
	50	80	Model B
	70	20	Model C

$$F_2 = \begin{array}{cc} \text{Manufacturing} & \text{Shipping} \\ \text{cost} & \text{cost} \end{array}$$

	Manufacturing cost	Shipping cost	
$F_2 =$	40	60	Model A
	50	50	Model B
	130	20	Model C

The matrix $F_1 + F_2$ gives the total manufacturing and shipping costs for each product. Thus the total manufacturing and shipping costs of a model C product are \$200 and \$40, respectively.   ■

Scalar Multiplication

DEFINITION   If $A = \begin{bmatrix} a_{ij} \end{bmatrix}$ is an $m \times n$ matrix and r is a real number, then the **scalar multiple** of A by r, rA, is the $m \times n$ matrix $B = \begin{bmatrix} b_{ij} \end{bmatrix}$, where

$$b_{ij} = ra_{ij} \qquad (1 \le i \le m,\ 1 \le j \le n).$$

That is, B is obtained by multiplying each element of A by r.

EXAMPLE 10   ■   If $r = -3$ and

$$A = \begin{bmatrix} 4 & -2 & 3 \\ 2 & -5 & 0 \\ 3 & 6 & -2 \end{bmatrix},$$

then

$$rA = -3 \begin{bmatrix} 4 & -2 & 3 \\ 2 & -5 & 0 \\ 3 & 6 & -2 \end{bmatrix} = \begin{bmatrix} (-3)(4) & (-3)(-2) & (-3)(3) \\ (-3)(2) & (-3)(-5) & (-3)(0) \\ (-3)(3) & (-3)(6) & (-3)(-2) \end{bmatrix}$$

$$= \begin{bmatrix} -12 & 6 & -9 \\ -6 & 15 & 0 \\ -9 & -18 & 6 \end{bmatrix}.$$ ■

If A and B are $m \times n$ matrices, we write $A + (-1)B$ as $A - B$ and call this the **difference** of A and B.

EXAMPLE 11   ■   Let

$$A = \begin{bmatrix} 2 & 3 & -5 \\ 4 & 2 & 1 \end{bmatrix} \quad \text{and} \quad B = \begin{bmatrix} 2 & -1 & 3 \\ 3 & 5 & -2 \end{bmatrix}.$$

Then

$$A - B = \begin{bmatrix} 2-2 & 3+1 & -5-3 \\ 4-3 & 2-5 & 1+2 \end{bmatrix} = \begin{bmatrix} 0 & 4 & -8 \\ 1 & -3 & 3 \end{bmatrix}.$$ ■

The Transpose of a Matrix

DEFINITION   If $A = [a_{ij}]$ is an $m \times n$ matrix, then the $n \times m$ matrix $A^T = [a_{ij}^T]$, where

$$a_{ij}^T = a_{ji} \qquad (1 \le i \le n, 1 \le j \le m)$$

is called the **transpose** of A. Thus the transpose of A is obtained by interchanging the rows and columns of A.

EXAMPLE 12 ■ Let

$$A = \begin{bmatrix} 4 & -2 & 3 \\ 0 & 5 & -2 \end{bmatrix}, \qquad B = \begin{bmatrix} 6 & 2 & -4 \\ 3 & -1 & 2 \\ 0 & 4 & 3 \end{bmatrix}, \qquad C = \begin{bmatrix} 5 & 4 \\ -3 & 2 \\ 2 & -3 \end{bmatrix},$$

$$D = \begin{bmatrix} 3 & -5 & 1 \end{bmatrix}, \qquad E = \begin{bmatrix} 2 \\ -1 \\ 3 \end{bmatrix}.$$

Then

$$A^T = \begin{bmatrix} 4 & 0 \\ -2 & 5 \\ 3 & -2 \end{bmatrix}, \qquad B^T = \begin{bmatrix} 6 & 3 & 0 \\ 2 & -1 & 4 \\ -4 & 2 & 3 \end{bmatrix},$$

$$C^T = \begin{bmatrix} 5 & -3 & 2 \\ 4 & 2 & -3 \end{bmatrix}, \qquad D^T = \begin{bmatrix} 3 \\ -5 \\ 1 \end{bmatrix}, \quad \text{and} \quad E^T = \begin{bmatrix} 2 & -1 & 3 \end{bmatrix}.$$

■

1.2 EXERCISES

1. Let

$$A = \begin{bmatrix} 2 & -3 & 5 \\ 6 & -5 & 4 \end{bmatrix}, \qquad B = \begin{bmatrix} 4 \\ -3 \\ 5 \end{bmatrix},$$

and

$$C = \begin{bmatrix} 7 & 3 & 2 \\ -4 & 3 & 5 \\ 6 & 1 & -1 \end{bmatrix}.$$

(a) What is a_{12}, a_{22}, a_{23}?

(b) What is b_{11}, b_{31}?

(c) What is c_{13}, c_{31}, c_{33}?

2. If

$$\begin{bmatrix} a+b & c+d \\ c-d & a-b \end{bmatrix} = \begin{bmatrix} 4 & 6 \\ 10 & 2 \end{bmatrix},$$

find a, b, c, and d.

3. If

$$\begin{bmatrix} a+2b & 2a-b \\ 2c+d & c-2d \end{bmatrix} = \begin{bmatrix} 4 & -2 \\ 4 & -3 \end{bmatrix},$$

find a, b, c, and d.

In Exercises 4 through 7, let

$$A = \begin{bmatrix} 1 & 2 & 3 \\ 2 & 1 & 4 \end{bmatrix}, \qquad B = \begin{bmatrix} 1 & 0 \\ 2 & 1 \\ 3 & 2 \end{bmatrix},$$

$$C = \begin{bmatrix} 3 & -1 & 3 \\ 4 & 1 & 5 \\ 2 & 1 & 3 \end{bmatrix}, \qquad D = \begin{bmatrix} 3 & -2 \\ 2 & 4 \end{bmatrix},$$

$$E = \begin{bmatrix} 2 & -4 & 5 \\ 0 & 1 & 4 \\ 3 & 2 & 1 \end{bmatrix}, \quad \text{and} \quad F = \begin{bmatrix} -4 & 5 \\ 2 & 3 \end{bmatrix}.$$

4. If possible, compute:

(a) $C + E$ and $E + C$.   (b) $A + B$.

(c) $D - F$.   (d) $-3C$.

(e) $2C - 3E$.   (f) $2B + F$.

5. If possible, compute:

(a) $3D + 2F$.

(b) $3(2A)$ and $6A$.

(c) $3A + 2A$ and $5A$.

(d) $2(D + F)$ and $2D + 2F$.

(e) $(2 + 3)D$ and $2D + 3D$.

(f) $3(B + D)$.

6. If possible, compute:

(a) A^T and $(A^T)^T$.

(b) $(C + E)^T$ and $C^T + E^T$.

(c) $(2D + 3F)^T$.

(d) $D - D^T$.

(e) $2A^T + B$.

(f) $(3D - 2F)^T$.

7. If possible, compute:

(a) $(2A)^T$.

(b) $(A - B)^T$.

(c) $(3B^T - 2A)^T$.

(d) $(3A^T - 5B^T)^T$.

(e) $(-A)^T$ and $-(A^T)$.

(f) $(C + E + F^T)^T$.

8. Let

$$A = \begin{bmatrix} 1 & 2 & 3 \\ 6 & -2 & 3 \\ 5 & 2 & 4 \end{bmatrix} \quad \text{and} \quad I_3 = \begin{bmatrix} 1 & 0 & 0 \\ 0 & 1 & 0 \\ 0 & 0 & 1 \end{bmatrix}.$$

If λ is a real number, compute $\lambda I_3 - A$.

THEORETICAL EXERCISES ▪

T.1. Show that the sum and difference of two diagonal matrices is a diagonal matrix.

T.2. Show that the sum and difference of two scalar matrices is a scalar matrix.

T.3. Let

$$A = \begin{bmatrix} a & b & c \\ c & d & e \\ e & e & f \end{bmatrix}.$$

(a) Compute $A - A^T$.

(b) Compute $A + A^T$.

(c) Compute $(A + A^T)^T$.

T.4. Let O be the $n \times n$ matrix all of whose entries are zero. Show that if k is a real number and A is an $n \times n$ matrix such that $kA = O$, then $k = 0$ or $A = O$.

T.5. A matrix $A = \begin{bmatrix} a_{ij} \end{bmatrix}$ is called **upper triangular** if $a_{ij} = 0$ for $i > j$. It is called **lower triangular** if $a_{ij} = 0$ for $i < j$.

$$\begin{bmatrix} a_{11} & a_{12} & \cdots & \cdots & \cdots & a_{1n} \\ 0 & a_{22} & \cdots & \cdots & \cdots & a_{2n} \\ 0 & 0 & a_{33} & \cdots & \cdots & a_{3n} \\ \vdots & \vdots & \vdots & \ddots & & \vdots \\ \vdots & \vdots & \vdots & & \ddots & \vdots \\ 0 & 0 & 0 & \cdots & 0 & a_{nn} \end{bmatrix}$$

Upper triangular matrix
(The elements below the main diagonal are zero.)

$$\begin{bmatrix} a_{11} & 0 & 0 & \cdots & \cdots & 0 \\ a_{21} & a_{22} & 0 & \cdots & \cdots & 0 \\ a_{31} & a_{32} & a_{33} & 0 & \cdots & 0 \\ \vdots & \vdots & \vdots & \ddots & & \vdots \\ \vdots & \vdots & \vdots & & \ddots & 0 \\ a_{n1} & a_{n2} & a_{n3} & \cdots & \cdots & a_{nn} \end{bmatrix}$$

Lower triangular matrix
(The elements above the main diagonal are zero.)

(a) Show that the sum and difference of two upper triangular matrices is upper triangular.

(b) Show that the sum and difference of two lower triangular matrices is lower triangular.

(c) Show that if a matrix is both upper and lower triangular, then it is a diagonal matrix.

T.6. (a) Show that if A is an upper triangular matrix, then A^T is lower triangular.

(b) Show that if A is a lower triangular matrix, then A^T is upper triangular.

MATLAB EXERCISES ▪

In order to use MATLAB in this section, you should first read Sections 10.1 and 10.2, which give basic information about MATLAB and about matrix operations in MATLAB. You are urged to do any examples or illustrations of MATLAB commands that appear in Sections 10.1 and 10.2 before trying these exercises.

ML.1. In MATLAB, enter the following matrices.

$$A = \begin{bmatrix} 5 & 1 & 2 \\ -3 & 0 & 1 \\ 2 & 4 & 1 \end{bmatrix},$$

$$B = \begin{bmatrix} 4*2 & 2/3 \\ 1/201 & 5-8.2 \\ 0.00001 & (9+4)/3 \end{bmatrix}.$$

Using MATLAB commands, display the following.

(a) a_{23}, b_{32}, b_{12}.

(b) $\text{row}_1(A)$, $\text{col}_3(A)$, $\text{row}_2(B)$.

(c) Type MATLAB command **format long** and display matrix B. Compare the elements of B from part (a) with the current display. Note that **format short** displays four decimal places rounded. Reset the format to **format short**.

ML.2. In MATLAB, type the command **H = hilb(5);** (Note that the last character is a semicolon, which suppresses the display of the contents of matrix H. See Section 10.1.) For more information on the **hilb** command, type **help hilb**. Using MATLAB commands, do the following:

(a) Determine the size of H.

(b) Display the contents of H.

(c) Display the contents of H as rational numbers.

(d) Extract as a matrix the first three columns.

(e) Extract as a matrix the last two rows.

1.3 ▼ Dot Product and Matrix Multiplication

In this section we introduce the operation of matrix multiplication. Unlike matrix addition, matrix multiplication has some properties that distinguish it from multiplication of real numbers.

DEFINITION   The **dot product** or **inner product** of the n-vectors

$$\mathbf{a} = \begin{bmatrix} a_1 & a_2 & \cdots & a_n \end{bmatrix} \quad \text{and} \quad \mathbf{b} = \begin{bmatrix} b_1 \\ b_2 \\ \vdots \\ b_n \end{bmatrix}$$

as

$$\mathbf{a} \cdot \mathbf{b} = a_1 b_1 + a_2 b_2 + \cdots + a_n b_n = \sum_{i=1}^{n} a_i b_i.^*$$

The dot product is an important operation that will be used here and in later sections.

---
*You may already be familiar with this useful notation, the summation notation. It is discussed in detail at the end of this section.

EXAMPLE 1 ■ The dot product of

$$\mathbf{u} = \begin{bmatrix} 1 & -2 & 3 & 4 \end{bmatrix} \quad \text{and} \quad \mathbf{v} = \begin{bmatrix} 2 \\ 3 \\ -2 \\ 1 \end{bmatrix}$$

is

$$\mathbf{u} \cdot \mathbf{v} = (1)(2) + (-2)(3) + (3)(-2) + (4)(1) = -6.$$ ■

EXAMPLE 2 ■ Let $\mathbf{a} = \begin{bmatrix} x & 2 & 3 \end{bmatrix}$ and $\mathbf{b} = \begin{bmatrix} 4 \\ 1 \\ 2 \end{bmatrix}$. If $\mathbf{a} \cdot \mathbf{b} = -4$, find x.

Solution   We have

$$\mathbf{a} \cdot \mathbf{b} = 4x + 2 + 6 = -4$$
$$4x + 8 = -4$$
$$x = -3.$$ ■

Matrix Multiplication

DEFINITION

If $A = \begin{bmatrix} a_{ij} \end{bmatrix}$ is an $m \times p$ matrix and $B = \begin{bmatrix} b_{ij} \end{bmatrix}$ is a $p \times n$ matrix, then the **product** of A and B, denoted AB, is the $m \times n$ matrix $C = \begin{bmatrix} c_{ij} \end{bmatrix}$, defined by

$$c_{ij} = a_{i1}b_{1j} + a_{i2}b_{2j} + \cdots + a_{ip}b_{pj} = \sum_{k=1}^{p} a_{ik}b_{kj} \quad (1 \le i \le m,\ 1 \le j \le n). \quad (1)$$

Equation (1) says that the i, jth element in the product matrix is the dot product of the ith row, $\text{row}_i(A)$, and the jth column, $\text{col}_j(B)$, of B; this is shown in Figure 1.2.

$$\text{row}_i(A) \begin{bmatrix} a_{11} & a_{12} & \cdots & a_{1p} \\ a_{21} & a_{22} & \cdots & a_{2p} \\ \vdots & \vdots & & \vdots \\ a_{i1} & a_{i2} & \cdots & a_{ip} \\ \vdots & \vdots & & \vdots \\ a_{m1} & a_{m2} & \cdots & a_{mp} \end{bmatrix} \begin{matrix} {}^{\cdot}\text{col}_j(B) \\ \begin{bmatrix} b_{11} & b_{12} & \cdots & b_{1j} & \cdots & b_{1n} \\ b_{21} & b_{22} & \cdots & b_{2j} & \cdots & b_{2n} \\ \vdots & \vdots & & \vdots & & \vdots \\ b_{p1} & b_{p2} & \cdots & b_{pj} & \cdots & b_{pn} \end{bmatrix} \end{matrix}$$

$$= \begin{bmatrix} c_{11} & c_{12} & \cdots & c_{1n} \\ c_{21} & c_{22} & \cdots & c_{2n} \\ \vdots & \vdots & c_{ij} & \vdots \\ c_{m1} & c_{m2} & \cdots & c_{mn} \end{bmatrix}.$$

$$\text{row}_i(A) \cdot \text{col}_j(B) = \sum_{k=1}^{p} a_{ik}b_{kj}$$

Observe that the product of A and B is defined only when the number of rows of B is exactly the same as the number of columns of A, as is indicated in Figure 1.3.

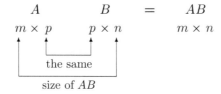

$$\begin{array}{ccccc} A & & B & = & AB \\ m \times p & & p \times n & & m \times n \end{array}$$

the same

size of AB

One might ask why matrix equality and matrix addition are defined in such a natural way while matrix multiplication appears to be much more complicated. Only a thorough understanding of the composition of functions and the relationship that exists between matrices and what shall be called linear transformations would show that the foregoing definition of multiplication is the natural one.

EXAMPLE 3 ■ Let

$$A = \begin{bmatrix} 1 & 2 & -1 \\ 3 & 1 & 4 \end{bmatrix} \quad \text{and} \quad B = \begin{bmatrix} -2 & 5 \\ 4 & -3 \\ 2 & 1 \end{bmatrix}.$$

Then

$$AB = \begin{bmatrix} (1)(-2)+(2)(4)+(-1)(2) & (1)(5)+(2)(-3)+(-1)(1) \\ (3)(-2)+(1)(4)+(4)(2) & (3)(5)+(1)(-3)+(4)(1) \end{bmatrix}$$
$$= \begin{bmatrix} 4 & -2 \\ 6 & 16 \end{bmatrix}.$$

■

EXAMPLE 4 ■ Let

$$A = \begin{bmatrix} 1 & -2 & 3 \\ 4 & 2 & 1 \\ 0 & 1 & -2 \end{bmatrix} \quad \text{and} \quad B = \begin{bmatrix} 1 & 4 \\ 3 & -1 \\ -2 & 2 \end{bmatrix}.$$

Compute the $(3,2)$ entry of AB.

Solution   If $AB = C$, then the $(3,2)$ entry of AB is c_{32}, which is $\text{row}_3(A) \cdot \text{col}_2(B)$. We now have

$$\text{row}_3(A) \cdot \text{col}_2(B) = \begin{bmatrix} 0 & 1 & -2 \end{bmatrix} \cdot \begin{bmatrix} 4 \\ -1 \\ 2 \end{bmatrix} = -5.$$

■

EXAMPLE 5 ■ Let

$$A = \begin{bmatrix} 1 & x & 3 \\ 2 & -1 & 1 \end{bmatrix} \quad \text{and} \quad B = \begin{bmatrix} 2 \\ 4 \\ y \end{bmatrix}.$$

If $AB = \begin{bmatrix} 12 \\ 6 \end{bmatrix}$, find x and y.

Solution   We have

$$AB = \begin{bmatrix} 1 & x & 3 \\ 2 & -1 & 1 \end{bmatrix} \begin{bmatrix} 2 \\ 4 \\ y \end{bmatrix} = \begin{bmatrix} 2 + 4x + 3y \\ 4 - 4 + y \end{bmatrix} = \begin{bmatrix} 12 \\ 6 \end{bmatrix}.$$

Then

$$2 + 4x + 3y = 12$$
$$y = 6,$$

so $x = -2$ and $y = 6$.   ■

The basic properties of matrix multiplication will be considered in the following section. However, multiplication of matrices requires much more care than their addition, since the algebraic properties of matrix multiplication differ from those satisfied by the real numbers. Part of the problem is due to the fact that AB is defined only when the number of columns of A is the same as the number of rows of B. Thus, if A is an $m \times p$ matrix and B is a $p \times n$ matrix, then AB is an $m \times n$ matrix. What about BA? Four different situations may occur:

1. BA may not be defined; this will take place if $n \neq m$.
2. If BA is defined, which means that $m = n$, then BA is $p \times p$ while AB is $m \times m$; thus, if $m \neq p$, AB and BA are of different sizes.
3. If AB and BA are both of the same size, they may be equal.
4. If AB and BA are both of the same size, they may be unequal.

EXAMPLE 6 ■ If A is a 2×3 matrix and B is a 3×4 matrix, then AB is a 2×4 matrix while BA is undefined.   ■

EXAMPLE 7 ■ Let A be 2×3 and let B be 3×2. Then AB is 2×2 while BA is 3×3.   ■

EXAMPLE 8 ■ Let

$$A = \begin{bmatrix} 1 & 2 \\ -1 & 3 \end{bmatrix} \quad \text{and} \quad B = \begin{bmatrix} 2 & 1 \\ 0 & 1 \end{bmatrix}.$$

Then

$$AB = \begin{bmatrix} 2 & 3 \\ -2 & 2 \end{bmatrix} \quad \text{while} \quad BA = \begin{bmatrix} 1 & 7 \\ -1 & 3 \end{bmatrix}.$$

Thus $AB \neq BA$.   ■

EXAMPLE 9 ■ Pesticides are sprayed on plants to eliminate harmful insects. However, some of
(*Ecology*)   the pesticide is absorbed by the plant. The pesticides are absorbed by herbivores when they eat the plants that have been sprayed. To determine the amount of pesticide absorbed by a herbivore, we proceed as follows. Suppose that we have three pesticides and four plants. Let a_{ij} denote the amount of pesticide

i (in milligrams) that has been absorbed by plant j. This information can be represented by the matrix

$$
A = \begin{array}{c} \\ \\ \left[\begin{array}{cccc} \overset{\text{Plant 1}}{2} & \overset{\text{Plant 2}}{3} & \overset{\text{Plant 3}}{4} & \overset{\text{Plant 4}}{3} \\ 3 & 2 & 2 & 5 \\ 4 & 1 & 6 & 4 \end{array} \right] \begin{array}{l} \text{Pesticide 1} \\ \text{Pesticide 2} \\ \text{Pesticide 3} \end{array} \end{array}
$$

Now suppose that we have three herbivores, and let b_{ij} denote the number of plants of type i that a herbivore of type j eats per month. This information can be represented by the matrix

$$
B = \left[\begin{array}{ccc} \overset{\text{Herbivore 1}}{20} & \overset{\text{Herbivore 2}}{12} & \overset{\text{Herbivore 3}}{8} \\ 28 & 15 & 15 \\ 30 & 12 & 10 \\ 40 & 16 & 20 \end{array} \right] \begin{array}{l} \text{Plant 1} \\ \text{Plant 2} \\ \text{Plant 3} \\ \text{Plant 4} \end{array}
$$

The (i, j) entry in AB gives the amount of pesticide of type i that animal j has absorbed. Thus, if $i = 2$ and $j = 3$, the $(2, 3)$ entry in AB is

$$
3(8) + 2(15) + 2(10) + 5(20)
$$
$$
= 174 \text{ mg of pesticide 2 absorbed by herbivore 3.}
$$

If we now have p carnivores (such as man) who eat the herbivores, we can repeat the analysis to find out how much of each pesticide has been absorbed by each carnivore. ∎

It is sometimes useful to be able to find a column in the matrix product AB without having to multiply the two matrices. It is not difficult to show (Exercise T.9) that the jth column of the matrix product AB is equal to the matrix product $A \cdot \mathrm{col}_j(B)$.

EXAMPLE 10 ∎ Let

$$
A = \begin{bmatrix} 1 & 2 \\ 3 & 4 \\ -1 & 5 \end{bmatrix} \quad \text{and} \quad B = \begin{bmatrix} -2 & 3 & 4 \\ 3 & 2 & 1 \end{bmatrix}.
$$

Then the second column of AB is

$$
A \cdot \mathrm{col}_2(B) = \begin{bmatrix} 1 & 2 \\ 3 & 4 \\ -1 & 5 \end{bmatrix} \begin{bmatrix} 3 \\ 2 \end{bmatrix} = \begin{bmatrix} 7 \\ 17 \\ 7 \end{bmatrix}.
$$

■

REMARK   If the n-vectors $\mathbf{u}$ and $\mathbf{v}$ are viewed as $n \times 1$ matrices, then it is easy to show (Exercise T.14) that

$$
\mathbf{u} \cdot \mathbf{v} = \mathbf{u}^T \mathbf{v}.
$$

This observation will be used in Chapter 3.

The Matrix-Vector Product Written in Terms of Columns

Let

$$A = \begin{bmatrix} a_{11} & a_{12} & \cdots & a_{1n} \\ a_{21} & a_{22} & \cdots & a_{2n} \\ \vdots & \vdots & & \vdots \\ a_{m1} & a_{m2} & \cdots & a_{mn} \end{bmatrix}$$

be an $m \times n$ matrix and let

$$\mathbf{c} = \begin{bmatrix} c_1 \\ c_2 \\ \vdots \\ c_n \end{bmatrix}$$

be an n-vector, that is, an $n \times 1$ matrix. Since A is $m \times n$ and $\mathbf{c}$ is $n \times 1$, the matrix product $A\mathbf{c}$ is the $m \times 1$ matrix

$$A\mathbf{c} = \begin{bmatrix} a_{11} & a_{12} & \cdots & a_{1n} \\ a_{21} & a_{22} & \cdots & a_{2n} \\ \vdots & \vdots & & \vdots \\ a_{m1} & a_{m2} & \cdots & a_{mn} \end{bmatrix} \begin{bmatrix} c_1 \\ c_2 \\ \vdots \\ c_n \end{bmatrix} = \begin{bmatrix} \mathrm{row}_1(A) \cdot \mathbf{c} \\ \mathrm{row}_2(A) \cdot \mathbf{c} \\ \vdots \\ \mathrm{row}_m(A) \cdot \mathbf{c} \end{bmatrix}$$

$$= \begin{bmatrix} a_{11}c_1 + a_{12}c_2 + \cdots + a_{1n}c_n \\ a_{21}c_1 + a_{22}c_2 + \cdots + a_{2n}c_n \\ \vdots \\ a_{m1}c_1 + a_{m2}c_2 + \cdots + a_{mn}c_n \end{bmatrix}. \tag{2}$$

The right side of this expression can be written as

$$c_1 \begin{bmatrix} a_{11} \\ a_{21} \\ \vdots \\ a_{m1} \end{bmatrix} + c_2 \begin{bmatrix} a_{12} \\ a_{22} \\ \vdots \\ a_{m2} \end{bmatrix} + \cdots + c_n \begin{bmatrix} a_{1n} \\ a_{2n} \\ \vdots \\ a_{mn} \end{bmatrix} \tag{3}$$

$$= c_1 \mathrm{col}_1(A) + c_2 \mathrm{col}_2(A) + \cdots + c_n \mathrm{col}_n(A).$$

The expression in (3) is called a **linear combination** of the columns of A. Thus the product $A\mathbf{c}$ of an $m \times n$ matrix A and an $n \times 1$ matrix $\mathbf{c}$ can be written as a linear combination of the columns of A, where the coefficients are the entries in $\mathbf{c}$.

EXAMPLE 11 ■ Let

$$A = \begin{bmatrix} 2 & -1 & -3 \\ 4 & 2 & -2 \end{bmatrix} \quad \text{and} \quad \mathbf{c} = \begin{bmatrix} 2 \\ -3 \\ 4 \end{bmatrix}.$$

Then the product $A\mathbf{c}$ written as a linear combination of the columns of A is

$$A\mathbf{c} = \begin{bmatrix} 2 & -1 & -3 \\ 4 & 2 & -2 \end{bmatrix} \begin{bmatrix} 2 \\ -3 \\ 4 \end{bmatrix} = \begin{bmatrix} -5 \\ -6 \end{bmatrix} = 2 \begin{bmatrix} 2 \\ 4 \end{bmatrix} - 3 \begin{bmatrix} -1 \\ 2 \end{bmatrix} + 4 \begin{bmatrix} -3 \\ -2 \end{bmatrix}.$$

∎

If A is an $m \times p$ matrix and B is a $p \times n$ matrix, we can then conclude that the jth column of the product AB can be written as a linear combination of the columns of matrix A, where the coefficients are the entries in the jth column of matrix B:

$$\text{col}_j(AB) = A\text{col}_j(B) = b_{1j}\text{col}_1(A) + b_{2j}\text{col}_2(A) + \cdots + b_{pj}\text{col}_p(A).$$

EXAMPLE 12 ∎ If A and B are the matrices defined in Example 10, then

$$AB = \begin{bmatrix} 1 & 2 \\ 3 & 4 \\ -1 & 5 \end{bmatrix} \begin{bmatrix} -2 & 3 & 4 \\ 3 & 2 & 1 \end{bmatrix} = \begin{bmatrix} 4 & 7 & 6 \\ 6 & 17 & 16 \\ 17 & 7 & 1 \end{bmatrix}.$$

The columns of AB as linear combinations of the columns of A are given by

$$\text{col}_1(AB) = \begin{bmatrix} 4 \\ 6 \\ 17 \end{bmatrix} = A\text{col}_1(B) = -2 \begin{bmatrix} 1 \\ 3 \\ -1 \end{bmatrix} + 3 \begin{bmatrix} 2 \\ 4 \\ 5 \end{bmatrix}$$

$$\text{col}_2(AB) = \begin{bmatrix} 7 \\ 17 \\ 7 \end{bmatrix} = A\text{col}_2(B) = 3 \begin{bmatrix} 1 \\ 3 \\ -1 \end{bmatrix} + 2 \begin{bmatrix} 2 \\ 4 \\ 5 \end{bmatrix}$$

$$\text{col}_3(AB) = \begin{bmatrix} 6 \\ 16 \\ 1 \end{bmatrix} = A\text{col}_3(B) = 4 \begin{bmatrix} 1 \\ 3 \\ -1 \end{bmatrix} + 1 \begin{bmatrix} 2 \\ 4 \\ 5 \end{bmatrix}.$$

Linear Systems

Consider the linear system of m equations in n unknowns,

$$
\begin{aligned}
a_{11}x_1 + a_{12}x_2 + \cdots + a_{1n}x_n &= b_1 \\
a_{21}x_1 + a_{22}x_2 + \cdots + a_{2n}x_n &= b_2 \\
&\vdots \\
a_{m1}x_1 + a_{m2}x_2 + \cdots + a_{mn}x_n &= b_m.
\end{aligned}
\tag{4}
$$

Now define the following matrices:

$$A = \begin{bmatrix} a_{11} & a_{12} & \cdots & a_{1n} \\ a_{21} & a_{22} & \cdots & a_{2n} \\ \vdots & \vdots & & \vdots \\ a_{m1} & a_{m2} & \cdots & a_{mn} \end{bmatrix}, \quad \mathbf{x} = \begin{bmatrix} x_1 \\ x_2 \\ \vdots \\ x_n \end{bmatrix}, \quad \mathbf{b} = \begin{bmatrix} b_1 \\ b_2 \\ \vdots \\ b_m \end{bmatrix}.$$

Then

$$
A\mathbf{x} =
\begin{bmatrix}
a_{11} & a_{12} & \cdots & a_{1n} \\
a_{21} & a_{22} & \cdots & a_{2n} \\
\vdots & \vdots & & \vdots \\
a_{m1} & a_{m2} & \cdots & a_{mn}
\end{bmatrix}
\begin{bmatrix}
x_1 \\ x_2 \\ \vdots \\ x_n
\end{bmatrix}
=
\begin{bmatrix}
a_{11}x_1 + a_{12}x_2 + \cdots + a_{1n}x_n \\
a_{21}x_1 + a_{22}x_2 + \cdots + a_{2n}x_n \\
\vdots \qquad \vdots \qquad\qquad \vdots \\
a_{m1}x_1 + a_{m2}x_2 + \cdots + a_{mn}x_n
\end{bmatrix}.
$$

The entries in the product $A\mathbf{x}$ are merely the left sides of the equations in (4). Hence the linear system (4) can be written in matrix form as

$$A\mathbf{x} = \mathbf{b}.$$

The matrix A is called the **coefficient matrix** of the linear system (4), and the matrix

$$
\begin{bmatrix}
a_{11} & a_{12} & \cdots & a_{1n} & \vdots & b_1 \\
a_{21} & a_{22} & \cdots & a_{2n} & \vdots & b_2 \\
\vdots & \vdots & & \vdots & \vdots & \vdots \\
a_{m1} & a_{m2} & \cdots & a_{mn} & \vdots & b_m
\end{bmatrix},
$$

obtained by adjoining $\mathbf{b}$ to A, is called the **augmented matrix** of the linear system (4). The augmented matrix of (4) will be written as $\begin{bmatrix} A \mathrel{\vdots} \mathbf{b} \end{bmatrix}$. Conversely, any matrix with more than one column can be thought of as the augmented matrix of a linear system. The coefficient and augmented matrices will play key roles in our method for solving linear systems.

EXAMPLE 13 ■ Consider the linear system

$$
\begin{aligned}
2x + 3y - 4z &= 7 \\
-2x \qquad\;\; + z &= 5 \\
3x + 2y + 2z &= 3.
\end{aligned}
$$

Letting

$$
A = \begin{bmatrix} 2 & 3 & -4 \\ -2 & 0 & 1 \\ 3 & 2 & 2 \end{bmatrix}, \quad
\mathbf{x} = \begin{bmatrix} x \\ y \\ z \end{bmatrix}, \quad \text{and} \quad
\mathbf{b} = \begin{bmatrix} 5 \\ 7 \\ 3 \end{bmatrix},
$$

we can write the given linear system in matrix form as

$$A\mathbf{x} = \mathbf{b}.$$

The coefficient matrix is A and the augmented matrix is

$$
\begin{bmatrix}
2 & 3 & -4 & \vdots & 5 \\
-2 & 0 & 1 & \vdots & 7 \\
3 & 2 & 2 & \vdots & 3
\end{bmatrix}.
$$
■

EXAMPLE 14 ■ The matrix

$$
\begin{bmatrix}
2 & -1 & 3 & \vdots & 4 \\
3 & 0 & 2 & \vdots & 5
\end{bmatrix}
$$

is the augmented matrix of the linear system

$$2x - y + 3z = 4$$
$$3x \quad\;\; + 2z = 5.$$ ∎

It follows from our discussion above that the linear system in (4) can be written as a linear combination of the columns of A as

$$x_1 \begin{bmatrix} a_{11} \\ a_{21} \\ \vdots \\ a_{m1} \end{bmatrix} + x_2 \begin{bmatrix} a_{12} \\ a_{22} \\ \vdots \\ a_{m2} \end{bmatrix} + \cdots + x_n \begin{bmatrix} a_{1n} \\ a_{2n} \\ \vdots \\ a_{mn} \end{bmatrix} = \begin{bmatrix} b_1 \\ b_2 \\ \vdots \\ b_m \end{bmatrix}. \tag{5}$$

Conversely, an equation as in (5) always describes a linear system as in (4).

Partitioned Matrices (Optional)

If we start out with an $m \times n$ matrix $A = \begin{bmatrix} a_{ij} \end{bmatrix}$ and cross out some, but not all, of its rows or columns, we obtain a **submatrix** of A.

EXAMPLE 15 ∎ Let

$$A = \begin{bmatrix} 1 & 2 & 3 & 4 \\ -2 & 4 & -3 & 5 \\ 3 & 0 & 5 & -3 \end{bmatrix}.$$

If we cross out the second row and third column, we obtain the submatrix

$$\begin{bmatrix} 1 & 2 & 4 \\ 3 & 0 & -3 \end{bmatrix}.$$ ∎

A matrix can be partitioned into submatrices by drawing horizontal lines between rows and vertical lines between columns. Of course, the partitioning can be carried out in many different ways.

EXAMPLE 16 ∎ The matrix

$$A = \left[\begin{array}{ccc:cc} a_{11} & a_{12} & a_{13} & a_{14} & a_{15} \\ a_{21} & a_{22} & a_{23} & a_{24} & a_{25} \\ \hdashline a_{31} & a_{32} & a_{33} & a_{34} & a_{35} \\ a_{41} & a_{42} & a_{43} & a_{44} & a_{45} \end{array} \right]$$

is partitioned as

$$A = \begin{bmatrix} A_{11} & A_{12} \\ A_{21} & A_{22} \end{bmatrix}.$$

We could also write

$$A = \left[\begin{array}{cc:ccc} a_{11} & a_{12} & a_{13} & a_{14} & a_{15} \\ a_{21} & a_{22} & a_{23} & a_{24} & a_{25} \\ \hdashline a_{31} & a_{32} & a_{33} & a_{34} & a_{35} \\ a_{41} & a_{42} & a_{43} & a_{44} & a_{45} \end{array} \right] = \begin{bmatrix} \widehat{A}_{11} & \widehat{A}_{12} & \widehat{A}_{13} \\ \widehat{A}_{21} & \widehat{A}_{22} & \widehat{A}_{23} \end{bmatrix}, \tag{6}$$

which gives another partitioning of A. We thus speak of **partitioned matrices**.

∎

EXAMPLE 17 ∎ The augmented matrix of a linear system is a partitioned matrix. Thus, if $A\mathbf{x} = \mathbf{b}$, we can write the augmented matrix of this system as $\begin{bmatrix} A \mathrel{\vdots} \mathbf{b} \end{bmatrix}$.   ∎

If A and B are both $m \times n$ matrices that are partitioned in the same way, then $A + B$ is obtained simply by adding the corresponding submatrices of A and B. Similarly, if A is a partitioned matrix, then the scalar multiple cA is obtained by forming the scalar multiple of each submatrix.

If A is partitioned as shown in (6) and

$$
B = \begin{bmatrix}
b_{11} & b_{12} & b_{13} & b_{14} \\
b_{21} & b_{22} & b_{23} & b_{24} \\
b_{31} & b_{32} & b_{33} & b_{34} \\
b_{41} & b_{42} & b_{43} & b_{44} \\
b_{51} & b_{52} & b_{53} & b_{54}
\end{bmatrix}
= \begin{bmatrix}
B_{11} & B_{12} \\
B_{21} & B_{22} \\
B_{31} & B_{32}
\end{bmatrix},
$$

then by straightforward computations we can show that

$$
AB = \begin{bmatrix}
(\widehat{A}_{11}B_{11} + \widehat{A}_{12}B_{21} + \widehat{A}_{13}B_{31}) & (\widehat{A}_{11}B_{12} + \widehat{A}_{12}B_{22} + \widehat{A}_{13}B_{32}) \\
(\widehat{A}_{21}B_{11} + \widehat{A}_{22}B_{21} + \widehat{A}_{23}B_{31}) & (\widehat{A}_{21}B_{12} + \widehat{A}_{22}B_{22} + \widehat{A}_{23}B_{32})
\end{bmatrix}.
$$

∎

EXAMPLE 18 ∎ Let

$$
A = \begin{bmatrix}
1 & 0 & 1 & 0 \\
0 & 2 & 3 & -1 \\
2 & 0 & -4 & 0 \\
0 & 1 & 0 & 3
\end{bmatrix}
= \begin{bmatrix}
A_{11} & A_{12} \\
A_{21} & A_{22}
\end{bmatrix}
$$

and let

$$
B = \begin{bmatrix}
2 & 0 & 0 & 1 & 1 & -1 \\
0 & 1 & 1 & -1 & 2 & 2 \\
1 & 3 & 0 & 0 & 1 & 0 \\
-3 & -1 & 2 & 1 & 0 & -1
\end{bmatrix}
= \begin{bmatrix}
B_{11} & B_{12} \\
B_{21} & B_{22}
\end{bmatrix}.
$$

Then

$$
AB = C = \begin{bmatrix}
3 & 3 & 0 & 1 & 2 & -1 \\
6 & 12 & 0 & -3 & 7 & 5 \\
0 & -12 & 0 & 2 & -2 & -2 \\
-9 & -2 & 7 & 2 & 2 & -1
\end{bmatrix}
= \begin{bmatrix}
C_{11} & C_{12} \\
C_{21} & C_{22}
\end{bmatrix},
$$

where C_{11} should be $A_{11}B_{11} + A_{12}B_{21}$. We verify that C_{11} is this expression as follows:

$$A_{11}B_{11} + A_{12}B_{21} = \begin{bmatrix} 1 & 0 \\ 0 & 2 \end{bmatrix} \begin{bmatrix} 2 & 0 & 0 \\ 0 & 1 & 1 \end{bmatrix} + \begin{bmatrix} 1 & 0 \\ 3 & -1 \end{bmatrix} \begin{bmatrix} 1 & 3 & 0 \\ -3 & -1 & 2 \end{bmatrix}$$

$$= \begin{bmatrix} 2 & 0 & 0 \\ 0 & 2 & 2 \end{bmatrix} + \begin{bmatrix} 1 & 3 & 0 \\ 6 & 10 & -2 \end{bmatrix}$$

$$= \begin{bmatrix} 3 & 3 & 0 \\ 6 & 12 & 0 \end{bmatrix} = C_{11}.$$ ■

This method of multiplying partitioned matrices is also known as **block multiplication**. Partitioned matrices can be used to great advantage in dealing with matrices that exceed the memory capacity of a computer. Thus, in multiplying two partitioned matrices, one can keep the matrices on disk and only bring into memory the submatrices required to form the submatrix products. The latter, of course, can be put out on disk as they are formed. The partitioning must be done so that the products of corresponding submatrices are defined. In contemporary computing technology, parallel-processing computers use partitioned matrices to perform matrix computations more rapidly.

Summation Notation (Optional)

We shall occasionally use the **summation notation** and we now review this useful and compact notation, which is widely used in mathematics.

By $\sum_{i=1}^{n} a_i$ we mean

$$a_1 + a_2 + \cdots + a_n.$$

The letter i is called the **index of summation**; it is a dummy variable that can be replaced by another letter. Hence we can write

$$\sum_{i=1}^{n} a_i = \sum_{j=1}^{n} a_j = \sum_{k=1}^{n} a_k.$$

EXAMPLE 19 ▶ If

$$a_1 = 3, \quad a_2 = 4, \quad a_3 = 5, \quad \text{and} \quad a_4 = 8,$$

then

$$\sum_{i=1}^{4} a_i = 3 + 4 + 5 + 8 = 20.$$ ■

EXAMPLE 20 ■ By $\sum_{i=1}^{n} r_i a_i$ we mean

$$r_1 a_1 + r_2 a_2 + \cdots + r_n a_n.$$

It is not difficult to show (Exercise T.11) that the summation notation satisfies the following properties:

(i) $$\sum_{i=1}^{n}(r_i + s_i)a_i = \sum_{i=1}^{n} r_i a_i + \sum_{i=1}^{n} s_i a_i.$$

(ii) $$\sum_{i=1}^{n} c(r_i a_i) = c\left(\sum_{i=1}^{n} r_i a_i\right).$$ ∎

EXAMPLE 21 ∎ If

$$\mathbf{a} = \begin{bmatrix} a_1 & a_2 & \cdots & a_n \end{bmatrix} \quad \text{and} \quad \mathbf{b} = \begin{bmatrix} b_1 \\ b_2 \\ \vdots \\ b_n \end{bmatrix},$$

then the dot product $\mathbf{a} \cdot \mathbf{b}$ can be expressed using summation notation as

$$\mathbf{a} \cdot \mathbf{b} = a_1 b_1 + a_2 b_2 + \cdots + a_n b_n = \sum_{i=1}^{n} a_i b_i.$$ ∎

EXAMPLE 22 ∎ We can write Equation (1), for the i, jth element in the product of the matrices A and B, in terms of the summation notation as

$$c_{ij} = \sum_{k=1}^{p} a_{ik} b_{kj} \qquad (1 \le i \le m,\ 1 \le j \le n).$$ ∎

It is also possible to form double sums. Thus by $\sum\limits_{j=1}^{m} \sum\limits_{i=1}^{n} a_i$ we mean that we first sum on i and then sum the resulting expression on j.

EXAMPLE 23 ∎ If $n = 2$ and $m = 3$, we have

$$\sum_{j=1}^{3} \sum_{i=1}^{2} a_{ij} = \sum_{j=1}^{3}(a_{1j} + a_{2j})$$
$$= (a_{11} + a_{21}) + (a_{12} + a_{22}) + (a_{13} + a_{23}) \qquad (7)$$

$$\sum_{i=1}^{2} \sum_{j=1}^{3} a_{ij} = \sum_{i=1}^{2}(a_{i1} + a_{i2} + a_{i3})$$
$$= (a_{11} + a_{12} + a_{13}) + (a_{21} + a_{22} + a_{23})$$
$$= \text{right side of } (7).$$ ∎

It is not difficult to show, in general (Exercise T.12), that

$$\sum_{i=1}^{m}\sum_{j=1}^{n} a_{ij} = \sum_{j=1}^{n}\sum_{i=1}^{m} a_{ij}. \tag{8}$$

Equation (8) can be interpreted as follows. Let A be the $m \times n$ matrix $[a_{ij}]$. If we add up the entries in each row of A and then add the resulting numbers, we obtain the same result as when we add up the entries in each column of A and then add the resulting numbers.

1.3 EXERCISES

In Exercises 1 and 2, compute $\mathbf{a} \cdot \mathbf{b}$.

1. (a) $\mathbf{a} = \begin{bmatrix} 1 & 2 \end{bmatrix}$, $\mathbf{b} = \begin{bmatrix} 4 \\ -1 \end{bmatrix}$.

(b) $\mathbf{a} = \begin{bmatrix} -3 & -2 \end{bmatrix}$, $\mathbf{b} = \begin{bmatrix} 1 \\ -2 \end{bmatrix}$.

(c) $\mathbf{a} = \begin{bmatrix} 4 & 2 & -1 \end{bmatrix}$, $\mathbf{b} = \begin{bmatrix} 1 \\ 3 \\ 6 \end{bmatrix}$.

(d) $\mathbf{a} = \begin{bmatrix} 1 & 1 & 0 \end{bmatrix}$, $\mathbf{b} = \begin{bmatrix} 1 \\ 0 \\ 1 \end{bmatrix}$.

2. (a) $\mathbf{a} = \begin{bmatrix} 2 & -1 \end{bmatrix}$, $\mathbf{b} = \begin{bmatrix} 3 \\ 2 \end{bmatrix}$.

(b) $\mathbf{a} = \begin{bmatrix} 1 & -1 \end{bmatrix}$, $\mathbf{b} = \begin{bmatrix} 1 \\ 1 \end{bmatrix}$.

(c) $\mathbf{a} = \begin{bmatrix} 1 & 2 & 3 \end{bmatrix}$, $\mathbf{b} = \begin{bmatrix} -2 \\ 0 \\ 1 \end{bmatrix}$.

(d) $\mathbf{a} = \begin{bmatrix} 1 & 0 & 0 \end{bmatrix}$, $\mathbf{b} = \begin{bmatrix} 1 \\ 0 \\ 0 \end{bmatrix}$.

3. Let

$$\mathbf{a} = \begin{bmatrix} -3 & 2 & x \end{bmatrix} \quad \text{and} \quad \mathbf{b} = \begin{bmatrix} -3 \\ 2 \\ x \end{bmatrix}.$$

If $\mathbf{a} \cdot \mathbf{b} = 17$, find x.

4. Let

$$A = \begin{bmatrix} 1 & 2 & x \\ 3 & -1 & 2 \end{bmatrix} \quad \text{and} \quad B = \begin{bmatrix} y \\ x \\ 1 \end{bmatrix}.$$

If $AB = \begin{bmatrix} 6 \\ 8 \end{bmatrix}$, find x and y.

In Exercises 5 and 6, let

$$A = \begin{bmatrix} 1 & 2 & -3 \\ 4 & 0 & -2 \end{bmatrix}, \quad B = \begin{bmatrix} 3 & 1 \\ 2 & 4 \\ -1 & 5 \end{bmatrix},$$

$$C = \begin{bmatrix} 2 & 3 & 1 \\ 3 & -4 & 5 \\ 1 & -1 & -2 \end{bmatrix}, \quad D = \begin{bmatrix} 2 & 3 \\ -1 & -2 \end{bmatrix},$$

$$E = \begin{bmatrix} 1 & 0 & -3 \\ -2 & 1 & 5 \\ 3 & 4 & 2 \end{bmatrix}, \quad \text{and} \quad F = \begin{bmatrix} 2 & -3 \\ 4 & 1 \end{bmatrix}.$$

5. If possible, compute:
   (a) AB.   (b) BA.   (c) $CB + D$.
   (d) $AB + DF$.   (e) $BA + FD$.

6. If possible, compute:
   (a) $A(BD)$.   (b) $(AB)D$.   (c) $A(C + E)$.
   (d) $AC + AE$.   (e) $(D + F)A$.

7. Let

$$A = \begin{bmatrix} 2 & 3 \\ -1 & 4 \\ 0 & 3 \end{bmatrix} \quad \text{and} \quad B = \begin{bmatrix} 3 & -1 & 3 \\ 1 & 2 & 4 \end{bmatrix}.$$

Compute the following entries of AB:
   (a) The $(1, 2)$ entry.   (b) The $(2, 3)$ entry.
   (c) The $(3, 1)$ entry.   (d) The $(3, 3)$ entry.

8. If $I_2 = \begin{bmatrix} 1 & 0 \\ 0 & 1 \end{bmatrix}$, compute DI_2 and I_2D.

9. Let

$$A = \begin{bmatrix} 1 & 2 \\ 3 & 2 \end{bmatrix} \quad \text{and} \quad B = \begin{bmatrix} 2 & -1 \\ -3 & 4 \end{bmatrix}.$$

Show that $AB \neq BA$.

10. If A is the matrix in Example 3 and O is the 3×2 matrix every one of whose entries is zero, compute AO.

In Exercises 11 and 12, let

$$A = \begin{bmatrix} 1 & -1 & 2 \\ 3 & 2 & 4 \\ 4 & -2 & 3 \\ 2 & 1 & 5 \end{bmatrix}$$

and

$$B = \begin{bmatrix} 1 & 0 & -1 & 2 \\ 3 & 3 & -3 & 4 \\ 4 & 2 & 5 & 1 \end{bmatrix}.$$

11. Using the method in Example 10, compute the following columns of AB:

(a) The first column.   (b) The third column.

12. Using the method in Example 10, compute the following columns of AB:

(a) The second column.   (b) The fourth column.

13. Let

$$A = \begin{bmatrix} 2 & -3 & 4 \\ -1 & 2 & 3 \\ 5 & -1 & -2 \end{bmatrix} \quad \text{and} \quad \mathbf{c} = \begin{bmatrix} 2 \\ 1 \\ 4 \end{bmatrix}.$$

Express $A\mathbf{c}$ as a linear combination of the columns of A.

14. Let

$$A = \begin{bmatrix} 1 & -2 & -1 \\ 2 & 4 & 3 \\ 3 & 0 & -2 \end{bmatrix} \quad \text{and} \quad B = \begin{bmatrix} 1 & -1 \\ 3 & 2 \\ 2 & 4 \end{bmatrix}.$$

Express the columns of AB as linear combinations of the columns of A.

15. Consider the following linear system:

$$\begin{aligned} 2x + \quad\quad w &= 7 \\ 3x + 2y + 3z \quad &= -2 \\ 2x + 3y - 4z \quad &= 3 \\ x + \quad 3z \quad &= 5. \end{aligned}$$

(a) Find the coefficient matrix.

(b) Write the linear system in matrix form.

(c) Find the augmented matrix.

16. Write the linear system with augmented matrix

$$\begin{bmatrix} -2 & -1 & 0 & 4 & \vdots & 5 \\ -3 & 2 & 7 & 8 & \vdots & 3 \\ 1 & 0 & 0 & 2 & \vdots & 4 \\ 3 & 0 & 1 & 3 & \vdots & 6 \end{bmatrix}.$$

17. Write the linear system with augmented matrix

$$\begin{bmatrix} 2 & 0 & -4 & \vdots & 3 \\ 0 & 1 & 2 & \vdots & 5 \\ 1 & 3 & 4 & \vdots & -1 \end{bmatrix}.$$

18. Consider the following linear system:

$$\begin{aligned} 3x - y + 2z &= 4 \\ 2x + y \quad\quad &= 2 \\ y + 3z &= 7 \\ 4x \quad\quad - z &= 4. \end{aligned}$$

(a) Find the coefficient matrix.

(b) Write the linear system in matrix form.

(c) Find the augmented matrix.

19. How are the linear systems whose augmented matrices are

$$\begin{bmatrix} 1 & 2 & 3 & \vdots & -1 \\ 2 & 3 & 6 & \vdots & 2 \end{bmatrix} \quad \text{and} \quad \begin{bmatrix} 1 & 2 & 3 & \vdots & -1 \\ 2 & 3 & 6 & \vdots & 2 \\ 0 & 0 & 0 & \vdots & 0 \end{bmatrix}$$

related?

20. Write each of the following as a linear system in matrix form.

(a) $x \begin{bmatrix} 1 \\ 2 \end{bmatrix} + y \begin{bmatrix} 2 \\ 5 \end{bmatrix} + z \begin{bmatrix} 0 \\ 3 \end{bmatrix} = \begin{bmatrix} 1 \\ 1 \end{bmatrix}.$

(b) $x \begin{bmatrix} 1 \\ 1 \\ 2 \end{bmatrix} + y \begin{bmatrix} 2 \\ 1 \\ 0 \end{bmatrix} + z \begin{bmatrix} 1 \\ 2 \\ 2 \end{bmatrix} = \begin{bmatrix} 0 \\ 0 \\ 0 \end{bmatrix}.$

21. Write each of the following linear systems as a linear combination of the columns of the coefficient matrix.

(a) $\begin{aligned} x + 2y &= 3 \\ 2x - y &= 5. \end{aligned}$

(b) $\begin{aligned} 2x - 3y + 5z &= -2 \\ x + 4y - z &= 3. \end{aligned}$

22. Let A be an $m \times n$ matrix and B an $n \times p$ matrix. What if anything can you say about the matrix product AB when:

(a) A has a column consisting entirely of zeros?

(b) B has a row consisting entirely of zeros?

23. (a) Find a value of r so that $AB^T = 0$, where

$$A = \begin{bmatrix} r & 1 & -2 \end{bmatrix} \quad \text{and} \quad B = \begin{bmatrix} 1 & 3 & -1 \end{bmatrix}.$$

(b) Give an alternate way to write this product.

24. Find a value of r and a value of s so that $AB^T = 0$, where

$$A = \begin{bmatrix} 1 & r & 1 \end{bmatrix} \quad \text{and} \quad B = \begin{bmatrix} -2 & 2 & s \end{bmatrix}.$$

25. Formulate the method for adding partitioned matrices and verify your method by partitioning the matrices

$$A = \begin{bmatrix} 1 & 3 & -1 \\ 2 & 1 & 0 \\ 2 & -3 & 1 \end{bmatrix} \quad \text{and} \quad B = \begin{bmatrix} 3 & 2 & 1 \\ -2 & 3 & 1 \\ 4 & 1 & 5 \end{bmatrix}$$

in two different ways and finding their sum.

26. Let A and B be the following matrices:

$$A = \begin{bmatrix} 2 & 1 & 3 & 4 & 2 \\ 1 & 2 & 3 & -1 & 4 \\ 2 & 3 & 2 & 1 & 4 \\ 5 & -1 & 3 & 2 & 6 \\ 3 & 1 & 2 & 4 & 6 \\ 2 & -1 & 3 & 5 & 7 \end{bmatrix}$$

and

$$B = \begin{bmatrix} 1 & 2 & 3 & 4 & 1 \\ 2 & 1 & 3 & 2 & -1 \\ 1 & 5 & 4 & 2 & 3 \\ 2 & 1 & 3 & 5 & 7 \\ 3 & 2 & 4 & 6 & 1 \end{bmatrix}.$$

Find AB by partitioning A and B in two different ways.

27. (*Manufacturing Costs*)   A furniture manufacturer makes chairs and tables, each of which must go through an assembly process and a finishing process. The times required for these processes are given (in hours) by the matrix

	Assembly process	Finishing process	
$A =$	2	2	Chair
	3	4	Table

The manufacturer has a plant in Salt Lake City and another in Chicago. The hourly rates for each of the processes are given (in dollars) by the matrix

	Salt Lake City	Chicago	
$B =$	9	10	Assembly process
	10	12	Finishing process

What do the entries in the matrix product AB tell the manufacturer?

28. (*Ecology—Pollution*)   A manufacturer makes two kinds of products, P and Q, at each of two plants, X and Y. In making these products, the pollutants sulfur dioxide, nitric oxide, and particulate matter are produced. The amounts of pollutants produced are given (in kilograms) by the matrix

	Sulfur dioxide	Nitric oxide	Particulate matter	
$A =$	300	100	150	Product P
	200	250	400	Product Q

State and federal ordinances require that these pollutants be removed. The daily cost of removing each kilogram of pollutant is given (in dollars) by the matrix

	Plant X	Plant Y	
$B =$	8	12	Sulfur dioxide
	7	9	Nitric oxide
	15	10	Particulate matter

What do the entries in the matrix product AB tell the manufacturer?

29. (*Medicine*)   A diet research project consists of adults and children of both sexes. The composition of the participants in the project is given by the matrix

	Adults	Children	
$A =$	80	120	Male
	100	200	Female

The number of daily grams of protein, fat, and carbohydrate consumed by each child and adult is given by the matrix

	Protein	Fat	Carbo-hydrate	
$B =$	20	20	20	Adult
	10	20	30	Child

(a) How many grams of protein are consumed daily by the males in the project?

(b) How many grams of fat are consumed daily by the females in the project?

30. (*Business*)   A photography business has a store in each of the following cities: New York, Denver, and Los Angeles. A particular make of camera is available in automatic, semiautomatic, and nonautomatic models. Moreover, each camera has a matched flash unit and a camera is usually sold together with the corresponding flash unit. The selling prices of the cameras and flash units are given (in dollars) by the matrix

	Automatic	Semi-automatic	Non-automatic	
$A =$	200	150	120	Camera
	50	40	25	Flash unit

The number of sets (camera and flash unit) available at each store is given by the matrix

$$B = \begin{bmatrix} 220 & 180 & 100 \\ 300 & 250 & 120 \\ 120 & 320 & 250 \end{bmatrix} \begin{array}{l} \text{Automatic} \\ \text{Semiautomatic} \\ \text{Nonautomatic} \end{array}$$

New York   Denver   Los Angeles

(a) What is the total value of the cameras in New York?

(b) What is the total value of the flash units in Los Angeles?

THEORETICAL EXERCISES

T.1. Let $\mathbf{x}$ be an n-vector.

(a) Is it possible for $\mathbf{x} \cdot \mathbf{x}$ to be negative? Explain.

(b) If $\mathbf{x} \cdot \mathbf{x} = 0$, what is $\mathbf{x}$?

T.2. Let $\mathbf{a}$, $\mathbf{b}$, and $\mathbf{c}$ be n-vectors and let k be a real number.

(a) Show that $\mathbf{a} \cdot \mathbf{b} = \mathbf{b} \cdot \mathbf{a}$.

(b) Show that $(\mathbf{a} + \mathbf{b}) \cdot \mathbf{c} = \mathbf{a} \cdot \mathbf{c} + \mathbf{b} \cdot \mathbf{c}$.

(c) Show that $(k\mathbf{a}) \cdot \mathbf{b} = \mathbf{a} \cdot (k\mathbf{b}) = k(\mathbf{a} \cdot \mathbf{b})$.

T.3. (a) Show that if A has a row of zeros, then AB has a row of zeros.

(b) Show that if B has a column of zeros, then AB has a column of zeros.

T.4. Show that the product of two diagonal matrices is a diagonal matrix.

T.5. Show that the product of two scalar matrices is a scalar matrix.

T.6. (a) Show that the product of two upper triangular matrices is upper triangular.

(b) Show that the product of two lower triangular matrices is lower triangular.

T.7. Let A and B be $n \times n$ diagonal matrices. Is $AB = BA$? Justify your answer.

T.8. (a) Let $\mathbf{a}$ be a $1 \times n$ matrix and B an $n \times p$ matrix. Show that the matrix product $\mathbf{a}B$ can be written as a linear combination of the rows of B, where the coefficients are the entries of $\mathbf{a}$.

(b) Let $\mathbf{a} = \begin{bmatrix} 1 & -2 & 3 \end{bmatrix}$ and

$$B = \begin{bmatrix} 2 & 1 & -4 \\ -3 & -2 & 3 \\ 4 & 5 & -2 \end{bmatrix}.$$

Write $\mathbf{a}B$ as a linear combination of the rows of B.

T.9. (a) Show that the jth column of the matrix product AB is equal to the matrix product $A\text{col}_j(B)$.

(b) Show that the ith row of the matrix product AB is equal to the matrix product $\text{row}_i(A)B$.

T.10. Let A be an $m \times n$ matrix whose entries are real numbers. Show that if $AA^T = O$ (the $m \times m$ matrix all of whose entries are zero), then $A = O$.

Exercises T.11 to T.13 depend upon material marked as optional.

T.11. Show that the summation notation satisfies the following properties:

(a) $\displaystyle\sum_{i=1}^{n}(r_i + s_i)a_i = \sum_{i=1}^{n}r_ia_i + \sum_{i=1}^{n}s_ia_i.$

(b) $\displaystyle\sum_{i=1}^{n}c(r_ia_i) = c\left(\sum_{i=1}^{n}r_ia_i\right).$

T.12. Show that $\displaystyle\sum_{i=1}^{n}\sum_{j=1}^{m}a_{ij} = \sum_{j=1}^{m}\sum_{i=1}^{n}a_{ij}.$

T.13. Answer the following as true or false. If true, prove the result; if false, give a counterexample.

(a) $\displaystyle\sum_{i=1}^{n}(a_i + 1) = \left(\sum_{i=1}^{n}a_i\right) + n.$

(b) $\displaystyle\sum_{i=1}^{n}\sum_{j=1}^{m}1 = mn.$

(c) $\displaystyle\sum_{j=1}^{m}\sum_{i=1}^{n}a_ib_j = \left[\sum_{i=1}^{n}a_i\right]\left[\sum_{j=1}^{m}b_j\right].$

T.14. Show that if the n-vectors $\mathbf{u}$ and $\mathbf{v}$ are viewed as $n \times 1$ matrices, then $\mathbf{u} \cdot \mathbf{v} = \mathbf{u}^T\mathbf{v}$.

MATLAB EXERCISES ■

ML.1. In MATLAB, type the command **clear**, then enter the following matrices:

$$A = \begin{bmatrix} 1 & 1/2 \\ 1/3 & 1/4 \\ 1/5 & 1/6 \end{bmatrix}, \quad B = \begin{bmatrix} 5 & -2 \end{bmatrix},$$

$$C = \begin{bmatrix} 4 & 5/4 & 9/4 \\ 1 & 2 & 3 \end{bmatrix}.$$

Using MATLAB commands, compute each of the following, if possible. Recall that a prime in MATLAB indicates transpose.

(a) $A * C$ (b) $A * B$
(c) $A + C'$ (d) $B * A - C' * A$
(e) $(2 * C - 6 * A') * B'$ (f) $A * C - C * A$
(g) $A * A' + C' * C$

ML.2. Enter the coefficient matrix of the system

$$\begin{aligned} 2x + 4y + 6z &= -12 \\ 2x - 3y - 4z &= 15 \\ 3x + 4y + 5z &= -8 \end{aligned}$$

into MATLAB and call it A. Enter the right-hand side of the system and call it **b**. Form the augmented matrix associated with this linear system using the MATLAB command [**A b**]. To give the augmented matrix a name, such as **aug**, use the command **aug** = [**A b**]. (Do not type the period!) Note that no bar appears between the coefficient matrix and the right-hand side in the MATLAB display.

ML.3. Repeat the preceding exercise with the following linear system:

$$\begin{aligned} 4x - 3y + 2z - w &= -5 \\ 2x + y - 3z &= 7 \\ -x + 4y + z + 2w &= 8. \end{aligned}$$

ML.4. Enter matrices

$$A = \begin{bmatrix} 1 & -1 & 2 \\ 3 & 2 & 4 \\ 4 & -2 & 3 \\ 2 & 1 & 5 \end{bmatrix}$$

and

$$B = \begin{bmatrix} 1 & 0 & -1 & 2 \\ 3 & 3 & -3 & 4 \\ 4 & 2 & 5 & 1 \end{bmatrix}$$

into MATLAB.

(a) Using MATLAB commands, assign $\text{row}_2(A)$ to **R** and $\text{col}_3(B)$ to **C**. Let **V** = **R** * **C**. What is **V** in terms of the entries of the product **A** * **B**?

(b) Using MATLAB commands, assign $\text{col}_2(B)$ to **C**, then compute **V** = **A** * **C**. What is **V** in terms of the entries of the product **A** * **B**?

(c) Using MATLAB commands, assign $\text{row}_3(A)$ to **R**, then compute **V** = **R** * **B**. What is **V** in terms of the entries of the product **A** * **B**?

ML.5. Use the MATLAB command **diag** to form each of the following diagonal matrices. Using **diag** we can form diagonal matrices without typing in all the entries. (To refresh your memory about command **diag**, use MATLAB's help feature.)

(a) The 4×4 diagonal matrix with main diagonal $\begin{bmatrix} 1 & 2 & 3 & 4 \end{bmatrix}$.

(b) The 5×5 diagonal matrix with main diagonal $\begin{bmatrix} 0 & 1 & \frac{1}{2} & \frac{1}{3} & \frac{1}{4} \end{bmatrix}$.

(c) The 5×5 scalar matrix with all 5's on the diagonal.

ML.6. In MATLAB the dot product of a pair of vectors can be computed using the **dot** command. If the vectors **v** and **w** have been entered into MATLAB as either rows or columns, their dot product is computed from the MATLAB command **dot(v, w)**. If the vectors do not have the same number of elements, an error message is displayed.

(a) Use **dot** to compute the dot product of each of the following vectors.

(i) $\mathbf{v} = \begin{bmatrix} 1 & 4 & -1 \end{bmatrix}, \mathbf{w} = \begin{bmatrix} 7 & 2 & 0 \end{bmatrix}.$

(ii) $\mathbf{v} = \begin{bmatrix} 2 \\ -1 \\ 0 \\ 6 \end{bmatrix}, \mathbf{w} = \begin{bmatrix} 4 \\ 2 \\ 3 \\ -1 \end{bmatrix}.$

(b) Let $\mathbf{a} = \begin{bmatrix} 3 & -2 & 1 \end{bmatrix}$. Find a value for k so that the dot product of **a** with $\mathbf{b} = \begin{bmatrix} k & 1 & 4 \end{bmatrix}$ is zero. Verify your results in MATLAB.

(c) For each of the following vectors $\mathbf{v}$, compute $\mathbf{dot(v,v)}$ in MATLAB.

(iii) $\mathbf{v} = \begin{bmatrix} 1 \\ 2 \\ -5 \\ -3 \end{bmatrix}$.

What sign is each of these dot products? Explain why this is true for almost all vectors $\mathbf{v}$. When is it not true?

(i) $\mathbf{v} = \begin{bmatrix} 4 & 2 & -3 \end{bmatrix}$.

(ii) $\mathbf{v} = \begin{bmatrix} -9 & 3 & 1 & 0 & 6 \end{bmatrix}$.

1.4 ▾ Properties of Matrix Operations

In this section we consider the algebraic properties of the matrix operations just defined. Many of these properties are similar to familiar properties of the real numbers. However, there will be striking differences between the set of real numbers and the set of matrices in their algebraic behavior under certain operations, for example, under multiplication (as seen in Section 1.3). Most of the properties will be stated as theorems, whose proofs will be left as exercises.

THEOREM 1.1 ■
(*Properties of Matrix Addition*)

Let A, B, C, and D be $m \times n$ matrices.

(a) $A + B = B + A$.

(b) $A + (B + C) = (A + B) + C$.

(c) *There is a unique $m \times n$ matrix O such that*

$$A + O = A \tag{1}$$

*for any $m \times n$ matrix A. The matrix O is called the $m \times n$ **additive identity** or **zero matrix**.*

(d) *For each $m \times n$ matrix A, there is a unique $m \times n$ matrix D such that*

$$A + D = O. \tag{2}$$

We shall write D as $-A$, so that (2) can be written as

$$A + (-A) = O.$$

*The matrix $-A$ is called the **additive inverse** or **negative** of A.*

Proof

(a) To establish (a), we must prove that the i, jth element of $A + B$ equals the i, jth element of $B + A$. The i, jth element of $A + B$ is $a_{ij} + b_{ij}$; the i, jth element of $B + A$ is $b_{ij} + a_{ij}$. Since the elements a_{ij} are real (or complex) numbers,

$$a_{ij} + b_{ij} = b_{ij} + a_{ij} \qquad (1 \le i \le m, 1 \le j \le n),$$

the result follows.

(b) Exercise T.1.

(c) Let $U = \begin{bmatrix} u_{ij} \end{bmatrix}$. Then

$$A + U = A$$

if and only if*

$$a_{ij} + u_{ij} = a_{ij},$$

which holds if and only if $u_{ij} = 0$. Thus U is the $m \times n$ matrix all of whose entries are zero; U is denoted by O.

(d) Exercise T.1. ∎

EXAMPLE 1 ∎ To illustrate (c) of Theorem 1.1, we note that the 2×2 zero matrix is

$$\begin{bmatrix} 0 & 0 \\ 0 & 0 \end{bmatrix}.$$

If

$$A = \begin{bmatrix} 4 & -1 \\ 2 & 3 \end{bmatrix},$$

we have

$$\begin{bmatrix} 4 & -1 \\ 2 & 3 \end{bmatrix} + \begin{bmatrix} 0 & 0 \\ 0 & 0 \end{bmatrix} = \begin{bmatrix} 4+0 & -1+0 \\ 2+0 & 3+0 \end{bmatrix} = \begin{bmatrix} 4 & -1 \\ 2 & 3 \end{bmatrix}.$$

The 2×3 zero matrix is

$$\begin{bmatrix} 0 & 0 & 0 \\ 0 & 0 & 0 \end{bmatrix}.$$ ∎

EXAMPLE 2 ∎ To illustrate (d) of Theorem 1.1, let

$$A = \begin{bmatrix} 2 & 3 & 4 \\ -4 & 5 & -2 \end{bmatrix}.$$

Then

$$-A = \begin{bmatrix} -2 & -3 & -4 \\ 4 & -5 & 2 \end{bmatrix}.$$

We now have $A + (-A) = O$. ∎

EXAMPLE 3 ∎ Let

$$A = \begin{bmatrix} 3 & -2 & 5 \\ -1 & 2 & 3 \end{bmatrix} \quad \text{and} \quad B = \begin{bmatrix} 2 & 3 & 2 \\ -3 & 4 & 6 \end{bmatrix}.$$

Then

$$A - B = \begin{bmatrix} 3-2 & -2-3 & 5-2 \\ -1+3 & 2-4 & 3-6 \end{bmatrix} = \begin{bmatrix} 1 & -5 & 3 \\ 2 & -2 & -3 \end{bmatrix}.$$ ∎

---

*The connector "if and only if" means that both statements are true or both statements are false. Thus (1) if $A + U = A$, then $a_{ij} + u_{ij} = a_{ij}$ and (2) if $a_{ij} + u_{ij} = a_{ij}$, then $A + U = A$.

THEOREM 1.2 ■ (a) *If A, B, and C are of the appropriate sizes, then*
(*Properties of*
Matrix Multiplication)
$$A(BC) = (AB)C.$$

(b) *If A, B, and C are of the appropriate sizes, then*
$$A(B + C) = AB + AC.$$

(c) *If A, B, and C are of the appropriate sizes, then*
$$(A + B)C = AC + BC.$$

Proof   (a) We omit a general proof here. Exercise T.2 asks the reader to prove the result for a specific case.

(b) Exercise T.3.

(c) Exercise T.3.   ■

EXAMPLE 4 ■ Let
$$A = \begin{bmatrix} 5 & 2 & 3 \\ 2 & -3 & 4 \end{bmatrix}, \quad B = \begin{bmatrix} 2 & -1 & 1 & 0 \\ 0 & 2 & 2 & 2 \\ 3 & 0 & -1 & 3 \end{bmatrix},$$

and
$$C = \begin{bmatrix} 1 & 0 & 2 \\ 2 & -3 & 0 \\ 0 & 0 & 3 \\ 2 & 1 & 0 \end{bmatrix}.$$

Then
$$A(BC) = \begin{bmatrix} 5 & 2 & 3 \\ 2 & -3 & 4 \end{bmatrix} \begin{bmatrix} 0 & 3 & 7 \\ 8 & -4 & 6 \\ 9 & 3 & 3 \end{bmatrix} = \begin{bmatrix} 43 & 16 & 56 \\ 12 & 30 & 8 \end{bmatrix}$$

and
$$(AB)C = \begin{bmatrix} 19 & -1 & 6 & 13 \\ 16 & -8 & -8 & 6 \end{bmatrix} \begin{bmatrix} 1 & 0 & 2 \\ 2 & -3 & 0 \\ 0 & 0 & 3 \\ 2 & 1 & 0 \end{bmatrix} = \begin{bmatrix} 43 & 16 & 56 \\ 12 & 30 & 8 \end{bmatrix}.$$ ■

EXAMPLE 5 ■ Let
$$A = \begin{bmatrix} 2 & 2 & 3 \\ 3 & -1 & 2 \end{bmatrix}, \quad B = \begin{bmatrix} 1 & 0 \\ 2 & 2 \\ 3 & -1 \end{bmatrix}, \quad \text{and} \quad C = \begin{bmatrix} -1 & 2 \\ 1 & 0 \\ 2 & -2 \end{bmatrix}.$$

Then
$$A(B + C) = \begin{bmatrix} 2 & 2 & 3 \\ 3 & -1 & 2 \end{bmatrix} \begin{bmatrix} 0 & 2 \\ 3 & 2 \\ 5 & -3 \end{bmatrix} = \begin{bmatrix} 21 & -1 \\ 7 & -2 \end{bmatrix}$$

and

$$AB + AC = \begin{bmatrix} 15 & 1 \\ 7 & -4 \end{bmatrix} + \begin{bmatrix} 6 & -2 \\ 0 & 2 \end{bmatrix} = \begin{bmatrix} 21 & -1 \\ 7 & -2 \end{bmatrix}.$$

■

The $n \times n$ scalar matrix

$$I_n = \begin{bmatrix} 1 & 0 & \cdots & 0 \\ 0 & 1 & \cdots & 0 \\ \vdots & \vdots & & \vdots \\ 0 & 0 & \cdots & 1 \end{bmatrix},$$

all of whose diagonal entries are 1, is called the **identity matrix of order n**. If A is an $m \times n$ matrix, then it is easy to verify (Exercise T.4) that

$$I_m A = A I_n = A.$$

It is also easy to see that every $n \times n$ scalar matrix can be written as $r I_n$ for some r.

EXAMPLE 6 ■ The identity matrix I_2 of order 2 is

$$I_2 = \begin{bmatrix} 1 & 0 \\ 0 & 1 \end{bmatrix}.$$

If

$$A = \begin{bmatrix} 4 & -2 & 3 \\ 5 & 0 & 2 \end{bmatrix},$$

then

$$I_2 A = A.$$

The identity matrix I_3 of order 3 is

$$I_3 = \begin{bmatrix} 1 & 0 & 0 \\ 0 & 1 & 0 \\ 0 & 0 & 1 \end{bmatrix}.$$

Hence

$$A I_3 = A.$$

■

Suppose that A is a square matrix. If p is a positive integer, then we define

$$A^p = \underbrace{A \cdot A \cdots A}_{p \text{ factors}}.$$

If A is $n \times n$, we also define

$$A^0 = I_n.$$

For nonnegative integers p and q, some of the familiar laws of exponents for the real numbers can also be proved for matrix multiplication of a square matrix A (Exercise T.5):

$$A^p A^q = A^{p+q}$$

and

$$(A^p)^q = A^{pq}.$$

It should be noted that

$$(AB)^p \neq A^p B^p$$

for square matrices in general. However, if $AB = BA$, then this rule does hold (Exercise T.6).

We now note two other peculiarities of matrix multiplication. If a and b are real numbers, then $ab = 0$ can hold only if a or b is zero. However, this is not true for matrices.

EXAMPLE 7 ■ If

$$A = \begin{bmatrix} 1 & 2 \\ 2 & 4 \end{bmatrix} \quad \text{and} \quad B = \begin{bmatrix} 4 & -6 \\ -2 & 3 \end{bmatrix},$$

then neither A nor B is the zero matrix, but

$$AB = \begin{bmatrix} 0 & 0 \\ 0 & 0 \end{bmatrix}.$$

■

If a, b, and c are real numbers for which $ab = ac$ and $a \neq 0$, it then follows that $b = c$. That is, we can cancel a out. However, the cancellation law does not hold for matrices, as the following example shows.

EXAMPLE 8 ■ If

$$A = \begin{bmatrix} 1 & 2 \\ 2 & 4 \end{bmatrix}, \quad B = \begin{bmatrix} 2 & 1 \\ 3 & 2 \end{bmatrix}, \quad \text{and} \quad C = \begin{bmatrix} -2 & 7 \\ 5 & -1 \end{bmatrix},$$

then

$$AB = AC = \begin{bmatrix} 8 & 5 \\ 16 & 10 \end{bmatrix},$$

but $B \neq C$.

■

EXAMPLE 9 ■ Suppose that only two rival companies, R and S, manufacture a certain product.
(*Business*)   Each year, company R keeps $\frac{1}{4}$ of its customers while $\frac{3}{4}$ switch to S. Each year, S keeps $\frac{2}{3}$ of its customers while $\frac{1}{3}$ switch to R. This information can be displayed in matrix form as

$$
\begin{array}{cc}
 & \begin{array}{cc} R & S \end{array} \\
A = & \begin{bmatrix} \frac{1}{4} & \frac{1}{3} \\ \frac{3}{4} & \frac{2}{3} \end{bmatrix} \begin{array}{c} R \\ S \end{array}
\end{array}
$$

When manufacture of the product first starts, R has $\frac{3}{5}$ of the market (the market is the total number of customers) while S has $\frac{2}{5}$ of the market. We denote the initial distribution of the market by

$$\mathbf{x}_0 = \begin{bmatrix} \frac{3}{5} \\ \frac{2}{5} \end{bmatrix}.$$

One year later, the distribution of the market is

$$\mathbf{x}_1 = A\mathbf{x}_0 = \begin{bmatrix} \frac{1}{4} & \frac{1}{3} \\ \frac{3}{4} & \frac{2}{3} \end{bmatrix} \begin{bmatrix} \frac{3}{5} \\ \frac{2}{5} \end{bmatrix} = \begin{bmatrix} \frac{1}{4}\left(\frac{3}{5}\right) + \frac{1}{3}\left(\frac{2}{5}\right) \\ \frac{3}{4}\left(\frac{3}{5}\right) + \frac{2}{3}\left(\frac{2}{5}\right) \end{bmatrix} = \begin{bmatrix} \frac{17}{60} \\ \frac{43}{60} \end{bmatrix}.$$

This can be readily seen as follows. Suppose that the initial market consists of k people, say $k = 12{,}000$, and no change in this number occurs with time. Then, initially, R has $\frac{3}{5}k$ customers, and S has $\frac{2}{5}k$ customers. At the end of the first year, R keeps $\frac{1}{4}$ of its customers and gains $\frac{1}{3}$ of S's customers. Thus R has

$$\tfrac{1}{4}\left(\tfrac{3}{5}k\right) + \tfrac{1}{3}\left(\tfrac{2}{5}k\right) = \left[\tfrac{1}{4}\left(\tfrac{3}{5}\right) + \tfrac{1}{3}\left(\tfrac{2}{5}\right)\right]k = \tfrac{17}{60}k \text{ customers.}$$

When $k = 12{,}000$, R has $\frac{17}{60}(12{,}000) = 3400$ customers. Similarly, at the end of the first year, S keeps $\frac{2}{3}$ of its customers and gains $\frac{3}{4}$ of R's customers. Thus S has

$$\tfrac{3}{4}\left(\tfrac{3}{5}k\right) + \tfrac{2}{3}\left(\tfrac{2}{5}k\right) = \left[\tfrac{3}{4}\left(\tfrac{3}{5}\right) + \tfrac{2}{3}\left(\tfrac{2}{5}\right)\right]k = \tfrac{43}{60}k \text{ customers.}$$

When $k = 12{,}000$, S has $\frac{43}{60}(12{,}000) = 8600$ customers. Similarly, at the end of 2 years, the distribution of the market will be given by

$$\mathbf{x}_2 = A\mathbf{x}_1 = A(A\mathbf{x}_0) = A^2\mathbf{x}_0.$$

If

$$\mathbf{x}_0 = \begin{bmatrix} a \\ b \end{bmatrix},$$

can we determine a and b so that the distribution will be the same from year to year? When this happens, the distribution of the market is said to be **stable**. We proceed as follows. Since R and S control the entire market, we must have

$$a + b = 1. \tag{3}$$

We also want the distribution after 1 year to be unchanged. Hence

$$A\mathbf{x}_0 = \mathbf{x}_0$$

or

$$\begin{bmatrix} \frac{1}{4} & \frac{1}{3} \\ \frac{3}{4} & \frac{2}{3} \end{bmatrix} \begin{bmatrix} a \\ b \end{bmatrix} = \begin{bmatrix} a \\ b \end{bmatrix}.$$

Then

$$\tfrac{1}{4}a + \tfrac{1}{3}b = a$$
$$\tfrac{3}{4}a + \tfrac{2}{3}b = b$$

or

$$-\tfrac{3}{4}a + \tfrac{1}{3}b = 0$$
$$\tfrac{3}{4}a - \tfrac{1}{3}b = 0. \tag{4}$$

Observe that the two equations in (4) are the same. Using Equation (3) and one of the equations in (4), we find (verify) that

$$a = \tfrac{4}{13} \quad \text{and} \quad b = \tfrac{9}{13}. \qquad \blacksquare$$

The problem just described is an example of a **Markov chain**. We shall return to this topic in Section 8.3.

THEOREM 1.3 ■ *If r and s are real numbers and A and B are matrices, then*
(*Properties of Scalar Multiplication*)

(a) $r(sA) = (rs)A$.

(b) $(r + s)A = rA + sA$.

(c) $r(A + B) = rA + rB$.

(d) $A(rB) = r(AB) = (rA)B$.

Proof   Exercise T.12.   ■

EXAMPLE 10 ■ Let $r = -2$,

$$A = \begin{bmatrix} 1 & 2 & 3 \\ -2 & 0 & 1 \end{bmatrix}, \quad \text{and} \quad B = \begin{bmatrix} 2 & -1 \\ 1 & 4 \\ 0 & -2 \end{bmatrix}.$$

Then

$$A(rB) = \begin{bmatrix} 1 & 2 & 3 \\ -2 & 0 & 1 \end{bmatrix} \begin{bmatrix} -4 & 2 \\ -2 & -8 \\ 0 & 4 \end{bmatrix} = \begin{bmatrix} -8 & -2 \\ 8 & 0 \end{bmatrix}$$

and

$$r(AB) = (-2) \begin{bmatrix} 4 & 1 \\ -4 & 0 \end{bmatrix} = \begin{bmatrix} -8 & -2 \\ 8 & 0 \end{bmatrix},$$

which illustrates (d) of Theorem 1.3.   ■

It is easy to show that $(-1)A = -A$ (Exercise T.13).

THEOREM 1.4 ■ *If r is a scalar and A and B are matrices, then*
(*Properties of Transpose*)

(a) $(A^T)^T = A$.

(b) $(A + B)^T = A^T + B^T$.

(c) $(AB)^T = B^T A^T$.

(d) $(rA)^T = rA^T$.

Proof   We leave the proofs of (a), (b), and (d) as an exercise (Exercise T.14) and prove only (c) here. Thus let $A = \begin{bmatrix} a_{ij} \end{bmatrix}$ be $m \times p$ and let $B = \begin{bmatrix} b_{ij} \end{bmatrix}$ be $p \times n$. The i, jth element of $(AB)^T$ is c_{ij}^T. Now

$$c_{ij}^T = c_{ji} = \text{row}_j(A) \cdot \text{col}_i(B)$$

$$= a_{j1}b_{1i} + a_{j2}b_{2i} + \cdots + a_{jp}b_{pi}$$

$$= a_{1j}^T b_{i1}^T + a_{2j}^T b_{i2}^T + \cdots + a_{pj}^T b_{ip}^T$$

$$= b_{i1}^T a_{1j}^T + b_{i2}^T a_{2j}^T + \cdots + b_{ip}^T a_{pj}^T$$

$$= \text{row}_i(B^T) \cdot \text{col}_j(A^T),$$

which is the i, jth element of $B^T A^T$.   ■

EXAMPLE 11 ■ Let

$$A = \begin{bmatrix} 1 & 3 & 2 \\ 2 & -1 & 3 \end{bmatrix} \quad \text{and} \quad B = \begin{bmatrix} 0 & 1 \\ 2 & 2 \\ 3 & -1 \end{bmatrix}.$$

Then

$$(AB)^T = \begin{bmatrix} 12 & 7 \\ 5 & -3 \end{bmatrix}$$

and

$$B^T A^T = \begin{bmatrix} 0 & 2 & 3 \\ 1 & 2 & -1 \end{bmatrix} \begin{bmatrix} 1 & 2 \\ 3 & -1 \\ 2 & 3 \end{bmatrix} = \begin{bmatrix} 12 & 7 \\ 5 & -3 \end{bmatrix}.$$

■

DEFINITION   A matrix $A = \begin{bmatrix} a_{ij} \end{bmatrix}$ is called **symmetric** if

$$A^T = A.$$

That is, A is symmetric if it is a square matrix for which

$$a_{ij} = a_{ji} \quad \text{(Exercise T.17)}.$$

If matrix A is symmetric, then the elements of A are symmetric with respect to the main diagonal of A.

EXAMPLE 12 ■ The matrices

$$A = \begin{bmatrix} 1 & 2 & 3 \\ 2 & 4 & 5 \\ 3 & 5 & 6 \end{bmatrix} \quad \text{and} \quad I_3 = \begin{bmatrix} 1 & 0 & 0 \\ 0 & 1 & 0 \\ 0 & 0 & 1 \end{bmatrix}$$

are symmetric.   ■

Section 8.1, which can be covered at this time, uses material from this section.

▼ *Preview of an Application*

GRAPH THEORY (SECTION 8.1)

In recent years, the need to solve problems dealing with communication among individuals, computers, and organizations has grown at an unprecedented rate. As an example, note the explosive growth of the Internet and the promises of using it to interact with all types of media. Graph Theory is an area of applied mathematics that deals with problems such as this one:

Consider a local area network consisting of six users denoted by $P_1, P_2, \ldots,$ P_6. We say that P_i has "access" to P_j if P_i can directly send a message to P_j. On the other hand, P_i may not be able to send a message directly to P_k, but can send it to P_j, who will then send it to P_k. In this way we say that P_i has "2-stage access" to P_k. In a similar way, we speak of "r-stage access." We may describe the access relation in the network shown in Figure A by defining the 6×6 matrix $A = \begin{bmatrix} a_{ij} \end{bmatrix}$, where $a_{ij} = 1$ if P_i has access to P_j and 0 otherwise. Thus A may be

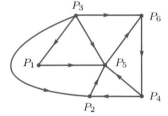

$$A = \begin{array}{c c} & \begin{array}{c c c c c c} P_1 & P_2 & P_3 & P_4 & P_5 & P_6 \end{array} \\ \begin{array}{c} P_1 \\ P_2 \\ P_3 \\ P_4 \\ P_5 \\ P_6 \end{array} & \left[\begin{array}{c c c c c c} 0 & 0 & 0 & 0 & 1 & 0 \\ 0 & 0 & 0 & 0 & 1 & 0 \\ 1 & 1 & 0 & 0 & 1 & 1 \\ 0 & 1 & 0 & 0 & 1 & 0 \\ 0 & 0 & 0 & 0 & 0 & 1 \\ 0 & 0 & 0 & 1 & 0 & 0 \end{array} \right] \end{array}.$$

Figure A

Using the matrix A and the techniques from Graph Theory discussed in Section 8.1, we can determine the number of ways that P_i has r-stage access to P_k, where $r = 1, 2, \ldots$. Many other problems involving communications can be solved using Graph Theory.

1.4 EXERCISES

1. Verify Theorem 1.1 for

$$A = \begin{bmatrix} 1 & 2 & -2 \\ 3 & 4 & 5 \end{bmatrix}, \quad B = \begin{bmatrix} 2 & 0 & 1 \\ 3 & -2 & 5 \end{bmatrix},$$

and

$$C = \begin{bmatrix} -4 & -6 & 1 \\ 2 & 3 & 0 \end{bmatrix}.$$

2. Verify (a) of Theorem 1.2 for

$$A = \begin{bmatrix} 1 & 3 \\ 2 & -1 \end{bmatrix}, \quad B = \begin{bmatrix} -1 & 3 & 2 \\ 1 & -3 & 4 \end{bmatrix},$$

and

$$C = \begin{bmatrix} 1 & 0 \\ 3 & -1 \\ 1 & 2 \end{bmatrix}.$$

3. Verify (b) of Theorem 1.2 for

$$A = \begin{bmatrix} 1 & -3 \\ -3 & 4 \end{bmatrix}, \quad B = \begin{bmatrix} 2 & -3 & 2 \\ 3 & -1 & -2 \end{bmatrix},$$

and

$$C = \begin{bmatrix} 0 & 1 & 2 \\ 1 & 3 & -2 \end{bmatrix}.$$

4. Verify (a), (b), and (c) of Theorem 1.3 for $r = 6$, $s = -2$, and

$$A = \begin{bmatrix} 4 & 2 \\ 1 & -3 \end{bmatrix}, \quad B = \begin{bmatrix} 0 & 2 \\ -4 & 3 \end{bmatrix}.$$

5. Verify (d) of Theorem 1.3 for $r = -3$ and

$$A = \begin{bmatrix} 1 & 3 \\ 2 & -1 \end{bmatrix}, \quad B = \begin{bmatrix} -1 & 3 & 2 \\ 1 & -3 & 4 \end{bmatrix}.$$

6. Verify (b) and (d) of Theorem 1.4 for $r = -4$ and

$$A = \begin{bmatrix} 1 & 3 & 2 \\ 2 & 1 & -3 \end{bmatrix}, \quad B = \begin{bmatrix} 4 & 2 & -1 \\ -2 & 1 & 5 \end{bmatrix}.$$

7. Verify (c) of Theorem 1.4 for

$$A = \begin{bmatrix} 1 & 3 & 2 \\ 2 & 1 & -3 \end{bmatrix}, \quad B = \begin{bmatrix} 3 & -1 \\ 2 & 4 \\ 1 & 2 \end{bmatrix}.$$

In Exercises 8 and 9, let

$$A = \begin{bmatrix} 2 & 1 & -2 \\ 3 & 2 & 5 \end{bmatrix}, \quad B = \begin{bmatrix} 2 & -1 \\ 3 & 4 \\ 1 & -2 \end{bmatrix},$$

$$C = \begin{bmatrix} 2 & 1 & 3 \\ -1 & 2 & 4 \\ 3 & 1 & 0 \end{bmatrix}, \quad D = \begin{bmatrix} 2 & -1 \\ -3 & 2 \end{bmatrix},$$

$$E = \begin{bmatrix} 1 & 1 & 2 \\ 2 & -1 & 3 \\ -3 & 2 & -1 \end{bmatrix}, \quad \text{and} \quad F = \begin{bmatrix} 1 & 0 \\ 2 & -3 \end{bmatrix}.$$

8. If possible, compute:
  (a) $(AB)^T$.
  (b) $B^T A^T$.
  (c) $A^T B^T$.
  (d) BB^T.
  (e) $B^T B$.

9. If possible, compute:
  (a) $(3C - 2E)^T B$.
  (b) $A^T (D + F)$.
  (c) $B^T C + A$.
  (d) $(2E) A^T$.
  (e) $(B^T + A) C$.

10. If

$$A = \begin{bmatrix} -2 & 3 \\ 2 & -3 \end{bmatrix} \quad \text{and} \quad B = \begin{bmatrix} 3 & 6 \\ 2 & 4 \end{bmatrix},$$

show that $AB = O$.

11. If

$$A = \begin{bmatrix} -2 & 3 \\ 2 & -3 \end{bmatrix}, \quad B = \begin{bmatrix} -1 & 3 \\ 2 & 0 \end{bmatrix},$$

and

$$C = \begin{bmatrix} -4 & -3 \\ 0 & -4 \end{bmatrix},$$

show that $AB = AC$.

12. If $A = \begin{bmatrix} 0 & 1 \\ 1 & 0 \end{bmatrix}$, show that $A^2 = I_2$.

13. Let $A = \begin{bmatrix} 4 & 2 \\ 1 & 3 \end{bmatrix}$. Find
  (a) $A^2 + 3A$.
  (b) $2A^3 + 3A^2 + 4A + 5I_2$.

14. Let $A = \begin{bmatrix} 1 & -1 \\ 2 & 3 \end{bmatrix}$. Find
  (a) $A^2 - 2A$.
  (b) $3A^3 - 2A^2 + 5A - 4I_2$.

15. Determine a scalar r such that $A\mathbf{x} = r\mathbf{x}$, where
$$A = \begin{bmatrix} 2 & 1 \\ 1 & 2 \end{bmatrix} \quad \text{and} \quad \mathbf{x} = \begin{bmatrix} 1 \\ 1 \end{bmatrix}.$$

16. Determine a constant k such that $(kA)^T(kA) = 1$, where
$$A = \begin{bmatrix} -2 \\ 1 \\ -1 \end{bmatrix}.$$
Is there more than one value of k that could be used?

17. Let
$$A = \begin{bmatrix} -3 & 2 & 1 \\ 4 & 5 & 0 \end{bmatrix}$$
and $\mathbf{a}_j = \text{col}_j(A)$, $j = 1, 2, 3$. Verify that
$$A^T A = \begin{bmatrix} \mathbf{a}_1 \cdot \mathbf{a}_1 & \mathbf{a}_1 \cdot \mathbf{a}_2 & \mathbf{a}_1 \cdot \mathbf{a}_3 \\ \mathbf{a}_2 \cdot \mathbf{a}_1 & \mathbf{a}_2 \cdot \mathbf{a}_2 & \mathbf{a}_2 \cdot \mathbf{a}_3 \\ \mathbf{a}_3 \cdot \mathbf{a}_1 & \mathbf{a}_3 \cdot \mathbf{a}_2 & \mathbf{a}_3 \cdot \mathbf{a}_3 \end{bmatrix}$$
$$= \begin{bmatrix} \mathbf{a}_1^T \mathbf{a}_1 & \mathbf{a}_1^T \mathbf{a}_2 & \mathbf{a}_1^T \mathbf{a}_3 \\ \mathbf{a}_2^T \mathbf{a}_1 & \mathbf{a}_2^T \mathbf{a}_2 & \mathbf{a}_2^T \mathbf{a}_3 \\ \mathbf{a}_3^T \mathbf{a}_1 & \mathbf{a}_3^T \mathbf{a}_2 & \mathbf{a}_3^T \mathbf{a}_3 \end{bmatrix}.$$

18. Suppose that the matrix A in Example 9 is
$$A = \begin{bmatrix} \frac{1}{3} & \frac{2}{5} \\ \frac{2}{3} & \frac{3}{5} \end{bmatrix} \quad \text{and} \quad \mathbf{x}_0 = \begin{bmatrix} \frac{2}{3} \\ \frac{1}{3} \end{bmatrix}.$$

(a) Find the distribution of the market after 1 year.

(b) Find the stable distribution of the market.

19. Consider two quick food companies, M and N. Each year, company M keeps $\frac{1}{3}$ of its customers, while $\frac{2}{3}$ switch to N. Each year, N keeps $\frac{1}{2}$ of its customers, while $\frac{1}{2}$ switch to M. Suppose that the initial distribution of the market is given by
$$\mathbf{x}_0 = \begin{bmatrix} \frac{1}{3} \\ \frac{2}{3} \end{bmatrix}.$$

(a) Find the distribution of the market after 1 year.

(b) Find the stable distribution of the market.

THEORETICAL EXERCISES

T.1. Prove properties (b) and (d) of Theorem 1.1.

T.2. If $A = \begin{bmatrix} a_{ij} \end{bmatrix}$ is a 2×3 matrix, $B = \begin{bmatrix} b_{ij} \end{bmatrix}$ is a 3×4 matrix, and $C = \begin{bmatrix} c_{ij} \end{bmatrix}$ is a 4×3 matrix, show that $A(BC) = (AB)C$.

T.3. Prove properties (b) and (c) of Theorem 1.2.

T.4. If A is an $m \times n$ matrix, show that
$$I_m A = A I_n = A.$$

T.5. Let p and q be nonnegative integers and let A be a square matrix. Show that
$$A^p A^q = A^{p+q} \quad \text{and} \quad (A^p)^q = A^{pq}.$$

T.6. If $AB = BA$, and p is a nonnegative integer, show that
$$(AB)^p = A^p B^p.$$

T.7. Show that if A and B are $n \times n$ diagonal matrices, then $AB = BA$.

T.8. Find a matrix $B \neq O_2$ and $B \neq I_2$ such that $AB = BA$, where
$$A = \begin{bmatrix} 1 & 2 \\ 2 & 1 \end{bmatrix}.$$
How many such matrices B are there?

T.9. Find a matrix $B \neq O_2$ and $B \neq I_2$ such that $AB = BA$, where
$$A = \begin{bmatrix} 1 & 2 \\ 0 & 1 \end{bmatrix}.$$
How many such matrices B are there?

T.10. Let $A = \begin{bmatrix} \cos\theta & \sin\theta \\ -\sin\theta & \cos\theta \end{bmatrix}$.

(a) Determine a simple expression for A^2.

(b) Determine a simple expression for A^3.

(c) Conjecture the form of a simple expression for A^k, k a positive integer.

(d) Prove or disprove your conjecture in part (c).

T.11. If p is a nonnegative integer and c is a scalar, show that
$$(cA)^p = c^p A^p.$$

T.12. Prove Theorem 1.3.

T.13. Show that $(-1)A = -A$.

T.14. Complete the proof of Theorem 1.4.

T.15. Show that $(A - B)^T = A^T - B^T$.

T.16. (a) Show that $(A^2)^T = (A^T)^2$.

(b) Show that $(A^3)^T = (A^T)^3$.

(c) Prove or disprove that, for $k = 4, 5, \ldots$, $(A^k)^T = (A^T)^k$.

T.17. Show that a square matrix A is symmetric if and only if $a_{ij} = a_{ji}$ for all i, j.

T.18. Show that if A is symmetric, then A^T is symmetric.

T.19. Let A be an $n \times n$ matrix. Show that if $A\mathbf{x} = \mathbf{0}$ for all $n \times 1$ matrices $\mathbf{x}$, then $A = O$.

T.20. Let A be an $n \times n$ matrix. Show that if $A\mathbf{x} = \mathbf{x}$ for all $n \times 1$ matrices $\mathbf{x}$, then $A = I_n$.

T.21. Show that if $AA^T = O$, then $A = O$.

T.22. Show that if A is a symmetric matrix, then A^k, $k = 2, 3, \ldots$, is symmetric.

T.23. Let A and B be symmetric matrices.

(a) Show that $A + B$ is symmetric.

(b) Show that AB is symmetric if and only if $AB = BA$.

T.24. A matrix $A = \begin{bmatrix} a_{ij} \end{bmatrix}$ is called **skew symmetric** if $A^T = -A$. Show that A is skew symmetric if and only if $a_{ij} = -a_{ji}$ for all i, j.

T.25. Describe all skew symmetric scalar matrices. (See Section 1.2 for the definition of scalar matrix.)

T.26. If A is an $n \times n$ matrix, show that AA^T and $A^T A$ are symmetric.

T.27. If A is an $n \times n$ matrix, show that

(a) $A + A^T$ is symmetric.

(b) $A - A^T$ is skew symmetric.

T.28. Show that if A is an $n \times n$ matrix, then A can be written uniquely as $A = S + K$, where S is symmetric and K is skew symmetric.

T.29. Show that if A is an $n \times n$ scalar matrix, then $A = rI_n$ for some real number r.

T.30. Show that $I_n^T = I_n$.

T.31. Let A be an $m \times n$ matrix. Show that if $rA = O$, then $r = 0$ or $A = O$.

T.32. Show that if $A\mathbf{x} = \mathbf{b}$ is a linear system that has more than one solution, then it has infinitely many solutions. (*Hint*: If $\mathbf{u}_1$ and $\mathbf{u}_2$ are solutions, consider $\mathbf{w} = r\mathbf{u}_1 + s\mathbf{u}_2$, where $r + s = 1$.)

T.33. Determine all 2×2 matrices A such that $AB = BA$ for any 2×2 matrix B.

MATLAB EXERCISES ▪

In order to use MATLAB in this section, you should first have read Chapter 10 through Section 10.3.

ML.1. Use MATLAB to find the smallest positive integer k in each of the following cases. (See also Exercise 12.)

(a) $A^k = I_3$ for $A = \begin{bmatrix} 0 & 0 & 1 \\ 1 & 0 & 0 \\ 0 & 1 & 0 \end{bmatrix}$.

(b) $A^k = A$ for $A = \begin{bmatrix} 0 & 1 & 0 & 0 \\ -1 & 0 & 0 & 0 \\ 0 & 0 & 0 & 1 \\ 0 & 0 & 1 & 0 \end{bmatrix}$.

ML.2. Use MATLAB to display the matrix A in each of the following cases. Find the smallest value of k such that A^k is a zero matrix. Here **tril**, **ones**, **triu**, **fix**, and **rand** are MATLAB commands. (To see a description, use **help**.)

(a) $A = \mathbf{tril}(\mathbf{ones}(5), -1)$

(b) $A = \mathbf{triu}(\mathbf{fix}(10 * \mathbf{rand}(7)), 2)$

ML.3. Let $A = \begin{bmatrix} 1 & -1 & 0 \\ 0 & 1 & -1 \\ -1 & 0 & 1 \end{bmatrix}$. Using command **polyvalm** in MATLAB, compute the following matrix polynomials:

(a) $A^4 - A^3 + A^2 + 2I_3$.

(b) $A^3 - 3A^2 + 3A$.

ML.4. Let $A = \begin{bmatrix} 0.1 & 0.3 & 0.6 \\ 0.2 & 0.2 & 0.6 \\ 0.3 & 0.3 & 0.4 \end{bmatrix}$. Using MATLAB, compute each of the following matrix expressions:

(a) $(A^2 - 7A)(A + 3I_3)$.

(b) $(A - I_3)^2 + (A^3 + A)$.

(c) Look at the sequence $A, A^2, A^3, \ldots, A^8$, $\ldots$. Does it appear to be converging to a matrix? If so, to what matrix?

ML.5. Let $A = \begin{bmatrix} 1 & \frac{1}{2} \\ 0 & \frac{1}{3} \end{bmatrix}$. Use MATLAB to compute members of the sequence A, A^2, A^3, ..., A^k, ... Write a description of the behavior of this matrix sequence.

ML.6. Let $A = \begin{bmatrix} \frac{1}{2} & \frac{1}{3} \\ 0 & -\frac{1}{5} \end{bmatrix}$. Repeat Exercise ML.5.

ML.7. Let $A = \begin{bmatrix} 1 & -2 & 1 \\ -1 & 1 & 2 \\ 0 & 2 & 1 \end{bmatrix}$. Use MATLAB to do the following:

(a) Compute $A^T A$ and AA^T. Are they equal?

(b) Compute $B = A + A^T$ and $C = A - A^T$. Show that B is symmetric and C is skew symmetric. (See Exercise T.24.)

(c) Determine a relationship between $B + C$ and A.

1.5 ▼ Solutions of Linear Systems of Equations

In this section we shall systematize the familiar method of elimination of unknowns (discussed in Section 1.1) for the solution of linear systems and thus obtain a useful method for solving such systems. This method starts with the augmented matrix of the given linear system and obtains a matrix of a certain form. This new matrix represents a linear system that has exactly the same solutions as the given system. For example, if

$$\begin{bmatrix} 1 & 0 & 0 & 2 & \vdots & 4 \\ 0 & 1 & 0 & -1 & \vdots & -5 \\ 0 & 0 & 1 & 3 & \vdots & 6 \end{bmatrix}$$

represents the augmented matrix of a linear system, then the solution is easily found from the corresponding equations

$$\begin{aligned} x_1 + \quad\quad\quad 2x_4 &= \ \ 4 \\ x_2 \quad - \ \ x_4 &= -5 \\ x_3 + 3x_4 &= \ \ 6. \end{aligned}$$

The task of this section is to manipulate the augmented matrix representing a given linear system into a form from which the solution can easily be found.

DEFINITION   An $m \times n$ matrix is said to be in **reduced row echelon form** when it satisfies the following properties:

(a) All rows consisting entirely of zeros, if any, are at the bottom of the matrix.

(b) Reading from left to right, the first nonzero entry in each row that does not consist entirely of zeros is a 1, called the **leading entry** of its row.

(c) If rows i and $i + 1$ are two successive rows that do not consist entirely of zeros, then the leading entry of row $i + 1$ is to the right of the leading entry of row i.

(d) If a column contains a leading entry of some row, then all other entries in that column are zero.

(Note that a matrix in reduced row echelon form might not have any rows that consist entirely of zeros.)

EXAMPLE 1   ■   The matrices

$$A = \begin{bmatrix} 1 & 0 & 0 & 4 \\ 0 & 1 & 0 & 5 \\ 0 & 0 & 1 & 2 \end{bmatrix}, \quad B = \begin{bmatrix} 1 & 2 & 0 & 0 & 2 \\ 0 & 0 & 1 & 0 & 1 \\ 0 & 0 & 0 & 1 & 0 \end{bmatrix},$$

and

$$C = \begin{bmatrix} 1 & 0 & 0 & 3 & 0 \\ 0 & 0 & 1 & 0 & 0 \\ 0 & 0 & 0 & 0 & 1 \\ 0 & 0 & 0 & 0 & 0 \\ 0 & 0 & 0 & 0 & 0 \end{bmatrix}$$

are in reduced row echelon form.   ■

EXAMPLE 2   ■   The matrices

$$A = \begin{bmatrix} 1 & 2 & 0 & 4 \\ 0 & 0 & 0 & 0 \\ 0 & 0 & 1 & -3 \end{bmatrix}, \quad B = \begin{bmatrix} 1 & 0 & 3 & 4 \\ 0 & 2 & -2 & 5 \\ 0 & 0 & 1 & 2 \end{bmatrix},$$

$$C = \begin{bmatrix} 1 & 0 & 3 & 4 \\ 0 & 1 & -2 & 5 \\ 0 & 1 & 2 & 2 \\ 0 & 0 & 0 & 0 \end{bmatrix}, \quad \text{and} \quad D = \begin{bmatrix} 1 & 2 & 3 & 4 \\ 0 & 1 & -2 & 5 \\ 0 & 0 & 1 & 2 \\ 0 & 0 & 0 & 0 \end{bmatrix}$$

are not in reduced row echelon form, since they fail to satisfy (a), (b), (c), and (d), respectively.   ■

Observe that when a matrix is in reduced row echelon form, the leading entries of its nonzero rows form a staircase ("echelon") pattern.

We shall now turn to the discussion of how to transform a given matrix to a matrix in reduced row echelon form.

DEFINITION

An **elementary row operation** on an $m \times n$ matrix $A = \begin{bmatrix} a_{ij} \end{bmatrix}$ is any of the following operations:

(a) Interchange rows r and s of A. That is, replace $a_{r1}, a_{r2}, \ldots, a_{rn}$ by $a_{s1}, a_{s2}, \ldots, a_{sn}$ and $a_{s1}, a_{s2}, \ldots, a_{sn}$ by $a_{r1}, a_{r2}, \ldots, a_{rn}$.

(b) Multiply row r of A by $c \neq 0$. That is, replace $a_{r1}, a_{r2}, \ldots, a_{rn}$ by $ca_{r1}, ca_{r2}, \ldots, ca_{rn}$.

(c) Add d times row r of A to row s of A, $r \neq s$. That is, replace $a_{s1}, a_{s2}, \ldots, a_{sn}$ by $a_{s1} + da_{r1}, a_{s2} + da_{r2}, \ldots, a_{sn} + da_{rn}$.

Observe that when a matrix is viewed as the augmented matrix of a linear system, the elementary row operations are equivalent, respectively, to interchanging two equations, multiplying an equation by a nonzero constant, and adding a multiple of one equation to another equation.

EXAMPLE 3 ■ Let

$$A = \begin{bmatrix} 0 & 0 & 1 & 2 \\ 2 & 3 & 0 & -2 \\ 3 & 3 & 6 & -9 \end{bmatrix}.$$

Interchanging rows 1 and 3 of A, we obtain

$$B = \begin{bmatrix} 3 & 3 & 6 & -9 \\ 2 & 3 & 0 & -2 \\ 0 & 0 & 1 & 2 \end{bmatrix}.$$

Multiplying the third row of A by $\frac{1}{3}$, we obtain

$$C = \begin{bmatrix} 0 & 0 & 1 & 2 \\ 2 & 3 & 0 & -2 \\ 1 & 1 & 2 & -3 \end{bmatrix}.$$

Adding (-2) times row 2 of A to row 3 of A, we obtain

$$D = \begin{bmatrix} 0 & 0 & 1 & 2 \\ 2 & 3 & 0 & -2 \\ -1 & -3 & 6 & -5 \end{bmatrix}.$$

■

DEFINITION An $m \times n$ matrix A is said to be **row equivalent** to an $m \times n$ matrix B if B can be obtained by applying a finite sequence of elementary row operations to A.

EXAMPLE 4 ■ The matrix

$$A = \begin{bmatrix} 1 & 2 & 4 & 3 \\ 2 & 1 & 3 & 2 \\ 1 & -1 & 2 & 3 \end{bmatrix}$$

is row equivalent to

$$D = \begin{bmatrix} 2 & 4 & 8 & 6 \\ 1 & -1 & 2 & 3 \\ 4 & -1 & 7 & 8 \end{bmatrix},$$

because if we add 2 times row 3 of A to its second row, we obtain

$$B = \begin{bmatrix} 1 & 2 & 4 & 3 \\ 4 & -1 & 7 & 8 \\ 1 & -1 & 2 & 3 \end{bmatrix}.$$

Interchanging rows 2 and 3 of B, we obtain

$$C = \begin{bmatrix} 1 & 2 & 4 & 3 \\ 1 & -1 & 2 & 3 \\ 4 & -1 & 7 & 8 \end{bmatrix}.$$

Multiplying row 1 of C by 2, we obtain D.   ■

It is easy to show (Exercise T.2) that (1) every matrix is row equivalent to itself; (2) if A is row equivalent to B, then B is row equivalent to A; and (3) if A is row equivalent to B and B is row equivalent to C, then A is row equivalent to C. In view of (2), both statements, "A is row equivalent to B" and "B is row equivalent to A," can be replaced by "A and B are row equivalent."

THEOREM 1.5   ■   *Every nonzero $m \times n$ matrix is row equivalent to a unique matrix in reduced row echelon form.*
                                                                                                                      ■

We shall illustrate the proof of the theorem by giving the steps that must be carried out on a specific matrix A to obtain a matrix in reduced row echelon form that is row equivalent to A. We omit the proof that the matrix thus obtained is unique. We use the following example to illustrate the steps involved.

EXAMPLE 5   ■   Let

$$A = \begin{bmatrix} 0 & 2 & 3 & -4 & 1 \\ 0 & 0 & 2 & 3 & 4 \\ 2 & 2 & -5 & 2 & 4 \\ 2 & 0 & -6 & 9 & 7 \end{bmatrix}.$$

The procedure for transforming a matrix to reduced row echelon form follows.

Procedure

Example

Step 1. Find the first (counting from left to right) column in A not all of whose entries are zero. This column is called the **pivotal column**.

$$A = \begin{bmatrix} 0 & 2 & 3 & -4 & 1 \\ 0 & 0 & 2 & 3 & 4 \\ 2 & 2 & -5 & 2 & 4 \\ 2 & 0 & -6 & 9 & 7 \end{bmatrix}$$

Pivotal column of A

Step 2. Identify the first (counting from top to bottom) nonzero entry in the pivotal column. This element is called the **pivot**, which we circle in A.

$$A = \begin{bmatrix} 0 & 2 & 3 & -4 & 1 \\ 0 & 0 & 2 & 3 & 4 \\ ② & 2 & -5 & 2 & 4 \\ 2 & 0 & -6 & 9 & 7 \end{bmatrix}$$

Pivot

Step 3. Interchange, if necessary, the first row with the row where the pivot occurs so that the pivot is now in the first row. Call the new matrix A_1.

$$A_1 = \begin{bmatrix} ② & 2 & -5 & 2 & 4 \\ 0 & 0 & 2 & 3 & 4 \\ 0 & 2 & 3 & -4 & 1 \\ 2 & 0 & -6 & 9 & 7 \end{bmatrix}$$

The first and third rows of A were interchanged.

Step 4. Multiply the first row of A_1 by the reciprocal of the pivot. Thus the entry in the first row and pivotal column (where the pivot was located) is now a 1. Call the new matrix A_2.

$$A_2 = \begin{bmatrix} 1 & 1 & -\frac{5}{2} & 1 & 2 \\ 0 & 0 & 2 & 3 & 4 \\ 0 & 2 & 3 & -4 & 1 \\ 2 & 0 & -6 & 9 & 7 \end{bmatrix}$$

The first row of A_1 was multiplied by $\frac{1}{2}$.

Step 5. Add multiples of the first row of A_2 to all other rows to make all entries in the pivotal column, except the entry where the pivot was located, equal to zero. Thus all entries in the pivotal column and rows 2, 3, ..., m are zero. Call the new matrix A_3.

$$A_3 = \begin{bmatrix} 1 & 1 & -\frac{5}{2} & 1 & 2 \\ 0 & 0 & 2 & 3 & 4 \\ 0 & 2 & 3 & -4 & 1 \\ 0 & -2 & -1 & 7 & 3 \end{bmatrix}$$

(-2) times the first row of A_2 was added to its fourth row.

Step 6. Identify B as the $(m-1) \times n$ submatrix of A_3 obtained by deleting the first row of A_3; do not erase the first row of A_3. Repeat steps 1 through 5 on B.

$$B = \begin{bmatrix} 1 & 1 & -\frac{5}{2} & 1 & 2 \\ 0 & 0 & 2 & 3 & 4 \\ 0 & ② & 3 & -4 & 1 \\ 0 & -2 & -1 & 7 & 3 \end{bmatrix}$$

Pivotal column of B ⟶ | **Pivot**

$$B_1 = \begin{bmatrix} 1 & 1 & -\frac{5}{2} & 1 & 2 \\ 0 & ② & 3 & -4 & 1 \\ 0 & 0 & 2 & 3 & 4 \\ 0 & -2 & -1 & 7 & 3 \end{bmatrix}$$

The first and second rows of B were interchanged.

$$B_2 = \begin{bmatrix} 1 & 1 & -\frac{5}{2} & 1 & 2 \\ 0 & 1 & \frac{3}{2} & -2 & \frac{1}{2} \\ 0 & 0 & 2 & 3 & 4 \\ 0 & -2 & -1 & 7 & 3 \end{bmatrix}$$

The first row of B_1 was multiplied by $\frac{1}{2}$.

$$B_3 = \begin{array}{ccccc} 1 & 1 & -\frac{5}{2} & 1 & 2 \\ \begin{bmatrix} 0 & 1 & \frac{3}{2} & -2 & \frac{1}{2} \\ 0 & 0 & 2 & 3 & 4 \\ 0 & 0 & 2 & 3 & 4 \end{bmatrix} \end{array}$$

2 times the first row of B_2 was added to its third row.

Step 7. Identify C as the $(m-2) \times n$ submatrix of B_3 obtained by deleting the first row of B_3; do not erase the first row of B_3. Repeat Steps 1–5 on C.

$$C = \begin{array}{ccccc} 1 & 1 & -\frac{5}{2} & 1 & 2 \\ 0 & 1 & \frac{3}{2} & -2 & \frac{1}{2} \\ \begin{bmatrix} 0 & 0 & ② & 3 & 4 \\ 0 & 0 & 2 & 3 & 4 \end{bmatrix} \end{array}$$

Pivotal column of C ——— **Pivot**

$$C_1 = C_2 = \begin{array}{ccccc} 1 & 1 & -\frac{5}{2} & 1 & 2 \\ 0 & 1 & \frac{3}{2} & -2 & \frac{1}{2} \\ \begin{bmatrix} 0 & 0 & 1 & \frac{3}{2} & 2 \\ 0 & 0 & 2 & 3 & 4 \end{bmatrix} \end{array}$$

No rows of C had to be interchanged. The first row of C was multiplied by $\frac{1}{2}$.

$$C_3 = \begin{array}{ccccc} 1 & 1 & -\frac{5}{2} & 1 & 2 \\ 0 & 1 & \frac{3}{2} & -2 & \frac{1}{2} \\ \begin{bmatrix} 0 & 0 & 1 & \frac{3}{2} & 2 \\ 0 & 0 & 0 & 0 & 0 \end{bmatrix} \end{array}$$

(-2) times the first row of C_2 was added to its second row.

Step 8. Let D be the matrix consisting of the shaded rows above matrix C_3 in Step 7 followed by C_3. Add multiples of each row of D having a leading 1 to zero out all entries above the leading 1. Denote the subsequent matrices as D_1, D_2, and so on.

$$D = \begin{bmatrix} 1 & 1 & -\frac{5}{2} & 1 & 2 \\ 0 & 1 & \frac{3}{2} & -2 & \frac{1}{2} \\ 0 & 0 & 1 & \frac{3}{2} & 2 \\ 0 & 0 & 0 & 0 & 0 \end{bmatrix}$$

$$D_1 = \begin{bmatrix} 1 & 1 & -\frac{5}{2} & 1 & 2 \\ 0 & 1 & 0 & -\frac{17}{4} & -\frac{5}{2} \\ 0 & 0 & 1 & \frac{3}{2} & 2 \\ 0 & 0 & 0 & 0 & 0 \end{bmatrix}$$

$\left(-\frac{3}{2}\right)$ times the third row of D was added to its second row.

$$D_2 = \begin{bmatrix} 1 & 1 & 0 & \frac{19}{4} & 7 \\ 0 & 1 & 0 & -\frac{17}{4} & -\frac{5}{2} \\ 0 & 0 & 1 & \frac{3}{2} & 2 \\ 0 & 0 & 0 & 0 & 0 \end{bmatrix}$$

$\frac{5}{2}$ times the third row of D_1 was added to its first row.

$$D_3 = \begin{bmatrix} 1 & 0 & 0 & 9 & \frac{19}{2} \\ 0 & 1 & 0 & -\frac{17}{4} & -\frac{5}{2} \\ 0 & 0 & 1 & \frac{3}{2} & 2 \\ 0 & 0 & 0 & 0 & 0 \end{bmatrix}$$

(-1) times the second row of D_2 was added to its first row.

The final matrix D_3 is in reduced row echelon form.   ■

The process just illustrated is used as many times as necessary until the matrix in reduced row echelon form is obtained.

REMARK   The procedure given here for finding a matrix in reduced row echelon form that is row equivalent to a given matrix is not the only one possible. As an alternate procedure, we could first zero out the entries below a leading 1 and then immediately zero out the entries above the leading 1. This procedure is not as efficient as the one described above. In actual practice, we do not take the time to identify the matrices $A_1, A_2, \ldots, B_1, B_2, \ldots, C_1, C_2, \ldots$, and so on. We merely start with the given matrix and transform it to reduced row echelon form.

Solving Linear Systems

We now apply these results to the solution of linear systems.

THEOREM 1.6 ■ *Let $A\mathbf{x} = \mathbf{b}$ and $C\mathbf{x} = \mathbf{d}$ be two linear systems each of m equations in n unknowns. If the augmented matrices $[A \mid \mathbf{b}]$ and $[C \mid \mathbf{d}]$ of these systems are row equivalent, then both linear systems have exactly the same solutions.*

Proof   This follows from the definition of row equivalence and from the fact that the three elementary row operations on the augmented matrix turn out to be the three manipulations on a linear system, discussed in Section 1.1, yielding a linear system having the same solutions as the given system. We also note that if one system has no solution, then the other system has no solution.   ■

COROLLARY 1.1 ■ *If A and C are row equivalent $m \times n$ matrices, then the linear system $A\mathbf{x} = \mathbf{0}$ and $C\mathbf{x} = \mathbf{0}$ have exactly the same solutions.*

Proof   Exercise T.3.   ■

The **Gauss–Jordan reduction** procedure for solving the linear system $A\mathbf{x} = \mathbf{b}$ is as follows.

Step 1. Form the augmented matrix $\left[A \mid \mathbf{b} \right]$.

Step 2. Transform the augmented matrix to reduced row echelon form by using elementary row operations.

Step 3. The linear system that corresponds to the matrix in reduced row echelon form that has been obtained in Step 2 has exactly the same solutions as the given linear system. For each nonzero row of the matrix in reduced row echelon form, solve the corresponding equation for the unknown that corresponds to the leading entry of the row. The rows consisting entirely of zeros can be ignored, since the corresponding equation will be satisfied for any values of the unknowns.

The following examples illustrate the Gauss*–Jordan** reduction procedure.

EXAMPLE 6 ■ Solve the linear system

$$
\begin{aligned}
x + 2y + 3z &= 9 \\
2x - y + z &= 8 \\
3x - z &= 3
\end{aligned}
\tag{1}
$$

by Gauss–Jordan reduction.

---

*Carl Friedrich Gauss (1777–1855) was born into a poor working-class family in Brunswick, Germany, and died in Göttingen, Germany, the most famous mathematician in the world. He was a child prodigy with a genius that did not impress his father, who called him a "stargazer." However, his teachers were impressed enough to arrange for the Duke of Brunswick to provide a scholarship for Gauss at the local secondary school. As a teenager there, he made original discoveries in number theory and began to speculate about non-Euclidean geometry. His scientific publications include important contributions in number theory, mathematical astronomy, mathematical geography, statistics, differential geometry, and magnetism. His diaries and private notes contain many other discoveries that he never published.

An austere, conservative man who had few friends and whose private life was generally unhappy, he was very concerned that proper credit be given for scientific discoveries. When he relied on the results of others, he was careful to acknowledge them; and when others independently discovered results in his private notes, he was quick to claim priority.

In his research he used a method of calculation that later generations generalized to row reduction of matrices and named in his honor although the method was used in China almost 2000 years earlier.

**Wilhelm Jordan (1842–1899) was born in southern Germany. He attended college in Stuttgart and in 1868 became full professor of geodesy at the technical college in Karlsruhe, Germany. He participated in surveying several regions of Germany. Jordan was a prolific writer whose major work, *Handbuch der Vermessungskunde* (*Handbook of Geodesy*), was translated into French, Italian, and Russian. He was considered a superb writer and an excellent teacher. Unfortunately, the Gauss–Jordan reduction method has been widely attributed to Camille Jordan (1838–1922), a well-known French mathematician. Moreover, it seems that the method was also discovered independently at the same time by B. I. Clasen, a priest who lived in Luxembourg. This biographical sketch is based on an excellent article: S. C. Althoen and R. McLaughlin, "Gauss–Jordan reduction: A brief history," *MAA Monthly*, 94 (1987), 130–142.

Solution   **Step 1.**   The augmented matrix of this linear system is

$$\begin{bmatrix} 1 & 2 & 3 & \vdots & 9 \\ 2 & -1 & 1 & \vdots & 8 \\ 3 & 0 & -1 & \vdots & 3 \end{bmatrix}.$$

Step 2.   We now transform the matrix in Step 1 to reduced row echelon form as follows:

$$\begin{bmatrix} 1 & 2 & 3 & \vdots & 9 \\ 2 & -1 & 1 & \vdots & 8 \\ 3 & 0 & -1 & \vdots & 3 \end{bmatrix}$$

$$\begin{bmatrix} 1 & 2 & 3 & \vdots & 9 \\ 0 & -5 & -5 & \vdots & -10 \\ 0 & -6 & -10 & \vdots & -24 \end{bmatrix}$$

(-2) times the first row was added to its second row.
(-3) times the first row was added to its third row.

$$\begin{bmatrix} 1 & 2 & 3 & \vdots & 9 \\ 0 & 1 & 1 & \vdots & 2 \\ 0 & -6 & -10 & \vdots & -24 \end{bmatrix}$$

The second row was multiplied by $\left(-\frac{1}{5}\right)$.

$$\begin{bmatrix} 1 & 2 & 3 & \vdots & 9 \\ 0 & 1 & 1 & \vdots & 2 \\ 0 & 0 & -4 & \vdots & -12 \end{bmatrix}$$

6 times the second row was added to its third row.

$$\begin{bmatrix} 1 & 2 & 3 & \vdots & 9 \\ 0 & 1 & 1 & \vdots & 2 \\ 0 & 0 & 1 & \vdots & 3 \end{bmatrix}$$

The third row was multiplied by $\left(-\frac{1}{4}\right)$.

$$\begin{bmatrix} 1 & 0 & 1 & \vdots & 5 \\ 0 & 1 & 1 & \vdots & 2 \\ 0 & 0 & 1 & \vdots & 3 \end{bmatrix}$$

(-2) times the second row was added to its first row.

$$\begin{bmatrix} 1 & 0 & 0 & \vdots & 2 \\ 0 & 1 & 0 & \vdots & -1 \\ 0 & 0 & 1 & \vdots & 3 \end{bmatrix}$$

(-1) times the third row was added to its second row and to its first row.

Thus, the augmented matrix is row equivalent to the matrix

$$\begin{bmatrix} 1 & 0 & 0 & \vdots & 2 \\ 0 & 1 & 0 & \vdots & -1 \\ 0 & 0 & 1 & \vdots & 3 \end{bmatrix} \tag{2}$$

in reduced row echelon form.

Step 3.   The linear system represented by (2) is

$$\begin{aligned} x & = 2 \\ y & = -1 \\ z & = 3 \end{aligned}$$

so that the unique solution to the given linear system (1) is

$$x = 2$$
$$y = -1$$
$$z = 3.$$

∎

EXAMPLE 7 ∎ Solve the linear system

$$
\begin{aligned}
x + y + 2z - 5w &= 3 \\
2x + 5y - z - 9w &= -3 \\
2x + y - z + 3w &= -11 \\
x - 3y + 2z + 7w &= -5
\end{aligned}
\tag{3}
$$

by Gauss–Jordan reduction.

Solution **Step 1.** The augmented matrix of this linear system is

$$
\begin{bmatrix}
1 & 1 & 2 & -5 & \vdots & 3 \\
2 & 5 & -1 & -9 & \vdots & -3 \\
2 & 1 & -1 & 3 & \vdots & -11 \\
1 & -3 & 2 & 7 & \vdots & -5
\end{bmatrix}.
$$

Step 2. The augmented matrix is row equivalent to the matrix (verify)

$$
\begin{bmatrix}
1 & 0 & 0 & 2 & \vdots & -5 \\
0 & 1 & 0 & -3 & \vdots & 2 \\
0 & 0 & 1 & -2 & \vdots & 3 \\
0 & 0 & 0 & 0 & \vdots & 0
\end{bmatrix}
\tag{4}
$$

in reduced row echelon form.

Step 3. The linear system represented by (4) is

$$
\begin{aligned}
x \phantom{{}+y} + 2w &= -5 \\
y - 3w &= 2 \\
z - 2w &= 3.
\end{aligned}
$$

The row in (4) consisting entirely of zeros has been ignored.

Solving each equation for the unknown that corresponds to the leading entry in each row of (4), we obtain

$$
\begin{aligned}
x &= -5 - 2w \\
y &= 2 + 3w \\
z &= 3 + 2w.
\end{aligned}
$$

Thus, a solution to the linear system (3) is

$$
\begin{aligned}
x &= -5 - 2r \\
y &= 2 + 3r \\
z &= 3 + 2r \\
w &= r,
\end{aligned}
\tag{5}
$$

where r is any real number. Since r can be assigned any real number in (5), the given linear system (3) has infinitely many solutions. ∎

EXAMPLE 8 ■ Solve the linear system

$$
\begin{aligned}
x_1 + 2x_2 &\quad -3x_4 + \ x_5 &\quad = 2 \\
x_1 + 2x_2 + x_3 - 3x_4 + \ x_5 + 2x_6 &= 3 \\
x_1 + 2x_2 &\quad -3x_4 + 2x_5 + \ x_6 &= 4 \\
3x_1 + 6x_2 + x_3 - 9x_4 + 4x_5 + 3x_6 &= 9
\end{aligned}
\tag{6}
$$

by Gauss–Jordan reduction.

Solution   **Step 1.**   The augmented matrix of this linear system is

$$
\left[\begin{array}{cccccc|c}
1 & 2 & 0 & -3 & 1 & 0 & 2 \\
1 & 2 & 1 & -3 & 1 & 2 & 3 \\
1 & 2 & 0 & -3 & 2 & 1 & 4 \\
3 & 6 & 1 & -9 & 4 & 3 & 9
\end{array}\right].
$$

Step 2.   The augmented matrix is row equivalent to the matrix (verify)

$$
\left[\begin{array}{cccccc|c}
1 & 2 & 0 & -3 & 0 & -1 & 0 \\
0 & 0 & 1 & 0 & 0 & 2 & 1 \\
0 & 0 & 0 & 0 & 1 & 1 & 2 \\
0 & 0 & 0 & 0 & 0 & 0 & 0
\end{array}\right].
\tag{7}
$$

Step 3.   The linear system represented by (7) is

$$
\begin{aligned}
x_1 + 2x_2 &\quad -3x_4 &\quad - \ x_6 &= 0 \\
&\quad x_3 &\quad + 2x_6 &= 1 \\
&\quad &\quad x_5 + \ x_6 &= 2.
\end{aligned}
$$

Solving each equation for the unknown that corresponds to the leading entry in each row of (7), we obtain

$$
\begin{aligned}
x_1 &= r + 3s - 2t \\
x_2 &= t \\
x_3 &= 1 - 2r \\
x_4 &= s \\
x_5 &= 2 - r \\
x_6 &= r,
\end{aligned}
\tag{8}
$$

where r, s, and t are any real numbers. Thus (8) is the solution to the given linear system (6). Since r, s, and t can be assigned any real numbers, the given linear system (6) has infinitely many solutions.   ■

EXAMPLE 9 ■ Solve the linear system

$$
\begin{aligned}
x + 2y + 3z + 4w &= \ \ 5 \\
x + 3y + 5z + 7w &= 11 \\
x \quad\quad - \ z - 2w &= -6
\end{aligned}
\tag{9}
$$

by Gauss–Jordan reduction.

Solution **Step 1.** The augmented matrix of this linear system is

$$\begin{bmatrix} 1 & 2 & 3 & 4 & \vdots & 5 \\ 1 & 3 & 5 & 7 & \vdots & 11 \\ 1 & 0 & -1 & -2 & \vdots & -6 \end{bmatrix}.$$

Step 2. The augmented matrix is row equivalent to the matrix (verify)

$$\begin{bmatrix} 1 & 0 & -1 & -2 & \vdots & 0 \\ 0 & 1 & 2 & 3 & \vdots & 0 \\ 0 & 0 & 0 & 0 & \vdots & 1 \end{bmatrix}. \tag{10}$$

Step 3. The last equation of the linear system represented by (10) is

$$0x + 0y + 0z + 0w = 1,$$

which has no solution for any x, y, z, and w. Consequently, the given linear system (9) has no solution. ∎

The last example is characteristic of the way in which a linear system has no solution. That is, a linear system $A\mathbf{x} = \mathbf{b}$ in n unknowns has no solution if and only if its augmented matrix is row equivalent to a matrix in reduced row echelon form, which has a row whose first n elements are zero and whose $(n + 1)$st element is 1 (Exercise T.4).

The linear systems of Examples 6, 7, and 8 each had at least one solution, while the system in Example 9 had no solution. Linear systems with at least one solution are called **consistent**, and linear systems with no solutions are called **inconsistent**. Every inconsistent linear system results in the situation illustrated in Example 9.

If a matrix satisfies the first three properties in the definition of a matrix in reduced row echelon form, it is said to be in **row echelon form**. A method called **Gaussian elimination**, which is used to solve the linear system $A\mathbf{x} = \mathbf{b}$, transforms the augmented matrix $\begin{bmatrix} A & \vdots & \mathbf{b} \end{bmatrix}$ to row echelon form and then uses **back substitution** to obtain the solution. This procedure is discussed in Section 9.2. Strictly speaking, the original Gauss–Jordan reduction procedure was more along the lines described in the Remark following Example 5 for transforming a matrix to reduced row echelon form. The version presented in this book is more efficient. As presented here, Gaussian elimination and Gauss–Jordan reduction are equally efficient. In actual practice, neither of these methods is used as much as the technique involving the LU-factorization of the coefficient matrix that is discussed in Section 9.3. However, Gaussian elimination and Gauss–Jordan reduction are fine for small problems and we use the latter heavily in this book.

Homogeneous Systems

A linear system of the form

$$a_{11}x_1 + a_{12}x_2 + \cdots + a_{1n}x_n = 0$$
$$a_{21}x_1 + a_{22}x_2 + \cdots + a_{2n}x_n = 0$$
$$\vdots \qquad \vdots \qquad \qquad \vdots \qquad \vdots \tag{11}$$
$$a_{m1}x_1 + a_{m2}x_2 + \cdots + a_{mn}x_n = 0$$

is called a **homogeneous system**. We can also write (11) in matrix form as

$$A\mathbf{x} = \mathbf{0}. \tag{12}$$

The solution

$$x_1 = x_2 = \cdots = x_n = 0$$

to the homogeneous system (12) is called the **trivial solution**. A solution $x_1, x_2, \ldots, x_n$ to a homogeneous system in which not all the x_i are zero is called a **nontrivial solution**. We see that a homogeneous system is always consistent, since it always has the trivial solution.

EXAMPLE 10 ■ Consider the homogeneous system

$$x + 2y + 3z = 0$$
$$-x + 3y + 2z = 0 \tag{13}$$
$$2x + \ y - 2z = 0.$$

The augmented matrix of this system,

$$\begin{bmatrix} 1 & 2 & 3 & \vdots & 0 \\ -1 & 3 & 2 & \vdots & 0 \\ 2 & 1 & -2 & \vdots & 0 \end{bmatrix},$$

is row equivalent to (verify)

$$\begin{bmatrix} 1 & 0 & 0 & \vdots & 0 \\ 0 & 1 & 0 & \vdots & 0 \\ 0 & 0 & 1 & \vdots & 0 \end{bmatrix},$$

which is in reduced row echelon form. Hence the solution to (13) is

$$x = y = z = 0,$$

which means that the given homogeneous system (13) has only the trivial solution.

■

EXAMPLE 11 ■ Consider the homogeneous system

$$x + \ y + z + w = 0$$
$$x \qquad \quad + w = 0 \tag{14}$$
$$x + 2y + z \qquad = 0.$$

The augmented matrix of this system,

$$\begin{bmatrix} 1 & 1 & 1 & 1 & \vdots & 0 \\ 1 & 0 & 0 & 1 & \vdots & 0 \\ 1 & 2 & 1 & 0 & \vdots & 0 \end{bmatrix},$$

is row equivalent to (verify)

$$\begin{bmatrix} 1 & 0 & 0 & 1 & \vdots & 0 \\ 0 & 1 & 0 & -1 & \vdots & 0 \\ 0 & 0 & 1 & 1 & \vdots & 0 \end{bmatrix},$$

which is in reduced row echelon form. Hence the solution to (14) is

$$\begin{aligned} x &= -r \\ y &= r \\ z &= -r \\ w &= r, \end{aligned}$$

where r is any real number. ■

Example 11 shows that a homogeneous system may have a nontrivial solution. The following theorem tells of one case when this occurs.

THEOREM 1.7 ■ *A homogeneous system of m equations in n unknowns always has a nontrivial solution if $m < n$, that is, if the number of unknowns exceeds the number of equations.*

Proof   Let C be the matrix in reduced row echelon form that is row equivalent to A. Then the homogeneous systems $A\mathbf{x} = \mathbf{0}$ and $C\mathbf{x} = \mathbf{0}$ are equivalent. If we let r be the number of nonzero rows of C, then $r \leq m$. If $m < n$, we conclude that $r < n$. We are then solving r equations in n unknowns and can solve for r unknowns in terms of the remaining $n - r$ unknowns, the latter being free to take on any values we please. Thus, by letting one of these $n - r$ unknowns be nonzero, we obtain a nontrivial solution to $C\mathbf{x} = \mathbf{0}$ and thus to $A\mathbf{x} = \mathbf{0}$. ■

We shall also use Theorem 1.7 in the following equivalent form: If A is $m \times n$ and $A\mathbf{x} = \mathbf{0}$ has only the trivial solution, then $m \geq n$.

The following result is important in the study of differential equations.

Let $A\mathbf{x} = \mathbf{b}$, $\mathbf{b} \neq \mathbf{0}$, be a consistent linear system. If $\mathbf{x}_p$ is a particular solution to the given nonhomogeneous system and $\mathbf{x}_h$ is a solution to the associated homogeneous system $A\mathbf{x} = \mathbf{0}$, then $\mathbf{x}_p + \mathbf{x}_h$ is a solution to the given system $A\mathbf{x} = \mathbf{b}$. Moreover, every solution $\mathbf{x}$ to the nonhomogeneous linear system $A\mathbf{x} = \mathbf{b}$ can be written as $\mathbf{x}_p + \mathbf{x}_h$, where $\mathbf{x}_p$ is a particular solution to the given nonhomogeneous system and $\mathbf{x}_h$ is a solution to the associated homogeneous system $A\mathbf{x} = \mathbf{0}$. For a proof, see Exercise T.13.

Chapter 7, Linear Programming, Section 8.2, Electrical Circuits, and Section 8.3, Markov Chains, which can be studied at this time, use material from this section.

▼ *Preview of an Application*

LINEAR PROGRAMMING (CHAPTER 7)

A problem that is typical in a manufacturing process is the following one:

A coffee packer uses Columbian and Kenyan coffee to prepare a regular blend and a deluxe blend. Each pound of regular blend consists of $\frac{1}{2}$ pound of Columbian coffee and $\frac{1}{2}$ pound of Kenyan coffee. Each pound of deluxe blend consists of $\frac{1}{4}$ pound of Columbian coffee and $\frac{3}{4}$ pound of Kenyan coffee. The packer will make a $2 profit on each pound of regular blend and a $3 profit on each pound of deluxe blend. If there are 100 pounds of Columbian coffee and 120 pounds of Kenyan coffee, how many pounds of each blend should be packed to make the largest possible profit?

We first translate this problem into mathematical form by letting x denote the number of pounds of regular blend and y the number of pounds of deluxe blend to be packed. Then our problem can be stated as follows:

Find values of x and y that will make the expression

$$z = 2x + 3y$$

as large as possible and satisfy the following restrictions:

$$\tfrac{1}{2}x + \tfrac{1}{4}y \leq 100$$

$$\tfrac{1}{2}x + \tfrac{3}{4}y \leq 120$$

$$x \geq 0$$

$$y \geq 0.$$

This problem can be readily solved by the techniques of linear programming, a recent area of applied mathematics, which is discussed in Chapter 7.

▼ *Preview of an Application*

ELECTRICAL CIRCUITS (SECTION 8.2)

An electrical circuit is a closed connection of batteries, resistors (such as light-bulbs) and wires connecting these. The batteries and resistors are denoted on paper by

Batteries   Resistors

An example of an electrical circuit is shown in Figure A.

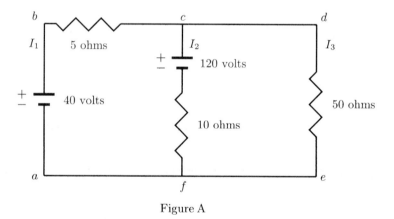

Figure A

For this circuit, the unknown currents I_1, I_2, and I_3 (in units of amps) are to be determined from the values of resistance (in units of ohms) across each resistor and electrostatic potential (in units of volts) across each battery (as shown in Figure A). Applying two basic laws of physics, to be discussed in Section 8.2, we find that I_1, I_2, and I_3 must satisfy the linear system

$$\begin{bmatrix} 1 & 1 & -1 \\ 1 & -2 & 0 \\ 0 & 1 & 5 \end{bmatrix} \begin{bmatrix} I_1 \\ I_2 \\ I_3 \end{bmatrix} = \begin{bmatrix} 0 \\ -16 \\ 12 \end{bmatrix} .$$

Section 8.2 gives a brief introduction to these types of electrical circuits.

▼ *Preview of an Application*

MARKOV CHAINS (SECTION 8.3)

Consider the following problem: A city that has just introduced a new public transportation system has predicted that each year 35% of those presently using public transportation to go to work will go back to driving their car, while 65% will stay with public transportation. They also expect that 45% of those presently driving to work will switch to public transportation, while 55% will continue to drive. Thus the probability that someone presently using public transportation will go back to driving is 0.35. We can display the expected behavior of the commuting population in terms of probabilities by the matrix

$$
A = \begin{array}{c} \textit{Mode of} \\ \textit{transportation} \\ \textit{this year} \end{array}
\begin{array}{c} \\ \text{Public transportation} \\ \text{Automobile} \end{array}
\begin{array}{c} \textit{Mode of transportation next year} \\ \begin{array}{cc} \text{Public} \\ \text{transportation} & \text{Automobile} \end{array} \\ \left[\begin{array}{cc} 0.65 & 0.35 \\ 0.45 & 0.55 \end{array} \right] \end{array}
$$

When the system becomes operational, 15% of the commuters use public transportation, while 85% drive to work. Assuming that the city's population remains constant for a long time, the management of the public transit system would like answers to the following questions:

▶ What is the percentage of commuters using each mode of transportation after, say, three years?

▶ What is the percentage of commuters using each mode of transportation in the long run?

This type of problem and the one described in Example 9 of Section 1.4 are Markov chains. The techniques discussed in Section 8.3 enable us to solve these and many other applied problems.

Polynomial Interpolation

Suppose we are given the n distinct points $(x_1, y_1), (x_2, y_2), \ldots, (x_n, y_n)$. Can we find a polynomial of degree $n-1$ or less that "interpolates" the data, that is, passes through the n points? Thus, the polynomial we seek has the form

$$y = a_{n-1}x^{n-1} + a_{n-2}x^{n-2} + \cdots + a_1 x + a_0.$$

The n given points can be used to obtain an $n \times n$ linear system whose unknowns are $a_0, a_1, \ldots, a_{n-1}$. It can be shown that this linear system has a unique solution. Thus, there is a unique interpolating polynomial.

We consider the case where $n = 3$ in detail. Here we are given the points (x_1, y_1), (x_2, y_2), (x_3, y_3), where $x_1 \neq x_2$, $x_1 \neq x_3$, and $x_2 \neq x_3$, and seek the polynomial

$$y = a_2 x^2 + a_1 x + a_0. \tag{15}$$

Substituting the given points in (15), we obtain the linear system

$$\begin{aligned} a_2 x_1^2 + a_1 x_1 + a_0 &= y_1 \\ a_2 x_2^2 + a_1 x_2 + a_0 &= y_2 \\ a_2 x_3^2 + a_1 x_3 + a_0 &= y_3. \end{aligned} \tag{16}$$

We show in Section 2.2 that the linear system (16) has a unique solution. Thus, there is a unique interpolating quadratic polynomial. In general, there is a unique interpolating polynomial of degree $n-1$ passing through n given points.

EXAMPLE 12 ■ Find the quadratic polynomial that interpolates the points $(1, 3)$, $(2, 4)$, $(3, 7)$.

Solution   Setting up the linear system (16) we have

$$\begin{aligned} a_2 + a_1 + a_0 &= 3 \\ 4a_2 + 2a_1 + a_0 &= 4 \\ 9a_2 + 3a_1 + a_0 &= 7 \end{aligned}$$

whose solution is (verify)

$$a_2 = 1, \qquad a_1 = -2, \qquad a_0 = 4.$$

Hence, the quadratic interpolating polynomial is

$$y = x^2 - 2x + 4.$$

■

1.5   EXERCISES

1. Which of the following matrices are in reduced row echelon form?

$$A = \begin{bmatrix} 1 & 0 & 0 & 0 & -3 \\ 0 & 0 & 1 & 0 & 4 \\ 0 & 0 & 0 & 1 & 2 \end{bmatrix}, \quad \text{Yes}$$

$$B = \begin{bmatrix} 0 & 1 & 0 & 0 & 5 \\ 0 & 0 & 1 & 0 & -4 \\ 0 & 0 & 0 & -1 & 3 \end{bmatrix}, \quad \text{No}$$

$$C = \begin{bmatrix} 0 & 1 & 0 & 0 & 5 \\ 0 & 0 & 1 & 0 & 4 \\ 0 & 1 & 0 & -2 & 3 \end{bmatrix}, \quad \text{No}$$

$$D = \begin{bmatrix} 0 & 1 & 0 & 0 & 2 \\ 0 & 0 & 0 & 0 & -1 \\ 0 & 0 & 0 & 1 & 4 \\ 0 & 0 & 0 & 0 & 0 \\ 0 & 0 & 0 & 0 & 1 \end{bmatrix}, \quad \text{No}$$

$$E = \begin{bmatrix} 1 & 0 & 0 & 0 & 2 \\ 0 & 0 & 1 & 0 & 0 \\ 0 & 0 & 0 & 1 & 3 \\ 0 & 0 & 0 & 0 & 0 \end{bmatrix}, \quad \text{Yes}$$

$$F = \begin{bmatrix} 0 & 0 & 0 & 0 & 0 \\ 0 & 0 & 1 & 2 & -3 \\ 0 & 0 & 0 & 1 & 0 \\ 0 & 0 & 0 & 0 & 0 \end{bmatrix}, \quad \text{No}$$

$$G = \begin{bmatrix} 1 & 0 & 0 & 0 & 1 \\ 0 & 1 & 0 & 0 & 2 \\ 0 & 0 & 0 & 1 & -1 \\ 0 & 0 & 0 & 0 & 0 \end{bmatrix}, \quad \text{Yes}$$

$$H = \begin{bmatrix} 1 & 0 & 0 & 1 \\ 0 & 1 & 0 & 2 \\ 0 & 0 & 0 & -1 \\ 0 & 0 & 0 & 0 \end{bmatrix} \quad \text{No}$$

2. Let

$$A = \begin{bmatrix} 1 & 0 & 3 \\ -3 & 1 & 4 \\ 4 & 2 & 2 \\ 5 & -1 & 5 \end{bmatrix}.$$

Find the matrices obtained by performing the following elementary row operations on A.

(a) Interchanging the second and fourth rows.

(b) Multiplying the third row by 3.

(c) Adding (-3) times the first row to the fourth row.

3. Let

$$A = \begin{bmatrix} 2 & 0 & 4 & 2 \\ 3 & -2 & 5 & 6 \\ -1 & 3 & 1 & 1 \end{bmatrix}.$$

Find the matrices obtained by performing the following elementary row operations on A.

(a) Interchanging the second and third rows.

(b) Multiplying the second row by (-4).

(c) Adding 2 times the third row to the first row.

4. Find three matrices that are row equivalent to

$$A = \begin{bmatrix} 2 & -1 & 3 & 4 \\ 0 & 1 & 2 & -1 \\ 5 & 2 & -3 & 4 \end{bmatrix}.$$

5. Find three matrices that are row equivalent to

$$\begin{bmatrix} 4 & 3 & 7 & 5 \\ -1 & 2 & -1 & 3 \\ 2 & 0 & 1 & 4 \end{bmatrix}.$$

6. If

$$A = \begin{bmatrix} 0 & 0 & -1 & 2 & 3 \\ 0 & 2 & 3 & 4 & 5 \\ 0 & 1 & 3 & -1 & 2 \\ 0 & 3 & 2 & 4 & 1 \end{bmatrix},$$

find a matrix C in reduced row echelon form that is row equivalent to A.

7. If

$$A = \begin{bmatrix} 1 & -2 & 0 & 2 \\ 2 & -3 & -1 & 5 \\ 1 & 3 & 2 & 5 \\ 1 & 1 & 0 & 2 \end{bmatrix},$$

find a matrix C in reduced row echelon form that is row equivalent to A.

In Exercises 8 through 10, find all solutions to the given linear system.

8. (a) $\begin{aligned} x + y + 2z &= -1 \\ x - 2y + z &= -5 \\ 3x + y + z &= 3. \end{aligned}$

(b) $\begin{aligned} x + y + 3z + 2w &= 7 \\ 2x - y + 4w &= 8 \\ 3y + 6z &= 8. \end{aligned}$

(c) $\begin{aligned} x + 2y - 4z &= 3 \\ x - 2y + 3z &= -1 \\ 2x + 3y - z &= 5 \\ 4x + 3y - 2z &= 7 \\ 5x + 2y - 6z &= 7. \end{aligned}$

9. (a) $\begin{aligned} x + y + 2z + 3w &= 13 \\ x - 2y + z + w &= 8 \\ 3x + y + z - w &= 1. \end{aligned}$

(b) $\begin{aligned} x + y + z &= 1 \\ x + y - 2z &= 3 \\ 2x + y + z &= 2. \end{aligned}$

(c) $\begin{aligned} 2x + y + z - 2w &= 1 \\ 3x - 2y + z - 6w &= -2 \\ x + y - z - w &= -1 \\ 6x + z - 9w &= -2 \\ 5x - y + 2z - 8w &= 3. \end{aligned}$

10. (a) $\begin{aligned} 2x - y + z &= 3 \\ x - 3y + z &= 4 \\ -5x - 2z &= -5. \end{aligned}$

(b) $\begin{aligned} x + y + z + w &= 6 \\ 2x + y - z &= 3 \\ 3x + y + 2w &= 6. \end{aligned}$

(c) $\begin{aligned} 2x - y + z &= 3 \\ 3x + y - 2z &= -2 \\ x - y + z &= 7 \\ x + 5y + 7z &= 13 \\ x - 7y - 5z &= 12. \end{aligned}$ _000 ⋮ -3 13_ _no solution_

In Exercises 11 through 14, find all values of a for which the resulting linear system has (a) no solution, (b) a unique solution, and (c) infinitely many solutions.

11. $\begin{aligned} x + y - z &= 2 \\ x + 2y + z &= 3 \\ x + y + (a^2 - 5)z &= a. \end{aligned}$

12. $\begin{aligned} x + y + z &= 2 \\ 2x + 3y + 2z &= 5 \\ 2x + 3y + (a^2 - 1)z &= a + 1. \end{aligned}$

13. $\begin{aligned} x + y + z &= 2 \\ x + 2y + z &= 3 \\ x + y + (a^2 - 5)z &= a. \end{aligned}$

14. $\begin{aligned} x + y &= 3 \\ x + (a^2 - 8)y &= a. \end{aligned}$

In Exercises 15 through 18, solve the linear system with the given augmented matrix.

15. (a) $\left[\begin{array}{ccc:c} 1 & 1 & 1 & 0 \\ 1 & 1 & 0 & 3 \\ 0 & 1 & 1 & 1 \end{array}\right].$

(b) $\left[\begin{array}{ccc:c} 1 & 2 & 3 & 0 \\ 1 & 1 & 1 & 0 \\ 1 & 1 & 2 & 0 \\ 1 & 3 & 3 & 0 \end{array}\right].$

16. (a) $\left[\begin{array}{ccc:c} 1 & 2 & 3 & 0 \\ 1 & 1 & 1 & 0 \\ 5 & 7 & 9 & 0 \end{array}\right].$

(b) $\left[\begin{array}{ccc:c} 1 & 2 & 1 & 7 \\ 2 & 0 & 1 & 4 \\ 1 & 0 & 2 & 5 \\ 1 & 2 & 3 & 11 \\ 2 & 1 & 4 & 12 \end{array}\right].$

17. (a) $\left[\begin{array}{cccc:c} 1 & 2 & 3 & 1 & 8 \\ 1 & 3 & 0 & 1 & 7 \\ 1 & 0 & 2 & 1 & 3 \end{array}\right].$

(b) $\left[\begin{array}{ccc:c} 1 & -2 & 3 & 4 \\ 2 & -1 & -3 & 5 \\ 3 & 0 & 1 & 2 \\ 3 & -3 & 0 & 7 \end{array}\right].$

18. (a) $\left[\begin{array}{ccc:c} 4 & 2 & -1 & 5 \\ 3 & 3 & 6 & 1 \\ 5 & 1 & -8 & 8 \end{array}\right].$

(b) $\left[\begin{array}{cccc:c} 1 & 1 & 3 & -3 & 0 \\ 0 & 2 & 1 & -3 & 3 \\ 1 & 0 & 2 & -1 & -1 \end{array}\right].$

In Exercises 19 and 20, let

$$A = \begin{bmatrix} 1 & 0 & 5 \\ 1 & 1 & 1 \\ 0 & 1 & -4 \end{bmatrix}.$$

19. Find a nontrivial solution to the homogeneous system $(-4I_3 - A)\mathbf{x} = \mathbf{0}$.

20. Find a nontrivial solution to the homogeneous system $(2I_3 - A)\mathbf{x} = \mathbf{0}$.

21. Find an equation relating a, b, and c so that the linear system

$$\begin{aligned} x + 2y - 3z &= a \\ 2x + 3y + 3z &= b \\ 5x + 9y - 6z &= c \end{aligned}$$

is consistent for any values of a, b, and c that satisfy that equation.

22. Find an equation relating a, b, and c so that the linear system

$$\begin{aligned} 2x + 2y + 3z &= a \\ 3x - y + 5z &= b \\ x - 3y + 2z &= c \end{aligned}$$

is consistent for any values of a, b, and c that satisfy that equation.

23. Find a 2×1 matrix $\mathbf{x}$ with entries not all zero such that $A\mathbf{x} = 4\mathbf{x}$, where

$$A = \begin{bmatrix} 4 & 1 \\ 0 & 2 \end{bmatrix}.$$

[*Hint*: Rewrite the matrix equation $A\mathbf{x} = 4\mathbf{x}$ as $4\mathbf{x} - A\mathbf{x} = (4I_2 - A)\mathbf{x} = \mathbf{0}$ and solve the homogeneous system.]

24. Find a 2×1 matrix $\mathbf{x}$ with entries not all zero such that $A\mathbf{x} = 3\mathbf{x}$, where

$$A = \begin{bmatrix} 2 & 1 \\ 1 & 2 \end{bmatrix}.$$

25. Find a 3×1 matrix with entries not all zero such that $A\mathbf{x} = 3\mathbf{x}$, where

$$A = \begin{bmatrix} 1 & 2 & -1 \\ 1 & 0 & 1 \\ 4 & -4 & 5 \end{bmatrix}.$$

26. Find a 3×1 matrix $\mathbf{x}$ with entries not all zero such that $A\mathbf{x} = 1\mathbf{x}$, where

$$A = \begin{bmatrix} 1 & 2 & -1 \\ 1 & 0 & 1 \\ 4 & -4 & 5 \end{bmatrix}.$$

In Exercises 27 and 28, find the quadratic polynomial that interpolates the given points.

27. $(1, 2), (3, 3), (5, 8)$.

28. $(1, 5), (2, 12), (3, 44)$.

In Exercises 29 and 30, find the cubic polynomial that interpolates the given points.

29. $(-1, -6), (1, 0), (2, 8), (3, 34)$.

30. $(-2, 2), (-1, 2), (1, 2), (2, 10)$.

31. A furniture manufacturer makes chairs, coffee tables, and dining-room tables. Each chair requires 10 minutes of sanding, 6 minutes of staining, and 12 minutes of varnishing. Each coffee table requires 12 minutes of sanding, 8 minutes of staining, and 12 minutes of varnishing. Each dining-room table requires 15 minutes of sanding, 12 minutes of staining, and 18 minutes of varnishing. The sanding bench is available 16 hours per week, the staining bench 11 hours per week, and the varnishing bench 18 hours per week. How many (per week) of each type of furniture should be made so that the benches are fully utilized?

32. A book publisher publishes a potential best seller in three different bindings: paperback, book club, and deluxe. Each paperback book requires 1 minute for sewing and 2 minutes for gluing. Each book club book requires 2 minutes for sewing and 4 minutes for gluing. Each deluxe book requires 3 minutes for sewing and 5 minutes for gluing. If the sewing plant is available 6 hours per day and the gluing plant is available 11 hours per day, how many books of each type can be produced per day so that the plants are fully utilized?

THEORETICAL EXERCISES ▪

T.1. Show that properties (a), (b), and (c) alone [excluding (d)] of the definition of the reduced row echelon form of a matrix A imply that if a column of A contains a leading entry of some row, then all other entries in that column *below the leading entry* are zero.

T.2. Show that:

(a) Every matrix is row equivalent to itself.

(b) If A is row equivalent to B, then B is row equivalent to A.

(c) If A is row equivalent to B and B is row equivalent to C, then A is row equivalent to C.

T.3. Prove Corollary 1.1.

T.4. Show that the linear system $A\mathbf{x} = \mathbf{b}$ has no solution if and only if its augmented matrix is row equivalent to a matrix in reduced echelon form that has a row whose first n elements are zero and whose $(n + 1)$st element is 1.

T.5. Let

$$A = \begin{bmatrix} a & b \\ c & d \end{bmatrix}.$$

Show that A is row equivalent to I_2 if and only if $ad - bc \neq 0$.

T.6. (a) Let

$$A = \begin{bmatrix} a & b \\ ka & kb \end{bmatrix}.$$

Use Exercise T.5 to determine if A is row equivalent to I_2.

(b) Let A be a 2×2 matrix with a row consisting entirely of zeros. Use Exercise T.5 to determine if A is row equivalent to I_2.

T.7. Determine the matrix in reduced row echelon form that is row equivalent to the matrix

$$\begin{bmatrix} \cos\theta & \sin\theta \\ -\sin\theta & \cos\theta \end{bmatrix}.$$

T.8. Let

$$A = \begin{bmatrix} a & b \\ c & d \end{bmatrix}.$$

Show that the homogeneous system $A\mathbf{x} = \mathbf{0}$ has only the trivial solution if and only if $ad - bc \neq 0$.

T.9. Let A be an $n \times n$ matrix in reduced row echelon form. Show that if $A \neq I_n$, then A has a row consisting entirely of zeros.

T.10. Show that the values of λ for which the homogeneous system

$$(a - \lambda)x + \qquad by = 0$$
$$cx + (d - \lambda)y = 0$$

has a nontrivial solution satisfy the equation $(a - \lambda)(d - \lambda) - bc = 0$. (*Hint*: See Exercise T.8.)

T.11. Let $\mathbf{u}$ and $\mathbf{v}$ be solutions to the homogeneous linear system $A\mathbf{x} = \mathbf{0}$.

(a) Show that $\mathbf{u} + \mathbf{v}$ is a solution.

(b) Show that $\mathbf{u} - \mathbf{v}$ is a solution.

(c) For any scalar r, show that $r\mathbf{u}$ is a solution.

(d) For any scalars r and s, show that $r\mathbf{u} + s\mathbf{v}$ is a solution.

T.12. Show that if $\mathbf{u}$ and $\mathbf{v}$ are solutions to the linear system $A\mathbf{x} = \mathbf{b}$, then $\mathbf{u} - \mathbf{v}$ is a solution to the associated homogeneous system $A\mathbf{x} = \mathbf{0}$.

T.13. Let $A\mathbf{x} = \mathbf{b}$, $\mathbf{b} \neq \mathbf{0}$, be a consistent linear system.

(a) Show that if $\mathbf{x}_p$ is a particular solution to the given nonhomogeneous system and $\mathbf{x}_h$ is a solution to the associated homogeneous system $A\mathbf{x} = \mathbf{0}$, then $\mathbf{x}_p + \mathbf{x}_h$ is a solution to the given system $A\mathbf{x} = \mathbf{b}$.

(b) Show that every solution $\mathbf{x}$ to the nonhomogeneous linear system $A\mathbf{x} = \mathbf{b}$ can be written as $\mathbf{x}_p + \mathbf{x}_h$, where $\mathbf{x}_p$ is a particular solution to the given nonhomogeneous system and $\mathbf{x}_h$ is a solution to the associated homogeneous system $A\mathbf{x} = \mathbf{0}$. [*Hint*: Let $\mathbf{x} = \mathbf{x}_p + (\mathbf{x} - \mathbf{x}_p)$.]

MATLAB EXERCISES ■

In order to use MATLAB in this section, you should first have read Chapter 10 through Section 10.4.

ML.1. Let

$$A = \begin{bmatrix} 4 & 2 & 2 \\ -3 & 1 & 4 \\ 1 & 0 & 3 \\ 5 & -1 & 5 \end{bmatrix}.$$

Find the matrices obtained by performing the following row operations in succession on matrix A. Do the row operations directly using the colon operator.

(a) Multiply row 1 by $\frac{1}{4}$.

(b) Add 3 times row 1 to row 2.

(c) Add (-1) times row 1 to row 3.

(d) Add (-5) times row 1 to row 4.

(e) Interchange rows 2 and 4.

ML.2. Let

$$A = \begin{bmatrix} \frac{1}{2} & \frac{1}{3} & \frac{1}{4} & \frac{1}{5} \\ \frac{1}{3} & \frac{1}{4} & \frac{1}{5} & \frac{1}{6} \\ 1 & \frac{1}{2} & \frac{1}{3} & \frac{1}{4} \end{bmatrix}.$$

Find the matrices obtained by performing the following row operations in succession on matrix A. Do the row operations directly using the colon operator.

(a) Multiply row 1 by 2.

(b) Add $\left(-\frac{1}{3}\right)$ times row 1 to row 2.

(c) Add (-1) times row 1 to row 3.

(d) Interchange rows 2 and 3.

ML.3. Use **reduce** to find the reduced row echelon form of matrix A in Exercise ML.1.

ML.4. Use **reduce** to find the reduced row echelon form of matrix A in Exercise ML.2.

ML.5. Use **reduce** to find all solutions to the linear system in Exercise 9(a).

ML.6. Use **reduce** to find all solutions to the linear system in Exercise 8(b).

ML.7. Use **reduce** to find all solutions to the linear system in Exercise 15(b).

ML.8. Use **reduce** to find all solutions to the linear system in Exercise 16(a).

ML.9. Let
$$A = \begin{bmatrix} 1 & 2 \\ 2 & 4 \end{bmatrix}.$$

Use **reduce** to find a nontrivial solution to the homogeneous system
$$(5I_2 - A)\mathbf{x} = \mathbf{0}.$$

[*Hint*: In MATLAB, enter matrix A, then use the command
reduce(5 * eye(size(A)) − A).]

ML.10. Let
$$A = \begin{bmatrix} 1 & 5 \\ 5 & 1 \end{bmatrix}.$$

Use **reduce** to find a nontrivial solution to the homogeneous system
$$(-4I_2 - A)\mathbf{x} = \mathbf{0}.$$

[*Hint*: In MATLAB, enter matrix A, then use the command
reduce(− 4 * eye(size(A)) − A).]

ML.11. Use **rref** in MATLAB to solve the linear systems in Exercises 15 and 16.

ML.12. MATLAB has an immediate command for solving square linear systems $A\mathbf{x} = \mathbf{b}$. Once the coefficient matrix A and right-hand side $\mathbf{b}$ are entered into MATLAB, command

$$\mathbf{x} = \mathbf{A} \backslash \mathbf{b}$$

displays the solution as long as A is considered nonsingular. (See the Definition at the beginning of Section 1.6.) The backslash command, $\backslash$, does not use reduced row echelon form, but does initiate numerical methods that are discussed briefly in Chapter 9. For more details on the command see D. R. Hill, *Experiments in Computational Matrix Algebra*, New York: Random House, 1988.

(a) Use $\backslash$ to solve Exercise 15(a).

(b) Use $\backslash$ to solve Exercise 9(b).

ML.13. The $\backslash$ command behaves differently than **rref**. Use both $\backslash$ and **rref** to solve $A\mathbf{x} = \mathbf{b}$, where

$$A = \begin{bmatrix} 1 & 2 & 3 \\ 4 & 5 & 6 \\ 7 & 8 & 9 \end{bmatrix}, \quad \mathbf{b} = \begin{bmatrix} 1 \\ 0 \\ 0 \end{bmatrix}.$$

1.6 ▾ The Inverse of a Matrix

In this section we restrict our attention to square matrices and formulate the notion corresponding to the reciprocal of a nonzero number.

DEFINITION
An $n \times n$ matrix A is called **nonsingular** (or **invertible**) if there exists an $n \times n$ matrix B such that
$$AB = BA = I_n.$$

The matrix B is called an **inverse** of A. If there exists no such matrix B, then A is called **singular** (or **noninvertible**).

EXAMPLE 1 ■ Let
$$A = \begin{bmatrix} 2 & 3 \\ 2 & 2 \end{bmatrix} \quad \text{and} \quad B = \begin{bmatrix} -1 & \frac{3}{2} \\ 1 & -1 \end{bmatrix}.$$

Since
$$AB = BA = I_2,$$
we conclude that B is an inverse of A and that A is nonsingular. ■

THEOREM 1.8 ■ *If a matrix has an inverse, then the inverse is unique.*

Proof   Let B and C be inverses of A. Then
$$BA = AC = I_n.$$
Therefore,
$$B = BI_n = B(AC) = (BA)C = I_nC = C,$$
which completes the proof. ■

We shall now write the inverse of A, if it exists, as A^{-1}. Thus
$$AA^{-1} = A^{-1}A = I_n.$$

EXAMPLE 2 ■ Let
$$A = \begin{bmatrix} 1 & 2 \\ 3 & 4 \end{bmatrix}.$$
To find A^{-1}, we let
$$A^{-1} = \begin{bmatrix} a & b \\ c & d \end{bmatrix}.$$
Then we must have
$$AA^{-1} = \begin{bmatrix} 1 & 2 \\ 3 & 4 \end{bmatrix} \begin{bmatrix} a & b \\ c & d \end{bmatrix} = I_2 = \begin{bmatrix} 1 & 0 \\ 0 & 1 \end{bmatrix}$$
so that
$$\begin{bmatrix} a + 2c & b + 2d \\ 3a + 4c & 3b + 4d \end{bmatrix} = \begin{bmatrix} 1 & 0 \\ 0 & 1 \end{bmatrix}.$$
Equating corresponding entries of these two matrices, we obtain the linear systems
$$\begin{matrix} a + 2c = 1 \\ 3a + 4c = 0 \end{matrix} \quad \text{and} \quad \begin{matrix} b + 2d = 0 \\ 3b + 4d = 1. \end{matrix}$$

The solutions are (verify) $a = -2$, $c = \frac{3}{2}$, $b = 1$, and $d = -\frac{1}{2}$. Moreover, since the matrix
$$\begin{bmatrix} a & b \\ c & d \end{bmatrix} = \begin{bmatrix} -2 & 1 \\ \frac{3}{2} & -\frac{1}{2} \end{bmatrix}$$
also satisfies the property that
$$\begin{bmatrix} -2 & 1 \\ \frac{3}{2} & -\frac{1}{2} \end{bmatrix} \begin{bmatrix} 1 & 2 \\ 3 & 4 \end{bmatrix} = \begin{bmatrix} 1 & 0 \\ 0 & 1 \end{bmatrix},$$

we conclude that A is nonsingular and that

$$A^{-1} = \begin{bmatrix} -2 & 1 \\ \frac{3}{2} & -\frac{1}{2} \end{bmatrix}.$$

■

Not every matrix has an inverse. For instance, consider the following example.

EXAMPLE 3 ■ Let

$$A = \begin{bmatrix} 1 & 2 \\ 2 & 4 \end{bmatrix}.$$

To find A^{-1}, we let

$$A^{-1} = \begin{bmatrix} a & b \\ c & d \end{bmatrix}.$$

Then we must have

$$AA^{-1} = \begin{bmatrix} 1 & 2 \\ 2 & 4 \end{bmatrix} \begin{bmatrix} a & b \\ c & d \end{bmatrix} = I_2 = \begin{bmatrix} 1 & 0 \\ 0 & 1 \end{bmatrix}$$

so that

$$\begin{bmatrix} a+2c & b+2d \\ 2a+4c & 2b+4d \end{bmatrix} = \begin{bmatrix} 1 & 0 \\ 0 & 1 \end{bmatrix}.$$

Equating corresponding entries of these two matrices, we obtain the linear systems

$$\begin{array}{ccc} a + 2c = 1 & & b + 2d = 0 \\ & \text{and} & \\ 2a + 4c = 0 & & 2b + 4d = 1. \end{array}$$

These linear systems have no solutions, so A has no inverse. Hence A is a singular matrix.

■

The method used in Example 2 to find the inverse of a matrix is not a very efficient one. We shall soon modify it and thereby obtain a much faster method. We first establish several properties of nonsingular matrices.

THEOREM 1.9 ■ (a) *If A is a nonsingular matrix, then A^{-1} is nonsingular and*
(*Properties of the Inverse*)
$$(A^{-1})^{-1} = A.$$

(b) *If A and B are nonsingular matrices, then AB is nonsingular and*
$$(AB)^{-1} = B^{-1}A^{-1}.$$

(c) *If A is a nonsingular matrix, then*
$$(A^T)^{-1} = (A^{-1})^T.$$

Proof (a) A^{-1} is nonsingular if we can find a matrix B such that

$$A^{-1}B = BA^{-1} = I_n.$$

Now

$$A^{-1}A = AA^{-1} = I_n.$$

Thus $B = A$ is an inverse of A^{-1}, and since inverses are unique, we conclude that

$$(A^{-1})^{-1} = A.$$

(b) We have

$$(AB)(B^{-1}A^{-1}) = A(BB^{-1})A^{-1} = AI_nA^{-1} = AA^{-1} = I_n$$

and

$$(B^{-1}A^{-1})(AB) = B^{-1}(A^{-1}A)B = B^{-1}I_nB = B^{-1}B = I_n.$$

Therefore, AB is nonsingular. Since the inverse of a matrix is unique, we conclude that

$$(AB)^{-1} = B^{-1}A^{-1}.$$

(c) We have

$$AA^{-1} = I_n \quad \text{and} \quad A^{-1}A = I_n.$$

Taking transposes, we obtain

$$(AA^{-1})^T = I_n^T = I_n \quad \text{and} \quad (A^{-1}A)^T = I_n^T = I_n.$$

Then

$$(A^{-1})^T A^T = I_n \quad \text{and} \quad A^T(A^{-1})^T = I_n.$$

These equations imply that

$$(A^T)^{-1} = (A^{-1})^T. \qquad \blacksquare$$

EXAMPLE 4 ■ If $A = \begin{bmatrix} 1 & 2 \\ 3 & 4 \end{bmatrix}$, then from Example 2

$$A^{-1} = \begin{bmatrix} -2 & 1 \\ \frac{3}{2} & -\frac{1}{2} \end{bmatrix} \quad \text{and} \quad (A^{-1})^T = \begin{bmatrix} -2 & \frac{3}{2} \\ 1 & -\frac{1}{2} \end{bmatrix}.$$

Also (verify)

$$A^T = \begin{bmatrix} 1 & 3 \\ 2 & 4 \end{bmatrix} \quad \text{and} \quad (A^T)^{-1} = \begin{bmatrix} -2 & \frac{3}{2} \\ 1 & -\frac{1}{2} \end{bmatrix}. \qquad \blacksquare$$

COROLLARY 1.2 ■ *If $A_1, A_2, \ldots, A_r$ are $n \times n$ nonsingular matrices, then $A_1 A_2 \cdots A_r$ is nonsingular and*

$$(A_1 A_2 \cdots A_r)^{-1} = A_r^{-1} A_{r-1}^{-1} \cdots A_1^{-1}.$$

Proof Exercise T.2. $\qquad \blacksquare$

Earlier, we defined a matrix B to be the inverse of A if $AB = BA = I_n$. The following theorem, whose proof we omit, shows that one of these equations follows from the other.

THEOREM 1.10 ■ *Suppose that A and B are $n \times n$ matrices.*

(a) *If $AB = I_n$, then $BA = I_n$.*

(b) *If $BA = I_n$, then $AB = I_n$.* ■

A Practical Method for Finding A^{-1}

We shall now develop a practical method for finding A^{-1}. If A is a given $n \times n$ matrix, we are looking for an $n \times n$ matrix $B = \begin{bmatrix} b_{ij} \end{bmatrix}$ such that

$$AB = BA = I_n.$$

Let the columns of B be denoted by the $n \times 1$ matrices $\mathbf{x}_1, \mathbf{x}_2, \ldots, \mathbf{x}_n$, where

$$\mathbf{x}_j = \begin{bmatrix} b_{1j} \\ b_{2j} \\ \vdots \\ b_{ij} \\ \vdots \\ b_{nj} \end{bmatrix} \qquad (1 \le j \le n).$$

Let the columns of I_n be denoted by the $n \times 1$ matrices $\mathbf{e}_1, \mathbf{e}_2, \ldots, \mathbf{e}_n$. Thus

$$\mathbf{e}_j = \begin{bmatrix} 0 \\ 0 \\ \vdots \\ 1 \\ 0 \\ \vdots \\ 0 \end{bmatrix} \longleftarrow j\text{th row.}$$

By Exercise T.9(a) of Section 1.3, the jth column of AB is the $n \times 1$ matrix $A\mathbf{x}_j$. Since equal matrices must agree column by column, it follows that the problem of finding an $n \times n$ matrix $B = A^{-1}$ such that

$$AB = I_n \tag{1}$$

is equivalent to the problem of finding n matrices (each $n \times 1$) $\mathbf{x}_1, \mathbf{x}_2, \ldots, \mathbf{x}_n$ such that

$$A\mathbf{x}_j = \mathbf{e}_j \qquad (1 \le j \le n). \tag{2}$$

Thus finding B is equivalent to solving n linear systems (each is n equations in n unknowns). This is precisely what we did in Example 2. Each of these systems can be solved by the Gauss–Jordan reduction method. To solve the first linear system, we form the augmented matrix $\begin{bmatrix} A & \vdots & \mathbf{e}_1 \end{bmatrix}$ and put it into reduced row echelon form. We do the same with

$$\begin{bmatrix} A & \vdots & \mathbf{e}_2 \end{bmatrix}, \ldots, \begin{bmatrix} A & \vdots & \mathbf{e}_n \end{bmatrix}.$$

However, if we observe that the coefficient matrix of each of these n linear systems is always A, we can solve all these systems simultaneously. We form the $n \times 2n$ matrix

$$\begin{bmatrix} A & \vdots & \mathbf{e}_1 & \mathbf{e}_2 & \cdots & \mathbf{e}_n \end{bmatrix} = \begin{bmatrix} A & \vdots & I_n \end{bmatrix}$$

and transform it to reduced row echelon form $\begin{bmatrix} C & \vdots & D \end{bmatrix}$. The $n \times n$ matrix C is the reduced row echelon form matrix that is row equivalent to A. Let $\mathbf{d}_1$, $\mathbf{d}_2$, $\ldots, \mathbf{d}_n$ be the n columns of D. Then the matrix $\begin{bmatrix} C & \vdots & D \end{bmatrix}$ gives rise to the n linear systems

$$C\mathbf{x}_j = \mathbf{d}_j \qquad (1 \leq j \leq n) \tag{3}$$

or to the matrix equation

$$CB = D. \tag{4}$$

There are now two possible cases.

Case 1. $C = I_n$. Then Equation (3) becomes

$$I_n\mathbf{x}_j = \mathbf{x}_j = \mathbf{d}_j,$$

and $B = D$, so we have obtained A^{-1}.

Case 2. $C \neq I_n$. It then follows from Exercise T.9 in Section 1.5 that C has a row consisting entirely of zeros. From Exercise T.3 in Section 1.3, we observe that the product CB in Equation (4) has a row of zeros. The matrix D in (4) arose from I_n by a sequence of elementary row operations, and it is intuitively clear that D cannot have a row of zeros. The statement that D cannot have a row of zeros can be rigorously established at this point, but we shall ask the reader to accept the argument now. In Section 2.2, an argument using determinants will show the validity of the result. Thus one of the equations $C\mathbf{x}_j = \mathbf{d}_j$ has no solution, so $A\mathbf{x}_j = \mathbf{e}_j$ has no solution and A is singular in this case.

The practical procedure for computing the inverse of matrix A is as follows.

Step 1. Form the $n \times 2n$ matrix $\begin{bmatrix} A & \vdots & I_n \end{bmatrix}$ obtained by adjoining the identity matrix I_n to the given matrix A.

Step 2. Transform the matrix obtained in Step 1 to reduced row echelon form by using elementary row operations. Remember that whatever we do to a row of A we also do to the corresponding row of I_n.

Step 3. Suppose that Step 2 has produced the matrix $\begin{bmatrix} C & \vdots & D \end{bmatrix}$ in reduced row echelon form.

(a) If $C = I_n$, then $D = A^{-1}$.

(b) If $C \neq I_n$, then C has a row of zeros. In this case A is singular and A^{-1} does not exist.

EXAMPLE 5 ■ Find the inverse of the matrix

$$A = \begin{bmatrix} 1 & 1 & 1 \\ 0 & 2 & 3 \\ 5 & 5 & 1 \end{bmatrix}.$$

Solution **Step 1.** The 3×6 matrix $\begin{bmatrix} A & \vdots & I_3 \end{bmatrix}$ is

$$[A \vdots I_3] = \begin{bmatrix} \overset{A}{} & & & \overset{I_3}{} & & \\ 1 & 1 & 1 & \vdots & 1 & 0 & 0 \\ 0 & 2 & 3 & \vdots & 0 & 1 & 0 \\ 5 & 5 & 1 & \vdots & 0 & 0 & 1 \end{bmatrix}.$$

Step 2. We transform the matrix obtained in Step 1 to reduced row echelon form:

$$\begin{array}{ccccccc} \overset{A}{} & & & & \overset{I_3}{} & & \\ 1 & 1 & 1 & \vdots & 1 & 0 & 0 \\ 0 & 2 & 3 & \vdots & 0 & 1 & 0 \\ 5 & 5 & 1 & \vdots & 0 & 0 & 1 \end{array}$$

Add (-5) times the first row to the third row to obtain

$$\begin{array}{ccccccc} 1 & 1 & 1 & \vdots & 1 & 0 & 0 \\ 0 & 2 & 3 & \vdots & 0 & 1 & 0 \\ 0 & 0 & -4 & \vdots & -5 & 0 & 1 \end{array}$$

Multiply the second row by $\frac{1}{2}$ to obtain

$$\begin{array}{ccccccc} 1 & 1 & 1 & \vdots & 1 & 0 & 0 \\ 0 & 1 & \frac{3}{2} & \vdots & 0 & \frac{1}{2} & 0 \\ 0 & 0 & -4 & \vdots & -5 & 0 & 1 \end{array}$$

Add (-1) times the second row to the first row to obtain

$$\begin{array}{ccccccc} 1 & 0 & -\frac{1}{2} & \vdots & 1 & -\frac{1}{2} & 0 \\ 0 & 1 & \frac{3}{2} & \vdots & 0 & \frac{1}{2} & 0 \\ 0 & 0 & -4 & \vdots & -5 & 0 & 1 \end{array}$$

Multiply the third row by $\left(-\frac{1}{4}\right)$ to obtain

$$\begin{array}{ccccccc} 1 & 0 & -\frac{1}{2} & \vdots & 1 & -\frac{1}{2} & 0 \\ 0 & 1 & \frac{3}{2} & \vdots & 0 & \frac{1}{2} & 0 \\ 0 & 0 & 1 & \vdots & \frac{5}{4} & 0 & -\frac{1}{4} \end{array}$$

Add $\left(-\frac{3}{2}\right)$ times the third row to the second row to obtain

$$\begin{array}{ccccccc} 1 & 0 & -\frac{1}{2} & \vdots & 1 & -\frac{1}{2} & 0 \\ 0 & 1 & 0 & \vdots & -\frac{15}{8} & \frac{1}{2} & \frac{3}{8} \\ 0 & 0 & 1 & \vdots & \frac{5}{4} & 0 & -\frac{1}{4} \end{array}$$

Add $\frac{1}{2}$ times the third row to the first row to obtain

$$\begin{array}{ccc|ccc} 1 & 0 & 0 & \frac{13}{8} & -\frac{1}{2} & -\frac{1}{8} \\ 0 & 1 & 0 & -\frac{15}{8} & \frac{1}{2} & \frac{3}{8} \\ 0 & 0 & 1 & \frac{5}{4} & 0 & -\frac{1}{4} \end{array}$$

Step 3. Since $C = I_3$, we conclude that $D = A^{-1}$. Hence

$$A^{-1} = \begin{bmatrix} \frac{13}{8} & -\frac{1}{2} & -\frac{1}{8} \\ -\frac{15}{8} & \frac{1}{2} & \frac{3}{8} \\ \frac{5}{4} & 0 & -\frac{1}{4} \end{bmatrix}.$$

It is easy to verify that $AA^{-1} = A^{-1}A = I_3$. ∎

If the reduced row echelon matrix under A has a row of zeros, then A is singular. Since each matrix under A is row equivalent to A, once a matrix under A has a row of zeros, every subsequent matrix that is row equivalent to A will have a row of zeros. Thus we can stop the procedure as soon as we find a matrix F that is row equivalent to A and has a row of zeros. In this case A^{-1} does not exist.

EXAMPLE 6 ■ Find the inverse of the matrix

$$A = \begin{bmatrix} 1 & 2 & -3 \\ 1 & -2 & 1 \\ 5 & -2 & -3 \end{bmatrix} \quad \text{if it exists.}$$

Solution   ***Step 1.*** The 3×6 matrix $\begin{bmatrix} A & \vdots & I_3 \end{bmatrix}$ is

$$\begin{bmatrix} A & \vdots & I_3 \end{bmatrix} = \begin{array}{c} \\ \end{array} \overset{\displaystyle A \qquad\qquad I_3}{\begin{bmatrix} 1 & 2 & -3 & \vdots & 1 & 0 & 0 \\ 1 & -2 & 1 & \vdots & 0 & 1 & 0 \\ 5 & -2 & -3 & \vdots & 0 & 0 & 1 \end{bmatrix}}.$$

Step 2. We transform the matrix obtained in Step 1 to reduced row echelon form. To find A^{-1}, we proceed as follows:

$$
\begin{array}{ccc}
A & & I_3
\end{array}
$$

$$
\left[\begin{array}{ccc:ccc}
1 & 2 & -3 & 1 & 0 & 0 \\
1 & -2 & 1 & 0 & 1 & 0 \\
5 & -2 & -3 & 0 & 0 & 1
\end{array}\right]
$$

Add (-1) times the first row to the second row to obtain

$$
\left[\begin{array}{ccc:ccc}
1 & 2 & -3 & 1 & 0 & 0 \\
0 & -4 & 4 & -1 & 1 & 0 \\
5 & -2 & -3 & 0 & 0 & 1
\end{array}\right]
$$

Add (-5) times the first row to the third row to obtain

$$
\left[\begin{array}{ccc:ccc}
1 & 2 & -3 & 1 & 0 & 0 \\
0 & -4 & 4 & -1 & 1 & 0 \\
0 & -12 & 12 & -5 & 0 & 1
\end{array}\right]
$$

Add (-3) times the second row to the third row to obtain

$$
\left[\begin{array}{ccc:ccc}
1 & 2 & -3 & 1 & 0 & 0 \\
0 & -4 & 4 & -1 & 1 & 0 \\
0 & 0 & 0 & -2 & -3 & 1
\end{array}\right]
$$

At this point A is row equivalent to

$$
F = \begin{bmatrix}
1 & 2 & -3 \\
0 & -4 & 4 \\
0 & 0 & 0
\end{bmatrix}.
$$

Since F has a row of zeros, we stop and conclude that A is a singular matrix. ■

Observe that to find A^{-1} we do not have to determine, in advance, whether or not it exists. We merely start the procedure given above and either obtain A^{-1} or find out that A is singular.

The foregoing discussion for the practical method of obtaining A^{-1} has actually established the following theorem.

THEOREM 1.11 ■ *An $n \times n$ matrix is nonsingular if and only if it is row equivalent to I_n.* ■

Linear Systems and Inverses

If A is an $n \times n$ matrix, then the linear system $A\mathbf{x} = \mathbf{b}$ is a system of n equations in n unknowns. Suppose that A is nonsingular. Then A^{-1} exists and we can multiply $A\mathbf{x} = \mathbf{b}$ by A^{-1} on both sides, obtaining

$$
A^{-1}(A\mathbf{x}) = A^{-1}\mathbf{b}
$$
$$
(A^{-1}A)\mathbf{x} = A^{-1}\mathbf{b}
$$
$$
I_n\mathbf{x} = A^{-1}\mathbf{b}
$$
$$
\mathbf{x} = A^{-1}\mathbf{b}.
$$

Moreover, $\mathbf{x} = A^{-1}\mathbf{b}$ is clearly a solution to the given linear system. Thus, if A is nonsingular, we have a unique solution.

Applications

This method is useful in industrial problems. Many physical models are described by linear systems. This means that if n values are used as inputs (which can be arranged as the $n \times 1$ matrix $\mathbf{x}$), then m values are obtained as outputs (which can be arranged as the $m \times 1$ matrix $\mathbf{b}$) by the rule $A\mathbf{x} = \mathbf{b}$. The matrix A is inherently tied to the process. Thus suppose that a chemical process has a certain matrix A associated with it. Any change in the process may result in a new matrix. In fact, we speak of a **black box**, meaning that the internal structure of the process does not interest us. The problem frequently encountered in systems analysis is that of determining the input to be used to obtain a desired output. That is, we want to solve the linear system $A\mathbf{x} = \mathbf{b}$ for $\mathbf{x}$ as we vary $\mathbf{b}$. If A is a nonsingular square matrix, an efficient way of handling this is as follows: Compute A^{-1} once; then whenever we change $\mathbf{b}$, we find the corresponding solution $\mathbf{x}$ by forming $A^{-1}\mathbf{b}$.

EXAMPLE 7 ■ Consider an industrial process whose matrix is the matrix A of Example 5. If $\mathbf{b}$
(Industrial Process)   is the output matrix

$$\begin{bmatrix} 8 \\ 24 \\ 8 \end{bmatrix},$$

then the input matrix $\mathbf{x}$ is the solution to the linear system $A\mathbf{x} = \mathbf{b}$. Then

$$\mathbf{x} = A^{-1}\mathbf{b} = \begin{bmatrix} \frac{13}{8} & -\frac{1}{2} & -\frac{1}{8} \\ -\frac{15}{8} & \frac{1}{2} & \frac{3}{8} \\ \frac{5}{4} & 0 & -\frac{1}{4} \end{bmatrix} \begin{bmatrix} 8 \\ 24 \\ 8 \end{bmatrix} = \begin{bmatrix} 0 \\ 0 \\ 8 \end{bmatrix}.$$

On the other hand, if $\mathbf{b}$ is the output matrix

$$\begin{bmatrix} 4 \\ 7 \\ 16 \end{bmatrix},$$

then (verify)

$$\mathbf{x} = A^{-1} \begin{bmatrix} 4 \\ 7 \\ 16 \end{bmatrix} = \begin{bmatrix} 1 \\ 2 \\ 1 \end{bmatrix}.$$

■

THEOREM 1.12 ■ *If A is an $n \times n$ matrix, the homogeneous system*

$$A\mathbf{x} = \mathbf{0} \tag{5}$$

has a nontrivial solution if and only if A is singular.

Proof   Suppose that A is nonsingular. Then A^{-1} exists, and multiplying both sides of (5) by A^{-1}, we have

$$A^{-1}(A\mathbf{x}) = A^{-1}\mathbf{0}$$
$$(A^{-1}A)\mathbf{x} = \mathbf{0}$$
$$I_n\mathbf{x} = \mathbf{0}$$
$$\mathbf{x} = \mathbf{0}.$$

Hence the only solution to (5) is $\mathbf{x} = \mathbf{0}$.

We leave the proof of the converse—if A is singular, then (5) has a nontrivial solution—as an exercise (Exercise T.3).   ■

EXAMPLE 8   ■   Consider the homogeneous system $A\mathbf{x} = \mathbf{0}$, where A is the matrix of Example 5. Since A is nonsingular,

$$\mathbf{x} = A^{-1}\mathbf{0} = \mathbf{0}.$$

We could also solve the given system by Gauss–Jordan reduction. In this case we find that the matrix in reduced row echelon form that is row equivalent to the augmented matrix of the given system,

$$\left[\begin{array}{ccc:c} 1 & 1 & 1 & 0 \\ 0 & 2 & 3 & 0 \\ 5 & 5 & 1 & 0 \end{array}\right],$$

is

$$\left[\begin{array}{ccc:c} 1 & 0 & 0 & 0 \\ 0 & 1 & 0 & 0 \\ 0 & 0 & 1 & 0 \end{array}\right],$$

which again shows that the solution is

$$\mathbf{x} = \mathbf{0}.$$ ■

EXAMPLE 9   ■   Consider the homogeneous system $A\mathbf{x} = \mathbf{0}$, where A is the singular matrix of Example 6. In this case the matrix in reduced row echelon form that is row equivalent to the augmented matrix of the given system,

$$\left[\begin{array}{ccc:c} 1 & 2 & -3 & 0 \\ 1 & -2 & 1 & 0 \\ 5 & -2 & -3 & 0 \end{array}\right],$$

is (verify)

$$\left[\begin{array}{ccc:c} 1 & 0 & -1 & 0 \\ 0 & 1 & -1 & 0 \\ 0 & 0 & 0 & 0 \end{array}\right],$$

which implies that

$$x = r$$
$$y = r$$
$$z = r,$$

where r is any real number. Thus the given system has a nontrivial solution.   ■

The proof of the following theorem is left as an exercise (Supplementary Exercise T.18).

THEOREM 1.13 ■ *If A is an $n \times n$ matrix, then A is nonsingular if and only if the linear system $A\mathbf{x} = \mathbf{b}$ has a unique solution for every $n \times 1$ matrix $\mathbf{b}$.* ■

The converse of this theorem is also true. (See Exercise T.8.)

We may summarize our results on homogeneous systems and nonsingular matrices in the following List of Nonsingular Equivalences.

List of Nonsingular Equivalences

The following statements are equivalent.

1. A is nonsingular.

2. $A\mathbf{x} = \mathbf{0}$ has only the trivial solution.

3. A is row equivalent to I_n.

4. The linear system $A\mathbf{x} = \mathbf{b}$ has a unique solution for every $n \times 1$ matrix $\mathbf{b}$.

This means that in solving a given problem we can use any of the four statements above; they are interchangeable. As you will see throughout this course, a given problem can often be solved in several alternative ways, and sometimes one solution procedure is easier to apply than another. This List of Nonsingular Equivalences will grow as we progress through the book. By the end of Appendix B it will have expanded to 12 equivalent statements.

Section 8.4, Least Squares, and Section 8.5, Linear Economic Models, which can be studied at this time, use material from this section.

▼ *Preview of an Application*

LEAST SQUARES (SECTION 8.4)

The gathering and analyzing of data is a problem that is widely encountered in the sciences, engineering, economics, and the social sciences. When some data are plotted we might obtain Figure A. The problem is that of drawing the straight line that "best fits" the given data. This line is shown in Figure B. The technique for solving this problem is called the method of least squares, which is covered in Section 8.4.

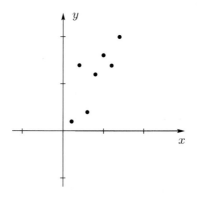

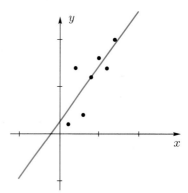

Figure A   Figure B

▼ *Preview of an Application*

LINEAR ECONOMIC MODELS (SECTION 8.5)

Economic analysis and forecasting has become increasingly important in our complex modern society. Suppose we have a simple society consisting of only three individuals: a farmer who produces all the food and nothing else, a carpenter who builds all the houses and makes nothing else, and a tailor who makes all the clothes and nothing else. We select units so that each person produces one unit of the commodity that he or she makes during the year. Suppose also that the portion of each commodity that is consumed by each person is given in Table 1.

Table 1

Goods Consumed by:	Goods Produced by:		
	Farmer	Carpenter	Tailor
Farmer	$\frac{7}{16}$	$\frac{1}{2}$	$\frac{3}{16}$
Carpenter	$\frac{5}{16}$	$\frac{1}{6}$	$\frac{5}{16}$
Tailor	$\frac{1}{4}$	$\frac{1}{3}$	$\frac{1}{2}$

Thus the farmer consumes $\frac{7}{16}$ of the food produced, $\frac{1}{2}$ of the shelter built by the carpenter, and $\frac{3}{16}$ of the clothes made by the tailor, and so on.

The economist has to determine relative prices p_1, p_2, and p_3 per unit of food, housing, and clothing, respectively, so that no one makes money or loses money. When this situation occurs we say that we have a state of equilibrium. Letting

$$\mathbf{p} = \begin{bmatrix} p_1 \\ p_2 \\ p_3 \end{bmatrix},$$

we find that $\mathbf{p}$ can be obtained by solving the linear system

$$A\mathbf{p} = \mathbf{p}.$$

Section 8.5 discusses this and several other economic models.

1.6 EXERCISES

In Exercises 1 through 4, use the method of Examples 2 and 3.

1. Show that $\begin{bmatrix} 2 & 1 \\ -2 & 3 \end{bmatrix}$ is nonsingular.

2. Show that $\begin{bmatrix} 2 & 1 \\ -4 & -2 \end{bmatrix}$ is singular.

3. Is the matrix

$$\begin{bmatrix} 1 & 1 \\ 3 & 4 \end{bmatrix}$$

singular or nonsingular? If it is nonsingular, find its inverse.

4. Is the matrix

$$\begin{bmatrix} 1 & 2 & -1 \\ 3 & 2 & 3 \\ 2 & 2 & 1 \end{bmatrix}$$

singular or nonsingular? If it is nonsingular, find its inverse.

In Exercises 5 through 10, find the inverses of the given matrices, if possible:

5. (a) $\begin{bmatrix} 1 & 3 \\ -2 & 6 \end{bmatrix}$. (b) $\begin{bmatrix} 1 & 2 & 3 \\ 1 & 1 & 2 \\ 0 & 1 & 2 \end{bmatrix}$.

(c) $\begin{bmatrix} 1 & 1 & 1 & 1 \\ 1 & 2 & -1 & 2 \\ 1 & -1 & 2 & 1 \\ 1 & 3 & 3 & 2 \end{bmatrix}$.

6. (a) $\begin{bmatrix} 1 & 3 \\ 2 & 6 \end{bmatrix}$. (b) $\begin{bmatrix} 1 & 2 & 3 \\ 0 & 2 & 3 \\ 1 & 2 & 4 \end{bmatrix}$.

(c) $\begin{bmatrix} 1 & 1 & 2 & 1 \\ 0 & -2 & 0 & 0 \\ 0 & 3 & 2 & 1 \\ 1 & 2 & 1 & -2 \end{bmatrix}$.

7. (a) $\begin{bmatrix} 1 & 3 \\ 2 & 4 \end{bmatrix}$. (b) $\begin{bmatrix} 1 & 1 & 1 & 1 \\ 1 & 3 & 1 & 2 \\ 1 & 2 & -1 & 1 \\ 5 & 9 & 1 & 6 \end{bmatrix}$.

(c) $\begin{bmatrix} 1 & 2 & 1 \\ 1 & 3 & 2 \\ 1 & 0 & 1 \end{bmatrix}$.

8. (a) $\begin{bmatrix} 1 & 1 & 1 \\ 1 & 2 & 3 \\ 0 & 1 & 1 \end{bmatrix}$. (b) $\begin{bmatrix} 1 & 2 & 2 \\ 1 & 3 & 1 \\ 1 & 3 & 2 \end{bmatrix}$.

(c) $\begin{bmatrix} 1 & 2 & 3 \\ 1 & 1 & 2 \\ 0 & 1 & 1 \end{bmatrix}$.

9. (a) $\begin{bmatrix} 1 & 2 & -3 & 1 \\ -1 & 3 & -3 & -2 \\ 2 & 0 & 1 & 5 \\ 3 & 1 & -2 & 5 \end{bmatrix}$.

(b) $\begin{bmatrix} 3 & 1 & 2 \\ 2 & 1 & 2 \\ 1 & 2 & 2 \end{bmatrix}$. (c) $\begin{bmatrix} 1 & 2 & 3 \\ 1 & 1 & 2 \\ 1 & 1 & 0 \end{bmatrix}$.

10. (a) $\begin{bmatrix} 2 & 1 & 3 \\ 0 & 1 & 2 \\ 1 & 0 & 3 \end{bmatrix}$. (b) $\begin{bmatrix} 1 & -1 & 2 & 3 \\ 4 & 1 & 2 & 0 \\ 2 & -1 & 3 & 1 \\ 4 & 2 & 1 & -5 \end{bmatrix}$.

(c) $\begin{bmatrix} 2 & 1 & -2 \\ 3 & 4 & 6 \\ 7 & 6 & 2 \end{bmatrix}$.

11. Which of the following linear systems have a nontrivial solution?

(a) $x + 2y + 3z = 0$ (b) $2x + y - z = 0$
$\quad\ \ 2y + 2z = 0$ $\qquad x - 2y - 3z = 0$
$x + 2y + 3z = 0.$ $\quad -3x - y + 2z = 0.$

12. Which of the following linear systems have a nontrivial solution?

(a) $x + y + 2z = 0$ (b) $x - y + z = 0$
$2x + y + z = 0$ $\quad 2x + y \qquad = 0$
$3x - y + z = 0.$ $\quad 2x - 2y + 2z = 0.$

(c) $2x - y + 5z = 0$
$3x + 2y - 3z = 0$
$\quad x - y + 4z = 0.$

13. If $A^{-1} = \begin{bmatrix} 2 & 3 \\ 1 & 4 \end{bmatrix}$, find A.

14. If $A^{-1} = \begin{bmatrix} 3 & 4 \\ -1 & -1 \end{bmatrix}$, find A.

15. Show that a matrix that has a row or column consisting entirely of zeros must be singular.

16. Find all values of a for which the inverse of

$$A = \begin{bmatrix} 1 & 1 & 0 \\ 1 & 0 & 0 \\ 1 & 2 & a \end{bmatrix}$$

exists. What is A^{-1}?

17. Consider an industrial process whose matrix is

$$A = \begin{bmatrix} 2 & 1 & 3 \\ 3 & 2 & -1 \\ 2 & 1 & 1 \end{bmatrix}.$$

Find the input matrix for each of the following output matrices:

(a) $\begin{bmatrix} 30 \\ 20 \\ 10 \end{bmatrix}$. (b) $\begin{bmatrix} 12 \\ 8 \\ 14 \end{bmatrix}$.

18. Suppose that $A = \begin{bmatrix} 1 & 3 \\ 2 & 7 \end{bmatrix}$.

(a) Find A^{-1}.

(b) Find $(A^T)^{-1}$. How do $(A^T)^{-1}$ and A^{-1} compare?

19. Is the inverse of a nonsingular symmetric matrix always symmetric? Explain.

20. (a) Is $(A + B)^{-1} = A^{-1} + B^{-1}$?

(b) Is $(cA)^{-1} = \dfrac{1}{c} A^{-1}$?

21. For what values of λ does the homogeneous system

$$\begin{aligned} (\lambda - 1)x + 2y &= 0 \\ 2x + (\lambda - 1)y &= 0 \end{aligned}$$

have a nontrivial solution?

22. If A and B are nonsingular, are $A + B$, $A - B$, and $-A$ nonsingular? Explain.

23. If

$$D = \begin{bmatrix} 4 & 0 & 0 \\ 0 & -2 & 0 \\ 0 & 0 & 3 \end{bmatrix},$$

find D^{-1}.

24. If

$$A^{-1} = \begin{bmatrix} 3 & 2 \\ 1 & 3 \end{bmatrix} \quad \text{and} \quad B^{-1} = \begin{bmatrix} 2 & 5 \\ 3 & -2 \end{bmatrix},$$

find $(AB)^{-1}$.

25. Solve $A\mathbf{x} = \mathbf{b}$ for $\mathbf{x}$ if

$$A^{-1} = \begin{bmatrix} 2 & 3 \\ 4 & 1 \end{bmatrix} \quad \text{and} \quad \mathbf{b} = \begin{bmatrix} 5 \\ 3 \end{bmatrix}.$$

26. Let A be a 3×3 matrix. Suppose that

$$\mathbf{x} = \begin{bmatrix} 1 \\ 2 \\ -3 \end{bmatrix}$$

is a solution to the homogeneous system $A\mathbf{x} = \mathbf{0}$. Is A singular or nonsingular? Justify your answer.

THEORETICAL EXERCISES ▪

T.1. Suppose that A and B are square matrices and $AB = O$. If B is nonsingular, find A.

T.2. Prove Corollary 1.2.

T.3. Let A be an $n \times n$ matrix. Show that if A is singular, then the homogeneous system $A\mathbf{x} = \mathbf{0}$ has a nontrivial solution. (*Hint*: Use Theorem 1.11.)

T.4. Show that the matrix

$$A = \begin{bmatrix} a & b \\ c & d \end{bmatrix}$$

is nonsingular if and only if $ad - bc \neq 0$. If this condition holds, show that

$$A^{-1} = \begin{bmatrix} \dfrac{d}{ad - bc} & \dfrac{-b}{ad - bc} \\[2mm] \dfrac{-c}{ad - bc} & \dfrac{a}{ad - bc} \end{bmatrix}.$$

T.5. Show that the matrix

$$\begin{bmatrix} \cos \theta & \sin \theta \\ -\sin \theta & \cos \theta \end{bmatrix}$$

is nonsingular, and compute its inverse.

T.6. Show that the inverse of a nonsingular upper (lower) triangular matrix is upper (lower) triangular.

T.7. Show that if A is singular and $A\mathbf{x} = \mathbf{b}$, $\mathbf{b} \neq \mathbf{0}$ has one solution, then it has infinitely many. (*Hint*: Use Exercise T.13 in Section 1.5.)

T.8. Show that if A is a nonsingular symmetric matrix, then A^{-1} is symmetric.

T.9. Let A be a diagonal matrix with nonzero diagonal entries $a_{11}, a_{22}, \ldots, a_{nn}$. Show that A^{-1} is nonsingular and that A^{-1} is a diagonal matrix with diagonal entries $1/a_{11}, 1/a_{22}, \ldots, 1/a_{nn}$.

MATLAB EXERCISES ▮

In order to use MATLAB in this section, you should first have read Chapter 10 through Section 10.5.

ML.1. Using MATLAB, determine which of the following matrices are nonsingular. Use command **rref**.

(a) $A = \begin{bmatrix} 1 & 2 \\ -2 & 1 \end{bmatrix}$.

(b) $A = \begin{bmatrix} 1 & 2 & 3 \\ 4 & 5 & 6 \\ 7 & 8 & 9 \end{bmatrix}$.

(c) $A = \begin{bmatrix} 1 & 2 & 3 \\ 4 & 5 & 6 \\ 7 & 8 & 0 \end{bmatrix}$.

ML.2. Using MATLAB, determine which of the following matrices are nonsingular. Use command **rref**.

(a) $A = \begin{bmatrix} 1 & 2 \\ 2 & 4 \end{bmatrix}$.   (b) $A = \begin{bmatrix} 1 & 0 & 0 \\ 0 & 1 & 0 \\ 1 & 1 & 1 \end{bmatrix}$.

(c) $A = \begin{bmatrix} 1 & 2 & 1 \\ 0 & 1 & 2 \\ 1 & 0 & 0 \end{bmatrix}$.

ML.3. Using MATLAB, determine the inverse of each of the following matrices. Use command **rref([A eye(size(A))])**.

(a) $A = \begin{bmatrix} 1 & 3 \\ 1 & 2 \end{bmatrix}$.   (b) $A = \begin{bmatrix} 1 & 1 & 2 \\ 2 & 1 & 1 \\ 1 & 2 & 1 \end{bmatrix}$.

ML.4. Using MATLAB, determine the inverse of each of the following matrices. Use command **rref([A eye(size(A))])**.

(a) $A = \begin{bmatrix} 2 & 1 \\ 2 & 3 \end{bmatrix}$.   (b) $A = \begin{bmatrix} 1 & -1 & 2 \\ 0 & 2 & 1 \\ 1 & 0 & 0 \end{bmatrix}$.

ML.5. Using MATLAB, determine a positive integer t so that $(tI - A)$ is singular.

(a) $A = \begin{bmatrix} 1 & 3 \\ 3 & 1 \end{bmatrix}$.   (b) $A = \begin{bmatrix} 4 & 1 & 2 \\ 1 & 4 & 1 \\ 0 & 0 & -4 \end{bmatrix}$.

KEY IDEAS FOR REVIEW ▮

☐ **Method of elimination.** To solve a linear system, repeatedly perform the following operations.

1. Interchange two equations.
2. Multiply an equation by a nonzero constant.
3. Add a multiple of one equation to another equation.

☐ **Matrix operations. Addition** (see page 14); **scalar multiplication** (see page 15); **transpose** (see page 16); **multiplication** (see page 19).

☐ **Theorem 1.1.** Properties of matrix addition. See page 35.

☐ **Theorem 1.2.** Properties of matrix multiplication. See page 37.

☐ **Theorem 1.3.** Properties of scalar multiplication. See page 41.

☐ **Theorem 1.4.** Properties of transpose. See page 41.

☐ **Reduced row echelon form.** See page 47.

☐ Procedure for transforming a matrix to reduced row echelon form. See pages 50–53.

☐ **Gauss–Jordan reduction procedure** (for solving the linear system $A\mathbf{x} = \mathbf{b}$). See page 54.

☐ **Theorem 1.7.** A homogeneous system of m equations in n unknowns always has a nontrivial solution if $m < n$.

☐ **Theorem 1.9.** Properties of the inverse. See page 71.

☐ Practical method for finding A^{-1}. See pages 73–74.

☐ **Theorem 1.11.** An $n \times n$ matrix is nonsingular if and only if it is row equivalent to I_n.

☐ **Theorem 1.12.** If A is an $n \times n$ matrix, the homogeneous system $A\mathbf{x} = \mathbf{0}$ has a nontrivial solution if and only if A is singular.

☐ **List of Nonsingular Equivalences.** The following statements are equivalent:

1. A is nonsingular.
2. $A\mathbf{x} = \mathbf{0}$ has only the trivial solution.
3. A is row equivalent to I_n.
4. The linear system $A\mathbf{x} = \mathbf{b}$ has a unique solution for every $n \times 1$ matrix $\mathbf{b}$.

SUPPLEMENTARY EXERCISES ■

In Exercises 1 through 3, let

$$A = \begin{bmatrix} 1 & 2 \\ 3 & -2 \end{bmatrix}, \quad B = \begin{bmatrix} 3 & -5 \\ 2 & 4 \end{bmatrix},$$

and

$$C = \begin{bmatrix} 4 & 1 \\ 3 & 2 \end{bmatrix}.$$

1. Compute $2A + BC$ if possible.

2. Compute $A^2 - 2A + 3I_2$ if possible.

3. Compute $A^T + B^T C$ if possible.

4. (a) If A and B are $n \times n$ matrices, when is
$$(A + B)(A - B) = A^2 - B^2?$$

   (b) Let A, B, and C be $n \times n$ matrices such that $AC = CA$ and $BC = CB$. Verify that $(AB)C = C(AB)$.

5. (a) Write the augmented matrix of the linear system

$$\begin{aligned} x_1 + 2x_2 - x_3 + \ x_4 &= \ \ 7 \\ 2x_1 - \ x_2 \qquad + 2x_4 &= -8. \end{aligned}$$

   (b) Write the linear system whose augmented matrix is
$$\begin{bmatrix} 3 & 2 & \vdots & -4 \\ 5 & 1 & \vdots & 2 \\ 3 & 2 & \vdots & 6 \end{bmatrix}.$$

6. If
$$A = \begin{bmatrix} 1 & 3 & -2 & 5 \\ 2 & 1 & 3 & 2 \\ 4 & 7 & -1 & -8 \\ -3 & 1 & -8 & 1 \end{bmatrix},$$

find a matrix C in reduced row echelon form that is row equivalent to A.

7. Find all solutions to the linear system

$$\begin{aligned} x + y - \ z &= 5 \\ 2x + y + \ z &= 2 \\ x - y - 2z &= 3. \end{aligned}$$

8. Find all solutions to the linear system

$$\begin{aligned} x + \ y - z + \ w &= \ \ 3 \\ 2x - \ y - z + 2w &= \ \ 4 \\ -3y + z \qquad &= -2 \\ -3x + 3y + z - 3w &= -5. \end{aligned}$$

9. Find all values of a for which the resulting system has (a) no solution, (b) a unique solution, and (c) infinitely many solutions.

$$\begin{aligned} x + y - \qquad z &= 3 \\ x - y + \qquad 3z &= 4 \\ x + y + (a^2 - 10)z &= a. \end{aligned}$$

10. Find an equation relating b_1, b_2, and b_3 so that the linear system with augmented matrix
$$\begin{bmatrix} 1 & 1 & -2 & \vdots & b_1 \\ 2 & -1 & -1 & \vdots & b_2 \\ 4 & 1 & -5 & \vdots & b_3 \end{bmatrix}$$
has a solution.

11. If
$$A = \begin{bmatrix} 5 & 3 & 1 \\ 0 & 4 & 2 \\ 0 & 0 & 4 \end{bmatrix}$$

and $\lambda = 4$, find all solutions to the homogeneous system $(\lambda I_3 - A)\mathbf{x} = \mathbf{0}$.

12. For what values of a is the linear system
$$\begin{aligned} x_1 \qquad + \ x_3 &= \ \ a^2 \\ 2x_1 + x_2 + 3x_3 &= -3a \\ 3x_1 + x_2 + 4x_3 &= -2 \end{aligned}$$
consistent?

13. If possible, find the inverse of the matrix
$$\begin{bmatrix} 1 & 2 & 3 \\ 2 & 5 & 3 \\ 1 & 0 & 8 \end{bmatrix}.$$

14. If possible, find the inverse of the matrix
$$\begin{bmatrix} -1 & 2 & 1 \\ 1 & 0 & 1 \\ -3 & 2 & 3 \end{bmatrix}.$$

15. Does the homogeneous system with coefficient matrix
$$\begin{bmatrix} 1 & -1 & 3 \\ 1 & 2 & -3 \\ 2 & 1 & 0 \end{bmatrix}$$
have a nontrivial solution?

16. If $A^{-1} = \begin{bmatrix} 1 & 2 & -1 \\ 3 & 4 & 2 \\ 0 & 1 & -2 \end{bmatrix}$ and
$$B^{-1} = \begin{bmatrix} 0 & 1 & 1 \\ 1 & 0 & 1 \\ -2 & 3 & 2 \end{bmatrix},$$
compute $(AB)^{-1}$.

17. Solve $A\mathbf{x} = \mathbf{b}$ for $\mathbf{x}$ if

$$A^{-1} = \begin{bmatrix} 1 & 2 & 0 \\ 0 & 1 & 0 \\ 3 & 1 & -1 \end{bmatrix} \quad \text{and} \quad \mathbf{b} = \begin{bmatrix} 2 \\ 1 \\ 3 \end{bmatrix}.$$

18. For what value(s) of λ does the homogeneous system

$$\begin{aligned} (\lambda - 2)x + \quad\quad 2y &= 0 \\ 2x + (\lambda - 2)y &= 0 \end{aligned}$$

have a nontrivial solution?

19. If A is an $n \times n$ matrix and $A^4 = O$, verify that $(I_n - A)^{-1} = I_n + A + A^2 + A^3$.

20. If A is a nonsingular $n \times n$ matrix and c is a nonzero scalar, what is $(cA)^{-1}$?

21. For what values of a does the linear system

$$\begin{aligned} x + \ y &= 3 \\ 5x + 5y &= a \end{aligned}$$

have (a) no solution; (b) exactly one solution; (c) infinitely many solutions?

22. Find all values of a for which the following linear systems have solutions.

(a) $\begin{aligned} x + 2y + \ z &= a^2 \\ x + \ y + 3z &= a \\ 3x + 4y + 7z &= 8. \end{aligned}$ (b) $\begin{aligned} x + 2y + \ z &= a^2 \\ x + \ y + 3z &= a \\ 3x + 4y + 8z &= 8. \end{aligned}$

23. Find all values of a for which the following homogeneous system has nontrivial solutions.

$$\begin{aligned} (1 - a)x \quad\quad + \ z &= 0 \\ -ay + \ z &= 0 \\ y - az &= 0. \end{aligned}$$

24. Determine the number of entries on or above the main diagonal of a $k \times k$ matrix when (a) $k = 2$; (b) $k = 3$; (c) $k = 4$; (d) $k = n$.

25. Let

$$A = \begin{bmatrix} 0 & 2 \\ 0 & 5 \end{bmatrix}.$$

(a) Find a $2 \times k$ matrix $B \neq O$ such that $AB = O$ for $k = 1, 2, 3, 4$.

(b) Are your answers to part (a) unique? Explain.

26. Find all 2×2 matrices with real entries of the form

$$A = \begin{bmatrix} a & b \\ 0 & c \end{bmatrix}$$

such that $A^2 = I_2$.

27. An $n \times n$ matrix A (with real entries) is called a **square root** of the $n \times n$ matrix B (with real entries) if $A^2 = B$.

(a) Find a square root of

$$B = \begin{bmatrix} 1 & 1 \\ 0 & 1 \end{bmatrix}.$$

(b) Find a square root of

$$B = \begin{bmatrix} 1 & 0 & 0 \\ 0 & 0 & 0 \\ 0 & 0 & 0 \end{bmatrix}.$$

(c) Find a square root of $B = I_4$.

(d) Show that there is no square root of

$$B = \begin{bmatrix} 0 & 1 \\ 0 & 0 \end{bmatrix}.$$

28. Compute the trace (see Supplementary Exercise T.1) of each of the following matrices.

(a) $\begin{bmatrix} 1 & 0 \\ 2 & 3 \end{bmatrix}.$ (b) $\begin{bmatrix} 2 & 2 & 3 \\ 2 & 4 & 4 \\ 3 & -2 & -5 \end{bmatrix}.$

(c) $\begin{bmatrix} 1 & 0 & 0 \\ 0 & 1 & 0 \\ 0 & 0 & 1 \end{bmatrix}.$

29. Develop a simple expression for the entries of A^n, where n is a positive integer and

$$A = \begin{bmatrix} 1 & \frac{1}{2} \\ 0 & \frac{1}{2} \end{bmatrix}.$$

THEORETICAL EXERCISES ▪

T.1. If $A = \begin{bmatrix} a_{ij} \end{bmatrix}$ is an $n \times n$ matrix, then the **trace** of A, $\mathrm{Tr}(A)$, is defined as the sum of all elements on the main diagonal of A,

$$\mathrm{Tr}(A) = a_{11} + a_{22} + \cdots + a_{nn}.$$

Show that

(a) $\mathrm{Tr}(cA) = c\,\mathrm{Tr}(A)$, where c is a real number.

(b) $\mathrm{Tr}(A + B) = \mathrm{Tr}(A) + \mathrm{Tr}(B)$.

(c) $\mathrm{Tr}(AB) = \mathrm{Tr}(BA)$.

(d) $\mathrm{Tr}(A^T) = \mathrm{Tr}(A)$.

(e) $\mathrm{Tr}(A^T A) \geq 0$.

T.2. Show that if $\operatorname{Tr}(AA^T) = 0$, then $A = O$.

T.3. Let A and B be $n \times n$ matrices. Show that if $A\mathbf{x} = B\mathbf{x}$ for all $n \times 1$ matrices $\mathbf{x}$, then $A = B$.

T.4. Show that there are no 2×2 matrices A and B such that
$$AB - BA = \begin{bmatrix} 1 & 0 \\ 0 & 1 \end{bmatrix}.$$

T.5. Show that if A is skew symmetric (see Exercise T.24 in Section 1.4), then A^k is skew symmetric for any positive integer k.

T.6. Let A be an $n \times n$ skew symmetric matrix and $\mathbf{x}$ an n-vector. Show that $\mathbf{x}^T A\mathbf{x} = 0$ for all $\mathbf{x}$ in R^n.

T.7. Show that every symmetric upper (or lower) triangular matrix (see Exercise T.5 in Section 1.2) is diagonal.

T.8. Let A be an upper triangular matrix. Show that A is nonsingular if and only if all the entries on the main diagonal of A are nonzero.

T.9. Let A be an $m \times n$ matrix. Show that A is row equivalent to O if and only if $A = O$.

T.10. Let A and B be row equivalent $n \times n$ matrices. Show that A is nonsingular if and only if B is nonsingular.

T.11. Show that if AB is nonsingular, then A and B are nonsingular. (*Hint:* First show that B is nonsingular by considering the homogeneous system $B\mathbf{x} = \mathbf{0}$ and using Theorem 1.12.)

T.12. Show that if A is skew symmetric (see Exercise T.24 in Section 1.4), then the elements on the main diagonal of A are all zero.

T.13. Show that if A is skew symmetric and nonsingular, then A^{-1} is skew symmetric.

T.14. If A is an $n \times n$ matrix, then A is called **idempotent** if $A^2 = A$.

    (a) Verify that I_n and O are idempotent.

    (b) Find an idempotent matrix that is not I_n or O.

    (c) Show that the only $n \times n$ nonsingular idempotent matrix is I_n.

T.15. If A is an $n \times n$ matrix, then A is called **nilpotent** if $A^k = O$ for some positive integer k.

    (a) Show that every nilpotent matrix is singular.

    (b) Verify that
$$A = \begin{bmatrix} 0 & 1 & 1 \\ 0 & 0 & 1 \\ 0 & 0 & 0 \end{bmatrix}$$
      is nilpotent.

    (c) If A is nilpotent, show that $I_n - A$ is nonsingular. [*Hint:* Find $(I_n - A)^{-1}$ in the cases $A^k = O$, $k = 1, 2, \ldots$, and look for a pattern.]

T.16. Let A and B be $n \times n$ idempotent matrices. (See Exercise T.14.)

    (a) Show that AB is idempotent if $AB = BA$.

    (b) Show that if A is idempotent, then A^T is idempotent.

    (c) Is $A + B$ idempotent? Justify your answer.

    (d) Find all values of k for which kA is also idempotent.

T.17. Show that the product of two 2×2 skew symmetric matrices is diagonal. Is this true for $n \times n$ skew symmetric matrices with $n > 2$?

T.18. Prove Theorem 1.13.

In Exercises T.19 through T.22, let $\mathbf{x}$ and $\mathbf{y}$ be column matrices with n elements. The **outer product** of $\mathbf{x}$ and $\mathbf{y}$ is the matrix product $\mathbf{x}\mathbf{y}^T$ that gives the $n \times n$ matrix
$$\begin{bmatrix} x_1 y_1 & x_1 y_2 & \cdots & x_1 y_n \\ x_2 y_1 & x_2 y_2 & \cdots & x_2 y_n \\ \vdots & \vdots & & \vdots \\ x_n y_1 & x_n y_2 & \cdots & x_n y_n \end{bmatrix}.$$

T.19. (a) Form the outer product of $\mathbf{x}$ and $\mathbf{y}$, where
$$\mathbf{x} = \begin{bmatrix} 1 \\ 2 \\ 3 \end{bmatrix} \quad \text{and} \quad \mathbf{y} = \begin{bmatrix} 4 \\ 5 \\ 0 \end{bmatrix}.$$

    (b) Form the outer product of $\mathbf{x}$ and $\mathbf{y}$, where
$$\mathbf{x} = \begin{bmatrix} 1 \\ 2 \\ 1 \\ 2 \end{bmatrix} \quad \text{and} \quad \mathbf{y} = \begin{bmatrix} -1 \\ 0 \\ 3 \\ 5 \end{bmatrix}.$$

T.20. Prove or disprove: The outer product of $\mathbf{x}$ and $\mathbf{y}$ equals the outer product of $\mathbf{y}$ and $\mathbf{x}$.

T.21. Show that $\operatorname{Tr}(\mathbf{x}\mathbf{y}^T) = \mathbf{x}^T\mathbf{y}$.

T.22. Show that the outer product of $\mathbf{x}$ and $\mathbf{y}$ is row equivalent to either O or to a matrix with $n - 1$ rows of zeros.

T.23. Let $\mathbf{w}$ be an $n \times 1$ matrix such that $\mathbf{w}^T\mathbf{w} = 1$. The $n \times n$ matrix

$$H = I_n - 2\mathbf{w}\mathbf{w}^T$$

is called a **Householder matrix**. (Note that a Householder matrix is the identity matrix plus a scalar multiple of an outer product.)

(a) Show that H is symmetric.

(b) Show that $H^{-1} = H^T$.

T.24. Let

$$A = \begin{bmatrix} 2 & 0 \\ -1 & 1 \end{bmatrix}.$$

Show that all 2×2 matrices B such that $AB = BA$ are of the form

$$\begin{bmatrix} r & 0 \\ s-r & s \end{bmatrix},$$

where r and s are any real numbers.

CHAPTER TEST ■

1. Find all solutions to the linear system

$$
\begin{aligned}
x_1 + x_2 + \ x_3 - 2x_4 &= 3 \\
2x_1 + x_2 + 3x_3 + 2x_4 &= 5 \\
- x_2 + \ x_3 + 6x_4 &= 3.
\end{aligned}
$$

2. Find all values of a for which the resulting linear system has (a) no solution, (b) a unique solution, and (c) infinitely many solutions.

$$
\begin{aligned}
x + \qquad\ \ z &= \quad 4 \\
2x + \ y + \qquad 3z &= \quad 5 \\
-3x - 3y + (a^2 - 5a)z &= a - 8.
\end{aligned}
$$

3. If possible, find the inverse of the following matrix:

$$\begin{bmatrix} 1 & 2 & -1 \\ 0 & 1 & 1 \\ 1 & 0 & -1 \end{bmatrix}.$$

4. If

$$A = \begin{bmatrix} -1 & -2 \\ -2 & 2 \end{bmatrix},$$

find all values of λ for which the homogeneous system $(\lambda I_2 - A)\mathbf{x} = \mathbf{0}$ has a nontrivial solution.

5. (a) If

$$A^{-1} = \begin{bmatrix} 1 & 3 & 0 \\ 0 & 1 & 1 \\ 1 & -1 & 4 \end{bmatrix}$$

and

$$B^{-1} = \begin{bmatrix} 2 & 1 & 1 \\ 0 & 0 & -2 \\ 1 & 1 & -1 \end{bmatrix},$$

compute $(AB)^{-1}$.

(b) Solve $A\mathbf{x} = \mathbf{b}$ for $\mathbf{x}$ if

$$A^{-1} = \begin{bmatrix} 1 & 0 & -2 \\ 2 & 1 & 3 \\ 4 & 2 & 5 \end{bmatrix} \quad \text{and} \quad \mathbf{b} = \begin{bmatrix} 2 \\ 1 \\ 3 \end{bmatrix}.$$

6. Answer each of the following as true or false.

(a) If A and B are $n \times n$ matrices, then

$$(A + B)(A + B) = A^2 + 2AB + B^2.$$

(b) If $\mathbf{u}_1$ and $\mathbf{u}_2$ are solutions to the linear system $A\mathbf{x} = \mathbf{b}$, then $\mathbf{w} = \frac{1}{4}\mathbf{u}_1 + \frac{3}{4}\mathbf{u}_2$ is also a solution to $A\mathbf{x} = \mathbf{b}$.

(c) If A is a nonsingular matrix, then the homogeneous system $A\mathbf{x} = \mathbf{0}$ has a nontrivial solution.

(d) A homogeneous system of three equations in four unknowns has a nontrivial solution.

(e) If A, B, and C are $n \times n$ nonsingular matrices, then $(ABC)^{-1} = C^{-1}A^{-1}B^{-1}$.

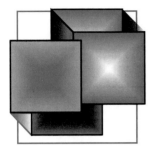

Determinants

2.1 ▼ Definition and Properties

In this section we define the notion of a determinant and study some of its properties. Determinants arose first in the solution of linear systems. Although the method given in Chapter 1 for solving such systems is much more efficient than those involving determinants, determinants are useful in other aspects of linear algebra; some of these areas will be considered in Chapter 5. First, we deal rather briefly with permutations, which are used in our definition of determinant. In this chapter, by matrix we mean a square matrix.

DEFINITION Let $S = \{1, 2, \ldots, n\}$ be the set of integers from 1 to n, arranged in ascending order. A rearrangement $j_1 j_2 \cdots j_n$ of the elements of S is called a **permutation** of S.

To illustrate the definition above, let $S = \{1, 2, 3, 4\}$. Then 4231 is a permutation of S. It corresponds to the function $f: S \to S$ defined by

$$f(1) = 4$$
$$f(2) = 2$$
$$f(3) = 3$$
$$f(4) = 1.$$

We can put any one of the n elements of S in first position, any one of the remaining $n - 1$ elements in second position, any one of the remaining $n - 2$ elements in third position, and so on, until the nth position can only be filled by the last remaining element. Thus there are

$$n(n - 1)(n - 2) \cdots 2 \cdot 1 \tag{1}$$

permutations of S; we denote the set of all permutations of S by S_n.

The expression in Equation (1) is denoted

$$n!, \quad \textbf{\textit{n}} \text{ \textbf{factorial}}.$$

We have

$$1! = 1$$
$$2! = 2 \cdot 1 = 2$$
$$3! = 3 \cdot 2 \cdot 1 = 6$$
$$4! = 4 \cdot 3 \cdot 2 \cdot 1 = 24$$
$$5! = 5 \cdot 4 \cdot 3 \cdot 2 \cdot 1 = 120$$
$$6! = 6 \cdot 5 \cdot 4 \cdot 3 \cdot 2 \cdot 1 = 720$$
$$7! = 7 \cdot 6 \cdot 5 \cdot 4 \cdot 3 \cdot 2 \cdot 1 = 5040$$
$$8! = 8 \cdot 7 \cdot 6 \cdot 5 \cdot 4 \cdot 3 \cdot 2 \cdot 1 = 40{,}320$$
$$9! = 9 \cdot 8 \cdot 7 \cdot 6 \cdot 5 \cdot 4 \cdot 3 \cdot 2 \cdot 1 = 362{,}880.$$

EXAMPLE 1 ■ S_1 consists of only $1! = 1$ permutation of the set $\{1\}$, namely, 1; S_2 consists of $2! = 2 \cdot 1 = 2$ permutations of the set $\{1, 2\}$, namely, 12 and 21; S_3 consists of $3! = 3 \cdot 2 \cdot 1 = 6$ permutations of the set $\{1, 2, 3\}$, namely, 123, 231, 312, 132, 213, and 321. ■

A permutation $j_1 j_2 \cdots j_n$ of $S = \{1, 2, \ldots, n\}$ is said to have an **inversion** if a larger integer j_r precedes a smaller one j_s. A permutation is called **even** or **odd** according to whether the total number of inversions in it is even or odd. Thus, the permutation 4132 of $S = \{1, 2, 3, 4\}$ has four inversions: 4 before 1, 4 before 3, 4 before 2, and 3 before 2. It is then an even permutation.

If $n \geq 2$, it can be shown that S_n has $n!/2$ even permutations and an equal number of odd permutations.

EXAMPLE 2 ■ In S_2, the permutation 12 is even, since it has no inversions; the permutation 21 is odd, since it has one inversion. ■

EXAMPLE 3 ■ The even permutations in S_3 are 123 (no inversions), 231 (two inversions: 21 and 31); and 312 (two inversions: 31 and 32). The odd permutations in S_3 are 132 (one inversion: 32); 213 (one inversion: 21); and 321 (three inversions: 32, 31, and 21).

■

DEFINITION   Let $A = \begin{bmatrix} a_{ij} \end{bmatrix}$ be an $n \times n$ matrix. We define the **determinant** of A (written $\det(A)$ or $|A|$) by

$$\det(A) = |A| = \sum (\pm) a_{1j_1} a_{2j_2} \cdots a_{nj_n}, \tag{2}$$

where the summation ranges over all permutations $j_1 j_2 \cdots j_n$ of the set $S = \{1, 2, \ldots, n\}$. The sign is taken as $+$ or $-$ according to whether the permutation $j_1 j_2 \cdots j_n$ is even or odd.

In each term $(\pm)a_{1j_1}a_{2j_2}\cdots a_{nj_n}$ of $\det(A)$, the row subscripts are in their natural order, whereas the column subscripts are in the order $j_1j_2\cdots j_n$. Since the permutation $j_1j_2\cdots j_n$ is merely a rearrangement of the numbers from 1 to n, it has no repeats. Thus each term in $\det(A)$ is a product of n elements of A each with its appropriate sign, with exactly one element from each row and exactly one element from each column. Since we sum over all the permutations of the set, $S = \{1, 2, \ldots, n\}$, $\det(A)$ has $n!$ terms in the sum.

EXAMPLE 4 ■ If $A = \begin{bmatrix} a_{11} \end{bmatrix}$ is a 1×1 matrix, then S_1 has only one permutation in it, the identity permutation 1, which is even. Thus $\det(A) = a_{11}$.   ■

EXAMPLE 5 ■ If

$$A = \begin{bmatrix} a_{11} & a_{12} \\ a_{21} & a_{22} \end{bmatrix}$$

is a 2×2 matrix, then to obtain $\det(A)$ we write down the terms

$$a_{1-}a_{2-} \quad \text{and} \quad a_{1-}a_{2-},$$

and fill in the blanks with all possible elements of S_2; these are 12 and 21. Since 12 is an even permutation, the term $a_{11}a_{22}$ has a $+$ sign associated with it; since 21 is an odd permutation, the term $a_{12}a_{21}$ has a $-$ sign associated with it. Hence

$$\det(A) = a_{11}a_{22} - a_{12}a_{21}.$$

We can also obtain $\det(A)$ by forming the product of the entries on the line from left to right in the following diagram and subtracting from this number the product of the entries on the line from right to left.

$$\begin{array}{cc} a_{11} & a_{12} \\ a_{21} & a_{22} \end{array} .$$

Thus, if

$$A = \begin{bmatrix} 2 & -3 \\ 4 & 5 \end{bmatrix},$$

then $\det(A) = (2)(5) - (-3)(4) = 22$.   ■

EXAMPLE 6 ■ If

$$A = \begin{bmatrix} a_{11} & a_{12} & a_{13} \\ a_{21} & a_{22} & a_{23} \\ a_{31} & a_{32} & a_{33} \end{bmatrix},$$

then, to compute $\det(A)$, we write down the six terms

$$a_{1-}a_{2-}a_{3-}, \quad a_{1-}a_{2-}a_{3-}, \quad a_{1-}a_{2-}a_{3-}, \quad a_{1-}a_{2-}a_{3-},$$
$$a_{1-}a_{2-}a_{3-}, \quad \text{and} \quad a_{1-}a_{2-}a_{3-}.$$

All the elements of S_3 are used to fill in the blanks, and if we prefix each term by $+$ or by $-$ according to whether the permutation used is even or odd, we find that

$$\det(A) = a_{11}a_{22}a_{33} + a_{12}a_{23}a_{31} + a_{13}a_{21}a_{32} - a_{11}a_{23}a_{32}$$
$$- a_{12}a_{21}a_{33} - a_{13}a_{22}a_{31}. \tag{3}$$

We can also obtain $\det(A)$ as follows. Repeat the first and second columns of A as shown below. Form the sum of the products of the entries on the lines from left to right, and subtract from this number the products of the entries on the lines from right to left (verify).

It should be emphasized that the methods given for evaluating $\det(A)$ in Examples 5 and 6 do not apply for $n \geq 4$.

EXAMPLE 7 ■ Let

$$A = \begin{bmatrix} 1 & 2 & 3 \\ 2 & 1 & 3 \\ 3 & 1 & 2 \end{bmatrix}.$$

Evaluate $\det(A)$.

Solution   Substituting in (3), we find that

$$\det(A) = (1)(1)(2) + (2)(3)(3) + (3)(2)(1)$$
$$- (1)(3)(1) - (2)(2)(2) - (3)(1)(3) = 6.$$

We could obtain the same result by using the easy method above (verify).   ■

It may already have struck the reader that this is an extremely tedious way of computing determinants for a sizable value of n. In fact, $10! = 3.6288 \times 10^6$ and $20! = 2.4329 \times 10^{18}$ are enormous numbers. We shall soon develop a number of properties satisfied by determinants, which will greatly reduce the computational effort.

Permutations are studied to some depth in abstract algebra courses and in courses dealing with group theory. We shall not make use of permutations in our methods for computing determinants. We require the following property of permutations. If we interchange two numbers in the permutation $j_1 j_2 \cdots j_n$, then the number of inversions is either increased or decreased by an odd number (Exercise T.1).

EXAMPLE 8 ■ The number of inversions in the permutation 54132 is 8. The number of inversions in the permutation 52134 is 5. The permutation 52134 was obtained from 54132 by interchanging 2 and 4. The number of inversions differs by 3, an odd number.   ■

Properties of Determinants

THEOREM 2.1 ■ *The determinants of a matrix and its transpose are equal, that is,* $\det(A^T) = \det(A)$.

Proof   Let $A = \left[a_{ij}\right]$ and let $A^T = \left[b_{ij}\right]$, where $b_{ij} = a_{ji}$ $(1 \leq i \leq n,\ 1 \leq j \leq n)$. Then from (2) we have

$$\det(A^T) = \sum(\pm)b_{1j_1}b_{2j_2}\cdots b_{nj_n} = \sum(\pm)a_{j_11}a_{j_22}\cdots a_{j_nn}. \qquad (4)$$

We can now rearrange the factors in the term $a_{j_11}a_{j_22}\cdots a_{j_nn}$ so that the row indices are in their natural order. Thus

$$b_{1j_1}b_{2j_2}\cdots b_{nj_n} = a_{j_11}a_{j_22}\cdots a_{j_nn} = a_{1k_1}a_{2k_2}\cdots a_{nk_n}.$$

It can be shown, by the properties of permutations discussed in an abstract algebra course,* that the permutation $k_1k_2\cdots k_n$, which determines the sign associated with $a_{1k_1}a_{2k_2}\cdots a_{nk_n}$, and the permutation $j_1j_2\cdots j_n$, which determines the sign associated with $a_{1j_1}a_{2j_2}\cdots a_{nj_n}$, are both even or both odd. As an example,

$$b_{13}b_{24}b_{35}b_{41}b_{52} = a_{31}a_{42}a_{53}a_{14}a_{25} = a_{14}a_{25}a_{31}a_{42}a_{53};$$

the number of inversions in the permutation 45123 is 6 and the number of inversions in the permutation 34512 is also 6. Since the terms and corresponding signs in (2) and (4) agree, we conclude that $\det(A) = \det(A^T)$. ∎

EXAMPLE 9 ■ Let A be the matrix of Example 7. Then

$$A^T = \begin{bmatrix} 1 & 2 & 3 \\ 2 & 1 & 1 \\ 3 & 3 & 2 \end{bmatrix}.$$

Substituting in (3), we find that

$$\det(A^T) = (1)(1)(2) + (2)(1)(3) + (3)(2)(3)$$
$$- (1)(1)(3) - (2)(2)(2) - (3)(1)(3)$$
$$= 6 = \det(A).$$

∎

Theorem 2.1 will enable us to replace "row" by "column" in many of the additional properties of determinants; we see how to do this in the following theorem.

THEOREM 2.2 ■ *If matrix B results from matrix A by interchanging two rows (columns) of A, then $\det(B) = -\det(A)$.*

Proof   Suppose that B arises from A by interchanging rows r and s of A, say $r < s$. Then we have $b_{rj} = a_{sj}$, $b_{sj} = a_{rj}$, and $b_{ij} = a_{ij}$ for $i \neq r$, $i \neq s$. Now

$$\det(B) = \sum(\pm)b_{1j_1}b_{2j_2}\cdots b_{rj_r}\cdots b_{sj_s}\cdots b_{nj_n}$$

$$= \sum(\pm)a_{1j_1}a_{2j_2}\cdots a_{sj_r}\cdots a_{rj_s}\cdots a_{nj_n}$$

$$= \sum(\pm)a_{1j_1}a_{2j_2}\cdots a_{rj_s}\cdots a_{sj_r}\cdots a_{nj_n}.$$

---

*See J. Fraleigh, *A First Course in Abstract Algebra*, 5th ed., Reading, Mass.: Addison-Wesley Publishing Company, Inc., 1994; and J. Gallian, *Contemporary Abstract Algebra*, 3rd ed., Lexington, Mass.: D. C. Heath and Company, 1994.

The permutation $j_1 j_2 \cdots j_s \cdots j_r \cdots j_n$ results from the permutation $j_1 j_2 \cdots j_r \cdots j_s \cdots j_n$ by an interchange of two numbers; the number of inversions in the former differs by an odd number from the number of inversions in the latter (see Exercise T.1). This means that the sign of each term in $\det(B)$ is the negative of the sign of the corresponding term in $\det(A)$. Hence $\det(B) = -\det(A)$.

Now suppose that B is obtained from A by interchanging two columns of A. Then B^T is obtained from A^T by interchanging two rows of A^T. So $\det(B^T) = -\det(A^T)$, but $\det(B^T) = \det(B)$ and $\det(A^T) = \det(A)$. Hence $\det(B) = -\det(A)$. ■

In the results to follow, proofs will be given only for the rows of A; the proofs of the corresponding column case proceed as at the end of the proof of Theorem 2.2.

EXAMPLE 10 ■ We have

$$\begin{vmatrix} 2 & -1 \\ 3 & 2 \end{vmatrix} = 7 \quad \text{and} \quad \begin{vmatrix} 3 & 2 \\ 2 & -1 \end{vmatrix} = -7.$$ ■

THEOREM 2.3 ■ *If two rows (columns) of A are equal, then $\det(A) = 0$.*

Proof   Suppose that rows r and s of A are equal. Interchange rows r and s of A to obtain a matrix B. Then $\det(B) = -\det(A)$. On the other hand, $B = A$, so $\det(B) = \det(A)$. Thus $\det(A) = -\det(A)$, so $\det(A) = 0$. ■

EXAMPLE 11 ■ Using Theorem 2.3, it follows that

$$\begin{vmatrix} 1 & 2 & 3 \\ -1 & 0 & 7 \\ 1 & 2 & 3 \end{vmatrix} = 0.$$ ■

THEOREM 2.4 ■ *If a row (column) of A consists entirely of zeros, then $\det(A) = 0$.*

Proof   Let the rth row of A consist entirely of zeros. Since each term in the definition for the determinant of A contains a factor from the rth row, each term in $\det(A)$ is zero. Hence $\det(A) = 0$. ■

EXAMPLE 12 ■ Using Theorem 2.4, it follows that

$$\begin{vmatrix} 1 & 2 & 3 \\ 4 & 5 & 6 \\ 0 & 0 & 0 \end{vmatrix} = 0.$$ ■

THEOREM 2.5 ■ *If B is obtained from A by multiplying a row (column) of A by a real number c, then $\det(B) = c \det(A)$.*

Proof   Suppose that the rth row of $A = \begin{bmatrix} a_{ij} \end{bmatrix}$ is multiplied by c to obtain $B = \begin{bmatrix} b_{ij} \end{bmatrix}$. Then $b_{ij} = a_{ij}$ if $i \neq r$ and $b_{rj} = ca_{rj}$. We obtain $\det(B)$ from Equation (2) as

$$\det(B) = \sum(\pm)b_{1j_1}b_{2j_2}\cdots b_{rj_r}\cdots b_{nj_n}$$

$$= \sum(\pm)a_{1j_1}a_{2j_2}\cdots(ca_{rj_r})\cdots a_{nj_n}$$

$$= c\left(\sum(\pm)a_{1j_1}a_{2j_2}\cdots a_{rj_r}\cdots a_{nj_n}\right) = c\det(A). \qquad \blacksquare$$

We can now use Theorem 2.5 to simplify the computation of $\det(A)$ by factoring out common factors from rows and columns of A.

EXAMPLE 13 ■ We have

$$\begin{vmatrix} 2 & 6 \\ 1 & 12 \end{vmatrix} = 2\begin{vmatrix} 1 & 3 \\ 1 & 12 \end{vmatrix} = (2)(3)\begin{vmatrix} 1 & 1 \\ 1 & 4 \end{vmatrix} = 6(4-1) = 18. \qquad \blacksquare$$

EXAMPLE 14 ■ We have

$$\begin{vmatrix} 1 & 2 & 3 \\ 1 & 5 & 3 \\ 2 & 8 & 6 \end{vmatrix} = 2\begin{vmatrix} 1 & 2 & 3 \\ 1 & 5 & 3 \\ 1 & 4 & 3 \end{vmatrix} = (2)(3)\begin{vmatrix} 1 & 2 & 1 \\ 1 & 5 & 1 \\ 1 & 4 & 1 \end{vmatrix} = (2)(3)(0) = 0.$$

Here we first factored out 2 from the third row, then 3 from the third column, and then used Theorem 2.3, since the first and third columns are equal.   ■

THEOREM 2.6 ■ *If $B = \begin{bmatrix} b_{ij} \end{bmatrix}$ is obtained from $A = \begin{bmatrix} a_{ij} \end{bmatrix}$ by adding to each element of the rth row (column) of A a constant c times the corresponding element of the sth row (column) $r \neq s$ of A, then $\det(B) = \det(A)$.*

Proof   We prove the theorem for rows. We have $b_{ij} = a_{ij}$ for $i \neq r$, and $b_{rj} = a_{rj} + ca_{sj}$, $r \neq s$, say $r < s$. Then

$$\det(B) = \sum(\pm)b_{1j_1}b_{2j_2}\cdots b_{rj_r}\cdots b_{nj_n}$$

$$= \sum(\pm)a_{1j_1}a_{2j_2}\cdots(a_{rj_r}+ca_{sj_r})\cdots a_{sj_s}\cdots a_{nj_n}$$

$$= \sum(\pm)a_{1j_1}a_{2j_2}\cdots a_{rj_r}\cdots a_{sj_s}\cdots a_{nj_n}$$

$$+ \sum(\pm)a_{1j_1}a_{2j_2}\cdots(ca_{sj_r})\cdots a_{sj_s}\cdots a_{nj_n}.$$

The first sum in this last expression is $\det(A)$; the second sum can be written as

$$c\left[\sum(\pm)a_{1j_1}a_{2j_2}\cdots a_{sj_r}\cdots a_{sj_s}\cdots a_{nj_n}\right].$$

Note that

$$\sum (\pm) a_{1j_1} a_{2j_2} \cdots a_{sj_r} \cdots a_{sj_s} \cdots a_{nj_n}$$

$$= \begin{vmatrix} a_{11} & a_{12} & \cdots & a_{1n} \\ a_{21} & a_{22} & \cdots & a_{2n} \\ \vdots & \vdots & & \vdots \\ a_{s1} & a_{s2} & \cdots & a_{sn} \\ \vdots & \vdots & & \vdots \\ a_{s1} & a_{s2} & \cdots & a_{sn} \\ \vdots & \vdots & & \vdots \\ a_{n1} & a_{n2} & \cdots & a_{nn} \end{vmatrix} \begin{matrix} \\ \\ \\ \leftarrow r\text{th row} \\ \\ \leftarrow s\text{th row} \\ \\ \\ \end{matrix}$$

$$= 0,$$

because there are two equal rows. Hence $\det(B) = \det(A) + 0 = \det(A)$. ∎

EXAMPLE 15 ■ We have

$$\begin{vmatrix} 1 & 2 & 3 \\ 2 & -1 & 3 \\ 1 & 0 & 1 \end{vmatrix} = \begin{vmatrix} 5 & 0 & 9 \\ 2 & -1 & 3 \\ 1 & 0 & 1 \end{vmatrix},$$

obtained by adding twice the second row to its first row. By applying the definition of determinant to the second determinant, we can see that both have the value 4. ∎

THEOREM 2.7 ■ *If a matrix $A = \begin{bmatrix} a_{ij} \end{bmatrix}$ is upper (lower) triangular (see Exercise T.5, Section 1.2), then*

$$\det(A) = a_{11} a_{22} \cdots a_{nn};$$

that is, the determinant of a triangular matrix is the product of the elements on the main diagonal.

Proof   Let $A = \begin{bmatrix} a_{ij} \end{bmatrix}$ be upper triangular (that is, $a_{ij} = 0$ for $i > j$). Then a term $a_{1j_1} a_{2j_2} \cdots a_{nj_n}$ in the expression for $\det(A)$ can be nonzero only for $1 \leq j_1$, $2 \leq j_2, \ldots, n \leq j_n$. Now $j_1 j_2 \cdots j_n$ must be a permutation, or rearrangement, of $\{1, 2, \ldots, n\}$. Hence we must have $j_n = n$, $j_{n-1} = n - 1, \ldots, j_2 = 2$, $j_1 = 1$. Thus the only term of $\det(A)$ that can be nonzero is the product of the elements on the main diagonal of A. Since the permutation $12 \cdots n$ has no inversions, the sign associated with it is $+$. Therefore, $\det(A) = a_{11} a_{22} \cdots a_{nn}$.

We leave the proof of the lower triangular case to the reader (Exercise T.2). ∎

Theorems 2.2, 2.5, 2.6, and 2.7 are very useful in the evaluation of $\det(A)$. What we do is transform A by means of our elementary row operations to a triangular matrix. Of course, we must keep track of how the determinant of the resulting matrices changes as we perform the elementary row operations.

EXAMPLE 16 ■ We have

$$
\begin{vmatrix} 4 & 3 & 2 \\ 3 & -2 & 5 \\ 2 & 4 & 6 \end{vmatrix}_{\frac{1}{2}\mathbf{r}_3 \to \mathbf{r}_3} = 2 \begin{vmatrix} 4 & 3 & 2 \\ 3 & -2 & 5 \\ 1 & 2 & 3 \end{vmatrix}_{\mathbf{r}_1 \leftrightarrow \mathbf{r}_3} = -2 \begin{vmatrix} 1 & 2 & 3 \\ 3 & -2 & 5 \\ 4 & 3 & 2 \end{vmatrix}_{-3\mathbf{r}_1 + \mathbf{r}_2 \to \mathbf{r}_2}
$$

$$
= -2 \begin{vmatrix} 1 & 2 & 3 \\ 0 & -8 & -4 \\ 4 & 3 & 2 \end{vmatrix}_{-4\mathbf{r}_1 + \mathbf{r}_3 \to \mathbf{r}_3} = -2 \begin{vmatrix} 1 & 2 & 3 \\ 0 & -8 & -4 \\ 0 & -5 & -10 \end{vmatrix}_{\frac{1}{4}\mathbf{r}_2 \to \mathbf{r}_2}
$$

$$
= (-2)(4) \begin{vmatrix} 1 & 2 & 3 \\ 0 & -2 & -1 \\ 0 & -5 & -10 \end{vmatrix}_{\frac{1}{5}\mathbf{r}_3 \to \mathbf{r}_3} = (-2)(4)(5) \begin{vmatrix} 1 & 2 & 3 \\ 0 & -2 & -1 \\ 0 & -1 & -2 \end{vmatrix}_{-\frac{1}{2}\mathbf{r}_2 + \mathbf{r}_3 \to \mathbf{r}_3}
$$

$$
= (-2)(4)(5) \begin{vmatrix} 1 & 2 & 3 \\ 0 & -2 & -1 \\ 0 & 0 & -\frac{3}{2} \end{vmatrix} = (-2)(4)(5)(1)(-2)\left(-\tfrac{3}{2}\right) = -120.
$$

Here $\frac{1}{2}\mathbf{r}_3 \to \mathbf{r}_3$ means that $\frac{1}{2}$ times the third row $\mathbf{r}_3$ replaces the third row; $\mathbf{r}_1 \leftrightarrow \mathbf{r}_3$ means that the first row $\mathbf{r}_1$ and the third row $\mathbf{r}_3$ are interchanged; $-3\mathbf{r}_1 + \mathbf{r}_2 \to \mathbf{r}_2$ means that (-3) times the first row added to the second row replaces the second row. ■

We shall omit the proof of the following important theorem.

THEOREM 2.8 ■ *The determinant of a product of two matrices is the product of their determinants; that is,*

$$
\det(AB) = \det(A)\det(B). \qquad ■
$$

EXAMPLE 17 ■ Let

$$
A = \begin{bmatrix} 1 & 2 \\ 3 & 4 \end{bmatrix} \quad \text{and} \quad B = \begin{bmatrix} 2 & -1 \\ 1 & 2 \end{bmatrix}.
$$

Then

$$
|A| = -2 \quad \text{and} \quad |B| = 5.
$$

Also,

$$
AB = \begin{bmatrix} 4 & 3 \\ 10 & 5 \end{bmatrix}
$$

and

$$
|AB| = -10 = |A||B|. \qquad ■
$$

As an immediate consequence of Theorem 2.1, we can readily compute $\det(A^{-1})$ from $\det(A)$, as the following corollary shows.

COROLLARY 2.1 ■ *If A is nonsingular, then $\det(A) \neq 0$ and*

$$\det(A^{-1}) = \frac{1}{\det(A)}.$$

Proof   Exercise T.4.   ■

EXAMPLE 18 ■ Let

$$A = \begin{bmatrix} 1 & 2 \\ 3 & 4 \end{bmatrix}.$$

Then $\det(A) = -2$ and

$$A^{-1} = \begin{bmatrix} -2 & 1 \\ \frac{3}{2} & -\frac{1}{2} \end{bmatrix}.$$

Now,

$$\det(A^{-1}) = -\frac{1}{2} = \frac{1}{\det(A)}.$$ ■

2.1 EXERCISES

1. Find the number of inversions in each of the following permutations of $S = \{1, 2, 3, 4, 5\}$.
   (a) 52134.   (b) 45213.   (c) 42135.
   (d) 13542.   (e) 35241.   (f) 12345.

2. Determine whether each of the following permutations of $S = \{1, 2, 3, 4\}$ is even or odd.
   (a) 4213.   (b) 1243.   (c) 1234.
   (d) 3214.   (e) 1423.   (f) 2431.

3. Determine the sign associated with each of the following permutations of $S = \{1, 2, 3, 4, 5\}$.
   (a) 25431.   (b) 31245.   (c) 21345.
   (d) 52341.   (e) 34125.   (f) 41253.

4. In each of the following pairs of permutations of $S = \{1, 2, 3, 4, 5, 6\}$, verify that the number of inversions differs by an odd number.
   (a) 436215 and 416235.
   (b) 623415 and 523416.
   (c) 321564 and 341562.
   (d) 123564 and 423561.

In Exercises 5 and 6, evaluate the determinants using Equation (2).

5. (a) $\begin{vmatrix} 2 & -1 \\ 3 & 2 \end{vmatrix}$.   (b) $\begin{vmatrix} 0 & 3 & 0 \\ 2 & 0 & 0 \\ 0 & 0 & -5 \end{vmatrix}$.

   (c) $\begin{vmatrix} 4 & 2 & 0 \\ 0 & -2 & 5 \\ 0 & 0 & 3 \end{vmatrix}$.   (d) $\begin{vmatrix} 4 & 2 & 2 & 0 \\ 2 & 0 & 0 & 0 \\ 3 & 0 & 0 & 1 \\ 0 & 0 & 1 & 0 \end{vmatrix}$.

6. (a) $\begin{vmatrix} 2 & 1 \\ 4 & 3 \end{vmatrix}$.   (b) $\begin{vmatrix} 0 & 0 & -2 \\ 0 & 3 & 0 \\ 4 & 0 & 0 \end{vmatrix}$.

   (c) $\begin{vmatrix} 3 & 4 & 2 \\ 2 & 5 & 0 \\ 3 & 0 & 0 \end{vmatrix}$.   (d) $\begin{vmatrix} -4 & 2 & 0 & 0 \\ 2 & 3 & 1 & 0 \\ 3 & 1 & 0 & 2 \\ 1 & 3 & 0 & 3 \end{vmatrix}$.

7. Let $A = \begin{bmatrix} a_{ij} \end{bmatrix}$ be a 4×4 matrix. Write the general expression for $\det(A)$ using Equation (2).

8. If

$$|A| = \begin{vmatrix} a_1 & a_2 & a_3 \\ b_1 & b_2 & b_3 \\ c_1 & c_2 & c_3 \end{vmatrix} = -4,$$

find the determinants of the following matrices:

$$B = \begin{bmatrix} a_3 & a_2 & a_1 \\ b_3 & b_2 & b_1 \\ c_3 & c_2 & c_1 \end{bmatrix},$$

$$C = \begin{bmatrix} a_1 & a_2 & a_3 \\ b_1 & b_2 & b_3 \\ 2c_1 & 2c_2 & 2c_3 \end{bmatrix},$$

and

$$D = \begin{bmatrix} a_1 & a_2 & a_3 \\ b_1 + 4c_1 & b_2 + 4c_2 & b_3 + 4c_3 \\ c_1 & c_2 & c_3 \end{bmatrix}.$$

9. If

$$|A| = \begin{vmatrix} a_1 & a_2 & a_3 \\ b_1 & b_2 & b_3 \\ c_1 & c_2 & c_3 \end{vmatrix} = 3,$$

find the determinants of the following matrices:

$$B = \begin{bmatrix} a_1 + 2b_1 - 3c_1 & a_2 + 2b_2 - 3c_2 & a_3 + 2b_3 - 3c_3 \\ b_1 & b_2 & b_3 \\ c_1 & c_2 & c_3 \end{bmatrix},$$

$$C = \begin{bmatrix} a_1 & 3a_2 & a_3 \\ b_1 & 3b_2 & b_3 \\ c_1 & 3c_2 & c_3 \end{bmatrix},$$

and

$$D = \begin{bmatrix} a_1 & a_2 & a_3 \\ c_1 & c_2 & c_3 \\ b_1 & b_2 & b_3 \end{bmatrix}.$$

10. If

$$A = \begin{bmatrix} 1 & -1 & 2 \\ 3 & 4 & 1 \\ 2 & 5 & 1 \end{bmatrix},$$

verify that $\det(A) = \det(A^T)$.

11. Evaluate:

(a) $\det\left(\begin{bmatrix} \lambda - 1 & 2 \\ 3 & \lambda - 2 \end{bmatrix} \right)$.

(b) $\det(\lambda I_2 - A)$, where $A = \begin{bmatrix} 4 & 2 \\ -1 & 1 \end{bmatrix}$.

12. Evaluate:

(a) $\det\left(\begin{bmatrix} \lambda - 1 & -1 & -2 \\ 0 & \lambda - 2 & 2 \\ 0 & 0 & \lambda - 3 \end{bmatrix} \right)$.

(b) $\det(\lambda I_3 - A)$, where $A = \begin{bmatrix} -1 & 0 & 1 \\ -2 & 0 & -1 \\ 0 & 0 & 1 \end{bmatrix}$.

13. For each of the matrices in Exercise 11, find all values of λ for which the determinant is 0.

14. For each of the matrices in Exercise 12, find all values of λ for which the determinant is 0.

In Exercises 15 through 18, evaluate the given determinant using the properties given in Theorems 2.1 through 2.7.

15. (a) $\begin{vmatrix} 4 & -3 & 5 \\ 5 & 2 & 0 \\ 2 & 0 & 4 \end{vmatrix}$. (b) $\begin{vmatrix} 2 & 0 & 1 & 4 \\ 3 & 2 & -4 & -2 \\ 2 & 3 & -1 & 0 \\ 11 & 8 & -4 & 6 \end{vmatrix}$.

(c) $\begin{vmatrix} 4 & 1 & 2 \\ 0 & 2 & 3 \\ 0 & 0 & -3 \end{vmatrix}$.

16. (a) $\begin{vmatrix} 4 & 0 & 0 & 0 \\ -1 & 2 & 0 & 0 \\ 1 & 2 & -3 & 0 \\ 1 & 5 & 3 & 5 \end{vmatrix}$. (b) $\begin{vmatrix} 4 & 1 & 3 \\ 2 & 3 & 0 \\ 1 & 3 & 2 \end{vmatrix}$.

(c) $\begin{vmatrix} 1 & 2 & 3 \\ 2 & 1 & 0 \\ -3 & 1 & 2 \end{vmatrix}$.

17. (a) $\begin{vmatrix} 4 & 2 & 3 & -4 \\ 3 & -2 & 1 & 5 \\ -2 & 0 & 1 & -3 \\ 8 & -2 & 6 & 4 \end{vmatrix}$.

(b) $\begin{vmatrix} 1 & 3 & -4 \\ -2 & 1 & 2 \\ -9 & 15 & 0 \end{vmatrix}$. (c) $\begin{vmatrix} 1 & 1 & 2 \\ 0 & 2 & -2 \\ 0 & 0 & 3 \end{vmatrix}$.

18. (a) $\begin{vmatrix} 1 & 0 & 1 \\ 1 & 1 & 0 \\ 2 & 1 & 0 \end{vmatrix}$. (b) $\begin{vmatrix} 2 & 0 & 0 & 0 \\ -5 & 3 & 0 & 0 \\ 3 & 2 & 4 & 0 \\ 4 & 2 & 1 & -5 \end{vmatrix}$.

(c) $\begin{vmatrix} 1 & 2 & -1 \\ 3 & 2 & 0 \\ 1 & 4 & 3 \end{vmatrix}$.

19. Verify that $\det(AB) = \det(A)\det(B)$ for the following:

(a) $A = \begin{bmatrix} 1 & -2 & 3 \\ -2 & 3 & 1 \\ 0 & 1 & 0 \end{bmatrix}$, $B = \begin{bmatrix} 1 & 0 & 2 \\ 3 & -2 & 5 \\ 2 & 1 & 3 \end{bmatrix}$.

(b) $A = \begin{bmatrix} 2 & 3 & 6 \\ 0 & 3 & 2 \\ 0 & 0 & -4 \end{bmatrix}$, $B = \begin{bmatrix} 3 & 0 & 0 \\ 4 & 5 & 0 \\ 2 & 1 & -2 \end{bmatrix}$.

20. If $|A| = -4$, find

(a) $|A^2|$. (b) $|A^4|$. (c) $|A^{-1}|$.

21. If A and B are $n \times n$ matrices with $|A| = 2$ and $|B| = -3$, calculate $|A^{-1}B^T|$.

THEORETICAL EXERCISES ▦

T.1. Show that if we interchange two numbers in the permutation $j_1 j_2 \cdots j_n$, then the number of inversions is either increased or decreased by an odd number. (*Hint:* First show that if two adjacent numbers are interchanged, the number of inversions is either increased or decreased by 1. Then show that an interchange of any two numbers can be achieved by an odd number of successive interchanges of adjacent numbers.)

T.2. Prove Theorem 2.7 for the lower triangular case.

T.3. Show that if c is a real number and A is $n \times n$, then $\det(cA) = c^n \det(A)$.

T.4. Prove Corollary 2.1.

T.5. Show that if $\det(AB) = 0$, then $\det(A) = 0$ or $\det(B) = 0$.

T.6. Is $\det(AB) = \det(BA)$? Justify your answer.

T.7. Show that if A is a matrix such that in each row and in each column one and only one element is $\neq 0$, then $\det(A) \neq 0$.

T.8. Show that if $AB = I_n$, then $\det(A) \neq 0$ and $\det(B) \neq 0$.

T.9. (a) Show that if $A = A^{-1}$, then $\det(A) = \pm 1$.
(b) Show that if $A^T = A^{-1}$, then $\det(A) = \pm 1$.

T.10. Show that if A is a nonsingular matrix such that $A^2 = A$, then $\det(A) = 1$.

T.11. Show that
$$\det(A^T B^T) = \det(A)\det(B^T)$$
$$= \det(A^T)\det(B).$$

T.12. Show that
$$\begin{vmatrix} a^2 & a & 1 \\ b^2 & b & 1 \\ c^2 & c & 1 \end{vmatrix} = (b-a)(c-a)(b-c).$$

This determinant is called a **Vandermonde**[*] **determinant**.

T.13. Let $A = \begin{bmatrix} a_{ij} \end{bmatrix}$ be an upper triangular matrix. Show that A is nonsingular if and only if $a_{ii} \neq 0$, $1 \leq i \leq n$.

T.14. Show that if $\det(A) = 0$, then $\det(AB) = 0$.

T.15. Show that if $A^n = O$, for some positive integer n, then $\det(A) = 0$.

T.16. Show that if A is $n \times n$, with A skew symmetric ($A^T = -A$, see Section 1.4, Exercise T.24), and n odd, then $\det(A) = 0$.

MATLAB EXERCISES ▦

In order to use MATLAB in this section, you should first have read Chapter 10 through Section 10.5.

ML.1. Use the routine **reduce** to perform row operations and keep track by hand of the changes in the determinant as in Example 16.

(a) $A = \begin{bmatrix} 2 & 1 & 3 \\ 1 & 3 & 2 \\ 3 & 2 & 1 \end{bmatrix}$.

(b) $A = \begin{bmatrix} 0 & 1 & 3 & -2 \\ -2 & 1 & 1 & 1 \\ 2 & 0 & 1 & 2 \\ 1 & 0 & 0 & 1 \end{bmatrix}$.

ML.2. Use routine **reduce** to perform row operations and keep track by hand of the changes in the determinant as in Example 16.

(a) $A = \begin{bmatrix} 1 & 0 & 2 \\ 0 & 2 & 1 \\ 2 & 1 & 0 \end{bmatrix}$.

---

[*]Alexandre-Théophile Vandermonde (1735–1796) was born in Paris. His father, a physician, tried to steer him toward a musical career. His published mathematical work consisted of four papers that were presented over a two-year period. He is generally considered the founder of the theory of determinants and also developed formulas for solving general quadratic, cubic, and quartic equations. Vandermonde was a cofounder of the Conservatoire des Arts et Métiers and was its director from 1782. In 1795 he helped to develop a course in political economy. He was an active revolutionary in the French Revolution and was a member of the Commune of Paris and the club of the Jacobins.

(b) $A = \begin{bmatrix} 1 & 2 & 0 & 0 \\ 2 & 1 & 2 & 0 \\ 0 & 2 & 1 & 2 \\ 0 & 0 & 2 & 1 \end{bmatrix}$.

ML.3. MATLAB has command **det**, which returns the value of the determinant of a matrix. Just type **det(A)**. Find the determinant of each of the following matrices using **det**.

(a) $A = \begin{bmatrix} 1 & -1 & 1 \\ 1 & 1 & -1 \\ -1 & 1 & 1 \end{bmatrix}$.

(b) $A = \begin{bmatrix} 1 & 2 & 3 & 4 \\ 2 & 3 & 4 & 5 \\ 3 & 4 & 5 & 6 \\ 4 & 5 & 6 & 7 \end{bmatrix}$.

ML.4. Use **det** (see Exercise ML.3) to compute the determinant of each of the following.

(a) $5 * \mathbf{eye(size(A))} - \mathbf{A}$, where

$$A = \begin{bmatrix} 2 & 3 & 0 \\ 4 & 1 & 0 \\ 0 & 0 & 5 \end{bmatrix}.$$

(b) $(3 * \mathbf{eye(size(A))} - \mathbf{A})\mathbf{\char`\^2}$, where

$$A = \begin{bmatrix} 1 & 1 \\ 5 & 2 \end{bmatrix}.$$

(c) $\mathbf{invert(A)} * \mathbf{A}$, where

$$A = \begin{bmatrix} 1 & 1 & 0 \\ 0 & 1 & 0 \\ 1 & 0 & 1 \end{bmatrix}.$$

ML.5. Determine a positive integer t so that $\mathbf{det(t * eye(size(A))} - \mathbf{A)} = \mathbf{0}$, where

$$A = \begin{bmatrix} 5 & 2 \\ -1 & 2 \end{bmatrix}.$$

2.2 ▾ Cofactor Expansion and Applications

So far, we have been evaluating determinants by using Equation (2) of the preceding section and have been aided by the properties established there. We now develop a different method for evaluating the determinant of an $n \times n$ matrix, which reduces the problem to the evaluation of determinants of matrices of order $n - 1$. We can then repeat the process for these $(n - 1) \times (n - 1)$ matrices until we get to 2×2 matrices.

DEFINITION

Let $A = \begin{bmatrix} a_{ij} \end{bmatrix}$ be an $n \times n$ matrix. Let M_{ij} be the $(n - 1) \times (n - 1)$ submatrix of A obtained by deleting the ith row and jth column of A. The determinant $\det(M_{ij})$ is called the **minor** of a_{ij}. The **cofactor** A_{ij} of a_{ij} is defined as

$$A_{ij} = (-1)^{i+j} \det(M_{ij}).$$

EXAMPLE 1 ■ Let

$$A = \begin{bmatrix} 3 & -1 & 2 \\ 4 & 5 & 6 \\ 7 & 1 & 2 \end{bmatrix}.$$

Then

$$\det(M_{12}) = \begin{vmatrix} 4 & 6 \\ 7 & 2 \end{vmatrix} = 8 - 42 = -34, \quad \det(M_{23}) = \begin{vmatrix} 3 & -1 \\ 7 & 1 \end{vmatrix} = 3 + 7 = 10,$$

and

$$\det(M_{31}) = \begin{vmatrix} -1 & 2 \\ 5 & 6 \end{vmatrix} = -6 - 10 = -16.$$

Also,

$$A_{12} = (-1)^{1+2} \det(M_{12}) = (-1)(-34) = 34,$$

$$A_{23} = (-1)^{2+3} \det(M_{23}) = (-1)(10) = -10,$$

and

$$A_{31} = (-1)^{3+1} \det(M_{31}) = (1)(-16) = -16. \qquad \blacksquare$$

If we think of the sign $(-1)^{i+j}$ as being located in position (i, j) of an $n \times n$ matrix, then the signs form a checkerboard pattern that has a $+$ in the $(1, 1)$ position. The patterns for $n = 3$ and $n = 4$ are as follows:

$$
\begin{array}{ccc}
+ & - & + \\
- & + & - \\
+ & - & +
\end{array}
\qquad\qquad
\begin{array}{cccc}
+ & - & + & - \\
- & + & - & + \\
+ & - & + & - \\
- & + & - & +
\end{array}
$$

$$n = 3 \qquad\qquad n = 4$$

The following theorem gives another method of evaluating determinants that is not as computationally efficient as reduction to triangular form.

THEOREM 2.9 ■ *Let $A = \begin{bmatrix} a_{ij} \end{bmatrix}$ be an $n \times n$ matrix. Then for each $1 \leq i \leq n$,*

$$\det(A) = a_{i1}A_{i1} + a_{i2}A_{i2} + \cdots + a_{in}A_{in}$$
$$(\text{expansion of } \det(A) \text{ about the } i\text{th row});$$

$$(1)$$

and for each $1 \leq j \leq n$,

$$\det(A) = a_{1j}A_{1j} + a_{2j}A_{2j} + \cdots + a_{nj}A_{nj}$$
$$(\text{expansion of } \det(A) \text{ about the } j\text{th column}).$$

$$(2)$$

Proof   The first formula follows from the second by Theorem 2.1, that is, from the fact that $\det(A^T) = \det(A)$. We omit the general proof and consider the 3×3 matrix $A = \begin{bmatrix} a_{ij} \end{bmatrix}$. From (3) in Section 2.1,

$$\det(A) = a_{11}a_{22}a_{33} + a_{12}a_{23}a_{31} + a_{13}a_{21}a_{32}$$
$$- a_{11}a_{23}a_{32} - a_{12}a_{21}a_{33} - a_{13}a_{22}a_{31}.$$

$$(3)$$

We can write this expression as

$$\det(A) = a_{11}(a_{22}a_{33} - a_{23}a_{32}) + a_{12}(a_{23}a_{31} - a_{21}a_{33})$$
$$+ a_{13}(a_{21}a_{32} - a_{22}a_{31}).$$

Now,

$$A_{11} = (-1)^{1+1} \begin{vmatrix} a_{22} & a_{23} \\ a_{32} & a_{33} \end{vmatrix} = (a_{22}a_{33} - a_{23}a_{32}),$$

$$A_{12} = (-1)^{1+2} \begin{vmatrix} a_{21} & a_{23} \\ a_{31} & a_{33} \end{vmatrix} = (a_{23}a_{31} - a_{21}a_{33}),$$

$$A_{13} = (-1)^{1+3} \begin{vmatrix} a_{21} & a_{22} \\ a_{31} & a_{32} \end{vmatrix} = (a_{21}a_{32} - a_{22}a_{31}).$$

Hence

$$\det(A) = a_{11}A_{11} + a_{12}A_{12} + a_{13}A_{13},$$

which is the expansion of $\det(A)$ about the first row.

If we now write (3) as

$$\det(A) = a_{13}(a_{21}a_{32} - a_{22}a_{31}) + a_{23}(a_{12}a_{31} - a_{11}a_{32})$$
$$+ a_{33}(a_{11}a_{22} - a_{12}a_{21}),$$

we can easily verify that

$$\det(A) = a_{13}A_{13} + a_{23}A_{23} + a_{33}A_{33},$$

which is the expansion of $\det(A)$ about the third column. ∎

EXAMPLE 2 ■ To evaluate the determinant

$$\begin{vmatrix} 1 & 2 & -3 & 4 \\ -4 & 2 & 1 & 3 \\ 3 & 0 & 0 & -3 \\ 2 & 0 & -2 & 3 \end{vmatrix},$$

we note that it is best to expand about either the second column or the third row because they each have two zeros. Obviously, the optimal course of action is to expand about a row or column having the largest number of zeros because in that case the cofactors A_{ij} of those a_{ij} which are zero do not have to be evaluated, since $a_{ij}A_{ij} = (0)(A_{ij}) = 0$. Thus, expanding about the third row, we have

$$\begin{vmatrix} 1 & 2 & -3 & 4 \\ -4 & 2 & 1 & 3 \\ 3 & 0 & 0 & -3 \\ 2 & 0 & -2 & 3 \end{vmatrix}$$

$$= (-1)^{3+1}(3)\begin{vmatrix} 2 & -3 & 4 \\ 2 & 1 & 3 \\ 0 & -2 & 3 \end{vmatrix} + (-1)^{3+2}(0)\begin{vmatrix} 1 & -3 & 4 \\ -4 & 1 & 3 \\ 2 & -2 & 3 \end{vmatrix}$$

$$+ (-1)^{3+3}(0)\begin{vmatrix} 1 & 2 & 4 \\ -4 & 2 & 3 \\ 2 & 0 & 3 \end{vmatrix} + (-1)^{3+4}(-3)\begin{vmatrix} 1 & 2 & -3 \\ -4 & 2 & 1 \\ 2 & 0 & -2 \end{vmatrix}.$$

(4)

We now evaluate

$$\begin{vmatrix} 2 & -3 & 4 \\ 2 & 1 & 3 \\ 0 & -2 & 3 \end{vmatrix}$$

by expanding about the first column, obtaining

$$(-1)^{1+1}(2)\begin{vmatrix} 1 & 3 \\ -2 & 3 \end{vmatrix} + (-1)^{2+1}(2)\begin{vmatrix} -3 & 4 \\ -2 & 3 \end{vmatrix} = (1)(2)(9) + (-1)(2)(-1) = 20.$$

Similarly, we evaluate

$$\begin{vmatrix} 1 & 2 & -3 \\ -4 & 2 & 1 \\ 2 & 0 & -2 \end{vmatrix}$$

by expanding about the third row, obtaining

$$(-1)^{3+1}(2)\begin{vmatrix} 2 & -3 \\ 2 & 1 \end{vmatrix} + (-1)^{3+3}(-2)\begin{vmatrix} 1 & 2 \\ -4 & 2 \end{vmatrix}$$

$$= (1)(2)(8) + (1)(-2)(10) = -4.$$

Substituting in Equation (4), we find the value of the given determinant as

$$(+1)(3)(20) + 0 + 0 + (-1)(-3)(-4) = 48.$$

On the other hand, evaluating the given determinant by expanding about the first column, we have

$$(-1)^{1+1}(1)\begin{vmatrix} 2 & 1 & 3 \\ 0 & 0 & -3 \\ 0 & -2 & 3 \end{vmatrix} + (-1)^{2+1}(-4)\begin{vmatrix} 2 & -3 & 4 \\ 0 & 0 & -3 \\ 0 & -2 & 3 \end{vmatrix}$$

$$+ (-1)^{3+1}(3)\begin{vmatrix} 2 & -3 & 4 \\ 2 & 1 & 3 \\ 0 & -2 & 3 \end{vmatrix} + (-1)^{4+1}(2)\begin{vmatrix} 2 & -3 & 4 \\ 2 & 1 & 3 \\ 0 & 0 & -3 \end{vmatrix}$$

$$= (1)(1)(-12) + (-1)(-4)(-12) + (1)(3)(20) + (-1)(2)(-24) = 48. \quad \blacksquare$$

We can use the properties of Section 2.1 to introduce many zeros in a given row or column and then expand about that row or column. This is illustrated in the following example.

EXAMPLE 3 ■ Consider the determinant of Example 2. We have

$$\begin{vmatrix} 1 & 2 & -3 & 4 \\ -4 & 2 & 1 & 3 \\ 3 & 0 & 0 & -3 \\ 2 & 0 & -2 & 3 \end{vmatrix}_{c_4 + c_1 \to c_4} = \begin{vmatrix} 1 & 2 & -3 & 5 \\ -4 & 2 & 1 & -1 \\ 3 & 0 & 0 & 0 \\ 2 & 0 & -2 & 5 \end{vmatrix}$$

$$= (-1)^{3+1}(3)\begin{vmatrix} 2 & -3 & 5 \\ 2 & 1 & -1 \\ 0 & -2 & 5 \end{vmatrix}_{r_1 - r_2 \to r_1}$$

$$= (-1)^4(3)\begin{vmatrix} 0 & -4 & 6 \\ 2 & 1 & -1 \\ 0 & -2 & 5 \end{vmatrix}$$

$$= (-1)^4(3)(-2)(-8) = 48.$$

Here $c_4 + c_1 \to c_4$ means that the fourth column c_4 of the matrix plus the first column c_1 replaces the fourth column. ■

The Inverse of a Matrix

It is interesting to ask what $a_{i1}A_{k1} + a_{i2}A_{k2} + \cdots + a_{in}A_{kn}$ is for $i \neq k$ because, as soon as we answer this question, we shall obtain another method for finding the inverse of a nonsingular matrix.

THEOREM 2.10 ■ *If* $A = \begin{bmatrix} a_{ij} \end{bmatrix}$ *is an* $n \times n$ *matrix, then*

$$a_{i1}A_{k1} + a_{i2}A_{k2} + \cdots + a_{in}A_{kn} = 0 \quad \text{for } i \neq k; \tag{5}$$

$$a_{1j}A_{1k} + a_{2j}A_{2k} + \cdots + a_{nj}A_{nk} = 0 \quad \text{for } j \neq k. \tag{6}$$

Proof   We prove only the first formula. The second follows from the first one by Theorem 2.1.

Consider the matrix B obtained from A by replacing the kth row of A by its ith row. Thus B is a matrix having two identical rows—the ith and kth rows. Then $\det(B) = 0$. Now expand $\det(B)$ about the kth row. The elements of the kth row of B are $a_{i1}, a_{i2}, \ldots, a_{in}$. The cofactors of the kth row are $A_{k1}, A_{k2}, \ldots, A_{kn}$. Thus from Equation (1) we have

$$0 = \det(B) = a_{i1}A_{k1} + a_{i2}A_{k2} + \cdots + a_{in}A_{kn},$$

which is what we wanted to show.   ■

This theorem says that if we sum the products of the elements of any row (column) times the corresponding cofactors of any other row (column), then we obtain zero.

EXAMPLE 4 ■ Let

$$A = \begin{bmatrix} 1 & 2 & 3 \\ -2 & 3 & 1 \\ 4 & 5 & -2 \end{bmatrix}.$$

Then

$$A_{21} = (-1)^{2+1} \begin{vmatrix} 2 & 3 \\ 5 & -2 \end{vmatrix} = 19, \qquad A_{22} = (-1)^{2+2} \begin{vmatrix} 1 & 3 \\ 4 & -2 \end{vmatrix} = -14,$$

$$A_{23} = (-1)^{2+3} \begin{vmatrix} 1 & 2 \\ 4 & 5 \end{vmatrix} = 3.$$

Now

$$a_{31}A_{21} + a_{32}A_{22} + a_{33}A_{23} = (4)(19) + (5)(-14) + (-2)(3) = 0$$

and

$$a_{11}A_{21} + a_{12}A_{22} + a_{13}A_{23} = (1)(19) + (2)(-14) + (3)(3) = 0. \qquad ■$$

We may combine (1) and (5) as

$$\begin{aligned} a_{i1}A_{k1} + a_{i2}A_{k2} + \cdots + a_{in}A_{kn} &= \det(A) \quad \text{if } i = k \\ &= 0 \qquad\quad \text{if } i \neq k. \end{aligned} \tag{7}$$

Similarly, we may combine (2) and (6) as

$$
\begin{aligned}
a_{1j}A_{1k} + a_{2j}A_{2k} + \cdots + a_{nj}A_{nk} &= \det(A) \quad \text{if } j = k \\
&= 0 \qquad\quad \text{if } j \neq k.
\end{aligned}
\tag{8}
$$

DEFINITION   Let $A = \begin{bmatrix} a_{ij} \end{bmatrix}$ be an $n \times n$ matrix. The $n \times n$ matrix adj A, called the **adjoint** of A, is the matrix whose i, jth element is the cofactor A_{ji} of a_{ji}. Thus

$$
\text{adj } A = \begin{bmatrix}
A_{11} & A_{21} & \cdots & A_{n1} \\
A_{12} & A_{22} & \cdots & A_{n2} \\
\vdots & \vdots & & \vdots \\
A_{1n} & A_{2n} & \cdots & A_{nn}
\end{bmatrix}.
$$

REMARK   It should be noted that the term *adjoint* has other meanings in linear algebra in addition to its use in the above definition.

EXAMPLE 5   ■   Let

$$
A = \begin{bmatrix}
3 & -2 & 1 \\
5 & 6 & 2 \\
1 & 0 & -3
\end{bmatrix}.
$$

Compute adj A.

Solution   The cofactors of A are

$$
A_{11} = (-1)^{1+1} \begin{vmatrix} 6 & 2 \\ 0 & -3 \end{vmatrix} = -18; \quad A_{12} = (-1)^{1+2} \begin{vmatrix} 5 & 2 \\ 1 & -3 \end{vmatrix} = 17;
$$

$$
A_{13} = (-1)^{1+3} \begin{vmatrix} 5 & 6 \\ 1 & 0 \end{vmatrix} = -6;
$$

$$
A_{21} = (-1)^{2+1} \begin{vmatrix} -2 & 1 \\ 0 & -3 \end{vmatrix} = -6; \quad A_{22} = (-1)^{2+2} \begin{vmatrix} 3 & 1 \\ 1 & -3 \end{vmatrix} = -10;
$$

$$
A_{23} = (-1)^{2+3} \begin{vmatrix} 3 & -2 \\ 1 & 0 \end{vmatrix} = -2;
$$

$$
A_{31} = (-1)^{3+1} \begin{vmatrix} -2 & 1 \\ 6 & 2 \end{vmatrix} = -10; \quad A_{32} = (-1)^{3+2} \begin{vmatrix} 3 & 1 \\ 5 & 2 \end{vmatrix} = -1;
$$

$$
A_{33} = (-1)^{3+3} \begin{vmatrix} 3 & -2 \\ 5 & 6 \end{vmatrix} = 28.
$$

Then

$$
\text{adj } A = \begin{bmatrix}
-18 & -6 & -10 \\
17 & -10 & -1 \\
-6 & -2 & 28
\end{bmatrix}.
$$

■

THEOREM 2.11   ■   *If $A = \begin{bmatrix} a_{ij} \end{bmatrix}$ is an $n \times n$ matrix, then*

$$
A(\text{adj } A) = (\text{adj } A)A = \det(A)I_n.
$$

Proof   We have

$$
A(\operatorname{adj} A) =
\begin{bmatrix}
a_{11} & a_{12} & \cdots & a_{1n} \\
a_{21} & a_{22} & \cdots & a_{2n} \\
\vdots & \vdots & & \vdots \\
a_{i1} & a_{i2} & \cdots & a_{in} \\
\vdots & \vdots & & \vdots \\
a_{n1} & a_{n2} & \cdots & a_{nn}
\end{bmatrix}
\begin{bmatrix}
A_{11} & A_{21} & \cdots & A_{j1} & \cdots & A_{n1} \\
A_{12} & A_{22} & \cdots & A_{j2} & \cdots & A_{n2} \\
\vdots & \vdots & & \vdots & & \vdots \\
A_{1n} & A_{2n} & \cdots & A_{jn} & \cdots & A_{nn}
\end{bmatrix}.
$$

The i, jth element in the product matrix $A(\operatorname{adj} A)$ is, by (7),

$$
\begin{aligned}
a_{i1}A_{j1} + a_{i2}A_{j2} + \cdots + a_{in}A_{jn} &= \det(A) \quad \text{if } i = j \\
&= 0 \qquad\quad \text{if } i \neq j.
\end{aligned}
$$

This means that

$$
A(\operatorname{adj} A) =
\begin{bmatrix}
\det(A) & 0 & \cdots & 0 \\
0 & \det(A) & & 0 \\
\vdots & \vdots & \vdots & \vdots \\
0 & & \cdots & 0 & \det(A)
\end{bmatrix}
= \det(A)I_n.
$$

The i, jth element in the product matrix $(\operatorname{adj} A)A$ is, by (8),

$$
\begin{aligned}
A_{1i}a_{1j} + A_{2i}a_{2j} + \cdots + A_{ni}a_{nj} &= \det(A) \quad \text{if } i = j \\
&= 0 \qquad\quad \text{if } i \neq j.
\end{aligned}
$$

Thus $(\operatorname{adj} A)A = \det(A)I_n$.   ■

EXAMPLE 6 ■ Consider the matrix of Example 5. Then

$$
\begin{bmatrix}
3 & -2 & 1 \\
5 & 6 & 2 \\
1 & 0 & -3
\end{bmatrix}
\begin{bmatrix}
-18 & -6 & -10 \\
17 & -10 & -1 \\
-6 & -2 & 28
\end{bmatrix}
=
\begin{bmatrix}
-94 & 0 & 0 \\
0 & -94 & 0 \\
0 & 0 & -94
\end{bmatrix}
$$

$$
= -94
\begin{bmatrix}
1 & 0 & 0 \\
0 & 1 & 0 \\
0 & 0 & 1
\end{bmatrix}
$$

and

$$
\begin{bmatrix}
-18 & -6 & -10 \\
17 & -10 & -1 \\
-6 & -2 & 28
\end{bmatrix}
\begin{bmatrix}
3 & -2 & 1 \\
5 & 6 & 2 \\
1 & 0 & -3
\end{bmatrix}
= -94
\begin{bmatrix}
1 & 0 & 0 \\
0 & 1 & 0 \\
0 & 0 & 1
\end{bmatrix}.
$$

■

We now have a new method for finding the inverse of a nonsingular matrix, and we state this result as the following corollary.

COROLLARY 2.2 ■ *If A is an $n \times n$ matrix and $\det(A) \neq 0$, then*

$$A^{-1} = \frac{1}{\det(A)}(\text{adj } A) = \begin{bmatrix} \dfrac{A_{11}}{\det(A)} & \dfrac{A_{21}}{\det(A)} & \cdots & \dfrac{A_{n1}}{\det(A)} \\[2mm] \dfrac{A_{12}}{\det(A)} & \dfrac{A_{22}}{\det(A)} & \cdots & \dfrac{A_{n2}}{\det(A)} \\[2mm] \vdots & \vdots & & \vdots \\[2mm] \dfrac{A_{1n}}{\det(A)} & \dfrac{A_{2n}}{\det(A)} & \cdots & \dfrac{A_{nn}}{\det(A)} \end{bmatrix}.$$

Proof   By Theorem 2.11, $A(\text{adj } A) = \det(A)I_n$, so if $\det(A) \neq 0$, then

$$A\frac{1}{\det(A)}(\text{adj } A) = \frac{1}{\det(A)}\left[A(\text{adj } A)\right] = \frac{1}{\det(A)}(\det(A)I_n) = I_n.$$

Hence

$$A^{-1} = \frac{1}{\det(A)}(\text{adj } A). \qquad ■$$

EXAMPLE 7 ■ Again consider the matrix of Example 5. Then $\det(A) = -94$, and

$$A^{-1} = \frac{1}{\det(A)}(\text{adj } A) = \begin{bmatrix} \frac{18}{94} & \frac{6}{94} & \frac{10}{94} \\[2mm] -\frac{17}{94} & \frac{10}{94} & \frac{1}{94} \\[2mm] \frac{6}{94} & \frac{2}{94} & -\frac{28}{94} \end{bmatrix}. \qquad ■$$

THEOREM 2.12 ■ *A matrix A is nonsingular if and only if $\det(A) \neq 0$.*

Proof   If $\det(A) \neq 0$, then Corollary 2.2 gives an expression for A^{-1}, so A is nonsingular. The converse has already been established in Corollary 2.1, whose proof was left to the reader as Exercise T.4 of Section 2.1. We now prove the converse here. Suppose that A is nonsingular. Then

$$AA^{-1} = I_n.$$

From Theorem 2.8 we have

$$\det(AA^{-1}) = \det(A)\det(A^{-1}) = \det(I_n) = 1,$$

which implies that $\det(A) \neq 0$. This completes the proof.   ■

COROLLARY 2.3 ■ *If A is an $n \times n$ matrix, then the homogeneous system $A\mathbf{x} = \mathbf{0}$ has a nontrivial solution if and only if $\det(A) = 0$.*

Proof   If $\det(A) \neq 0$, then, by Theorem 2.12, A is nonsingular and thus $A\mathbf{x} = \mathbf{0}$ has only the trivial solution (Theorem 1.12 in Section 1.6).

Conversely, if $\det(A) = 0$, then A is singular (Theorem 2.12). Suppose that A is row equivalent to a matrix B in reduced row echelon form. It then follows from Theorem 1.11 in Section 1.6 and Exercise T.9 in Section 1.5 that B has a row of zeros. The system $B\mathbf{x} = \mathbf{0}$ has the same solutions as the system $A\mathbf{x} = \mathbf{0}$. Let C_1 be the matrix obtained by deleting the zero rows of B. Then the system $B\mathbf{x} = \mathbf{0}$ has the same solutions as the system $C_1\mathbf{x} = \mathbf{0}$. Since the latter is a homogeneous system of at most $n - 1$ equations in n unknowns, it has a nontrivial solution (Theorem 1.7 in Section 1.5). Hence the given system $A\mathbf{x} = \mathbf{0}$ has a nontrivial solution. We might note that the proof of the converse is essentially the proof of Exercise T.3 in Section 1.6.   ■

In Section 1.6 we developed a practical method for finding A^{-1}. In describing the situation showing that a matrix is singular and has no inverse, we used the fact that if we start with the identity matrix I_n and use only elementary row operations on I_n, we can never obtain a matrix having a row consisting entirely of zeros. We can now justify this statement as follows. If matrix B results from I_n by interchanging two rows of I_n, then $\det(B) = -\det(I_n) = -1$ (Theorem 2.2); if C results from I_n by multiplying a row of I_n by $c \neq 0$, then $\det(C) = c\det(I_n) = c$ (Theorem 2.5), and if D results from I_n by adding a multiple of a row of I_n to another row of I_n, then $\det(D) = \det(I_n) = 1$ (Theorem 2.6). Performing elementary row operations on I_n will thus never yield a matrix with zero determinant. Now suppose that F is obtained from I_n by a sequence of elementary row operations and F has a row of zeros. Then $\det(F) = 0$ (Theorem 2.4). This contradiction justifies the statement used in Section 1.6.

We might note that the method of inverting a nonsingular matrix given in Corollary 2.2 is much less efficient than the method given in Chapter 1. In fact, the computation of A^{-1} using determinants, as given in Corollary 2.2, becomes too expensive for $n > 4$ from a computing point of view. We discuss these matters in Section 2.3, where we deal with determinants from a computational point of view. However, Corollary 2.2 is still a useful result on other grounds.

We may summarize our results on determinants, homogeneous systems, and nonsingular matrices in the following List of Nonsingular Equivalences.

List of Nonsingular Equivalences

The following statements are equivalent.

1. A is nonsingular.

2. $A\mathbf{x} = \mathbf{0}$ has only the trivial solution.

3. A is row equivalent to I_n.

4. The linear system $A\mathbf{x} = \mathbf{b}$ has a unique solution for every $n \times 1$ matrix $\mathbf{b}$.

5. $\det(A) \neq 0$.

Cramer's* Rule

We can use the result of Theorem 2.11 to obtain another method, known as Cramer's rule, for solving a linear system of n equations in n unknowns with a nonsingular coefficient matrix.

THEOREM 2.13 ■ *Let*
(*Cramer's Rule*)

$$a_{11}x_1 + a_{12}x_2 + \cdots + a_{1n}x_n = b_1$$
$$a_{21}x_1 + a_{22}x_2 + \cdots + a_{2n}x_n = b_2$$
$$\vdots \qquad \vdots \qquad \vdots \qquad \vdots \qquad \vdots$$
$$a_{n1}x_1 + a_{n2}x_2 + \cdots + a_{nn}x_n = b_n$$

be a linear system of n equations in n unknowns and let $A = \begin{bmatrix} a_{ij} \end{bmatrix}$ be the coefficient matrix so that we can write the given system as $A\mathbf{x} = \mathbf{b}$, where

$$\mathbf{b} = \begin{bmatrix} b_1 \\ b_2 \\ \vdots \\ b_n \end{bmatrix}.$$

If $\det(A) \neq 0$, then the system has the unique solution

$$x_1 = \frac{\det(A_1)}{\det(A)}, \quad x_2 = \frac{\det(A_2)}{\det(A)}, \dots, \quad x_n = \frac{\det(A_n)}{\det(A)},$$

where A_i is the matrix obtained from A by replacing the ith column of A by $\mathbf{b}$.

Proof   If $\det(A) \neq 0$, then, by Theorem 2.12, A is nonsingular. Hence

$$\mathbf{x} = \begin{bmatrix} x_1 \\ x_2 \\ \vdots \\ x_n \end{bmatrix} = A^{-1}\mathbf{b} = \begin{bmatrix} \dfrac{A_{11}}{\det(A)} & \dfrac{A_{21}}{\det(A)} & \cdots & \dfrac{A_{n1}}{\det(A)} \\[2mm] \dfrac{A_{12}}{\det(A)} & \dfrac{A_{22}}{\det(A)} & \cdots & \dfrac{A_{n2}}{\det(A)} \\[2mm] \vdots & \vdots & & \vdots \\[2mm] \dfrac{A_{1i}}{\det(A)} & \dfrac{A_{2i}}{\det(A)} & \cdots & \dfrac{A_{ni}}{\det(A)} \\[2mm] \vdots & \vdots & & \vdots \\[2mm] \dfrac{A_{1n}}{\det(A)} & \dfrac{A_{2n}}{\det(A)} & \cdots & \dfrac{A_{nn}}{\det(A)} \end{bmatrix} \begin{bmatrix} b_1 \\ b_2 \\ \vdots \\ b_n \end{bmatrix}.$$

---

*Gabriel Cramer (1704–1752) was born in Geneva, Switzerland, and lived there all his life. Remaining single, he traveled extensively, taught at the Académie de Calvin, and participated actively in civic affairs.

The rule for solving systems of linear equations appeared in an appendix to his 1750 book, *Introduction à l'analyse des lignes courbes algébriques*. It was known previously by other mathematicians but was not widely known or very clearly explained until its appearance in Cramer's influential work.

This means that

$$x_i = \frac{A_{1i}}{\det(A)}b_1 + \frac{A_{2i}}{\det(A)}b_2 + \cdots + \frac{A_{ni}}{\det(A)}b_n \qquad (1 \le i \le n).$$

Now let

$$A_i = \begin{bmatrix} a_{11} & a_{12} & \cdots & a_{1\,i-1} & b_1 & a_{1\,i+1} & \cdots & a_{1n} \\ a_{21} & a_{22} & \cdots & a_{2\,i-1} & b_2 & a_{2\,i+1} & \cdots & a_{2n} \\ \vdots & \vdots & & \vdots & \vdots & \vdots & & \vdots \\ a_{n1} & a_{n2} & \cdots & a_{n\,i-1} & b_n & a_{n\,i+1} & \cdots & a_{nn} \end{bmatrix}.$$

If we evaluate $\det(A_i)$ by expanding about the ith column, we find that

$$\det(A_i) = A_{1i}b_1 + A_{2i}b_2 + \cdots + A_{ni}b_n.$$

Hence

$$x_i = \frac{\det(A_i)}{\det(A)}$$

for $i = 1, 2, \ldots, n$. In this expression for x_i, the determinant of A_i, $\det(A_i)$, can be calculated by any method. It was only in the derivation of the expression for x_i that we had to evaluate it by expanding about the ith column. ■

EXAMPLE 8 ■ Consider the following linear system:

$$\begin{aligned} -2x_1 + 3x_2 - x_3 &= 1 \\ x_1 + 2x_2 - x_3 &= 4 \\ -2x_1 - x_2 + x_3 &= -3. \end{aligned}$$

Then

$$|A| = \begin{vmatrix} -2 & 3 & -1 \\ 1 & 2 & -1 \\ -2 & -1 & 1 \end{vmatrix} = -2.$$

Hence

$$x_1 = \frac{\begin{vmatrix} 1 & 3 & -1 \\ 4 & 2 & -1 \\ -3 & -1 & 1 \end{vmatrix}}{|A|} = \frac{-4}{-2} = 2;$$

$$x_2 = \frac{\begin{vmatrix} -2 & 1 & -1 \\ 1 & 4 & -1 \\ -2 & -3 & 1 \end{vmatrix}}{|A|} = \frac{-6}{-2} = 3;$$

$$x_3 = \frac{\begin{vmatrix} -2 & 3 & 1 \\ 1 & 2 & 4 \\ -2 & -1 & -3 \end{vmatrix}}{|A|} = \frac{-8}{-2} = 4.$$

■

Cramer's rule for solving the linear system $A\mathbf{x} = \mathbf{b}$, where A is $n \times n$, is as follows:

Step 1. Compute $\det(A)$. If $\det(A) = 0$, Cramer's rule is not applicable. Use Gauss–Jordan reduction.

Step 2. If $\det(A) \neq 0$, for each i,

$$x_i = \frac{\det(A_i)}{\det(A)},$$

where A_i is the matrix obtained from A by replacing the ith column of A by $\mathbf{b}$.

We note that Cramer's rule is applicable only to the case where we have n equations in n unknowns and where the coefficient matrix A is nonsingular. Cramer's rule becomes computationally inefficient for $n > 4$, and it is better to use the Gauss–Jordan reduction method discussed in Section 1.5.

Polynomial Interpolation Revisited

At the end of Section 1.5 we discussed the problem of finding a quadratic polynomial that interpolates the points (x_1, y_1), (x_2, y_2), (x_3, y_3), $x_1 \neq x_2$, $x_1 \neq x_3$, and $x_2 \neq x_3$. Thus, the polynomial has the form

$$y = a_2 x^2 + a_1 x + a_0 \tag{9}$$

[this was Equation (15) in Section 1.5]. Substituting the given points in (9), we obtain the linear system

$$\begin{aligned} a_2 x_1^2 + a_1 x_1 + a_0 &= y_1 \\ a_2 x_2^2 + a_1 x_2 + a_0 &= y_2 \\ a_2 x_3^2 + a_1 x_3 + a_0 &= y_3. \end{aligned} \tag{10}$$

The coefficient matrix of this linear system is

$$\begin{bmatrix} x_1^2 & x_1 & 1 \\ x_2^2 & x_2 & 1 \\ x_3^2 & x_3 & 1 \end{bmatrix}$$

whose determinant is the Vandermonde determinant (see Exercise T.12 in Section 2.1), which has the value

$$(x_2 - x_1)(x_3 - x_1)(x_2 - x_3).$$

Since the three given points are *distinct*, the Vandermonde determinant is not zero. Hence, the coefficient matrix of the linear system in (10) is nonsingular, which implies that the linear system has a unique solution. Thus there is a unique interpolating quadratic polynomial. The general proof for n points is similar.

2.2   EXERCISES

1. Let
$$A = \begin{bmatrix} 1 & 0 & -2 \\ 3 & 1 & 4 \\ 5 & 2 & -3 \end{bmatrix}.$$

Compute all the cofactors.

2. Let
$$A = \begin{bmatrix} 1 & 0 & 3 & 0 \\ 2 & 1 & 4 & -1 \\ 3 & 2 & 4 & 0 \\ 0 & 3 & -1 & 0 \end{bmatrix}.$$

Compute all the cofactors of the elements in the second row and all the cofactors of the elements in the third column.

In Exercises 3 through 6, evaluate the determinants using Theorem 2.9.

3. (a) $\begin{vmatrix} 1 & 2 & 3 \\ -1 & 5 & 2 \\ 3 & 2 & 0 \end{vmatrix}$.

(b) $\begin{vmatrix} 4 & -4 & 2 & 1 \\ 1 & 2 & 0 & 3 \\ 2 & 0 & 3 & 4 \\ 0 & -3 & 2 & 1 \end{vmatrix}$.

(c) $\begin{vmatrix} 4 & -2 & 0 \\ 0 & 2 & 4 \\ -1 & -1 & -3 \end{vmatrix}$.

4. (a) $\begin{vmatrix} 2 & 2 & -3 & 1 \\ 0 & 1 & 2 & -1 \\ 3 & -1 & 4 & 1 \\ 2 & 3 & 0 & 0 \end{vmatrix}$.

(b) $\begin{vmatrix} 0 & 1 & -2 \\ -1 & 3 & 1 \\ 2 & -2 & 3 \end{vmatrix}$.

(c) $\begin{vmatrix} 2 & 1 & -3 \\ 0 & 1 & 2 \\ -4 & 2 & 1 \end{vmatrix}$.

5. (a) $\begin{vmatrix} 3 & 1 & 2 & -1 \\ 2 & 0 & 3 & -7 \\ 1 & 3 & 4 & -5 \\ 0 & -1 & 1 & -5 \end{vmatrix}$.

(b) $\begin{vmatrix} 3 & 1 & 0 \\ 3 & 2 & 1 \\ 0 & 1 & -1 \end{vmatrix}$.

(c) $\begin{vmatrix} 3 & -3 & 0 \\ 2 & 0 & 2 \\ 2 & 1 & -3 \end{vmatrix}$.

6. (a) $\begin{vmatrix} 0 & 0 & -1 & 3 \\ 0 & 1 & 2 & 1 \\ 2 & -2 & 5 & 2 \\ 3 & 3 & 0 & 0 \end{vmatrix}$.

(b) $\begin{vmatrix} 4 & 2 & 0 \\ 1 & 1 & 2 \\ -1 & 3 & 4 \end{vmatrix}$.

(c) $\begin{vmatrix} -1 & 2 & -1 \\ 3 & 2 & 1 \\ 1 & 4 & 2 \end{vmatrix}$.

7. Verify Theorem 2.10 for the matrix
$$A = \begin{bmatrix} -2 & 3 & 0 \\ 4 & 1 & -3 \\ 2 & 0 & 1 \end{bmatrix}$$

by computing $a_{11}A_{12} + a_{21}A_{22} + a_{31}A_{32}$.

8. Let
$$A = \begin{bmatrix} 2 & 1 & 3 \\ -1 & 2 & 0 \\ 3 & -2 & 1 \end{bmatrix}.$$

(a) Find adj A.

(b) Compute $\det(A)$.

(c) Verify Theorem 2.11; that is, show that
$$A(\text{adj } A) = (\text{adj } A)A = \det(A)I_3.$$

9. Let
$$A = \begin{bmatrix} 6 & 2 & 8 \\ -3 & 4 & 1 \\ 4 & -4 & 5 \end{bmatrix}.$$

(a) Find adj A.

(b) Compute $\det(A)$.

(c) Verify Theorem 2.11; that is, show that
$$A(\text{adj } A) = (\text{adj } A)A = \det(A)I_3.$$

In Exercises 10 through 13, compute the inverses of the given matrices, if they exist, using Corollary 2.2.

10. (a) $\begin{bmatrix} 3 & 2 \\ -3 & 4 \end{bmatrix}$. **(b)** $\begin{bmatrix} 4 & 2 & 2 \\ 0 & 1 & 2 \\ 1 & 0 & 3 \end{bmatrix}$.

(c) $\begin{bmatrix} 2 & 0 & -1 \\ 3 & 7 & 2 \\ 1 & 1 & 0 \end{bmatrix}$.

11. (a) $\begin{bmatrix} 1 & 2 & -3 \\ -4 & -5 & 2 \\ -1 & 1 & -7 \end{bmatrix}$.   (b) $\begin{bmatrix} 2 & 3 \\ -1 & 2 \end{bmatrix}$.

(c) $\begin{bmatrix} 4 & 0 & 2 \\ 0 & 3 & 4 \\ 0 & 1 & -2 \end{bmatrix}$.

12. (a) $\begin{bmatrix} 2 & 0 & 1 \\ 3 & 2 & -1 \\ 1 & 0 & 1 \end{bmatrix}$.   (b) $\begin{bmatrix} 5 & -1 \\ 2 & -1 \end{bmatrix}$.

(c) $\begin{bmatrix} 1 & 2 & 4 \\ 1 & -5 & 6 \\ 3 & -1 & 2 \end{bmatrix}$.

13. (a) $\begin{bmatrix} -3 & 1 \\ 2 & 0 \end{bmatrix}$.   (b) $\begin{bmatrix} 4 & 0 & 0 \\ 0 & -3 & 0 \\ 0 & 0 & 2 \end{bmatrix}$.

(c) $\begin{bmatrix} 0 & 2 & 1 & 3 \\ 2 & -1 & 3 & 4 \\ -2 & 1 & 5 & 2 \\ 0 & 1 & 0 & 2 \end{bmatrix}$.

14. Use Theorem 2.12 to determine which of the following matrices are nonsingular.

(a) $\begin{bmatrix} 1 & 2 & 3 \\ 0 & 1 & 2 \\ 2 & -3 & 1 \end{bmatrix}$.   (b) $\begin{bmatrix} 1 & 2 \\ 3 & 4 \end{bmatrix}$.

(c) $\begin{bmatrix} 1 & 3 & 2 \\ 2 & 1 & 4 \\ 1 & -7 & 2 \end{bmatrix}$.

(d) $\begin{bmatrix} 1 & 2 & 0 & 5 \\ 3 & 4 & 1 & 7 \\ -2 & 5 & 2 & 0 \\ 0 & 1 & 2 & -7 \end{bmatrix}$.

15. Use Theorem 2.12 to determine which of the following matrices are nonsingular.

(a) $\begin{bmatrix} 4 & 3 & -5 \\ -2 & -1 & 3 \\ 4 & 6 & -2 \end{bmatrix}$.

(b) $\begin{bmatrix} 1 & 3 & -1 & 2 \\ 2 & -6 & 4 & 1 \\ 3 & 5 & -1 & 3 \\ 4 & -6 & 5 & 2 \end{bmatrix}$.

(c) $\begin{bmatrix} 2 & 2 & -4 \\ 1 & 5 & 2 \\ 3 & 7 & -2 \end{bmatrix}$.   (d) $\begin{bmatrix} 0 & 1 & 2 \\ 1 & 2 & 0 \\ 1 & 3 & 4 \end{bmatrix}$.

16. Find all values of λ for which

(a) $\det\left(\begin{bmatrix} \lambda - 2 & 2 \\ 3 & \lambda - 3 \end{bmatrix} \right) = 0.$

(b) $\det(\lambda I_3 - A) = 0$, where $A = \begin{bmatrix} 1 & 0 & -1 \\ 2 & 0 & 1 \\ 0 & 0 & -1 \end{bmatrix}$.

17. Find all values of λ for which

(a) $\det\left(\begin{bmatrix} \lambda - 1 & -4 \\ 0 & \lambda - 4 \end{bmatrix} \right) = 0.$

(b) $\det(\lambda I_3 - A) = 0$, where

$A = \begin{bmatrix} -3 & -1 & -3 \\ 0 & 3 & 0 \\ -2 & -1 & -2 \end{bmatrix}.$

18. Use Corollary 2.3 to find whether the following homogeneous systems have nontrivial solutions.

(a) $x - 2y + z = 0$
$2x + 3y + z = 0$
$3x + y + 2z = 0.$

(b) $x + 2y + w = 0$
$ + 2y + 3z = 0$
$ z + 2w = 0$
$y + 2z - w = 0.$

19. Repeat Exercise 18 for the following homogeneous systems.

(a) $x + 2y - z = 0$
$2x + y + 2z = 0$
$3x - y + z = 0.$

(b) $x + y + 2z + w = 0$
$2x - y + z - w = 0$
$3x + y + 2z + 3w = 0$
$2x - y - z + w = 0.$

In Exercises 20 through 23, if possible, solve the given linear system by Cramer's rule.

20. $2x + 4y + 6z = 2$
$x + 2z = 0$
$2x + 3y - z = -5.$

21. $x + y + z - 2w = -4$
$2y + z + 3w = 4$
$2x + y - z + 2w = 5$
$x - y + w = 4.$

22. $2x + y + z = 6$
$3x + 2y - 2z = -2$
$x + y + 2z = 4.$

23. $2x + 3y + 7z = 2$
$-2x - 4z = 0$
$x + 2y + 4z = 0.$

THEORETICAL EXERCISES ▪

T.1. Show by a column (row) expansion that if $A = \begin{bmatrix} a_{ij} \end{bmatrix}$ is upper (lower) triangular, then $\det(A) = a_{11}a_{22} \cdots a_{nn}$.

T.2. If $A = \begin{bmatrix} a_{ij} \end{bmatrix}$ is a 3×3 matrix, develop the general expression for $\det(A)$ by expanding (a) about the second column, and (b) about the third row. Compare these answers with those obtained for Example 6 in Section 2.1.

T.3. Show that if A is symmetric, then adj A is also symmetric.

T.4. Show that if A is a nonsingular upper triangular matrix, then A^{-1} is also upper triangular.

T.5. Show that
$$A = \begin{bmatrix} a & b \\ c & d \end{bmatrix}$$
is nonsingular if and only if $ad - bc \neq 0$. If this condition is satisfied, use Corollary 2.2 to find A^{-1}.

T.6. Using Corollary 2.2, find the inverse of
$$A = \begin{bmatrix} 1 & a & a^2 \\ 1 & b & b^2 \\ 1 & c & c^2 \end{bmatrix}.$$

[*Hint*: See Exercise T.12 in Section 2.1, where $\det(A)$ was computed.]

T.7. Show that if A is singular, then adj A is singular. [*Hint*: Show that if A is singular, then $A(\text{adj } A) = O$.]

T.8. Show that if A is an $n \times n$ matrix, then $\det(\text{adj } A) = [\det(A)]^{n-1}$.

T.9. Do Exercise T.10 in Section 1.5 using determinants.

T.10. Let $AB = AC$. Show that if $\det(A) \neq 0$, then $B = C$.

T.11. Let A be an $n \times n$ matrix all of whose entries are integers. Show that if $\det(A) = \pm 1$, then all entries of A^{-1} are integers.

T.12. Show that if A is nonsingular, then adj A is nonsingular and
$$(\text{adj } A)^{-1} = \frac{1}{\det(A)} A = \text{adj}\,(A^{-1}).$$

MATLAB EXERCISES ▪

ML.1. In MATLAB there is a routine **cofactor** that computes the (i, j) cofactor of a matrix. For directions on using this routine, type **help cofactor**. Use **cofactor** to check your hand computations for the matrix A in Exercise 1.

ML.2. Use the **cofactor** routine (see Exercise ML.1) to compute the cofactor of the elements in the second row of
$$A = \begin{bmatrix} 1 & 5 & 0 \\ 2 & -1 & 3 \\ 3 & 2 & 1 \end{bmatrix}.$$

ML.3. Use the **cofactor** routine to evaluate the determinant of A using Theorem 2.9.
$$A = \begin{bmatrix} 4 & 0 & -1 \\ -2 & 2 & -1 \\ 0 & 4 & -3 \end{bmatrix}.$$

ML.4. Use the **cofactor** routine to evaluate the determinant of A using Theorem 2.9.
$$A = \begin{bmatrix} -1 & 2 & 0 & 0 \\ 2 & -1 & 2 & 0 \\ 0 & 2 & -1 & 2 \\ 0 & 0 & 2 & -1 \end{bmatrix}.$$

ML.5. In MATLAB there is a routine **adjoint**, which computes the adjoint of a matrix. For directions on using this routine, type **help adjoint**. Use **adjoint** to aid in computing the inverses of the matrices in Exercise 11.

2.3 ▼ Determinants from a Computational Point of View

In this book we have, by now, developed two methods for solving a linear system of n equations in n unknowns: Gauss–Jordan reduction and Cramer's rule. We also have two methods for finding the inverse of a nonsingular matrix: the method involving determinants and the method discussed in Section 1.6. In this section we discuss criteria to be considered when selecting one of these methods instead of another one.

Most sizable problems in linear algebra are solved on computers so that it is natural to compare two methods by estimating their computing time for the same problem. Since addition is so much faster than multiplication, the number of multiplications is often used as a basis of comparison for two numerical procedures.

Consider the linear system $A\mathbf{x} = \mathbf{b}$, where A is 25×25. If we find $\mathbf{x}$ by Cramer's rule, we must first obtain $\det(A)$. We can find $\det(A)$ by cofactor expansion, say $\det(A) = a_{11}A_{11} + a_{21}A_{21} + \cdots + a_{n1}A_{n1}$, where we have expanded $\det(A)$ about the first column. Note that if each cofactor is available, we require 25 multiplications. Now each cofactor A_{ij} is plus or minus the determinant of a 24×24 matrix, which can be expanded about a given row or column, requiring 24 multiplications. Thus the computation of $\det(A)$ requires more than $25 \times 24 \times \cdots \times 2 \times 1 = 25!$ multiplications. Even if we were to use a futuristic (not very far into the future) computer capable of performing 10 trillion (1×10^{12}) multiplications *per second* $(3.15 \times 10^{19}$ *per year*$)$, it would take *about 49,000 years* to evaluate $\det(A)$. However, Gauss–Jordan reduction takes about $25^3/3$ multiplications, and we would find the solution in less than *one second*. Of course, we can compute $\det(A)$ in a much more efficient way by using elementary row operations to reduce A to triangular form and then use Theorem 2.7 (see Example 16 in Section 2.1). When implemented this way for an $n \times n$ matrix, Cramer's rule will require approximately n^4 multiplications, compared to $n^3/3$ multiplications for Gauss–Jordan reduction. Thus Gauss–Jordan reduction is still much faster.

In general, if we are seeking numerical answers, then any method involving determinants can be used for $n \leq 4$. For $n > 5$, determinant-dependent methods are much less efficient than Gauss–Jordan reduction or the method of Section 1.6. for inverting a matrix.

The importance of determinants obviously does not lie in their computational use. Note that methods involving determinants enable one to express the inverse of a matrix and the solution to a linear system of n equations in n unknowns by means of expressions or formulas. Gauss–Jordan reduction and the method for finding A^{-1} given in Section 1.6 do not yield a *formula* for the answer; we must proceed numerically to obtain the answer. Sometimes we do not need numerical answers but an expression for the answer because we may wish to further manipulate the answer. Another important reason for studying determinants is that they play a key role in the study of eigenvalues and eigenvectors, which will be undertaken in Chapter 5.

KEY IDEAS FOR REVIEW ▪

☐ **Theorem 2.1.** $\det(A^T) = \det(A)$.

☐ **Theorem 2.2.** If B results from A by interchanging two rows (columns) of A, then $\det(B) = -\det(A)$.

☐ **Theorem 2.3.** If two rows (columns) of A are equal, then $\det(A) = 0$.

☐ **Theorem 2.4.** If a row (column) of A consists entirely of zeros, then $\det(A) = 0$.

☐ **Theorem 2.5.** If B is obtained from A by multiplying a row (column) of A by a real number c, then $\det(B) = c \det(A)$.

☐ **Theorem 2.6.** If B is obtained from A by adding a multiple of a row (column) of A to another row (column) of A, then $\det(B) = \det(A)$.

☐ **Theorem 2.7.** If $A = \begin{bmatrix} a_{ij} \end{bmatrix}$ is upper (lower) triangular, then $\det(A) = a_{11}a_{22}\cdots a_{nn}$.

☐ **Theorem 2.8.** $\det(AB) = \det(A)\det(B)$.

☐ **Theorem 2.9 (Cofactor Expansion).** If $A = \begin{bmatrix} a_{ij} \end{bmatrix}$, then
$$\det(A) = a_{i1}A_{i1} + a_{i2}A_{i2} + \cdots + a_{in}A_{in}$$
and
$$\det(A) = a_{1j}A_{1j} + a_{2j}A_{2j} + \cdots + a_{nj}A_{nj}.$$

☐ **Corollary 2.2.** If $\det(A) \neq 0$, then
$$A^{-1} = \begin{bmatrix} \dfrac{A_{11}}{\det(A)} & \dfrac{A_{21}}{\det(A)} & \cdots & \dfrac{A_{n1}}{\det(A)} \\ \dfrac{A_{12}}{\det(A)} & \dfrac{A_{22}}{\det(A)} & \cdots & \dfrac{A_{n2}}{\det(A)} \\ \vdots & \vdots & & \vdots \\ \dfrac{A_{1n}}{\det(A)} & \dfrac{A_{2n}}{\det(A)} & \cdots & \dfrac{A_{nn}}{\det(A)} \end{bmatrix}.$$

☐ **Theorem 2.12.** A is nonsingular if and only if $\det(A) \neq 0$.

☐ **Corollary 2.3.** If A is an $n \times n$ matrix, then the homogeneous system $A\mathbf{x} = \mathbf{0}$ has a nontrivial solution if and only if $\det(A) = 0$.

☐ **Theorem 2.13 (Cramer's Rule).** Let $A\mathbf{x} = \mathbf{b}$ be a linear system of n equations in n unknowns. If $\det(A) \neq 0$, then the system has the unique solution
$$x_1 = \frac{\det(A_1)}{\det(A)}, \quad x_2 = \frac{\det(A_2)}{\det(A)}, \dots,$$
$$x_n = \frac{\det(A_n)}{\det(A)},$$
where A_i is the matrix obtained from A by replacing the ith column of A by $\mathbf{b}$.

☐ **List of Nonsingular Equivalences.** The following statements are equivalent:

1. A is nonsingular.
2. $A\mathbf{x} = \mathbf{0}$ has only the trivial solution.
3. A is row equivalent to I_n.
4. The linear system $A\mathbf{x} = \mathbf{b}$ has a unique solution for every $n \times 1$ matrix $\mathbf{b}$.
5. $\det(A) \neq 0$.

SUPPLEMENTARY EXERCISES ▪

1. Evaluate the following determinants using Equation (2) of Section 2.1.

(a) $\begin{vmatrix} 0 & 2 & 0 \\ 0 & 0 & -3 \\ 4 & 0 & 0 \end{vmatrix}$.   (b) $\begin{vmatrix} 3 & 0 & 0 & 0 \\ 0 & -2 & 0 & 0 \\ 0 & 4 & 1 & 0 \\ 3 & 2 & -1 & -4 \end{vmatrix}$.

2. If
$$\begin{vmatrix} a_1 & a_2 & a_3 \\ b_1 & b_2 & b_3 \\ c_1 & c_2 & c_3 \end{vmatrix} = 5,$$

find the determinants of the following matrices:

(a) $B = \begin{bmatrix} \frac{1}{2}a_1 & \frac{1}{2}a_2 & \frac{1}{2}a_3 \\ b_1 & b_2 & b_3 \\ c_1 & c_2 & c_3 \end{bmatrix}$.

(b) $C = \begin{bmatrix} a_1 - b_1 & a_2 - b_2 & a_3 - b_3 \\ 3b_1 & 3b_2 & 3b_3 \\ 2c_1 & 2c_2 & 2c_3 \end{bmatrix}$.

3. Let A be 4×4 and suppose that $|A| = 5$. Compute
(a) $|A^{-1}|$.   (b) $|2A|$.   (c) $|2A^{-1}|$.
(d) $|(2A)^{-1}|$.

4. Let $|A| = 3$ and $|B| = 4$. Compute
(a) $|AB|$.   (b) $|ABA^T|$.   (c) $|B^{-1}AB|$.

5. Find all values of λ for which

$$\det\left(\begin{bmatrix} \lambda+2 & -1 & 3 \\ 2 & \lambda-1 & 2 \\ 0 & 0 & \lambda+4 \end{bmatrix}\right) = 0.$$

6. Evaluate

$$\begin{vmatrix} 2 & -1 & 3 \\ 4 & 1 & 5 \\ -2 & -3 & -2 \end{vmatrix}.$$

7. Evaluate

$$\begin{vmatrix} 3 & 2 & -1 & 1 \\ 4 & 1 & 1 & 0 \\ -1 & 2 & 3 & 4 \\ -2 & 3 & 5 & 1 \end{vmatrix}.$$

8. Compute all the cofactors of

$$A = \begin{bmatrix} 2 & -1 & 3 \\ 1 & 4 & 5 \\ -3 & -4 & 6 \end{bmatrix}.$$

9. Evaluate

$$\begin{vmatrix} 3 & 2 & -1 & 0 \\ -1 & 0 & 3 & 2 \\ 4 & 1 & 5 & -2 \\ 1 & 3 & 2 & -3 \end{vmatrix}$$

by cofactor expansion.

10. Let

$$A = \begin{bmatrix} 3 & -1 & 2 \\ 0 & 4 & 5 \\ 1 & 3 & 2 \end{bmatrix}.$$

(a) Find adj A.
(b) Compute $\det(A)$.
(c) Show that $A(\text{adj } A) = \det(A)I_3$.

11. Compute the inverse of the following matrix, if it exists, using Corollary 2.2:

$$\begin{bmatrix} 2 & -1 & 3 \\ 0 & 1 & 2 \\ -1 & 1 & 2 \end{bmatrix}.$$

12. Find all values of λ for which

$$\begin{bmatrix} \lambda-3 & 0 & 3 \\ 0 & \lambda+2 & 0 \\ -5 & 0 & \lambda+5 \end{bmatrix}$$

is singular.

13. If

$$A = \begin{bmatrix} \lambda & 0 & 1 \\ 1 & \lambda-1 & 0 \\ 0 & 0 & \lambda+1 \end{bmatrix},$$

find all values of λ for which the homogeneous system $A\mathbf{x} = \mathbf{0}$ has only the trivial solution.

14. If possible, solve the following linear system by Cramer's rule:

$$\begin{aligned} 3x + 2y - z &= -1 \\ x - y - z &= 0 \\ 2x + y - 2z &= 3. \end{aligned}$$

15. Using only elementary row or elementary column operations and Theorems 2.2, 2.5, and 2.6 (do not expand the determinants), verify the following.

(a) $\begin{vmatrix} a-b & 1 & a \\ b-c & 1 & b \\ c-a & 1 & c \end{vmatrix} = \begin{vmatrix} a & 1 & b \\ b & 1 & c \\ c & 1 & a \end{vmatrix}.$

(b) $\begin{vmatrix} 1 & a & bc \\ 1 & b & ca \\ 1 & c & ab \end{vmatrix} = \begin{vmatrix} 1 & a & a^2 \\ 1 & b & b^2 \\ 1 & c & c^2 \end{vmatrix}.$

16. Find all values of a for which the linear system

$$\begin{aligned} 2x + ay &= 0 \\ ax + 2y &= 0 \end{aligned}$$

has (a) a unique solution; (b) infinitely many solutions.

17. Find all values of a for which the matrix

$$\begin{bmatrix} a-2 & 2 \\ a-2 & a+2 \end{bmatrix}$$

is nonsingular.

18. Use Cramer's rule to find all values of a for which the linear system

$$\begin{aligned} x - 2y + 2z &= 9 \\ 2x + y &= a \\ 3x - y - z &= -10 \end{aligned}$$

has the solution in which $y = 1$.

THEORETICAL EXERCISES ■

T.1. Show that if two rows (columns) of the $n \times n$ matrix A are proportional, then $\det(A) = 0$.

T.2. Show that if A is an $n \times n$ matrix, then $\det(AA^T) \geq 0$.

T.3. Let Q be an $n \times n$ matrix in which each entry is 1. Show that $\det(Q - nI_n) = 0$.

T.4. Let P be an invertible matrix. Show that if $B = PAP^{-1}$, then $\det(B) = \det(A)$.

T.5. Show that if A is a singular $n \times n$ matrix, then $A(\operatorname{adj} A) = O$. (*Hint*: See Theorem 2.11.)

T.6. Show that if A is a singular $n \times n$ matrix, then AB is singular for any $n \times n$ matrix B.

T.7. Show that if A and B are square matrices, then
$$\det\left(\begin{bmatrix} A & O \\ O & B \end{bmatrix} \right) = \det(A)\det(B).$$

T.8. Show that if A, B, and C are square matrices, then
$$\det\left(\begin{bmatrix} A & O \\ C & B \end{bmatrix} \right) = \det(A)\det(B).$$

T.9. Let A be an $n \times n$ matrix with integer entries and $\det(A) = \pm 1$. Show that if **b** has all integer entries, then every solution to $A\mathbf{x} = \mathbf{b}$ consists of integers.

CHAPTER TEST ■

1. Evaluate
$$\begin{vmatrix} 1 & 1 & 2 & -1 \\ 0 & 1 & 0 & 3 \\ -1 & 2 & -3 & 4 \\ 0 & 5 & 0 & -2 \end{vmatrix}.$$

2. Let A be 3×3 and suppose that $|A| = 2$. Compute

(a) $|3A|$.   (b) $|3A^{-1}|$.   (c) $|(3A)^{-1}|$.

3. For what value of a is
$$\begin{vmatrix} 2 & 1 & 0 \\ 0 & -1 & 3 \\ 0 & 1 & a \end{vmatrix} + \begin{vmatrix} 0 & a & 1 \\ 1 & 3a & 0 \\ -2 & a & 2 \end{vmatrix} = 14?$$

4. Find all values of a for which the matrix
$$\begin{bmatrix} a^2 & 0 & 3 \\ 5 & a & 2 \\ 3 & 0 & 1 \end{bmatrix}$$
is singular.

5. Solve the following linear system by Cramer's rule.
$$\begin{aligned} x - y + z &= -1 \\ 2x + y - 3z &= 8 \\ x - 2y + 3z &= -5. \end{aligned}$$

6. Answer each of the following as true or false.

(a) $\det(AA^T) = \det(A^2)$.

(b) $\det(-A) = -\det(A)$.

(c) If $A^T = A^{-1}$, then $\det(A) = 1$.

(d) If $\det(A) = 0$, then $A = O$.

(e) If $\det(A) = 7$, then $A\mathbf{x} = \mathbf{0}$ has only the trivial solution.

(f) The sign of the term $a_{15}a_{23}a_{31}a_{42}a_{54}$ in the expansion of the determinant of a 5×5 matrix is $+$.

(g) If $\det(A) = 0$, then $\det(\operatorname{adj} A) = 0$.

(h) If $B = PAP^{-1}$, and P is nonsingular, then $\det(B) = \det(A)$.

(i) If $A^4 = I_n$, then $\det(A) = 1$.

(j) If $A^2 = A$ and $A \neq I_n$, then $\det(A) = 0$.

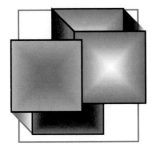

Vectors in R^2 and R^n

3.1 ▾ Vectors in the Plane

Coordinate Systems

In many applications we deal with measurable quantities, such as pressure, mass, and speed, which can be described completely by giving their magnitude. There are also many other measurable quantities, such as velocity, force, and acceleration, that require for their description not only magnitude, but also a direction. These are called **vectors**, and their study comprises this chapter. Vectors will be denoted by lowercase boldface letters such as **u**, **v**, **w**, **x**, **y**, and **z**. The real numbers will be called **scalars** and will be denoted by lowercase italic letters.

We recall that the real number system may be visualized as a straight line L, which is usually taken in a horizontal position. A point O, called the **origin**, is chosen on L; O corresponds to the number 0. A point A is chosen to the right of O, thereby fixing the length of OA as 1 and specifying a positive direction. Thus the positive real numbers lie to the right of O; the negative real numbers lie to the left of O (Figure 3.1).

The **absolute value** $|x|$ of the real number x is defined by

$$|x| = \begin{cases} x & \text{if } x \geq 0 \\ -x & \text{if } x < 0. \end{cases}$$

Thus $|3| = 3$, $|-2| = 2$, $|0| = 0$, $\left|-\frac{2}{3}\right| = \frac{2}{3}$, and $|-1.82| = 1.82$.

FIGURE 3.1

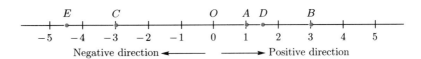

The real number x corresponding to the point P is called the **coordinate** of P and the point P whose coordinate is x is denoted by $P(x)$. The line L is called a **coordinate axis**. If P is to the right of O, then its coordinate is the length of the segment OP. If Q is to the left of O, then its coordinate is the negative of the length of the segment OQ. The distance between the points P and Q with respective coordinates a and b is $|b - a|$.

EXAMPLE 1 ■ Referring to Figure 3.1, we see that the coordinates of the points B, C, D, and E are, respectively, 3, -3, 1.5, and -4.5. The distance between B and C is $|-3 - 3| = 6$. The distance between A and B is $|3 - 1| = 2$. The distance between C and E is $|-4.5 - (-3)| = 1.5$. ■

We shall now turn to the analogous situation for the plane. We draw a pair of perpendicular lines intersecting at a point O, called the **origin**. One of the lines, the **x-axis**, is usually taken in a horizontal position. The other line, the **y-axis**, is then taken in a vertical position. We now choose a point on the x-axis to the right of O and a point on the y-axis above O to fix the units of length and positive directions on the x- and y-axes. Frequently, but not always, these points are chosen so that they are both equidistant from O, that is, so that the same unit of length is used for both axes. The x- and y-axes together are called **coordinate axes** (Figure 3.2). The **projection** of a point P in the plane on a line L is the point Q obtained by intersecting L with the line L' passing through P and perpendicular to L [Figure 3.3(a) and (b)].

FIGURE 3.2

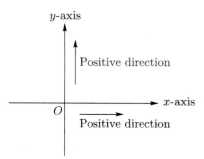

FIGURE 3.3

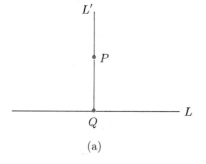

(a)

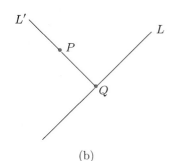

(b)

Let P be a point in the plane and let Q be its projection onto the x-axis. The coordinate of Q on the x-axis is called the **x-coordinate** of P. Similarly, let Q' be the projection of P onto the y-axis. The coordinate of Q' is called the **y-coordinate** of P. Thus with every point in the plane we associate an ordered pair (x, y) of real numbers, its coordinates. The point P with coordinates x and y is denoted by $P(x, y)$. Conversely, it is easy to see (Exercise T.1) how we can associate a point in the plane with each ordered pair (x, y) of real numbers (Figure 3.4). The correspondence given above between points in the plane and ordered pairs of real numbers is called a **rectangular coordinate system** or a **Cartesian coordinate system** (after René Descartes*). The set of points in the plane is denoted by R^2. It is also called **2-space**.

FIGURE 3.4

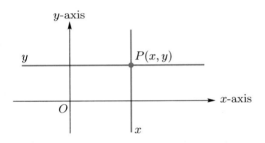

EXAMPLE 2 ■ In Figure 3.5 we show a number of points and their coordinates.

The coordinates of the origin are $(0, 0)$. The coordinates of the projection of the point $P(x, y)$ on the x-axis are $(x, 0)$ and the coordinates of its projection on the y-axis are $(0, y)$. ■

Vectors

Recall that at the beginning of Section 1.3 we briefly introduced vectors algebraically to better understand matrix multiplication. In this section we view 2-vectors geometrically and in the next section we do the same with n-vectors.

---

*René Descartes (1596–1650) was one of the best-known scientists and philosophers of his day; he was considered by some to be the founder of modern philosophy. After completing a university degree in law, he turned to the private study of mathematics, simultaneously pursuing interests in Parisian night life and in the military, volunteering for brief periods in the Dutch, Bavarian, and French armies. The most productive period of his life was 1628–1648, when he lived in Holland. In 1649 he accepted an invitation from Queen Christina of Sweden to be her private tutor and to establish an Academy of Sciences there. Unfortunately, he had no time for this project, for he died of pneumonia in 1650.

In 1619 Descartes had a dream in which he realized that the method of mathematics is the best way for obtaining truth. However, his only mathematical publication was *La Géométrie*, which appeared as an appendix to his major philosophical work *Discours de la méthode pour bien conduire sa raison, et chercher la vérité dans les sciences*. In *La Géométrie* he proposes the radical idea of doing geometry algebraically. To express a curve algebraically, one chooses any convenient line of reference and, on the line, a point of reference. If y represents the distance from any point of the curve to the reference line and x represents the distance along the line to the reference point, there is an equation relating x and y that represents the curve. The systematic use of "Cartesian" coordinates described above was introduced later in the seventeenth century by authors following up on Descartes' work.

FIGURE 3.5

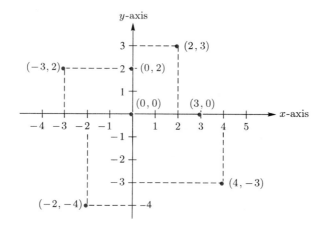

Consider the 2-vector

$$\mathbf{u} = \begin{bmatrix} x_1 \\ y_1 \end{bmatrix},$$

where x_1 and y_1 are real numbers. With $\mathbf{u}$ we associate the directed line segment with initial point at the origin $O(0,0)$ and terminal point at $P(x_1, y_1)$. The directed line segment from O to P is denoted by $\overrightarrow{OP}$; O is called its **tail** and P its **head**. We distinguish tail and head by placing an arrow at the head (Figure 3.6).

FIGURE 3.6

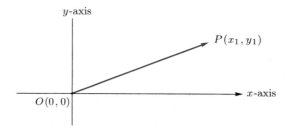

A directed line segment has a **direction**, which is the angle made with the positive x-axis, indicated by the arrow at its head. The **magnitude** of a directed line segment is its length.

EXAMPLE 3 ■ Let

$$\mathbf{u} = \begin{bmatrix} 2 \\ 3 \end{bmatrix}.$$

With $\mathbf{u}$ we can associate the directed line segment with tail $O(0,0)$ and head $P(2,3)$, shown in Figure 3.7. ■

Conversely, with a directed line segment $\overrightarrow{OP}$ with tail $O(0,0)$ and head $P(x_1, y_1)$, we can associate the 2-vector

$$\begin{bmatrix} x_1 \\ y_1 \end{bmatrix}.$$

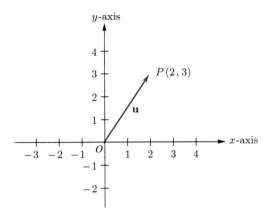

FIGURE 3.7

EXAMPLE 4 ■ With the directed line segment $\overrightarrow{OP}$ with head $P(4,5)$, we can associate the 2-vector

$$\begin{bmatrix} 4 \\ 5 \end{bmatrix}.$$

■

DEFINITION   A **vector in the plane** is a 2-vector

$$\mathbf{u} = \begin{bmatrix} x_1 \\ y_1 \end{bmatrix},$$

where x_1 and y_1 are real numbers, called the **components of u**. We refer to a vector in the plane merely as a **vector**.

Thus with every vector we can associate a directed line segment, and conversely, with every directed line segment emanating from the origin we can associate a vector. As we have seen, a coordinate system is needed to set up this correspondence. The magnitude and direction of a vector are the magnitude and direction of its associated directed line segment. Frequently, the notions of directed line segment and vector are used interchangeably and a directed line segment is called a **vector**.

Since a vector is a matrix, the vectors

$$\mathbf{u} = \begin{bmatrix} x_1 \\ y_1 \end{bmatrix} \quad \text{and} \quad \mathbf{v} = \begin{bmatrix} x_2 \\ y_2 \end{bmatrix}$$

are said to be **equal** if $x_1 = x_2$ and $y_1 = y_2$. That is, two vectors are equal if their respective components are equal.

EXAMPLE 5 ■ The vectors

$$\begin{bmatrix} 1 \\ 0 \end{bmatrix} \quad \text{and} \quad \begin{bmatrix} 1 \\ -2 \end{bmatrix}$$

are not equal, since their respective components are not equal.   ■

With each vector

$$\mathbf{u} = \begin{bmatrix} x_1 \\ y_1 \end{bmatrix}$$

we can also associate in a unique manner the point $P(x_1, y_1)$; conversely, with each point $P(x_1, y_1)$ we can associate in a unique manner the vector

$$\begin{bmatrix} x_1 \\ y_1 \end{bmatrix}.$$

Thus we also write the vector $\mathbf{u}$ as

$$\mathbf{u} = (x_1, y_1).$$

Of course, this association is obtained by means of the directed line segment $\overrightarrow{OP}$, where O is the origin (Figure 3.6).

Thus the plane may be viewed both as the set of all points or as the set of all vectors. For this reason and, depending upon the context, we sometimes take R^2 as the set of all ordered pairs (x_1, y_1) and sometimes as the set of all 2-vectors

$$\begin{bmatrix} x_1 \\ y_1 \end{bmatrix}.$$

Frequently, in physical applications it is necessary to deal with a directed line segment $\overrightarrow{PQ}$, from the point $P(x_1, y_1)$ (not the origin) to the point $Q(x_2, y_2)$, as shown in Figure 3.8(a). Such a directed line segment will also be called a **vector in the plane**, or simply a **vector** with **tail** $P(x_1, y_1)$ and **head** $Q(x_2, y_2)$. The **components** of such a vector are $x_2 - x_1$ and $y_2 - y_1$. Thus the vector PQ in Figure 3.8(a) can also be represented by the vector $(x_2 - x_1, y_2 - y_1)$ with tail O and head $P''(x_2 - x_1, y_2 - y_1)$. Two such vectors in the plane will be called **equal** if their components are equal. Consider the vectors $\overrightarrow{P_1Q_1}$, $\overrightarrow{P_2Q_2}$, and $\overrightarrow{P_3Q_3}$ joining the points $P_1(3, 2)$ and $Q_1(5, 5)$, $P_2(0, 0)$ and $Q_2(2, 3)$, $P_3(-3, 1)$

FIGURE 3.8

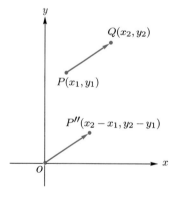

(a) Different directed line segments representing the same vector.

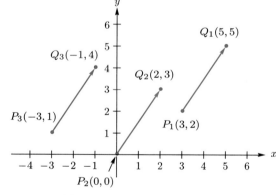

(b) Vectors in the plane.

and $Q_3(-1, 4)$, respectively, as shown in Figure 3.8(b). Since they all have the same components, they are equal.

To find the head $Q_4(a, b)$ of the vector

$$\overrightarrow{P_4 Q_4} = \begin{bmatrix} 2 \\ 3 \end{bmatrix} = \overrightarrow{P_2 Q_2},$$

with tail $P_4(-5, 2)$, we proceed as follows. We must have $a - (-5) = 2$ and $b - 2 = 3$ so that $a = 2 - 5 = -3$ and $b = 3 + 2 = 5$, so the coordinates of Q_4 are $(-3, 5)$. Similarly, to find the tail $P_5(c, d)$ of the vector

$$\overrightarrow{P_5 Q_5} = \begin{bmatrix} 2 \\ 3 \end{bmatrix},$$

with head $Q_5(8, 6)$, we must have $8 - c = 2$ and $6 - d = 3$ so that $c = 8 - 2 = 6$ and $d = 6 - 3 = 3$. Hence, the coordinates of P_5 are $(6, 3)$.

Length

By the Pythagorean theorem (Figure 3.9), the **length** or **magnitude** of the vector $\mathbf{u} = (x_1, y_1)$ is

$$\|\mathbf{u}\| = \sqrt{x_1^2 + y_1^2}. \tag{1}$$

It also follows, by the Pythagorean theorem, that the length of the directed line segment with initial point $P_1(x_1, y_1)$ and terminal point $P_2(x_2, y_2)$ is (Figure 3.10)

$$\|\overrightarrow{P_1 P_2}\| = \sqrt{(x_2 - x_1)^2 + (y_2 - y_1)^2}. \tag{2}$$

FIGURE 3.9

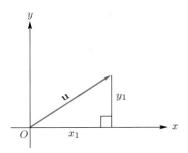

FIGURE 3.10

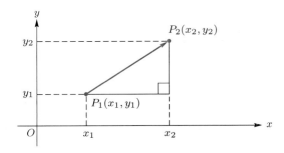

Equation (2) also gives the distance between the points P_1 and P_2.

EXAMPLE 6 ■ If $\mathbf{u} = (2, -5)$, then, by Equation (1),

$$\|\mathbf{u}\| = \sqrt{(2)^2 + (-5)^2} = \sqrt{4 + 25} = \sqrt{29}.$$

■

EXAMPLE 7 ■ The distance between $P(3, 2)$ and $Q(-1, 5)$, or the length of the directed line segment $\overrightarrow{PQ}$, is, by Equation (2),

$$\|\overrightarrow{PQ}\| = \sqrt{(-1 - 3)^2 + (5 - 2)^2} = \sqrt{(-4)^2 + 3^2} = \sqrt{25} = 5.$$

■

The length of each vector (directed line segment) $\overrightarrow{P_1Q_1}$, $\overrightarrow{P_2Q_2}$, and $\overrightarrow{P_3Q_3}$ in Figure 3.8(b) is $\sqrt{13}$ (verify).

Two nonzero vectors

$$\mathbf{u} = \begin{bmatrix} x_1 \\ y_1 \end{bmatrix} \quad \text{and} \quad \mathbf{v} = \begin{bmatrix} x_2 \\ y_2 \end{bmatrix}$$

are said to be **parallel** if one is a multiple of the other. They are parallel if the lines on which they lie are both vertical or have the same slopes. Thus the vectors (directed line segments) $\overrightarrow{P_1Q_1}$, $\overrightarrow{P_2Q_2}$, and $\overrightarrow{P_3Q_3}$ in Figure 3.8(b) are parallel.

Using Determinants to Compute Area

Consider the triangle with vertices (x_1, y_1), (x_2, y_2), and (x_3, y_3), which we take as shown in Figure 3.11.

FIGURE 3.11

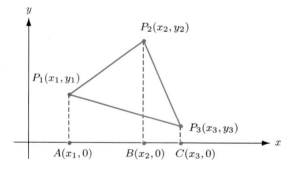

We may compute the area of this triangle as

$$\text{area of trapezoid } AP_1P_2B + \text{area of trapezoid } BP_2P_3C$$
$$- \text{area of trapezoid } AP_1P_3C.$$

Now recall that the area of a trapezoid is $\frac{1}{2}$ the distance between the parallel sides of the trapezoid times the sum of the lengths of the parallel sides. Thus

area of triangle $P_1 P_2 P_3$

$$= \tfrac{1}{2}(x_2 - x_1)(y_1 + y_2) + \tfrac{1}{2}(x_3 - x_2)(y_2 + y_3) - \tfrac{1}{2}(x_3 - x_1)(y_1 + y_3)$$

$$= \tfrac{1}{2}x_2 y_1 - \tfrac{1}{2}x_1 y_2 + \tfrac{1}{2}x_3 y_2 - \tfrac{1}{2}x_2 y_3 - \tfrac{1}{2}x_3 y_1 + \tfrac{1}{2}x_1 y_3$$

It turns out that this expression is

$$\tfrac{1}{2} \det \left(\begin{bmatrix} x_1 & y_1 & 1 \\ x_2 & y_2 & 1 \\ x_3 & y_3 & 1 \end{bmatrix} \right).$$

When the points are in the other quadrants or the points are labeled in a different order, the formula just obtained will yield the negative of the area of the triangle. Thus, for a triangle with vertices (x_1, y_1), (x_2, y_2), and (x_3, y_3), we have

$$\text{area of triangle} = \tfrac{1}{2} \left| \det \left(\begin{bmatrix} x_1 & y_1 & 1 \\ x_2 & y_2 & 1 \\ x_3 & y_3 & 1 \end{bmatrix} \right) \right| \tag{3}$$

(the area is $\frac{1}{2}$ the absolute value of the determinant).

EXAMPLE 8 ■ Compute the area of the triangle T shown in Figure 3.12 with vertices $(-1, 4)$, $(3, 1)$, and $(2, 6)$.

Solution   By Equation (3), the area of T is

$$\tfrac{1}{2} \left| \det \left(\begin{bmatrix} -1 & 4 & 1 \\ 3 & 1 & 1 \\ 2 & 6 & 1 \end{bmatrix} \right) \right| = \tfrac{1}{2} |17| = 8.5.$$

■

FIGURE 3.12

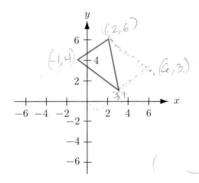

Suppose we now have the parallelogram shown in Figure 3.13. Since a diagonal divides the parallelogram into two equal triangles, it follows from Equation (3) that

$$\text{area of parallelogram} = \left| \det \left(\begin{bmatrix} x_1 & y_1 & 1 \\ x_2 & y_2 & 1 \\ x_3 & y_3 & 1 \end{bmatrix} \right) \right|.$$

FIGURE 3.13

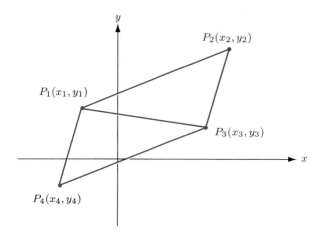

FIGURE 3.13

Vector Operations

DEFINITION

Let $\mathbf{u} = (x_1, y_1)$ and $\mathbf{v} = (x_2, y_2)$ be two vectors in the plane. The **sum** of the vectors $\mathbf{u}$ and $\mathbf{v}$ is the vector

$$(x_1 + x_2, y_1 + y_2)$$

and is denoted by $\mathbf{u} + \mathbf{v}$. Thus vectors are added by adding their components.

EXAMPLE 9 ■ Let $\mathbf{u} = (1, 2)$ and $\mathbf{v} = (3, -4)$. Then

$$\mathbf{u} + \mathbf{v} = (1 + 3, 2 + (-4)) = (4, -2).$$ ■

We can interpret vector addition geometrically as follows. In Figure 3.14, the vector from (x_1, y_1) to $(x_1 + x_2, y_1 + y_2)$ is also $\mathbf{v}$. Thus the vector with tail O and head $(x_1 + x_2, y_1 + y_2)$ is $\mathbf{u} + \mathbf{v}$.

FIGURE 3.14
Vector addition

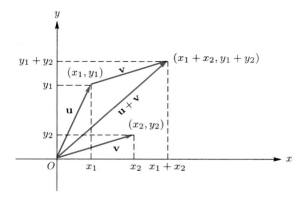

We can also describe $\mathbf{u} + \mathbf{v}$ as the diagonal of the parallelogram defined by $\mathbf{u}$ and $\mathbf{v}$, as shown in Figure 3.15.

FIGURE 3.15
Vector addition

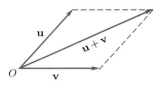

FIGURE 3.16

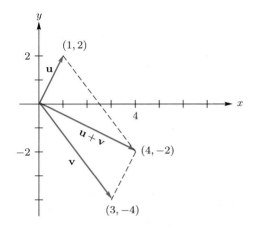

EXAMPLE 10 ■ If **u** and **v** are as in Example 9, then **u** + **v** is as shown in Figure 3.16.   ■

DEFINITION ─ If $\mathbf{u} = (x_1, y_1)$ and c is a scalar (a real number), then the **scalar multiple** $c\mathbf{u}$ of **u** by c is the vector (cx_1, cy_1). Thus the scalar multiple $c\mathbf{u}$ of **u** by c is obtained by multiplying each component of **u** by c.

If $c > 0$, then $c\mathbf{u}$ is in the same direction as **u**, whereas if $d < 0$, then $d\mathbf{u}$ is in the opposite direction (Figure 3.17).

FIGURE 3.17
Scalar multiplication

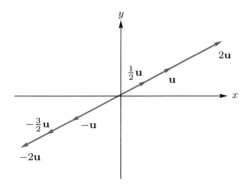

EXAMPLE 11 ■ If $c = 2$, $d = -3$, and $\mathbf{u} = (1, -2)$, then

$$c\mathbf{u} = 2(1, -2) = (2, -4) \quad \text{and} \quad d\mathbf{u} = -3(1, -2) = (-3, 6),$$

FIGURE 3.18

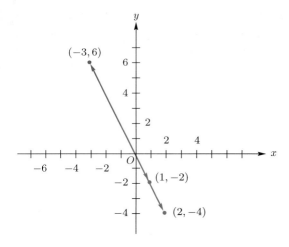

which are shown in Figure 3.18.   ■

The vector $(0,0)$ is called the **zero vector** and is denoted by **0**. If **u** is any vector, it follows that (Exercise T.2)

$$\mathbf{u} + \mathbf{0} = \mathbf{u}. \tag{4}$$

We can also show (Exercise T.3) that

$$\mathbf{u} + (-1)\mathbf{u} = \mathbf{0}, \tag{5}$$

and we write $(-1)\mathbf{u}$ as $-\mathbf{u}$ and call it the **negative** of **u**. Moreover, we write $\mathbf{u} + (-1)\mathbf{v}$ as $\mathbf{u} - \mathbf{v}$ and call it the **difference** of **u** and **v**. The vector $\mathbf{u} - \mathbf{v}$ is shown in Figure 3.19(a).

Observe that while vector addition gives one diagonal of a parallelogram, vector subtraction gives the other diagonal. See Figure 3.19(b).

FIGURE 3.19

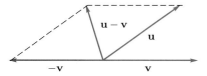

(a) Difference between vectors.

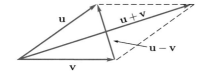

(b) Vector sum and vector difference.

Application (Vectors in Physics)

When several forces act on a body, we can find a single force, called the **resultant force**, having the equivalent effect. The resultant force can be determined using vectors. The following example illustrates the method.

EXAMPLE 12 ■ Suppose that a force of 12 pounds is applied to an object along the negative x-axis and a force of 5 pounds is applied to the object along the positive y-axis. Find the magnitude and direction of the resultant force.

Solution

In Figure 3.20 we have represented the force along the negative x-axis by the vector $\overrightarrow{OA}$ and the force along the positive y-axis by the vector $\overrightarrow{OB}$. The resultant force is the vector $\overrightarrow{OC} = \overrightarrow{OA} + \overrightarrow{OB}$. Thus the magnitude of the resultant force is 13 pounds and its direction is as indicated in the figure.   ■

FIGURE 3.20

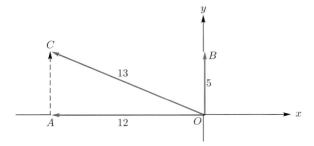

Vectors are also used in physics to deal with velocity problems, as the following example illustrates.

EXAMPLE 13   ■   Suppose that a boat is traveling east across a river at the rate of 4 miles per hour while the river's current is flowing south at a rate of 3 miles per hour. Find the resultant velocity of the boat.

Solution

In Figure 3.21, we have represented the velocity of the boat by the vector $\overrightarrow{OA}$ and the velocity of the river's current by the vector $\overrightarrow{OB}$. The resultant velocity is the vector $\overrightarrow{OC} = \overrightarrow{OA} + \overrightarrow{OB}$. Thus the magnitude of the resultant velocity is 5 miles per hour and its direction is as indicated in the figure.   ■

FIGURE 3.21

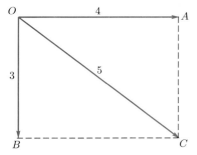

Angle Between Two Vectors

The angle between the nonzero vectors $\mathbf{u} = (x_1, y_1)$ and $\mathbf{v} = (x_2, y_2)$ is the angle θ, $0 \le \theta \le \pi$, shown in Figure 3.22. Applying the law of cosines to the triangle in that figure, we obtain

$$\|\mathbf{u} - \mathbf{v}\|^2 = \|\mathbf{u}\|^2 + \|\mathbf{v}\|^2 - 2\|\mathbf{u}\|\|\mathbf{v}\| \cos \theta. \tag{6}$$

From (2),

$$\|\mathbf{u} - \mathbf{v}\|^2 = (x_1 - x_2)^2 + (y_1 - y_2)^2$$
$$= x_1^2 + x_2^2 + y_1^2 + y_2^2 - 2(x_1 x_2 + y_1 y_2)$$
$$= \|\mathbf{u}\|^2 + \|\mathbf{v}\|^2 - 2(x_1 x_2 + y_1 y_2).$$

If we substitute this expression in (6) and solve for $\cos\theta$ (recall that since $\mathbf{u}$ and $\mathbf{v}$ are nonzero vectors, then $\|\mathbf{u}\| \neq 0$ and $\|\mathbf{v}\| \neq 0$), we obtain

$$\cos\theta = \frac{x_1 x_2 + y_1 y_2}{\|\mathbf{u}\|\|\mathbf{v}\|}. \tag{7}$$

FIGURE 3.22

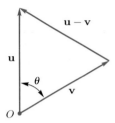

Recall from the first part of Section 1.3 that the **dot product** of the vectors $\mathbf{u} = (x_1, y_1)$ and $\mathbf{v} = (x_2, y_2)$ is defined to be

$$\mathbf{u} \cdot \mathbf{v} = x_1 x_2 + y_1 y_2.$$

Thus we can rewrite (7) as

$$\cos\theta = \frac{\mathbf{u} \cdot \mathbf{v}}{\|\mathbf{u}\|\|\mathbf{v}\|} \qquad (0 \leq \theta \leq \pi). \tag{8}$$

EXAMPLE 14 ■ If $\mathbf{u} = (2, 4)$ and $\mathbf{v} = (-1, 2)$, then

$$\mathbf{u} \cdot \mathbf{v} = (2)(-1) + (4)(2) = 6.$$

Also,

$$\|\mathbf{u}\| = \sqrt{2^2 + 4^2} = \sqrt{20}$$

and

$$\|\mathbf{v}\| = \sqrt{(-1)^2 + 2^2} = \sqrt{5}.$$

Hence

$$\cos\theta = \frac{6}{\sqrt{20}\,\sqrt{5}} = 0.6.$$

We can obtain the approximate angle by using a calculator or a table of cosines; we find that θ is approximately $53°8'$ or 0.93 radian. ■

If $\mathbf{u}$ is a vector in R^2, then we can use the definition of dot product to write

$$\|\mathbf{u}\| = \sqrt{\mathbf{u} \cdot \mathbf{u}}.$$

FIGURE 3.23
Orthogonal vectors

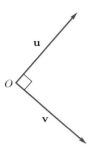

FIGURE 3.23
Orthogonal vectors

If the nonzero vectors **u** and **v** are at right angles (Figure 3.23), then the cosine of the angle θ between them is zero. Hence, from (8), we have $\mathbf{u} \cdot \mathbf{v} = 0$. Conversely, if $\mathbf{u} \cdot \mathbf{v} = 0$, then $\cos \theta = 0$ and the vectors are at right angles. Thus the nonzero vectors **u** and **v** are **perpendicular** or **orthogonal** if and only if $\mathbf{u} \cdot \mathbf{v} = 0$. We shall also say that two vectors are orthogonal if at least one of them is zero. Thus, we can now say that two vectors **u** and **v** are orthogonal if and only if $\mathbf{u} \cdot \mathbf{v} = 0$.

EXAMPLE 15 ■ The vectors $\mathbf{u} = (2, -4)$ and $\mathbf{v} = (4, 2)$ are orthogonal, since

$$\mathbf{u} \cdot \mathbf{v} = (2)(4) + (-4)(2) = 0$$

(Figure 3.24). ■

FIGURE 3.24
Orthogonal vectors

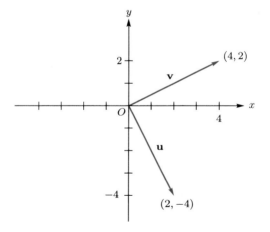

THEOREM 3.1 ■ *If* **u**, **v**, *and* **w** *are vectors and c is a scalar, then:*
(*Properties of Dot Product*)

(a) $\mathbf{u} \cdot \mathbf{u} > 0$ *if* $\mathbf{u} \neq \mathbf{0}$; $\mathbf{u} \cdot \mathbf{u} = 0$ *if and only if* $\mathbf{u} = \mathbf{0}$.

(b) $\mathbf{u} \cdot \mathbf{v} = \mathbf{v} \cdot \mathbf{u}$.

(c) $(\mathbf{u} + \mathbf{v}) \cdot \mathbf{w} = \mathbf{u} \cdot \mathbf{w} + \mathbf{v} \cdot \mathbf{w}$.

(d) $(c\mathbf{u}) \cdot \mathbf{v} = \mathbf{u} \cdot (c\mathbf{v}) = c(\mathbf{u} \cdot \mathbf{v})$.

Proof   Exercise T.7. ■

Unit Vectors

A **unit vector** is a vector whose length is 1. If $\mathbf{x}$ is any nonzero vector, then the vector

$$\mathbf{u} = \frac{1}{\|\mathbf{x}\|}\,\mathbf{x}$$

is a unit vector in the direction of $\mathbf{x}$ (Exercise T.5).

EXAMPLE 16 ■ Let $\mathbf{x} = (-3, 4)$. Then

$$\|\mathbf{x}\| = \sqrt{(-3)^2 + 4^4} = 5.$$

Hence the vector $\mathbf{u} = \frac{1}{5}(-3, 4) = \left(-\frac{3}{5}, \frac{4}{5}\right)$ is a unit vector, since

$$\|\mathbf{u}\| = \sqrt{\left(-\frac{3}{5}\right)^2 + \left(\frac{4}{5}\right)^2} = \sqrt{\frac{9+16}{25}} = 1.$$

Also, $\mathbf{u}$ lies in the direction of $\mathbf{x}$ (Figure 3.25). ■

FIGURE 3.25

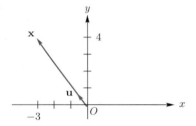

There are two unit vectors in R^2 that are of special importance. They are $\mathbf{i} = (1, 0)$ and $\mathbf{j} = (0, 1)$, the unit vectors along the positive x- and y-axes, respectively (Figure 3.26).

FIGURE 3.26

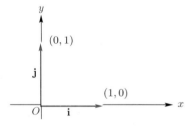

If $\mathbf{u} = (x_1, y_1)$ is any vector in R^2, then we can write $\mathbf{u}$ as a linear combination of $\mathbf{i}$ and $\mathbf{j}$ as

$$\mathbf{u} = x_1 \mathbf{i} + y_1 \mathbf{j}.$$

EXAMPLE 17 ■ If

$$\mathbf{u} = (4, -5),$$

then

$$\mathbf{u} = 4\mathbf{i} - 5\mathbf{j}. \qquad ■$$

EXAMPLE 18 ■ The vectors **i** and **j** are orthogonal [see Exercise 23(b)]. ■

3.1 EXERCISES

1. Plot the following points in R^2.
   (a) $(2, -1)$. (b) $(-1, 2)$. (c) $(3, 4)$
   (d) $(-3, -2)$. (e) $(0, 2)$. (f) $(0, -3)$.

2. Sketch a directed line segment in R^2 representing each of the following vectors.
   (a) $\mathbf{u}_1 = \begin{bmatrix} -2 \\ 3 \end{bmatrix}$. (b) $\mathbf{u}_2 = \begin{bmatrix} 3 \\ 4 \end{bmatrix}$.
   (c) $\mathbf{u}_3 = \begin{bmatrix} -3 \\ -3 \end{bmatrix}$. (d) $\mathbf{u}_4 = \begin{bmatrix} 0 \\ -3 \end{bmatrix}$.

3. Determine the head of the vector $\begin{bmatrix} -2 \\ 5 \end{bmatrix}$ whose tail is at $(3, 2)$. Make a sketch.

4. Determine the head of the vector $\begin{bmatrix} 2 \\ 5 \end{bmatrix}$ whose tail is at $(1, 2)$. Make a sketch.

5. Find $\mathbf{u} + \mathbf{v}$, $\mathbf{u} - \mathbf{v}$, $2\mathbf{u}$, and $3\mathbf{u} - 2\mathbf{v}$ if
   (a) $\mathbf{u} = (2, 3)$, $\mathbf{v} = (-2, 5)$.
   (b) $\mathbf{u} = (0, 3)$, $\mathbf{v} = (3, 2)$.
   (c) $\mathbf{u} = (2, 6)$, $\mathbf{v} = (3, 2)$.

6. Repeat Exercise 5 for
   (a) $\mathbf{u} = (-1, 3)$, $\mathbf{v} = (2, 4)$.
   (b) $\mathbf{u} = (-4, -3)$, $\mathbf{v} = (5, 2)$.
   (c) $\mathbf{u} = (3, 2)$, $\mathbf{v} = (-2, 0)$.

7. Let $\mathbf{u} = (1, 2)$, $\mathbf{v} = (-3, 4)$, $\mathbf{w} = (w_1, 4)$, and $\mathbf{x} = (-2, x_2)$. Find w_1 and x_2 so that
   (a) $\mathbf{w} = 2\mathbf{u}$. (b) $\frac{3}{2}\mathbf{x} = \mathbf{v}$.
   (c) $\mathbf{w} + \mathbf{x} = \mathbf{u}$.

8. Let $\mathbf{u} = (-4, 3)$, $\mathbf{v} = (2, -5)$, and $\mathbf{w} = (w_1, w_2)$. Find w_1 and w_2 so that
   (a) $\mathbf{w} = 2\mathbf{u} + 3\mathbf{v}$. (b) $\mathbf{u} + \mathbf{w} = 2\mathbf{u} - \mathbf{v}$.
   (c) $\mathbf{w} = \frac{5}{2}\mathbf{v}$.

9. Find the length of the following vectors.
   (a) $(1, 2)$. (b) $(-3, -4)$.
   (c) $(0, 2)$. (d) $(-4, 3)$.

10. Find the length of the following vectors.
    (a) $(-2, 3)$. (b) $(3, 0)$.
    (c) $(-4, -5)$. (d) $(3, 2)$.

11. Find the distance between the following pairs of points.
    (a) $(2, 3)$, $(3, 4)$. (b) $(0, 0)$, $(3, 4)$.
    (c) $(-3, 2)$, $(0, 1)$. (d) $(0, 3)$, $(2, 0)$.

12. Find the distance between the following pairs of points.
    (a) $(4, 2)$, $(1, 2)$. (b) $(-2, -3)$, $(0, 1)$.
    (c) $(2, 4)$, $(-1, 1)$. (d) $(2, 0)$, $(3, 2)$.

13. Is it possible to write the vector $(-5, 6)$ as a linear combination (defined before Example 11 in Section 1.3) of the vectors $(1, 2)$ and $(3, 4)$?

14. If possible, find scalars c_1 and c_2, not both zero, so that
    $$c_1 \begin{bmatrix} 1 \\ 2 \end{bmatrix} + c_2 \begin{bmatrix} 3 \\ 4 \end{bmatrix} = \begin{bmatrix} 0 \\ 0 \end{bmatrix}.$$

15. Find the area of the triangle with vertices $(3, 3)$, $(-1, -1)$, $(4, 1)$.

16. Find the area of the right triangle with vertices $(0, 0)$, $(0, 3)$, $(4, 0)$. Verify by using the formula $A = \frac{1}{2}(\text{base})(\text{height})$.

17. Find the area of the parallelogram with vertices $(2, 3)$, $(5, 3)$, $(4, 5)$, $(7, 5)$.

18. Let Q be the quadrilateral with vertices $(-2, 3)$, $(1, 4)$, $(3, 0)$, and $(-1, -3)$. Find the area of Q.

19. Find a unit vector in the direction of $\mathbf{x}$.
    (a) $\mathbf{x} = (3, 4)$. (b) $\mathbf{x} = (-2, -3)$. (c) $\mathbf{x} = (5, 0)$.

20. Find a unit vector in the direction of $\mathbf{x}$.
    (a) $\mathbf{x} = (2, 4)$. (b) $\mathbf{x} = (0, -2)$.
    (c) $\mathbf{x} = (-1, -3)$.

21. Find the cosine of the angle between each pair of vectors $\mathbf{u}$ and $\mathbf{v}$.
    (a) $\mathbf{u} = (1, 2)$, $\mathbf{v} = (2, -3)$.
    (b) $\mathbf{u} = (1, 0)$, $\mathbf{v} = (0, 1)$.
    (c) $\mathbf{u} = (-3, -4)$, $\mathbf{v} = (4, -3)$.
    (d) $\mathbf{u} = (2, 1)$, $\mathbf{v} = (-2, -1)$.

22. Find the cosine of the angle between each pair of vectors $\mathbf{u}$ and $\mathbf{v}$.
    (a) $\mathbf{u} = (0, -1)$, $\mathbf{v} = (1, 0)$.
    (b) $\mathbf{u} = (2, 2)$, $\mathbf{v} = (4, -5)$.

(c) $\mathbf{u} = (2, -1)$, $\mathbf{v} = (-3, -2)$.

(d) $\mathbf{u} = (0, 2)$, $\mathbf{v} = (3, -3)$.

23. Show that

(a) $\mathbf{i} \cdot \mathbf{i} = \mathbf{j} \cdot \mathbf{j} = 1$.   (b) $\mathbf{i} \cdot \mathbf{j} = 0$.

24. Which of the vectors $\mathbf{u}_1 = (1, 2)$, $\mathbf{u}_2 = (0, 1)$, $\mathbf{u}_3 = (-2, -4)$, $\mathbf{u}_4 = (-2, 1)$, $\mathbf{u}_5 = (2, 4)$, $\mathbf{u}_6 = (-6, 3)$ are

(a) Orthogonal?

(b) In the same direction?

(c) In opposite directions?

25. Find all constants a such that the vectors $(a, 4)$ and $(2, 5)$ are parallel.

26. Find all constants a such that the vectors $(a, 2)$ and $(a, -2)$ are orthogonal.

27. Write each of the following vectors in terms of $\mathbf{i}$ and $\mathbf{j}$.

(a) $(1, 3)$.   (b) $(-2, -3)$.

(c) $(-2, 0)$.   (d) $(0, 3)$.

28. Write each of the following vectors as a 2×1 matrix.

(a) $3\mathbf{i} - 2\mathbf{j}$.   (b) $2\mathbf{i}$.   (c) $-2\mathbf{i} - 3\mathbf{j}$.

29. A ship is being pushed by a tugboat with a force of 300 pounds along the negative y-axis while another tugboat is pushing along the negative x-axis with a force of 400 pounds. Find the magnitude and sketch the direction of the resultant force.

30. Suppose that an airplane is flying with an airspeed of 260 kilometers per hour while a wind is blowing to the west at 100 kilometers per hour. Indicate on a figure the approximate direction that the plane must follow to result in a flight directly south. What will be the resultant speed?

THEORETICAL EXERCISES

T.1. Show how we can associate a point in the plane with each ordered pair (x, y) of real numbers.

T.2. Show that $\mathbf{u} + \mathbf{0} = \mathbf{u}$.

T.3. Show that $\mathbf{u} + (-1)\mathbf{u} = \mathbf{0}$.

T.4. Show that if c is a scalar, then $\|c\mathbf{u}\| = |c|\|\mathbf{u}\|$.

T.5. Show that if $\mathbf{x}$ is a nonzero vector, then

$$\mathbf{u} = \frac{1}{\|\mathbf{x}\|} \mathbf{x}$$

is a unit vector in the direction of $\mathbf{x}$.

T.6. Show that

(a) $1\mathbf{u} = \mathbf{u}$.

(b) $(rs)\mathbf{u} = r(s\mathbf{u})$, where r and s are scalars.

T.7. Prove Theorem 3.1.

T.8. Show that if $\mathbf{w}$ is orthogonal to $\mathbf{u}$ and $\mathbf{v}$, then $\mathbf{w}$ is orthogonal to $r\mathbf{u} + s\mathbf{v}$, where r and s are scalars.

T.9. Let θ be the angle between the nonzero vectors $\mathbf{u} = (x_1, y_1)$ and $\mathbf{v} = (x_2, y_2)$ in the plane. Show that if $\mathbf{u}$ and $\mathbf{v}$ are parallel, then $\cos \theta = \pm 1$.

MATLAB EXERCISES

The following exercises use the routine **vec2demo**, which provides a graphical display of vectors in the plane. For a pair of vectors $\mathbf{u} = (x_1, y_1)$ and $\mathbf{v} = (x_2, y_2)$, routine **vec2demo** graphs $\mathbf{u}$ and $\mathbf{v}$, $\mathbf{u} + \mathbf{v}$, $\mathbf{u} - \mathbf{v}$, and a scalar multiple. Once the vectors $\mathbf{u}$ and $\mathbf{v}$ are entered into MATLAB, type

$$\mathbf{vec2demo(u, v)}$$

For further information, use **help vec2demo**.

ML.1. Use the routine **vec2demo** with each of the following pairs of vectors. (Square brackets are used in MATLAB.)

(a) $\mathbf{u} = \begin{bmatrix} 2 & 0 \end{bmatrix}$, $\mathbf{v} = \begin{bmatrix} 0 & 3 \end{bmatrix}$

(b) $\mathbf{u} = \begin{bmatrix} -3 & 1 \end{bmatrix}$, $\mathbf{v} = \begin{bmatrix} 2 & 2 \end{bmatrix}$

(c) $\mathbf{u} = \begin{bmatrix} 5 & 2 \end{bmatrix}$, $\mathbf{v} = \begin{bmatrix} -3 & 3 \end{bmatrix}$

ML.2. Use the routine **vec2demo** with each of the following pairs of vectors. (Square brackets are used in MATLAB.)

(a) $\mathbf{u} = \begin{bmatrix} 2 & -2 \end{bmatrix}$, $\mathbf{v} = \begin{bmatrix} 1 & 3 \end{bmatrix}$

(b) $\mathbf{u} = \begin{bmatrix} 0 & 3 \end{bmatrix}$, $\mathbf{v} = \begin{bmatrix} -2 & 0 \end{bmatrix}$

(c) $\mathbf{u} = \begin{bmatrix} 4 & -1 \end{bmatrix}$, $\mathbf{v} = \begin{bmatrix} -3 & 5 \end{bmatrix}$

ML.3. Choose pairs of vectors $\mathbf{u}$ and $\mathbf{v}$ to use with **vec2demo**.

3.2 ▾ *n*-Vectors

In this section we look at *n*-vectors from a geometric point of view by generalizing the notions discussed in the preceding section. The case of $n = 3$ will be of special interest, and we shall discuss it in some detail.

As we have already seen in the first part of Section 1.3, an *n*-vector is an $n \times 1$ matrix

$$\mathbf{u} = \begin{bmatrix} u_1 \\ u_2 \\ \vdots \\ u_n \end{bmatrix},$$

where $u_1, u_2, \ldots, u_n$ are real numbers, which are called the **components** of **u**. Since an *n*-vector is a matrix, the *n*-vectors

$$\mathbf{u} = \begin{bmatrix} u_1 \\ u_2 \\ \vdots \\ u_n \end{bmatrix} \quad \text{and} \quad \mathbf{v} = \begin{bmatrix} v_1 \\ v_2 \\ \vdots \\ v_n \end{bmatrix}$$

are said to be **equal** if $u_i = v_i \ (1 \le i \le n)$.

EXAMPLE 1   ■   The 4-vectors $\begin{bmatrix} 1 \\ -2 \\ 3 \\ 4 \end{bmatrix}$ and $\begin{bmatrix} 1 \\ -2 \\ 3 \\ -4 \end{bmatrix}$ are not equal, since their fourth components are not the same.   ■

The set of all *n*-vectors is denoted by R^n and is called ***n*-space**. When the actual value of n need not be specified, we refer to *n*-vectors simply as **vectors**. The real numbers are called **scalars**. The components of a vector are real numbers and hence the components of a vector are scalars.

Vector Operations

DEFINITION   Let

$$\mathbf{u} = \begin{bmatrix} u_1 \\ u_2 \\ \vdots \\ u_n \end{bmatrix} \quad \text{and} \quad \mathbf{v} = \begin{bmatrix} v_1 \\ v_2 \\ \vdots \\ v_n \end{bmatrix}$$

be two vectors in R^n. The **sum** of the vectors **u** and **v** is the vector

$$\begin{bmatrix} u_1 + v_1 \\ u_2 + v_2 \\ \vdots \\ u_n + v_n \end{bmatrix},$$

and it is denoted by $\mathbf{u} + \mathbf{v}$.

EXAMPLE 2 ■ If $\mathbf{u} = \begin{bmatrix} 1 \\ -2 \\ 3 \end{bmatrix}$ and $\begin{bmatrix} 2 \\ 3 \\ -3 \end{bmatrix}$ are vectors in R^3, then

$$\mathbf{u} + \mathbf{v} = \begin{bmatrix} 1+2 \\ -2+3 \\ 3+(-3) \end{bmatrix} = \begin{bmatrix} 3 \\ 1 \\ 0 \end{bmatrix}.$$

■

DEFINITION   If

$$\mathbf{u} = \begin{bmatrix} u_1 \\ u_2 \\ \vdots \\ u_n \end{bmatrix}$$

is a vector in R^n and c is a scalar, then the **scalar multiple** $c\mathbf{u}$ of $\mathbf{u}$ by c is the vector

$$\begin{bmatrix} cu_1 \\ cu_2 \\ \vdots \\ cu_n \end{bmatrix}.$$

EXAMPLE 3 ■ If $\mathbf{u} = \begin{bmatrix} 2 \\ 3 \\ -1 \\ 2 \end{bmatrix}$ is a vector in R^4 and $c = -2$, then

$$c\mathbf{u} = (-2) \begin{bmatrix} 2 \\ 3 \\ -1 \\ 2 \end{bmatrix} = \begin{bmatrix} -4 \\ -6 \\ 2 \\ -4 \end{bmatrix}.$$

■

The operations of vector addition and scalar multiplication satisfy the following properties.

THEOREM 3.2 ■ *Let $\mathbf{u}$, $\mathbf{v}$, and $\mathbf{w}$ be any vectors in R^n; let c and d be any scalars. Then*

(α) $\mathbf{u} + \mathbf{v}$ *is a vector in R^n (i.e., R^n is closed under the operation of vector addition).*

(a) $\mathbf{u} + \mathbf{v} = \mathbf{v} + \mathbf{u}$.

(b) $\mathbf{u} + (\mathbf{v} + \mathbf{w}) = (\mathbf{u} + \mathbf{v}) + \mathbf{w}$.

(c) *There is a vector $\mathbf{0}$ in R^n such that $\mathbf{u} + \mathbf{0} = \mathbf{0} + \mathbf{u} = \mathbf{u}$ for all $\mathbf{u}$ in R^n.*

(d) *For each vector $\mathbf{u}$ in R^n, there is a vector $-\mathbf{u}$ in R^n such that $\mathbf{u} + (-\mathbf{u}) = \mathbf{0}$.*

(β) $c\mathbf{u}$ *is a vector in R^n (i.e., R^n is closed under the operation of scalar multiplication).*

(e) $c(\mathbf{u} + \mathbf{v}) = c\mathbf{u} + c\mathbf{v}$.

(f) $(c + d)\mathbf{u} = c\mathbf{u} + d\mathbf{u}$.

(g) $c(d\mathbf{u}) = (cd)\mathbf{u}$.

(h) $1\mathbf{u} = \mathbf{u}$.

Proof (α) and (β) are immediate from the definitions for vector sum and scalar multiple. We verify (f) here and leave the rest of the proof to the reader (Exercise T.1). Thus

$$(c + d)\mathbf{u} = (c + d) \begin{bmatrix} u_1 \\ u_2 \\ \vdots \\ u_n \end{bmatrix} = \begin{bmatrix} (c + d)u_1 \\ (c + d)u_2 \\ \vdots \\ (c + d)u_n \end{bmatrix} = \begin{bmatrix} cu_1 + du_1 \\ cu_2 + du_2 \\ \vdots \\ cu_n + du_n \end{bmatrix}$$

$$= \begin{bmatrix} cu_1 \\ cu_2 \\ \vdots \\ cu_n \end{bmatrix} + \begin{bmatrix} du_1 \\ du_2 \\ \vdots \\ du_n \end{bmatrix} = c \begin{bmatrix} u_1 \\ u_2 \\ \vdots \\ u_n \end{bmatrix} + d \begin{bmatrix} u_1 \\ u_2 \\ \vdots \\ u_n \end{bmatrix} = c\mathbf{u} + d\mathbf{u}. \quad \blacksquare$$

It is easy to show that the vectors $\mathbf{0}$ and $-\mathbf{u}$ in Properties (c) and (d) are unique. Moreover,

$$\mathbf{0} = \begin{bmatrix} 0 \\ 0 \\ \vdots \\ 0 \end{bmatrix}$$

and if $\mathbf{u} = \begin{bmatrix} u_1 \\ u_2 \\ \vdots \\ u_n \end{bmatrix}$, then $-\mathbf{u} = \begin{bmatrix} -u_1 \\ -u_2 \\ \vdots \\ -u_n \end{bmatrix}$. The vector $\mathbf{0}$ is called the **zero vector** and $-\mathbf{u}$ is called the **negative** of $\mathbf{u}$. It is easy to verify (Exercise T.2) that

$$-\mathbf{u} = (-1)\mathbf{u}.$$

We shall also write $\mathbf{u} + (-\mathbf{v})$ as $\mathbf{u} - \mathbf{v}$ and call it the **difference** of $\mathbf{u}$ and $\mathbf{v}$.

EXAMPLE 4 ■ If $\mathbf{u}$ and $\mathbf{v}$ are as in Example 2, then

$$\mathbf{u} - \mathbf{v} = \begin{bmatrix} 1 - 2 \\ -2 - 3 \\ 3 - (-3) \end{bmatrix} = \begin{bmatrix} -1 \\ -5 \\ 6 \end{bmatrix}.$$

■

As in the case of R^2, we shall identify the vector

$$\begin{bmatrix} u_1 \\ u_2 \\ \vdots \\ u_n \end{bmatrix}$$

with the point $(u_1, u_2, \ldots, u_n)$ so that points and vectors can be used interchangeably. Thus we may view R^n as consisting of vectors or of points, and we also write

$$\mathbf{u} = (u_1, u_2, \ldots, u_n).$$

Moreover, an n-vector is an $n \times 1$ matrix, vector addition is matrix addition, and scalar multiplication is merely the operation of multiplication of a matrix by a real number. Thus R^n can be viewed as the set of all $n \times 1$ matrices with the operations of matrix addition and scalar multiplication. *The important point here is that no matter how we view R^n—as n-vectors, points, or $n \times 1$ matrices—the algebraic behavior is always the same.*

Application

Vectors in R^n can be used to handle large amounts of data. Indeed, a number of computer software products, notably MATLAB, make extensive use of vectors. The following example illustrates these ideas.

EXAMPLE 5 ■
(*Application: Inventory Control*)

Suppose that a store handles 100 different items. The inventory on hand at the beginning of the week can be described by the inventory vector $\mathbf{u}$ in R^{100}. The number of items sold at the end of the week can be described by the vector $\mathbf{v}$, and the vector

$$\mathbf{u} - \mathbf{v}$$

represents the inventory at the end of the week. If the store receives a new shipment of goods, represented by the vector $\mathbf{w}$, then its new inventory would be

$$\mathbf{u} - \mathbf{v} + \mathbf{w}. \qquad ■$$

Visualizing R^3

We cannot draw pictures of R^n for $n > 3$. However, since R^3 is the world we live in, we can visualize it in a manner similar to that used for R^2.

We first fix a **coordinate system** by choosing a point, called the **origin**, and three lines, called the **coordinate axes**, each passing through the origin, so that each line is perpendicular to the other two. These lines are individually called the x-, y-, and z-**axes**. On each of these coordinate axes we choose a point fixing the units of length and positive directions. Frequently, but not always, the same unit of length is used for all the coordinate axes. In Figure 3.27(a) and (b) we show two of the many possible coordinate systems. The coordinate system shown in Figure 3.27(a) is called a **right-handed coordinate system**; the one shown in Figure 3.27(b) is called **left-handed**. A right-handed system is characterized by the following property. If we curl the fingers of the right hand in the direction of a 90° rotation from the positive x-axis to the positive y-axis, then the thumb will point in the direction of the positive z-axis (Figure 3.28).

FIGURE 3.27

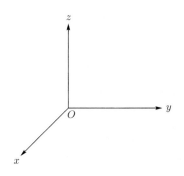

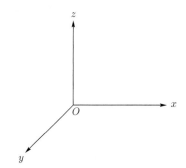

(a) Right-handed coordinate system.   (b) Left-handed coordinate system.

FIGURE 3.28

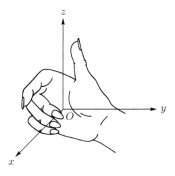

The projection of a point P in space on a line L is the point Q obtained by intersecting L with the line L' passing through P and perpendicular to L (Figure 3.29).

FIGURE 3.29

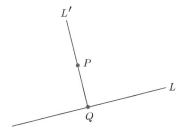

The **x-coordinate** of the point P is the number associated with the projection of P on the x-axis; similarly, for the **y-** and **z-coordinates**. These three numbers are called the **coordinates** of P. Thus with each point in space we associate an ordered triple (x, y, z) of real numbers, and conversely, with each ordered triple of real numbers, we associate a point in space. This correspondence is called a **rectangular coordinate system**. We write $P(x, y, z)$, or simply (x, y, z).

EXAMPLE 6 ■ In Figure 3.30(a) and (b) we show two points and their coordinates. ■

FIGURE 3.30

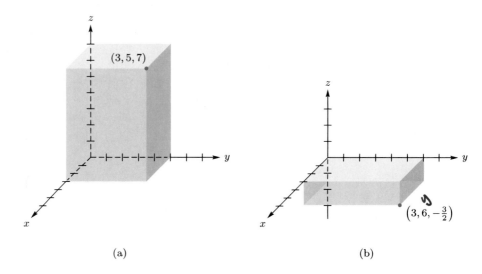

(a)   (b)

The **xy-plane** is the plane determined by the x- and y-axes. Similarly, we have the **xz-** and **yz-planes**.

In R^3, the components of a vector **u** are denoted by x_1, y_1, and z_1. Thus $\mathbf{u} = (x_1, y_1, z_1)$.

As in the plane, with the vector $\mathbf{u} = (x_1, y_1, z_1)$ we associate the directed line segment $\overrightarrow{OP}$, whose tail is $O(0, 0, 0)$ and whose head is $P(x_1, y_1, z_1)$ [Figure 3.31(a)]. Again as in the plane, in physical applications we often deal with a directed line segment $\overrightarrow{PQ}$, from the point $P(x_1, y_1, z_1)$ (not the origin) to the point $Q(x_2, y_2, z_2)$, as shown in Figure 3.31(b). Such a directed line segment will also be called a **vector in R^3**, or simply a **vector** with tail $P(x_1, y_1, z_1)$ and head $Q(x_2, y_2, z_2)$. The components of such a vector are $x_2 - x_1$, $y_2 - y_1$, and $z_2 - z_1$. Two such vectors in R^3 will be called **equal** if their components are

FIGURE 3.31

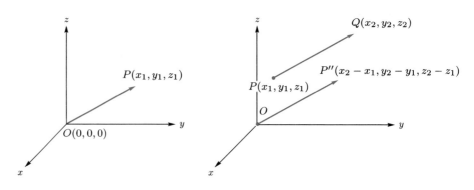

(a) A vector in R^3.   (b) Different directed line segments representing the same vector.

equal. Thus the vector $\overrightarrow{PQ}$ in Figure 3.31(b) can also be represented by the vector $(x_2 - x_1, y_2 - y_1, z_2 - z_1)$ with tail O and head $P''(x_2 - x_1, y_2 - y_1, z_2 - z_1)$.

We shall soon define the length of a vector in R^n and the angle between two nonzero vectors in R^n. Once this has been done, it can be shown that two vectors in R^3 are equal if and only if any directed line segments representing them are parallel, have the same direction, and are of the same length.

The sum $\mathbf{u} + \mathbf{v}$ of the vectors $\mathbf{u} = (x_1, y_1, z_1)$ and $\mathbf{v} = (x_2, y_2, z_2)$ in R^3 is the diagonal of the parallelogram determined by $\mathbf{u}$ and $\mathbf{v}$, as shown in Figure 3.32.

FIGURE 3.32

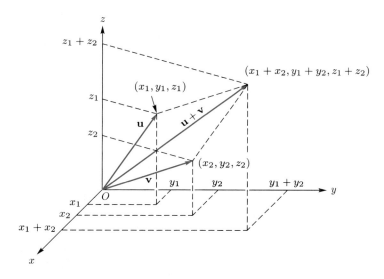

The reader will note that Figure 3.32 looks very much like Figure 3.15 in Section 3.1 in that in both R^2 and R^3 the vector $\mathbf{u} + \mathbf{v}$ is the diagonal of the parallelogram determined by $\mathbf{u}$ and $\mathbf{v}$.

The scalar multiple in R^3 is shown in Figure 3.33, which looks very much like Figure 3.17 in Section 3.1.

FIGURE 3.33
Scalar multiplication

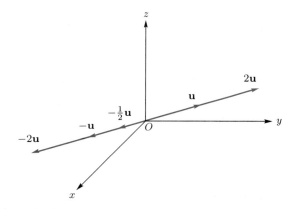

Dot Product on R^n

We shall now define the notion of the length of a vector in R^n by generalizing the corresponding idea for R^2.

DEFINITION

The **length** (also called **magnitude** or **norm**) of the vector $\mathbf{u} = (u_1, u_2, \ldots, u_n)$ in R^n is

$$\|\mathbf{u}\| = \sqrt{u_1^2 + u_2^2 + \cdots + u_n^2}. \tag{1}$$

We also define the distance from the point $(u_1, u_2, \ldots, u_n)$ to the origin by (1). The **distance** between the points $(u_1, u_2, \ldots, u_n)$ and $(v_1, v_2, \ldots, v_n)$ is then defined as the length of the vector $\mathbf{u} - \mathbf{v}$, where

$$\mathbf{u} = (u_1, u_2, \ldots, u_n) \quad \text{and} \quad \mathbf{v} = (v_1, v_2, \ldots, v_n).$$

Thus this distance is given by

$$\|\mathbf{u} - \mathbf{v}\| = \sqrt{(u_1 - v_1)^2 + (u_2 - v_2)^2 + \cdots + (u_n - v_n)^2}. \tag{2}$$

EXAMPLE 7 ■ Let $\mathbf{u} = (2, 3, 2, -1)$ and $\mathbf{v} = (4, 2, 1, 3)$. Then

$$\|\mathbf{u}\| = \sqrt{2^2 + 3^2 + 2^2 + (-1)^2} = \sqrt{18},$$
$$\|\mathbf{v}\| = \sqrt{4^2 + 2^2 + 1^2 + 3^2} = \sqrt{30}.$$

The distance between the points $(2, 3, 2, -1)$ and $(4, 2, 1, 3)$ is the length of the vector $\mathbf{u} - \mathbf{v}$. Thus, from Equation (2),

$$\|\mathbf{u} - \mathbf{v}\| = \sqrt{(2-4)^2 + (3-2)^2 + (2-1)^2 + (-1-3)^2} = \sqrt{22}. \qquad ■$$

It should be noted that in R^3, Equations (1) and (2), for the length of a vector and the distance between two points, do not have to be defined. They can easily be established by means of two applications of the Pythagorean theorem (Exercise T.3).

We shall define the cosine of the angle between two vectors in R^n by generalizing the corresponding formula in R^2. However, we first recall the notion of dot product in R^n as defined in the first part of Section 1.3.

If $\mathbf{u} = (u_1, u_2, \ldots, u_n)$ and $\mathbf{v} = (v_1, v_2, \ldots, v_n)$ are vectors in R^n, then their **dot product** is defined by

$$\mathbf{u} \cdot \mathbf{v} = u_1 v_1 + u_2 v_2 + \cdots + u_n v_n.$$

This is exactly how the dot product was defined in R^2. The dot product in R^n is also called the **standard inner product**.

EXAMPLE 8 ■ If $\mathbf{u}$ and $\mathbf{v}$ are as in Example 7, then

$$\mathbf{u} \cdot \mathbf{v} = (2)(4) + (3)(2) + (2)(1) + (-1)(3) = 13. \qquad ■$$

EXAMPLE 9
(*Application:*
Revenue Monitoring)

■ Consider the store in Example 5. If the vector **p** denotes the price of each of the 100 items, then the dot product

$$\mathbf{v} \cdot \mathbf{p}$$

gives the total revenue received at the end of the week.   ■

If **u** is a vector in R^n, then we can use the definition of dot product in R^n to write

$$\|\mathbf{u}\| = \sqrt{\mathbf{u} \cdot \mathbf{u}}.$$

The dot product in R^n satisfies the same properties as in R^2. We state these as the following theorem, which reads just like Theorem 3.1.

THEOREM 3.3
(*Properties of*
Dot Product

■ *If* **u**, **v**, *and* **w** *are vectors in* R^n *and* *c* *is a scalar, then:*

(a) $\mathbf{u} \cdot \mathbf{u} > 0$ *for* $\mathbf{u} \neq \mathbf{0}$; $\mathbf{u} \cdot \mathbf{u} = 0$ *if and only if* $\mathbf{u} = \mathbf{0}$.

(b) $\mathbf{u} \cdot \mathbf{v} = \mathbf{v} \cdot \mathbf{u}$.

(c) $(\mathbf{u} + \mathbf{v}) \cdot \mathbf{w} = \mathbf{u} \cdot \mathbf{w} + \mathbf{v} \cdot \mathbf{w}$.

(d) $(c\mathbf{u}) \cdot \mathbf{v} = \mathbf{u} \cdot (c\mathbf{v}) = c(\mathbf{u} \cdot \mathbf{v})$.

Proof

Exercise T.4.   ■

We shall now prove a result that will enable us to give a worthwhile definition for the cosine of an angle between two nonzero vectors. This result, called the **Cauchy–Schwarz inequality**, has many important applications in mathematics. The proof of this result, although not difficult, is one that is not too natural but does call for a clever start.

THEOREM 3.4
(*Cauchy*–Schwarz***
Inequality)

■ *If* **u** *and* **v** *are vectors in* R^n, *then*

$$|\mathbf{u} \cdot \mathbf{v}| \leq \|\mathbf{u}\|\|\mathbf{v}\|. \tag{3}$$

(*Observe that* | | *on the left stands for the absolute value of a real number;* ‖ ‖ *on the right denotes the length of a vector.*)

---

*Augustin-Louis Cauchy (1789–1857) grew up in a suburb of Paris as a neighbor of several leading mathematicians of the day, attended the École Polytechnique and the École des Ponts et Chaussées, and was for a time a practicing engineer. He was a devout Roman Catholic, with an abiding interest in Catholic charities. He was also strongly devoted to royalty, especially to the Bourbon kings who ruled France after Napoleon's defeat. When Charles X was deposed in 1830, Cauchy voluntarily followed him into exile in Prague.

Cauchy wrote 7 books and more than 700 papers of varying quality, touching all branches of mathematics. He made important contributions to the early theory of determinants, the theory of eigenvalues, the study of ordinary and partial differential equations, the theory of permutation groups, and the foundations of calculus, and he founded the theory of functions of a complex variable.

**Hermann Amandus Schwarz (1843–1921) was born in Poland but was educated and taught in Germany. He was a protégé of Karl Weierstrass and Ernst Eduard Kummer, whose daughter he married. His main contributions to mathematics were in the geometric aspects of analysis, such as conformal mappings and minimal surfaces. In connection with the latter he sought certain numbers associated with differential equations, numbers that have since come to be called eigenvalues. The inequality given above was used in the search for these numbers.

Proof   If $\mathbf{u} = 0$, then $\|\mathbf{u}\| = 0$ and $\mathbf{u} \cdot \mathbf{v} = 0$, so (3) holds. Now suppose that $\mathbf{u}$ is nonzero. Let r be a scalar and consider the vector $r\mathbf{u} + \mathbf{v}$. By Theorem 3.3,

$$0 \le (r\mathbf{u} + \mathbf{v}) \cdot (r\mathbf{u} + \mathbf{v}) = r^2 \mathbf{u} \cdot \mathbf{u} + 2r\mathbf{u} \cdot \mathbf{v} + \mathbf{v} \cdot \mathbf{v}$$
$$= ar^2 + 2br + c,$$

where

$$a = \mathbf{u} \cdot \mathbf{u}, \quad b = \mathbf{u} \cdot \mathbf{v}, \quad \text{and} \quad c = \mathbf{v} \cdot \mathbf{v}.$$

Now $p(r) = ar^2 + 2br + c$ is a quadratic polynomial in r (whose graph is a parabola, opening upward, since $a > 0$) that is nonnegative for all values of r. This means that either this polynomial has no real roots, or if it has real roots, then both roots are equal. [If $p(r)$ had two distinct roots r_1 and r_2, then it would be negative for some value of r between r_1 and r_2, as seen in Figure 3.34.]

Recall that the roots of $p(r)$ are given by the quadratic formula as

$$\frac{-2b + \sqrt{4b^2 - 4ac}}{2a} \quad \text{and} \quad \frac{-2b - \sqrt{4b^2 - 4ac}}{2a}$$

($a \ne 0$ since $\mathbf{u} \ne \mathbf{0}$). Both roots are equal or there are no real roots when

$$4b^2 - 4ac \le 0,$$

which means that

$$b^2 \le ac.$$

Upon taking square roots of both sides and observing that $\sqrt{a} = \sqrt{\mathbf{u} \cdot \mathbf{u}} = \|\mathbf{u}\|$, $\sqrt{c} = \sqrt{\mathbf{v} \cdot \mathbf{v}} = \|\mathbf{v}\|$, we obtain (3). ∎

FIGURE 3.34

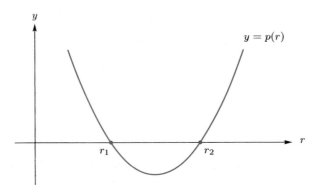

$y = p(r)$

EXAMPLE 10 ■ If $\mathbf{u}$ and $\mathbf{v}$ are as in Example 7, then from Example 8, $\mathbf{u} \cdot \mathbf{v} = 13$. Hence

$$|\mathbf{u} \cdot \mathbf{v}| = 13 \le \|\mathbf{u}\|\|\mathbf{v}\| = \sqrt{18}\sqrt{30}.$$

■

FIGURE 3.35

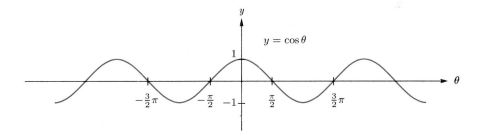

We shall now use the Cauchy–Schwarz inequality to define the angle between two nonzero vectors in R^n.

If $\mathbf{u}$ and $\mathbf{v}$ are nonzero vectors, then it follows from the Cauchy–Schwarz inequality that

$$\left|\frac{\mathbf{u}\cdot\mathbf{v}}{\|\mathbf{u}\|\|\mathbf{v}\|}\right|\leq 1$$

or

$$-1\leq\frac{\mathbf{u}\cdot\mathbf{v}}{\|\mathbf{u}\|\|\mathbf{v}\|}\leq 1.$$

Examining the portion of the graph of $y=\cos\theta$ (see Figure 3.35) for $0\leq\theta\leq\pi$, we see that for any number r in the interval $[-1,1]$, there is a unique real number θ such that $\cos\theta=r$. This implies that there is a unique real number θ such that

$$\cos\theta=\frac{\mathbf{u}\cdot\mathbf{v}}{\|\mathbf{u}\|\|\mathbf{v}\|},\qquad 0\leq\theta\leq\pi. \tag{4}$$

The angle θ is called the **angle between $\mathbf{u}$ and $\mathbf{v}$**.

In the case of R^3, we can establish by the law of cosines, as was done in R^2, that the cosine of the angle between $\mathbf{u}$ and $\mathbf{v}$ is given by (4). However, for R^n, $n>3$, we have to define it by (4).

EXAMPLE 11 ■ Let $\mathbf{u}=(1,0,0,1)$ and $\mathbf{v}=(0,1,0,1)$. Then

$$\|\mathbf{u}\|=\sqrt{2},\quad \|\mathbf{v}\|=\sqrt{2},\quad\text{and}\quad \mathbf{u}\cdot\mathbf{v}=1.$$

Thus

$$\cos\theta=\tfrac{1}{2}$$

and $\theta=60°$ or $\frac{\pi}{3}$ radians. ■

It is very useful to talk about orthogonality and parallelism in R^n, and we accordingly formulate the following definitions.

DEFINITION | Two nonzero vectors $\mathbf{u}$ and $\mathbf{v}$ in R^n are said to be **orthogonal** if $\mathbf{u}\cdot\mathbf{v}=0$. If one of the vectors is the zero vector, we agree to say that the vectors are orthogonal. They are said to be **parallel** if $|\mathbf{u}\cdot\mathbf{v}|=\|\mathbf{u}\|\|\mathbf{v}\|$. They are in the **same direction** if $\mathbf{u}\cdot\mathbf{v}=\|\mathbf{u}\|\|\mathbf{v}\|$. That is, they are orthogonal if $\cos\theta=0$, parallel if $\cos\theta=\pm1$, and in the same direction if $\cos\theta=1$.

EXAMPLE 12 ■ Consider the vectors $\mathbf{u} = (1, 0, 0, 1)$, $\mathbf{v} = (0, 1, 1, 0)$, and $\mathbf{w} = (3, 0, 0, 3)$. Then

$$\mathbf{u} \cdot \mathbf{v} = 0 \quad \text{and} \quad \mathbf{v} \cdot \mathbf{w} = 0,$$

which implies that $\mathbf{u}$ and $\mathbf{v}$ are orthogonal and $\mathbf{v}$ and $\mathbf{w}$ are orthogonal. Since

$$\mathbf{u} \cdot \mathbf{w} = 6, \quad \|\mathbf{u}\| = \sqrt{2}, \quad \|\mathbf{w}\| = \sqrt{18}, \quad \text{and} \quad \mathbf{u} \cdot \mathbf{w} = \|\mathbf{u}\|\|\mathbf{w}\|,$$

we conclude that $\mathbf{u}$ and $\mathbf{w}$ are in the same direction. ■

An easy consequence of the Cauchy–Schwarz inequality is the triangle inequality, which we prove next.

THEOREM 3.5 ■ *If $\mathbf{u}$ and $\mathbf{v}$ are vectors in R^n, then*
(*Triangle Inequality*)

$$\|\mathbf{u} + \mathbf{v}\| \le \|\mathbf{u}\| + \|\mathbf{v}\|.$$

Proof We have, from the definition of the length of a vector,

$$\|\mathbf{u} + \mathbf{v}\|^2 = (\mathbf{u} + \mathbf{v}) \cdot (\mathbf{u} + \mathbf{v})$$
$$= \mathbf{u} \cdot \mathbf{u} + 2(\mathbf{u} \cdot \mathbf{v}) + \mathbf{v} \cdot \mathbf{v}$$
$$= \|\mathbf{u}\|^2 + 2(\mathbf{u} \cdot \mathbf{v}) + \|\mathbf{v}\|^2.$$

By the Cauchy–Schwarz inequality we have

$$\|\mathbf{u}\|^2 + 2(\mathbf{u} \cdot \mathbf{v}) + \|\mathbf{v}\|^2 \le \|\mathbf{u}\|^2 + 2\|\mathbf{u}\|\|\mathbf{v}\| + \|\mathbf{v}\|^2$$
$$= (\|\mathbf{u}\| + \|\mathbf{v}\|)^2.$$

Taking square roots, we obtain the desired result. ■

The triangle inequality in R^2 and R^3 merely states that the length of a side of a triangle does not exceed the sum of the lengths of the other two sides (Figure 3.36).

FIGURE 3.36
The triangle
inequality

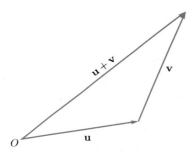

EXAMPLE 13 ■ For $\mathbf{u}$ and $\mathbf{v}$ as in Example 12, we have

$$\|\mathbf{u} + \mathbf{v}\| = \sqrt{4} = 2 < \sqrt{2} + \sqrt{2} = \|\mathbf{u}\| + \|\mathbf{v}\|. \qquad ■$$

Another useful result is the Pythagorean theorem in R^n: If $\mathbf{u}$ and $\mathbf{v}$ are vectors in R^n, then

$$\|\mathbf{u} + \mathbf{v}\|^2 = \|\mathbf{u}\|^2 + \|\mathbf{v}\|^2$$

if and only if $\mathbf{u}$ and $\mathbf{v}$ are orthogonal (Exercise T.10).

DEFINITION   A **unit vector** $\mathbf{u}$ in R^n is a vector of length 1. If $\mathbf{x}$ is a nonzero vector, then the vector

$$\mathbf{u} = \frac{1}{\|\mathbf{x}\|}\,\mathbf{x}$$

is a unit vector in the direction of $\mathbf{x}$.

EXAMPLE 14 ■ If $\mathbf{x} = (1, 0, 0, 1)$, then since $\|\mathbf{x}\| = \sqrt{2}$, the vector $\mathbf{u} = \frac{1}{\sqrt{2}}(1, 0, 0, 1)$ is a unit vector in the direction of $\mathbf{x}$.   ■

In the case of R^3 the unit vectors in the positive directions of the x-, y-, and z-axes are denoted by $\mathbf{i} = (1, 0, 0)$, $\mathbf{j} = (0, 1, 0)$, and $\mathbf{k} = (0, 0, 1)$ and are shown in Figure 3.37. If $\mathbf{u} = (x_1, y_1, z_1)$ is any vector in R^3, then we can write $\mathbf{u}$ as a linear combination (see Section 1.3) of $\mathbf{i}$, $\mathbf{j}$, and $\mathbf{k}$ as

$$\mathbf{u} = x_1\mathbf{i} + y_1\mathbf{j} + z_1\mathbf{k}.$$

FIGURE 3.37
Unit vectors
in R^3

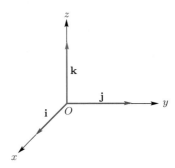

EXAMPLE 15 ■ If $\mathbf{u} = (2, -1, 3)$, then $\mathbf{u} = 2\mathbf{i} - \mathbf{j} + 3\mathbf{k}$.   ■

The n-vectors

$$\mathbf{e}_1 = (1, 0, \ldots, 0), \quad \mathbf{e}_2 = (0, 1, \ldots, 0), \ldots, \quad \mathbf{e}_n = (0, 0, \ldots, 1)$$

are unit vectors in R^n that are mutually orthogonal. If $\mathbf{u} = (u_1, u_2, \ldots, u_n)$ is any vector in R^n, then $\mathbf{u}$ can be written as a linear combination of $\mathbf{e}_1, \mathbf{e}_2, \ldots, \mathbf{e}_n$ as

$$\mathbf{u} = u_1\mathbf{e}_1 + u_2\mathbf{e}_2 + \cdots + u_n\mathbf{e}_n.$$

The vector $\mathbf{e}_i$, $1 \leq i \leq n$, can be viewed as the ith column of the identity matrix I_n. Thus we see that the columns of I_n form a set of n vectors that are mutually orthogonal. We shall discuss such sets of vectors in more detail in Section 4.8.

Section 3.6, which can be covered at this time, uses material from this section.

▼ *Preview of an Application*

LINES AND PLANES (SECTION 3.6)

The simplest curve in R^2 is a straight line. It is also one of the most useful curves, since it enables us to approximate any curve.

For example, when a computer or graphing calculator sketches a curve, it does it by piecing together a great many line segments, which approximate the curve. In Figure A below we show $y = \sin x$ and then its approximations with 5 and 15 equally spaced line segments, respectively. In general, the shorter the line segments, the better the approximation.

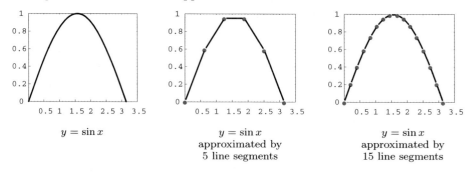

$y = \sin x$ $y = \sin x$ approximated by 5 line segments $y = \sin x$ approximated by 15 line segments

Figure A

The line L in Figure B is described by the vector equation

$$\mathbf{x} = \mathbf{w}_0 + t\mathbf{u} \qquad (-\infty < t < \infty).$$

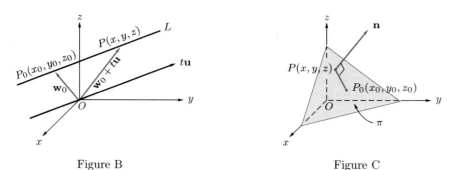

Figure B   Figure C

Similarly, the simplest surface in R^3 is a plane. Planes can be used to approximate a more complicated surface. The plane π shown in Figure C is described by the set of points $P(x, y, z)$ that satisfy the equation

$$\mathbf{n} \cdot \overrightarrow{P_0P} = 0.$$

Section 3.6 gives a brief introduction to lines and planes from a linear algebra point of view.

3.2 EXERCISES

1. Find $\mathbf{u} + \mathbf{v}$, $\mathbf{u} - \mathbf{v}$, $2\mathbf{u}$, and $3\mathbf{u} - 2\mathbf{v}$ if
   (a) $\mathbf{u} = (1, 2, -3)$, $\mathbf{v} = (0, 1, -2)$.
   (b) $\mathbf{u} = (4, -2, 1, 3)$, $\mathbf{v} = (-1, 2, 5, -4)$.

2. Repeat Exercise 1 for
   (a) $\mathbf{u} = \begin{bmatrix} 2 \\ 0 \\ -4 \end{bmatrix}$, $\mathbf{v} = \begin{bmatrix} 3 \\ 2 \\ 1 \end{bmatrix}$.

   (b) $\mathbf{u} = \begin{bmatrix} -3 \\ 5 \\ -3 \\ 0 \end{bmatrix}$, $\mathbf{v} = \begin{bmatrix} 2 \\ 1 \\ 5 \\ -2 \end{bmatrix}$.

3. Let
   $$\mathbf{u} = \begin{bmatrix} 1 \\ -2 \\ 3 \end{bmatrix}, \qquad \mathbf{v} = \begin{bmatrix} -3 \\ -1 \\ 3 \end{bmatrix},$$

   $$\mathbf{w} = \begin{bmatrix} a \\ -1 \\ b \end{bmatrix}, \qquad \mathbf{x} = \begin{bmatrix} 3 \\ c \\ 2 \end{bmatrix}.$$

   Find a, b, and c so that
   (a) $\mathbf{w} = \frac{1}{2}\mathbf{u}$.   (b) $\mathbf{w} + \mathbf{v} = \mathbf{u}$.
   (c) $\mathbf{w} + \mathbf{x} = \mathbf{v}$.

4. Let $\mathbf{u} = (4, -1, -2, 3)$, $\mathbf{v} = (3, -2, -4, 1)$, $\mathbf{w} = (a, -3, -6, b)$, and $\mathbf{x} = (2, c, d, 4)$. Find a, b, c, and d so that
   (a) $\mathbf{w} = 3\mathbf{u}$.   (b) $\mathbf{w} + \mathbf{x} = \mathbf{u}$.
   (c) $\mathbf{w} - \mathbf{u} = \mathbf{v}$.

5. Let $\mathbf{u} = (4, 5, -2, 3)$, $\mathbf{v} = (3, -2, 0, 1)$, $\mathbf{w} = (-3, 2, -5, 3)$, $c = 2$, and $d = 3$. Verify properties (a) through (h) in Theorem 3.2.

6. Plot the following points in R^3.
   (a) $(3, -1, 2)$.   (b) $(1, 0, 2)$.
   (c) $(0, 0, -4)$.   (d) $(1, 0, 0)$.
   (e) $(0, -2, 0)$.

7. Sketch a directed line segment in R^3 representing each of the following vectors.
   (a) $\mathbf{u}_1 = (2, -3, -1)$.   (b) $\mathbf{u}_2 = (0, 1, 4)$.
   (c) $\mathbf{u}_3 = (0, 0, -1)$.

8. For each of the following pairs of points in R^3, determine the vector that is associated with the directed line segment whose tail is the first point and whose head is the second point.
   (a) $(2, 3, -1)$, $(0, 0, 2)$.

   (b) $(1, 1, 0)$, $(0, 1, 1)$.
   (c) $(-1, -2, -3)$, $(3, 4, 5)$.
   (d) $(1, 1, 3)$, $(0, 0, 1)$.

9. Determine the head of the vector $(3, 4, -1)$ whose tail is $(1, -2, 3)$.

10. Find the length of the following vectors.
    (a) $(1, 2, -3)$.   (b) $(2, 3, -1, 4)$.
    (c) $(1, 0, 3)$.   (d) $(0, 0, 3, 4)$.

11. Find the length of the following vectors.
    (a) $(2, 3, 4)$.   (b) $(0, -1, 2, 3)$.
    (c) $(-1, -2, 0)$.   (d) $(1, 2, -3, -4)$.

12. Find the distance between the following pairs of points.
    (a) $(1, -1, 2)$, $(3, 0, 2)$.
    (b) $(4, 2, -1, 5)$, $(2, 3, -1, 4)$.
    (c) $(0, 0, 2)$, $(-3, 0, 0)$.
    (d) $(1, 0, 0, 2)$, $(3, -1, 5, 2)$.

13. Find the distance between the following pairs of points.
    (a) $(1, 1, 0)$, $(2, -3, 1)$.
    (b) $(4, 2, -1, 6)$, $(4, 3, 1, 5)$.
    (c) $(0, 2, 3)$, $(1, 2, -4)$.
    (d) $(3, 4, 0, 1)$, $(2, 2, 1, -1)$.

14. Is the vector $(2, -2, 3)$ a linear combination of the vectors $(1, 2, -3)$, $(-1, 1, 1)$, and $(-1, 4, -1)$?

15. If possible, find scalars c_1, c_2, and c_3, not all zero, so that
    $$c_1 \begin{bmatrix} 1 \\ 2 \\ -1 \end{bmatrix} + c_2 \begin{bmatrix} 1 \\ 3 \\ -2 \end{bmatrix} + c_3 \begin{bmatrix} 3 \\ 7 \\ -4 \end{bmatrix} = \begin{bmatrix} 0 \\ 0 \\ 0 \end{bmatrix}.$$

16. Find all constants a such that $\|(1, a, -3, 2)\| = 5$.

17. Find all constants a such that $\mathbf{u} \cdot \mathbf{v} = 0$, where $\mathbf{u} = (a, 2, 1, a)$ and $\mathbf{v} = (a, -1, -2, -3)$.

18. Verify Theorem 3.3 for $c = 3$ and $\mathbf{u} = (1, 2, 3)$, $\mathbf{v} = (1, 2, -4)$, and $\mathbf{w} = (1, 0, 2)$.

19. Verify Theorem 3.4 for $\mathbf{u}$ and $\mathbf{v}$ as in Exercise 18.

20. Find the cosine of the angle between each pair of vectors $\mathbf{u}$ and $\mathbf{v}$.
    (a) $\mathbf{u} = (1, 2, 3)$, $\mathbf{v} = (-4, 4, 5)$.
    (b) $\mathbf{u} = (0, 2, 3, 1)$, $\mathbf{v} = (-3, 1, -2, 0)$.

(c) $\mathbf{u} = (0, 0, 1)$, $\mathbf{v} = (2, 2, 0)$.

(d) $\mathbf{u} = (2, 0, -1, 3)$, $\mathbf{v} = (-3, -5, 2, -1)$.

21. Find the cosine of the angle between each pair of vectors $\mathbf{u}$ and $\mathbf{v}$.

(a) $\mathbf{u} = (2, 3, 1)$, $\mathbf{v} = (3, -2, 0)$.

(b) $\mathbf{u} = (1, 2, -1, 3)$, $\mathbf{v} = (0, 0, -1, -2)$.

(c) $\mathbf{u} = (2, 0, 1)$, $\mathbf{v} = (2, 2, -1)$.

(d) $\mathbf{u} = (0, 4, 2, 3)$, $\mathbf{v} = (0, -1, 2, 0)$.

22. Show that

(a) $\mathbf{i} \cdot \mathbf{i} = \mathbf{j} \cdot \mathbf{j} = \mathbf{k} \cdot \mathbf{k} = 1$.

(b) $\mathbf{i} \cdot \mathbf{j} = \mathbf{i} \cdot \mathbf{k} = \mathbf{j} \cdot \mathbf{k} = 0$.

23. Which of the vectors

$$\mathbf{u}_1 = (4, 2, 6, -8), \qquad \mathbf{u}_2 = (-2, 3, -1, -1),$$

$$\mathbf{u}_3 = (-2, -1, -3, 4), \qquad \mathbf{u}_4 = (1, 0, 0, 2),$$

$$\mathbf{u}_5 = (1, 2, 3, -4), \quad \text{and } \mathbf{u}_6 = (0, -3, 1, 0)$$

are

(a) Orthogonal?   (b) Parallel?

(c) In the same direction?

24. Find c so that the vector $\mathbf{v} = (2, c, 3)$ is orthogonal to $\mathbf{w} = (1, -2, 1)$.

25. If possible, find a, b, and c so that $\mathbf{v} = (a, b, c)$ is orthogonal to both

$$\mathbf{w} = (1, 2, 1) \quad \text{and} \quad \mathbf{x} = (1, -1, 1).$$

26. Verify the triangle inequality for $\mathbf{u} = (1, 2, 3, -1)$ and $\mathbf{v} = (1, 0, -2, 3)$.

27. Find a unit vector in the direction of $\mathbf{x}$.

(a) $\mathbf{x} = (2, -1, 3)$.   (b) $\mathbf{x} = (1, 2, 3, 4)$.

(c) $\mathbf{x} = (0, 1, -1)$.   (d) $\mathbf{x} = (0, -1, 2, -1)$.

28. Find a unit vector in the direction of $\mathbf{x}$.

(a) $\mathbf{x} = (1, 2, -1)$.   (b) $\mathbf{x} = (0, 0, 2, 0)$.

(c) $\mathbf{x} = (-1, 0, -2)$.   (d) $\mathbf{x} = (0, 0, 3, 4)$.

29. Write each of the following vectors in R^3 in terms of $\mathbf{i}$, $\mathbf{j}$, and $\mathbf{k}$.

(a) $(1, 2, -3)$.   (b) $(2, 3, -1)$.

(c) $(0, 1, 2)$.   (d) $(0, 0, -2)$.

30. Write each of the following vectors in R^3 as a 3×1 matrix.

(a) $2\mathbf{i} + 3\mathbf{j} - 4\mathbf{k}$.   (b) $\mathbf{i} + 2\mathbf{j}$.

(c) $-3\mathbf{i}$.   (d) $3\mathbf{i} - 2\mathbf{k}$.

31. Verify that the triangle with vertices $P_1(2, 3, -4)$, $P_2(3, 1, 2)$, and $P_3(-3, 0, 4)$ is isosceles.

32. Verify that the triangle with vertices $P_1(2, 3, -4)$, $P_2(3, 1, 2)$, and $P_3(7, 0, 1)$ is a right triangle.

33. A large steel manufacturer, who has 2000 employees, lists each employee's salary as a component of a vector $\mathbf{u}$ in R^{2000}. If an 8 percent across-the-board salary increase has been approved, find an expression involving $\mathbf{u}$ giving all the new salaries.

34. The vector $\mathbf{u} = (20, 30, 80, 10)$ gives the number of receivers, CD players, speakers, and cassette recorders that are on hand in a stereo shop. The vector $\mathbf{v} = (200, 120, 80, 70)$ gives the price (in dollars) of each receiver, CD player, speaker, and cassette recorder, respectively. What does the dot product $\mathbf{u} \cdot \mathbf{v}$ tell the shop owner?

35. A brokerage firm records the high and low values of the price of IBM stock each day. The information for a given week is presented in two vectors, $\mathbf{t}$ and $\mathbf{b}$, in R^5, giving the high and low values, respectively. What expression gives the average daily values of the price of IBM stock for the entire 5-day week?

THEORETICAL EXERCISES ■

T.1. Prove the rest of Theorem 3.2.

T.2. Show that $-\mathbf{u} = (-1)\mathbf{u}$.

T.3. Establish Equations (1) and (2) in R^3, for the length of a vector and the distance between two points, by using the Pythagorean theorem.

T.4. Prove Theorem 3.3.

T.5. Suppose that $\mathbf{u}$ is orthogonal to both $\mathbf{v}$ and $\mathbf{w}$. Show that $\mathbf{u}$ is orthogonal to any vector of the

form $r\mathbf{v} + s\mathbf{w}$, where r and s are scalars.

T.6. Show that if $\mathbf{u} \cdot \mathbf{v} = 0$ for all vectors $\mathbf{v}$, then $\mathbf{u} = \mathbf{0}$.

T.7. Show that $\mathbf{u} \cdot (\mathbf{v} + \mathbf{w}) = \mathbf{u} \cdot \mathbf{v} + \mathbf{u} \cdot \mathbf{w}$.

T.8. Show that if $\mathbf{u} \cdot \mathbf{v} = \mathbf{u} \cdot \mathbf{w}$ for all $\mathbf{u}$, then $\mathbf{v} = \mathbf{w}$.

T.9. Show that if c is a scalar, then $\|c\mathbf{u}\| = |c|\|\mathbf{u}\|$, where $|c|$ is the absolute value of c.

T.10. (Pythagorean Theorem in R^n). Show that $\|\mathbf{u} + \mathbf{v}\|^2 = \|\mathbf{u}\|^2 + \|\mathbf{v}\|^2$ if and only if $\mathbf{u} \cdot \mathbf{v} = 0$.

T.11. Let A be an $n \times n$ matrix and let $\mathbf{x}$ and $\mathbf{y}$ be vectors in R^n. Show that $A\mathbf{x} \cdot \mathbf{y} = \mathbf{x} \cdot A^T\mathbf{y}$.

T.12. Define the **distance** between two vectors $\mathbf{u}$ and $\mathbf{v}$ in R^n as $d(\mathbf{u}, \mathbf{v}) = \|\mathbf{u} - \mathbf{v}\|$. Show that
(a) $d(\mathbf{u}, \mathbf{v}) \geq 0$.
(b) $d(\mathbf{u}, \mathbf{v}) = 0$ if and only if $\mathbf{u} = \mathbf{v}$.
(c) $d(\mathbf{u}, \mathbf{v}) = d(\mathbf{v}, \mathbf{u})$.

(d) $d(\mathbf{u}, \mathbf{w}) \leq d(\mathbf{u}, \mathbf{v}) + d(\mathbf{v}, \mathbf{w})$.

T.13. Prove the **parallelogram law**:
$$\|\mathbf{u} + \mathbf{v}\|^2 + \|\mathbf{u} - \mathbf{v}\|^2 = 2\|\mathbf{u}\|^2 + 2\|\mathbf{v}\|^2.$$

T.14. If $\mathbf{x}$ is a nonzero vector, show that
$$\mathbf{u} = \frac{1}{\|\mathbf{x}\|}\mathbf{x}$$
is a unit vector in the direction of $\mathbf{x}$.

T.15. Show that $\mathbf{u} \cdot \mathbf{v} = \frac{1}{4}\|\mathbf{u} + \mathbf{v}\|^2 - \frac{1}{4}\|\mathbf{u} - \mathbf{v}\|^2$.

MATLAB EXERCISES

In order to use MATLAB in this section, you should first have read Section 10.6.

ML.1. As aid for visualizing vector operations in R^3, we have **vec3demo**. This routine provides a graphical display of vectors in 3-space. For a pair of vectors $\mathbf{u}$ and $\mathbf{v}$, routine **vec3demo** graphs $\mathbf{u}$ and $\mathbf{v}$, $\mathbf{u} + \mathbf{v}$, $\mathbf{u} - \mathbf{v}$, and a scalar multiple. Once the pair of vectors from R^3 are entered into MATLAB, type

$$\textbf{vec3demo}(\mathbf{u}, \mathbf{v})$$

Use **vec3demo** on each of the following pairs from R^3.
(a) $\mathbf{u} = (2, 6, 4)$, $\mathbf{v} = (6, 2, -5)$.
(b) $\mathbf{u} = (3, -5, 4)$, $\mathbf{v} = (7, -1, -2)$.
(c) $\mathbf{u} = (4, 0, -5)$, $\mathbf{v} = (0, 6, 3)$.

ML.2. Determine the norm or length of each of the following vectors using MATLAB.

(a) $\mathbf{u} = \begin{bmatrix} 2 \\ 2 \\ -1 \end{bmatrix}$.   (b) $\mathbf{v} = \begin{bmatrix} 0 \\ 4 \\ -3 \\ 0 \end{bmatrix}$.

(c) $\mathbf{w} = \begin{bmatrix} 1 \\ 0 \\ 1 \\ 0 \\ 3 \end{bmatrix}$.

ML.3. Determine the distance between each of the following pairs of vectors using MATLAB.

(a) $\mathbf{u} = \begin{bmatrix} 2 \\ 0 \\ 3 \end{bmatrix}$, $\mathbf{v} = \begin{bmatrix} 2 \\ -1 \\ 1 \end{bmatrix}$.
(b) $\mathbf{u} = (2, 0, 0, 1)$, $\mathbf{v} = (2, 5, -1, 3)$.
(c) $\mathbf{u} = (1, 0, 4, 3)$, $\mathbf{v} = (-1, 1, 2, 2)$.

ML.4. Determine the lengths of the sides of the triangle ABC, which has vertices in R^3, given

by $\mathbf{A}(1, 3, -2)$, $\mathbf{B}(4, -1, 0)$, $\mathbf{C}(1, 1, 2)$. (*Hint*: Determine a vector for each side and compute its length.)

ML.5. Determine the dot product of each one of the following pairs of vectors using MATLAB.
(a) $\mathbf{u} = (5, 4, -4)$, $\mathbf{v} = (3, 2, 1)$.
(b) $\mathbf{u} = (3, -1, 0, 2)$, $\mathbf{v} = (-1, 2, -5, -3)$.
(c) $\mathbf{u} = (1, 2, 3, 4, 5)$, $\mathbf{v} = -\mathbf{u}$.

ML.6. The norm or length of a vector can be computed using dot products as follows:

$$\|\mathbf{u}\| = \sqrt{\mathbf{u} \cdot \mathbf{u}}.$$

In MATLAB, the right side of the preceding expression is computed as

$$\textbf{sqrt}(\textbf{dot}(\mathbf{u}, \mathbf{u}))$$

Verify this alternative procedure on the vectors in Exercise ML.2.

ML.7. In MATLAB, if the n-vectors $\mathbf{u}$ and $\mathbf{v}$ are entered as columns, then

$$\mathbf{u}' * \mathbf{v} \quad \text{or} \quad \mathbf{v}' * \mathbf{u}$$

gives the dot product of vectors $\mathbf{u}$ and $\mathbf{v}$. Verify this using the vectors in Exercise ML.5.

ML.8. Use MATLAB to find the angle between each of the following pairs of vectors. (To convert the angle from radians to degrees, multiply by 180/pi.)
(a) $\mathbf{u} = (3, 2, 4, 0)$, $\mathbf{v} = (0, 2, -1, 0)$.
(b) $\mathbf{u} = (2, 2, -1)$, $\mathbf{v} = (2, 0, 1)$.
(c) $\mathbf{u} = (1, 0, 0, 2)$, $\mathbf{v} = (0, 3, -4, 0)$.

ML.9. Use MATLAB to find a unit vector in the direction of the vectors in Exercise ML.2.

3.3 ▾ Introduction to Linear Transformations

Functions occur in almost every application of mathematics. In this section we shall give a brief introduction from a geometric point of view to certain functions mapping R^n into R^m. Since we wish to picture this special class of functions, called linear transformations, we limit most of our discussion in this section to the situation where m and n have the values 2 and 3. In the next section we give an application of these functions to computer graphics in the plane, that is, for m and n equal to 2. In Chapter 6 we consider linear transformations from a much more general point of view and we study their properties in some detail.

Linear transformations play an important role in many areas of mathematics, as well as in numerous applied problems in the physical sciences, the social sciences, and economics.

DEFINITION

A **linear transformation** L of R^n into R^m is a function assigning a unique vector $L(\mathbf{u})$ in R^m to each $\mathbf{u}$ in R^n such that:

(a) $L(\mathbf{u} + \mathbf{v}) = L(\mathbf{u}) + L(\mathbf{v})$, for every $\mathbf{u}$ and $\mathbf{v}$ in R^n.
(b) $L(k\mathbf{u}) = kL(\mathbf{u})$, for every $\mathbf{u}$ in R^n and every scalar k.

The vector $L(\mathbf{u})$ is called the **image** of $\mathbf{u}$ and the set of all images of the vectors in R^n is called the **range** of L. Since R^n can be viewed as consisting of points or of vectors, $L(\mathbf{u})$, for $\mathbf{u}$ in R^n, can be considered as a point or a vector in R^m.

We shall write the fact that L maps R^n into R^m, even if it is not a linear transformation, as

$$L\colon R^n \to R^m.$$

If $n = m$, a linear transformation $L\colon R^n \to R^n$ is also called a **linear operator** on R^n.

EXAMPLE 1 ■ Let $L\colon R^3 \to R^2$ be defined by

$$L(x, y, z) = (x, y).$$

To verify that L is a linear transformation, we let

$$\mathbf{u} = (x_1, y_1, z_1) \quad \text{and} \quad \mathbf{v} = (x_2, y_2, z_2).$$

Then

$$
\begin{aligned}
L(\mathbf{u} + \mathbf{v}) &= L((x_1, y_1, z_1) + (x_2, y_2, z_2)) \\
&= L(x_1 + x_2, y_1 + y_2, z_1 + z_2) = (x_1 + x_2, y_1 + y_2) \\
&= (x_1, y_1) + (x_2, y_2) = L(\mathbf{u}) + L(\mathbf{v}).
\end{aligned}
$$

Also, if k is a real number, then

$$L(k\mathbf{u}) = L(kx_1, ky_1, kz_1) = (kx_1, ky_1) = k(x_1, y_1) = kL(\mathbf{u}).$$

Hence L is a linear transformation, which is called a **projection**. The image of the vector (or point) $(3, 5, 7)$ is the vector (or point) $(3, 5)$. See Figure 3.38.

FIGURE 3.38

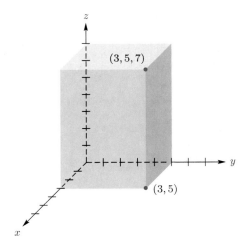

It is simple and helpful to describe geometrically the effect of L. The image under L of a vector $\mathbf{u} = (a, b, c)$ in R^3 is found by drawing a line through the endpoint $P(a, b, c)$ of $\mathbf{u}$ and perpendicular to R^2, the xy-plane. We obtain the point of intersection $Q(a, b)$ of this line with the xy-plane. The vector $\mathbf{v} = (a, b)$ in R^2 with endpoint Q is the image of $\mathbf{u}$ under L (Figure 3.39).   ∎

FIGURE 3.39
Projection

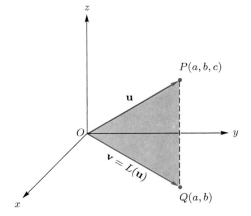

EXAMPLE 2   ∎   Let $L\colon R^3 \to R^2$ be defined by

$$L\left(\begin{bmatrix} u_1 \\ u_2 \\ u_3 \end{bmatrix}\right) = \begin{bmatrix} u_1 + 1 \\ u_2 - u_3 \end{bmatrix}.$$

To determine whether L is a linear transformation, let

$$\mathbf{u} = \begin{bmatrix} u_1 \\ u_2 \\ u_3 \end{bmatrix} \quad \text{and} \quad \mathbf{v} = \begin{bmatrix} v_1 \\ v_2 \\ v_3 \end{bmatrix}.$$

Then

$$L(\mathbf{u} + \mathbf{v}) = L\left(\begin{bmatrix} u_1 \\ u_2 \\ u_3 \end{bmatrix} + \begin{bmatrix} v_1 \\ v_2 \\ v_3 \end{bmatrix}\right) = L\left(\begin{bmatrix} u_1 + v_1 \\ u_2 + v_2 \\ u_3 + v_3 \end{bmatrix}\right)$$

$$= \begin{bmatrix} (u_1 + v_1) + 1 \\ (u_2 + v_2) - (u_3 + v_3) \end{bmatrix}.$$

On the other hand,

$$L(\mathbf{u}) + L(\mathbf{v}) = \begin{bmatrix} u_1 + 1 \\ u_2 - u_3 \end{bmatrix} + \begin{bmatrix} v_1 + 1 \\ v_2 - v_3 \end{bmatrix} = \begin{bmatrix} (u_1 + v_1) + 2 \\ (u_2 - u_3) + (v_2 - v_3) \end{bmatrix}.$$

Since the first coordinates of $L(\mathbf{u} + \mathbf{v})$ and $L(\mathbf{u}) + L(\mathbf{v})$ are different, $L(\mathbf{u} + \mathbf{v}) \neq L(\mathbf{u}) + L(\mathbf{v})$, so we conclude that the function L is not a linear transformation. ∎

EXAMPLE 3 ■ Let $L\colon R^3 \to R^3$ be defined by

$$L(\mathbf{u}) = r\mathbf{u},$$

where r is a real number. It is easy to verify that L is a linear operator. If $r > 1$, L is called a **dilation**; if $0 < r < 1$, L is called a **contraction**. In Figure 3.40(a) we show the vector $L_1(\mathbf{u}) = 2\mathbf{u}$, and in Figure 3.40(b) the vector $L_2(\mathbf{u}) = \frac{1}{2}\mathbf{u}$. Thus dilation stretches a vector, and contraction shrinks it. ∎

FIGURE 3.40

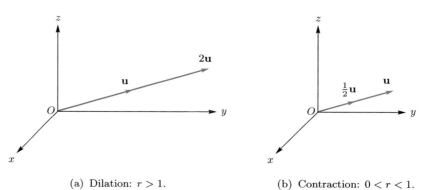

(a) Dilation: $r > 1$.   (b) Contraction: $0 < r < 1$.

EXAMPLE 4 ■ Let $L\colon R^2 \to R^2$ be defined by

$$L(x, y) = (x, -y).$$

Then L is a linear operator, which is shown in Figure 3.41. This linear transformation is called a **reflection with respect to the x-axis**. The reader may consider a **reflection with respect to the y-axis**. ∎

The following two theorems give some additional basic properties of linear transformations from R^n to R^m. The proofs will be left as exercises. Moreover, a more general version of the second theorem below will be proved in Section 6.1.

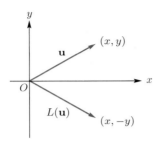

FIGURE 3.41
Reflection

THEOREM 3.6 ■ *If $L: R^n \to R^m$ is a linear transformation, then*

$$L(c_1\mathbf{u}_1 + c_2\mathbf{u}_2 + \cdots + c_k\mathbf{u}_k) = c_1 L(\mathbf{u}_1) + c_2 L(\mathbf{u}_2) + \cdots + c_k L(\mathbf{u}_k)$$

for any vectors $\mathbf{u}_1, \mathbf{u}_2, \ldots, \mathbf{u}_k$ in R^n and any scalars $c_1, c_2, \ldots, c_k$.

Proof   Exercise T.1.   ■

THEOREM 3.7 ■ *Let $L: R^n \to R^m$ be a linear transformation. Then*

(a) $L(\mathbf{0}_{R^n}) = \mathbf{0}_{R^m}$.
(b) $L(\mathbf{u} - \mathbf{v}) = L(\mathbf{u}) - L(\mathbf{v})$, *for $\mathbf{u}$ and $\mathbf{v}$ in R^n.*

Proof   Exercise T.2.   ■

These theorems can be used to compute the image of a vector $\mathbf{u}$ in R^2 or R^3 under a linear transformation $L: R^2 \to R^n$ once we know $L(\mathbf{i})$ and $L(\mathbf{j})$, where $\mathbf{i} = (1,0)$ and $\mathbf{j} = (0,1)$. Similarly, we can compute $L(\mathbf{u})$ for $\mathbf{u}$ in R^3 under the linear transformation $L: R^3 \to R^n$ if we know $L(\mathbf{i})$, $L(\mathbf{j})$, and $L(\mathbf{k})$, where $\mathbf{i} = (1,0,0)$, $\mathbf{j} = (0,1,0)$, and $\mathbf{k} = (0,0,1)$. It follows from the observations in Sections 3.1 and 3.2 that if $\mathbf{v} = (v_1, v_2)$ is any vector in R^2 and $\mathbf{u} = (u_1, u_2, u_3)$ is any vector in R^3, then

$$\mathbf{v} = v_1\mathbf{i} + v_2\mathbf{j} \quad \text{and} \quad \mathbf{u} = u_1\mathbf{i} + u_2\mathbf{j} + u_3\mathbf{k}.$$

EXAMPLE 5 ■ Let $L: R^3 \to R^2$ be a linear transformation for which we know that

$$L(1,0,0) = (2,-1), \quad L(0,1,0) = (3,1), \quad \text{and} \quad L(0,0,1) = (-1,2).$$

Find $L(-3, 4, 2)$.

Solution   Since

$$(-3, 4, 2) = -3\mathbf{i} + 4\mathbf{j} + 2\mathbf{k},$$

we have

$$\begin{aligned}
L(-3, 4, 2) &= L(-3\mathbf{i} + 4\mathbf{j} + 2\mathbf{k}) \\
&= -3L(\mathbf{i}) + 4L(\mathbf{j}) + 2L(\mathbf{k}) \\
&= -3(2,-1) + 4(3,1) + 2(-1,2) \\
&= (4, 11).
\end{aligned}$$
                                                                                 ■

More generally, we pointed out after Example 15 in Section 3.2, that if $\mathbf{u} = (u_1, u_2, \ldots, u_n)$ is any vector in R^n, then

$$\mathbf{u} = u_1\mathbf{e}_1 + u_2\mathbf{e}_2 + \cdots + u_n\mathbf{e}_n,$$

where

$$\mathbf{e}_1 = (1, 0, \ldots, 0), \quad \mathbf{e}_2 = (0, 1, \ldots, 0), \ldots, \quad \mathbf{e}_n = (0, 0, \ldots, 1).$$

This implies that if $L\colon R^n \to R^m$ is a linear transformation for which we know $L(\mathbf{e}_1), L(\mathbf{e}_2), \ldots, L(\mathbf{e}_n)$, then we can compute $L(\mathbf{u})$.

EXAMPLE 6 ■ Let $L\colon R^2 \to R^3$ be defined by

$$L\left(\begin{bmatrix} x \\ y \end{bmatrix}\right) = \begin{bmatrix} 1 & 0 \\ 0 & 1 \\ 1 & -1 \end{bmatrix} \begin{bmatrix} x \\ y \end{bmatrix}.$$

Then L is a linear transformation (verify). Observe that the vector

$$\mathbf{v} = \begin{bmatrix} 2 \\ 3 \\ -1 \end{bmatrix}$$

lies in range L because

$$L\left(\begin{bmatrix} 2 \\ 3 \end{bmatrix}\right) = \begin{bmatrix} 1 & 0 \\ 0 & 1 \\ 1 & -1 \end{bmatrix} \begin{bmatrix} 2 \\ 3 \end{bmatrix} = \begin{bmatrix} 2 \\ 3 \\ 2-3 \end{bmatrix} = \begin{bmatrix} 2 \\ 3 \\ -1 \end{bmatrix} = \mathbf{v}.$$

To find out whether the vector

$$\mathbf{w} = \begin{bmatrix} 3 \\ 5 \\ 2 \end{bmatrix}$$

lies in range L we need to determine whether there is a vector

$$\mathbf{u} = \begin{bmatrix} x \\ y \end{bmatrix}$$

so that $L(\mathbf{u}) = \mathbf{w}$. We have

$$L(\mathbf{u}) = L\left(\begin{bmatrix} x \\ y \end{bmatrix}\right) = \begin{bmatrix} 1 & 0 \\ 0 & 1 \\ 1 & -1 \end{bmatrix} \begin{bmatrix} x \\ y \end{bmatrix} = \begin{bmatrix} x \\ y \\ x-y \end{bmatrix}.$$

Since we want $L(\mathbf{u}) = \mathbf{w}$, we have

$$\begin{bmatrix} x \\ y \\ x-y \end{bmatrix} = \begin{bmatrix} 3 \\ 5 \\ 2 \end{bmatrix}.$$

This means that

$$x = 3$$
$$y = 5$$
$$x - y = 2.$$

This linear system of three equations in two unknowns has no solution. Hence, **w** is not in range L. ∎

EXAMPLE 7 ∎ The linear transformation given in Example 6 can be generalized as follows. Let $L: R^n \to R^m$ be defined by

$$L(\mathbf{x}) = A\mathbf{x} \tag{1}$$

for **x** in R^n, where A is a fixed $m \times n$ matrix. Using the properties of the matrix operations discussed in Section 1.4, we can show that L is a linear transformation. (See Exercise T.4.)

EXAMPLE 8 ∎
(*Production*)
A book publisher publishes a book in three different editions: trade, book club, and deluxe. Each book requires a certain amount of paper and canvas (for the cover). The requirements are given (in grams) by the matrix

$$A = \begin{array}{c} \\ \\ \end{array} \begin{matrix} \text{Trade} & \overset{\text{Book}}{\text{Club}} & \text{Deluxe} \\ \begin{bmatrix} 300 & 500 & 800 \\ 40 & 50 & 60 \end{bmatrix} & \begin{array}{l} \text{Paper} \\ \text{Canvas} \end{array} \end{matrix}$$

Let

$$\mathbf{x} = \begin{bmatrix} x_1 \\ x_2 \\ x_3 \end{bmatrix}$$

denote the production vector, where x_1, x_2, and x_3 are the number of trade, book club, and deluxe books, respectively, that are published. The linear transformation $L: R^3 \to R^2$ defined by $L(\mathbf{x}) = A\mathbf{x}$ gives the vector

$$\begin{bmatrix} y_1 \\ y_2 \end{bmatrix},$$

where y_1 is the total amount of paper required and y_2 is the total amount of canvas required. ∎

EXAMPLE 9 ∎
(*Cryptography*)
Cryptography is the technique of coding and decoding messages; it goes back to the time of the ancient Greeks. A simple code is constructed by associating a different number with every letter in the alphabet. For example,

$$\begin{matrix} \text{A} & \text{B} & \text{C} & \text{D} & \cdots & \text{X} & \text{Y} & \text{Z} \\ \updownarrow & \updownarrow & \updownarrow & \updownarrow & & \updownarrow & \updownarrow & \updownarrow \\ 1 & 2 & 3 & 4 & \cdots & 24 & 25 & 26 \end{matrix}$$

Suppose that Mark S. and Susan J. are two undercover agents who want to communicate with each other by using a code because they suspect that their phone is being tapped and their mail is being intercepted. In particular, Mark wants to send Susan the message

MEET TOMORROW

Using the substitution scheme given above, Mark sends the message

13   5   5   20   20   15   13   15   18   18   15   23

A code of this type could be cracked without too much difficulty by a number of techniques, including the analysis of frequency of letters. To make it difficult to crack the code, the agents proceed as follows. First, when they undertook the mission they agreed on a 3×3 nonsingular matrix such as

$$A = \begin{bmatrix} 1 & 2 & 3 \\ 1 & 1 & 2 \\ 0 & 1 & 2 \end{bmatrix}.$$

Mark then breaks the message into four vectors in R^3 (if this cannot be done, we can add extra letters). Thus, we have the vectors

$$\begin{bmatrix} 13 \\ 5 \\ 5 \end{bmatrix}, \qquad \begin{bmatrix} 20 \\ 20 \\ 15 \end{bmatrix}, \qquad \begin{bmatrix} 13 \\ 15 \\ 18 \end{bmatrix}, \qquad \begin{bmatrix} 18 \\ 15 \\ 23 \end{bmatrix}.$$

Mark now defines the linear transformation $L \colon R^3 \to R^3$ by $L(\mathbf{x}) = A\mathbf{x}$, so the message becomes

$$A \begin{bmatrix} 13 \\ 5 \\ 5 \end{bmatrix} = \begin{bmatrix} 38 \\ 28 \\ 15 \end{bmatrix}, \qquad A \begin{bmatrix} 20 \\ 20 \\ 15 \end{bmatrix} = \begin{bmatrix} 105 \\ 70 \\ 50 \end{bmatrix},$$

$$A \begin{bmatrix} 13 \\ 15 \\ 18 \end{bmatrix} = \begin{bmatrix} 97 \\ 64 \\ 51 \end{bmatrix}, \qquad A \begin{bmatrix} 18 \\ 15 \\ 23 \end{bmatrix} = \begin{bmatrix} 117 \\ 79 \\ 61 \end{bmatrix}.$$

Thus Mark transmits the message

$$38 \quad 28 \quad 15 \quad 105 \quad 70 \quad 50 \quad 97 \quad 64 \quad 51 \quad 117 \quad 79 \quad 61$$

Suppose now that Mark receives the following message from Susan,

$$77 \quad 54 \quad 38 \quad 71 \quad 49 \quad 29 \quad 68 \quad 51 \quad 33 \quad 76 \quad 48 \quad 40 \quad 86 \quad 53 \quad 52$$

which he wants to decode with the same key matrix A above. To decode it, Mark breaks the message into five vectors in R^3:

$$\begin{bmatrix} 77 \\ 54 \\ 38 \end{bmatrix}, \qquad \begin{bmatrix} 71 \\ 49 \\ 29 \end{bmatrix}, \qquad \begin{bmatrix} 68 \\ 51 \\ 33 \end{bmatrix}, \qquad \begin{bmatrix} 76 \\ 48 \\ 40 \end{bmatrix}, \qquad \begin{bmatrix} 86 \\ 53 \\ 52 \end{bmatrix}$$

and solves the equation

$$L(\mathbf{x}_1) = \begin{bmatrix} 77 \\ 54 \\ 38 \end{bmatrix} = A\mathbf{x}_1$$

for $\mathbf{x}_1$. Since A is nonsingular,

$$\mathbf{x}_1 = A^{-1} \begin{bmatrix} 77 \\ 54 \\ 38 \end{bmatrix} = \begin{bmatrix} 0 & 1 & -1 \\ 2 & -2 & -1 \\ -1 & 1 & 1 \end{bmatrix} \begin{bmatrix} 77 \\ 54 \\ 38 \end{bmatrix} = \begin{bmatrix} 16 \\ 8 \\ 15 \end{bmatrix}.$$

Similarly,

$$\mathbf{x}_2 = A^{-1}\begin{bmatrix} 71 \\ 49 \\ 29 \end{bmatrix} = \begin{bmatrix} 20 \\ 15 \\ 7 \end{bmatrix}, \qquad \mathbf{x}_3 = A^{-1}\begin{bmatrix} 68 \\ 51 \\ 33 \end{bmatrix} = \begin{bmatrix} 18 \\ 1 \\ 16 \end{bmatrix},$$

$$\mathbf{x}_4 = A^{-1}\begin{bmatrix} 76 \\ 48 \\ 40 \end{bmatrix} = \begin{bmatrix} 8 \\ 16 \\ 12 \end{bmatrix}, \qquad \mathbf{x}_5 = A^{-1}\begin{bmatrix} 86 \\ 53 \\ 52 \end{bmatrix} = \begin{bmatrix} 1 \\ 14 \\ 19 \end{bmatrix}.$$

Using our correspondence between letters and numbers, Mark has received the message

PHOTOGRAPH PLANS

Additional material on cryptography may be found in the references given in Further Readings at the end of this section.   ■

EXAMPLE 10 ■ Suppose that we rotate every point in R^2 counterclockwise through an angle ϕ about the origin of a Cartesian coordinate system. Thus, if P has coordinates (x, y), then after rotating, we get P' with coordinates (x', y'). To obtain a relationship between the coordinates of P' and those of P, we let $\mathbf{u}$ be the vector from the origin to $P(x, y)$. See Figure 3.42(a). Also, let θ be the angle made by $\mathbf{u}$ with the positive x-axis.

FIGURE 3.42

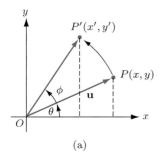

(a)

(b) Rotation.

Letting r denote the length of vector $\mathbf{u}$, we see from Figure 3.42(a) that

$$x = r\cos\theta, \qquad y = r\sin\theta \tag{2}$$

and

$$x' = r\cos(\theta + \phi), \qquad y' = r\sin(\theta + \phi). \tag{3}$$

Using the formulas for the sine and cosine of a sum of angles (3) becomes

$$x' = r\cos\theta\cos\phi - r\sin\theta\sin\phi$$
$$y' = r\sin\theta\cos\phi + r\cos\theta\sin\phi.$$

Substituting the expression in (2) into the last pair of equations, we obtain

$$x' = x\cos\phi - y\sin\phi, \qquad y' = x\sin\phi + y\cos\phi. \tag{4}$$

Solving (4) for x and y, we have

$$x = x' \cos\phi + y' \sin\phi \quad \text{and} \quad y = -x' \sin\phi + y' \cos\phi. \tag{5}$$

Equation (4) gives the coordinates of P' in terms of those of P and (5) expresses the coordinates of P in terms of those of P'. This type of rotation is used to simplify the general equation of second degree $ax^2 + bxy + cy^2 + dx + ey + f = 0$. Substituting for x and y in terms of x' and y', we obtain

$$a'x'^2 + b'x'y' + c'y'^2 + d'x' + e'y' + f' = 0.$$

The key point is to choose ϕ so that $b' = 0$. Once this is done (we might now have to perform a translation of coordinates), we identify the general equation of second degree as a circle, ellipse, hyperbola, parabola, or a degenerate form of these. This topic will be treated from a linear algebra point of view in Section 8.3.

We may also perform this change of coordinates by considering the function $L\colon R^2 \to R^2$ defined by

$$L\left(\begin{bmatrix} x \\ y \end{bmatrix}\right) = \begin{bmatrix} \cos\phi & -\sin\phi \\ \sin\phi & \cos\phi \end{bmatrix} \begin{bmatrix} x \\ y \end{bmatrix}. \tag{6}$$

If $\mathbf{u} = \begin{bmatrix} x \\ y \end{bmatrix}$, then (6) can be written as

$$L(\mathbf{u}) = \begin{bmatrix} x\cos\phi - y\sin\phi \\ x\sin\phi + y\cos\phi \end{bmatrix},$$

so $L(\mathbf{u})$ is the vector from the origin to $P'(x', y')$. See Figure 3.42(b). Then L is a linear transformation (verify) called a **rotation**. ■

REMARK Observe that if we let

$$A = \begin{bmatrix} \cos\phi & -\sin\phi \\ \sin\phi & \cos\phi \end{bmatrix} \quad \text{and} \quad \mathbf{u} = \begin{bmatrix} x \\ y \end{bmatrix},$$

then Equation (6) can be written as

$$L(\mathbf{u}) = A\mathbf{u},$$

which is of the form in Equation (1).

In Example 7, we have seen, in general, that if A is an $m \times n$ matrix, then the function $L\colon R^n \to R^m$ defined by $L(\mathbf{x}) = A\mathbf{x}$ for $\mathbf{x}$ in R^n is a linear transformation. Examples 8 and 9 give specific illustrations of this result. In the following theorem, we show that if $L\colon R^n \to R^m$ is a linear transformation, then L must be of this form.

THEOREM 3.8 ■ *Let $L\colon R^n \to R^m$ be a linear transformation. Then there exists a unique $m \times n$ matrix A such that*

$$L(\mathbf{x}) = A\mathbf{x} \tag{7}$$

for $\mathbf{x}$ in R^n.

Proof   If

$$\mathbf{x} = \begin{bmatrix} c_1 \\ c_2 \\ \vdots \\ c_n \end{bmatrix}$$

is any vector in R^n, then

$$\mathbf{x} = c_1\mathbf{e}_1 + c_2\mathbf{e}_2 + \cdots + c_n\mathbf{e}_n,$$

so by Theorem 3.6,

$$L(\mathbf{x}) = c_1 L(\mathbf{e}_1) + c_2 L(\mathbf{e}_2) + \cdots + c_n L(\mathbf{e}_n). \tag{8}$$

If we let A be the $m \times n$ matrix whose jth column is $L(\mathbf{e}_j)$ and

$$\mathbf{x} = \begin{bmatrix} c_1 \\ c_2 \\ \vdots \\ c_n \end{bmatrix},$$

then Equation (8) can be written as

$$L(\mathbf{x}) = A\mathbf{x}.$$

We now show that the matrix A is unique. Suppose that we also have

$$L(\mathbf{x}) = B\mathbf{x} \quad \text{for } \mathbf{x} \text{ in } R^n.$$

Letting $\mathbf{x} = \mathbf{e}_j$, $j = 1, \ldots, n$, we obtain

$$L(\mathbf{e}_j) = A\mathbf{e}_j = \text{col}_j(A)$$

and

$$L(\mathbf{e}_j) = B\mathbf{e}_j = \text{col}_j(B).$$

Thus the columns of A and B agree, so $A = B$.   ∎

The matrix $A = \begin{bmatrix} L(\mathbf{e}_1) & L(\mathbf{e}_2) & \cdots & L(\mathbf{e}_n) \end{bmatrix}$ in Equation (7) is called the **standard matrix** representing L.

EXAMPLE 11   ■   Let $L \colon R^3 \to R^3$ be the linear operator defined by

$$L\left(\begin{bmatrix} x \\ y \\ z \end{bmatrix}\right) = \begin{bmatrix} x + y \\ y - z \\ x + z \end{bmatrix}.$$

Find the standard matrix representing L and verify Equation (7).

Solution   The standard matrix A representing L is the 3×3 matrix whose columns are $L(\mathbf{e}_1)$, $L(\mathbf{e}_2)$, and $L(\mathbf{e}_3)$, respectively. Thus

$$L(\mathbf{e}_1) = L\left(\begin{bmatrix} 1 \\ 0 \\ 0 \end{bmatrix}\right) = \begin{bmatrix} 1+0 \\ 0-0 \\ 1+0 \end{bmatrix} = \begin{bmatrix} 1 \\ 0 \\ 1 \end{bmatrix} = \mathrm{col}_1(A)$$

$$L(\mathbf{e}_2) = L\left(\begin{bmatrix} 0 \\ 1 \\ 0 \end{bmatrix}\right) = \begin{bmatrix} 0+1 \\ 1-0 \\ 0+0 \end{bmatrix} = \begin{bmatrix} 1 \\ 1 \\ 0 \end{bmatrix} = \mathrm{col}_2(A)$$

$$L(\mathbf{e}_3) = L\left(\begin{bmatrix} 0 \\ 0 \\ 1 \end{bmatrix}\right) = \begin{bmatrix} 0+0 \\ 0-1 \\ 0+1 \end{bmatrix} = \begin{bmatrix} 0 \\ -1 \\ 1 \end{bmatrix} = \mathrm{col}_3(A).$$

Hence

$$A = \begin{bmatrix} 1 & 1 & 0 \\ 0 & 1 & -1 \\ 1 & 0 & 1 \end{bmatrix}.$$

Thus we have

$$A\mathbf{x} = \begin{bmatrix} 1 & 1 & 0 \\ 0 & 1 & -1 \\ 1 & 0 & 1 \end{bmatrix} \begin{bmatrix} x \\ y \\ z \end{bmatrix} = \begin{bmatrix} x+y \\ y-z \\ x+z \end{bmatrix} = L(\mathbf{x}),$$

so Equation (7) holds. ∎

Further Readings in Cryptography ▶

Elementary presentation

KOHN, BERNICE. *Secret Codes and Ciphers*. Englewood Cliffs, N.J.: Prentice Hall, Inc., 1968 (63 pages).

Advanced presentation

FISHER, JAMES L. *Applications-Oriented Algebra*. New York: T. Harper & Row, Publishers, 1977 (Chapter 9, "Coding Theory").

KAHN, DAVID. *The Codebreakers*. New York: The New American Library Inc., 1973.

3.3   EXERCISES

1. Which of the following are linear transformations?

(a) $L(x, y) = (x + 1, y, x + y)$.

(b) $L\left(\begin{bmatrix} x \\ y \\ z \end{bmatrix} \right) = \begin{bmatrix} x + y \\ y \\ x - z \end{bmatrix}$.

(c) $L(x, y) = (x^2 + x, y - y^2)$.

2. Which of the following are linear transformations?

(a) $L(x, y, z) = (x - y, x^2, 2z)$.

(b) $L\left(\begin{bmatrix} x \\ y \\ z \end{bmatrix} \right) = \begin{bmatrix} 2x - 3y \\ 3y - 2z \\ 2z \end{bmatrix}$.

(c) $L(x, y) = (x - y, 2x + 2)$.

3. Which of the following are linear transformations?

(a) $L(x, y, z) = (x + y, 0, 2x - z)$.

(b) $L\left(\begin{bmatrix} x \\ y \end{bmatrix} \right) = \begin{bmatrix} x^2 - y^2 \\ x^2 + y^2 \end{bmatrix}$.

(c) $L(x, y) = (x - y, 0, 2x + 3)$.

4. Which of the following are linear transformations?

(a) $L\left(\begin{bmatrix} u_1 \\ u_2 \\ u_3 \\ u_4 \end{bmatrix} \right) = \begin{bmatrix} u_1 \\ u_1^2 + u_2 \\ u_1 - u_3 \end{bmatrix}$.

(b) $L\left(\begin{bmatrix} x \\ y \\ z \end{bmatrix} \right) = \begin{bmatrix} 1 & 1 & 0 \\ 0 & -1 & 2 \\ 1 & 1 & -1 \end{bmatrix} \begin{bmatrix} x \\ y \\ z \end{bmatrix}$.

(c) $L\left(\begin{bmatrix} x \\ y \\ z \end{bmatrix} \right) = \begin{bmatrix} 0 \\ 0 \\ 0 \end{bmatrix}$.

In Exercises 5–12, sketch the image of the given point P or vector $\mathbf{u}$ under the given linear transformation L.

5. $L\colon R^2 \to R^2$ is defined by
$$L(x, y) = (x, -y); P = (2, 3).$$

6. $L\colon R^2 \to R^2$ is defined by
$$L\left(\begin{bmatrix} x \\ y \end{bmatrix} \right) = \begin{bmatrix} 1 & -1 \\ 2 & 1 \end{bmatrix} \begin{bmatrix} x \\ y \end{bmatrix}; \mathbf{u} = (1, -2).$$

7. $L\colon R^2 \to R^2$ is a counterclockwise rotation through $30°$; $P = (-1, 3)$.

8. $L\colon R^2 \to R^2$ is a counterclockwise rotation through $\frac{2}{3}\pi$ radians; $\mathbf{u} = (-2, -3)$.

9. $L\colon R^2 \to R^2$ is defined by $L(\mathbf{u}) = -\mathbf{u}$; $\mathbf{u} = (3, 2)$.

10. $L\colon R^2 \to R^2$ is defined by $L(\mathbf{u}) = 2\mathbf{u}$; $\mathbf{u} = (-3, 3)$.

11. $L\colon R^3 \to R^2$ is defined by
$$L\left(\begin{bmatrix} x \\ y \\ z \end{bmatrix} \right) = \begin{bmatrix} x \\ x - y \end{bmatrix}; \mathbf{u} = (2, -1, 3).$$

12. $L\colon R^3 \to R^3$ is defined by
$$L\left(\begin{bmatrix} x \\ y \\ z \end{bmatrix} \right) = \begin{bmatrix} 1 & 0 & 1 \\ -1 & 1 & 0 \\ 0 & 0 & 1 \end{bmatrix} \begin{bmatrix} x \\ y \\ z \end{bmatrix}; \mathbf{u} = (0, -2, 4).$$

13. Let $L\colon R^3 \to R^3$ be the linear transformation defined by
$$L\left(\begin{bmatrix} x \\ y \\ z \end{bmatrix} \right) = \begin{bmatrix} x + z \\ y + z \\ x + 2y + 2z \end{bmatrix}.$$

Is $\mathbf{w}$ in range L?

(a) $\mathbf{w} = \begin{bmatrix} 1 \\ -1 \\ 0 \end{bmatrix}$.

(b) $\mathbf{w} = \begin{bmatrix} 2 \\ -1 \\ 3 \end{bmatrix}$.

14. Let $L\colon R^3 \to R^3$ be the linear transformation defined by
$$L\left(\begin{bmatrix} x \\ y \\ z \end{bmatrix} \right) = \begin{bmatrix} -1 & 2 & 0 \\ 1 & 1 & 1 \\ 2 & -1 & 1 \end{bmatrix} \begin{bmatrix} x \\ y \\ z \end{bmatrix}.$$

Is $\mathbf{w}$ in range L?

(a) $\mathbf{w} = \begin{bmatrix} 1 \\ 2 \\ -1 \end{bmatrix}$.

(b) $\mathbf{w} = \begin{bmatrix} 1 \\ 3 \\ 2 \end{bmatrix}$.

15. Let $L\colon R^3 \to R^3$ be defined by
$$L\left(\begin{bmatrix} x \\ y \\ z \end{bmatrix} \right) = \begin{bmatrix} 4 & 1 & 3 \\ 2 & -1 & 3 \\ 2 & 2 & 0 \end{bmatrix} \begin{bmatrix} x \\ y \\ z \end{bmatrix}.$$

Find an equation relating a, b, and c so that
$$\mathbf{w} = \begin{bmatrix} a \\ b \\ c \end{bmatrix}$$

will lie in range L.

16. Repeat Exercise 15 if $L: R^3 \rightarrow R^3$ is defined by

$$L\left(\begin{bmatrix} x \\ y \\ z \end{bmatrix}\right) = \begin{bmatrix} x + 2y + 3z \\ -3x - 2y - z \\ -2x + 2z \end{bmatrix}.$$

17. Let $L: R^2 \rightarrow R^2$ be a linear transformation such that

$$L(\mathbf{i}) = \begin{bmatrix} 2 \\ 3 \end{bmatrix} \quad \text{and} \quad L(\mathbf{j}) = \begin{bmatrix} -1 \\ 2 \end{bmatrix}.$$

Find $L\left(\begin{bmatrix} 4 \\ -3 \end{bmatrix}\right)$.

18. Let $L: R^3 \rightarrow R^3$ be a linear transformation such that

$$L(\mathbf{i}) = \begin{bmatrix} 1 \\ 2 \\ -1 \end{bmatrix}, \ L(\mathbf{j}) = \begin{bmatrix} 1 \\ 0 \\ 2 \end{bmatrix}, \ \text{and} \ L(\mathbf{k}) = \begin{bmatrix} 1 \\ 1 \\ 3 \end{bmatrix}.$$

Find $L\left(\begin{bmatrix} 2 \\ -1 \\ 3 \end{bmatrix}\right)$.

19. Let L be the linear transformation defined in Exercise 11. Find all vectors $\mathbf{x}$ in R^3 such that $L(\mathbf{x}) = \mathbf{0}$.

20. Repeat Exercise 19, where L is the linear transformation defined in Exercise 12.

21. Describe the following linear transformations geometrically.
(a) $L(x, y) = (-x, y)$.
(b) $L(x, y) = (-x, -y)$.
(c) $L(x, y) = (-y, x)$.

22. Describe the following linear transformations geometrically.
(a) $L(x, y) = (y, x)$.

(b) $L(x, y) = (-y, -x)$.
(c) $L(x, y) = (2x, 2y)$.

In Exercises 23–28, find the standard matrix representing L.

23. $L: R^2 \rightarrow R^2$ is reflection with respect to the y-axis.

24. $L: R^2 \rightarrow R^2$ is defined by

$$L\left(\begin{bmatrix} x \\ y \end{bmatrix}\right) = \begin{bmatrix} x - y \\ x + y \end{bmatrix}.$$

25. $L: R^2 \rightarrow R^2$ is counterclockwise rotation through $\pi/4$ radians.

26. $L: R^2 \rightarrow R^2$ is counterclockwise rotation through $\pi/3$ radians.

27. $L: R^3 \rightarrow R^3$ is defined by

$$L\left(\begin{bmatrix} x \\ y \\ z \end{bmatrix}\right) = \begin{bmatrix} x - y \\ x + z \\ y - z \end{bmatrix}.$$

28. $L: R^3 \rightarrow R^3$ is defined by $L(\mathbf{u}) = -2\mathbf{u}$.

29. Use the substitution and matrix A of Example 9.
(a) Code the message SEND HIM MONEY.
(b) Decode the message 67 44 41 49 39 19 113 76 62 104 69 55.

30. Use the substitution scheme of Example 9 and the matrix

$$A = \begin{bmatrix} 5 & 3 \\ 2 & 1 \end{bmatrix}.$$

(a) Code the message WORK HARD.
(b) Decode the message

93 36 60 21 159 60 110 43

THEORETICAL EXERCISES

T.1. Prove Theorem 3.6.

T.2. Prove Theorem 3.7.

T.3. Show that $L: R^n \rightarrow R^n$ defined by $L(\mathbf{u}) = r\mathbf{u}$, where r is a scalar, is a linear operator on R^n.

T.4. Let A be an $m \times n$ matrix, and suppose that $L: R^n \rightarrow R^m$ is defined by $L(\mathbf{x}) = A\mathbf{x}$ for $\mathbf{x}$ in R^n. Show that L is a linear transformation.

T.5. Let $L: R^1 \rightarrow R^1$ be defined by $L(\mathbf{u}) = a\mathbf{u} + b$,

where a and b are real numbers (of course, $\mathbf{u}$ is a vector in R^1, which in this case means that $\mathbf{u}$ is also a real number). Find all values of a and b such that L is a linear transformation.

T.6. Show that the function $O: R^n \rightarrow R^m$ defined by $O(\mathbf{u}) = \mathbf{0}_{R^m}$ is a linear transformation, which is called the **zero** linear transformation.

T.7. Let $I: R^n \rightarrow R^n$ be defined by $I(\mathbf{u}) = \mathbf{u}$, for $\mathbf{u}$ in R^n. Show that I is a linear transformation, which is called the **identity operator** on R^n.

T.8. Let $L\colon R^n \to R^m$ be a linear transformation. Show that if **u** and **v** are vectors in R^n such that $L(\mathbf{u}) = \mathbf{0}$ and $L(\mathbf{v}) = \mathbf{0}$, then $L(a\mathbf{u} + b\mathbf{v}) = \mathbf{0}$ for any scalars a and b.

MATLAB EXERCISES ▦

ML.1. Let $L\colon R^n \to R^1$ be defined by $L(\mathbf{u}) = \|\mathbf{u}\|$.

(a) Find a pair of vectors **u** and **v** in R^2 such that
$$L(\mathbf{u} + \mathbf{v}) \neq L(\mathbf{u}) + L(\mathbf{v}).$$
Use MATLAB to do the computations. It follows that L is not a linear transformation.

(b) Find a pair of vectors **u** and **v** in R^3 such that
$$L(\mathbf{u} + \mathbf{v}) \neq L(\mathbf{u}) + L(\mathbf{v}).$$
Use MATLAB to do the computations.

3.4 ▾ Computer Graphics (Optional)

We are all familiar with the astounding results being developed with computer graphics in the areas of video games and special effects in the film industry. Computer graphics also play a major role in the manufacturing world. *Computer-aided design* (CAD) is used to design a computer model of a product and then by subjecting the computer model to a variety of tests (carried out on the computer) changes to the current design can be implemented to obtain an improved design. One of the most notable successes of this approach has been in the automobile industry, where the computer model can be viewed from different angles to obtain a most pleasing and popular style; can be tested for strength of components, for roadability, for seating comfort, and for safety in a crash; and so on.

In this section we give illustrations of linear operators $L\colon R^2 \to R^2$ that are useful in two-dimensional graphics. In each of the following four examples, we describe a geometric transformation on a set of points in R^2 and show the construction of the standard matrix representing the corresponding linear operator $L\colon R^2 \to R^2$.

EXAMPLE 1 ■ A reflection with respect to the x-axis of a vector **u** in R^2 is defined by the linear operator

$$L(\mathbf{u}) = L\left(\begin{bmatrix} a_1 \\ a_2 \end{bmatrix} \right) = \begin{bmatrix} a_1 \\ -a_2 \end{bmatrix}$$

(see Example 4 in Section 3.3). Then

$$L(\mathbf{e}_1) = L\left(\begin{bmatrix} 1 \\ 0 \end{bmatrix} \right) = \begin{bmatrix} 1 \\ 0 \end{bmatrix} \quad \text{and} \quad L(\mathbf{e}_2) = L\left(\begin{bmatrix} 0 \\ 1 \end{bmatrix} \right) = \begin{bmatrix} 0 \\ -1 \end{bmatrix}.$$

Hence the standard matrix representing L is

$$A = \begin{bmatrix} 1 & 0 \\ 0 & -1 \end{bmatrix}.$$

FIGURE 3.43

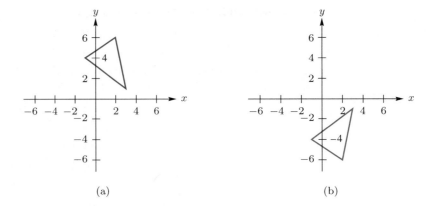

(a)   (b)

Thus we have

$$L(\mathbf{u}) = A\mathbf{u} = \begin{bmatrix} 1 & 0 \\ 0 & -1 \end{bmatrix} \begin{bmatrix} a_1 \\ a_2 \end{bmatrix} = \begin{bmatrix} a_1 \\ -a_2 \end{bmatrix}.$$

To illustrate a reflection with respect to the x-axis in computer graphics, let the triangle T in Figure 3.43(a) have vertices

$$(-1, 4), \quad (3, 1), \quad \text{and} \quad (2, 6).$$

To reflect T with respect to the x-axis, we let

$$\mathbf{u}_1 = \begin{bmatrix} -1 \\ 4 \end{bmatrix}, \qquad \mathbf{u}_2 = \begin{bmatrix} 3 \\ 1 \end{bmatrix}, \qquad \mathbf{u}_3 = \begin{bmatrix} 2 \\ 6 \end{bmatrix}$$

and compute the images $L(\mathbf{u}_1)$, $L(\mathbf{u}_2)$, and $L(\mathbf{u}_3)$ by forming the products

$$A\mathbf{u}_1 = \begin{bmatrix} 1 & 0 \\ 0 & -1 \end{bmatrix} \begin{bmatrix} -1 \\ 4 \end{bmatrix} = \begin{bmatrix} -1 \\ -4 \end{bmatrix},$$

$$A\mathbf{u}_2 = \begin{bmatrix} 1 & 0 \\ 0 & -1 \end{bmatrix} \begin{bmatrix} 3 \\ 1 \end{bmatrix} = \begin{bmatrix} 3 \\ -1 \end{bmatrix},$$

$$A\mathbf{u}_3 = \begin{bmatrix} 1 & 0 \\ 0 & -1 \end{bmatrix} \begin{bmatrix} 2 \\ 6 \end{bmatrix} = \begin{bmatrix} 2 \\ -6 \end{bmatrix}.$$

Thus the image of T has vertices

$$(-1, -4), \quad (3, -1), \quad \text{and} \quad (2, -6)$$

and is displayed in Figure 3.43(b). ■

EXAMPLE 2 ■ The reflection with respect to the line $y = -x$ of a vector $\mathbf{u}$ in R^2 is defined by the linear operator

$$L(\mathbf{u}) = L\left(\begin{bmatrix} a_1 \\ a_2 \end{bmatrix} \right) = \begin{bmatrix} -a_2 \\ -a_1 \end{bmatrix}.$$

Then

$$L(\mathbf{e}_1) = L\left(\begin{bmatrix} 1 \\ 0 \end{bmatrix} \right) = \begin{bmatrix} 0 \\ -1 \end{bmatrix} \quad \text{and} \quad L(\mathbf{e}_2) = L\left(\begin{bmatrix} 0 \\ 1 \end{bmatrix} \right) = \begin{bmatrix} -1 \\ 0 \end{bmatrix}.$$

Hence the standard matrix representing L is

$$B = \begin{bmatrix} 0 & -1 \\ -1 & 0 \end{bmatrix}.$$

To illustrate reflection with respect to the line $y = -x$, we use triangle T as defined in Example 1 and compute the products

$$B\mathbf{u}_1 = \begin{bmatrix} 0 & -1 \\ -1 & 0 \end{bmatrix} \begin{bmatrix} -1 \\ 4 \end{bmatrix} = \begin{bmatrix} -4 \\ 1 \end{bmatrix},$$

$$B\mathbf{u}_2 = \begin{bmatrix} 0 & -1 \\ -1 & 0 \end{bmatrix} \begin{bmatrix} 3 \\ 1 \end{bmatrix} = \begin{bmatrix} -1 \\ -3 \end{bmatrix},$$

$$B\mathbf{u}_3 = \begin{bmatrix} 0 & -1 \\ -1 & 0 \end{bmatrix} \begin{bmatrix} 2 \\ 6 \end{bmatrix} = \begin{bmatrix} -6 \\ -2 \end{bmatrix}.$$

Thus the image of T has vertices

$$(-4, 1), \quad (-1, -3), \quad \text{and} \quad (-6, -2)$$

and is displayed in Figure 3.44.   ■

FIGURE 3.44

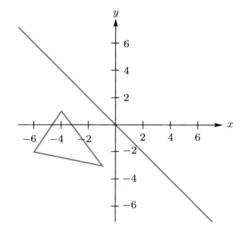

To perform a reflection with respect to the x-axis on triangle T of Example 1 followed by a reflection with respect to the line $y = -x$, we compute

$$B(A\mathbf{u}_1), \quad B(A\mathbf{u}_2), \quad \text{and} \quad B(A\mathbf{u}_3).$$

It is easy to show that reversing the order of these linear operators produces a different image (verify). Thus the order in which graphics transformations are performed is important.

EXAMPLE 3   ■   Rotations in a plane have been defined in Example 10 of Section 3.3. A plane figure is rotated counterclockwise through an angle ϕ using the linear operator $L\colon R^2 \to R^2$ whose standard matrix is

$$A = \begin{bmatrix} \cos\phi & -\sin\phi \\ \sin\phi & \cos\phi \end{bmatrix}.$$

Now suppose that we wish to rotate the parabola $y = x^2$ counterclockwise through $\frac{5}{18}\pi$ radians. We start by choosing a sample of points from the parabola, say

$$(-2, 4), \quad (-1, 1), \quad (0, 0), \quad \left(\tfrac{1}{2}, \tfrac{1}{4}\right), \quad \text{and} \quad (3, 9)$$

[see Figure 3.45(a)]. We then compute the images of these points. Thus letting

$$\mathbf{u}_1 = \begin{bmatrix} -2 \\ 4 \end{bmatrix}, \quad \mathbf{u}_2 = \begin{bmatrix} -1 \\ 1 \end{bmatrix}, \quad \mathbf{u}_3 = \begin{bmatrix} 0 \\ 0 \end{bmatrix}, \quad \mathbf{u}_4 = \begin{bmatrix} \frac{1}{2} \\ \frac{1}{4} \end{bmatrix}, \quad \mathbf{u}_5 = \begin{bmatrix} 3 \\ 9 \end{bmatrix},$$

we compute the products (to four decimal places) (verify)

$$A\mathbf{u}_1 = \begin{bmatrix} -4.3498 \\ 1.0391 \end{bmatrix}, \quad A\mathbf{u}_2 = \begin{bmatrix} -1.4088 \\ -0.1233 \end{bmatrix}, \quad A\mathbf{u}_3 = \begin{bmatrix} 0 \\ 0 \end{bmatrix},$$

$$A\mathbf{u}_4 = \begin{bmatrix} 0.1299 \\ 0.5437 \end{bmatrix}, \quad \text{and} \quad A\mathbf{u}_5 = \begin{bmatrix} -4.9660 \\ 8.0832 \end{bmatrix}.$$

The image points

$$(-4.3498, 1.0391), \quad (-1.4088, -0.1233), \quad (0, 0),$$
$$(0.1299, 0.5437), \quad \text{and} \quad (-4.9660, 8.0832)$$

are plotted as shown in Figure 3.45(b) and successive points are connected showing the image of the parabola. ∎

FIGURE 3.45

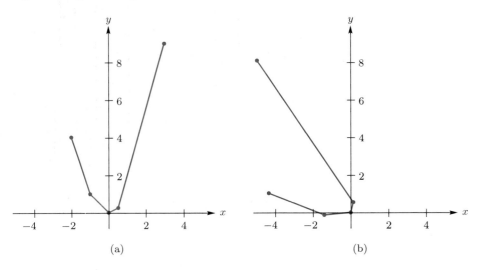

(a)　　　　　　　(b)

EXAMPLE 4 ∎ **A shear in the x-direction** is defined by the linear operator

$$L(\mathbf{u}) = L\left(\begin{bmatrix} a_1 \\ a_2 \end{bmatrix} \right) = \begin{bmatrix} a_1 + ka_2 \\ a_2 \end{bmatrix},$$

where k is a scalar. Hence the standard matrix representing L is

$$A = \begin{bmatrix} 1 & k \\ 0 & 1 \end{bmatrix}.$$

A shear in the x-direction takes the point (a_1, a_2) and moves it to the point $(a_1 + ka_2, a_2)$. That is, the point (a_1, a_2) is moved parallel to the x-axis by the amount ka_2 (see Figure 3.46).

FIGURE 3.46

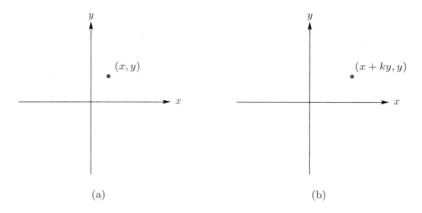

(a)   (b)

Consider now the rectangle R shown in Figure 3.47(a) with vertices

$$(0,0), \quad (0,2), \quad (4,0), \quad \text{and} \quad (4,2).$$

If we apply the shear in the x-direction with $k = 2$, then the image of R is the parallelogram with vertices

$$(0,0), \quad (4,2), \quad (4,0), \quad \text{and} \quad (8,2),$$

shown in Figure 3.47(b). If we apply the shear in the x-direction with $k = -3$, then the image of R is the parallelogram with vertices

$$(0,0), \quad (-6,2), \quad (4,0), \quad \text{and} \quad (-2,2),$$

shown in Figure 3.47(c).

In the exercises we consider shears in the y-direction. ■

Other linear operators used in two-dimensional computer graphics are considered in the exercises at the end of this section. For a detailed discussion of computer graphics, the reader is referred to the books listed in the Further Readings section that follows.

An arbitrary 2×2 matrix A defines a linear operator $L\colon R^2 \to R^2$ by

$$L(\mathbf{x}) = A\mathbf{x}$$

for a vector $\mathbf{x}$ in R^2. Using computer graphics, it is interesting to view the image of a closed figure, like the triangle T in Figure 3.43(a), that results from using such a linear operator. The original figure and its image can be compared in a number of ways. We present one such comparison in the next example.

EXAMPLE 5 ■ The area of a triangle with vertices (x_1, y_1), (x_2, y_2), and (x_3, y_3) has been given by Equation (3) in Section 3.1 as

$$\text{area of triangle} = \tfrac{1}{2} \left| \det \left(\begin{bmatrix} x_1 & y_1 & 1 \\ x_2 & y_2 & 1 \\ x_3 & y_3 & 1 \end{bmatrix} \right) \right| \tag{1}$$

FIGURE 3.47

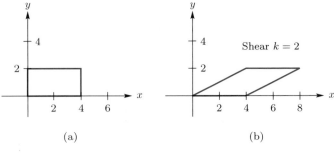

(a)   (b)

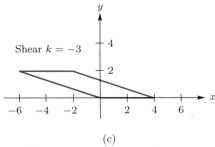

(c)

(the area is $\frac{1}{2}$ the absolute value of the determinant).

The triangle T in Figure 3.43(a) has vertices $(-1, 4)$, $(3, 1)$, and $(2, 6)$. Thus the area of T is, by (1),

$$\frac{1}{2} \left| \det \left(\begin{bmatrix} -1 & 4 & 1 \\ 3 & 1 & 1 \\ 2 & 6 & 1 \end{bmatrix} \right) \right| = \frac{1}{2}|17| = 8.5.$$

The image of T using the linear operator defined by the matrix

$$A = \begin{bmatrix} 6 & -3 \\ -1 & 1 \end{bmatrix}$$

has vertices $(-18, 5)$, $(15, -2)$, and $(-6, 4)$ (verify). See Figure 3.48. The area of the image of triangle T obtained from the linear operator using matrix A is obtained by (1):

$$\frac{1}{2} \left| \det \left(\begin{bmatrix} -18 & 5 & 1 \\ 15 & -2 & 1 \\ -6 & 4 & 1 \end{bmatrix} \right) \right| = \frac{1}{2}|51| = 25.5.$$

We have $\det(A) = 3$ and we see that

$$3 \times \text{area of triangle } T = \text{area of the image.}$$

This result holds in general. That is, the

$$\text{area of the image} = |\det(A)| \times \text{the area of the original figure.} \qquad \blacksquare$$

FIGURE 3.48

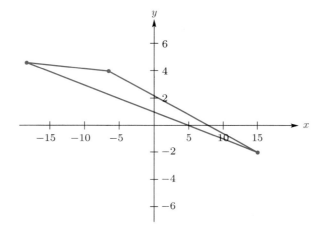

Further Readings ▶

FOLEY, J. D., A. VAN DAM, S. K. FEINER, J. F. HUGHES, and R. L. PHILLIPS. *Introduction to Computer Graphics*. Reading, Mass.: Addison-Wesley, 1994.

ROGERS, D. F., and J. A. ADAMS. *Mathematical Elements for Computer Graphics*, 2nd ed. New York: McGraw-Hill, 1989.

3.4   EXERCISES

1. Determine the standard matrix representing each of the following linear operators $L\colon R^2 \to R^2$.

(a) Reflection with respect to the y-axis.

(b) Reflection with respect to the line $y = x$.

(c) Rotation of $3\pi/4$ radians counterclockwise.

(d) Rotations of $\pi/6$ radians clockwise.

2. Let R be the rectangle with vertices $(1,1)$, $(1,4)$, $(3,1)$, and $(3,4)$ and let L be the shear in the x-direction with $k = 3$. Determine and sketch the image of R.

3. A **shear in the y-direction** is defined by the linear operator

$$L(\mathbf{u}) = L\left(\begin{bmatrix} u_1 \\ u_2 \end{bmatrix}\right) = \begin{bmatrix} u_1 \\ u_2 + ku_1 \end{bmatrix},$$

where k is a scalar.

(a) Determine the standard matrix representing L.

(b) Let R be the rectangle defined in Exercise 2 and let L be the shear in the y-direction with $k = -2$. Determine and sketch the image of R.

4. Consider the linear operator $L\colon R^2 \to R^2$ defined by

$$L(\mathbf{u}) = L\left(\begin{bmatrix} u_1 \\ u_2 \end{bmatrix}\right) = k\begin{bmatrix} u_1 \\ u_2 \end{bmatrix},$$

where k is a scalar, which has been defined in Example 3 in Section 3.3. Recall that L is called a **dilation** if $k > 1$ and it is called a **contraction** if $0 < k < 1$. Thus dilation stretches a vector, while contraction shrinks it.

(a) Determine the standard matrix representing L.

(b) Let R be the rectangle defined in Exercise 2 and let L be the dilation with $k = 3$. Find and sketch the image of R.

5. A linear operator $L: R^2 \to R^2$ defined by

$$L(\mathbf{u}) = L\left(\begin{bmatrix} u_1 \\ u_2 \end{bmatrix}\right) = \begin{bmatrix} ku_1 \\ u_2 \end{bmatrix},$$

where k is a scalar, is called a **dilation in the x-direction** if $k > 1$; it is called a **contraction in the x-direction** if $0 < k < 1$.

(a) Determine the standard matrix representing L.

(b) Let R be the unit square and let L be the dilation in the x-direction with $k = 2$. Find and sketch the image of R.

6. A linear operator $L: R^2 \to R^2$ defined by

$$L(\mathbf{u}) = L\left(\begin{bmatrix} u_1 \\ u_2 \end{bmatrix}\right) = \begin{bmatrix} u_1 \\ ku_2 \end{bmatrix},$$

where k is a scalar, is called a **dilation in the y-direction** if $k > 1$; it is called a **contraction in the y-direction** if $0 < k < 1$.

(a) Determine the standard matrix representing L.

(b) Let R be the unit square and let L be the dilation in the y-direction with $k = \frac{1}{2}$. Find and sketch the image of R.

7. Let T be the triangle with vertices $(5, 0)$, $(0, 3)$, and $(-2, 1)$.

(a) Find the area of triangle T.

(b) Find the coordinates of the vertices of the image of T under the linear operator $L: R^2 \to R^2$ defined by $L(\mathbf{x}) = A\mathbf{x}$, where

$$A = \begin{bmatrix} -2 & 1 \\ 3 & 4 \end{bmatrix}.$$

(c) Find the area of the triangle whose vertices are obtained in part (b). [*Hint*: Use Equation (1).]

8. Let T be the triangle with vertices $(1, 1)$, $(-3, -3)$, and $(2, -1)$.

(a) Find the area of the triangle T. [*Hint*: Use Equation (1).]

(b) Find the coordinates of the vertices of the image of T under the linear operator $L: R^2 \to R^2$ defined by $L(\mathbf{x}) = A\mathbf{x}$, where

$$A = \begin{bmatrix} 4 & -3 \\ -4 & 2 \end{bmatrix}.$$

(c) Find the area of the triangle whose vertices are obtained in part (b).

9. Let L be the counterclockwise rotation through $60°$. If T is the triangle defined in Exercise 8, find and sketch the image of T under L.

THEORETICAL EXERCISE ▨

T.1. Show that a rotation leaves the area of a triangle unchanged.

MATLAB EXERCISES ▨

The routine **planelt** in MATLAB provides a geometric visualization for the standard algebraic approach to linear transformations by illustrating **plane linear transformations**, which are linear transformations from R^2 to R^2. The name, of course, follows from the fact that we are mapping points in the plane into points in a corresponding plane.
In MATLAB, type the command

planelt

A description of the routine will be displayed. Read it, then press ENTER. You will then see a set of General Directions. Read them carefully, then press ENTER. A set of Figure Choices will then be displayed. As you do the following exercises, record the figures requested on a separate sheet of paper. (After each menu choice, press ENTER. To return to a menu when the graphics are visible, press ENTER.)

ML.1. In **planelt**, choose figure #4 (the triangle). Then choose option #1 (see the triangle). Return to the menu and choose option #2 (use this figure). Next perform the following linear transformations.

(a) Reflect the triangle about the y-axis. Record the Current Figure.

(b) Now rotate the result from part (a) $60°$. Record the figure.

(c) Return to the menu, restore the original figure, reflect it about the line $y = x$, then dilate it in the x-direction by a factor of 2. Record the figure.

(d) Repeat the experiment in part (c) but interchange the order of the linear transformations. Record the figure.

(e) Are the results from parts (c) and (d) the same? Compare the two figures you recorded.

(f) What does your answer in part (e) imply about the order of the linear transformations as applied to the triangle?

ML.2. Restore the original triangle selected in Exercise ML.1.

(a) Reflect it about the x-axis. Predict the result before pressing ENTER. (Call this linear transformation L_1.)

(b) Then reflect the figure resulting in part (a) about the y-axis. Predict the result before pressing ENTER. (Call this linear transformation L_2.)

(c) Record the figure that resulted from parts (a) and (b).

(d) Inspect the relationship between the Current Figure and the Original Figure. What (single) transformation do you think will accomplish the same result? (Use names that appear in the transformation menu. Call the linear transformation you select L_3.)

(e) Write a formula involving reflection L_1 about the x-axis, reflection L_2 about the y-axis, and the transformation L_3 that expresses the relationship you saw in part (d).

(f) Experiment with the formula in part (e) on several other figures until you can determine whether or not the formula in part (e) is correct in general. Write a brief summary of your experiments, observations, and conclusions.

ML.3. Choose the unit square as the figure.

(a) Reflect it about the x-axis, reflect the resulting figure about the y-axis, and then reflect that figure about the line $y = -x$. Record the figure.

(b) Compare the Current Figure with the Original Figure. Denote the reflection about the x-axis as L_1, the reflection about the y-axis as L_2, and the reflection about the line $y = -x$ as L_3. What formula relating these linear transformations does your comparison suggest when L_1 is followed by L_2, and then by L_3 on this figure?

(c) If M_i denotes the standard matrix representing the transformation L_i in part (b), to what matrix is $M_3 * M_2 * M_1$ equal? Does this result agree with your conclusion in (b)?

(d) Experiment with the successive application of these three linear transformations on other figures.

ML.4. In routine **planelt** you can enter any figure you like and perform linear transformations on it. Follow the directions on entering your own figure and experiment with various linear transformations. It is recommended that you draw the figure on graph paper first and assign the coordinates of its vertices.

ML.5. On a figure, **planelt** allows you to select any 2×2 matrix to use as a linear transformation. Perform the following experiment. Choose a singular matrix and apply it to each of the stored figures. Write a brief summary of your experiments, observations, and conclusions about the behavior of "singular" linear transformations.

3.5 ▾ Cross Product in R^3 (Optional)

In this section we discuss an operation that is meaningful only in R^3. Despite this limitation, it has many important applications in a number of different situations. We shall consider several of these applications in this section.

DEFINITION   If

$$\mathbf{u} = u_1\mathbf{i} + u_2\mathbf{j} + u_3\mathbf{k} \quad \text{and} \quad \mathbf{v} = v_1\mathbf{i} + v_2\mathbf{j} + v_3\mathbf{k}$$

are two vectors in R^3, then their **cross product** is the vector $\mathbf{u} \times \mathbf{v}$ defined by

$$\mathbf{u} \times \mathbf{v} = (u_2v_3 - u_3v_2)\mathbf{i} + (u_3v_1 - u_1v_3)\mathbf{j} + (u_1v_2 - u_2v_1)\mathbf{k}. \tag{1}$$

The cross product $\mathbf{u} \times \mathbf{v}$ can also be written as a "determinant,"

$$\mathbf{u} \times \mathbf{v} = \begin{vmatrix} \mathbf{i} & \mathbf{j} & \mathbf{k} \\ u_1 & u_2 & u_3 \\ v_1 & v_2 & v_3 \end{vmatrix}. \tag{2}$$

The right side of (2) is not really a determinant, but it is convenient to think of the computation in this manner. If we expand (2) about the first row, we obtain

$$\mathbf{u} \times \mathbf{v} = \begin{vmatrix} u_2 & u_3 \\ v_2 & v_3 \end{vmatrix}\mathbf{i} - \begin{vmatrix} u_1 & u_3 \\ v_1 & v_3 \end{vmatrix}\mathbf{j} + \begin{vmatrix} u_1 & u_2 \\ v_1 & v_2 \end{vmatrix}\mathbf{k},$$

which is the right side of (1). Observe that the cross product $\mathbf{u} \times \mathbf{v}$ is a vector while the dot product $\mathbf{u} \cdot \mathbf{v}$ is a number.

EXAMPLE 1 ■ Let $\mathbf{u} = 2\mathbf{i} + \mathbf{j} + 2\mathbf{k}$ and $\mathbf{v} = 3\mathbf{i} - \mathbf{j} - 3\mathbf{k}$. Then

$$\mathbf{u} \times \mathbf{v} = \begin{vmatrix} \mathbf{i} & \mathbf{j} & \mathbf{k} \\ 2 & 1 & 2 \\ 3 & -1 & -3 \end{vmatrix} = -\mathbf{i} + 12\mathbf{j} - 5\mathbf{k}.$$

■

Some of the algebraic properties of cross product are described in the following theorem. The proof, which follows easily from the properties of determinants, is left to the reader (Exercise T.1).

THEOREM 3.9 ■ *If* $\mathbf{u}$, $\mathbf{v}$, *and* $\mathbf{w}$ *are vectors in* R^3 *and* c *is a scalar, then:*
(Properties of
Cross Product)   (a) $\mathbf{u} \times \mathbf{v} = -(\mathbf{v} \times \mathbf{u})$.

(b) $\mathbf{u} \times (\mathbf{v} + \mathbf{w}) = \mathbf{u} \times \mathbf{v} + \mathbf{u} \times \mathbf{w}$.

(c) $(\mathbf{u} + \mathbf{v}) \times \mathbf{w} = \mathbf{u} \times \mathbf{w} + \mathbf{v} \times \mathbf{w}$.

(d) $c(\mathbf{u} \times \mathbf{v}) = (c\mathbf{u}) \times \mathbf{v} = \mathbf{u} \times (c\mathbf{v})$.

(e) $\mathbf{u} \times \mathbf{u} = \mathbf{0}$.

(f) $\mathbf{0} \times \mathbf{u} = \mathbf{u} \times \mathbf{0} = \mathbf{0}$.

(g) $\mathbf{u} \times (\mathbf{v} \times \mathbf{w}) = (\mathbf{u} \cdot \mathbf{w})\mathbf{v} - (\mathbf{u} \cdot \mathbf{v})\mathbf{w}$.

(h) $(\mathbf{u} \times \mathbf{v}) \times \mathbf{w} = (\mathbf{w} \cdot \mathbf{u})\mathbf{v} - (\mathbf{w} \cdot \mathbf{v})\mathbf{u}$.

EXAMPLE 2 ■ It follows from (1) that

$$\mathbf{i} \times \mathbf{i} = \mathbf{j} \times \mathbf{j} = \mathbf{k} \times \mathbf{k} = \mathbf{0}; \qquad \mathbf{i} \times \mathbf{j} = \mathbf{k}, \quad \mathbf{j} \times \mathbf{k} = \mathbf{i}, \quad \mathbf{k} \times \mathbf{i} = \mathbf{j}.$$

Also,

$$\mathbf{j} \times \mathbf{i} = -\mathbf{k}, \qquad \mathbf{k} \times \mathbf{j} = -\mathbf{i}, \qquad \mathbf{i} \times \mathbf{k} = -\mathbf{j}.$$

These rules can be remembered by the method indicated in Figure 3.49. Moving around the circle in a clockwise direction, we see that the cross product of two vectors taken in the indicated order is the third vector; moving in a counter-clockwise direction, we see that the cross product taken in the indicated order is the negative of the third vector. The cross product of a vector with itself is the zero vector. ■

FIGURE 3.49

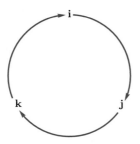

Although many of the familiar properties of the real numbers hold for the cross product, it should be noted that two important properties do not hold. The commutative law does not hold, since $\mathbf{u} \times \mathbf{v} = -(\mathbf{v} \times \mathbf{u})$. Also, the associative law does not hold, since $\mathbf{i} \times (\mathbf{i} \times \mathbf{j}) = \mathbf{i} \times \mathbf{k} = -\mathbf{j}$ while $(\mathbf{i} \times \mathbf{i}) \times \mathbf{j} = \mathbf{0} \times \mathbf{j} = \mathbf{0}$.

We shall now take a closer look at the geometric properties of the cross product. First, we observe the following additional property of the cross product, whose proof we leave to the reader:

$$(\mathbf{u} \times \mathbf{v}) \cdot \mathbf{w} = \mathbf{u} \cdot (\mathbf{v} \times \mathbf{w}) \qquad \qquad \text{Exercise T.2.} \qquad (3)$$

It is also easy to show (Exercise T.4) that

$$(\mathbf{u} \times \mathbf{v}) \cdot \mathbf{w} = \begin{vmatrix} u_1 & u_2 & u_3 \\ v_1 & v_2 & v_3 \\ w_1 & w_2 & w_3 \end{vmatrix}. \qquad (4)$$

EXAMPLE 3 ■ Let $\mathbf{u}$ and $\mathbf{v}$ be as in Example 1, and let $\mathbf{w} = \mathbf{i} + 2\mathbf{j} + 3\mathbf{k}$. Then

$$\mathbf{u} \times \mathbf{v} = -\mathbf{i} + 12\mathbf{j} - 5\mathbf{k} \quad \text{and} \quad (\mathbf{u} \times \mathbf{v}) \cdot \mathbf{w} = 8,$$
$$\mathbf{v} \times \mathbf{w} = 3\mathbf{i} - 12\mathbf{j} + 7\mathbf{k} \quad \text{and} \quad \mathbf{u} \cdot (\mathbf{v} \times \mathbf{w}) = 8,$$

which illustrates Equation (3). ■

From Equation (3) and (a) and (e) of Theorem 3.9, we have

$$(\mathbf{u} \times \mathbf{v}) \cdot \mathbf{v} = \mathbf{u} \cdot (\mathbf{v} \times \mathbf{v}) = \mathbf{u} \cdot \mathbf{0} = 0, \qquad (5)$$
$$(\mathbf{u} \times \mathbf{v}) \cdot \mathbf{u} = -(\mathbf{v} \times \mathbf{u}) \cdot \mathbf{u} = -\mathbf{v} \cdot (\mathbf{u} \times \mathbf{u}) = -\mathbf{v} \cdot \mathbf{0} = 0. \qquad (6)$$

From (5) and (6) and the definition for orthogonality of vectors given in Section 3.2, it follows that if $\mathbf{u} \times \mathbf{v} \neq \mathbf{0}$, then $\mathbf{u} \times \mathbf{v}$ is orthogonal to both $\mathbf{u}$ and $\mathbf{v}$ and to the plane determined by them. It can be shown that if θ is the angle between $\mathbf{u}$ and $\mathbf{v}$, then the direction of $\mathbf{u} \times \mathbf{v}$ is that in which a right-hand screw perpendicular to the plane of $\mathbf{u}$ and $\mathbf{v}$ will move if we rotate it through the angle θ from $\mathbf{u}$ to $\mathbf{v}$ (Figure 3.50).

FIGURE 3.50

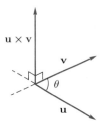

The magnitude of $\mathbf{u} \times \mathbf{v}$ can be determined as follows. From the definition of the length of a vector we have

$$
\begin{aligned}
\|\mathbf{u} \times \mathbf{v}\|^2 &= (\mathbf{u} \times \mathbf{v}) \cdot (\mathbf{u} \times \mathbf{v}) \\
&= \mathbf{u} \cdot [\mathbf{v} \times (\mathbf{u} \times \mathbf{v})] && \text{by (3)} \\
&= \mathbf{u} \cdot [(\mathbf{v} \cdot \mathbf{v})\mathbf{u} - (\mathbf{v} \cdot \mathbf{u})\mathbf{v}] && \text{by (g) of Theorem 3.9} \\
&= (\mathbf{u} \cdot \mathbf{u})(\mathbf{v} \cdot \mathbf{v}) - (\mathbf{v} \cdot \mathbf{u})(\mathbf{v} \cdot \mathbf{u}) && \text{by (b), (c) and (d) of Theorem 3.3} \\
&= \|\mathbf{u}\|^2 \|\mathbf{v}\|^2 - (\mathbf{u} \cdot \mathbf{v})^2 && \text{by (b) of Theorem 3.3 and the}
\end{aligned}
$$
definition of length of a vector.

From Equation (4) of Section 3.2 it follows that

$$
\mathbf{u} \cdot \mathbf{v} = \|\mathbf{u}\| \|\mathbf{v}\| \cos \theta,
$$

where θ is the angle between $\mathbf{u}$ and $\mathbf{v}$. Hence

$$
\begin{aligned}
\|\mathbf{u} \times \mathbf{v}\|^2 &= \|\mathbf{u}\|^2 \|\mathbf{v}\|^2 - \|\mathbf{u}\|^2 \|\mathbf{v}\|^2 \cos^2 \theta \\
&= \|\mathbf{u}\|^2 \|\mathbf{v}\|^2 (1 - \cos^2 \theta) \\
&= \|\mathbf{u}\|^2 \|\mathbf{v}\|^2 \sin^2 \theta.
\end{aligned}
$$

Taking square roots, we obtain

$$
\|\mathbf{u} \times \mathbf{v}\| = \|\mathbf{u}\| \|\mathbf{v}\| \sin \theta. \tag{7}
$$

Note that in (7) we do not have to write $|\sin \theta|$, since $\sin \theta$ is nonnegative for $0 \leq \theta \leq \pi$. It follows that vectors $\mathbf{u}$ and $\mathbf{v}$ are parallel if and only if $\mathbf{u} \times \mathbf{v} = \mathbf{0}$ (Exercise T.5).

We now consider several applications of cross product.

Area of a Triangle

Consider the triangle with vertices

$$
P_1(x_1, y_1, z_1), \qquad P_2(x_2, y_2, z_2), \qquad P_3(x_3, y_3, z_3)
$$

FIGURE 3.51

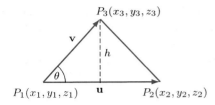

(Figure 3.51). The area of this triangle is $\frac{1}{2}bh$, where b is the base and h is the height. If we take the segment between $P_1(x_1, y_1, z_1)$ and $P_2(x_2, y_2, z_2)$ to be the base and denote $\overrightarrow{P_1P_2}$ by the vector $\mathbf{u}$, then

$$b = \|\mathbf{u}\|.$$

Letting $\overrightarrow{P_1P_3} = \mathbf{v}$, we find that the height h is given by

$$h = \|\mathbf{v}\| \sin\theta.$$

Hence, by (7), the area A_T of the triangle is

$$A_T = \tfrac{1}{2}\|\mathbf{u}\|\,\|\mathbf{v}\| \sin\theta = \tfrac{1}{2}\|\mathbf{u} \times \mathbf{v}\|.$$

EXAMPLE 4 ■ Find the area of the triangle with vertices $P_1(2, 2, 4)$, $P_2(-1, 0, 5)$, and $P_3(3, 4, 3)$.

Solution We have

$$\mathbf{u} = \overrightarrow{P_1P_2} = -3\mathbf{i} - 2\mathbf{j} + \mathbf{k}$$
$$\mathbf{v} = \overrightarrow{P_1P_3} = \mathbf{i} + 2\mathbf{j} - \mathbf{k}.$$

Then

$$A_T = \tfrac{1}{2}\|(-3\mathbf{i} - 2\mathbf{j} + \mathbf{k}) \times (\mathbf{i} + 2\mathbf{j} - \mathbf{k})\|$$
$$= \tfrac{1}{2}\|-2\mathbf{j} - 4\mathbf{k}\| = \|-\mathbf{j} - 2\mathbf{k}\| = \sqrt{5}. \qquad ■$$

Area of a Parallelogram

The area A_P of the parallelogram with adjacent sides $\mathbf{u}$ and $\mathbf{v}$ (Figure 3.52) is $2A_T$, so

$$A_P = \|\mathbf{u} \times \mathbf{v}\|.$$

FIGURE 3.52

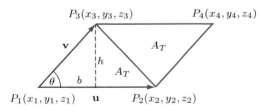

EXAMPLE 5 ■ If P_1, P_2, and P_3 are as in Example 4, then the area of the parallelogram with adjacent sides $\overrightarrow{P_1P_2}$ and $\overrightarrow{P_1P_3}$ is $2\sqrt{5}$. (Verify.) ■

FIGURE 3.53

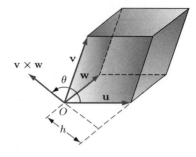

Volume of a Parallelepiped

Consider the parallelepiped with a vertex at the origin and edges **u**, **v**, and **w** (Figure 3.53). The volume V of the parallelepiped is the product of the area of the face containing **v** and **w** and the distance h from this face to the face parallel to it. Now

$$h = \|\mathbf{u}\| \, |\cos\theta| \,,$$

where θ is the angle between **u** and $\mathbf{v} \times \mathbf{w}$, and the area of the face determined by **v** and **w** is $\|\mathbf{v} \times \mathbf{w}\|$. Hence

$$V = \|\mathbf{v} \times \mathbf{w}\| \|\mathbf{u}\| \, |\cos\theta| = |\mathbf{u} \cdot (\mathbf{v} \times \mathbf{w})|. \tag{8}$$

From Equations (3) and (4), we also have

$$V = \left| \det \left(\begin{bmatrix} u_1 & u_2 & u_3 \\ v_1 & v_2 & v_3 \\ w_1 & w_2 & w_3 \end{bmatrix} \right) \right|. \tag{9}$$

EXAMPLE 6 ■ Consider the parallelepiped with a vertex at the origin and edges $\mathbf{u} = \mathbf{i} - 2\mathbf{j} + 3\mathbf{k}$, $\mathbf{v} = \mathbf{i} + 3\mathbf{j} + \mathbf{k}$, and $\mathbf{w} = 2\mathbf{i} + \mathbf{j} + 2\mathbf{k}$. Then

$$\mathbf{v} \times \mathbf{w} = 5\mathbf{i} - 5\mathbf{k}.$$

Hence $\mathbf{u} \cdot (\mathbf{v} \times \mathbf{w}) = -10$. Thus the volume V is given by (8) as

$$V = |\mathbf{u} \cdot (\mathbf{v} \times \mathbf{w})| = |-10| = 10.$$

We can also compute the volume by Equation (9) as

$$V = \left| \det \left(\begin{bmatrix} 1 & -2 & 3 \\ 1 & 3 & 1 \\ 2 & 1 & 2 \end{bmatrix} \right) \right| = |-10| = 10.$$

■

3.5 EXERCISES

In Exercises 1 and 2, compute $\mathbf{u} \times \mathbf{v}$.

1. (a) $\mathbf{u} = 2\mathbf{i} + 3\mathbf{j} + 4\mathbf{k}$, $\mathbf{v} = -\mathbf{i} + 3\mathbf{j} - \mathbf{k}$.
   (b) $\mathbf{u} = (1, 0, 1)$, $\mathbf{v} = (2, 3, -1)$.
   (c) $\mathbf{u} = \mathbf{i} - \mathbf{j} + 2\mathbf{k}$, $\mathbf{v} = 3\mathbf{i} - 4\mathbf{j} + \mathbf{k}$.
   (d) $\mathbf{u} = (2, -1, 1)$, $\mathbf{v} = -2\mathbf{u}$.

2. (a) $\mathbf{u} = (1, -1, 2)$, $\mathbf{v} = (3, 1, 2)$.
   (b) $\mathbf{u} = 2\mathbf{i} + \mathbf{j} - 2\mathbf{k}$, $\mathbf{v} = \mathbf{i} + 3\mathbf{k}$.
   (c) $\mathbf{u} = 2\mathbf{j} + \mathbf{k}$, $\mathbf{v} = 3\mathbf{u}$.
   (d) $\mathbf{u} = (4, 0, -2)$, $\mathbf{v} = (0, 2, -1)$.

3. Let $\mathbf{u} = \mathbf{i} + 2\mathbf{j} - 3\mathbf{k}$, $\mathbf{v} = 2\mathbf{i} + 3\mathbf{j} + \mathbf{k}$, $\mathbf{w} = 2\mathbf{i} - \mathbf{j} + 2\mathbf{k}$, and $c = -3$. Verify properties (a) through (d) of Theorem 3.9.

4. Let $\mathbf{u} = 2\mathbf{i} - \mathbf{j} + 3\mathbf{k}$, $\mathbf{v} = 3\mathbf{i} + \mathbf{j} - \mathbf{k}$, and $\mathbf{w} = 3\mathbf{i} + \mathbf{j} + 2\mathbf{k}$.
   (a) Verify Equation (3).
   (b) Verify Equation (4).

5. Let $\mathbf{u} = \mathbf{i} - \mathbf{j} + 2\mathbf{k}$, $\mathbf{v} = 2\mathbf{i} + 2\mathbf{j} - \mathbf{k}$, and $\mathbf{w} = \mathbf{i} + \mathbf{j} - \mathbf{k}$.
   (a) Verify Equation (3).
   (b) Verify Equation (4).

6. Verify that each of the cross products $\mathbf{u} \times \mathbf{v}$ in Exercise 1 is orthogonal to both $\mathbf{u}$ and $\mathbf{v}$.

7. Verify that each of the cross products $\mathbf{u} \times \mathbf{v}$ in Exercise 2 is orthogonal to both $\mathbf{u}$ and $\mathbf{v}$.

8. Verify Equation (7) for the pairs of vectors in Exercise 1.

9. Find the area of the triangle with vertices $P_1(1, -2, 3)$, $P_2(-3, 1, 4)$, $P_3(0, 4, 3)$.

10. Find the area of the triangle with vertices P_1, P_2, and P_3, where $\overrightarrow{P_1 P_2} = 2\mathbf{i} + 3\mathbf{j} - \mathbf{k}$ and $\overrightarrow{P_1 P_3} = \mathbf{i} + 2\mathbf{j} + 2\mathbf{k}$.

11. Find the area of the parallelogram with adjacent sides $\mathbf{u} = \mathbf{i} + 3\mathbf{j} - 2\mathbf{k}$ and $\mathbf{v} = 3\mathbf{i} - \mathbf{j} - \mathbf{k}$.

12. Find the volume of the parallelepiped with a vertex at the origin and edges $\mathbf{u} = 2\mathbf{i} - \mathbf{j}$, $\mathbf{v} = \mathbf{i} - 2\mathbf{j} - 2\mathbf{k}$, and $\mathbf{w} = 3\mathbf{i} - \mathbf{j} + \mathbf{k}$.

13. Repeat Exercise 12 for $\mathbf{u} = \mathbf{i} - 2\mathbf{j} + 4\mathbf{k}$, $\mathbf{v} = 3\mathbf{i} + 4\mathbf{j} + \mathbf{k}$, and $\mathbf{w} = -\mathbf{i} + \mathbf{j} + \mathbf{k}$.

THEORETICAL EXERCISES ▣

T.1. Prove Theorem 3.9.

T.2. Show that $(\mathbf{u} \times \mathbf{v}) \cdot \mathbf{w} = \mathbf{u} \cdot (\mathbf{v} \times \mathbf{w})$.

T.3. Show that $\mathbf{j} \times \mathbf{i} = -\mathbf{k}$, $\mathbf{k} \times \mathbf{j} = -\mathbf{i}$, $\mathbf{i} \times \mathbf{k} = -\mathbf{j}$.

T.4. Show that
$$(\mathbf{u} \times \mathbf{v}) \cdot \mathbf{w} = \begin{vmatrix} u_1 & u_2 & u_3 \\ v_1 & v_2 & v_3 \\ w_1 & w_2 & w_3 \end{vmatrix}.$$

T.5. Show that $\mathbf{u}$ and $\mathbf{v}$ are parallel if and only if $\mathbf{u} \times \mathbf{v} = \mathbf{0}$.

T.6. Show that $\|\mathbf{u} \times \mathbf{v}\|^2 + (\mathbf{u} \cdot \mathbf{v})^2 = \|\mathbf{u}\|^2 \|\mathbf{v}\|^2$.

T.7. Prove the **Jacobi identity**:
$$(\mathbf{u} \times \mathbf{v}) \times \mathbf{w} + (\mathbf{v} \times \mathbf{w}) \times \mathbf{u} + (\mathbf{w} \times \mathbf{u}) \times \mathbf{v} = \mathbf{0}.$$

MATLAB EXERCISES ▣

There are two MATLAB routines that apply to the material in this section. They are **cross**, which computes the cross product of a pair of 3-vectors; and **crossdemo**, which displays graphically a pair of vectors and their cross product. Using routine **dot** with **cross**, we can carry out the computations in Example 6. (For directions on the use of MATLAB routines, use **help** followed by a space and the name of the routine.)

ML.1. Use **cross** in MATLAB to find the cross product of each of the following pairs of vectors.
   (a) $\mathbf{u} = \mathbf{i} - 2\mathbf{j} + 3\mathbf{k}$, $\mathbf{v} = \mathbf{i} + 3\mathbf{j} + \mathbf{k}$.
   (b) $\mathbf{u} = (1, 0, 3)$, $\mathbf{v} = (1, -1, 2)$.
   (c) $\mathbf{u} = (1, 2, -3)$, $\mathbf{v} = (2, -1, 2)$.

ML.2. Use routine **cross** to find the cross product of each of the following pairs of vectors.
  (a) $\mathbf{u} = (2, 3, -1)$, $\mathbf{v} = (2, 3, 1)$.
  (b) $\mathbf{u} = 3\mathbf{i} - \mathbf{j} + \mathbf{k}$, $\mathbf{v} = 2\mathbf{u}$.
  (c) $\mathbf{u} = (1, -2, 1)$, $\mathbf{v} = (3, 1, -1)$.

ML.3. Use **crossdemo** in MATLAB to display the vectors $\mathbf{u}$, $\mathbf{v}$, and their cross product.
  (a) $\mathbf{u} = \mathbf{i} + 2\mathbf{j} + 4\mathbf{k}$, $\mathbf{v} = -2\mathbf{i} + 4\mathbf{j} + 3\mathbf{k}$.
  (b) $\mathbf{u} = (-2, 4, 5)$, $\mathbf{v} = (0, 1, -3)$.
  (c) $\mathbf{u} = (2, 2, 2)$, $\mathbf{v} = (3, -3, 3)$.

ML.4. Use **cross** in MATLAB to check your answers to Exercises 1 and 2.

ML.5. Use MATLAB to find the volume of the parallelepiped with vertex at the origin and edges $\mathbf{u} = (3, -2, 1)$, $\mathbf{v} = (1, 2, 3)$, and $\mathbf{w} = (2, -1, 2)$.

ML.6. The angle of intersection of two planes in 3-space is the same as the angle of intersection of perpendiculars to the planes. Find the angle of intersection of plane P_1 determined by $\mathbf{x}$ and $\mathbf{y}$ and plane P_2 determined by $\mathbf{v}$, $\mathbf{w}$, where

$$\mathbf{x} = (2, -1, 2), \quad \mathbf{y} = (3, -2, 1),$$
$$\mathbf{v} = (1, 3, 1), \quad \mathbf{w} = (0, 2, -1).$$

3.6 ▾ Lines and Planes

Lines in R^2

Any two distinct points $P_1(x_1, y_1)$ and $P_2(x_2, y_2)$ in R^2 (Figure 3.54) determine a straight line whose equation is

$$ax + by + c = 0, \tag{1}$$

where a, b, and c are real numbers, and a and b are not both zero. Since P_1 and P_2 lie on the line, their coordinates satisfy Equation (1):

$$ax_1 + by_1 + c = 0 \tag{2}$$
$$ax_2 + by_2 + c = 0. \tag{3}$$

We now write (1), (2), and (3) as a linear system in the unknowns a, b, and c, obtaining

$$xa + yb + c = 0$$
$$x_1 a + y_1 b + c = 0 \tag{4}$$
$$x_2 a + y_2 b + c = 0.$$

FIGURE 3.54

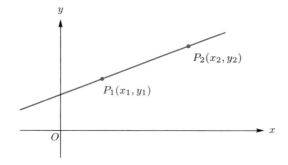

We seek a condition on the values x and y that allow (4) to have a nontrivial solution a, b, and c. Since (4) is a homogeneous system, it has a nontrivial solution if and only if the determinant of the coefficient matrix is zero, that is, if and only if

$$\begin{vmatrix} x & y & 1 \\ x_1 & y_1 & 1 \\ x_2 & y_2 & 1 \end{vmatrix} = 0. \tag{5}$$

Thus every point $P(x, y)$ on the line satisfies (5) and, conversely, a point satisfying (5) lies on the line.

EXAMPLE 1 ■ Find an equation of the line determined by the points $P_1(-1, 3)$ and $P_2(4, 6)$.

Solution   Substituting in (5), we obtain

$$\begin{vmatrix} x & y & 1 \\ -1 & 3 & 1 \\ 4 & 6 & 1 \end{vmatrix} = 0.$$

Expanding this determinant in cofactors about the first row, we have (verify)

$$-3x + 5y - 18 = 0. \qquad \blacksquare$$

Lines in R^3

We may recall that in R^2 a line is determined by specifying its slope and one of its points. In R^3 a line is determined by specifying its direction and one of its points. Let $\mathbf{u} = (a, b, c)$ be a nonzero vector in R^3, and let $P_0 = (x_0, y_0, z_0)$ be a point in R^3. Let $\mathbf{w}_0$ be the vector associated with P_0 and let $\mathbf{x}$ be the vector associated with the point $P(x, y, z)$. The line L through P_0 and parallel to $\mathbf{u}$ consists of the points $P(x, y, z)$ (Figure 3.55) such that

$$\mathbf{x} = \mathbf{w}_0 + t\mathbf{u} \qquad (-\infty < t < \infty). \tag{6}$$

FIGURE 3.55

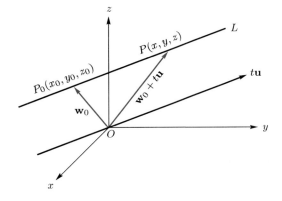

Equation (6) is called a **parametric equation** of L, since it contains the parameter t, which can be assigned any real number. Equation (6) can also be written in terms of the components as

$$
\begin{aligned}
x &= x_0 + ta \\
y &= y_0 + tb \qquad (-\infty < t < \infty) \\
z &= z_0 + tc,
\end{aligned}
\tag{7}
$$

which are called **parametric equations** of L.

EXAMPLE 2 ■ Parametric equations of the line through the point $P_0(-3, 2, 1)$, which is parallel to the vector $\mathbf{u} = (2, -3, 4)$, are

$$
\begin{aligned}
x &= -3 + 2t \\
y &= 2 - 3t \qquad (-\infty < t < \infty) \\
z &= 1 + 4t.
\end{aligned}
$$

■

EXAMPLE 3 ■ Find parametric equations of the line L through the points $P_0(2, 3, -4)$ and $P_1(3, -2, 5)$.

Solution   The desired line is parallel to the vector $\mathbf{u} = \overrightarrow{P_0 P_1}$. Now

$$
\mathbf{u} = (3 - 2, -2 - 3, 5 - (-4)) = (1, -5, 9).
$$

Since P_0 lies on the line, we can write parametric equations of L as

$$
\begin{aligned}
x &= 2 + t \\
y &= 3 - 5t \qquad (-\infty < t < \infty) \\
z &= -4 + 9t.
\end{aligned}
$$

■

In Example 3 we could have used the point P_2 instead of P_1. In fact, we could use any point on the line in the parametric equations of L. Thus a line can be represented in infinitely many ways in parametric form. If a, b, and c are all nonzero in (7), we can solve each equation for t and equate the results to obtain the equations in **symmetric form** of the line through P_0 and parallel to $\mathbf{u}$:

$$
\frac{x - x_0}{a} = \frac{y - y_0}{b} = \frac{z - z_0}{c}.
$$

The equations in symmetric form of a line are useful in some analytic geometry applications.

EXAMPLE 4 ■ The equations in symmetric form of the line in Example 3 are

$$
\frac{x - 2}{1} = \frac{y - 3}{-5} = \frac{z + 4}{9}.
$$

■

Exercises T.2 and T.3 consider the intersection of two lines in R^3.

Planes in R^3

A plane in R^3 can be determined by specifying a point in the plane and a vector perpendicular to it. A vector perpendicular to a plane is called a **normal** to the plane.

To obtain an equation of the plane passing through the point $P_0(x_0, y_0, z_0)$ and having the nonzero vector $\mathbf{n} = (a, b, c)$ as a normal, we proceed as follows. A point $P(x, y, z)$ lies in the plane if and only if the vector $\overrightarrow{P_0 P}$ is perpendicular to $\mathbf{n}$ (Figure 3.56). Thus $P(x, y, z)$ lies in the plane if and only if

$$\mathbf{n} \cdot \overrightarrow{P_0 P} = 0. \tag{8}$$

Since

$$\overrightarrow{P_0 P} = (x - x_0, y - y_0, z - z_0),$$

we can write (8) as

$$a(x - x_0) + b(y - y_0) + c(z - z_0) = 0. \tag{9}$$

FIGURE 3.56

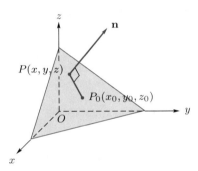

EXAMPLE 5 ■ Find an equation of the plane passing through the point $(3, 4, -3)$ and perpendicular to the vector $\mathbf{n} = (5, -2, 4)$.

Solution   Substituting in (9), we obtain the equation of the plane as

$$5(x - 3) - 2(y - 4) + 4(z + 3) = 0. \qquad ■$$

If we multiply out and simplify, (9) can be rewritten as

$$ax + by + cz + d = 0. \tag{10}$$

It is not difficult to show (Exercise T.1) that the graph of an equation of the form given in (10), where a, b, c, and d are constants, is a plane with normal $\mathbf{n} = (a, b, c)$ provided a, b, and c are not all zero.

EXAMPLE 6 ■ Find an equation of the plane passing through the points $P_1(2, -2, 1), P_2(-1, 0, 3)$, and $P_3(5, -3, 4)$.

Solution   Let an equation of the desired plane be as given by (10). Since P_1, P_2, and P_3 lie in the plane, their coordinates satisfy (10). Thus we obtain the linear system (verify)

$$\begin{aligned} 2a - 2b + c + d &= 0 \\ -a + 3c + d &= 0 \\ 5a - 3b + 4c + d &= 0. \end{aligned}$$

Solving this system, we have (verify)

$$a = \tfrac{8}{17}r, \qquad b = \tfrac{15}{17}r, \qquad c = -\tfrac{3}{17}r, \qquad d = r,$$

where r is any real number. Letting $r = 17$, we obtain

$$a = 8, \qquad b = 15, \qquad c = -3, \qquad d = 17.$$

Thus, an equation for the desired plane is

$$8x + 15y - 3z + 17 = 0. \tag{11} \quad \blacksquare$$

EXAMPLE 7 ■ A second solution to Example 6 is as follows. Proceeding as in the case of a line in R^2 determined by two distinct points P_1 and P_2, it is not difficult to show (Exercise T.7) that an equation of the plane through the noncollinear points $P_1(x_1, y_1, z_1)$, $P_2(x_2, y_2, z_2)$, and $P_3(x_3, y_3, z_3)$ is

$$\begin{vmatrix} x & y & z & 1 \\ x_1 & y_1 & z_1 & 1 \\ x_2 & y_2 & z_2 & 1 \\ x_3 & y_3 & z_3 & 1 \end{vmatrix} = 0.$$

In our example, the equation of the desired plane is

$$\begin{vmatrix} x & y & z & 1 \\ 2 & -2 & 1 & 1 \\ -1 & 0 & 3 & 1 \\ 5 & -3 & 4 & 1 \end{vmatrix} = 0.$$

Expanding this determinant in cofactors about the first row, we obtain (verify) Equation (11).   ■

EXAMPLE 8 ■ A third solution to Example 6 using optional Section 3.5, Cross Product in R^3, is as follows. The nonparallel vectors $\overrightarrow{P_1P_2} = (-3, 2, 2)$ and $\overrightarrow{P_1P_3} = (3, -1, 3)$ lie in the plane, since the points P_1, P_2, and P_3 lie in the plane. The vector

$$\mathbf{n} = \overrightarrow{P_1P_2} \times \overrightarrow{P_1P_3} = (8, 15, -3)$$

is then perpendicular to both $\overrightarrow{P_1P_2}$ and $\overrightarrow{P_1P_3}$ and is thus a normal to the plane. Using the vector $\mathbf{n}$ and the point $P_1(2, -2, 1)$ in (9), we obtain an equation of the plane as

$$8(x - 2) + 15(y + 2) - 3(z - 1) = 0,$$

which when expanded agrees with Equation (11).   ■

The equations of a line in symmetric form can be used to determine two planes whose intersection is the given line.

EXAMPLE 9 ■ Find two planes whose intersection is the line

$$x = -2 + 3t$$
$$y = 3 - 2t \qquad (-\infty < t < \infty)$$
$$z = 5 + 4t.$$

Solution   First, find equations of the line in symmetric form as

$$\frac{x+2}{3} = \frac{y-3}{-2} = \frac{z-5}{4}.$$

The given line is then the intersection of the planes

$$\frac{x+2}{3} = \frac{y-3}{-2} \quad \text{and} \quad \frac{x+2}{3} = \frac{z-5}{4}.$$

Thus the given line is the intersection of the planes

$$2x + 3y - 5 = 0 \quad \text{and} \quad 4x - 3z + 23 = 0. \qquad \blacksquare$$

Two planes are either parallel or they intersect in a straight line. They are parallel if their normals are parallel. In the following example we determine the line of intersection of two planes.

EXAMPLE 10 ■ Find parametric equations of the line of intersection of the planes

$$\pi_1 \colon 2x + 3y - 2z + 4 = 0 \quad \text{and} \quad \pi_2 \colon x - y + 2z + 3 = 0.$$

Solution   Solving the linear system consisting of the equations of π_1 and π_2, we obtain (verify)

$$x = -\tfrac{13}{5} - \tfrac{4}{5}t$$
$$y = \tfrac{2}{5} + \tfrac{6}{5}t \qquad (-\infty < t < \infty)$$
$$z = 0 + t$$

as parametric equations of the line L of intersection of the planes (see Figure 3.57). ■

FIGURE 3.57

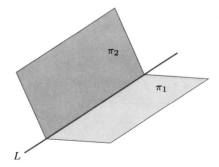

Three planes in R^3 may intersect in a plane, in a line, in a unique point, or have no set of points in common. These possibilities can be detected by solving the linear system consisting of their equations.

3.6 EXERCISES

1. In each of the following, find an equation of the line in R^2 determined by the given points.
   (a) $P_1(-2, -3)$, $P_2(3, 4)$.
   (b) $P_1(2, -5)$, $P_2(-3, 4)$.
   (c) $P_1(0, 0)$, $P_2(-3, 5)$.
   (d) $P_1(-3, -5)$, $P_2(0, 2)$.

2. In each of the following, find the equation of the line in R^2 determined by the given points.
   (a) $P_1(1, 1)$, $P_2(2, 2)$.
   (b) $P_1(1, 2)$, $P_2(1, 3)$.
   (c) $P_1(2, -4)$, $P_2(-3, -4)$.
   (d) $P_1(2, -3)$, $P_2(3, -2)$.

3. State which of the following points are on the line

$$x = 3 + 2t$$
$$y = -2 + 3t \qquad (-\infty < t < \infty)$$
$$z = 4 - 3t.$$

   (a) $(1, 1, 1)$.   (b) $(1, -1, 0)$.
   (c) $(1, 0, -2)$.   (d) $\left(4, -\frac{1}{2}, \frac{5}{2}\right)$.

4. State which of the following points are on the line
$$\frac{x - 4}{-2} = \frac{y + 3}{2} = \frac{z - 4}{-5}.$$

   (a) $(0, 1, -6)$.   (b) $(1, 2, 3)$.
   (c) $(4, -3, 4)$.   (d) $(0, 1, -1)$.

5. In each of the following, find the parametric equations of the line through the point $P_0(x_0, y_0, z_0)$, which is parallel to the vector $\mathbf{u}$.
   (a) $P_0 = (3, 4, -2)$, $\mathbf{u} = (4, -5, 2)$.
   (b) $P_0 = (3, 2, 4)$, $\mathbf{u} = (-2, 5, 1)$.
   (c) $P_0 = (0, 0, 0)$, $\mathbf{u} = (2, 2, 2)$.
   (d) $P_0 = (-2, -3, 1)$, $\mathbf{u} = (2, 3, 4)$.

6. In each of the following, find the parametric equations of the line through the given points.
   (a) $(2, -3, 1)$, $(4, 2, 5)$.   (b) $(-3, -2, -2)$, $(5, 5, 4)$.
   (c) $(-2, 3, 4)$, $(2, -3, 5)$.   (d) $(0, 0, 0)$, $(4, 5, 2)$.

7. For each of the lines in Exercise 6, find the equations in symmetric form.

8. State which of the following points are on the plane $3(x - 2) + 2(y + 3) - 4(z - 4) = 0$.
   (a) $(0, -2, 3)$.   (b) $(1, -2, 3)$.
   (c) $(1, -1, 3)$.   (d) $(0, 0, 4)$.

9. In each of the following, find an equation of the plane passing through the given point and perpendicular to the given vector $\mathbf{n}$.
   (a) $(0, 2, -3)$, $\mathbf{n} = (3, -2, 4)$.
   (b) $(-1, 3, 2)$, $\mathbf{n} = (0, 1, -3)$.
   (c) $(-2, 3, 4)$, $\mathbf{n} = (0, 0, -4)$.
   (d) $(5, 2, 3)$, $\mathbf{n} = (-1, -2, 4)$.

10. In each of the following, find an equation of the plane passing through the given three points.
    (a) $(0, 1, 2)$, $(3, -2, 5)$, $(2, 3, 4)$.
    (b) $(2, 3, 4)$, $(-1, -2, 3)$, $(-5, -4, 2)$.
    (c) $(1, 2, 3)$, $(0, 0, 0)$, $(-2, 3, 4)$.
    (d) $(1, 1, 1)$, $(2, 3, 4)$, $(-5, 3, 2)$.

11. In each of the following, find parametric equations of the line of intersection of the given planes.
    (a) $2x + 3y - 4z + 5 = 0$ and $-3x + 2y + 5z + 6 = 0$.
    (b) $3x - 2y - 5z + 4 = 0$ and $2x + 3y + 4z + 8 = 0$.
    (c) $-x + 2y + z = 0$ and $2x - y + 2z + 8 = 0$.

12. In each of the following, find a pair of planes whose intersection is the given line.
    (a) $x = 2 - 3t$
    $\ y = 3 + t$
    $\ z = 2 - 4t.$
    (b) $\dfrac{x - 2}{-2} = \dfrac{y - 3}{4} = \dfrac{z + 4}{3}.$
    (c) $x = 4t$
    $\ y = 1 + 5t$
    $\ z = 2 - t.$

13. Are the points $(2, 3, -2)$, $(4, -2, -3)$, and $(0, 8, -1)$ on the same line?

14. Are the points $(-2, 4, 2)$, $(3, 5, 1)$, and $(4, 2, -1)$ on the same line?

15. Find the point of intersection of the lines

$$
\begin{array}{ll}
x = 2 - 3s & x = 5 + 2t \\
y = 3 + 2s \quad \text{and} & y = 1 - 3t \\
z = 4 + 2s & z = 2 + \ t.
\end{array}
$$

16. Which of the following pairs of lines are perpendicular?

   (a) $\begin{array}{ll} x = \ \ 2 + 2t & x = 2 + t \\ y = -3 - 3t \quad \text{and} & y = 4 - t \\ z = \ \ \ 4 + 4t & z = 5 - t. \end{array}$

   (b) $\begin{array}{ll} x = 3 - \ t & x = \ \ \ \ 2t \\ y = 4 + \ t \quad \text{and} & y = 3 - 2t \\ z = 2 + 2t & z = 4 + 2t. \end{array}$

17. Show that the following parametric equations define the same line.

$$
\begin{array}{ll}
x = \ \ 2 + 3t & x = -1 - \ \ 9t \\
y = \ \ 3 - 2t \quad \text{and} & y = \ \ 5 + \ \ 6t \\
z = -1 + 4t & z = -5 - 12t.
\end{array}
$$

18. Find parametric equations of the line passing through the point $(3, -1, -3)$ and perpendicular to the line passing through the points $(3, -2, 4)$ and $(0, 3, 5)$.

19. Find an equation of the plane passing through the point $(-2, 3, 4)$ and perpendicular to the line passing through the points $(4, -2, 5)$ and $(0, 2, 4)$.

20. Find the point of intersection of the line

$$
\begin{array}{l}
x = 2 - 3t \\
y = 4 + 2t \\
z = 3 - 5t
\end{array}
$$

   and the plane $2x + 3y + 4z + 8 = 0$.

21. Find a plane containing the lines

$$
\begin{array}{ll}
x = 3 + 2t & x = 1 - 2t \\
y = 4 - 3t \quad \text{and} & y = 7 + 4t \\
z = 5 + 4t & z = 1 - 3t.
\end{array}
$$

22. Find a plane that passes through the point $(2, 4, -3)$ and is parallel to the plane $-2x + 4y - 5z + 6 = 0$.

23. Find a line that passes through the point $(-2, 5, -3)$ and is perpendicular to the plane $2x - 3y + 4z + 7 = 0$.

THEORETICAL EXERCISES

T.1. Show that the graph of the equation $ax + by + cz + d = 0$, where a, b, c, and d are constants with a, b, and c not all zero, is a plane with normal $\mathbf{n} = (a, b, c)$.

T.2. Let the lines L_1 and L_2 be given parametrically by

$$L_1 \colon \mathbf{x} = \mathbf{w}_0 + s\mathbf{u} \quad \text{and} \quad L_2 \colon \mathbf{x} = \mathbf{w}_1 + t\mathbf{v}.$$

   Show that
   (a) L_1 and L_2 are parallel if and only if $\mathbf{u} = k\mathbf{v}$ for some scalar k.
   (b) L_1 and L_2 are identical if and only if $\mathbf{w}_1 - \mathbf{w}_0$ and $\mathbf{u}$ are both parallel to $\mathbf{v}$.
   (c) L_1 and L_2 are perpendicular if and only if $\mathbf{u} \cdot \mathbf{v} = 0$.
   (d) L_1 and L_2 intersect if and only if $\mathbf{w}_1 - \mathbf{w}_0$ is a linear combination of $\mathbf{u}$ and $\mathbf{v}$.

T.3. The lines L_1 and L_2 in R^3 are said to be **skew** if they are not parallel and do not intersect. Give an example of skew lines L_1 and L_2.

T.4. Consider the planes $a_1 x + b_1 y + c_1 z + d_1 = 0$ and $a_2 x + b_2 y + c_2 z + d_2 = 0$ with normals $\mathbf{n}_1$ and $\mathbf{n}_2$, respectively. Show that if the planes are identical, then $\mathbf{n}_2 = a\mathbf{n}_1$ for some scalar a.

T.5. Show that an equation of the plane through the noncollinear points $P_1(a_1, b_1, c_1)$, $P_2(a_2, b_2, c_2)$, and $P_3(a_3, b_3, c_3)$ is

$$
\begin{vmatrix}
x & y & z & 1 \\
a_1 & b_1 & c_1 & 1 \\
a_2 & b_2 & c_2 & 1 \\
a_3 & b_3 & c_3 & 1
\end{vmatrix} = 0.
$$

KEY IDEAS FOR REVIEW ■

☐ **Theorem 3.2.** Properties of vector addition and scalar multiplication in R^n (see page 142).

☐ **Theorem 3.3 (Properties of Dot Product).** See page 149.

☐ **Theorem 3.4 (Cauchy–Schwarz Inequality)**
$$|\mathbf{u} \cdot \mathbf{v}| \le \|\mathbf{u}\|\|\mathbf{v}\|$$

☐ **Theorem 3.6.** If $L: R^n \to R^m$ is a linear transformation, then
$$L(c_1\mathbf{u}_1 + c_2\mathbf{u}_2 + \cdots + c_k\mathbf{u}_k)$$
$$= c_1 L(\mathbf{u}_1) + c_2 L(\mathbf{u}_2) + \cdots + c_k L(\mathbf{u}_k).$$

☐ **Theorem 3.8.** Let $L: R^n \to R^m$ be a linear transformation. Then there exists a unique $m \times n$ matrix A such that $L(\mathbf{x}) = A\mathbf{x}$ for $\mathbf{x}$ in R^n.

☐ **Theorem 3.9 (Properties of Cross Product).** See page 180.

☐ A **parametric equation** of the line through P_0 and parallel to $\mathbf{u}$ is
$$\mathbf{x} = \mathbf{w}_0 + t\mathbf{u} \qquad (-\infty < t < \infty),$$
where $\mathbf{w}_0$ is the vector associated with P_0.

☐ An equation of the plane with normal $\mathbf{n} = (a, b, c)$ and passing through the point $P_0 = (x_0, y_0, z_0)$ is
$$a(x - x_0) + b(y - y_0) + c(z - z_0) = 0.$$

SUPPLEMENTARY EXERCISES ■

In Exercises 1 through 3, let
$$\mathbf{u} = (2, -1), \quad \mathbf{v} = (1, 3), \quad \text{and} \quad \mathbf{w} = (4, 1).$$

1. Sketch a diagram showing that $\mathbf{u} + \mathbf{v} = \mathbf{v} + \mathbf{u}$.

2. Sketch a diagram showing that $(\mathbf{u} + \mathbf{v}) + \mathbf{w} = \mathbf{u} + (\mathbf{v} + \mathbf{w})$.

3. Sketch a diagram showing that $2(\mathbf{u} + \mathbf{v}) = 2\mathbf{u} + 2\mathbf{v}$.

In Exercises 4 and 5, let $\mathbf{u} = (1, 2, -3)$, $\mathbf{v} = (3, 0, 1)$, and $\mathbf{w} = (-2, 1, 1)$.

4. Find $\mathbf{x}$ so that $\mathbf{u} + \mathbf{x} = \mathbf{v} - \mathbf{w}$.

5. Find $\mathbf{x}$ so that $2\mathbf{u} + 3\mathbf{x} = \mathbf{w} - 5\mathbf{x}$.

6. Write the vector $(1, 2)$ as a linear combination of the vectors $(-2, 3)$ and $(1, -1)$.

7. Let $\mathbf{u} = (1, -1, 2, 3)$, and $\mathbf{v} = (2, 3, 1, -2)$. Compute
   (a) $\|\mathbf{u}\|$.   (b) $\|\mathbf{v}\|$.
   (c) $\|\mathbf{u} - \mathbf{v}\|$.   (d) $\mathbf{u} \cdot \mathbf{v}$.
   (e) Cosine of the angle between $\mathbf{u}$ and $\mathbf{v}$.

8. If $\mathbf{u} = (x, y)$ is any vector in R^2, show that the vector $\mathbf{v} = (-y, x)$ is orthogonal to $\mathbf{u}$.

9. Find all values of c for which $\|c(1, -2, 2, 0)\| = 9$.

10. Let $\mathbf{u} = (a, 2, a)$ and $\mathbf{v} = (4, -3, 2)$. For what values of a are $\mathbf{u}$ and $\mathbf{v}$ orthogonal?

11. Is $L: R^2 \to R^2$ defined by
$$L(x, y) = (x - 1, y - x)$$
a linear transformation?

12. Let $\mathbf{u}_0$ be a fixed vector in R^n. Let $L: R^n \to R^1$ be defined by $L(\mathbf{u}) = \mathbf{u} \cdot \mathbf{u}_0$. Show that L is a linear transformation.

13. Find a unit vector parallel to the vector $(-1, 2, 3)$.

14. Show that a parallelogram is a rhombus, a parallelogram with four equal sides, if and only if its diagonals are orthogonal.

15. If possible, find a and b so that
$$\mathbf{v} = \begin{bmatrix} a \\ b \\ 2 \end{bmatrix}$$
is orthogonal to both
$$\mathbf{w} = \begin{bmatrix} 2 \\ 1 \\ 1 \end{bmatrix} \quad \text{and} \quad \mathbf{x} = \begin{bmatrix} 1 \\ 0 \\ 1 \end{bmatrix}.$$

16. Find a unit vector that is orthogonal to the vector $(1, 2)$.

17. Find the area of the quadrilateral with vertices $(-3, 1)$, $(-2, -2)$, $(2, 4)$, and $(5, 0)$.

18. Find all values of a so that the vector $(a, -5, -2)$ is orthogonal to the vector $(a, a, -3)$.

19. Find the standard matrix representing a clockwise rotation of R^2 through $\pi/6$ radians.

20. Let $L: R^2 \to R^1$ be the linear transformation defined by $L(\mathbf{u}) = \mathbf{u} \cdot \mathbf{u}_0$, where $\mathbf{u}_0 = (1, 2)$. (See Exercise 12.) Find the standard matrix representing L.

21. (a) Write the vector $(1, 3, -2)$ as a linear combination of the vectors $(1, 1, 0)$, $(0, 1, 1)$, and $(0, 0, 1)$.

(b) If $L\colon R^3 \to R^2$ is the linear transformation for which $L(1, 1, 0) = (2, -1)$, $L(0, 1, 1) = (3, 2)$, and $L(0, 0, 1) = (1, -1)$, find $L(1, 3, -2)$.

22. A river flows south at a rate of 2 miles per hour. If a swimmer is trying to swim westward at a rate of 8 miles per hour, draw a figure showing the magnitude and direction of the resultant velocity.

23. If the 5×3 matrix A is the standard matrix representing the linear transformation $L\colon R^n \to R^m$, what are the values of n and m?

24. Let $L\colon R^2 \to R^2$ be the linear transformation defined by

$$L\left(\begin{bmatrix} x \\ y \end{bmatrix}\right) = \begin{bmatrix} x - y \\ x + y \end{bmatrix}.$$

Is $\begin{bmatrix} 2 \\ 3 \end{bmatrix}$ in range L?

25. Let $L\colon R^3 \to R^3$ be the linear transformation defined by $L(\mathbf{x}) = A\mathbf{x}$, where

$$A = \begin{bmatrix} 1 & 2 & 4 \\ 2 & 3 & 5 \\ -1 & -3 & -7 \end{bmatrix}.$$

Find an equation relating a, b, and c so that $\begin{bmatrix} a \\ b \\ c \end{bmatrix}$ will lie in range L.

26. Which of the following points are on the line

$$\frac{x-3}{2} = \frac{y+3}{4} = \frac{z+5}{-4}?$$

(a) $(1, 2, 3)$.   (b) $(5, 1, -9)$.   (c) $(1, -7, -1)$.

27. Find parametric equations of the line of intersection of the planes $x - 2y + z + 3 = 0$ and $2x - y + 3z + 4 = 0$.

THEORETICAL EXERCISES

T.1. Let $\mathbf{u}$ and $\mathbf{v}$ be vectors in R^n. Show that $\mathbf{u} \cdot \mathbf{v} = 0$ if and only if $\|\mathbf{u} + \mathbf{v}\| = \|\mathbf{u} - \mathbf{v}\|$.

T.2. Show that for any vectors $\mathbf{u}$, $\mathbf{v}$, and $\mathbf{w}$ in R^2 or R^3 and any scalar c, we have:
(a) $(\mathbf{u} + c\mathbf{v}) \cdot \mathbf{w} = \mathbf{u} \cdot \mathbf{w} + c(\mathbf{v} \cdot \mathbf{w})$.
(b) $\mathbf{u} \cdot (c\mathbf{v}) = c(\mathbf{u} \cdot \mathbf{v})$.
(c) $(\mathbf{u} + \mathbf{v}) \cdot (c\mathbf{w}) = c(\mathbf{u} \cdot \mathbf{w}) + c(\mathbf{v} \cdot \mathbf{w})$.

T.3. Show that the only vector $\mathbf{x}$ in R^2 or R^3 that is orthogonal to every other vector is the zero vector.

T.4. Show that $L\colon R^n \to R^m$ is a linear transformation if and only if

$$L(a\mathbf{u} + b\mathbf{v}) = aL(\mathbf{u}) + bL(\mathbf{v})$$

for any scalars a and b and any vectors $\mathbf{u}$ and $\mathbf{v}$ in R^n.

T.5. Show that if $\|\mathbf{u}\| = 0$ in R^n, then $\mathbf{u} = \mathbf{0}$.

T.6. Give an example in R^4 showing that $\mathbf{u}$ is orthogonal to $\mathbf{v}$ and $\mathbf{v}$ is orthogonal to $\mathbf{w}$, but $\mathbf{u}$ is not orthogonal to $\mathbf{w}$.

CHAPTER TEST

1. Find the cosine of the angle between the vectors $(1, 2, -1, 4)$ and $(3, -2, 4, 1)$.

2. Find a unit vector in the direction of $(2, -1, 1, 3)$.

3. Is the vector $(1, 2, 3)$ a linear combination of the vectors $(1, 3, 2)$, $(2, 2, -1)$, and $(3, 7, 0)$?

4. Let $L\colon R^3 \to R^3$ be the linear transformation defined by $L(\mathbf{x}) = A\mathbf{x}$, where

$$A = \begin{bmatrix} 1 & 2 & 0 \\ 2 & -1 & 5 \\ 3 & 2 & 4 \end{bmatrix}.$$

Is $\begin{bmatrix} 1 \\ 2 \\ 3 \end{bmatrix}$ in range L?

5. Let $L\colon R^2 \to R^3$ be defined by $L(x, y) = (2x + 3y, -2x + 3y, x + y)$. Find the standard matrix representing L.

6. Find parametric equations of the line through the point $(5, -2, 1)$ that is parallel to the vector $\mathbf{u} = (3, -2, 5)$.

7. Find an equation of the plane passing through the points $(1, 2, -1)$, $(3, 4, 5)$, $(0, 1, 1)$.

8. Answer each of the following as true or false.

    (a) In R^n, if $\mathbf{u} \cdot \mathbf{v} = 0$, then $\mathbf{u} = \mathbf{0}$ or $\mathbf{v} = \mathbf{0}$.

    (b) In R^n, if $\mathbf{u} \cdot \mathbf{v} = \mathbf{u} \cdot \mathbf{w}$, then $\mathbf{v} = \mathbf{w}$.

    (c) In R^n, if $c\mathbf{u} = \mathbf{0}$, then $c = 0$ or $\mathbf{u} = \mathbf{0}$.

    (d) In R^n, $\|c\mathbf{u}\| = c\|\mathbf{u}\|$.

    (e) In R^n, $\|\mathbf{u} + \mathbf{v}\| = \|\mathbf{u}\| + \|\mathbf{v}\|$.

    (f) If $L\colon R^4 \to R^3$ is a linear transformation defined by $L(\mathbf{x}) = A\mathbf{x}$, then A is 3×4.

    (g) The vectors $(1, 0, 1)$ and $(-1, 1, 0)$ are orthogonal.

    (h) In R^n, if $\|\mathbf{u}\| = 0$, then $\mathbf{u} = \mathbf{0}$.

    (i) In R^n, if $\mathbf{u}$ is orthogonal to $\mathbf{v}$ and $\mathbf{w}$, then $\mathbf{u}$ is orthogonal to $2\mathbf{v} + 3\mathbf{w}$.

    (j) If $L\colon R^n \to R^m$ is a linear transformation, then $L(\mathbf{u}) = L(\mathbf{v})$ implies that $\mathbf{u} = \mathbf{v}$.

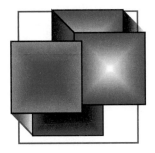

4

Real Vector Spaces

4.1 ▾ Vector Spaces

We have already defined R^n and examined some of its basic properties. We must now study the fundamental structure of R^n. In many applications in mathematics, the sciences, and engineering, the notion of a vector space arises. This idea is merely a carefully constructed generalization of R^n. In studying the properties and structure of a vector space, we can study not only R^n, in particular, but many other important vector spaces. In this section we define the notion of a vector space in general and in later sections we study their structure.

DEFINITION 1*   A **real vector space** is a set of elements V together with two operations $\oplus$ and $\odot$ satisfying the following properties:

(α) If **u** and **v** are any elements of V, then $\mathbf{u} \oplus \mathbf{v}$ is in V (i.e., V is closed under the operation $\oplus$).

   (a) $\mathbf{u} \oplus \mathbf{v} = \mathbf{v} \oplus \mathbf{u}$, for **u** and **v** in V.

   (b) $\mathbf{u} \oplus (\mathbf{v} \oplus \mathbf{w}) = (\mathbf{u} \oplus \mathbf{v}) \oplus \mathbf{w}$, for **u**, **v**, and **w** in V.

   (c) There is an element **0** in V such that

$$\mathbf{u} \oplus \mathbf{0} = \mathbf{0} \oplus \mathbf{u} = \mathbf{u}, \quad \text{for all } \mathbf{u} \text{ in } V.$$

   (d) For each **u** in V, there is an element $-\mathbf{u}$ in V such that

$$\mathbf{u} \oplus -\mathbf{u} = \mathbf{0}.$$

(β) If **u** is any element of V and c is any real number, then $c \odot \mathbf{u}$ is in V (i.e., V is closed under the operation $\odot$).

   (e) $c \odot (\mathbf{u} \oplus \mathbf{v}) = c \odot \mathbf{u} \oplus c \odot \mathbf{v}$, for all real numbers c and all **u** and **v** in V.

---

*Although the definitions in this book are not numbered, *this* definition is numbered because it will be referred to a number of times in this chapter.

197

(f) $(c+d) \odot \mathbf{u} = c \odot \mathbf{u} \oplus d \odot \mathbf{u}$, for all real numbers c and d, and all $\mathbf{u}$ in V.

(g) $c \odot (d \odot \mathbf{u}) = (cd) \odot \mathbf{u}$, for all real numbers c and d and all $\mathbf{u}$ in V.

(h) $1 \odot \mathbf{u} = \mathbf{u}$, for all $\mathbf{u}$ in V.

The elements of V are called **vectors**; the real numbers are called **scalars**. The operation $\oplus$ is called **vector addition**; the operation $\odot$ is called **scalar multiplication**. The vector $\mathbf{0}$ in property (c) is called a **zero vector**. The vector $-\mathbf{u}$ in property (d) is called a **negative** of $\mathbf{u}$. It can be shown (see Exercises T.5 and T.6) that the vectors $\mathbf{0}$ and $-\mathbf{u}$ are unique.

REMARK   If we allow the scalars to be complex numbers, we obtain a **complex vector space**. Such spaces are important in many applications in mathematics and the physical sciences. We provide a brief introduction to complex vector spaces in Appendix A. However, in this book we limit our study to real vector spaces.

EXAMPLE 1 ■ Consider the set R^n together with the operations of vector addition and scalar multiplication as defined in Section 3.2. Theorem 3.2 in Section 3.2 established the fact that R^n is a vector space under the operations of addition and scalar multiplication of n-vectors.   ■

EXAMPLE 2 ■ Consider the set V of all ordered triples of real numbers of the form $(x, y, 0)$ and define the operations $\oplus$ and $\odot$ by

$$(x, y, 0) \oplus (x', y', 0) = (x + x', y + y', 0)$$
$$c \odot (x, y, 0) = (cx, cy, 0).$$

It is then easy to show (Exercise 7) that V is a vector space, since it satisfies all the properties of Definition 1.   ■

EXAMPLE 3 ■ Consider the set V of all ordered triples of real numbers (x, y, z) and define the operations $\oplus$ and $\odot$ by

$$(x, y, z) \oplus (x', y', z') = (x + x', y + y', z + z')$$
$$c \odot (x, y, z) = (cx, y, z).$$

It is then easy to verify (Exercise 8) that properties (α), (β), (a), (b), (c), (d), and (e) of Definition 1 hold. Here $\mathbf{0} = (0, 0, 0)$ and the negative of the vector (x, y, z) is the vector $(-x, -y, -z)$. For example, to verify property (e) we proceed as follows. First,

$$c \odot [(x, y, z) \oplus (x', y', z')] = c \odot (x + x', y + y', z + z')$$
$$= (c(x + x'), y + y', z + z').$$

Also,

$$c \odot (x, y, z) \oplus c \odot (x', y', z') = (cx, y, z) \oplus (cx', y', z')$$
$$= (cx + cx', y + y', z + z')$$
$$= (c(x + x'), y + y', z + z').$$

However, we now show that property (f) fails to hold. Thus

$$(c + d) \odot (x, y, z) = ((c + d)x, y, z).$$

On the other hand,

$$\begin{aligned}
c \odot (x, y, z) \oplus d \odot (x, y, z) &= (cx, y, z) \oplus (dx, y, z) \\
&= (cx + dx, y + y, z + z) \\
&= ((c + d)x, 2y, 2z).
\end{aligned}$$

Thus V is not a vector space under the prescribed operations. Incidentally, properties (g) and (h) *do* hold for this example. ■

EXAMPLE 4 ■ Consider the set M_{23} of all 2×3 matrices under the usual operations of matrix addition and scalar multiplication. In Section 1.4 (Theorems 1.1 and 1.3) we have established that the properties in Definition 1 hold, thereby making M_{23} into a vector space. Similarly, the set of all $m \times n$ matrices under the usual operations of matrix addition and scalar multiplication is a vector space. This vector space will be denoted by M_{mn}. ■

EXAMPLE 5 ■ Let V be the set of all real-valued functions that are defined on the interval $[a, b]$. If f and g are in V, we define $f \oplus g$ by

$$(f \oplus g)(t) = f(t) + g(t).$$

If f is in V and c is a scalar, we define $c \odot f$ by

$$(c \odot f)(t) = cf(t).$$

Then V is a vector space (Exercise 9). Similarly, the set of all real-valued functions defined for all real numbers is a vector space. ■

Another source of examples of vector spaces will be sets of polynomials; therefore, we recall some well-known facts about such functions. A **polynomial** (in t) is a function that is expressible as

$$p(t) = a_n t^n + a_{n-1} t^{n-1} + \cdots + a_1 t + a_0, \tag{1}$$

where $a_0, a_1, \ldots, a_n$ are real numbers.

EXAMPLE 6 ■ The following functions are polynomials:

$$\begin{aligned}
p_1(t) &= 3t^4 - 2t^2 + 5t - 1 \\
p_2(t) &= 2t + 1 \\
p_3(t) &= 4.
\end{aligned}$$

The following functions are not polynomials (explain why):

$$f_4(t) = 2\sqrt{t} - 6 \quad \text{and} \quad f_5(t) = \frac{1}{t^2} - 2t + 1.$$ ■

The polynomial $p(t)$ in (1) is said to have **degree n** if $a_n \neq 0$. Thus the degree of a polynomial is the highest power having a nonzero coefficient.

EXAMPLE 7 ■ The polynomials defined in Example 6 have the following degrees:

$$p_1(t): \text{degree 4}$$
$$p_2(t): \text{degree 1}$$
$$p_3(t): \text{degree 0.}$$
■

The **zero polynomial** is defined as

$$0t^n + 0t^{n-1} + \cdots + 0t + 0.$$

Note that, by definition, the zero polynomial has no degree.

We now let P_n be the set of all polynomials of degree $\leq n$ together with the zero polynomial. Thus $2t^2 - 3t + 5$ is a member of P_2.

EXAMPLE 8 ■ If

$$p(t) = a_n t^n + a_{n-1} t^{n-1} + \cdots + a_1 t + a_0$$

and

$$q(t) = b_n t^n + b_{n-1} t^{n-1} + \cdots + b_1 t + b_0$$

we define $p(t) \oplus q(t)$ as

$$p(t) \oplus q(t) = (a_n + b_n) t^n + (a_{n-1} + b_{n-1}) t^{n-1} + \cdots + (a_1 + b_1) t + (a_0 + b_0),$$

(i.e., add coefficients of like-power terms). If c is a scalar, we also define $c \odot p(t)$ as

$$c \odot p(t) = (ca_n) t^n + (ca_{n-1}) t^{n-1} + \cdots + (ca_1) t + (ca_0)$$

(i.e., multiply each coefficient by c). We now show that P_n is a vector space.

Let $p(t)$ and $q(t)$ as above be elements of P_n; that is, they are polynomials of degree $\leq n$ or the zero polynomial. Then the definitions above of the operations $\oplus$ and $\odot$ show that $p(t) \oplus q(t)$ and $c \odot p(t)$, for any scalar c, are polynomials of degree $\leq n$ or the zero polynomial. That is, $p(t) \oplus q(t)$ and $c \odot p(t)$ are in P_n so that (α) and (β) in Definition 1 hold. To verify property (a), we observe that

$$q(t) \oplus p(t) = (b_n + a_n) t^n + (b_{n-1} + a_{n-1}) t^{n-1} + \cdots + (b_1 + a_1) t + (a_0 + b_0),$$

and since $a_i + b_i = b_i + a_i$ holds for the real numbers, we conclude that $p(t) \oplus q(t) = q(t) \oplus p(t)$. Similarly, we verify property (b). The zero polynomial is the element **0** needed in property (c). If $p(t)$ is as given above, then its negative, $-p(t)$, is

$$-a_n t^n - a_{n-1} t^{n-1} - \cdots - a_1 t - a_0.$$

We shall now verify property (f) and will leave the verification of the remaining properties to the reader. Thus

$$\begin{aligned}
(c+d) \odot p(t) &= (c+d) a_n t^n + (c+d) a_{n-1} t^{n-1} + \cdots + (c+d) a_1 t \\
&\quad + (c+d) a_0 \\
&= ca_n t^n + da_n t^n + ca_{n-1} t^{n-1} + da_{n-1} t^{n-1} + \cdots + ca_1 t \\
&\quad + da_1 t + ca_0 + da_0 \\
&= c(a_n t^n + a_{n-1} t^{n-1} + \cdots + a_1 t + a_0) \\
&\quad + d(a_n t^n + a_{n-1} t^{n-1} + \cdots + a_1 t + a_0) \\
&= c \odot p(t) \oplus d \odot p(t).
\end{aligned}$$
■

EXAMPLE 9 ■ Let V be the set of all real numbers with the operations $\mathbf{u} \oplus \mathbf{v} = \mathbf{u} - \mathbf{v}$ ($\oplus$ is ordinary subtraction) and $c \odot \mathbf{u} = c\mathbf{u}$ ($\odot$ is ordinary multiplication). Is V a vector space? If it is not, which properties in Definition 1 fail to hold?

Solution   If $\mathbf{u}$ and $\mathbf{v}$ are in V, and c is a scalar, then $\mathbf{u} \oplus \mathbf{v}$ and $c \odot \mathbf{u}$ are in V so that (α) and (β) in Definition 1 hold. However, property (a) fails to hold, as we can see by taking $\mathbf{u} = 2$ and $\mathbf{v} = 3$:

$$\mathbf{u} \oplus \mathbf{v} = 2 \oplus 3 = -1$$

and

$$\mathbf{v} \oplus \mathbf{u} = 3 \oplus 2 = 1.$$

Also, properties (b), (c) and (d) fail to hold (verify). Properties (e), (g), and (h) hold, but property (f) does not hold, as we can see by taking $c = 2$, $d = 3$, and $\mathbf{u} = 4$:

$$(c + d) \odot \mathbf{u} = (2 + 3) \odot 4 = 5 \odot 4 = 20$$

while

$$c \odot \mathbf{u} \oplus d \odot \mathbf{u} = 2 \odot 4 \oplus 3 \odot 4 = 8 \oplus 12 = -4.$$

Thus V is not a vector space.   ■

For each natural number n, we have just defined the vector space P_n of all polynomials of degree $\leq n$ together with the zero polynomial. We could also consider the space P of *all* polynomials (of any degree), together with the zero polynomial. Here P is the mathematical union of all the vector spaces P_n. Two polynomials $p(t)$ of degree n and $g(t)$ of degree m are added in P in the same way as they would be added in P_r, where r is the maximum of the two numbers m and n. Then P is a vector space (Exercise 10).

> To verify that a given set V with two operations $\oplus$ and $\odot$ is a real vector space, we must show that it satisfies all the properties of Definition 1. The first thing to check is whether (α) and (β) hold, for, if either of these fails, we do not have a vector space. If both (α) and (β) hold, it is recommended that (c), the existence of a zero element, be verified next. Naturally, if (c) fails to hold, we do not have a vector space and do not have to check the remaining properties.

We frequently refer to a real vector space simply as a **vector space**. We also write $\mathbf{u} \oplus \mathbf{v}$ simply as $\mathbf{u} + \mathbf{v}$ and $c \odot \mathbf{u}$ simply as $c\mathbf{u}$, being careful to keep the particular operation in mind.

Many other important examples of vector spaces occur in numerous areas of mathematics.

The advantage of Definition 1 is that it is not concerned with the question of what a vector is. For example, is a vector in R^3 a point, a directed line segment, or a 3×1 matrix? Definition 1 deals only with the algebraic behavior of the elements in a vector space. In the case of R^3, whichever point of view we take, the algebraic behavior is the same. The mathematician abstracts those features that all such objects have in common (i.e., those properties that make them all behave alike) and defines a new structure, called a real vector space. We can now talk about properties of all vector spaces without having to refer to any one vector space in particular. Thus a "vector" is now merely an element of a vector space, and it no longer needs to be associated with a directed line segment. The following theorem presents several useful properties common to all vector spaces.

THEOREM 4.1 ■ *If V is a vector space, then*:

(a) $0\mathbf{u} = \mathbf{0}$, *for every* $\mathbf{u}$ *in* V.

(b) $c\mathbf{0} = \mathbf{0}$, *for every scalar* c.

(c) *If* $c\mathbf{u} = \mathbf{0}$, *then* $c = 0$ *or* $\mathbf{u} = \mathbf{0}$.

(d) $(-1)\mathbf{u} = -\mathbf{u}$, *for every* $\mathbf{u}$ *in* V.

Proof   (a) We have

$$0\mathbf{u} = (0 + 0)\mathbf{u} = 0\mathbf{u} + 0\mathbf{u}, \tag{2}$$

by (f) of Definition 1. Adding $-0\mathbf{u}$ to both sides of (2), we obtain by (b), (c), and (d) of Definition 1,

$$\mathbf{0} = 0\mathbf{u} + (-0\mathbf{u}) = (0\mathbf{u} + 0\mathbf{u}) + (-0\mathbf{u})$$
$$= 0\mathbf{u} + [0\mathbf{u} + (-0\mathbf{u})] = 0\mathbf{u} + \mathbf{0} = 0\mathbf{u}.$$

(b) Exercise T.1.

(c) Suppose that $c\mathbf{u} = \mathbf{0}$ and $c \neq 0$. We have

$$\mathbf{u} = 1\mathbf{u} = \left(\frac{1}{c}c\right)\mathbf{u} = \frac{1}{c}(c\mathbf{u}) = \frac{1}{c}\mathbf{0} = \mathbf{0}$$

by (b) of this theorem and (g) and (h) of Definition 1.

(d) $(-1)\mathbf{u} + \mathbf{u} = (-1)\mathbf{u} + (1)\mathbf{u} = (-1 + 1)\mathbf{u} = 0\mathbf{u} = \mathbf{0}$. Since $-\mathbf{u}$ is unique, we conclude that $(-1)\mathbf{u} = -\mathbf{u}$.   ■

4.1 EXERCISES

In Exercises 1 through 4, determine whether the given set V is closed under the operations $\oplus$ and $\odot$.

1. V is the set of all ordered pairs of real numbers (x, y), where $x > 0$ and $y > 0$;

$$(x, y) \oplus (x', y') = (x + x', y + y')$$

and

$$c \odot (x, y) = (cx, cy).$$

2. V is the set of all ordered triples of real numbers of the form $(0, y, z)$;

$$(0, y, z) \oplus (0, y', z') = (0, y + y', z + z')$$

and

$$c \odot (0, y, z) = (0, 0, cz).$$

3. V is the set of all polynomials of the form $at^2 + bt + c$, where a, b, and c are real numbers with $b = a + 1$;

$$(a_1t^2 + b_1t + c_1) \oplus (a_2t^2 + b_2t + c_2)$$
$$= (a_1 + a_2)t^2 + (b_1 + b_2)t + (c_1 + c_2)$$

and

$$r \odot (at^2 + bt + c) = (ra)t^2 + (rb)t + rc.$$

4. V is the set of all 2×2 matrices

$$\begin{bmatrix} a & b \\ c & d \end{bmatrix},$$

where $a = d$; $\oplus$ is matrix addition and $\odot$ is scalar multiplication.

5. Verify in detail that R^2 is a vector space.

6. Verify in detail that R^3 is a vector space.

7. Verify that the set in Example 2 is a vector space.

8. Verify that all the properties of Definition 1, except property (f), hold for the set in Example 3.

9. Show that the set in Example 5 is a vector space.

10. Show that P is a vector space.

In Exercises 11 through 17, determine whether the given set together with the given operations is a vector space. If it is not a vector space, list the properties of Definition 1 that fail to hold.

11. The set of all ordered triples of real numbers (x, y, z) with the operations

$$(x, y, z) \oplus (x', y', z') = (x', y + y', z')$$

and

$$c \odot (x, y, z) = (cx, cy, cz).$$

12. The set of all ordered triples of real numbers (x, y, z) with the operations

$$(x, y, z) \oplus (x', y', z') = (x + x', y + y', z + z')$$

and

$$c \odot (x, y, z) = (x, 1, z).$$

13. The set of all ordered triples of real numbers of the form $(0, 0, z)$ with the operations

$$(0, 0, z) \oplus (0, 0, z') = (0, 0, z + z')$$

and

$$c \odot (0, 0, z) = (0, 0, cz).$$

14. The set of all real numbers with the usual operations of addition and multiplication.

15. The set of all ordered pairs of real numbers (x, y), where $x \leq 0$, with the usual operations in R^2.

16. The set of all ordered pairs of real numbers (x, y) with the operations
$(x, y) \oplus (x', y') = (x + x', y + y')$ and
$c \odot (x, y) = (0, 0)$.

17. The set of all positive real numbers $\mathbf{u}$ with the operations $\mathbf{u} \oplus \mathbf{v} = \mathbf{uv}$ and $c \odot \mathbf{u} = \mathbf{u}^c$.

18. Let V be the set of all real numbers; define $\oplus$ by $\mathbf{u} \oplus \mathbf{v} = 2\mathbf{u} - \mathbf{v}$ and $\odot$ by $c \odot \mathbf{u} = c\mathbf{u}$. Is V a vector space?

19. Let V be the set consisting of a single element $\mathbf{0}$. Let $\mathbf{0} \oplus \mathbf{0} = \mathbf{0}$ and $c \odot \mathbf{0} = \mathbf{0}$. Show that V is a vector space.

20. (a) If V is a vector space that has a nonzero vector, how many vectors are in V?

    (b) Describe all vector spaces having a finite number of vectors.

THEORETICAL EXERCISES ▪

In Exercises T.1 through T.4, establish the indicated result for a real vector space V.

T.1. Show that $c\mathbf{0} = \mathbf{0}$ for every scalar c.

T.2. Show that $-(-\mathbf{u}) = \mathbf{u}$.

T.3. Show that if $\mathbf{u} + \mathbf{v} = \mathbf{u} + \mathbf{w}$, then $\mathbf{v} = \mathbf{w}$.

T.4. Show that if $\mathbf{u} \neq \mathbf{0}$ and $a\mathbf{u} = b\mathbf{u}$, then $a = b$.

T.5. Show that a vector space has only one zero vector.

T.6. Show that a vector $\mathbf{u}$ in a vector space has only one negative $-\mathbf{u}$.

MATLAB EXERCISES ▪

The concepts discussed in this section are not easily implemented in MATLAB routines. The items in Definition 1 must hold for *all* vectors. Just because we demonstrate in MATLAB that a property of Definition 1 holds for a few vectors it does not imply that it holds for all such vectors. You must guard against such faulty reasoning. However, if, for a particular choice of vectors, we show that a property fails in MATLAB, then we have established that the property does not hold in all possible cases. Hence the property is considered to be false. In this way we might be able to show that a set is not a vector space.

ML.1. Let V be the set of all 2×2 matrices with operations given by the following MATLAB commands:

$$A \oplus B \quad \text{is} \quad A.*B$$
$$k \odot A \quad \text{is} \quad k + A$$

Is V a vector space? (*Hint*: Enter some 2×2 matrices and experiment with the MATLAB commands to understand their behavior before checking the conditions in Definition 1.)

ML.2. Following Example 8, we discuss the vector space P_n of polynomials of degree n or less.

Operations on polynomials can be performed in linear algebra software by associating a row matrix of size $n + 1$ with polynomial $p(t)$ of P_n. The row matrix consists of the coefficients of $p(t)$ using the association

$$p(t) = a_n t^n + a_{n-1} t^{n-1} + \cdots + a_1 t + a_0$$
$$\rightarrow \begin{bmatrix} a_n & a_{n-1} & \cdots & a_1 & a_0 \end{bmatrix}.$$

If any term of $p(t)$ is explicitly missing, a zero is used for its coefficient. Then the addition of polynomials corresponds to matrix addition and multiplication of a polynomial by a scalar corresponds to scalar multiplication of matrices. Use MATLAB to perform the following operations on polynomials, using the matrix association described above. Let $n = 3$ and

$$p(t) = 2t^3 + 5t^2 + t - 2$$
$$q(t) = t^3 + 3t + 5.$$

(a) $p(t) + q(t)$.   (b) $5p(t)$.

(c) $3p(t) - 4q(t)$.

4.2 ▼ Subspaces

In this section we begin to analyze the structure of a vector space. First, it is convenient to have a name for a subset of a given vector space that is itself a vector space with respect to the same operations as those in V. Thus we have the following.

DEFINITION   Let V be a vector space and W a nonempty subset of V. If W is a vector space with respect to the operations in V, then W is called a **subspace** of V.

EXAMPLE 1 ▪ Every vector space has at least two subspaces, itself and the subspace $\{\mathbf{0}\}$ consisting only of the zero vector [recall that $\mathbf{0} \oplus \mathbf{0} = \mathbf{0}$ and $c \odot \mathbf{0} = \mathbf{0}$ in any vector space (see Exercise 19 in Section 4.1)]. The subspace $\{\mathbf{0}\}$ is called the **zero subspace**. ▪

EXAMPLE 2 ▪ Let W be the subset of R^3 consisting of all vectors of the form $(a, b, 0)$, where a and b are any real numbers. To check if W is a subspace of R^3, we first see whether properties (α) and (β) of Definition 1 hold. Thus let $\mathbf{u} = (a_1, b_1, 0)$ and $\mathbf{v} = (a_2, b_2, 0)$ be vectors in W. Then $\mathbf{u} + \mathbf{v} = (a_1, b_1, 0) + (a_2, b_2, 0) = (a_1 + a_2, b_1 + b_2, 0)$ is in W, since the third component is zero. Also, if c is a scalar, then

$c\mathbf{u} = c(a_1, b_1, 0) = (ca_1, cb_1, 0)$ is in W. Thus properties (α) and (β) of Definition 1 hold. We can easily verify that properties (a) – (h) hold. Hence W is a subspace of R^3. ∎

Before listing other subspaces we pause to develop a labor-saving result. We just noted that to verify that a nonempty subset W of a vector space V is a subspace, we must check that (α), (β), and (a) – (h) of Definition 1 hold. However, the following theorem says that it is enough to merely check that (α) and (β) hold. Property (α) is called the **closure** property for $\oplus$, and (β) is called the **closure** property for $\odot$.

THEOREM 4.2 ■ *Let V be a vector space with operations $\oplus$ and $\odot$ and let W be a nonempty subset of V. Then W is a subspace of V if and only if the following conditions hold:*

(α) *If $\mathbf{u}$ and $\mathbf{v}$ are any vectors in W, then $\mathbf{u} \oplus \mathbf{v}$ is in W.*
(β) *If c is any real number and $\mathbf{u}$ is any vector in W, then $c \odot \mathbf{u}$ is in W.*

Proof Exercise T.1. ∎

EXAMPLE 3 ■ Consider the set W consisting of all 2×3 matrices of the form

$$\begin{bmatrix} a & b & 0 \\ 0 & c & d \end{bmatrix},$$

where a, b, c, and d are arbitrary real numbers. Then W is a subset of the vector space M_{23} defined in Example 4 of Section 4.1. Show that W is a subspace of M_{23}. Note that a 2×3 matrix is in W provided its $(1, 3)$ and $(2, 1)$ entries are zero.

Solution Consider

$$\mathbf{u} = \begin{bmatrix} a_1 & b_1 & 0 \\ 0 & c_1 & d_1 \end{bmatrix} \quad \text{and} \quad \mathbf{v} = \begin{bmatrix} a_2 & b_2 & 0 \\ 0 & c_2 & d_2 \end{bmatrix} \quad \text{in } W.$$

Then

$$\mathbf{u} + \mathbf{v} = \begin{bmatrix} a_1 + a_2 & b_1 + b_2 & 0 \\ 0 & c_1 + c_2 & d_1 + d_2 \end{bmatrix} \quad \text{is in } W$$

so that (α) of Theorem 4.2 is satisfied. Also, if k is a scalar, then

$$k\mathbf{u} = \begin{bmatrix} ka_1 & kb_1 & 0 \\ 0 & kc_1 & kd_1 \end{bmatrix} \quad \text{is in } W$$

so that (β) of Theorem 4.2 is satisfied. Hence W is a subspace of M_{23}. ∎

We can also show that a nonempty subset W of a vector space V is a subspace of V if and only if $a\mathbf{u} + b\mathbf{v}$ is in W for any vectors $\mathbf{u}$ and $\mathbf{v}$ in W and any scalars a and b (Exercise T.2).

EXAMPLE 4 ■ Let W be the subset of R^3 consisting of all vectors of the form $(a, b, 1)$, where a and b are any real numbers. To check whether properties (α) and (β) of

Theorem 4.2 hold, we let $\mathbf{u} = (a_1, b_1, 1)$ and $\mathbf{v} = (a_2, b_2, 1)$ be vectors in W. Then $\mathbf{u} + \mathbf{v} = (a_1, b_1, 1) + (a_2, b_2, 1) = (a_1 + a_2, b_1 + b_2, 2)$, which is not in W, since the third component is 2 and not 1. Since (α) of Theorem 4.2 does not hold, W is not a subspace of R^3. ■

EXAMPLE 5 ■ In Section 4.1, we let P_n denote the vector space consisting of all polynomials of degree $\leq n$ and the zero polynomial, and P the vector space of all polynomials. It is easy to verify that P_2 is a subspace of P_3 and, in general, that P_n is a subspace of P_{n+1} (Exercise 7). Also, P_n is a subspace of P (Exercise 8). ■

EXAMPLE 6 ■ Let V be the set of all polynomials of degree exactly $= 2$; V is a *subset* of P_2, but it is not a *subspace* of P_2, since the sum of the polynomials $2t^2 + 3t + 1$ and $-2t^2 + t + 2$, a polynomial of degree 1, is not in V. ■

EXAMPLE 7 ■ Let $C[a, b]$ denote the set of all real-valued continuous functions that are defined
(*Calculus Required*) on the interval $[a, b]$. If f and g are in $C[a, b]$, then $f + g$ is in $C[a, b]$, since the sum of two continuous functions is continuous. Similarly, if c is a scalar, then cf is in $C[a, b]$. Hence $C[a, b]$ is a subspace of the vector space of all real-valued functions that are defined on $[a, b]$, which was introduced in Example 5 of Section 4.1. If the functions are defined for all real numbers, the vector space is denoted by $C(-\infty, \infty)$. ■

We now come to a very important example of a subspace.

EXAMPLE 8 ■ Consider the homogeneous system $A\mathbf{x} = \mathbf{0}$, where A is an $m \times n$ matrix. A solution consists of a vector $\mathbf{x}$ in R^n. Let W be the subset of R^n consisting of all solutions to the homogeneous system. Since $A\mathbf{0} = \mathbf{0}$, we conclude that W is not empty. To check that W is a subspace of R^n, we verify properties (α) and (β) of Theorem 4.2. Thus let $\mathbf{x}$ and $\mathbf{y}$ be solutions. Then

$$A\mathbf{x} = \mathbf{0} \quad \text{and} \quad A\mathbf{y} = \mathbf{0}.$$

Now

$$A(\mathbf{x} + \mathbf{y}) = A\mathbf{x} + A\mathbf{y} = \mathbf{0} + \mathbf{0} = \mathbf{0},$$

so $\mathbf{x} + \mathbf{y}$ is a solution. Also, if c is a scalar, then

$$A(c\mathbf{x}) = c(A\mathbf{x}) = c\mathbf{0} = \mathbf{0},$$

so $c\mathbf{x}$ is also a solution. Hence W is a subspace of R^n, called the **solution space** of the homogeneous system, or the **null space** of A. ■

It should be noted that the set of all solutions to the linear system $A\mathbf{x} = \mathbf{b}$, where A is $m \times n$, is not a subspace of R^n if $\mathbf{b} \neq \mathbf{0}$ (Exercise T.3).

EXAMPLE 9 ■ A simple way of constructing subspaces in a vector space is as follows. Let $\mathbf{v}_1$ and $\mathbf{v}_2$ be fixed vectors in a vector space V and let W be the set of all linear combinations (see Section 1.3) of $\mathbf{v}_1$ and $\mathbf{v}_2$, that is, W consists of all vectors in

V of the form $a_1\mathbf{v}_1 + a_2\mathbf{v}_2$, where a_1 and a_2 are any real numbers. To show that W is a subspace of V, we verify properties (α) and (β) of Theorem 4.2. Thus let

$$\mathbf{w}_1 = a_1\mathbf{v}_1 + a_2\mathbf{v}_2 \quad \text{and} \quad \mathbf{w}_2 = b_1\mathbf{v}_1 + b_2\mathbf{v}_2$$

be vectors in W. Then

$$\mathbf{w}_1 + \mathbf{w}_2 = (a_1\mathbf{v}_1 + a_2\mathbf{v}_2) + (b_1\mathbf{v}_1 + b_2\mathbf{v}_2) = (a_1 + b_1)\mathbf{v}_1 + (a_2 + b_2)\mathbf{v}_2,$$

which is in W. Also, if c is a scalar, then

$$c\mathbf{w}_1 = (ca_1)\mathbf{v}_1 + (ca_2)\mathbf{v}_2$$

is in W. Hence W is a subspace of V. ∎

The construction carried out in Example 9 for two vectors can easily be performed for more than two vectors. We now give a formal definition.

DEFINITION Let $\mathbf{v}_1, \mathbf{v}_2, \ldots, \mathbf{v}_k$ be vectors in a vector space V. A vector $\mathbf{v}$ in V is called a **linear combination** of $\mathbf{v}_1, \mathbf{v}_2, \ldots, \mathbf{v}_k$ if

$$\mathbf{v} = c_1\mathbf{v}_1 + c_2\mathbf{v}_2 + \cdots + c_k\mathbf{v}_k$$

for some real numbers $c_1, c_2, \ldots, c_k$. (See also Section 1.3.)

In Figure 4.1 we show the vector $\mathbf{v}$ in R^2 or R^3 as a linear combination of the vectors $\mathbf{v}_1$ and $\mathbf{v}_2$.

FIGURE 4.1
Linear combination
of two vectors

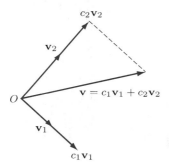

EXAMPLE 10 ∎ In R^3 let
$$\mathbf{v}_1 = (1, 2, 1), \quad \mathbf{v}_2 = (1, 0, 2), \quad \text{and} \quad \mathbf{v}_3 = (1, 1, 0).$$

The vector
$$\mathbf{v} = (2, 1, 5)$$

is a linear combination of $\mathbf{v}_1$, $\mathbf{v}_2$, and $\mathbf{v}_3$ if we can find real numbers c_1, c_2, and c_3 so that

$$c_1\mathbf{v}_1 + c_2\mathbf{v}_2 + c_3\mathbf{v}_3 = \mathbf{v}.$$

Substituting for $\mathbf{v}$, $\mathbf{v}_1$, $\mathbf{v}_2$, and $\mathbf{v}_3$, we have

$$c_1(1, 2, 1) + c_2(1, 0, 2) + c_3(1, 1, 0) = (2, 1, 5).$$

Combining terms on the left and equating corresponding entries leads to the linear system (verify)

$$\begin{aligned} c_1 + c_2 + c_3 &= 2 \\ 2c_1 + c_3 &= 1 \\ c_1 + 2c_2 &= 5. \end{aligned}$$

Solving this linear system by the methods of Chapter 1 gives (verify) $c_1 = 1$, $c_2 = 2$, and $c_3 = -1$, which means that $\mathbf{v}$ is a linear combination of $\mathbf{v}_1$, $\mathbf{v}_2$, and $\mathbf{v}_3$. Thus

$$\mathbf{v} = \mathbf{v}_1 + 2\mathbf{v}_2 - \mathbf{v}_3. \qquad \blacksquare$$

DEFINITION   If $S = \{\mathbf{v}_1, \mathbf{v}_2, \ldots, \mathbf{v}_k\}$ is a set of vectors in a vector space V, then the set of all vectors in V that are linear combinations of the vectors in S is denoted by

$$\operatorname{span} S \quad \text{or} \quad \operatorname{span} \{\mathbf{v}_1, \mathbf{v}_2, \ldots, \mathbf{v}_k\}.$$

In Figure 4.2 we show a portion of span $\{\mathbf{v}_1, \mathbf{v}_2\}$, where $\mathbf{v}_1$ and $\mathbf{v}_2$ are noncollinear vectors in R^3; span $\{\mathbf{v}_1, \mathbf{v}_2\}$ is a plane that passes through the origin and contains the vectors $\mathbf{v}_1$ and $\mathbf{v}_2$.

FIGURE 4.2

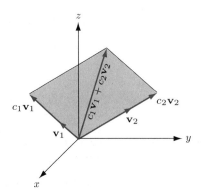

EXAMPLE 11 ■ Consider the set S of 2×3 matrices given by

$$S = \left\{ \begin{bmatrix} 1 & 0 & 0 \\ 0 & 0 & 0 \end{bmatrix}, \begin{bmatrix} 0 & 1 & 0 \\ 0 & 0 & 0 \end{bmatrix}, \begin{bmatrix} 0 & 0 & 0 \\ 0 & 1 & 0 \end{bmatrix}, \begin{bmatrix} 0 & 0 & 0 \\ 0 & 0 & 1 \end{bmatrix} \right\}.$$

Then span S is the set in M_{23} consisting of all vectors of the form

$$a \begin{bmatrix} 1 & 0 & 0 \\ 0 & 0 & 0 \end{bmatrix} + b \begin{bmatrix} 0 & 1 & 0 \\ 0 & 0 & 0 \end{bmatrix} + c \begin{bmatrix} 0 & 0 & 0 \\ 0 & 1 & 0 \end{bmatrix} + d \begin{bmatrix} 0 & 0 & 0 \\ 0 & 0 & 1 \end{bmatrix}$$

$$= \begin{bmatrix} a & b & 0 \\ 0 & c & d \end{bmatrix}, \quad \text{where } a, b, c, \text{ and } d \text{ are real numbers.}$$

That is, span S is the subset of M_{23} consisting of all matrices of the form

$$\begin{bmatrix} a & b & 0 \\ 0 & c & d \end{bmatrix},$$

where a, b, c, and d are real numbers. ■

THEOREM 4.3 ■ *Let $S = \{\mathbf{v}_1, \mathbf{v}_2, \ldots, \mathbf{v}_k\}$ be a set of vectors in a vector space V. Then span S is a subspace of V.*

Proof   See Exercise T.4. ■

EXAMPLE 12 ■ In P_2 let

$$\mathbf{v}_1 = 2t^2 + t + 2, \quad \mathbf{v}_2 = t^2 - 2t, \quad \mathbf{v}_3 = 5t^2 - 5t + 2, \quad \mathbf{v}_4 = -t^2 - 3t - 2.$$

Determine if the vector

$$\mathbf{u} = t^2 + t + 2$$

belongs to span $\{\mathbf{v}_1, \mathbf{v}_2, \mathbf{v}_3, \mathbf{v}_4\}$.

Solution   If we can find scalars c_1, c_2, c_3, and c_4 so that

$$c_1\mathbf{v}_1 + c_2\mathbf{v}_2 + c_3\mathbf{v}_3 + c_4\mathbf{v}_4 = \mathbf{u},$$

then $\mathbf{u}$ belongs to span $\{\mathbf{v}_1, \mathbf{v}_2, \mathbf{v}_3, \mathbf{v}_4\}$. Substituting for $\mathbf{u}$, $\mathbf{v}_1$, $\mathbf{v}_2$, $\mathbf{v}_3$, and $\mathbf{v}_4$, we have

$$c_1(2t^2 + t + 2) + c_2(t^2 - 2t) + c_3(5t^2 - 5t + 2) + c_4(-t^2 - 3t - 2)$$
$$= t^2 + t + 2$$

or

$$(2c_1 + c_2 + 5c_3 - c_4)t^2 + (c_1 - 2c_2 - 5c_3 - 3c_4)t + (2c_1 + 2c_3 - 2c_4)$$
$$= t^2 + t + 2.$$

Now two polynomials agree for all values of t only if the coefficients of respective powers of t agree. Thus we get the linear system

$$
\begin{aligned}
2c_1 + c_2 + 5c_3 - c_4 &= 1 \\
c_1 - 2c_2 - 5c_3 - 3c_4 &= 1 \\
2c_1 \phantom{{}- 2c_2} + 2c_3 - 2c_4 &= 2.
\end{aligned}
$$

To investigate whether or not this system of linear equations is consistent, we form the augmented matrix and transform it to reduced row echelon form, obtaining (verify)

$$
\begin{bmatrix}
1 & 0 & 1 & -1 & \vdots & 0 \\
0 & 1 & 3 & 1 & \vdots & 0 \\
0 & 0 & 0 & 0 & \vdots & 1
\end{bmatrix},
$$

which indicates that the system is inconsistent, that is, it has no solution. Hence $\mathbf{u}$ does not belong to span $\{\mathbf{v}_1, \mathbf{v}_2, \mathbf{v}_3, \mathbf{v}_4\}$. ■

REMARK   In general, to determine if a specific vector $\mathbf{v}$ belongs to span S, we investigate the consistency of an appropriate linear system.

4.2   EXERCISES

1. Which of the following subsets of R^3 are subspaces of R^3? The set of all vectors of the form
   (a) $(a, b, 2)$.
   (b) (a, b, c), where $c = a + b$.
   (c) (a, b, c), where $c > 0$.

2. Which of the following subsets of R^3 are subspaces of R^3? The set of all vectors of the form
   (a) (a, b, c), where $a = c = 0$.
   (b) (a, b, c), where $a = -c$.
   (c) (a, b, c), where $b = 2a + 1$.

3. Which of the following subsets of R^4 are subspaces of R^4? The set of all vectors of the form
   (a) (a, b, c, d), where $a - b = 2$.
   (b) (a, b, c, d), where $c = a + 2b$ and $d = a - 3b$.
   (c) (a, b, c, d), where $a = 0$ and $b = -d$.

4. Which of the following subsets of R^4 are subspaces of R^4? The set of all vectors of the form
   (a) (a, b, c, d), where $a = b = 0$.
   (b) (a, b, c, d), where $a = 1$, $b = 0$, and $a + d = 1$.
   (c) (a, b, c, d), where $a > 0$ and $b < 0$.

5. Which of the following subsets of P_2 are subspaces? The set of all polynomials of the form
   (a) $a_2 t^2 + a_1 t + a_0$, where $a_0 = 0$.
   (b) $a_2 t^2 + a_1 t + a_0$, where $a_0 = 2$.
   (c) $a_2 t^2 + a_1 t + a_0$, where $a_2 + a_1 = a_0$.

6. Which of the following subsets of P_2 are subspaces? The set of all polynomials of the form
   (a) $a_2 t^2 + a_1 t + a_0$, where $a_1 = 0$ and $a_0 = 0$.
   (b) $a_2 t^2 + a_1 t + a_0$, where $a_1 = 2a_0$.
   (c) $a_2 t^2 + a_1 t + a_0$, where $a_2 + a_1 + a_0 = 2$.

7. (a) Show that P_2 is a subspace of P_3.
   (b) Show that P_n is a subspace of P_{n+1}.

8. Show that P_n is a subspace of P.

9. Show that P is a subspace of the vector space defined in Example 5 of Section 4.1.

10. Let $\mathbf{u} = (1, 2, -3)$ and $\mathbf{v} = (-2, 3, 0)$ be two vectors in R^3 and let W be the subset of R^3 consisting of all vectors of the form $a\mathbf{u} + b\mathbf{v}$, where a and b are any real numbers. Give an argument to show that W is a subspace of R^3.

11. Let $\mathbf{u} = (2, 0, 3, -4)$ and $\mathbf{v} = (4, 2, -5, 1)$ be two vectors in R^4 and let W be the subset of R^4 consisting of all vectors of the form $a\mathbf{u} + b\mathbf{v}$, where a and b are any real numbers. Give an argument to show that W is a subspace of R^4.

12. Which of the following subsets of the vector space M_{23} defined in Example 4 of Section 4.1 are subspaces? The set of all matrices of the form
   (a) $\begin{bmatrix} a & b & c \\ d & 0 & 0 \end{bmatrix}$, where $b = a + c$.
   (b) $\begin{bmatrix} a & b & c \\ d & 0 & 0 \end{bmatrix}$, where $c > 0$.
   (c) $\begin{bmatrix} a & b & c \\ d & e & f \end{bmatrix}$, where $a = -2c$ and $f = 2e + d$.

13. Which of the following subsets of the vector space M_{23} defined in Example 4 of Section 4.1 are subspaces? The set of all matrices of the form
   (a) $\begin{bmatrix} a & b & c \\ d & e & f \end{bmatrix}$, where $a = 2c + 1$.
   (b) $\begin{bmatrix} 0 & 1 & a \\ b & c & 0 \end{bmatrix}$.
   (c) $\begin{bmatrix} a & b & c \\ d & e & f \end{bmatrix}$, where $a + c = 0$ and $b + d + f = 0$.

14. Which of the following subsets of the vector space M_{nn} are subspaces?
   (a) The set of all $n \times n$ symmetric matrices.
   (b) The set of all $n \times n$ nonsingular matrices.
   (c) The set of all $n \times n$ diagonal matrices.

15. Which of the following subsets of the vector space M_{nn} are subspaces?
   (a) The set of all $n \times n$ singular matrices.
   (b) The set of all $n \times n$ upper triangular matrices.
   (c) The set of all $n \times n$ matrices whose determinant is 1.

16. (**Calculus Required**). Which of the following subsets are subspaces of the vector space $C(-\infty, \infty)$ defined in Example 7?
   (a) All nonnegative functions.
   (b) All constant functions.
   (c) All functions f such that $f(0) = 0$.
   (d) All functions f such that $f(0) = 5$.
   (e) All differentiable functions.

17. (**Calculus Required**). Which of the following subsets of the vector space $C(-\infty, \infty)$ defined in Example 7 are subspaces?

(a) All integrable functions.

(b) All bounded functions

(c) All functions that are integrable on $[a, b]$.

(d) All functions that are bounded on $[a, b]$.

18. (**Calculus Required**). Consider the differential equation

$$y'' - y' + 2y = 0.$$

A solution is a real-valued function f satisfying the equation. Let V be the set of all solutions to the given differential equation; define $\oplus$ and $\odot$ as in Example 5 in Section 4.1. Show that V is a subspace of the vector space of all real-valued functions defined on $(-\infty, \infty)$. (See also Section 8.11.)

19. Determine which of the following subsets of R^2 are subspaces.

(a)

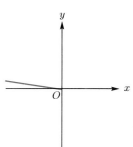

(b)

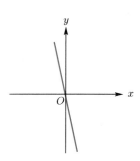

20. Determine which of the following subsets of R^2 are subspaces.

(a)

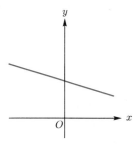

(b)

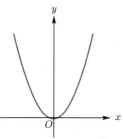

21. In each part, determine whether the given vector $\mathbf{v}$ belongs to span $\{\mathbf{v_1}, \mathbf{v_2}, \mathbf{v_3}\}$, where

$$\mathbf{v_1} = (1, 0, 0, 1), \qquad \mathbf{v_2} = (1, -1, 0, 0),$$

and

$$\mathbf{v_3} = (0, 1, 2, 1).$$

(a) $\mathbf{v} = (-1, 4, 2, 2)$.   (b) $\mathbf{v} = (1, 2, 0, 1)$.

(c) $\mathbf{v} = (-1, 1, 4, 3)$.   (d) $\mathbf{v} = (0, 1, 1, 0)$.

22. Which of the following vectors are linear combinations of

$$A_1 = \begin{bmatrix} 1 & -1 \\ 0 & 3 \end{bmatrix}, \qquad A_2 = \begin{bmatrix} 1 & 1 \\ 0 & 2 \end{bmatrix},$$

$$A_3 = \begin{bmatrix} 2 & 2 \\ -1 & 1 \end{bmatrix}?$$

(a) $\begin{bmatrix} 5 & 1 \\ -1 & 9 \end{bmatrix}$.   (b) $\begin{bmatrix} -3 & -1 \\ 3 & 2 \end{bmatrix}$.

(c) $\begin{bmatrix} 3 & -2 \\ 3 & 2 \end{bmatrix}$.   (d) $\begin{bmatrix} 1 & 0 \\ 2 & 1 \end{bmatrix}$.

23. In each part, determine whether the given vector $p(t)$ belongs to span $\{p_1(t), p_2(t), p_3(t)\}$, where $p_1(t) = t^2 - t$, $p_2(t) = t^2 - 2t + 1$, $p_3(t) = -t^2 + 1$.

(a) $p(t) = 3t^2 - 3t + 1$.   (b) $p(t) = t^2 - t + 1$.

(c) $p(t) = t + 1$.   (d) $p(t) = 2t^2 - t - 1$.

THEORETICAL EXERCISES ▩

T.1. Prove Theorem 4.2.

T.2. Show that a subset W of a vector space V is a subspace of V if and only if the following condition holds: If $\mathbf{u}$ and $\mathbf{v}$ are any vectors in W and a and b are any scalars, then $a\mathbf{u} + b\mathbf{v}$ is in W.

T.3. Show that the set of all solutions to $A\mathbf{x} = \mathbf{b}$, where A is $m \times n$, is not a subspace of R^n if $\mathbf{b} \neq \mathbf{0}$.

T.4. Prove Theorem 4.3.

T.5. Let $S = \{\mathbf{v}_1, \mathbf{v}_2, \ldots, \mathbf{v}_k\}$ be a set of vectors in a vector space V, and let W be a subspace of V containing S. Show that W contains span S.

T.6. If A is a nonsingular matrix, what is the null space of A? Justify your answer.

T.7. Let $\mathbf{x}_0$ be a fixed vector in a vector space V. Show that the set W consisting of all scalar multiples $c\mathbf{x}_0$ of $\mathbf{x}_0$ is a subspace of V.

T.8. Let A be an $m \times n$ matrix. Is the set W of all vectors $\mathbf{x}$ in R^n such that $A\mathbf{x} \neq \mathbf{0}$ a subspace of R^n? Justify your answer.

T.9. Show that the only subspaces of R^1 are $\{\mathbf{0}\}$ and R^1 itself.

T.10. Let W_1 and W_2 be subspaces of a vector space V. Let $W_1 + W_2$ be the set of all vectors $\mathbf{v}$ in V such that $\mathbf{v} = \mathbf{w}_1 + \mathbf{w}_2$, where $\mathbf{w}_1$ is in W_1 and $\mathbf{w}_2$ is in W_2. Show that $W_1 + W_2$ is a subspace of V.

T.11. Let W_1 and W_2 be subspaces of a vector space V with $W_1 \cap W_2 = \{\mathbf{0}\}$. Let $W_1 + W_2$ be as defined in Exercise T.10. Suppose that $V = W_1 + W_2$. Show that every vector in V can be uniquely written as $\mathbf{w}_1 + \mathbf{w}_2$, where $\mathbf{w}_1$ is in W_1 and $\mathbf{w}_2$ is in W_2. In this case we write $V = W_1 \oplus W_2$ and say that V is the **direct sum** of the subspaces W_1 and W_2.

T.12. (a) Show that the set of all points in the plane $ax + by + cz = 0$ is a subspace of R^3.

(b) Find a basis for the plane $2x - 3y + 4z = 0$.

MATLAB EXERCISES ▩

ML.1. Let V be R^3 and let W be the subset of V of vectors of the form $(2, a, b)$, where a and b are any real numbers. Is W a subspace of V? Use the following MATLAB commands to help you determine the answer.

$$\mathbf{a1} = \mathbf{fix}(10 * \mathbf{randn});$$
$$\mathbf{a2} = \mathbf{fix}(10 * \mathbf{randn});$$
$$\mathbf{b1} = \mathbf{fix}(10 * \mathbf{randn});$$
$$\mathbf{b2} = \mathbf{fix}(10 * \mathbf{randn});$$
$$\mathbf{v} = [\mathbf{2\ a1\ b1}]$$
$$\mathbf{w} = [\mathbf{2\ a2\ b2}]$$
$$\mathbf{v} + \mathbf{w}$$
$$\mathbf{3 * v}$$

ML.2. Let V be P_2 and let W be the subset of V of vectors of the form $ax^2 + bx + 5$, where a and b are arbitrary real numbers. With each such polynomial in W we associate a vector $(a, b, 5)$ in R^3. Construct commands like those in Exercise ML.1 to show that W is not a subspace of V.

Before solving the following MATLAB exercises, you should have read Section 10.7.

ML.3. Use MATLAB to determine if vector $\mathbf{v}$ is a linear combination of the members of set S.

(a) $S = \{\mathbf{v}_1, \mathbf{v}_2, \mathbf{v}_3\}$
$= \{(1, 0, 0, 1), (0, 1, 1, 0), (1, 1, 1, 1)\}$
$\mathbf{v} = (0, 1, 1, 1)$.

(b) $S = \{\mathbf{v}_1, \mathbf{v}_2, \mathbf{v}_3\}$

$$= \left\{ \begin{bmatrix} 1 \\ 2 \\ -1 \end{bmatrix}, \begin{bmatrix} 2 \\ -1 \\ 0 \end{bmatrix}, \begin{bmatrix} -1 \\ 8 \\ -3 \end{bmatrix} \right\}$$

$$\mathbf{v} = \begin{bmatrix} 0 \\ 5 \\ -2 \end{bmatrix}.$$

ML.4. Use MATLAB to determine if $\mathbf{v}$ is a linear combination of the members of set S. If it is, express $\mathbf{v}$ in terms of the members of S.

(a) $S = \{\mathbf{v}_1, \mathbf{v}_2, \mathbf{v}_3\}$
$= \{(1, 2, 1), (3, 0, 1), (1, 8, 3)\}$
$\mathbf{v} = (-2, 14, 4)$.

(b) $S = \{A_1, A_2, A_3\}$

$$= \left\{ \begin{bmatrix} 1 & 2 \\ 1 & 0 \end{bmatrix}, \begin{bmatrix} 2 & -1 \\ 1 & 2 \end{bmatrix}, \begin{bmatrix} -3 & 1 \\ 0 & 1 \end{bmatrix} \right\}$$

$\mathbf{v} = I_2$.

ML.5. Use MATLAB to determine if $\mathbf{v}$ is a linear combination of the members of set S. If it is, express $\mathbf{v}$ in terms of the members of S.

(a) $S = \{\mathbf{v}_1, \mathbf{v}_2, \mathbf{v}_3, \mathbf{v}_4\}$

$$= \left\{ \begin{bmatrix} 1 \\ 2 \\ 1 \\ 0 \\ 1 \end{bmatrix}, \begin{bmatrix} 0 \\ 1 \\ 2 \\ -1 \\ 1 \end{bmatrix}, \begin{bmatrix} 2 \\ 1 \\ 0 \\ 0 \\ -1 \end{bmatrix}, \begin{bmatrix} -2 \\ 1 \\ 1 \\ 1 \\ 1 \end{bmatrix} \right\}$$

$$\mathbf{v} = \begin{bmatrix} 0 \\ -1 \\ 1 \\ -2 \\ 1 \end{bmatrix}.$$

(b) $S = \{p_1(t), p_2(t), p_3(t)\}$

$\qquad = \{2t^2 - t + 1, t^2 - 2, t - 1\}$

$\qquad \mathbf{v} = p(t) = 4t^2 + t - 5.$

ML.6. In each part, determine whether $\mathbf{v}$ belongs to span S, where

$\qquad S = \{\mathbf{v}_1, \mathbf{v}_2, \mathbf{v}_3\}$

$\qquad = \{(1, 1, 0, 1), (1, -1, 0, 1), (0, 1, 2, 1)\}.$

(a) $\mathbf{v} = (2, 3, 2, 3).$

(b) $\mathbf{v} = (2, -3, -2, 3).$

(c) $\mathbf{v} = (0, 1, 2, 3).$

ML.7. In each part, determine whether $p(t)$ belongs to span S, where

$\qquad S = \{p_1(t), p_2(t), p_3(t)\}$

$\qquad = \{t - 1, t + 1, t^2 + t + 1\}.$

(a) $p(t) = t^2 + 2t + 4.$

(b) $p(t) = 2t^2 + t - 2.$

(c) $p(t) = -2t^2 + 1.$

4.3 ▾ Linear Independence

So far we have defined a mathematical system called a real vector space and noted some of its properties. We further observe that the only real vector space having a finite number of vectors in it is the vector space whose only vector is $\mathbf{0}$, for if $\mathbf{v} \neq \mathbf{0}$ is in a vector space V, then $c\mathbf{v}$ is in V, where c is any real number, and so we can show that V has infinitely many vectors in it. However, in this section and the following one we show that each vector space V studied here has a set composed of a finite number of vectors that completely describe V. It should be noted that, in general, there is more than one such set describing V. We now turn to a formulation of these ideas.

DEFINITION   The vectors $\mathbf{v}_1, \mathbf{v}_2, \ldots, \mathbf{v}_k$ in a vector space V are said to **span** V if every vector in V is a linear combination of $\mathbf{v}_1, \mathbf{v}_2, \ldots, \mathbf{v}_k$. Moreover, if these vectors are distinct and we denote them as a set $S = \{\mathbf{v}_1, \mathbf{v}_2, \ldots, \mathbf{v}_k\}$, then we also say that the set S **spans** V, or that $\{\mathbf{v}_1, \mathbf{v}_2, \ldots, \mathbf{v}_k\}$ **spans** V, or that V is **spanned by** S, or in the language of Section 4.2, span $S = V$.

The procedure to check if the vectors $\mathbf{v}_1, \mathbf{v}_2, \ldots, \mathbf{v}_k$ span the vector space V is as follows.

Step 1. Choose an arbitrary vector $\mathbf{v}$ in V.

Step 2. Determine if $\mathbf{v}$ is a linear combination of the given vectors. If it is, then the given vectors span V. If it is not, they do not span V.

Again we investigate the consistency of a linear system, but this time for a right side that represents an arbitrary vector in a vector space V.

EXAMPLE 1 ■ Let V be the vector space R^3 and let

$$\mathbf{v}_1 = (1, 2, 1), \quad \mathbf{v}_2 = (1, 0, 2), \quad \text{and} \quad \mathbf{v}_3 = (1, 1, 0).$$

Do $\mathbf{v}_1$, $\mathbf{v}_2$, and $\mathbf{v}_3$ span V?

Solution **Step 1.** Let $\mathbf{v} = (a, b, c)$ be any vector in R^3, where a, b, and c are arbitrary real numbers.

Step 2. We must find out whether there are constants c_1, c_2, and c_3 such that

$$c_1 \mathbf{v}_1 + c_2 \mathbf{v}_2 + c_3 \mathbf{v}_3 = \mathbf{v}.$$

This leads to the linear system (verify)

$$\begin{array}{rcl} c_1 + c_2 + c_3 &=& a \\ 2c_1 + c_3 &=& b \\ c_1 + 2c_2 &=& c. \end{array}$$

A solution is (verify)

$$c_1 = \frac{-2a + 2b + c}{3}, \qquad c_2 = \frac{a - b + c}{3}, \qquad c_3 = \frac{4a - b - 2c}{3}.$$

Since we have obtained a solution for every choice of a, b, and c, we conclude that $\mathbf{v}_1$, $\mathbf{v}_2$, $\mathbf{v}_3$ span R^3. This is equivalent to saying that span $\{\mathbf{v}_1, \mathbf{v}_2, \mathbf{v}_3\} = R^3$. ■

EXAMPLE 2 ■ Show that

$$S = \left\{ \begin{bmatrix} 1 & 0 \\ 0 & 0 \end{bmatrix}, \begin{bmatrix} 0 & 1 \\ 1 & 0 \end{bmatrix}, \begin{bmatrix} 0 & 0 \\ 0 & 1 \end{bmatrix} \right\}$$

spans the subspace of M_{22} consisting of all symmetric matrices.

Solution **Step 1.** An arbitrary symmetric matrix has the form

$$A = \begin{bmatrix} a & b \\ b & c \end{bmatrix},$$

where a, b, and c are any real numbers.

Step 2. We must find constants d_1, d_2, and d_3 such that

$$d_1 \begin{bmatrix} 1 & 0 \\ 0 & 0 \end{bmatrix} + d_2 \begin{bmatrix} 0 & 1 \\ 1 & 0 \end{bmatrix} + d_3 \begin{bmatrix} 0 & 0 \\ 0 & 1 \end{bmatrix} = A = \begin{bmatrix} a & b \\ b & c \end{bmatrix},$$

which leads to a linear system whose solution is (verify)

$$d_1 = a, \qquad d_2 = b, \qquad d_3 = c.$$

Since we have found a solution for every choice of a, b, and c, we conclude that S spans the given subspace. ■

EXAMPLE 3 ■   Let V be P_2, the vector space consisting of all polynomials of degree ≤ 2 and the zero polynomial. Let $S = \{p_1(t), p_2(t)\}$, where $p_1(t) = t^2 + 2t + 1$ and $p_2(t) = t^2 + 2$. Does S span P_2?

Solution   ***Step 1.*** Let $p(t) = at^2 + bt + c$ be any polynomial in P_2, where a, b, and c are any real numbers.

Step 2. We must find out whether there are constants c_1 and c_2 such that

$$p(t) = c_1 p_1(t) + c_2 p_2(t)$$

or

$$at^2 + bt + c = c_1(t^2 + 2t + 1) + c_2(t^2 + 2).$$

Thus

$$(c_1 + c_2)t^2 + (2c_1)t + (c_1 + 2c_2) = at^2 + bt + c.$$

Since two polynomials agree for all values of t only if the coefficients of respective powers of t agree. Thus we obtain the linear system

$$
\begin{aligned}
c_1 + \ c_2 &= a \\
2c_1 \quad\ \ &= b \\
c_1 + 2c_2 &= c.
\end{aligned}
$$

Using elementary row operations on the augmented matrix of this linear system, we obtain (verify)

$$
\begin{bmatrix}
1 & 0 & 2a - c \\
0 & 1 & c - a \\
0 & 0 & b - 4a + 2c
\end{bmatrix}.
$$

If $b - 4a + 2c \neq 0$, then there is no solution. Hence $S = \{p_1(t), p_2(t)\}$ does not span P_2. For example, the polynomial $3t^2 + 2t - 1$ cannot be written as a linear combination of $p_1(t)$ and $p_2(t)$.   ■

EXAMPLE 4 ■   The vectors $\mathbf{e}_1 = \mathbf{i} = (1, 0)$ and $\mathbf{e}_2 = \mathbf{j} = (0, 1)$ span R^2, for as was observed in Section 3.1, if $\mathbf{u} = (u_1, u_2)$ is any vector in R^2, then $\mathbf{u} = u_1 \mathbf{e}_1 + u_2 \mathbf{e}_2$. As was noted in Section 3.2, every vector $\mathbf{u}$ in R^3 can be written as a linear combination of the vectors $\mathbf{e}_1 = \mathbf{i} = (1, 0, 0)$, $\mathbf{e}_2 = \mathbf{j} = (0, 1, 0)$, and $\mathbf{e}_3 = \mathbf{k} = (0, 0, 1)$. Thus $\mathbf{e}_1$, $\mathbf{e}_2$, and $\mathbf{e}_3$ span R^3. Similarly, the vectors $\mathbf{e}_1 = (1, 0, \ldots, 0)$, $\mathbf{e}_2 = (0, 1, 0, \ldots, 0)$, $\ldots$, $\mathbf{e}_n = (0, 0, \ldots, 1)$ span R^n, since any vector $\mathbf{u} = (u_1, u_2, \ldots, u_n)$ in R^n can be written as

$$\mathbf{u} = u_1 \mathbf{e}_1 + u_2 \mathbf{e}_2 + \cdots + u_n \mathbf{e}_n.$$ ■

EXAMPLE 5 ■   The set $S = \{t^n, t^{n-1}, \ldots, t, 1\}$ spans P_n, since every polynomial in P_n is of the form

$$a_0 t^n + a_1 t^{n-1} + \cdots + a_{n-1} t + a_n,$$

which is a linear combination of the elements in S.   ■

EXAMPLE 6 ■ Consider the homogeneous linear system $A\mathbf{x} = \mathbf{0}$, where

$$A = \begin{bmatrix} 1 & 1 & 0 & 2 \\ -2 & -2 & 1 & -5 \\ 1 & 1 & -1 & 3 \\ 4 & 4 & -1 & 9 \end{bmatrix}.$$

From Example 8 in Section 4.2, the set of all solutions to $A\mathbf{x} = \mathbf{0}$ forms a subspace of R^4. To determine a spanning set for the solution space of this homogeneous system, we find that the reduced row echelon form of the augmented matrix is (verify)

$$\begin{bmatrix} 1 & 1 & 0 & 2 & \vdots & 0 \\ 0 & 0 & 1 & -1 & \vdots & 0 \\ 0 & 0 & 0 & 0 & \vdots & 0 \\ 0 & 0 & 0 & 0 & \vdots & 0 \end{bmatrix}.$$

The general solution is then given by

$$x_1 = -r - 2s$$
$$x_2 = r$$
$$x_3 = s$$
$$x_4 = s,$$

where r and s are any real numbers. In matrix form we have that any member of the solution space is given by

$$\mathbf{x} = r \begin{bmatrix} -1 \\ 1 \\ 0 \\ 0 \end{bmatrix} + s \begin{bmatrix} -2 \\ 0 \\ 1 \\ 1 \end{bmatrix}.$$

Hence the vectors $\begin{bmatrix} -1 \\ 1 \\ 0 \\ 0 \end{bmatrix}$ and $\begin{bmatrix} -2 \\ 0 \\ 1 \\ 1 \end{bmatrix}$ span the solution space.

■

Linear Independence

DEFINITION The vectors $\mathbf{v}_1, \mathbf{v}_2, \ldots, \mathbf{v}_k$ in a vector space V are said to be **linearly dependent** if there exist constants $c_1, c_2, \ldots, c_k$, not all zero, such that

$$c_1 \mathbf{v}_1 + c_2 \mathbf{v}_2 + \cdots + c_k \mathbf{v}_k = \mathbf{0}. \tag{1}$$

Otherwise, $\mathbf{v}_1, \mathbf{v}_2, \ldots, \mathbf{v}_k$ are called **linearly independent**. That is, $\mathbf{v}_1, \mathbf{v}_2, \ldots, \mathbf{v}_k$ are linearly independent if whenever $c_1 \mathbf{v}_1 + c_2 \mathbf{v}_2 + \cdots + c_k \mathbf{v}_k = \mathbf{0}$, we must have

$$c_1 = c_2 = \cdots = c_k = 0.$$

That is, the *only* linear combination of $\mathbf{v}_1, \mathbf{v}_2, \ldots, \mathbf{v}_k$ that yields the zero vector is that in which all the coefficients are zero. If the vectors $\mathbf{v}_1, \mathbf{v}_2, \ldots, \mathbf{v}_k$ are distinct and we denote them as a set $S = \{\mathbf{v}_1, \mathbf{v}_2, \ldots, \mathbf{v}_k\}$, then we also say that the set S is **linearly dependent** or **linearly independent**.

It should be emphasized that for any vectors $\mathbf{v}_1, \mathbf{v}_2, \ldots, \mathbf{v}_k$, Equation (1) always holds if we choose all the scalars $c_1, c_2, \ldots, c_k$ equal to zero. The important point in this definition is whether or not it is possible to satisfy (1) with at least one of the scalars different from zero.

The procedure to determine if the vectors $\mathbf{v}_1, \mathbf{v}_2, \ldots, \mathbf{v}_k$ are linearly dependent or linearly independent is as follows.

Step 1. Form Equation (1), which leads to a homogeneous system.

Step 2. If the homogeneous system obtained in Step 1 has only the trivial solution, then the given vectors are linearly independent; if it has a nontrivial solution, then the vectors are linearly dependent.

EXAMPLE 7 ■ Determine whether the vectors

$$\begin{bmatrix} -1 \\ 1 \\ 0 \\ 0 \end{bmatrix} \quad \text{and} \quad \begin{bmatrix} -2 \\ 0 \\ 1 \\ 1 \end{bmatrix}$$

found in Example 6 as spanning the solution space of $A\mathbf{x} = \mathbf{0}$ are linearly dependent or linearly independent.

Solution   Forming Equation (1),

$$c_1 \begin{bmatrix} -1 \\ 1 \\ 0 \\ 0 \end{bmatrix} + c_2 \begin{bmatrix} -2 \\ 0 \\ 1 \\ 1 \end{bmatrix} = \begin{bmatrix} 0 \\ 0 \\ 0 \\ 0 \end{bmatrix},$$

we obtain the homogeneous system

$$\begin{aligned} -c_1 - 2c_2 &= 0 \\ c_1 + 0c_2 &= 0 \\ 0c_1 + c_2 &= 0 \\ 0c_1 + c_2 &= 0, \end{aligned}$$

whose only solution is $c_1 = c_2 = 0$. Hence the given vectors are linearly independent. ■

EXAMPLE 8 ■ Are the vectors $\mathbf{v}_1 = (1, 0, 1, 2)$, $\mathbf{v}_2 = (0, 1, 1, 2)$, and $\mathbf{v}_3 = (1, 1, 1, 3)$ in R^4 linearly dependent or linearly independent?

Solution   We form Equation (1),

$$c_1 \mathbf{v}_1 + c_2 \mathbf{v}_2 + c_3 \mathbf{v}_3 = \mathbf{0},$$

and solve for c_1, c_2, and c_3. The resulting homogeneous system is (verify)

$$
\begin{aligned}
c_1 \qquad\quad +\ c_3 &= 0 \\
c_2 +\ c_3 &= 0 \\
c_1 +\ c_2 +\ c_3 &= 0 \\
2c_1 + 2c_2 + 3c_3 &= 0,
\end{aligned}
$$

which has as its only solution $c_1 = c_2 = c_3 = 0$ (verify), showing that the given vectors are linearly independent. ∎

EXAMPLE 9 ∎ Consider the vectors

$$\mathbf{v}_1 = (1, 2, -1), \qquad \mathbf{v}_2 = (1, -2, 1), \qquad \mathbf{v}_3 = (-3, 2-1),$$

and

$$\mathbf{v}_4 = (2, 0, 0) \quad \text{in } R^3.$$

Is $S = \{\mathbf{v}_1, \mathbf{v}_2, \mathbf{v}_3, \mathbf{v}_4\}$ linearly dependent or linearly independent?

Solution   Setting up Equation (1), we are led to the homogeneous system (verify)

$$
\begin{aligned}
c_1 +\ c_2 - 3c_3 + 2c_4 &= 0 \\
2c_1 - 2c_2 + 2c_3 \qquad &= 0 \\
-c_1 +\ c_2 -\ c_3 \qquad &= 0,
\end{aligned}
$$

a homogeneous system of three equations in four unknowns. By Theorem 1.7, Section 1.5, we are assured of the existence of a nontrivial solution. Hence, S is linearly dependent. In fact, two of the infinitely many solutions are

$$
\begin{aligned}
c_1 = 1, \qquad c_2 = 2, \qquad c_3 = 1, \qquad c_4 = 0; \\
c_1 = 1, \qquad c_2 = 1, \qquad c_3 = 0, \qquad c_4 = -1.
\end{aligned}
$$
∎

EXAMPLE 10 ∎ The vectors $\mathbf{e}_1$ and $\mathbf{e}_2$ in R^2, defined in Example 4, are linearly independent, since

$$c_1(1, 0) + c_2(0, 1) = (0, 0)$$

can hold only if $c_1 = c_2 = 0$. Similarly, the vectors $\mathbf{e}_1$, $\mathbf{e}_2$, and $\mathbf{e}_3$ in R^3, and more generally, the vectors $\mathbf{e}_1, \mathbf{e}_2, \ldots, \mathbf{e}_n$ in R^n are linearly independent (Exercise T.1). ∎

Corollary 4.4 in Section 4.6, to follow, gives another way of testing whether n given vectors in R^n are linearly dependent or linearly independent. We form the matrix A, whose columns are the given n vectors. Then the given vectors are linearly independent if and only if $\det(A) \neq 0$. Thus, in Example 10,

$$A = \begin{bmatrix} 1 & 0 \\ 0 & 1 \end{bmatrix}$$

and $\det(A) = 1$ so that $\mathbf{e}_1$ and $\mathbf{e}_2$ are linearly independent.

EXAMPLE 11 ■ Consider the vectors

$$p_1(t) = t^2 + t + 2, \qquad p_2(t) = 2t^2 + t, \qquad p_3(t) = 3t^2 + 2t + 2.$$

To find out whether $S = \{p_1(t), p_2(t), p_3(t)\}$ is linearly dependent or linearly independent, we set up Equation (1) and solve for c_1, c_2, and c_3. The resulting homogeneous system is (verify)

$$c_1 + 2c_2 + 3c_3 = 0$$
$$c_1 + c_2 + 2c_3 = 0$$
$$2c_1 + 2c_3 = 0,$$

which has infinitely many solutions (verify). A particular solution is $c_1 = 1$, $c_2 = 1$, $c_3 = -1$, so

$$p_1(t) + p_2(t) - p_3(t) = 0.$$

Hence S is linearly dependent.   ■

EXAMPLE 12 ■ If $\mathbf{v}_1, \mathbf{v}_2, \ldots, \mathbf{v}_k$ are k vectors in any vector space and $\mathbf{v}_i$ is the zero vector, Equation (1) holds by letting $c_i = 1$ and $c_j = 0$ for $j \neq i$. Thus $S = \{\mathbf{v}_1, \mathbf{v}_2, \ldots, \mathbf{v}_k\}$ is linearly dependent. *Hence every set of vectors containing the zero vector is linearly dependent.*   ■

Let S_1 and S_2 be finite subsets of a vector space and let S_1 be a subset of S_2. Then (a) if S_1 is linearly dependent, so is S_2; and (b) if S_2 is linearly independent, so is S_1 (Exercise T.2).

We consider next the meaning of linear independence in R^2 and R^3. Suppose that $\mathbf{v}_1$ and $\mathbf{v}_2$ are linearly dependent in R^2. Then there exist scalars c_1 and c_2, not both zero, such that

$$c_1 \mathbf{v}_1 + c_2 \mathbf{v}_2 = \mathbf{0}.$$

If $c_1 \neq 0$, then

$$\mathbf{v}_1 = \left(-\frac{c_2}{c_1} \right) \mathbf{v}_2.$$

If $c_2 \neq 0$, then

$$\mathbf{v}_2 = \left(-\frac{c_1}{c_2} \right) \mathbf{v}_1.$$

Thus one of the vectors is a multiple of the other. Conversely, suppose that $\mathbf{v}_1 = c\mathbf{v}_2$. Then

$$1\mathbf{v}_1 - c\mathbf{v}_2 = \mathbf{0},$$

and since the coefficients of $\mathbf{v}_1$ and $\mathbf{v}_2$ are not both zero, it follows that $\mathbf{v}_1$ and $\mathbf{v}_2$ are linearly dependent. Thus $\mathbf{v}_1$ and $\mathbf{v}_2$ are linearly dependent in R^2 if and only if one of the vectors is a multiple of the other (Figure 4.3). Hence two vectors in R^2 are linearly dependent if and only if they both lie on the same line passing through the origin [Figure 4.3(a)].

Suppose now that $\mathbf{v}_1$, $\mathbf{v}_2$, and $\mathbf{v}_3$ are linearly dependent in R^3. Then we can write

$$c_1 \mathbf{v}_1 + c_2 \mathbf{v}_2 + c_3 \mathbf{v}_3 = \mathbf{0},$$

FIGURE 4.3

(a) Linearly dependent vectors in R^2.   (b) Linearly independent vectors in R^2.

where c_1, c_2, and c_3 are not all zero, say $c_2 \neq 0$. Then

$$\mathbf{v}_2 = \left(-\frac{c_1}{c_2}\right)\mathbf{v}_1 - \left(\frac{c_3}{c_2}\right)\mathbf{v}_3,$$

which means that $\mathbf{v}_2$ is in the subspace W spanned by $\mathbf{v}_1$ and $\mathbf{v}_3$.

    Now W is either a plane through the origin (when $\mathbf{v}_1$ and $\mathbf{v}_3$ are linearly independent), or a line through the origin (when $\mathbf{v}_1$ and $\mathbf{v}_3$ are linearly dependent), or the origin (when $\mathbf{v}_1 = \mathbf{v}_2 = \mathbf{v}_3 = \mathbf{0}$). Since a line through the origin always lies in a plane through the origin, we conclude that $\mathbf{v}_1$, $\mathbf{v}_2$, and $\mathbf{v}_3$ all lie in the same plane through the origin. Conversely, suppose that $\mathbf{v}_1$, $\mathbf{v}_2$, and $\mathbf{v}_3$ all lie in the same plane through the origin. Then either all three vectors are the zero vector, or all three vectors lie on the same line through the origin, or all three vectors lie in a plane through the origin spanned by two vectors, say $\mathbf{v}_1$ and $\mathbf{v}_3$. Thus, in all these cases, $\mathbf{v}_2$ is a linear combination of $\mathbf{v}_1$ and $\mathbf{v}_3$:

$$\mathbf{v}_2 = a_1\mathbf{v}_1 + a_3\mathbf{v}_3.$$

Then

$$a_1\mathbf{v}_1 - 1\mathbf{v}_2 + a_3\mathbf{v}_3 = \mathbf{0},$$

which means that $\mathbf{v}_1$, $\mathbf{v}_2$, and $\mathbf{v}_3$ are linearly dependent. Hence three vectors in R^3 are linearly dependent if and only if they all lie in the same plane passing through the origin (Figure 4.4).

FIGURE 4.4

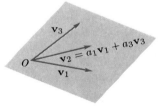

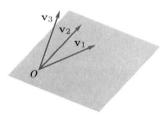

(a) Linearly dependent vectors in R^3.   (b) Linearly independent vectors in R^3.

    More generally, let $\mathbf{u}$ and $\mathbf{v}$ be nonzero vectors in a vector space V. We can show (Exercise T.13) that $\mathbf{u}$ and $\mathbf{v}$ are linearly dependent if and only if there is a scalar k such that $\mathbf{v} = k\mathbf{u}$. Equivalently, $\mathbf{u}$ and $\mathbf{v}$ are linearly independent if and only if neither vector is a multiple of the other. This approach will not work with sets having three or more vectors. Instead, we use the result given by the following theorem.

THEOREM 4.4 ■ *The nonzero vectors $\mathbf{v}_1, \mathbf{v}_2, \ldots, \mathbf{v}_n$ in a vector space V are linearly dependent if and only if one of the vectors $\mathbf{v}_j$ is a linear combination of the preceding vectors $\mathbf{v}_1, \mathbf{v}_2, \ldots, \mathbf{v}_{j-1}$.*

Proof   If $\mathbf{v}_j$ is a linear combination of $\mathbf{v}_1, \mathbf{v}_2, \ldots, \mathbf{v}_{j-1}$,

$$\mathbf{v}_j = c_1\mathbf{v}_1 + c_2\mathbf{v}_2 + \cdots + c_{j-1}\mathbf{v}_{j-1},$$

then

$$c_1\mathbf{v}_1 + c_2\mathbf{v}_2 + \cdots + c_{j-1}\mathbf{v}_{j-1} + (-1)\mathbf{v}_j + 0\mathbf{v}_{j+1} + \cdots + 0\mathbf{v}_n = \mathbf{0}.$$

Since at least one coefficient, -1, is nonzero, we conclude that $\mathbf{v}_1, \mathbf{v}_2, \ldots, \mathbf{v}_n$ are linearly dependent.

Conversely, suppose that $\mathbf{v}_1, \mathbf{v}_2, \ldots, \mathbf{v}_n$ are linearly dependent. Then there exist scalars $c_1, c_2, \ldots, c_n$, not all zero, such that

$$c_1\mathbf{v}_1 + c_2\mathbf{v}_2 + \cdots + c_n\mathbf{v}_n = \mathbf{0}.$$

Now let j be the largest subscript for which $c_j \neq 0$. If $j > 1$, then

$$\mathbf{v}_j = -\left(\frac{c_1}{c_j}\right)\mathbf{v}_1 - \left(\frac{c_2}{c_j}\right)\mathbf{v}_2 - \cdots - \left(\frac{c_{j-1}}{c_j}\right)\mathbf{v}_{j-1}.$$

If $j = 1$, then $c_1\mathbf{v}_1 = \mathbf{0}$, which implies that $\mathbf{v}_1 = \mathbf{0}$, a contradiction to the hypothesis that none of the vectors are the zero vector. Thus one of the vectors $\mathbf{v}_j$ is a linear combination of the preceding vectors $\mathbf{v}_1, \mathbf{v}_2, \ldots, \mathbf{v}_{j-1}$.   ∎

EXAMPLE 13 ■ If $\mathbf{v}_1$, $\mathbf{v}_2$, $\mathbf{v}_3$, and $\mathbf{v}_4$ are as in Example 9, then we find (verify) that

$$\mathbf{v}_1 + 2\mathbf{v}_2 + \mathbf{v}_3 + 0\mathbf{v}_4 = \mathbf{0},$$

so $\mathbf{v}_3 = -\mathbf{v}_1 - 2\mathbf{v}_2$. Also,

$$\mathbf{v}_1 + \mathbf{v}_2 + 0\mathbf{v}_3 - \mathbf{v}_4 = \mathbf{0},$$

so $\mathbf{v}_4 = \mathbf{v}_1 + \mathbf{v}_2$.   ■

REMARKS   1. We observe that Theorem 4.4 does not say that *every* vector $\mathbf{v}$ is a linear combination of the preceding vectors. Thus, in Example 9, we have the equation $\mathbf{v}_1 + 2\mathbf{v}_2 + \mathbf{v}_3 + 0\mathbf{v}_4 = \mathbf{0}$. We cannot solve, in this equation, for $\mathbf{v}_4$ as a linear combination of $\mathbf{v}_1$, $\mathbf{v}_2$, and $\mathbf{v}_3$, since its coefficient is zero. However, in this same example we have $\mathbf{v}_1 + \mathbf{v}_2 + 0\mathbf{v}_3 - \mathbf{v}_4 = \mathbf{0}$. In this equation we *can* solve for $\mathbf{v}_4$, since its coefficient is nonzero.

2. We can also prove that if $S = \{\mathbf{v}_1, \mathbf{v}_2, \ldots, \mathbf{v}_k\}$ is a set of vectors in a vector space V, then S is linearly dependent if and only if one of the vectors in S is a linear combination of all the other vectors in S (see Exercise T.3). Thus, in Example 13,

$$\mathbf{v}_1 = -\mathbf{v}_3 - 2\mathbf{v}_2, \qquad \mathbf{v}_2 = -\tfrac{1}{2}\mathbf{v}_1 - \tfrac{1}{2}\mathbf{v}_3, \qquad \mathbf{v}_1 = \mathbf{v}_4 - \mathbf{v}_2,$$

and

$$\mathbf{v}_2 = \mathbf{v}_4 - \mathbf{v}_1.$$

3. Observe that if $\mathbf{v}_1, \mathbf{v}_2, \ldots, \mathbf{v}_k$ are linearly independent vectors in a vector space, then they must be distinct and none can be the zero vector.

The following result will be used in Section 4.4 as well as in several other places. Suppose that $S = \{\mathbf{v}_1, \mathbf{v}_2, \ldots, \mathbf{v}_n\}$ spans a vector space V and $\mathbf{v}_j$ is a linear combination of the preceding vectors in S. Then the set

$$S_1 = \{\mathbf{v}_1, \mathbf{v}_2, \ldots, \mathbf{v}_{j-1}, \mathbf{v}_{j+1}, \ldots, \mathbf{v}_n\},$$

consisting of S with $\mathbf{v}_j$ deleted, also spans V. To show this result, observe that if $\mathbf{v}$ is any vector in V, then, since S spans V, we can find scalars $a_1, a_2, \ldots, a_n$ such that

$$\mathbf{v} = a_1 \mathbf{v}_1 + a_2 \mathbf{v}_2 + \cdots + a_{j-1} \mathbf{v}_{j-1} + a_j \mathbf{v}_j + a_{j+1} \mathbf{v}_{j+1} + \cdots + a_n \mathbf{v}_n.$$

Now if

$$\mathbf{v}_j = b_1 \mathbf{v}_1 + b_2 \mathbf{v}_2 + \cdots + b_{j-1} \mathbf{v}_{j-1},$$

then

$$\begin{aligned}
\mathbf{v} &= a_1 \mathbf{v}_1 + a_2 \mathbf{v}_2 + \cdots + a_{j-1} \mathbf{v}_{j-1} + a_j (b_1 \mathbf{v}_1 + b_2 \mathbf{v}_2 + \cdots + b_{j-1} \mathbf{v}_{j-1}) \\
&\quad + a_{j+1} \mathbf{v}_{j+1} + \cdots + a_n \mathbf{v}_n \\
&= c_1 \mathbf{v}_1 + c_2 \mathbf{v}_2 + \cdots + c_{j-1} \mathbf{v}_{j-1} + c_{j+1} \mathbf{v}_{j+1} + \cdots + c_n \mathbf{v}_n,
\end{aligned}$$

which means that span $S_1 = V$.

EXAMPLE 14 ■ Consider the set of vectors $S = \{\mathbf{v}_1, \mathbf{v}_2, \mathbf{v}_3, \mathbf{v}_4\}$ in R^4, where

$$\mathbf{v}_1 = \begin{bmatrix} 1 \\ 1 \\ 0 \\ 0 \end{bmatrix}, \quad \mathbf{v}_2 = \begin{bmatrix} 1 \\ 0 \\ 1 \\ 0 \end{bmatrix}, \quad \mathbf{v}_3 = \begin{bmatrix} 0 \\ 1 \\ 1 \\ 0 \end{bmatrix}, \quad \text{and} \quad \mathbf{v}_4 = \begin{bmatrix} 2 \\ 1 \\ 1 \\ 0 \end{bmatrix},$$

and let $W = \text{span } S$. Since $\mathbf{v}_4 = \mathbf{v}_1 + \mathbf{v}_2$, we conclude that $W = \text{span } S_1$, where $S_1 = \{\mathbf{v}_1, \mathbf{v}_2, \mathbf{v}_3\}$. ■

4.3 EXERCISES

1. Which of the following vectors span R^2?
   (a) $(1, 2)$, $(-1, 1)$.
   (b) $(0, 0)$, $(1, 1)$, $(-2, -2)$.
   (c) $(1, 3)$, $(2, -3)$, $(0, 2)$.
   (d) $(2, 4)$, $(-1, 2)$.

2. Which of the following sets of vectors span R^3?
   (a) $\{(1, -1, 2), (0, 1, 1)\}$.
   (b) $\{(1, 2, -1), (6, 3, 0), (4, -1, 2), (2, -5, 4)\}$.
   (c) $\{(2, 2, 3), (-1, -2, 1), (0, 1, 0)\}$.
   (d) $\{(1, 0, 0), (0, 1, 0), (0, 0, 1), (1, 1, 1)\}$.

3. Which of the following vectors span R^4?

(a) $(1, 0, 0, 1)$, $(0, 1, 0, 0)$, $(1, 1, 1, 1)$, $(1, 1, 1, 0)$.
(b) $(1, 2, 1, 0)$, $(1, 1, -1, 0)$, $(0, 0, 0, 1)$.
(c) $(6, 4, -2, 4)$, $(2, 0, 0, 1)$, $(3, 2, -1, 2)$, $(5, 6, -3, 2)$, $(0, 4, -2, -1)$.
(d) $(1, 1, 0, 0)$, $(1, 2, -1, 1)$, $(0, 0, 1, 1)$, $(2, 1, 2, 1)$.

4. Which of the following sets of polynomials span P_2?
   (a) $\{t^2 + 1, t^2 + t, t + 1\}$.
   (b) $\{t^2 + 1, t - 1, t^2 + t\}$.
   (c) $\{t^2 + 2, 2t^2 - t + 1, t + 2, t^2 + t + 4\}$.
   (d) $\{t^2 + 2t - 1, t^2 - 1\}$.

5. Do the polynomials $t^3 + 2t + 1$, $t^2 - t + 2$, $t^3 + 2$, $-t^3 + t^2 - 5t + 2$ span P_3?

6. Find a set of vectors spanning the solution space of $A\mathbf{x} = \mathbf{0}$, where

$$A = \begin{bmatrix} 1 & 0 & 1 & 0 \\ 1 & 2 & 3 & 1 \\ 2 & 1 & 3 & 1 \\ 1 & 1 & 2 & 1 \end{bmatrix}.$$

7. Find a set of vectors spanning the null space of

$$A = \begin{bmatrix} 1 & 1 & 2 & -1 \\ 2 & 3 & 6 & -2 \\ -2 & 1 & 2 & 2 \\ 0 & -2 & -4 & 0 \end{bmatrix}.$$

8. Let

$$\mathbf{x}_1 = \begin{bmatrix} 2 \\ -1 \\ 1 \end{bmatrix}, \qquad \mathbf{x}_2 = \begin{bmatrix} 4 \\ -7 \\ -1 \end{bmatrix}, \qquad \mathbf{x}_3 = \begin{bmatrix} 1 \\ 2 \\ 2 \end{bmatrix}$$

belong to the solution space of $A\mathbf{x} = \mathbf{0}$. Is $\{\mathbf{x}_1, \mathbf{x}_2, \mathbf{x}_3\}$ linearly independent?

9. Let

$$\mathbf{x}_1 = \begin{bmatrix} 1 \\ 2 \\ 0 \\ 1 \end{bmatrix}, \qquad \mathbf{x}_2 = \begin{bmatrix} 1 \\ 0 \\ -1 \\ 1 \end{bmatrix}, \qquad \mathbf{x}_3 = \begin{bmatrix} 1 \\ 6 \\ 2 \\ 0 \end{bmatrix}$$

belong to the null space of A. Is $\{\mathbf{x}_1, \mathbf{x}_2, \mathbf{x}_3\}$ linearly independent?

10. Which of the following sets of vectors in R^3 are linearly dependent? For those that are, express one vector as a linear combination of the rest.

(a) $\{(1, 2, -1), (3, 2, 5)\}$.

(b) $\{(4, 2, 1), (2, 6, -5), (1, -2, 3)\}$.

(c) $\{(1, 1, 0), (0, 2, 3), (1, 2, 3), (3, 6, 6)\}$.

(d) $\{(1, 2, 3), (1, 1, 1), (1, 0, 1)\}$.

11. Consider the vector space R^4. Follow the directions of Exercise 10.

(a) $\{(1, 1, 2, 1), (1, 0, 0, 2), (4, 6, 8, 6), (0, 3, 2, 1)\}$.

(b) $\{(1, -2, 3, -1), (-2, 4, -6, 2)\}$.

(c) $\{(1, 1, 1, 1), (2, 3, 1, 2), (3, 1, 2, 1), (2, 2, 1, 1)\}$.

(d) $\{(4, 2, -1, 3), (6, 5, -5, 1), (2, -1, 3, 5)\}$.

12. Consider the vector space P_2. Follow the directions of Exercise 10.

(a) $\{t^2 + 1, t - 2, t + 3\}$.

(b) $\{2t^2 + 1, t^2 + 3, t\}$.

(c) $\{3t + 1, 3t^2 + 1, 2t^2 + t + 1\}$.

(d) $\{t^2 - 4, 5t^2 - 5t - 6, 3t^2 - 5t + 2\}$.

13. Consider the vector space M_{22}. Follow the directions of Exercise 10.

(a) $\left\{ \begin{bmatrix} 1 & 1 \\ 1 & 2 \end{bmatrix}, \begin{bmatrix} 1 & 0 \\ 0 & 2 \end{bmatrix}, \begin{bmatrix} 0 & 3 \\ 1 & 2 \end{bmatrix}, \begin{bmatrix} 2 & 6 \\ 4 & 6 \end{bmatrix} \right\}$.

(b) $\left\{ \begin{bmatrix} 1 & 1 \\ 1 & 1 \end{bmatrix}, \begin{bmatrix} 1 & 0 \\ 0 & 2 \end{bmatrix}, \begin{bmatrix} 0 & 1 \\ 0 & 2 \end{bmatrix} \right\}$.

(c) $\left\{ \begin{bmatrix} 1 & 1 \\ 1 & 1 \end{bmatrix}, \begin{bmatrix} 2 & 3 \\ 1 & 2 \end{bmatrix}, \begin{bmatrix} 3 & 1 \\ 2 & 1 \end{bmatrix}, \begin{bmatrix} 2 & 2 \\ 1 & 1 \end{bmatrix} \right\}$.

14. Let V be the vector space of all real-valued continuous functions. Follow the directions of Exercise 10.

(a) $\{\cos t, \sin t, e^t\}$.

(b) $\{t, e^t, \sin t\}$.

(c) $\{t^2, t, e^t\}$.

(d) $\{\cos^2 t, \sin^2 t, \cos 2t\}$.

15. For what values of c are the vectors $(-1, 0, -1)$, $(2, 1, 2)$, and $(1, 1, c)$ in R^3 linearly dependent?

16. For what values of λ are the vectors $t + 3$ and $2t + \lambda^2 + 2$ in P_1 linearly dependent?

THEORETICAL EXERCISES ■

T.1. Show that the vectors $\mathbf{e}_1, \mathbf{e}_2, \ldots, \mathbf{e}_n$ in R^n are linearly independent.

T.2. Let S_1 and S_2 be finite subsets of a vector space and let S_1 be a subset of S_2. Show that:

(a) If S_1 is linearly dependent, so is S_2.

(b) If S_2 is linearly independent, so is S_1.

T.3. Let $S = \{\mathbf{v}_1, \mathbf{v}_2, \ldots, \mathbf{v}_k\}$ be a set of vectors in a vector space. Show that S is linearly dependent if and only if one of the vectors in S is a linear combination of all the other vectors in S.

T.4. Suppose that $S = \{\mathbf{v}_1, \mathbf{v}_2, \mathbf{v}_3\}$ is a linearly independent set of vectors in a vector space V. Show that $T = \{\mathbf{w}_1, \mathbf{w}_2, \mathbf{w}_3\}$ is also linearly independent, where $\mathbf{w}_1 = \mathbf{v}_1 + \mathbf{v}_2 + \mathbf{v}_3$, $\mathbf{w}_2 = \mathbf{v}_2 + \mathbf{v}_3$, and $\mathbf{w}_3 = \mathbf{v}_3$.

T.5. Suppose that $S = \{\mathbf{v}_1, \mathbf{v}_2, \mathbf{v}_3\}$ is a linearly independent set of vectors in a vector space V. Is $T = \{\mathbf{w}_1, \mathbf{w}_2, \mathbf{w}_3\}$, where $\mathbf{w}_1 = \mathbf{v}_1 + \mathbf{v}_2$, $\mathbf{w}_2 = \mathbf{v}_1 + \mathbf{v}_3$, $\mathbf{w}_3 = \mathbf{v}_2 + \mathbf{v}_3$, linearly dependent or linearly independent? Justify your answer.

T.6. Suppose that $S = \{\mathbf{v}_1, \mathbf{v}_2, \mathbf{v}_3\}$ is a linearly dependent set of vectors in a vector space V. Is $T = \{\mathbf{w}_1, \mathbf{w}_2, \mathbf{w}_3\}$, where $\mathbf{w}_1 = \mathbf{v}_1$, $\mathbf{w}_2 = \mathbf{v}_1 + \mathbf{v}_2$, $\mathbf{w}_3 = \mathbf{v}_1 + \mathbf{v}_2 + \mathbf{v}_3$, linearly dependent or linearly independent? Justify your answer.

T.7. Let $\mathbf{v}_1$, $\mathbf{v}_2$, and $\mathbf{v}_3$ be vectors in a vector space such that $\{\mathbf{v}_1, \mathbf{v}_2\}$ is linearly independent. Show that if $\mathbf{v}_3$ does not belong to span $\{\mathbf{v}_1, \mathbf{v}_2\}$, then $\{\mathbf{v}_1, \mathbf{v}_2, \mathbf{v}_3\}$ is linearly independent.

T.8. Let A be an $m \times n$ matrix in reduced row echelon form. Show that the nonzero rows of A, viewed as vectors in R^n, form a linearly independent set of vectors.

T.9. Let $S = \{\mathbf{u}_1, \mathbf{u}_2, \ldots, \mathbf{u}_k\}$ be a set of vectors in a vector space, and let $T = \{\mathbf{v}_1, \mathbf{v}_2, \ldots, \mathbf{v}_m\}$, where each $\mathbf{v}_i$, $i = 1, 2, \ldots, m$, is a linear combination of the vectors in S. Show that

$$\mathbf{w} = b_1 \mathbf{v}_1 + b_2 \mathbf{v}_2 + \cdots + b_m \mathbf{v}_m$$

is a linear combination of the vectors in S.

T.10. Suppose that $\{\mathbf{v}_1, \mathbf{v}_2, \ldots, \mathbf{v}_n\}$ is a linearly independent set of vectors in R^n. Show that if A is an $n \times n$ nonsingular matrix, then $\{A\mathbf{v}_1, A\mathbf{v}_2, \ldots, A\mathbf{v}_n\}$ is linearly independent.

T.11. Let S_1 and S_2 be finite subsets of a vector space V and let S_1 be a subset of S_2. If S_2 is linearly dependent, show by examples that S_1 may be either linearly dependent or linearly independent.

T.12. Let S_1 and S_2 be finite subsets of a vector space and let S_1 be a subset of S_2. If S_1 is linearly independent, show by examples that S_2 may be either linearly dependent or linearly independent.

T.13. Let $\mathbf{u}$ and $\mathbf{v}$ be nonzero vectors in a vector space V. Show that $\{\mathbf{u}, \mathbf{v}\}$ is linearly dependent if and only if there is a scalar k such that $\mathbf{v} = k\mathbf{u}$. Equivalently, $\{\mathbf{u}, \mathbf{v}\}$ is linearly independent if and only if one of the vectors is not a multiple of the other.

MATLAB EXERCISES ◼

ML.1. Determine if S is linearly independent or linearly dependent.

(a) $S = \{(1, 0, 0, 1), (0, 1, 1, 0), (1, 1, 1, 1)\}$.

(b) $S = \left\{ \begin{bmatrix} 1 & 2 \\ 1 & 0 \end{bmatrix}, \begin{bmatrix} 2 & -1 \\ 1 & 2 \end{bmatrix}, \begin{bmatrix} -3 & 1 \\ 0 & 1 \end{bmatrix} \right\}$.

(c) $S = \left\{ \begin{bmatrix} 1 \\ 2 \\ 1 \\ 0 \\ 1 \end{bmatrix}, \begin{bmatrix} 0 \\ 1 \\ 2 \\ -1 \\ 1 \end{bmatrix}, \begin{bmatrix} 2 \\ 1 \\ 0 \\ 0 \\ -1 \end{bmatrix}, \begin{bmatrix} -2 \\ 1 \\ 1 \\ 1 \\ 1 \end{bmatrix} \right\}$.

ML.2. Find a spanning set of the solution space of $A\mathbf{x} = \mathbf{0}$, where

$$A = \begin{bmatrix} 1 & 2 & 0 & 1 \\ 1 & 1 & 1 & 2 \\ 2 & -1 & 5 & 7 \\ 0 & 2 & -2 & -2 \end{bmatrix}.$$

4.4 ▾ Basis and Dimension

In this section we continue our study of the structure of a vector space V by determining a smallest set of vectors in V that completely describes V.

Basis

| DEFINITION | The vectors $\mathbf{v}_1, \mathbf{v}_2, \ldots, \mathbf{v}_k$ in a vector space V are said to form a **basis** for V if (a) $\mathbf{v}_1, \mathbf{v}_2, \ldots, \mathbf{v}_k$ span V and (b) $\mathbf{v}_1, \mathbf{v}_2, \ldots, \mathbf{v}_k$ are linearly independent. |

REMARK If $\mathbf{v}_1, \mathbf{v}_2, \ldots, \mathbf{v}_k$ form a basis for a vector space V, then they must be distinct and nonzero, so we write them as a set $\{\mathbf{v}_1, \mathbf{v}_2, \ldots, \mathbf{v}_k\}$.

EXAMPLE 1 ■ The vectors $\mathbf{e}_1 = (1, 0)$ and $\mathbf{e}_2 = (0, 1)$ form a basis for R^2, the vectors $\mathbf{e}_1$, $\mathbf{e}_2$, and $\mathbf{e}_3$ form a basis for R^3 and, in general, the vectors $\mathbf{e}_1, \mathbf{e}_2, \ldots, \mathbf{e}_n$ form a basis for R^n. Each of these sets of vectors is called the **natural basis** or **standard basis** for R^2, R^3, and R^n, respectively. ■

EXAMPLE 2 ■ Show that the set $S = \{\mathbf{v}_1, \mathbf{v}_2, \mathbf{v}_3, \mathbf{v}_4\}$, where $\mathbf{v}_1 = (1, 0, 1, 0)$, $\mathbf{v}_2 = (0, 1, -1, 2)$, $\mathbf{v}_3 = (0, 2, 2, 1)$, and $\mathbf{v}_4 = (1, 0, 0, 1)$ is a basis for R^4.

Solution To show that S is linearly independent, we form the equation

$$c_1\mathbf{v}_1 + c_2\mathbf{v}_2 + c_3\mathbf{v}_3 + c_4\mathbf{v}_4 = \mathbf{0}$$

and solve for c_1, c_2, c_3, and c_4. Substituting for $\mathbf{v}_1$, $\mathbf{v}_2$, $\mathbf{v}_3$, and $\mathbf{v}_4$, we obtain the linear system (verify)

$$\begin{aligned} c_1 \quad\quad\quad + c_4 &= 0 \\ c_2 + 2c_3 \quad\quad &= 0 \\ c_1 - c_2 + 2c_3 \quad\quad &= 0 \\ 2c_2 + c_3 + c_4 &= 0, \end{aligned}$$

which has as its only solution $c_1 = c_2 = c_3 = c_4 = 0$ (verify), showing that S is linearly independent.

To show that S spans R^4, we let $\mathbf{v} = (a, b, c, d)$ be any vector in R^4. We now seek constants k_1, k_2, k_3, and k_4 such that

$$k_1\mathbf{v}_1 + k_2\mathbf{v}_2 + k_3\mathbf{v}_3 + k_4\mathbf{v}_4 = \mathbf{v}.$$

Substituting for $\mathbf{v}_1$, $\mathbf{v}_2$, $\mathbf{v}_3$, $\mathbf{v}_4$, and $\mathbf{v}$, we find a solution (verify) for k_1, k_2, k_3, and k_4 to the resulting linear system for any a, b, c, and d. Hence S spans R^4 and is a basis for R^4. ■

EXAMPLE 3 ■ In Example 5 of Section 4.3 we observed that the set of polynomials

$$S = \{t^n, t^{n-1}, \ldots, t, 1\}$$

spans P_n. Moreover, it is easy to show that S is linearly independent as follows. Form a linear combination of the vectors in S and set it equal to zero:

$$c_1 t^n + c_2 t^{n-1} + \cdots + c_n t + c_{n+1} = 0, \tag{1}$$

which is Equation (1) in Section 4.3. In order for this equation to be true for all values of t, the polynomial on the left side of Equation (1) must have infinitely many roots. But the only polynomial that has infinitely many roots is the zero polynomial. Thus Equation (1) can hold only if

$$c_1 = c_2 = \cdots = c_n = c_{n+1} = 0.$$

Hence S is a linearly independent set of vectors. Therefore, S is a basis for P_n, called the **natural basis** or **standard basis** for P_n. ∎

EXAMPLE 4 ■ The set

$$S = \left\{ \begin{bmatrix} 1 & 0 \\ 0 & 0 \end{bmatrix}, \begin{bmatrix} 0 & 1 \\ 0 & 0 \end{bmatrix}, \begin{bmatrix} 0 & 0 \\ 1 & 0 \end{bmatrix}, \begin{bmatrix} 0 & 0 \\ 0 & 1 \end{bmatrix} \right\}$$

is a basis for the vector space V of all 2×2 matrices. To verify that S is linearly independent, we form a linear combination of the vectors in S and set it equal to zero:

$$c_1 \begin{bmatrix} 1 & 0 \\ 0 & 0 \end{bmatrix} + c_2 \begin{bmatrix} 0 & 1 \\ 0 & 0 \end{bmatrix} + c_3 \begin{bmatrix} 0 & 0 \\ 1 & 0 \end{bmatrix} + c_4 \begin{bmatrix} 0 & 0 \\ 0 & 1 \end{bmatrix} = \begin{bmatrix} 0 & 0 \\ 0 & 0 \end{bmatrix}.$$

This gives

$$\begin{bmatrix} c_1 & c_2 \\ c_3 & c_4 \end{bmatrix} = \begin{bmatrix} 0 & 0 \\ 0 & 0 \end{bmatrix},$$

which implies that $c_1 = c_2 = c_3 = c_4 = 0$. Hence S is linearly independent. To verify that S spans V, we take any vector

$$\begin{bmatrix} a & b \\ c & d \end{bmatrix}$$

in V and we must find scalars c_1, c_2, c_3, and c_4 such that

$$\begin{bmatrix} a & b \\ c & d \end{bmatrix} = c_1 \begin{bmatrix} 1 & 0 \\ 0 & 0 \end{bmatrix} + c_2 \begin{bmatrix} 0 & 1 \\ 0 & 0 \end{bmatrix} + c_3 \begin{bmatrix} 0 & 0 \\ 1 & 0 \end{bmatrix} + c_4 \begin{bmatrix} 0 & 0 \\ 0 & 1 \end{bmatrix}.$$

We find that $c_1 = a$, $c_2 = b$, $c_3 = c$, and $c_4 = d$ so that S spans V. ∎

THEOREM 4.5 ■ *If $S = \{\mathbf{v}_1, \mathbf{v}_2, \ldots, \mathbf{v}_n\}$ is a basis for a vector space V, then every vector in V can be written in one and only one way as a linear combination of the vectors in S.*

Proof   First, every vector $\mathbf{v}$ in V can be written as a linear combination of the vectors in S because S spans V. Now let

$$\mathbf{v} = c_1 \mathbf{v}_1 + c_2 \mathbf{v}_2 + \cdots + c_n \mathbf{v}_n \tag{2}$$

and

$$\mathbf{v} = d_1 \mathbf{v}_1 + d_2 \mathbf{v}_2 + \cdots + d_n \mathbf{v}_n. \tag{3}$$

Subtracting (3) from (2), we obtain

$$0 = (c_1 - d_1)\mathbf{v}_1 + (c_2 - d_2)\mathbf{v}_2 + \cdots + (c_n - d_n)\mathbf{v}_n.$$

Since S is linearly independent, it follows that $c_i - d_i = 0$, $1 \leq i \leq n$, so $c_i = d_i$, $1 \leq i \leq n$. Hence there is only one way to express $\mathbf{v}$ as a linear combination of the vectors in S. ∎

We can also show (Exercise T.11) that if $S = \{\mathbf{v}_1, \mathbf{v}_2, \ldots, \mathbf{v}_n\}$ is a set of non-zero vectors in a vector space V such that every vector in V can be written in one and only one way as a linear combination of the vectors in S, then S is a basis for V.

THEOREM 4.6 ■ *Let $S = \{\mathbf{v}_1, \mathbf{v}_2, \ldots, \mathbf{v}_n\}$ be a set of nonzero vectors in a vector space V and let $W = span\ S$. Then some subset of S is a basis for W.*

Proof **Case I.** If S is linearly independent, then since S already spans W, we conclude that S is a basis for W.

Case II. If S is linearly dependent, then

$$c_1\mathbf{v}_1 + c_2\mathbf{v}_2 + \cdots + c_n\mathbf{v}_n = \mathbf{0}, \tag{4}$$

where $c_1, c_2, \ldots, c_n$ are not all zero. Thus, some $\mathbf{v}_j$ is a linear combination of the preceding vectors in S (Theorem 4.4). We now delete $\mathbf{v}_j$ from S, getting a subset S_1 of S. Then, by the observation made just before Example 14 in Section 4.3, we conclude that $S_1 = \{\mathbf{v}_1, \mathbf{v}_2, \ldots, \mathbf{v}_{j-1}, \mathbf{v}_{j+1}, \ldots, \mathbf{v}_n\}$ also spans W.

If S_1 is linearly independent, then S_1 is a basis. If S_1 is linearly dependent, delete a vector of S_1 that is a linear combination of the preceding vectors of S_1 and get a new set S_2 which spans W. Continuing, since S is a finite set, we will eventually find a subset T of S that is linearly independent and spans W. The set T is a basis for W.

Alternate Constructive Proof When V Is R^m, $n \geq m$. We take the vectors in S as $m \times 1$ matrices and form Equation (4) above. This equation leads to a homogeneous system in the n unknowns $c_1, c_2, \ldots, c_n$; the columns of its $m \times n$ coefficient matrix A are $\mathbf{v}_1, \mathbf{v}_2, \ldots, \mathbf{v}_n$. We now transform A to a matrix B in reduced row echelon form, having r nonzero rows, $1 \leq r \leq m$. Without loss of generality, we may assume that the r leading 1's in the r nonzero rows of B occur in the first r columns. Thus we have

$$B = \begin{bmatrix} 1 & 0 & 0 & \cdots & 0 & b_{1\,r+1} & \cdots & b_{1\,n} \\ 0 & 1 & 0 & \cdots & 0 & b_{2\,r+1} & \cdots & b_{2\,n} \\ 0 & 0 & 1 & \cdots & 0 & b_{3\,r+1} & \cdots & b_{3\,n} \\ \vdots & \vdots & \vdots & \ddots & \vdots & \vdots & & \vdots \\ 0 & 0 & 0 & \cdots & 1 & b_{r\,r+1} & \cdots & b_{r\,n} \\ 0 & 0 & 0 & \cdots & 0 & 0 & \cdots & 0 \\ \vdots & \vdots & \vdots & & \vdots & \vdots & \ddots & \vdots \\ 0 & 0 & 0 & \cdots & 0 & 0 & \cdots & 0 \end{bmatrix}.$$

Solving for the unknowns corresponding to the leading 1's, we see that $c_1, c_2, \ldots, c_r$ can be solved for in terms of the other unknowns $c_{r+1}, c_{r+2}, \ldots, c_n$.

Thus

$$c_1 = -b_{1\,r+1}c_{r+1} - b_{1\,r+2}c_{r+2} - \cdots - b_{1\,n}c_n,$$
$$c_2 = -b_{2\,r+1}c_{r+1} - b_{2\,r+2}c_{r+2} - \cdots - b_{2\,n}c_n, \tag{5}$$
$$\vdots$$
$$c_r = -b_{r\,r+1}c_{r+1} - b_{r\,r+2}c_{r+2} - \cdots - b_{r\,n}c_n,$$

where $c_{r+1}, c_{r+2}, \ldots, c_n$ can be assigned arbitrary real values. Letting

$$c_{r+1} = 1, \quad c_{r+2} = 0, \ldots, \quad c_n = 0$$

in Equation (5) and using these values in Equation (4), we have

$$-b_{1\,r+1}\mathbf{v}_1 - b_{2\,r+1}\mathbf{v}_2 - \cdots - b_{r\,r+1}\mathbf{v}_r + \mathbf{v}_{r+1} = \mathbf{0},$$

which implies that $\mathbf{v}_{r+1}$ is a linear combination of $\mathbf{v}_1, \mathbf{v}_2, \ldots, \mathbf{v}_r$. By the observation made just before Example 14 in Section 4.3, the set of vectors obtained from S by deleting $\mathbf{v}_{r+1}$ spans W. Similarly, letting

$$c_{r+1} = 0, \quad c_{r+2} = 1, \quad c_{r+3} = 0, \ldots, \quad c_n = 0,$$

we find that $\mathbf{v}_{r+2}$ is a linear combination of $\mathbf{v}_1, \mathbf{v}_2, \ldots, \mathbf{v}_r$ and the set of vectors obtained from S by deleting $\mathbf{v}_{r+1}$ and $\mathbf{v}_{r+2}$ spans W. Continuing in this manner, $\mathbf{v}_{r+3}, \mathbf{v}_{r+4}, \ldots, \mathbf{v}_n$ are linear combinations of $\mathbf{v}_1, \mathbf{v}_2, \ldots, \mathbf{v}_r$, so it follows that $\{\mathbf{v}_1, \mathbf{v}_2, \ldots, \mathbf{v}_r\}$ spans W.

We next show that $\{\mathbf{v}_1, \mathbf{v}_2, \ldots, \mathbf{v}_r\}$ is linearly independent. Consider the matrix B_D obtained by deleting from B all columns not containing a leading 1. In this case, B_D consists of the first r columns of B. Thus,

$$B_D = \begin{bmatrix} 1 & 0 & 0 & \cdots & 0 \\ 0 & 1 & 0 & \cdots & 0 \\ 0 & 0 & 1 & & \\ & & & \ddots & \vdots \\ 0 & 0 & & \cdots & 1 \\ 0 & 0 & & \cdots & 0 \\ \vdots & \vdots & & & \vdots \\ 0 & 0 & & \cdots & 0 \end{bmatrix}.$$

Let A_D be the matrix obtained from A by deleting the columns corresponding to the columns that were deleted in B to obtain B_D. In this case, the columns of A_D are $\mathbf{v}_1, \mathbf{v}_2, \ldots, \mathbf{v}_r$, the first r columns of A. Since A and B are row equivalent, so are A_D and B_D. Then the homogeneous systems

$$A_D\mathbf{x} = \mathbf{0} \quad \text{and} \quad B_D\mathbf{x} = \mathbf{0}$$

are equivalent. Recall now that the homogeneous system $B_D\mathbf{x} = \mathbf{0}$ can be written equivalently as

$$x_1\mathbf{w}_1 + x_2\mathbf{w}_2 + \cdots + x_r\mathbf{w}_r = \mathbf{0}, \tag{6}$$

where

$$\mathbf{x} = \begin{bmatrix} x_1 \\ x_2 \\ \vdots \\ x_r \end{bmatrix}$$

and $\mathbf{w}_1, \mathbf{w}_2, \ldots, \mathbf{w}_r$ are the columns of B_D. Since the columns of B_D form a linearly independent set of vectors in R^m, Equation (6) has only the trivial solution. Hence, $A_D\mathbf{x} = \mathbf{0}$ also has only the trivial solution. Thus the columns of A_D are linearly independent. That is, $\{\mathbf{v}_1, \mathbf{v}_2, \ldots, \mathbf{v}_r\}$ is linearly independent. ∎

The first proof of Theorem 4.6 leads to a simple procedure for finding a subset of a spanning set that is a basis.

Let $S = \{\mathbf{v}_1, \mathbf{v}_2, \ldots, \mathbf{v}_n\}$ be a set of nonzero vectors in a vector space V. The procedure for finding a subset of S that is a basis for $W = \text{span } S$ is as follows:

Step 1. Form Equation (4),

$$c_1\mathbf{v}_1 + c_2\mathbf{v}_2 + \cdots + c_n\mathbf{v}_n = \mathbf{0},$$

which we solve for $c_1, c_2, \ldots, c_n$. If these are all zero, then S is linearly independent and is then a basis for W.

Step 2. If $c_1, c_2, \ldots, c_n$ are not all zero, then S is linearly dependent, so one of the vectors in S — say, $\mathbf{v}_j$ — is a linear combination of the preceding vectors in S. Delete $\mathbf{v}_j$ from S, getting the subset S_1, which also spans W.

Step 3. Repeat Step 1, using S_1 instead of S. By repeatedly deleting vectors of S we obtain a subset T that spans W and is linearly independent. Thus T is a basis for W.

This procedure can be rather tedious, since *every time* we delete a vector from S, we must solve a linear system. In Section 4.6 we shall present a much more efficient procedure for finding a basis for $W = \text{span } S$, but the basis is *not* guaranteed to be a subset of S. In many cases this is not a cause for concern, since one basis for $W = \text{span } S$ is as good as any other basis. However, there are cases when the vectors in S have some special properties and we want the basis for $W = \text{span } S$ to have the same properties, so we want the basis to be a subset of S. If $V = R^m$, the alternate proof of Theorem 4.6 yields a very efficient procedure (see Example 5 below) for finding a basis for $W = \text{span } S$ consisting of vectors from S.

Let $V = R^m$ and let $S = \{\mathbf{v}_1, \mathbf{v}_2, \ldots, \mathbf{v}_n\}$ be a set of nonzero vectors in V. The procedure for finding a subset of S that is a basis for $W = \operatorname{span} S$ is as follows:

Step 1. Form Equation (4),

$$c_1\mathbf{v}_1 + c_2\mathbf{v}_2 + \cdots + c_n\mathbf{v}_n = \mathbf{0}.$$

Step 2. Construct the augmented matrix associated with the homogeneous system of Equation (4), and transform it to reduced row echelon form.

Step 3. The vectors corresponding to the columns containing the leading 1's form a basis for $W = \operatorname{span} S$. Recall that in the alternate proof of the theorem we assumed without loss of generality that the r leading 1's in the r nonzero rows of B occur in the first r columns. Thus, if $S = \{\mathbf{v}_1, \mathbf{v}_2, \ldots, \mathbf{v}_6\}$ and the leading 1's occur in columns 1, 3, and 4, then $\{\mathbf{v}_1, \mathbf{v}_3, \mathbf{v}_4\}$ is a basis for span S.

REMARK   In Step 2 of the procedure above, it is sufficient to transform the augmented matrix to row echelon form (see Section 1.5).

EXAMPLE 5 ■ Let $S = \{\mathbf{v}_1, \mathbf{v}_2, \mathbf{v}_3, \mathbf{v}_4, \mathbf{v}_5\}$ be a set of vectors in R^4, where

$$\mathbf{v}_1 = (1, 2, -2, 1), \qquad \mathbf{v}_2 = (-3, 0, -4, 3),$$

$$\mathbf{v}_3 = (2, 1, 1, -1), \quad \mathbf{v}_4 = (-3, 3, -9, 6), \quad \text{and} \quad \mathbf{v}_5 = (9, 3, 7, -6).$$

Find a subset of S that is a basis for $W = \operatorname{span} S$.

Solution   ***Step 1.*** Form Equation (4),

$$c_1(1, 2, -2, 1) + c_2(-3, 0, -4, 3) + c_3(2, 1, 1, -1)$$
$$+ c_4(-3, 3, -9, 6) + c_5(9, 3, 7, -6) = (0, 0, 0, 0).$$

Step 2. Equating corresponding components, we obtain the homogeneous system

$$\begin{aligned}
c_1 - 3c_2 + 2c_3 - 3c_4 + 9c_5 &= 0 \\
2c_1 \quad\quad + c_3 + 3c_4 + 3c_5 &= 0 \\
-2c_1 - 4c_2 + c_3 - 9c_4 + 7c_5 &= 0 \\
c_1 + 3c_2 - c_3 + 6c_4 - 6c_5 &= 0.
\end{aligned}$$

The reduced row echelon form of the associated augmented matrix is (verify)

$$\begin{bmatrix}
1 & 0 & \frac{1}{2} & \frac{3}{2} & \frac{3}{2} & \vdots & 0 \\
0 & 1 & -\frac{1}{2} & \frac{3}{2} & -\frac{5}{2} & \vdots & 0 \\
0 & 0 & 0 & 0 & 0 & \vdots & 0 \\
0 & 0 & 0 & 0 & 0 & \vdots & 0
\end{bmatrix}.$$

Step 3. The leading 1's appear in columns 1 and 2, so $\{\mathbf{v}_1, \mathbf{v}_2\}$ is a basis for $W = \operatorname{span} S$. ■

REMARK

In the alternate proof of Theorem 4.6 when $V = R^n$, the order of the vectors in the original spanning set S determines which basis for W is obtained. If, for example, we consider Example 5, where $S = \{\mathbf{w}_1, \mathbf{w}_2, \mathbf{w}_3, \mathbf{w}_4\}$ with $\mathbf{w}_1 = \mathbf{v}_4$, $\mathbf{w}_2 = \mathbf{v}_3$, $\mathbf{w}_3 = \mathbf{v}_2$, $\mathbf{w}_4 = \mathbf{v}_1$, and $\mathbf{w}_5 = \mathbf{v}_5$, then the reduced row echelon form of the augmented matrix is (verify)

$$\begin{bmatrix} 1 & 0 & \frac{1}{3} & \frac{1}{3} & -\frac{1}{3} & \vdots & 0 \\ 0 & 1 & -1 & 1 & 4 & \vdots & 0 \\ 0 & 0 & 0 & 0 & 0 & \vdots & 0 \\ 0 & 0 & 0 & 0 & 0 & \vdots & 0 \end{bmatrix}.$$

It now follows that $\{\mathbf{w}_1, \mathbf{w}_2\} = \{\mathbf{v}_4, \mathbf{v}_3\}$ is a basis for $W = \text{span } S$.

We will soon establish a main result (Corollary 4.1) of this section, which will tell us about the number of vectors in two different bases. First, observe that if $\{\mathbf{v}_1, \mathbf{v}_2, \dots, \mathbf{v}_n\}$ is a basis for a vector space V, then $\{c\mathbf{v}_1, \mathbf{v}_2, \dots, \mathbf{v}_n\}$ is also a basis if $c \neq 0$. Thus a vector space always has infinitely many bases.

THEOREM 4.7 ■ *If $S = \{\mathbf{v}_1, \mathbf{v}_2, \dots, \mathbf{v}_n\}$ is a basis for a vector space V and $T = \{\mathbf{w}_1, \mathbf{w}_2, \dots, \mathbf{w}_r\}$ is a linearly independent set of vectors in V, then $r \leq n$.*

Proof

Let $T_1 = \{\mathbf{w}_1, \mathbf{v}_1, \dots, \mathbf{v}_n\}$. Since S spans V, so does T_1. Since $\mathbf{w}_1$ is a linear combination of the vectors in S, we find that T_1 is linearly dependent. Then, by Theorem 4.4, some $\mathbf{v}_j$ is a linear combination of the preceding vectors in T_1. Delete that particular vector $\mathbf{v}_j$.

Let $S_1 = \{\mathbf{w}_1, \mathbf{v}_1, \dots, \mathbf{v}_{j-1}, \mathbf{v}_{j+1}, \dots, \mathbf{v}_n\}$. Note that S_1 spans V. Next, let $T_2 = \{\mathbf{w}_2, \mathbf{w}_1, \mathbf{v}_1, \dots, \mathbf{v}_{j-1}, \mathbf{v}_{j+1}, \dots, \mathbf{v}_n\}$. Then T_2 is linearly dependent and some vector in T_2 is a linear combination of the preceding vectors in T_2. Since T is linearly independent, this vector cannot be $\mathbf{w}_1$, so it is $\mathbf{v}_i$, $i \neq j$. Repeat this process over and over. If the $\mathbf{v}$ vectors are all eliminated before we can run out of $\mathbf{w}$ vectors, then the resulting set of $\mathbf{w}$ vectors, a subset of T, is linearly dependent, which implies that T is also linearly dependent. Since we have reached a contradiction, we conclude that the number r of $\mathbf{w}$ vectors must be no greater than the number n of $\mathbf{v}$ vectors. That is, $r \leq n$. ■

COROLLARY 4.1 ■ *If $S = \{\mathbf{v}_1, \mathbf{v}_2, \dots, \mathbf{v}_n\}$ and $T = \{\mathbf{w}_1, \mathbf{w}_2, \dots, \mathbf{w}_m\}$ are bases for a vector space, then $n = m$.*

Proof

Since T is a linearly independent set of vectors, Theorem 4.7 implies that $m \leq n$. Similarly, $n \leq m$ because S is linearly independent. Hence $n = m$. ■

Thus, although a vector space has many bases, we have just shown that for a particular vector space V, all bases have the same number of vectors. We can then make the following definition.

Dimension

DEFINITION	The **dimension** of a nonzero vector space V is the number of vectors in a basis for V. We often write **dim** V for the dimension of V. Since the set $\{\mathbf{0}\}$ is linearly dependent, it is natural to say that the vector space $\{\mathbf{0}\}$ has dimension **zero**.

EXAMPLE 6 ■ The dimension of R^2 is 2; the dimension of R^3 is 3; and in general, the dimension of R^n is n. ■

EXAMPLE 7 ■ The dimension of P_2 is 3; the dimension of P_3 is 4; and in general, the dimension of P_n is $n + 1$. ■

Almost all the vector spaces considered in this book are **finite dimensional**; that is, their dimension is a finite number. However, we point out that there are many vector spaces of infinite dimension that are extremely important in mathematics and physics; their study lies beyond the scope of this book. The vector space P, consisting of all polynomials, and the vector space $C(-\infty, \infty)$ defined in Example 7 of Section 4.2 are not finite dimensional. All finite-dimensional vector spaces of the same dimension differ only in the nature of their elements; their algebraic properties are identical.

It can be shown that if V is a finite-dimensional vector space, then every nonzero subspace W of V has a finite basis and dim $W \leq$ dim V (Exercise T.2).

EXAMPLE 8 ■ The subspace W of R^4 considered in Example 5 has dimension 2. ■

We might also consider the subspaces of R^2 [recall that R^2 can be viewed as the (x, y)-plane]. First, we have $\{\mathbf{0}\}$ and R^2, the trivial subspaces of dimension 0 and 2, respectively. Now the subspace V of R^2 spanned by a vector $\mathbf{v} \neq \mathbf{0}$ is a one-dimensional subspace of R^2; V is represented by a line through the origin. Thus the subspaces of R^2 are $\{\mathbf{0}\}$, R^2, and all the lines through the origin. Similarly, we ask you to show (Exercise T.8) that the subspaces of R^3 are $\{\mathbf{0}\}$, R^3, all lines through the origin, and all planes through the origin.

It can be shown that if a vector space V has dimension n, then any set of $n + 1$ vectors in V is necessarily linearly dependent (Exercise T.3). Any set of more than n vectors in R^n is linearly dependent. Thus the four vectors in R^3 considered in Example 9 of Section 4.3 are linearly dependent. Also, if a vector space V is of dimension n, then no set of $n - 1$ vectors in V can span V (Exercise T.4). Thus in Example 3 of Section 4.3, polynomials $p_1(t)$ and $p_2(t)$ do not span P_2, which is of dimension 3.

We now come to a theorem that we shall have occasion to use several times in constructing a basis containing a given set of linearly independent vectors. We shall leave the proof as an exercise (Exercise T.5). The example following the theorem completely imitates the proof.

THEOREM 4.8 ■ *If S is a linearly independent set of vectors in a finite-dimensional vector space V, then there is a basis T for V, which contains S.* ■

Theorem 4.8 says that a linearly independent set of vectors in a vector space V can be extended to a basis for V.

EXAMPLE 9 ■ Suppose that we wish to find a basis for R^4 that contains the vectors $\mathbf{v}_1 = (1, 0, 1, 0)$ and $\mathbf{v}_2 = (-1, 1, -1, 0)$.

We use Theorem 4.8 as follows. First, let $\{\mathbf{e}_1, \mathbf{e}_2, \mathbf{e}_3, \mathbf{e}_4\}$ be the natural basis for R^4, where

$$\mathbf{e}_1 = (1, 0, 0, 0), \qquad \mathbf{e}_2 = (0, 1, 0, 0), \qquad \mathbf{e}_3 = (0, 0, 1, 0),$$

and

$$\mathbf{e}_4 = (0, 0, 0, 1).$$

Form the set $S = \{\mathbf{v}_1, \mathbf{v}_2, \mathbf{e}_1, \mathbf{e}_2, \mathbf{e}_3, \mathbf{e}_4\}$. Since $\{\mathbf{e}_1, \mathbf{e}_2, \mathbf{e}_3, \mathbf{e}_4\}$ spans R^4, so does S. We now use the alternate proof of Theorem 4.6 to find a subset of S that is a basis for R^4. Thus, we form Equation (4),

$$c_1\mathbf{v}_1 + c_2\mathbf{v}_2 + c_3\mathbf{e}_1 + c_4\mathbf{e}_2 + c_5\mathbf{e}_3 + c_6\mathbf{e}_4 = \mathbf{0},$$

which leads to the homogeneous system

$$
\begin{aligned}
c_1 - c_2 + c_3 & & & = 0 \\
- c_2 & + c_4 & & = 0 \\
c_1 - c_2 & & + c_5 & = 0 \\
& & c_6 & = 0.
\end{aligned}
$$

Transforming the augmented matrix to reduced row echelon form, we obtain (verify)

$$
\left[\begin{array}{cccccc|c}
1 & 0 & 0 & 1 & 1 & 0 & 0 \\
0 & 1 & 0 & 1 & 0 & 0 & 0 \\
0 & 0 & 1 & 0 & -1 & 0 & 0 \\
0 & 0 & 0 & 0 & 0 & 1 & 0
\end{array}\right].
$$

Since the leading 1's appear in columns 1, 2, 3, and 6, we conclude that $\{\mathbf{v}_1, \mathbf{v}_2, \mathbf{e}_1, \mathbf{e}_4\}$ is a basis for R^4 containing $\mathbf{v}_1$ and $\mathbf{v}_2$. ■

From the definition of a basis, a set of vectors in a vector space V is a basis for V if it spans V and is linearly independent. However, if we are given the additional information that the dimension of V is n, we need only verify one of the two conditions. This is the content of the following theorem.

THEOREM 4.9 ■ *Let V be an n-dimensional vector space, and let $S = \{\mathbf{v}_1, \mathbf{v}_2, \dots, \mathbf{v}_n\}$ be a set of n vectors in V.*

(a) *If S is linearly independent, then it is a basis for V.*

(b) *If S spans V, then it is a basis for V.*

Proof   Exercise T.6. ■

As a particular application of Theorem 4.9, we have the following. To determine if a subset S of R^n is a basis for R^n, first count the number of elements in S. If S has n elements, we can use either part (a) or part (b) of Theorem 4.9 to determine whether S is or is not a basis. If S does not have n elements, it is not a basis for R^n. (Why?) The same line of reasoning applies to any vector space or subspace whose *dimension is known*.

EXAMPLE 10 ■ In Example 5, $W = \text{span } S$ is a subspace of R^4, so $\dim W \leq 4$. Since S contains five vectors, we conclude by Corollary 4.1 that S is not a basis for W. In Example 2, since $\dim R^4 = 4$ and the set S contains four vectors, it is possible for S to be a basis for R^4. If S is linearly independent *or* spans R^4, it is a basis; otherwise it is not a basis. Thus, we need only check one of the conditions in Theorem 4.9, not both. ■

4.4   EXERCISES

1. Which of the following sets of vectors are bases for R^2?

(a) $\{(1,3),(1,-1)\}$.

(b) $\{(0,0),(1,2),(2,4)\}$.

(c) $\{(1,2),(2,-3),(3,2)\}$.

(d) $\{(1,3),(-2,6)\}$.

2. Which of the following sets of vectors are bases for R^3?

(a) $\{(1,2,0),(0,1,-1)\}$.

(b) $\{(1,1,-1),(2,3,4),(4,1,-1),(0,1,-1)\}$.

(c) $\{(3,2,2),(-1,2,1),(0,1,0)\}$.

(d) $\{(1,0,0),(0,2,-1),(3,4,1),(0,1,0)\}$.

3. Which of the following sets of vectors are bases for R^4?

(a) $\{(1,0,0,1),(0,1,0,0),(1,1,1,1),(0,1,1,1)\}$.

(b) $\{(1,-1,0,2),(3,-1,2,1),(1,0,0,1)\}$.

(c) $\{(-2,4,6,4),(0,1,2,0),(-1,2,3,2),$
$(-3,2,5,6),(-2,-1,0,4)\}$.

(d) $\{(0,0,1,1),(-1,1,1,2),(1,1,0,0),(2,1,2,1)\}$.

4. Which of the following sets of vectors are bases for P_2?

(a) $\{-t^2+t+2,2t^2+2t+3,4t^2-1\}$.

(b) $\{t^2+2t-1,2t^2+3t-2\}$.

(c) $\{t^2+1,3t^2+2t,3t^2+2t+1,6t^2+6t+3\}$.

(d) $\{3t^2+2t+1,t^2+t+1,t^2+1\}$.

5. Which of the following sets of vectors are bases for P_3?

(a) $\{t^3+2t^2+3t,2t^3+1,6t^3+8t^2+6t+4,$
$t^3+2t^2+t+1\}$.

(b) $\{t^3+t^2+1,t^3-1,t^3+t^2+t\}$.

(c) $\{t^3+t^2+t+1,t^3+2t^2+t+3,$
$2t^3+t^2+3t+2,t^3+t^2+2t+2\}$.

(d) $\{t^3-t,t^3+t^2+1,t-1\}$.

6. Show that the matrices

$$\begin{bmatrix} 1 & 1 \\ 0 & 0 \end{bmatrix}, \quad \begin{bmatrix} 0 & 0 \\ 1 & 1 \end{bmatrix}, \quad \begin{bmatrix} 1 & 0 \\ 0 & 1 \end{bmatrix}, \quad \begin{bmatrix} 0 & 1 \\ 1 & 1 \end{bmatrix}$$

form a basis for the vector space M_{22}.

In Exercises 7 and 8, determine which of the given subsets forms a basis for R^3. Express the vector $(2,1,3)$ as a linear combination of the vectors in each subset that is a basis.

7. (a) $\{(1,1,1),(1,2,3),(0,1,0)\}$.

(b) $\{(1,2,3),(2,1,3),(0,0,0)\}$.

8. (a) $\{(2,1,3),(1,2,1),(1,1,4),(1,5,1)\}$.

(b) $\{(1,1,2),(2,2,0),(3,4,-1)\}$.

In Exercises 9 and 10, determine which of the given subsets forms a basis for P_2. Express $5t^2-3t+8$ as a linear combination of the vectors in each subset that is a basis.

9. (a) $\{t^2+t,t-1,t+1\}$.

(b) $\{t^2+1,t-1\}$.

10. (a) $\{t^2 + t, t^2, t^2 + 1\}$.

(b) $\{t^2 + 1, t^2 - t + 1\}$.

11. Let $S = \{\mathbf{v}_1, \mathbf{v}_2, \mathbf{v}_3, \mathbf{v}_4\}$, where

$$\mathbf{v}_1 = (1, 2, 2), \qquad \mathbf{v}_2 = (3, 2, 1),$$
$$\mathbf{v}_3 = (11, 10, 7), \quad \text{and} \quad \mathbf{v}_4 = (7, 6, 4).$$

Find a basis for the subspace $W = \text{span } S$ of R^3. What is $\dim W$?

12. Let $S = \{\mathbf{v}_1, \mathbf{v}_2, \mathbf{v}_3, \mathbf{v}_4, \mathbf{v}_5\}$, where

$$\mathbf{v}_1 = (1, 1, 0, -1), \quad \mathbf{v}_2 = (0, 1, 2, 1),$$
$$\mathbf{v}_3 = (1, 0, 1, -1), \quad \mathbf{v}_4 = (1, 1, -6, -3),$$

and $\mathbf{v}_5 = (-1, -5, 1, 0)$. Find a basis for the subspace $W = \text{span } S$ of R^4. What is $\dim W$?

13. Consider the following subset of P_3:

$$S = \{t^3 + t^2 - 2t + 1, t^2 + 1, t^3 - 2t,$$
$$2t^3 + 3t^2 - 4t + 3\}.$$

Find a basis for the subspace $W = \text{span } S$. What is $\dim W$?

14. Let

$$S = \left\{ \begin{bmatrix} 1 & 0 \\ 0 & 1 \end{bmatrix}, \begin{bmatrix} 0 & 1 \\ 1 & 0 \end{bmatrix}, \begin{bmatrix} 1 & 1 \\ 1 & 1 \end{bmatrix}, \begin{bmatrix} -1 & 1 \\ 1 & -1 \end{bmatrix} \right\}.$$

Find a basis for the subspace $W = \text{span } S$ of M_{22}.

15. Find a basis for M_{23}. What is the dimension of M_{23}? Generalize to M_{mn}.

16. Consider the following subset of the vector space of all real-valued functions

$$S = \{\cos^2 t, \sin^2 t, \cos 2t\}.$$

Find a basis for the subspace $W = \text{span } S$. What is $\dim W$?

In Exercises 17 and 18, find a basis for the given subspaces of R^3 and R^4.

17. (a) All vectors of the form (a, b, c), where $b = a + c$.

(b) All vectors of the form (a, b, c), where $b = a$.

(c) All vectors of the form (a, b, c), where $2a + b - c = 0$.

18. (a) All vectors of the form (a, b, c), where $a = 0$.

(b) All vectors of the form
$(a + c, a - b, b + c, -a + b)$.

(c) All vectors of the form (a, b, c), where $a - b + 5c = 0$.

In Exercises 19 and 20, find the dimensions of the given subspaces of R^4.

19. (a) All vectors of the form (a, b, c, d), where $d = a + b$.

(b) All vectors of the form (a, b, c, d), where $c = a - b$ and $d = a + b$.

20. (a) All vectors of the form (a, b, c, d), where $a = b$.

(b) All vectors of the form
$(a + c, -a + b, -b - c, a + b + 2c)$.

21. Find a basis for the subspace of P_2 consisting of all vectors of the form $at^2 + bt + c$, where $c = a - b$.

22. Find a basis for the subspace of P_3 consisting of all vectors of the form $at^3 + bt^2 + ct + d$, where $a = b$ and $c = d$.

23. Find the dimensions of the subspaces of R^2 spanned by the vectors in Exercise 1.

24. Find the dimensions of the subspaces of R^3 spanned by the vectors in Exercise 2.

25. Find the dimensions of the subspaces of R^4 spanned by the vectors in Exercise 3.

26. Find a basis for R^3 that includes the vectors

(a) $(1, 0, 2)$.

(b) $(1, 0, 2)$ and $(0, 1, 3)$.

27. Find a basis for R^4 that includes the vectors $(1, 0, 1, 0)$ and $(0, 1, -1, 0)$.

28. Find all values of a for which $\{(a^2, 0, 1), (0, a, 2), (1, 0, 1)\}$ is a basis for R^3.

29. Find a basis for the subspace W of M_{33} consisting of all symmetric matrices.

30. Find a basis for the subspace of M_{33} consisting of all diagonal matrices.

31. Give an example of a two-dimensional subspace of R^4.

32. Give an example of a two-dimensional subspace of P_3.

THEORETICAL EXERCISES ■

T.1. Suppose that in the nonzero vector space V, the largest number of vectors in a linearly independent set is m. Show that any set of m linearly independent vectors in V is a basis for V.

T.2. Show that if V is a finite-dimensional vector space, then every nonzero subspace W of V has a finite basis and $\dim W \leq \dim V$.

T.3. Show that if $\dim V = n$, then any $n + 1$ vectors in V are linearly dependent.

T.4. Show that if $\dim V = n$, then no set of $n - 1$ vectors in V can span V.

T.5. Prove Theorem 4.8.

T.6. Prove Theorem 4.9.

T.7. Show that if W is a subspace of a finite-dimensional vector space V and $\dim W = \dim V$, then $W = V$.

T.8. Show that the subspaces of R^3 are $\{0\}$, R^3, all lines through the origin, and all planes through the origin.

T.9. Show that if $\{\mathbf{v}_1, \mathbf{v}_2, \ldots, \mathbf{v}_n\}$ is a basis for a vector space V and $c \neq 0$, then $\{c\mathbf{v}_1, \mathbf{v}_2, \ldots, \mathbf{v}_n\}$ is also a basis for V.

T.10. Let $S = \{\mathbf{v}_1, \mathbf{v}_2, \mathbf{v}_3\}$ be a basis for vector space V. Then show that $T = \{\mathbf{w}_1, \mathbf{w}_2, \mathbf{w}_3\}$, where $\mathbf{w}_1 = \mathbf{v}_1 + \mathbf{v}_2 + \mathbf{v}_3$, $\mathbf{w}_2 = \mathbf{v}_2 + \mathbf{v}_3$, and $\mathbf{w}_3 = \mathbf{v}_3$, is also a basis for V.

T.11. Let $S = \{\mathbf{v}_1, \mathbf{v}_2, \ldots, \mathbf{v}_n\}$ be a set of nonzero vectors in a vector space V such that every vector in V can be written in one and only one way as a linear combination of the vectors in S. Show that S is a basis for V.

T.12. Suppose that $\{\mathbf{v}_1, \mathbf{v}_2, \ldots, \mathbf{v}_n\}$ is a basis for R^n. Show that if A is an $n \times n$ nonsingular matrix, then $\{A\mathbf{v}_1, A\mathbf{v}_2, \ldots, A\mathbf{v}_n\}$ is also a basis for R^n. (*Hint*: See Exercise T.10 in Section 4.3.)

T.13. Suppose that $\{\mathbf{v}_1, \mathbf{v}_2, \ldots, \mathbf{v}_n\}$ is a linearly independent set of vectors in R^n and let A be a singular matrix. Prove or disprove that $\{A\mathbf{v}_1, A\mathbf{v}_2, \ldots, A\mathbf{v}_n\}$ is linearly independent.

T.14. Show that the vector space P of all polynomials is not finite dimensional. [*Hint*: Suppose that $\{p_1(t), p_2(t), \ldots, p_k(t)\}$ is a finite basis for P. Let $d_j = \text{degree } p_j(t)$. Establish a contradiction.]

MATLAB EXERCISES ■

In order to use MATLAB in this section, you should have read Section 10.7. In the following exercises we relate the theory developed in the section to computational procedures in MATLAB that aid in analyzing the situation.

To determine if a set $S = \{\mathbf{v}_1, \mathbf{v}_2, \ldots, \mathbf{v}_k\}$ is a basis for a vector space V, the definition requires that we show span $S = V$ and S is linearly independent. However, if we know that $\dim V = k$, then Theorem 4.9 implies that we need only show that either span $S = V$ or S is linearly independent. The linear independence, in this special case, is easily analyzed using MATLAB's **rref** command. Construct the homogeneous system $A\mathbf{c} = \mathbf{0}$ associated with the linear independence/dependence question. Then S is linearly independent if and only if

$$\text{rref}(\mathbf{A}) = \begin{bmatrix} \mathbf{I}_k \\ \mathbf{0} \end{bmatrix}.$$

In Exercises ML.1 through ML.6, if this special case can be applied, do so; otherwise, determine if S is a basis for V in the conventional manner.

ML.1. $S = \{(1, 2, 1), (2, 1, 1), (2, 2, 1)\}$ in $V = R^3$.

ML.2. $S = \{2t - 2, t^2 - 3t + 1, 2t^2 - 8t + 4\}$ in $V = P_2$.

ML.3. $S = \{(1, 1, 0, 0), (2, 1, 1, -1), (0, 0, 1, 1), (1, 2, 1, 2)\}$ in $V = R^4$.

ML.4. $S = \{(1, 2, 1, 0), (2, 1, 3, 1), (2, -2, 4, 2)\}$ in $V = \text{span } S$.

ML.5. $S = \{(1, 2, 1, 0), (2, 1, 3, 1), (2, 2, 1, 2)\}$ in $V = \text{span } S$.

ML.6. $V =$ the subspace of R^3 of all vectors of the form (a, b, c), where $b = 2a - c$ and $S = \{(0, 1, -1), (1, 1, 1)\}$.

In Exercises ML.7 through ML.9, use MATLAB's **rref** command to determine a subset of S that is a basis for span S. See Example 5.

ML.7. $S = \{(1, 1, 0, 0), (-2, -2, 0, 0), (1, 0, 2, 1),$
$(2, 1, 2, 1), (0, 1, 1, 1)\}.$

What is dim span S? Does span $S = R^4$?

ML.8. $S = \left\{ \begin{bmatrix} 1 & 2 \\ 1 & 2 \end{bmatrix}, \begin{bmatrix} 1 & 0 \\ 1 & 1 \end{bmatrix}, \begin{bmatrix} 0 & 2 \\ 0 & 1 \end{bmatrix}, \right.$
$\left. \begin{bmatrix} 2 & 4 \\ 2 & 4 \end{bmatrix}, \begin{bmatrix} 1 & 0 \\ 0 & 1 \end{bmatrix} \right\}.$

What is dim span S? Does span $S = M_{22}$?

ML.9. $S = \{t - 2, 2t - 1, 4t - 2, t^2 - t + 1, t^2 + 2t + 1\}.$
What is dim span S? Does span $S = P_2$?

An interpretation of Theorem 4.8 is that any linearly independent subset S of vector space V can be extended to a basis for V. Following the ideas in Example 9, use MATLAB's **rref** command to extend S to a basis for V in Exercises ML.10 through ML.12.

ML.10. $S = \{(1, 1, 0, 0), (1, 0, 1, 0)\}, V = R^4$.

ML.11. $S = \{t^3 - t + 1, t^3 + 2\}, V = P_3$.

ML.12. $S = \{(0, 3, 0, 2, -1)\},$
$V =$ the subspace of R^5 consisting of all vectors of the form (a, b, c, d, e), where $c = a$, $b = 2d + e$.

4.5 ▼ Homogeneous Systems

Homogeneous systems play a central role in linear algebra. This will be seen in Chapter 5, where the foundations of the subject are all integrated to solve one of the major problems occurring in a wide variety of applications. In this section we deal with several problems involving homogeneous systems that will be fundamental in Chapter 5. Here we are able to focus our attention on these problems without being distracted by the additional material in Chapter 5.

Consider the homogeneous system

$$Ax = 0,$$

where A is an $m \times n$ matrix. As we have already observed in Example 8 of Section 4.2, the set of all solutions to this homogeneous system is a subspace of R^n. An extremely important problem, which will occur repeatedly in Chapter 5, is that of finding a basis for this solution space. To find such a basis, we use the method of Gauss–Jordan reduction presented in Section 1.5. Thus we transform the augmented matrix $\begin{bmatrix} A & \vdots & 0 \end{bmatrix}$ of the system to a matrix $\begin{bmatrix} B & \vdots & 0 \end{bmatrix}$ in reduced row echelon form, where B has r nonzero rows, $1 \le r \le m$. Without loss of generality we may assume that the leading 1's in the r nonzero rows of B occur in the first r columns. If $r = n$, then

$$B = \begin{bmatrix} 1 & 0 & \cdots & 0 & 0 & \vdots & 0 \\ 0 & 1 & \cdots & 0 & 0 & \vdots & 0 \\ \vdots & \vdots & \ddots & \vdots & \vdots & \vdots & \vdots \\ 0 & 0 & \cdots & 0 & 1 & \vdots & 0 \\ 0 & 0 & \cdots & & 0 & \vdots & 0 \\ \vdots & \vdots & & & \vdots & \vdots & \\ 0 & 0 & \cdots & & 0 & \vdots & 0 \end{bmatrix} \begin{matrix} \left. \vphantom{\begin{matrix} 1 \\ 0 \\ \vdots \\ 0 \end{matrix}} \right\} r = n \\ \\ \\ \\ \end{matrix}$$

and the only solution to $Ax = 0$ is the trivial one. The solution space has no basis and its dimension is zero.

If $r < n$, then

$$
B = \begin{bmatrix}
1 & 0 & 0 & \cdots & 0 & b_{1\,r+1} & \cdots & b_{1n} & \vdots & 0 \\
0 & 1 & 0 & \cdots & 0 & b_{2\,r+1} & \cdots & b_{2n} & \vdots & 0 \\
0 & 0 & 1 & \cdots & 0 & & & \vdots & \vdots & \vdots \\
\vdots & \vdots & \vdots & \ddots & \vdots & & & & & \\
0 & 0 & 0 & \cdots & 1 & b_{r\,r+1} & \cdots & b_{rn} & \vdots & 0 \\
0 & 0 & 0 & \cdots & 0 & 0 & \cdots & 0 & \vdots & 0 \\
\vdots & \vdots & \vdots & & \vdots & \vdots & & \vdots & \vdots & \\
0 & 0 & 0 & \cdots & 0 & \cdots & & 0 & \vdots & 0
\end{bmatrix}
$$

(overbrace n over the columns; brace $\}r$ grouping the first r rows; brace $\}m$ grouping all rows)

Solving for the unknowns corresponding to the leading 1's, we have

$$x_1 = -b_{1\,r+1}x_{r+1} - b_{1\,r+2}x_{r+2} - \cdots - b_{1n}x_n$$

$$x_2 = -b_{2\,r+1}x_{r+1} - b_{2\,r+2}x_{r+2} - \cdots - b_{2n}x_n$$

$$\vdots$$

$$x_r = -b_{r\,r+1}x_{r+1} - b_{r\,r+2}x_{r+2} - \cdots - b_{rn}x_n,$$

where $x_{r+1}, x_{r+2}, \ldots, x_n$ can be assigned arbitrary real values s_j, $j = 1, 2, \ldots, p$, and $p = n - r$. Thus

$$
\mathbf{x} = \begin{bmatrix} x_1 \\ x_2 \\ \vdots \\ x_r \\ x_{r+1} \\ x_{r+2} \\ \vdots \\ x_n \end{bmatrix} = \begin{bmatrix} -b_{1\,r+1}s_1 - b_{1\,r+2}s_2 - \cdots - b_{1n}s_p \\ -b_{2\,r+1}s_1 - b_{2\,r+2}s_2 - \cdots - b_{2n}s_p \\ \vdots \\ -b_{r\,r+1}s_1 - b_{r\,r+2}s_2 - \cdots - b_{rn}s_p \\ s_1 \\ s_2 \\ \vdots \\ s_p \end{bmatrix}
$$

$$
= s_1 \begin{bmatrix} -b_{1\,r+1} \\ -b_{2\,r+1} \\ \vdots \\ -b_{r\,r+1} \\ 1 \\ 0 \\ 0 \\ \vdots \\ 0 \\ 0 \end{bmatrix} + s_2 \begin{bmatrix} -b_{1\,r+2} \\ -b_{2\,r+2} \\ \vdots \\ -b_{r\,r+2} \\ 0 \\ 1 \\ 0 \\ \vdots \\ 0 \\ 0 \end{bmatrix} + \cdots + s_p \begin{bmatrix} -b_{1n} \\ -b_{2n} \\ \vdots \\ -b_{rn} \\ 0 \\ 0 \\ 0 \\ \vdots \\ 0 \\ 1 \end{bmatrix}.
$$

Since $s_1, s_2, \ldots, s_p$ can be assigned arbitrary real values, we make the following choices for these values:

$$
\begin{array}{llll}
s_1 = 1, & s_2 = 0, & \ldots, & s_p = 0 \\
s_1 = 0, & s_2 = 1, & \ldots, & s_p = 0 \\
& & \vdots & \\
s_1 = 0, & s_2 = 0, & \ldots, \quad s_{p-1} = 0, & s_p = 1,
\end{array}
$$

obtaining the solutions

$$
\mathbf{x}_1 = \begin{bmatrix} -b_{1\,r+1} \\ -b_{2\,r+1} \\ \vdots \\ -b_{r\,r+1} \\ 1 \\ 0 \\ 0 \\ \vdots \\ 0 \\ 0 \end{bmatrix}, \quad
\mathbf{x}_2 = \begin{bmatrix} -b_{1\,r+2} \\ -b_{2\,r+2} \\ \vdots \\ -b_{r\,r+2} \\ 0 \\ 1 \\ 0 \\ \vdots \\ 0 \\ 0 \end{bmatrix}, \ldots, \quad
\mathbf{x}_p = \begin{bmatrix} -b_{1n} \\ -b_{2n} \\ \vdots \\ -b_{rn} \\ 0 \\ 0 \\ 0 \\ \vdots \\ 0 \\ 1 \end{bmatrix}.
$$

Since

$$\mathbf{x} = s_1\mathbf{x}_1 + s_2\mathbf{x}_2 + \cdots + s_p\mathbf{x}_p,$$

we see that $\{\mathbf{x}_1, \mathbf{x}_2, \ldots, \mathbf{x}_p\}$ spans the solution space of $A\mathbf{x} = \mathbf{0}$. Moreover, if we form the equation

$$c_1\mathbf{x}_1 + c_2\mathbf{x}_2 + \cdots + c_p\mathbf{x}_p = \mathbf{0},$$

its coefficient matrix is the matrix whose columns are $\mathbf{x}_1, \mathbf{x}_2, \ldots, \mathbf{x}_p$. If we look at rows $r+1, r+2, \ldots, n$ of this matrix, we readily see that

$$c_1 = c_2 = \cdots = c_p = 0.$$

Hence $\{\mathbf{x}_1, \mathbf{x}_2, \ldots, \mathbf{x}_p\}$ is linearly independent and forms a basis for the solution space of $A\mathbf{x} = \mathbf{0}$, the null space of A.

The procedure for finding a basis for the solution space of a homogeneous system $A\mathbf{x} = \mathbf{0}$, or the null space of A, where A is $m \times n$, is as follows.

Step 1. Solve the given homogeneous system by Gauss–Jordan reduction. If the solution contains no arbitrary constants, then the solution space is $\{\mathbf{0}\}$, which has no basis; the dimension of the solution space is zero.

Step 2. If the solution $\mathbf{x}$ contains arbitrary constants, write $\mathbf{x}$ as a linear combination of vectors $\mathbf{x}_1, \mathbf{x}_2, \ldots, \mathbf{x}_p$ with $s_1, s_2, \ldots, s_p$ as coefficients:

$$\mathbf{x} = s_1\mathbf{x}_1 + s_2\mathbf{x}_2 + \cdots + s_p\mathbf{x}_p.$$

Step 3. The set of vectors $\{\mathbf{x}_1, \mathbf{x}_2, \ldots, \mathbf{x}_p\}$ is a basis for the solution space of $A\mathbf{x} = \mathbf{0}$; the dimension of the solution space is p.

REMARK   In Step 1, suppose that the matrix in reduced row echelon form to which $\begin{bmatrix} A & \vdots & \mathbf{0} \end{bmatrix}$ has been transformed has r nonzero rows (also r leading 1's). Then $p = n - r$. That is, the dimension of the solution space is $n - r$. Moreover, a solution $\mathbf{x}$ to $A\mathbf{x} = \mathbf{0}$ has $n - r$ arbitrary constants.

If A is an $m \times n$ matrix, we refer to the dimension of the null space of A as the **nullity** of A, denoted by nullity A.

EXAMPLE 1 ■ Find a basis for and the dimension of the solution space W of the homogeneous system

$$\begin{bmatrix} 1 & 1 & 4 & 1 & 2 \\ 0 & 1 & 2 & 1 & 1 \\ 0 & 0 & 0 & 1 & 2 \\ 1 & -1 & 0 & 0 & 2 \\ 2 & 1 & 6 & 0 & 1 \end{bmatrix} \begin{bmatrix} x_1 \\ x_2 \\ x_3 \\ x_4 \\ x_5 \end{bmatrix} = \begin{bmatrix} 0 \\ 0 \\ 0 \\ 0 \\ 0 \end{bmatrix}.$$

Solution   **Step 1.** To solve the given system by the Gauss–Jordan reduction method, we transform the augmented matrix to reduced row echelon form, obtaining (verify)

$$\begin{bmatrix} 1 & 0 & 2 & 0 & 1 & \vdots & 0 \\ 0 & 1 & 2 & 0 & -1 & \vdots & 0 \\ 0 & 0 & 0 & 1 & 2 & \vdots & 0 \\ 0 & 0 & 0 & 0 & 0 & \vdots & 0 \\ 0 & 0 & 0 & 0 & 0 & \vdots & 0 \end{bmatrix}.$$

Every solution is of the form (verify)

$$\mathbf{x} = \begin{bmatrix} -2s - t \\ -2s + t \\ s \\ -2t \\ t \end{bmatrix}, \tag{1}$$

where s and t are any real numbers.

Step 2. Every vector in W is a solution and is then of the form given by Equation (1). We can then write every vector in W as

$$\mathbf{x} = s \begin{bmatrix} -2 \\ -2 \\ 1 \\ 0 \\ 0 \end{bmatrix} + t \begin{bmatrix} -1 \\ 1 \\ 0 \\ -2 \\ 1 \end{bmatrix}. \tag{2}$$

Since s and t can take on any values, we first let $s = 1$, $t = 0$, and then let $s = 0$, $t = 1$, in Equation (2), obtaining as solutions

$$\mathbf{x}_1 = \begin{bmatrix} -2 \\ -2 \\ 1 \\ 0 \\ 0 \end{bmatrix} \quad \text{and} \quad \mathbf{x}_2 = \begin{bmatrix} -1 \\ 1 \\ 0 \\ -2 \\ 1 \end{bmatrix}.$$

Step 3. The set $\{\mathbf{x}_1, \mathbf{x}_2\}$ is a basis for W. Moreover, $\dim W = 2$.   ■

The following example illustrates a type of problem that we will be solving frequently in Chapter 5.

EXAMPLE 2 ■ Find a basis for the solution space of the homogeneous system $(\lambda I_3 - A)\mathbf{x} = \mathbf{0}$ for $\lambda = -2$ and

$$A = \begin{bmatrix} -3 & 0 & -1 \\ 2 & 1 & 0 \\ 0 & 0 & -2 \end{bmatrix}.$$

Solution   We form $-2I_3 - A$:

$$-2\begin{bmatrix} 1 & 0 & 0 \\ 0 & 1 & 0 \\ 0 & 0 & 1 \end{bmatrix} - \begin{bmatrix} -3 & 0 & -1 \\ 2 & 1 & 0 \\ 0 & 0 & -2 \end{bmatrix} = \begin{bmatrix} 1 & 0 & 1 \\ -2 & -3 & 0 \\ 0 & 0 & 0 \end{bmatrix}.$$

This last matrix is the coefficient matrix of the homogeneous system, so we transform the augmented matrix

$$\begin{bmatrix} 1 & 0 & 1 & \vdots & 0 \\ -2 & -3 & 0 & \vdots & 0 \\ 0 & 0 & 0 & \vdots & 0 \end{bmatrix}$$

to reduced row echelon form, obtaining (verify)

$$\begin{bmatrix} 1 & 0 & 1 & \vdots & 0 \\ 0 & 1 & -\frac{2}{3} & \vdots & 0 \\ 0 & 0 & 0 & \vdots & 0 \end{bmatrix}.$$

Every solution is then of the form (verify)

$$\mathbf{x} = \begin{bmatrix} -s \\ \frac{2}{3}s \\ s \end{bmatrix},$$

where s is any real number. Then every vector in the solution can be written as

$$\mathbf{x} = s\begin{bmatrix} -1 \\ \frac{2}{3} \\ 1 \end{bmatrix},$$

so

$$\left\{ \begin{bmatrix} -1 \\ \frac{2}{3} \\ 1 \end{bmatrix} \right\}$$

is a basis for the solution space.   ■

Another important problem that we have to solve often in Chapter 5 is illustrated in the following example.

EXAMPLE 3 ■ Find all real numbers λ such that the homogeneous system $(\lambda I_2 - A)\mathbf{x} = \mathbf{0}$ has a nontrivial solution for

$$A = \begin{bmatrix} 1 & 5 \\ 3 & -1 \end{bmatrix}.$$

Solution   We form $\lambda I_2 - A$:

$$\lambda \begin{bmatrix} 1 & 0 \\ 0 & 1 \end{bmatrix} - \begin{bmatrix} 1 & 5 \\ 3 & -1 \end{bmatrix} = \begin{bmatrix} \lambda - 1 & -5 \\ -3 & \lambda + 1 \end{bmatrix}.$$

The homogeneous system $(\lambda I_2 - A)\mathbf{x} = \mathbf{0}$ is then

$$\begin{bmatrix} \lambda - 1 & -5 \\ -3 & \lambda + 1 \end{bmatrix} \begin{bmatrix} x_1 \\ x_2 \end{bmatrix} = \begin{bmatrix} 0 \\ 0 \end{bmatrix}.$$

By Corollary 2.3, this homogeneous system has a nontrivial solution if and only if the determinant of the coefficient matrix is zero; that is, if and only if

$$\det\left(\begin{bmatrix} \lambda - 1 & -5 \\ -3 & \lambda + 1 \end{bmatrix} \right) = 0,$$

or if and only if

$$(\lambda - 1)(\lambda + 1) - 15 = 0$$
$$\lambda^2 - 16 = 0$$
$$\lambda = 4 \quad \text{or} \quad \lambda = -4.$$

Thus, when $\lambda = 4$ or -4, the homogeneous system $(\lambda I_2 - A)\mathbf{x} = \mathbf{0}$ for the given matrix A has a nontrivial solution.   ■

Relationship Between Nonhomogeneous Linear Systems and Homogeneous Systems

The following result, which is important in the study of differential equations, was presented in Section 1.5 and its proof was left to Exercise T.13 of that section.

If $\mathbf{x}_p$ is a particular solution to the nonhomogeneous system $A\mathbf{x} = \mathbf{b}$, $\mathbf{b} \neq \mathbf{0}$, and $\mathbf{x}_h$ is a solution to the associated homogeneous system $A\mathbf{x} = \mathbf{0}$, then $\mathbf{x}_p + \mathbf{x}_h$ is a solution to the given system $A\mathbf{x} = \mathbf{b}$. Moreover, every solution $\mathbf{x}$ to the nonhomogeneous linear system $A\mathbf{x} = \mathbf{b}$ can be written as $\mathbf{x}_p + \mathbf{x}_h$, where $\mathbf{x}_p$ is a particular solution to the given nonhomogeneous system and $\mathbf{x}_h$ is a solution to the associated homogeneous system $A\mathbf{x} = \mathbf{0}$.

4.5 EXERCISES

In Exercises 1 through 8, find a basis for and the dimension of the solution space of the given homogeneous system.

1. $x_1 + x_2 + x_3 + x_4 = 0$
$2x_1 + x_2 - x_3 + x_4 = 0.$

2. $\begin{bmatrix} 1 & -1 & 1 & -2 & 1 \\ 3 & -3 & 2 & 0 & 2 \end{bmatrix} \begin{bmatrix} x_1 \\ x_2 \\ x_3 \\ x_4 \\ x_5 \end{bmatrix} = \begin{bmatrix} 0 \\ 0 \end{bmatrix}.$

3. $x_1 + 2x_2 - x_3 + 3x_4 = 0$
$2x_1 + 2x_2 - x_3 + 2x_4 = 0$
$x_1 \qquad + 3x_3 + 3x_4 = 0.$

4. $x_1 - x_2 + 2x_3 + 3x_4 + 4x_5 = 0$
$-x_1 + 2x_2 + 3x_3 + 4x_4 + 5x_5 = 0$
$x_1 - x_2 + 3x_3 + 5x_4 + 6x_5 = 0$
$3x_1 - 4x_2 + x_3 + 2x_4 + 3x_5 = 0.$

5. $\begin{bmatrix} 1 & 2 & 1 & 2 & 1 \\ 1 & 2 & 2 & 1 & 2 \\ 2 & 4 & 3 & 3 & 3 \\ 0 & 0 & 1 & -1 & -1 \end{bmatrix} \begin{bmatrix} x_1 \\ x_2 \\ x_3 \\ x_4 \\ x_5 \end{bmatrix} = \begin{bmatrix} 0 \\ 0 \\ 0 \\ 0 \end{bmatrix}.$

6. $\begin{bmatrix} 1 & 0 & 2 \\ 2 & 1 & 3 \\ 3 & 1 & 2 \end{bmatrix} \begin{bmatrix} x_1 \\ x_2 \\ x_3 \end{bmatrix} = \begin{bmatrix} 0 \\ 0 \\ 0 \end{bmatrix}.$

7. $\begin{bmatrix} 1 & 2 & 2 & -1 & 1 \\ 0 & 2 & 2 & -2 & -1 \\ 2 & 6 & 2 & -4 & 1 \\ 1 & 4 & 0 & -3 & 0 \end{bmatrix} \begin{bmatrix} x_1 \\ x_2 \\ x_3 \\ x_4 \\ x_5 \end{bmatrix} = \begin{bmatrix} 0 \\ 0 \\ 0 \\ 0 \end{bmatrix}.$

8. $\begin{bmatrix} 1 & 2 & -3 & -2 & 1 & 3 \\ 1 & 2 & -4 & 3 & 3 & 4 \\ -2 & -4 & 6 & 4 & -3 & 2 \\ 0 & 0 & -1 & 5 & 1 & 9 \\ 1 & 2 & -3 & -2 & 0 & 7 \end{bmatrix} \begin{bmatrix} x_1 \\ x_2 \\ x_3 \\ x_4 \\ x_5 \\ x_6 \end{bmatrix} = \begin{bmatrix} 0 \\ 0 \\ 0 \\ 0 \\ 0 \end{bmatrix}.$

In Exercises 9 and 10, find a basis for the null space of the given matrix A.

9. $A = \begin{bmatrix} 1 & 2 & 3 & -1 \\ 2 & 3 & 2 & 0 \\ 3 & 4 & 1 & 1 \\ 1 & 1 & -1 & 1 \end{bmatrix}.$

10. $A = \begin{bmatrix} 1 & -1 & 2 & 1 & 0 \\ 2 & 0 & 1 & -1 & 3 \\ 5 & -1 & 3 & 0 & 3 \\ 4 & -2 & 5 & 1 & 3 \\ 1 & 3 & -4 & -5 & 6 \end{bmatrix}.$

In Exercises 11 through 14, find a basis for the solution space of the homogeneous system $(\lambda I_n - A)\mathbf{x} = \mathbf{0}$ for the given scalar λ and given matrix A.

11. $\lambda = 1; A = \begin{bmatrix} 3 & 2 \\ 1 & 2 \end{bmatrix}.$

12. $\lambda = -3, A = \begin{bmatrix} -4 & -3 \\ 2 & 3 \end{bmatrix}.$

13. $\lambda = 1, A = \begin{bmatrix} 0 & 0 & 1 \\ 1 & 0 & -3 \\ 0 & 1 & 3 \end{bmatrix}.$

14. $\lambda = 3, A = \begin{bmatrix} 1 & 1 & -2 \\ -1 & 2 & 1 \\ 0 & 1 & -1 \end{bmatrix}.$

In Exercises 15 through 18, find all real numbers λ such that the homogeneous system $(\lambda I_n - A)\mathbf{x} = \mathbf{0}$ has a nontrivial solution.

15. $A = \begin{bmatrix} 2 & 3 \\ 2 & -3 \end{bmatrix}.$ **16.** $A = \begin{bmatrix} 3 & 0 \\ 2 & -2 \end{bmatrix}.$

17. $A = \begin{bmatrix} 0 & 0 & 0 \\ 0 & 1 & -1 \\ 1 & 0 & 0 \end{bmatrix}.$

18. $A = \begin{bmatrix} -2 & 0 & 0 \\ 0 & -2 & -3 \\ 0 & 4 & 5 \end{bmatrix}.$

THEORETICAL EXERCISES ◼

T.1. Let $S = \{\mathbf{x}_1, \mathbf{x}_2, \ldots, \mathbf{x}_k\}$ be a set of solutions to a homogeneous system $A\mathbf{x} = \mathbf{0}$. Show that every vector in span S is a solution to $A\mathbf{x} = \mathbf{0}$.

T.2. Show that if the $n \times n$ coefficient matrix A of the homogeneous system $A\mathbf{x} = \mathbf{0}$ has a row or column of zeros, then $A\mathbf{x} = \mathbf{0}$ has a nontrivial solution.

T.3. (a) Show that the zero matrix is the only 3×3 matrix whose null space has dimension 3.

(b) Let A be a nonzero 3×3 matrix and suppose that $A\mathbf{x} = \mathbf{0}$ has a nontrivial solution. Show that the dimension of the null space of A is either 1 or 2.

T.4. A and B are $m \times n$ and their reduced row echelon forms are the same. What is the relationship between the null space of A and the null space of B?

MATLAB EXERCISES ◼

In Exercises ML.1 through ML.3, use MATLAB's **rref** command to aid in finding a basis for the null space of A. You may also use routine **homsoln**. For directions, use **help**.

ML.1. $A = \begin{bmatrix} 1 & 1 & 2 & 2 & 1 \\ 2 & 0 & 4 & 2 & 4 \\ 1 & 1 & 2 & 2 & 1 \end{bmatrix}$.

ML.2. $A = \begin{bmatrix} 2 & 2 & 2 \\ 1 & 2 & 1 \\ 3 & 1 & 0 \\ 0 & 0 & 1 \\ 1 & 0 & 0 \end{bmatrix}$.

ML.3. $A = \begin{bmatrix} 1 & 4 & 7 & 0 \\ 2 & 5 & 8 & -1 \\ 3 & 6 & 9 & -2 \end{bmatrix}$.

ML.4. For the matrix

$$A = \begin{bmatrix} 1 & 2 \\ 2 & 1 \end{bmatrix}$$

and $\lambda = 3$, the homogeneous system $(\lambda I_2 - A)\mathbf{x} = \mathbf{0}$ has a nontrivial solution. Find such a solution using MATLAB commands.

ML.5. For the matrix

$$A = \begin{bmatrix} 1 & 2 & 3 \\ 3 & 2 & 1 \\ 2 & 1 & 3 \end{bmatrix}$$

and $\lambda = 6$, the homogeneous linear system $(\lambda I_3 - A)\mathbf{x} = \mathbf{0}$ has a nontrivial solution. Find such a solution using MATLAB commands.

4.6 ▾ The Rank of a Matrix and Applications

In this section we obtain another effective method for finding a basis for a vector space V spanned by a given set of vectors $S = \{\mathbf{v}_1, \mathbf{v}_2, \ldots, \mathbf{v}_n\}$. In the proof of Theorem 4.6 we developed a technique for choosing a basis for V that is a *subset* of S. The method to be developed in this section produces a basis for V that is *not* guaranteed to be a subset of S. We shall also attach a unique number to a matrix A that we later show gives us information about the dimension of the solution space of a homogeneous system with coefficient matrix A.

DEFINITION   Let

$$A = \begin{bmatrix} a_{11} & a_{12} & \cdots & a_{1n} \\ a_{21} & a_{22} & \cdots & a_{2n} \\ \vdots & \vdots & & \vdots \\ a_{m1} & a_{m2} & \cdots & a_{mn} \end{bmatrix}$$

be an $m \times n$ matrix. The rows of A,

$$\mathbf{v}_1 = (a_{11}, a_{12}, \ldots, a_{1n})$$
$$\mathbf{v}_2 = (a_{21}, a_{22}, \ldots, a_{2n})$$
$$\vdots$$
$$\mathbf{v}_m = (a_{m1}, a_{m2}, \ldots, a_{mn}),$$

considered as vectors in R^n, span a subspace of R^n, called the **row space** of A. Similarly, the columns of A,

$$\mathbf{w}_1 = \begin{bmatrix} a_{11} \\ a_{21} \\ \vdots \\ a_{m1} \end{bmatrix}, \qquad \mathbf{w}_2 = \begin{bmatrix} a_{12} \\ a_{22} \\ \vdots \\ a_{m2} \end{bmatrix}, \ldots, \qquad \mathbf{w}_n = \begin{bmatrix} a_{1n} \\ a_{2n} \\ \vdots \\ a_{mn} \end{bmatrix},$$

considered as vectors in R^m, span a subspace of R^m, called the **column space** of A.

THEOREM 4.10 ■ *If A and B are two $m \times n$ row equivalent matrices, then the row spaces of A and B are equal.*

Proof   If A and B are row equivalent, then the rows of B are obtained from those of A by a finite number of the three elementary row operations. Thus each row of B is a linear combination of the rows of A. Hence the row space of B is contained in the row space of A. Similarly, A can be obtained from B by a finite number of the three elementary row operations, so the row space of A is contained in the row space of B. Hence the row spaces of A and B are equal.   ■

REMARK   For a related result see Theorem 1.6 in Section 1.5.

It follows from Theorem 4.10 that if we take a given matrix A and transform it to reduced row echelon form B, then the row spaces of A and B are equal. Furthermore, recall from Exercise T.8 in Section 4.3 that the nonzero rows of a matrix that is in reduced row echelon form are linearly independent and thus form a basis for its row space. We can use this method to find a basis for a vector space spanned by a given set of vectors in R^n, as illustrated in Example 1.

EXAMPLE 1 ■ Let $S = \{\mathbf{v}_1, \mathbf{v}_2, \mathbf{v}_3, \mathbf{v}_4\}$, where

$$\mathbf{v}_1 = (1, -2, 0, 3, -4), \qquad \mathbf{v}_2 = (3, 2, 8, 1, 4),$$
$$\mathbf{v}_3 = (2, 3, 7, 2, 3), \qquad \mathbf{v}_4 = (-1, 2, 0, 4, -3),$$

and let V be the subspace of R^5 given by $V = \text{span } S$. Find a basis for V.

Solution   Note that V is the row space of the matrix A whose rows are the given vectors

$$A = \begin{bmatrix} 1 & -2 & 0 & 3 & -4 \\ 3 & 2 & 8 & 1 & 4 \\ 2 & 3 & 7 & 2 & 3 \\ -1 & 2 & 0 & 4 & -3 \end{bmatrix}.$$

Using elementary row operations, we find that A is row equivalent to the matrix (verify)

$$B = \begin{bmatrix} 1 & 0 & 2 & 0 & 1 \\ 0 & 1 & 1 & 0 & 1 \\ 0 & 0 & 0 & 1 & -1 \\ 0 & 0 & 0 & 0 & 0 \end{bmatrix},$$

which is in reduced row echelon form. The row spaces of A and B are identical, and a basis for the row space of B consists of the nonzero rows of B. Hence

$$\mathbf{w}_1 = (1, 0, 2, 0, 1), \quad \mathbf{w}_2 = (0, 1, 1, 0, 1), \quad \text{and} \quad \mathbf{w}_3 = (0, 0, 0, 1, -1)$$

form a basis for V.   ∎

We may now summarize the method used in Example 1 to find a basis for the subspace V of R^n given by $V = \text{span } S$, where S is a set of vectors in R^n.

The procedure for finding a basis for the subspace V of R^n given by $V = \text{span } S$, where $S = \{\mathbf{v}_1, \mathbf{v}_2, \ldots, \mathbf{v}_k\}$, is a set of vectors in R^n that are given in row form, is as follows.

Step 1. Form the matrix

$$A = \begin{bmatrix} \mathbf{v}_1 \\ \mathbf{v}_2 \\ \vdots \\ \mathbf{v}_k \end{bmatrix}$$

whose rows are the given vectors in S.

Step 2. Transform A to reduced row echelon form, obtaining the matrix B.

Step 3. The nonzero rows of B form a basis for V.

Of course, the basis that has been obtained by the procedure used in Example 1 produced a basis that is not a subset of the spanning set. However, the basis for a subspace V of R^n that is obtained in this manner is analogous, in its simplicity, to the standard basis for R^n. Thus if $\mathbf{v} = (a_1, a_2, \ldots, a_n)$ is a vector in V and $\{\mathbf{v}_1, \mathbf{v}_2, \ldots, \mathbf{v}_k\}$ is a basis for V obtained by the method of Example 1 where the leading 1's occur in columns $j_1, j_2, \ldots, j_k$, then it can be shown that

$$\mathbf{v} = a_{j_1}\mathbf{v}_1 + a_{j_2}\mathbf{v}_2 + \cdots + a_{j_k}\mathbf{v}_k.$$

EXAMPLE 2 ■ Let V be the subspace of Example 1. Given that the vector

$$\mathbf{v} = (5, 4, 14, 6, 3)$$

is in V, write $\mathbf{v}$ as a linear combination of the basis determined in Example 1.

Solution   We have $j_1 = 1$, $j_2 = 2$, and $j_3 = 4$, so

$$\mathbf{v} = 5\mathbf{w}_1 + 4\mathbf{w}_2 + 6\mathbf{w}_3.$$ ■

REMARK   The solution to Example 1 also yields a basis for the row space of matrix A in that example. Observe that the vectors in this basis are not rows of matrix A.

DEFINITION   The dimension of the row space of A is called the **row rank** of A, and the dimension of the column space of A is called the **column rank** of A.

EXAMPLE 3 ■   Find a basis for the row space of the matrix A defined in the solution of Example 1 that contains only row vectors from A. Also compute the row rank of A.

Solution   Using the procedure in the alternate proof of Theorem 4.6, we form the equation

$$c_1(1, -2, 0, 3, -4) + c_2(3, 2, 8, 1, 4)$$
$$+ c_3(2, 3, 7, 2, 3) + c_4(-1, 2, 0, 4, -3) = (0, 0, 0, 0, 0),$$

whose augmented matrix is

$$\begin{bmatrix} 1 & 3 & 2 & -1 & \vdots & 0 \\ -2 & 2 & 3 & 2 & \vdots & 0 \\ 0 & 8 & 7 & 0 & \vdots & 0 \\ 3 & 1 & 2 & 4 & \vdots & 0 \\ -4 & 4 & 3 & -3 & \vdots & 0 \end{bmatrix} = \begin{bmatrix} A^T & \vdots & \mathbf{0} \end{bmatrix}; \tag{1}$$

that is, the coefficient matrix is A^T. Transforming the augmented matrix $\begin{bmatrix} A^T & \vdots & \mathbf{0} \end{bmatrix}$ in (1) to reduced row echelon form, we obtain (verify)

$$\begin{bmatrix} 1 & 0 & 0 & \frac{11}{24} & \vdots & 0 \\ 0 & 1 & 0 & -\frac{49}{24} & \vdots & 0 \\ 0 & 0 & 1 & \frac{7}{3} & \vdots & 0 \\ 0 & 0 & 0 & 0 & \vdots & 0 \\ 0 & 0 & 0 & 0 & \vdots & 0 \end{bmatrix}. \tag{2}$$

Since the leading 1's in (2) occur in columns 1, 2, and 3, we conclude that the first three rows of A form a basis for the row space of A. That is,

$$\{(1, -2, 0, 3, -4), (3, 2, 8, 1, 4), (2, 3, 7, 2, 3)\}$$

is a basis for the row space of A. The row rank of A is 3.   ■

EXAMPLE 4 ■   Find a basis for the column space of the matrix A defined in the solution of Example 1 and compute the column rank of A.

Solution 1   Writing the columns of A as row vectors, we obtain the matrix A^T, which when transformed to reduced row echelon form is (as we saw in Example 3)

$$\begin{bmatrix} 1 & 0 & 0 & \frac{11}{24} \\ 0 & 1 & 0 & -\frac{49}{24} \\ 0 & 0 & 1 & \frac{7}{3} \\ 0 & 0 & 0 & 0 \\ 0 & 0 & 0 & 0 \end{bmatrix}.$$

Thus the vectors $\left(1, 0, 0, \frac{11}{24}\right)$, $\left(0, 1, 0, -\frac{49}{24}\right)$, and $\left(0, 0, 1, \frac{7}{3}\right)$ form a basis for the row space of A^T. Hence the vectors

$$\begin{bmatrix} 1 \\ 0 \\ 0 \\ \frac{11}{24} \end{bmatrix}, \quad \begin{bmatrix} 0 \\ 1 \\ 0 \\ -\frac{49}{24} \end{bmatrix}, \quad \text{and} \quad \begin{bmatrix} 0 \\ 0 \\ 1 \\ \frac{7}{3} \end{bmatrix}$$

form a basis for the column space of A and we conclude that the column rank of A is 3.

Solution 2   If we want to find a basis for the column space of A that contains only the column vectors from A, we follow the procedure developed in the proof of Theorem 4.6, forming the equation

$$c_1 \begin{bmatrix} 1 \\ 3 \\ 2 \\ -1 \end{bmatrix} + c_2 \begin{bmatrix} -2 \\ 2 \\ 3 \\ 2 \end{bmatrix} + c_3 \begin{bmatrix} 0 \\ 8 \\ 7 \\ 0 \end{bmatrix} + c_4 \begin{bmatrix} 3 \\ 1 \\ 2 \\ 4 \end{bmatrix} + c_5 \begin{bmatrix} -4 \\ 4 \\ 3 \\ -3 \end{bmatrix} = \begin{bmatrix} 0 \\ 0 \\ 0 \\ 0 \end{bmatrix}$$

whose augmented matrix is $\begin{bmatrix} A & \vdots & \mathbf{0} \end{bmatrix}$. Transforming this matrix to reduced row echelon form, we obtain (as in Example 1)

$$\begin{bmatrix} 1 & 0 & 2 & 0 & 1 & \vdots & 0 \\ 0 & 1 & 1 & 0 & 1 & \vdots & 0 \\ 0 & 0 & 0 & 1 & -1 & \vdots & 0 \\ 0 & 0 & 0 & 0 & 0 & \vdots & 0 \end{bmatrix}.$$

Since the leading 1's occur in columns 1, 2, and 4, we conclude that the first, second, and fourth columns of A form a basis for the column space of A. That is,

$$\left\{ \begin{bmatrix} 1 \\ 3 \\ 2 \\ -1 \end{bmatrix}, \begin{bmatrix} -2 \\ 2 \\ 3 \\ 2 \end{bmatrix}, \begin{bmatrix} 3 \\ 1 \\ 2 \\ 4 \end{bmatrix} \right\}$$

is a basis for the column space of A. The column rank of A is 3.   ■

From Examples 3 and 4 we observe that the row and column ranks of A are equal. This is always true, and is a very important result in linear algebra. We now turn to the proof of this theorem.

THEOREM 4.11 ■ *The row rank and column rank of the $m \times n$ matrix $A = \left[a_{ij} \right]$ are equal.*

Proof   Let $\mathbf{w}_1, \mathbf{w}_2, \ldots, \mathbf{w}_n$ denote the columns of A. To determine the dimension of the column space of A, we use the procedure in the alternate proof of Theorem 4.6. Thus we consider the equation

$$c_1 \mathbf{w}_1 + c_2 \mathbf{w}_2 + \cdots + c_n \mathbf{w}_n = \mathbf{0}.$$

We now transform the augmented matrix, $\left[A \mathrel{\vdots} \mathbf{0} \right]$, of this homogeneous system to reduced row echelon form. The vectors corresponding to the columns containing the leading 1's form a basis for the column space of A. Thus the column rank of A is the number of leading 1's. But this number is also the number of nonzero rows in the reduced row echelon form matrix that is row equivalent to A, so it is the row rank of A. Thus row rank A = column rank A. ■

DEFINITION   Since row rank A = column rank A, we now merely refer to the **rank** of an $m \times n$ matrix and write rank A.

We next summarize the procedure for computing the rank of a matrix.

> The procedure for computing the rank of the matrix A is as follows.
>
> **Step 1.** Using elementary row operations, transform A to a matrix B in reduced row echelon form.
>
> **Step 2.** Rank A = the number of nonzero rows of B.

If A is an $m \times n$ matrix, we have defined (see Section 4.5) the nullity of A as the dimension of the null space of A, that is, the dimension of the solution space of $A\mathbf{x} = \mathbf{0}$. If A is transformed to a matrix B in reduced row echelon form having r nonzero rows, then we know that the dimension of the solution space of $A\mathbf{x} = \mathbf{0}$ is $n - r$. Since r is also the rank of A, we have obtained a fundamental relationship between the rank and nullity of A, which we state in the following theorem.

THEOREM 4.12 ■ *If A is an $m \times n$ matrix, then rank A + nullity $A = n$.* ■

EXAMPLE 5 ■ Let

$$A = \begin{bmatrix} 1 & 1 & 4 & 1 & 2 \\ 0 & 1 & 2 & 1 & 1 \\ 0 & 0 & 0 & 1 & 2 \\ 1 & -1 & 0 & 0 & 2 \\ 2 & 1 & 6 & 0 & 1 \end{bmatrix},$$

which was defined in Example 1 of Section 4.5. When A is transformed to reduced row echelon form, we obtain

$$\begin{bmatrix} 1 & 0 & 2 & 0 & 1 \\ 0 & 1 & 2 & 0 & -1 \\ 0 & 0 & 0 & 1 & 2 \\ 0 & 0 & 0 & 0 & 0 \\ 0 & 0 & 0 & 0 & 0 \end{bmatrix}.$$

Then rank $A = 3$ and nullity $A = 2$. This agrees with the result obtained in the solution of Example 1 of Section 4.5, where we found that the dimension of the solution space of $A\mathbf{x} = \mathbf{0}$ is 2. ∎

Rank and Singularity

For a square matrix, its rank can be used to determine whether the matrix is singular or nonsingular, as the following theorem shows.

THEOREM 4.13 ■ *An $n \times n$ matrix is nonsingular if and only if rank $A = n$.*

Proof   Suppose that A is nonsingular. Then A is row equivalent to I_n (Theorem 1.11, Section 1.6), and so rank $A = n$.

Conversely, let rank $A = n$. Suppose that A is row equivalent to a matrix B in reduced row echelon form. Then rank $B = n$, so the rows of A must be linearly independent. Hence B has no zero rows and since it is in reduced row echelon form, it must be I_n. Thus A is row equivalent to I_n and so A is nonsingular (Theorem 1.11, Section 1.6). ∎

An immediate consequence of Theorem 4.13 is the following corollary, which is a criterion for the rank of an $n \times n$ matrix to be n.

COROLLARY 4.2 ■ *If A is an $n \times n$ matrix, then rank $A = n$ if and only if $\det(A) \neq 0$.*

Proof   Exercise T.1. ∎

Another result easily obtained from Theorem 4.13 is contained in the following corollary.

COROLLARY 4.3 ■ *Let A be an $n \times n$ matrix. The linear system $A\mathbf{x} = \mathbf{b}$ has a unique solution for every $n \times 1$ matrix $\mathbf{b}$ if and only if rank $A = n$.*

Proof   Exercise T.2. ∎

The following corollary gives another method of testing whether n given vectors in R^n are linearly dependent or linearly independent.

COROLLARY 4.4 ■ *Let $S = \{\mathbf{v}_1, \mathbf{v}_2, \ldots, \mathbf{v}_n\}$ be a set of n vectors in R^n and let A be the matrix whose rows (columns) are the vectors in S. Then S is linearly independent if and only if $\det(A) \neq 0$.*

Proof   Exercise T.3.   ■

For $n > 4$, the method of Corollary 4.4 for testing linear dependence is not as efficient as the direct method of Section 4.3, calling for the solution of a homogeneous system.

COROLLARY 4.5 ■   *The homogeneous system $A\mathbf{x} = \mathbf{0}$ of n linear equations in n unknowns has a nontrivial solution if and only if rank $A < n$.*

Proof   This follows from Corollary 4.2 and from the fact that $A\mathbf{x} = \mathbf{0}$ has a nontrivial solution if and only if A is singular (Theorem 1.12, Section 1.6).   ■

EXAMPLE 6 ■   Let

$$A = \begin{bmatrix} 1 & 2 & 0 \\ 0 & 1 & 3 \\ 2 & 1 & 3 \end{bmatrix}.$$

If we transform A to reduced row echelon form B, we find that $B = I_3$ (verify). Thus rank $A = 3$ and matrix A is nonsingular. Moreover, the homogeneous system $A\mathbf{x} = \mathbf{0}$ has only the trivial solution.   ■

EXAMPLE 7 ■   Let

$$A = \begin{bmatrix} 1 & 2 & 0 \\ 1 & 1 & -3 \\ 1 & 3 & 3 \end{bmatrix}.$$

Then A is row equivalent to

$$\begin{bmatrix} 1 & 0 & -6 \\ 0 & 1 & 3 \\ 0 & 0 & 0 \end{bmatrix},$$

a matrix in reduced row echelon form. Hence rank $A < 3$, and A is singular. Moreover, $A\mathbf{x} = \mathbf{0}$ has a nontrivial solution.   ■

Applications of Rank to the Linear System $A\mathbf{x} = \mathbf{b}$

In Corollary 4.5 we have seen that the rank of A provides us with information about the existence of a nontrivial solution to the homogeneous system $A\mathbf{x} = \mathbf{0}$. We shall now obtain some results that use the rank of A to provide information about the solutions to the linear system $A\mathbf{x} = \mathbf{b}$, where $\mathbf{b}$ is an arbitrary $n \times 1$ matrix. When $\mathbf{b} \neq \mathbf{0}$, the linear system is said to be **nonhomogeneous**.

THEOREM 4.14 ■   *The linear system $A\mathbf{x} = \mathbf{b}$ has a solution if and only if rank $A = $ rank $\begin{bmatrix} A & \vdots & \mathbf{b} \end{bmatrix}$; that is, if and only if the ranks of the coefficient and augmented matrices are equal.*

Proof First, observe that if $A = \begin{bmatrix} a_{ij} \end{bmatrix}$ is $m \times n$, then the given linear system may be written as

$$x_1 \begin{bmatrix} a_{11} \\ a_{21} \\ \vdots \\ a_{m1} \end{bmatrix} + x_2 \begin{bmatrix} a_{12} \\ a_{22} \\ \vdots \\ a_{m2} \end{bmatrix} + \cdots + x_n \begin{bmatrix} a_{1n} \\ a_{2n} \\ \vdots \\ a_{mn} \end{bmatrix} = \begin{bmatrix} b_1 \\ b_2 \\ \vdots \\ b_m \end{bmatrix}. \tag{3}$$

Suppose now that $A\mathbf{x} = \mathbf{b}$ has a solution. Then there exist values of $x_1, x_2, \ldots, x_n$ that satisfy Equation (3). Thus, $\mathbf{b}$ is a linear combination of the columns of A and so belongs to the column space of A. Hence, rank $A = $ rank $\begin{bmatrix} A \vdots \mathbf{b} \end{bmatrix}$.

Conversely, suppose that rank $A = $ rank $\begin{bmatrix} A \vdots \mathbf{b} \end{bmatrix}$. Then $\mathbf{b}$ is in the column space of A, which means that we can find values of $x_1, x_2, \ldots, x_n$ that satisfy Equation (3). Hence $A\mathbf{x} = \mathbf{b}$ has a solution. ■

REMARK The result given in Theorem 4.14, while of interest, is not of great computational value, since we usually are interested in finding a solution rather than in knowing whether or not a solution exists.

EXAMPLE 8 ■ Consider the linear system

$$\begin{bmatrix} 2 & 1 & 3 \\ 1 & -2 & 2 \\ 0 & 1 & 3 \end{bmatrix} \begin{bmatrix} x_1 \\ x_2 \\ x_3 \end{bmatrix} = \begin{bmatrix} 1 \\ 2 \\ 3 \end{bmatrix}.$$

Since rank $A = $ rank $\begin{bmatrix} A \vdots \mathbf{b} \end{bmatrix} = 3$, the linear system has a solution. ■

EXAMPLE 9 ■ The linear system

$$\begin{bmatrix} 1 & 2 & 3 \\ 1 & -3 & 4 \\ 2 & -1 & 7 \end{bmatrix} \begin{bmatrix} x_1 \\ x_2 \\ x_3 \end{bmatrix} = \begin{bmatrix} 4 \\ 5 \\ 6 \end{bmatrix}$$

has no solution because rank $A = 2$ and rank $\begin{bmatrix} A \vdots \mathbf{b} \end{bmatrix} = 3$ (verify). ■

We now extend our List of Nonsingular Equivalences.

List of Nonsingular Equivalences

The following statements are equivalent for an $n \times n$ matrix A.

1. A is nonsingular.
2. $A\mathbf{x} = \mathbf{0}$ has only the trivial solution.
3. A is row equivalent to I_n.
4. The linear system $A\mathbf{x} = \mathbf{b}$ has a unique solution for every $n \times 1$ matrix $\mathbf{b}$.
5. $\det(A) \neq 0$.
6. A has rank n.
7. A has nullity 0.
8. The rows of A form a linearly independent set of n vectors in R^n.
9. The columns of A form a linearly independent set of n vectors in R^n.

4.6   EXERCISES

1. Let $S = \{\mathbf{v}_1, \mathbf{v}_2, \mathbf{v}_3, \mathbf{v}_4, \mathbf{v}_5\}$, where

$$\mathbf{v}_1 = (1, 2, 3), \qquad \mathbf{v}_2 = (2, 1, 4),$$
$$\mathbf{v}_3 = (-1, -1, 2), \qquad \mathbf{v}_4 = (0, 1, 2),$$

and $\mathbf{v}_5 = (1, 1, 1)$. Find a basis for the subspace $V = \text{span } S$ of R^3.

2. Let $S = \{\mathbf{v}_1, \mathbf{v}_2, \mathbf{v}_3, \mathbf{v}_4, \mathbf{v}_5\}$, where

$$\mathbf{v}_1 = (1, 1, 2, 1), \qquad \mathbf{v}_2 = (1, 0, -3, 1),$$
$$\mathbf{v}_3 = (0, 1, 1, 2), \qquad \mathbf{v}_4 = (0, 0, 1, 1),$$

and $\mathbf{v}_5 = (1, 0, 0, 1)$. Find a basis for the subspace $V = \text{span } S$ of R^4.

3. Let $S = \{\mathbf{v}_1, \mathbf{v}_2, \mathbf{v}_3, \mathbf{v}_4, \mathbf{v}_5\}$, where

$$\mathbf{v}_1 = \begin{bmatrix} 1 \\ 2 \\ 1 \\ 2 \end{bmatrix}, \qquad \mathbf{v}_2 = \begin{bmatrix} 2 \\ 1 \\ 2 \\ 1 \end{bmatrix},$$

$$\mathbf{v}_3 = \begin{bmatrix} 3 \\ 2 \\ 3 \\ 2 \end{bmatrix}, \qquad \mathbf{v}_4 = \begin{bmatrix} 3 \\ 3 \\ 3 \\ 3 \end{bmatrix}, \quad \text{and} \quad \mathbf{v}_5 = \begin{bmatrix} 5 \\ 3 \\ 5 \\ 3 \end{bmatrix}.$$

Find a basis for the subspace $V = \text{span } S$ of R^4.

4. Let $S = \{\mathbf{v}_1, \mathbf{v}_2, \mathbf{v}_3, \mathbf{v}_4, \mathbf{v}_5\}$, where

$$\mathbf{v}_1 = \begin{bmatrix} 1 \\ 2 \\ 1 \\ 1 \end{bmatrix}, \qquad \mathbf{v}_2 = \begin{bmatrix} 2 \\ 1 \\ 3 \\ 1 \end{bmatrix},$$

$$\mathbf{v}_3 = \begin{bmatrix} 0 \\ 2 \\ 1 \\ 2 \end{bmatrix}, \qquad \mathbf{v}_4 = \begin{bmatrix} 3 \\ 2 \\ 1 \\ 4 \end{bmatrix}, \quad \text{and} \quad \mathbf{v}_5 = \begin{bmatrix} 5 \\ 0 \\ 0 \\ -1 \end{bmatrix}.$$

Find a basis for the subspace $V = \text{span } S$ of R^4.

In Exercises 5 and 6, find a basis for the row space of A
(a) consisting of vectors that are *not* row vectors of A;
(b) consisting of vectors that *are* row vectors of A.

5. $A = \begin{bmatrix} 1 & 2 & -1 \\ 1 & 9 & -1 \\ -3 & 8 & 3 \\ -2 & 3 & 2 \end{bmatrix}.$

6. $A = \begin{bmatrix} 1 & 2 & -1 & 3 \\ 3 & 5 & 2 & 0 \\ 0 & 1 & 2 & 1 \\ -1 & 0 & -2 & 7 \end{bmatrix}.$

In Exercises 7 and 8, find a basis for the column space of A
(a) consisting of vectors that are *not* column vectors of A;
(b) consisting of vectors that *are* column vectors of A.

7. $A = \begin{bmatrix} 1 & -2 & 7 & 0 \\ 1 & -1 & 4 & 0 \\ 3 & 2 & -3 & 5 \\ 2 & 1 & -1 & 3 \end{bmatrix}.$

8. $A = \begin{bmatrix} -2 & 2 & 3 & 7 & 1 \\ -2 & 2 & 4 & 8 & 0 \\ -3 & 3 & 2 & 8 & 4 \\ 4 & -2 & 1 & -5 & -7 \end{bmatrix}.$

In Exercises 9 and 10, compute the row and column ranks of A, verifying Theorem 4.11.

9. $A = \begin{bmatrix} 1 & 2 & 3 & 2 & 1 \\ 3 & 1 & -5 & -2 & 1 \\ 7 & 8 & -1 & 2 & 5 \end{bmatrix}.$

10. $A = \begin{bmatrix} 1 & 3 & 2 & 0 & 0 & 1 \\ 2 & 1 & -5 & 1 & 2 & 0 \\ 3 & 2 & 5 & 1 & -2 & 1 \\ 5 & 8 & 9 & 1 & -2 & 2 \\ 9 & 9 & 4 & 2 & 0 & 2 \end{bmatrix}.$

In Exercises 11 through 15, compute the rank and nullity of A and verify Theorem 4.12.

11. $\begin{bmatrix} 1 & 2 & 1 & 3 \\ 2 & 1 & -4 & -5 \\ 7 & 8 & -5 & -1 \\ 10 & 14 & -2 & 8 \end{bmatrix}.$

12. $\begin{bmatrix} 1 & 2 & 1 & 3 \\ 2 & 1 & -4 & -5 \\ 1 & 1 & 0 & 0 \\ 0 & 0 & 1 & 1 \end{bmatrix}.$

13. $\begin{bmatrix} 1 & 2 & 3 \\ -1 & 2 & 1 \\ 3 & 1 & 2 \end{bmatrix}.$

14. $\begin{bmatrix} 1 & -2 & -1 \\ 2 & -1 & 3 \\ 7 & -8 & 3 \end{bmatrix}.$

15. $\begin{bmatrix} 1 & -2 & -1 \\ 2 & -1 & 3 \\ 7 & -8 & 3 \\ 5 & -7 & 0 \end{bmatrix}.$

16. If A is a 3×4 matrix, what is the largest possible value for rank A?

17. If A is a 4×6 matrix, show that the columns of A are linearly dependent.

18. If A is a 5×3 matrix, show that the rows of A are linearly dependent.

In Exercises 19 and 20, let A be a 7×3 matrix whose rank is 3.

19. Are the rows of A linearly dependent or linearly independent? Justify your answer.

20. Are the columns of A linearly dependent or linearly independent? Justify your answer.

In Exercises 21 through 23, use Theorem 4.13 to determine whether each matrix is singular or nonsingular.

21. $\begin{bmatrix} 1 & 2 & -3 \\ -1 & 2 & 3 \\ 0 & 8 & 0 \end{bmatrix}$. **22.** $\begin{bmatrix} 1 & 2 & -3 \\ -1 & 2 & 3 \\ 0 & 1 & 1 \end{bmatrix}$.

23. $\begin{bmatrix} 1 & 1 & 4 & -1 \\ 1 & 2 & 3 & 2 \\ -1 & 3 & 2 & 1 \\ -2 & 6 & 12 & -4 \end{bmatrix}$.

In Exercises 24 and 25, use Corollary 4.3 to determine whether the linear system $A\mathbf{x} = \mathbf{b}$ has a unique solution for every 3×1 matrix $\mathbf{b}$.

24. $A = \begin{bmatrix} 1 & 2 & -2 \\ 0 & 8 & -7 \\ 3 & -2 & 1 \end{bmatrix}$. **25.** $A = \begin{bmatrix} 1 & -1 & 2 \\ 3 & 2 & 3 \\ 1 & -2 & 1 \end{bmatrix}$.

Use Corollary 4.4 to do Exercises 26 and 27.

26. Is
$$S = \left\{ \begin{bmatrix} 2 \\ 2 \\ 3 \end{bmatrix}, \begin{bmatrix} 1 \\ 0 \\ 2 \end{bmatrix}, \begin{bmatrix} 0 \\ 1 \\ 3 \end{bmatrix} \right\}$$
a linearly independent set of vectors in R^3?

27. Is
$$S = \left\{ \begin{bmatrix} 4 \\ 1 \\ 2 \end{bmatrix}, \begin{bmatrix} 2 \\ 5 \\ -5 \end{bmatrix}, \begin{bmatrix} 2 \\ -1 \\ 3 \end{bmatrix} \right\}$$
a linearly independent set of vectors in R^3?

In Exercises 28 through 30, find which homogeneous systems have a nontrivial solution for the given matrix A by using Corollary 4.5.

28. $A = \begin{bmatrix} 1 & 1 & 2 & -1 \\ 1 & 3 & -1 & 2 \\ 1 & 1 & 1 & 3 \\ 1 & 2 & 1 & 1 \end{bmatrix}$.

29. $A = \begin{bmatrix} 1 & 2 & 3 \\ 0 & 1 & 0 \\ 1 & 0 & 3 \end{bmatrix}$. **30.** $A = \begin{bmatrix} 1 & 2 & -1 \\ 2 & -1 & 3 \\ 5 & -4 & 3 \end{bmatrix}$.

In Exercises 31 through 34, determine which of the linear systems have a solution by using Theorem 4.14.

31. $\begin{bmatrix} 1 & 2 & 5 & -2 \\ 2 & 3 & -2 & 4 \\ 5 & 1 & 0 & 2 \end{bmatrix} \begin{bmatrix} x_1 \\ x_2 \\ x_3 \\ x_4 \end{bmatrix} = \begin{bmatrix} 0 \\ 0 \\ 0 \end{bmatrix}$.

32. $\begin{bmatrix} 1 & 2 & 5 & -2 \\ 2 & 3 & -2 & 4 \\ 5 & 1 & 0 & 2 \end{bmatrix} \begin{bmatrix} x_1 \\ x_2 \\ x_3 \\ x_4 \end{bmatrix} = \begin{bmatrix} 1 \\ -13 \\ 3 \end{bmatrix}$.

33. $\begin{bmatrix} 1 & -2 & -3 & 4 \\ 4 & -1 & -5 & 6 \\ 2 & 3 & 1 & -2 \end{bmatrix} \begin{bmatrix} x_1 \\ x_2 \\ x_3 \\ x_4 \end{bmatrix} = \begin{bmatrix} 1 \\ 2 \\ 2 \end{bmatrix}$.

34. $\begin{bmatrix} 1 & 1 & 1 \\ 1 & -1 & 1 \\ 5 & 1 & 5 \end{bmatrix} \begin{bmatrix} x_1 \\ x_2 \\ x_3 \end{bmatrix} = \begin{bmatrix} 6 \\ 2 \\ 5 \end{bmatrix}$.

THEORETICAL EXERCISES ■

T.1. Prove Corollary 4.2.

T.2. Prove Corollary 4.3.

T.3. Prove Corollary 4.4.

T.4. Let A be an $n \times n$ matrix. Show that the homogeneous system $A\mathbf{x} = \mathbf{0}$ has a nontrivial solution if and only if the columns of A are linearly dependent.

T.5. Let A be an $n \times n$ matrix. Show that rank $A = n$ if and only if the columns of A are linearly independent.

T.6. Let A be an $n \times n$ matrix. Show that the rows of A are linearly independent if and only if the columns of A span R^n.

T.7. Let A be an $m \times n$ matrix. Show that the linear system $A\mathbf{x} = \mathbf{b}$ has a solution for every $m \times 1$

matrix **b** if and only if rank $A = m$.

T.8. Let A be an $m \times n$ matrix. Show that the columns of A are linearly independent if and only if the homogeneous system $A\mathbf{x} = \mathbf{0}$ has only the trivial solution.

T.9. Let A be an $m \times n$ matrix. Show that the linear system $A\mathbf{x} = \mathbf{b}$ has at most one solution for every $m \times 1$ matrix **b** if and only if the associated homogeneous system $A\mathbf{x} = \mathbf{0}$ has only the trivial solution.

T.10. Let A be an $m \times n$ matrix with $m \neq n$. Show that either the rows or the columns of A are linearly dependent.

T.11. Suppose that the linear system $A\mathbf{x} = \mathbf{b}$, where A is $m \times n$, is consistent (has a solution). Show that the solution is unique if and only if rank $A = n$.

T.12. Show that if A is an $m \times n$ matrix such that AA^T is nonsingular, then rank $A = m$.

MATLAB EXERCISES ▪

Given a matrix A, the nonzero rows of **rref(A)** form a basis for the row space of A and the nonzero rows of **rref(A′)** transformed to columns give a basis for the column space of A.

ML.1. Solve Exercises 1 through 4 using MATLAB.

To find a basis for the row space of A that consists of rows of A, we compute **rref(A′)**. The leading 1's point to the original rows of A that give us a basis for the row space. See Example 3.

ML.2. Determine two bases for the row space of A that have no vectors in common.

(a) $A = \begin{bmatrix} 1 & 3 & 1 \\ 2 & 5 & 0 \\ 4 & 11 & 2 \\ 6 & 9 & 1 \end{bmatrix}$.

(b) $A = \begin{bmatrix} 2 & 1 & 2 & 0 \\ 0 & 0 & 0 & 0 \\ 1 & 2 & 2 & 1 \\ 4 & 5 & 6 & 2 \\ 3 & 3 & 4 & 1 \end{bmatrix}$.

ML.3. Repeat Exercise ML.2 for column spaces.

To compute the rank of a matrix A in MATLAB, use the command **rank(A)**.

ML.4. Compute the rank and nullity of each of the following matrices.

(a) $\begin{bmatrix} 3 & 2 & 1 \\ 1 & 2 & -1 \\ 2 & 1 & 3 \end{bmatrix}$.

(b) $\begin{bmatrix} 1 & 2 & 1 & 2 & 1 \\ 2 & 1 & 0 & 0 & 2 \\ 1 & -1 & -1 & -2 & 1 \\ 3 & 0 & -1 & -2 & 3 \end{bmatrix}$.

ML.5. Using only the rank command, determine which of the following linear systems is consistent.

(a) $\begin{bmatrix} 1 & 2 & 4 & -1 \\ 0 & 1 & 2 & 0 \\ 3 & 1 & 1 & -2 \end{bmatrix} \begin{bmatrix} x_1 \\ x_2 \\ x_3 \\ x_4 \end{bmatrix} = \begin{bmatrix} 21 \\ 8 \\ 16 \end{bmatrix}$.

(b) $\begin{bmatrix} 1 & 2 & 1 \\ 1 & 1 & 0 \\ 2 & 1 & -1 \end{bmatrix} \begin{bmatrix} x_1 \\ x_2 \\ x_3 \end{bmatrix} = \begin{bmatrix} 3 \\ 3 \\ 3 \end{bmatrix}$.

(c) $\begin{bmatrix} 1 & 2 \\ 2 & 0 \\ 2 & 1 \\ -1 & 2 \end{bmatrix} \begin{bmatrix} x_1 \\ x_2 \end{bmatrix} = \begin{bmatrix} 3 \\ 2 \\ 3 \\ 2 \end{bmatrix}$.

4.7 ▾ Coordinates and Change of Basis

Coordinates

If V is an n-dimensional vector space, we know that V has a basis S with n vectors in it; so far we have not paid much attention to the order of the vectors in S. However, in the discussion of this section we speak of an **ordered basis** $S = \{\mathbf{v}_1, \mathbf{v}_2, \ldots, \mathbf{v}_n\}$ for V; thus $S_1 = \{\mathbf{v}_2, \mathbf{v}_1, \ldots, \mathbf{v}_n\}$ is a different ordered basis for V.

If $S = \{\mathbf{v}_1, \mathbf{v}_2, \ldots, \mathbf{v}_n\}$ is an ordered basis for the n-dimensional vector space V, then every vector $\mathbf{v}$ in V can be uniquely expressed in the form

$$\mathbf{v} = c_1 \mathbf{v}_1 + c_2 \mathbf{v}_2 + \cdots + c_n \mathbf{v}_n,$$

where $c_1, c_2, \ldots, c_n$ are real numbers. We shall refer to

$$\left[\mathbf{v}\right]_S = \begin{bmatrix} c_1 \\ c_2 \\ \vdots \\ c_n \end{bmatrix}$$

as the **coordinate vector of v with respect to the ordered basis** S. The entries of $\left[\mathbf{v}\right]_S$ are called the **coordinates of v with respect to** S. Theorem 4.5 implies that $\left[\mathbf{v}\right]_S$ is unique.

Observe that the coordinate vector $\left[\mathbf{v}\right]_S$ depends on the order in which the vectors in S are listed; a change in the order of this listing may change the coordinates of $\mathbf{v}$ with respect to S. All bases considered in this section are assumed to be ordered bases.

EXAMPLE 1 ■ Let $S = \{\mathbf{v}_1, \mathbf{v}_2, \mathbf{v}_3, \mathbf{v}_4\}$ be a basis for R^4, where

$$\mathbf{v}_1 = (1, 1, 0, 0), \quad \mathbf{v}_2 = (2, 0, 1, 0), \quad \mathbf{v}_3 = (0, 1, 2, -1), \quad \mathbf{v}_4 = (0, 1, -1, 0).$$

If

$$\mathbf{v} = (-1, 2, -6, 5),$$

compute $\left[\mathbf{v}\right]_S$.

Solution   To find $\left[\mathbf{v}\right]_S$ we need to compute constants c_1, c_2, c_3, and c_4 such that

$$c_1 \mathbf{v}_1 + c_2 \mathbf{v}_2 + c_3 \mathbf{v}_3 + c_4 \mathbf{v}_4 = \mathbf{v},$$

which is just a linear combination problem. The previous equation leads to the linear system whose augmented matrix is (verify)

$$\begin{bmatrix} 1 & 2 & 0 & 0 & \vdots & 1 \\ 1 & 0 & 1 & 1 & \vdots & 2 \\ 0 & 1 & 2 & -1 & \vdots & -6 \\ 0 & 0 & -1 & 0 & \vdots & 2 \end{bmatrix}, \tag{1}$$

or, equivalently,

$$\begin{bmatrix} \mathbf{v}_1^T & \mathbf{v}_2^T & \mathbf{v}_3^T & \mathbf{v}_4^T & \vdots & \mathbf{v}^T \end{bmatrix}.$$

Transforming the matrix in (1) to reduced row echelon form, we obtain the solution (verify)

$$c_1 = 3, \qquad c_2 = -1, \qquad c_3 = -2, \qquad c_4 = 1,$$

so the coordinate vector of $\mathbf{v}$ with respect to the basis S is

$$\left[\mathbf{v}\right]_S = \begin{bmatrix} 3 \\ -1 \\ -2 \\ 1 \end{bmatrix}.$$

■

EXAMPLE 2 ■ Let $S = \{\mathbf{e}_1, \mathbf{e}_2, \mathbf{e}_3\}$ be the natural basis for R^3 and let
$$\mathbf{v} = (2, -1, 3).$$
Compute $[\mathbf{v}]_S$.

Solution   Since S is the natural basis,
$$\mathbf{v} = 2\mathbf{e}_1 - 1\mathbf{e}_2 + 3\mathbf{e}_3,$$
so
$$[\mathbf{v}]_S = \begin{bmatrix} 2 \\ -1 \\ 3 \end{bmatrix}.$$

■

REMARK   In Example 2 the coordinate vector $[\mathbf{v}]_S$ of $\mathbf{v}$ with respect to S agrees with $\mathbf{v}$ because S is the natural basis for R^3.

EXAMPLE 3 ■ Let V be P_1, the vector space of all polynomials of degree ≤ 1, and let $S = \{\mathbf{v}_1, \mathbf{v}_2\}$ and $T = \{\mathbf{w}_1, \mathbf{w}_2\}$ be bases for P_1, where
$$\mathbf{v}_1 = t, \qquad \mathbf{v}_2 = 1, \qquad \mathbf{w}_1 = t + 1, \qquad \mathbf{w}_2 = t - 1.$$
Let $\mathbf{v} = p(t) = 5t - 2$.

(a) Compute $[\mathbf{v}]_S$.

(b) Compute $[\mathbf{v}]_T$.

Solution   (a) Since S is the standard or natural basis for P_1, we have
$$5t - 2 = 5t + (-2)(1).$$
Hence
$$[\mathbf{v}]_S = \begin{bmatrix} 5 \\ -2 \end{bmatrix}.$$

(b) To compute $[\mathbf{v}]_T$, we must write $\mathbf{v}$ as a linear combination of $\mathbf{w}_1$ and $\mathbf{w}_2$. Thus,
$$5t - 2 = c_1(t + 1) + c_2(t - 1),$$
or
$$5t - 2 = (c_1 + c_2)t + (c_1 - c_2).$$
Equating coefficients of like powers of t, we obtain the linear system
$$\begin{aligned} c_1 + c_2 &= 5 \\ c_1 - c_2 &= -2, \end{aligned}$$
whose solution is (verify)
$$c_1 = \tfrac{3}{2} \quad \text{and} \quad c_2 = \tfrac{7}{2}.$$
Hence
$$[\mathbf{v}]_T = \begin{bmatrix} \frac{3}{2} \\ \frac{7}{2} \end{bmatrix}.$$

■

In some important ways the coordinate vectors of elements in a vector space behave algebraically in ways that are similar to the way the vectors themselves behave. For example, it is not difficult to show (see Exercise T.2) that if S is a basis for an n-dimensional vector space V, $\mathbf{v}$ and $\mathbf{w}$ are vectors in V, and k is a scalar, then

$$\left[\mathbf{v} + \mathbf{w}\right]_S = \left[\mathbf{v}\right]_S + \left[\mathbf{w}\right]_S \tag{2}$$

and

$$\left[k\mathbf{v}\right]_S = k\left[\mathbf{v}\right]_S. \tag{3}$$

That is, the coordinate vector of a sum of two vectors is the sum of the coordinate vectors, and the coordinate vector of a scalar multiple of a vector is the scalar multiple of the coordinate vector. Moreover, the results in Equations (2) and (3) can be generalized to show that

$$\left[k_1\mathbf{v}_1 + k_2\mathbf{v}_2 + \cdots + k_n\mathbf{v}_n\right]_S = k_1\left[\mathbf{v}_1\right]_S + k_2\left[\mathbf{v}_2\right]_S + \cdots + k_n\left[\mathbf{v}_n\right]_S.$$

That is, the coordinate vector of a linear combination of vectors is the same linear combination of the individual coordinate vectors.

Picturing a Vector Space

The choice of an ordered basis and the consequent assignment of a coordinate vector for every $\mathbf{v}$ in V enables us to "picture" the vector space. We illustrate this notion by using Example 3. Choose a fixed point O in the plane R^2 and draw any two arrows $\mathbf{w}_1$ and $\mathbf{w}_2$ from O that depict the basis vectors t and 1 in the ordered basis $S = \{t, 1\}$ for P_1 (see Figure 4.5). The directions of $\mathbf{w}_1$ and $\mathbf{w}_2$ determine two lines, which we call the $\boldsymbol{x_1}$- and $\boldsymbol{x_2}$-**axes**, respectively. The positive direction on the x_1-axis is in the direction of $\mathbf{w}_1$; the negative direction on the x_1-axis is along $-\mathbf{w}_1$. Similarly, the positive direction on the x_2-axis is in the direction of $\mathbf{w}_2$; the negative direction on the x_2-axis is along $-\mathbf{w}_2$. The lengths of $\mathbf{w}_1$ and $\mathbf{w}_2$ determine the scales on the x_1- and x_2-axes, respectively. If $\mathbf{v}$ is a vector in P_1, we can write $\mathbf{v}$, uniquely, as $\mathbf{v} = c_1\mathbf{w}_1 + c_2\mathbf{w}_2$. We now mark off a segment of length $|c_1|$ on the x_1-axis (in the positive direction if c_1 is

FIGURE 4.5

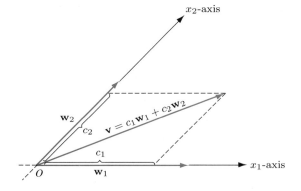

positive and in the negative direction if c_1 is negative) and draw a line through the endpoint of this segment parallel to $\mathbf{w}_2$. Similarly, mark off a segment of length $|c_2|$ on the x_2-axis (in the positive direction if c_2 is positive and in the negative direction if c_2 is negative) and draw a line through the endpoint of this segment parallel to $\mathbf{w}_1$. We draw a directed line segment from O to the point of intersection of these two lines. This directed line segment represents $\mathbf{v}$.

Transition Matrices

Suppose now that $S = \{\mathbf{v}_1, \mathbf{v}_2, \ldots, \mathbf{v}_n\}$ and $T = \{\mathbf{w}_1, \mathbf{w}_2, \ldots, \mathbf{w}_n\}$ are bases for the n-dimensional vector space V. We shall examine the relationship between the coordinate vectors $\begin{bmatrix} \mathbf{v} \end{bmatrix}_S$ and $\begin{bmatrix} \mathbf{v} \end{bmatrix}_T$ of the vector $\mathbf{v}$ in V with respect to the bases S and T, respectively.

If $\mathbf{v}$ is any vector in V, then

$$\mathbf{v} = c_1 \mathbf{w}_1 + c_2 \mathbf{w}_2 + \cdots + c_n \mathbf{w}_n \tag{4}$$

so that

$$\begin{bmatrix} \mathbf{v} \end{bmatrix}_T = \begin{bmatrix} c_1 \\ c_2 \\ \vdots \\ c_n \end{bmatrix}.$$

Then

$$\begin{aligned} \begin{bmatrix} \mathbf{v} \end{bmatrix}_S &= \begin{bmatrix} c_1 \mathbf{w}_1 + c_2 \mathbf{w}_2 + \cdots + c_n \mathbf{w}_n \end{bmatrix}_S \\ &= \begin{bmatrix} c_1 \mathbf{w}_1 \end{bmatrix}_S + \begin{bmatrix} c_2 \mathbf{w}_2 \end{bmatrix}_S + \cdots + \begin{bmatrix} c_n \mathbf{w}_n \end{bmatrix}_S \\ &= c_1 \begin{bmatrix} \mathbf{w}_1 \end{bmatrix}_S + c_2 \begin{bmatrix} \mathbf{w}_2 \end{bmatrix}_S + \cdots + c_n \begin{bmatrix} \mathbf{w}_n \end{bmatrix}_S. \end{aligned}$$

Let the coordinate vector of $\mathbf{w}_j$ with respect to S be denoted by

$$\begin{bmatrix} \mathbf{w}_j \end{bmatrix}_S = \begin{bmatrix} a_{1j} \\ a_{2j} \\ \vdots \\ a_{nj} \end{bmatrix}.$$

Then

$$\begin{bmatrix} \mathbf{v} \end{bmatrix}_S = c_1 \begin{bmatrix} a_{11} \\ a_{21} \\ \vdots \\ a_{n1} \end{bmatrix} + c_2 \begin{bmatrix} a_{12} \\ a_{22} \\ \vdots \\ a_{n2} \end{bmatrix} + \cdots + c_n \begin{bmatrix} a_{1n} \\ a_{2n} \\ \vdots \\ a_{nn} \end{bmatrix} = \begin{bmatrix} a_{11} & a_{12} & \cdots & a_{1n} \\ a_{21} & a_{22} & \cdots & a_{2n} \\ \vdots & \vdots & & \vdots \\ a_{n1} & a_{n2} & \cdots & a_{nn} \end{bmatrix} \begin{bmatrix} c_1 \\ c_2 \\ \vdots \\ c_n \end{bmatrix}$$

or

$$\begin{bmatrix} \mathbf{v} \end{bmatrix}_S = P_{S \leftarrow T} \begin{bmatrix} \mathbf{v} \end{bmatrix}_T, \tag{5}$$

where

$$P_{S \leftarrow T} = \begin{bmatrix} a_{11} & a_{12} & \cdots & a_{1n} \\ a_{21} & a_{22} & \cdots & a_{2n} \\ \vdots & \vdots & & \vdots \\ a_{n1} & a_{n2} & \cdots & a_{nn} \end{bmatrix} = \begin{bmatrix} \begin{bmatrix} \mathbf{w}_1 \end{bmatrix}_S & \begin{bmatrix} \mathbf{w}_2 \end{bmatrix}_S & \cdots & \begin{bmatrix} \mathbf{w}_n \end{bmatrix}_S \end{bmatrix}$$

is called the **transition matrix from the T-basis to the S-basis**. Equation (5) says that the coordinate vector of $\mathbf{v}$ with respect to the basis S is the transition matrix $P_{S \leftarrow T}$ times the coordinate vector of $\mathbf{v}$ with respect to the basis T. Figure 4.6 illustrates Equation (5).

FIGURE 4.6

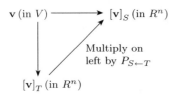

We may now summarize the procedure just developed for computing the transition matrix from the T-basis to the S-basis.

The procedure for computing the transition matrix $P_{S \leftarrow T}$ from the basis $T = \{\mathbf{w}_1, \mathbf{w}_2, \ldots, \mathbf{w}_n\}$ for V to the basis $S = \{\mathbf{v}_1, \mathbf{v}_2, \ldots, \mathbf{v}_n\}$ for V is as follows.

Step 1. Compute the coordinate vector of $\mathbf{w}_j$, $j = 1, 2, \ldots, n$, with respect to the basis S. This means that we first have to express $\mathbf{w}_j$ as a linear combination of the vectors in S:

$$a_{1j}\mathbf{v}_1 + a_{2j}\mathbf{v}_2 + \cdots + a_{nj}\mathbf{v}_n = \mathbf{w}_j, \qquad j = 1, 2, \ldots, n.$$

We now solve for $a_{1j}, a_{2j}, \ldots, a_{nj}$ by Gauss–Jordan reduction, transforming the augmented matrix of this linear system to reduced row echelon form.

Step 2. The transition matrix $P_{S \leftarrow T}$ from the T-basis to the S-basis is formed by choosing $\left[\mathbf{w}_j\right]_S$ as the jth column of $P_{S \leftarrow T}$.

EXAMPLE 4 ■ Let V be R^3 and let $S = \{\mathbf{v}_1, \mathbf{v}_2, \mathbf{v}_3\}$ and $T = \{\mathbf{w}_1, \mathbf{w}_2, \mathbf{w}_3\}$ be bases for R^3, where

$$\mathbf{v}_1 = \begin{bmatrix} 2 \\ 0 \\ 1 \end{bmatrix}, \qquad \mathbf{v}_2 = \begin{bmatrix} 1 \\ 2 \\ 0 \end{bmatrix}, \qquad \mathbf{v}_3 = \begin{bmatrix} 1 \\ 1 \\ 1 \end{bmatrix}$$

and

$$\mathbf{w}_1 = \begin{bmatrix} 6 \\ 3 \\ 3 \end{bmatrix}, \qquad \mathbf{w}_2 = \begin{bmatrix} 4 \\ -1 \\ 3 \end{bmatrix}, \qquad \mathbf{w}_3 = \begin{bmatrix} 5 \\ 5 \\ 2 \end{bmatrix}.$$

(a) Compute the transition matrix $P_{S \leftarrow T}$ from the T-basis to the S-basis.

(b) Verify Equation (5) for $\mathbf{v} = \begin{bmatrix} 4 \\ -9 \\ 5 \end{bmatrix}$.

Solution   (a) To compute $P_{S \leftarrow T}$, we find a_1, a_2, a_3 such that

$$a_1 \mathbf{v}_1 + a_2 \mathbf{v}_2 + a_3 \mathbf{v}_3 = \mathbf{w}_1.$$

In this case we are led to a linear system of three equations in three unknowns, whose augmented matrix is

$$\begin{bmatrix} \mathbf{v}_1 & \mathbf{v}_2 & \mathbf{v}_3 & \vdots & \mathbf{w}_1 \end{bmatrix}.$$

That is, the augmented matrix is

$$\begin{bmatrix} 2 & 1 & 1 & \vdots & 6 \\ 0 & 2 & 1 & \vdots & 3 \\ 1 & 0 & 1 & \vdots & 3 \end{bmatrix}.$$

Similarly, to find b_1, b_2, b_3 and c_1, c_2, c_3 such that

$$b_1 \mathbf{v}_1 + b_2 \mathbf{v}_2 + b_3 \mathbf{v}_3 = \mathbf{w}_2$$
$$c_1 \mathbf{v}_1 + c_2 \mathbf{v}_2 + c_3 \mathbf{v}_3 = \mathbf{w}_3,$$

we are led to two linear systems, each of three equations in three unknowns, whose augmented matrices are

$$\begin{bmatrix} \mathbf{v}_1 & \mathbf{v}_2 & \mathbf{v}_3 & \vdots & \mathbf{w}_2 \end{bmatrix} \quad \text{and} \quad \begin{bmatrix} \mathbf{v}_1 & \mathbf{v}_2 & \mathbf{v}_3 & \vdots & \mathbf{w}_3 \end{bmatrix},$$

or, specifically,

$$\begin{bmatrix} 2 & 1 & 1 & \vdots & 4 \\ 0 & 2 & 1 & \vdots & -1 \\ 1 & 0 & 1 & \vdots & 3 \end{bmatrix} \quad \text{and} \quad \begin{bmatrix} 2 & 1 & 1 & \vdots & 5 \\ 0 & 2 & 1 & \vdots & 5 \\ 1 & 0 & 1 & \vdots & 2 \end{bmatrix}.$$

Since the coefficient matrix of all three linear systems is $\begin{bmatrix} \mathbf{v}_1 & \mathbf{v}_2 & \mathbf{v}_3 \end{bmatrix}$, we can transform the three augmented matrices to reduced row echelon form simultaneously by transforming the partitioned matrix

$$\begin{bmatrix} \mathbf{v}_1 & \mathbf{v}_2 & \mathbf{v}_3 & \vdots & \mathbf{w}_1 & \vdots & \mathbf{w}_2 & \vdots & \mathbf{w}_3 \end{bmatrix}$$

to reduced row echelon form. Thus we transform

$$\begin{bmatrix} 2 & 1 & 1 & \vdots & 6 & \vdots & 4 & \vdots & 5 \\ 0 & 2 & 1 & \vdots & 3 & \vdots & -1 & \vdots & 5 \\ 1 & 0 & 1 & \vdots & 3 & \vdots & 3 & \vdots & 2 \end{bmatrix}$$

to reduced row echelon form, obtaining (verify)

$$\begin{bmatrix} 1 & 0 & 0 & \vdots & 2 & \vdots & 2 & \vdots & 1 \\ 0 & 1 & 0 & \vdots & 1 & \vdots & -1 & \vdots & 2 \\ 0 & 0 & 1 & \vdots & 1 & \vdots & 1 & \vdots & 1 \end{bmatrix},$$

which implies that the transition matrix from the T-basis to the S-basis is

$$P_{S \leftarrow T} = \begin{bmatrix} 2 & 2 & 1 \\ 1 & -1 & 2 \\ 1 & 1 & 1 \end{bmatrix}.$$

(b) If

$$\mathbf{v} = \begin{bmatrix} 4 \\ -9 \\ 5 \end{bmatrix},$$

then to express $\mathbf{v}$ in terms of the T-basis, we use Equation (4). From the associated linear system we find that (verify)

$$\mathbf{v} = \begin{bmatrix} 4 \\ -9 \\ 5 \end{bmatrix} = 1 \begin{bmatrix} 6 \\ 3 \\ 3 \end{bmatrix} + 2 \begin{bmatrix} 4 \\ -1 \\ 3 \end{bmatrix} - 2 \begin{bmatrix} 5 \\ 5 \\ 2 \end{bmatrix} = 1\mathbf{w}_1 + 2\mathbf{w}_2 - 2\mathbf{w}_3,$$

so

$$\begin{bmatrix} \mathbf{v} \end{bmatrix}_T = \begin{bmatrix} 1 \\ 2 \\ -2 \end{bmatrix}.$$

Then by Equation (5) we find that $\begin{bmatrix} \mathbf{v} \end{bmatrix}_S$ is

$$P_{S \leftarrow T} \begin{bmatrix} \mathbf{v} \end{bmatrix}_T = \begin{bmatrix} 2 & 2 & 1 \\ 1 & -1 & 2 \\ 1 & 1 & 1 \end{bmatrix} \begin{bmatrix} 1 \\ 2 \\ -2 \end{bmatrix} = \begin{bmatrix} 4 \\ -5 \\ 1 \end{bmatrix}.$$

If we compute $\begin{bmatrix} \mathbf{v} \end{bmatrix}_S$ directly by setting up and solving the associated linear system, we find that (verify)

$$\mathbf{v} = \begin{bmatrix} 4 \\ -9 \\ 5 \end{bmatrix} = 4 \begin{bmatrix} 2 \\ 0 \\ 1 \end{bmatrix} - 5 \begin{bmatrix} 1 \\ 2 \\ 0 \end{bmatrix} + 1 \begin{bmatrix} 1 \\ 1 \\ 1 \end{bmatrix} = 4\mathbf{v}_1 - 5\mathbf{v}_2 + 1\mathbf{v}_3,$$

so

$$\begin{bmatrix} \mathbf{v} \end{bmatrix}_S = \begin{bmatrix} 4 \\ -5 \\ 1 \end{bmatrix}.$$

Hence

$$\begin{bmatrix} \mathbf{v} \end{bmatrix}_S = P_{S \leftarrow T} \begin{bmatrix} \mathbf{v} \end{bmatrix}_T. \qquad \blacksquare$$

We next show that the transition matrix $P_{S \leftarrow T}$ from the T-basis to the S-basis is nonsingular and that $P_{S \leftarrow T}^{-1}$ is the transition matrix from the S-basis to the T-basis.

THEOREM 4.15 ■ *Let $S = \{\mathbf{v}_1, \mathbf{v}_2, \ldots, \mathbf{v}_n\}$ and $T = \{\mathbf{w}_1, \mathbf{w}_2, \ldots, \mathbf{w}_n\}$ be bases for the n-dimensional vector space V. Let $P_{S \leftarrow T}$ be the transition matrix from the T-basis to the S-basis. Then $P_{S \leftarrow T}$ is nonsingular and $P_{S \leftarrow T}^{-1}$ is the transition matrix from the S-basis to the T-basis.*

Proof   We proceed by showing that the null space of $P_{S \leftarrow T}$ contains only the zero vector. Suppose that $P_{S \leftarrow T} \begin{bmatrix} \mathbf{v} \end{bmatrix}_T = \mathbf{0}_{R^n}$ for some $\mathbf{v}$ in V. From Equation (5) we have

$$P_{S \leftarrow T} \begin{bmatrix} \mathbf{v} \end{bmatrix}_T = \begin{bmatrix} \mathbf{v} \end{bmatrix}_S = \mathbf{0}_{R^n}.$$

If $\mathbf{v} = b_1\mathbf{v}_1 + b_2\mathbf{v}_2 + \cdots + b_n\mathbf{v}_n$, then

$$\begin{bmatrix} b_1 \\ b_2 \\ \vdots \\ b_n \end{bmatrix} = \begin{bmatrix} \mathbf{v} \end{bmatrix}_S = \mathbf{0}_{R^n} = \begin{bmatrix} 0 \\ 0 \\ \vdots \\ 0 \end{bmatrix},$$

so

$$\mathbf{v} = 0\mathbf{v}_1 + 0\mathbf{v}_2 + \cdots + 0\mathbf{v}_n = \mathbf{0}_V.$$

Hence $\begin{bmatrix} \mathbf{v} \end{bmatrix}_T = \mathbf{0}_{R^n}$. Thus the homogeneous system $P_{S \leftarrow T}\mathbf{x} = \mathbf{0}$ has only the trivial solution; it then follows from Theorem 1.12 that $P_{S \leftarrow T}$ is nonsingular. Multiplying both sides of Equation (5) on the left by $P_{S \leftarrow T}^{-1}$, we have

$$\begin{bmatrix} \mathbf{v} \end{bmatrix}_T = P_{S \leftarrow T}^{-1} \begin{bmatrix} \mathbf{v} \end{bmatrix}_S.$$

That is, $P_{S \leftarrow T}^{-1}$ is then the transition matrix from the S-basis to the T-basis; the jth column of $P_{S \leftarrow T}^{-1}$ is $\begin{bmatrix} \mathbf{v}_j \end{bmatrix}_T$. ■

REMARK   In Exercises T.5 through T.7 we ask you to show that if S and T are bases for the vector space R^n, then

$$P_{S \leftarrow T} = M_S^{-1} M_T,$$

where M_S is the $n \times n$ matrix whose jth column is $\mathbf{v}_j$ and M_T is the $n \times n$ matrix whose jth column is $\mathbf{w}_j$. This formula implies that $P_{S \leftarrow T}$ is nonsingular and it is helpful in solving some of the exercises in this section.

EXAMPLE 5 ■   Let S and T be the bases for R^3 defined in Example 4. Compute the transition matrix $Q_{T \leftarrow S}$ from the S-basis to the T-basis directly and show that $Q_{T \leftarrow S} = P_{S \leftarrow T}^{-1}$.

Solution $Q_{T \leftarrow S}$ is the matrix whose columns are the solution vectors to the linear system obtained from the vector equations

$$a_1\mathbf{w}_1 + a_2\mathbf{w}_2 + a_3\mathbf{w}_3 = \mathbf{v}_1$$
$$b_1\mathbf{w}_1 + b_2\mathbf{w}_2 + b_3\mathbf{w}_3 = \mathbf{v}_2$$
$$c_1\mathbf{w}_1 + c_2\mathbf{w}_2 + c_3\mathbf{w}_3 = \mathbf{v}_3.$$

As in Example 4, we can solve these linear systems simultaneously by transforming the partitioned matrix

$$\begin{bmatrix} \mathbf{w}_1 & \mathbf{w}_2 & \mathbf{w}_3 & \vdots & \mathbf{v}_1 & \vdots & \mathbf{v}_2 & \vdots & \mathbf{v}_3 \end{bmatrix}$$

to reduced row echelon form. That is, we transform

$$\begin{bmatrix} 6 & 4 & 5 & \vdots & 2 & \vdots & 1 & \vdots & 1 \\ 3 & -1 & 5 & \vdots & 0 & \vdots & 2 & \vdots & 1 \\ 3 & 3 & 2 & \vdots & 1 & \vdots & 0 & \vdots & 1 \end{bmatrix}$$

to reduced row echelon form, obtaining (verify)

$$\begin{bmatrix} 1 & 0 & 0 & \vdots & \frac{3}{2} & \vdots & \frac{1}{2} & \vdots & -\frac{5}{2} \\ 0 & 1 & 0 & \vdots & -\frac{1}{2} & \vdots & -\frac{1}{2} & \vdots & \frac{3}{2} \\ 0 & 0 & 1 & \vdots & -1 & \vdots & 0 & \vdots & 2 \end{bmatrix},$$

so

$$Q_{T \leftarrow S} = \begin{bmatrix} \frac{3}{2} & \frac{1}{2} & -\frac{5}{2} \\ -\frac{1}{2} & -\frac{1}{2} & \frac{3}{2} \\ -1 & 0 & 2 \end{bmatrix}.$$

Multiplying $Q_{T \leftarrow S}$ by $P_{S \leftarrow T}$, we find (verify) that $Q_{T \leftarrow S} P_{S \leftarrow T} = I_3$, so we conclude that $Q_{T \leftarrow S} = P_{S \leftarrow T}^{-1}$. ■

EXAMPLE 6 ■ Let V be P_1 and let $S = \{\mathbf{v}_1, \mathbf{v}_2\}$ and $T = \{\mathbf{w}_1, \mathbf{w}_2\}$ be bases for P_1, where

$$\mathbf{v}_1 = t, \qquad \mathbf{v}_2 = t - 3, \qquad \mathbf{w}_1 = t - 1, \qquad \mathbf{w}_2 = t + 1.$$

(a) Compute the transition matrix $P_{S \leftarrow T}$ from the T-basis to the S-basis.
(b) Verify Equation (5) for $\mathbf{v} = 5t + 1$.
(c) Compute the transition matrix $Q_{T \leftarrow S}$ from the S-basis to the T-basis and show that $Q_{T \leftarrow S} = P_{S \leftarrow T}^{-1}$.

Solution (a) To compute $P_{S \leftarrow T}$, we need to solve the vector equations

$$a_1 \mathbf{v}_1 + a_2 \mathbf{v}_2 = \mathbf{w}_1$$
$$b_1 \mathbf{v}_1 + b_2 \mathbf{v}_2 = \mathbf{w}_2$$

simultaneously by transforming the resulting partitioned matrix (verify)

$$\begin{bmatrix} 1 & 1 & \vdots & 1 & \vdots & 1 \\ 0 & -3 & \vdots & -1 & \vdots & 1 \end{bmatrix}$$

to reduced row echelon form. The result is (verify)

$$\begin{bmatrix} 1 & 0 & \vdots & \frac{2}{3} & \vdots & \frac{4}{3} \\ 0 & 1 & \vdots & \frac{1}{3} & \vdots & -\frac{1}{3} \end{bmatrix},$$

so

$$P_{S \leftarrow T} = \begin{bmatrix} \frac{2}{3} & \frac{4}{3} \\ \frac{1}{3} & -\frac{1}{3} \end{bmatrix}.$$

(b) If $\mathbf{v} = 5t + 1$, then expressing $\mathbf{v}$ in terms of the T-basis, we have (verify)

$$\mathbf{v} = 5t + 1 = 2(t - 1) + 3(t + 1),$$

so

$$[\mathbf{v}]_T = \begin{bmatrix} 2 \\ 3 \end{bmatrix}.$$

To verify this, set up and solve the associated linear system resulting from the vector equation

$$\mathbf{v} = a_1 \mathbf{w}_1 + a_2 \mathbf{w}_2.$$

Then

$$\left[\mathbf{v}\right]_S = P_{S\leftarrow T}\left[\mathbf{v}\right]_T = \begin{bmatrix} \frac{2}{3} & \frac{4}{3} \\ \frac{1}{3} & -\frac{1}{3} \end{bmatrix}\begin{bmatrix} 2 \\ 3 \end{bmatrix} = \begin{bmatrix} \frac{16}{3} \\ -\frac{1}{3} \end{bmatrix}.$$

Computing $\left[\mathbf{v}\right]_S$ directly from the associated linear system arising from the vector equation

$$\mathbf{v} = b_1\mathbf{w}_1 + b_2\mathbf{w}_2$$

we find that (verify)

$$\mathbf{v} = 5t + 1 = \tfrac{16}{3}t - \tfrac{1}{3}(t-3),$$

so

$$\left[\mathbf{v}\right]_S = \begin{bmatrix} \frac{16}{3} \\ -\frac{1}{3} \end{bmatrix}.$$

Hence

$$\left[\mathbf{v}\right]_S = P_{S\leftarrow T}\left[\mathbf{v}\right]_T,$$

which is Equation (5).

(c) The transition matrix $Q_{T\leftarrow S}$ from the S-basis to the T-basis is obtained (verify) by transforming the partitioned matrix

$$\begin{bmatrix} 1 & 1 & \vdots & 1 & \vdots & 1 \\ -1 & 1 & \vdots & 0 & \vdots & -3 \end{bmatrix}$$

to reduced row echelon form, obtaining (verify)

$$\begin{bmatrix} 1 & 0 & \vdots & \frac{1}{2} & \vdots & 2 \\ 0 & 1 & \vdots & \frac{1}{2} & \vdots & -1 \end{bmatrix}.$$

Hence

$$Q_{T\leftarrow S} = \begin{bmatrix} \frac{1}{2} & 2 \\ \frac{1}{2} & -1 \end{bmatrix}.$$

Multiplying $Q_{T\leftarrow S}$ by $P_{S\leftarrow T}$, we find (verify) that $Q_{T\leftarrow S}P_{S\leftarrow T} = I_2$, so $Q_{T\leftarrow S} = P_{S\leftarrow T}^{-1}$. ■

4.7   EXERCISES

All bases considered in these exercises are assumed to be ordered bases. In Exercises 1 through 6, compute the coordinate vector of $\mathbf{v}$ with respect to the given basis S for V.

1. V is R^2, $S = \left\{\begin{bmatrix} 1 \\ 0 \end{bmatrix}, \begin{bmatrix} 0 \\ 1 \end{bmatrix}\right\}$, $\mathbf{v} = \begin{bmatrix} 3 \\ -2 \end{bmatrix}$.

2. V is R^3, $S = \{(1,-1,0),(0,1,0),(1,0,2)\}$, $\mathbf{v} = (2,-1,-2)$.

3. V is P_1, $S = \{t+1, t-2\}$, $\mathbf{v} = t+4$.

4. V is P_2, $S = \{t^2 - t + 1, t + 1, t^2 + 1\}$, $\mathbf{v} = 4t^2 - 2t + 3$.

5. V is M_{22}, $S = \left\{ \begin{bmatrix} 1 & 0 \\ 0 & 0 \end{bmatrix}, \begin{bmatrix} 0 & 0 \\ 1 & 0 \end{bmatrix}, \begin{bmatrix} 0 & 1 \\ 0 & 0 \end{bmatrix}, \right.$

$\left. \begin{bmatrix} 0 & 0 \\ 0 & 1 \end{bmatrix} \right\}$, $\mathbf{v} = \begin{bmatrix} 1 & 0 \\ -1 & 2 \end{bmatrix}$.

6. V is M_{22}, $S = \left\{ \begin{bmatrix} 1 & -1 \\ 0 & 0 \end{bmatrix}, \begin{bmatrix} 0 & 1 \\ 1 & 0 \end{bmatrix}, \begin{bmatrix} 1 & 0 \\ 0 & -1 \end{bmatrix}, \right.$

$\left. \begin{bmatrix} 1 & 0 \\ -1 & 0 \end{bmatrix} \right\}$, $\mathbf{v} = \begin{bmatrix} 1 & 3 \\ -2 & 2 \end{bmatrix}$.

In Exercises 7 through 12, compute the vector $\mathbf{v}$ if the coordinate vector $\begin{bmatrix} \mathbf{v} \end{bmatrix}_S$ is given with respect to the basis S for V.

7. V is R^2, $S = \left\{ \begin{bmatrix} 2 \\ 1 \end{bmatrix}, \begin{bmatrix} -1 \\ 1 \end{bmatrix} \right\}$, $\begin{bmatrix} \mathbf{v} \end{bmatrix}_S = \begin{bmatrix} 1 \\ 2 \end{bmatrix}$.

8. V is R^3, $S = \{(0, 1, -1), (1, 0, 0), (1, 1, 1)\}$,

$\begin{bmatrix} \mathbf{v} \end{bmatrix}_S = \begin{bmatrix} -1 \\ 1 \\ 2 \end{bmatrix}$.

9. V is P_1, $S = \{t, 2t - 1\}$, $\begin{bmatrix} \mathbf{v} \end{bmatrix}_S = \begin{bmatrix} -2 \\ 3 \end{bmatrix}$.

10. V is P_2, $S = \{t^2 + 1, t + 1, t^2 + t\}$, $\begin{bmatrix} \mathbf{v} \end{bmatrix}_S = \begin{bmatrix} 3 \\ -1 \\ -2 \end{bmatrix}$.

11. V is M_{22}, $S = \left\{ \begin{bmatrix} -1 & 0 \\ 1 & 0 \end{bmatrix}, \begin{bmatrix} 2 & 2 \\ 0 & 1 \end{bmatrix}, \begin{bmatrix} 1 & 2 \\ -1 & 3 \end{bmatrix}, \right.$

$\left. \begin{bmatrix} 0 & 0 \\ 2 & 3 \end{bmatrix} \right\}$, $\begin{bmatrix} \mathbf{v} \end{bmatrix}_S = \begin{bmatrix} 2 \\ 1 \\ -1 \\ 3 \end{bmatrix}$.

12. V is M_{22}, $S = \left\{ \begin{bmatrix} 1 & -2 \\ 0 & 0 \end{bmatrix}, \begin{bmatrix} -1 & 3 \\ 0 & 1 \end{bmatrix}, \begin{bmatrix} 1 & 0 \\ 0 & 0 \end{bmatrix}, \right.$

$\left. \begin{bmatrix} 0 & -1 \\ 1 & 0 \end{bmatrix} \right\}$, $\begin{bmatrix} \mathbf{v} \end{bmatrix}_S = \begin{bmatrix} 0 \\ 1 \\ 0 \\ 2 \end{bmatrix}$.

13. Let $S = \{(1, 2), (0, 1)\}$ and $T = \{(1, 1), (2, 3)\}$ be bases for R^2. Let $\mathbf{v} = (1, 5)$ and $\mathbf{w} = (5, 4)$.

(a) Find the coordinate vectors of $\mathbf{v}$ and $\mathbf{w}$ with respect to the basis T.

(b) What is the transition matrix $P_{S \leftarrow T}$ from the T- to the S-basis?

(c) Find the coordinate vectors of $\mathbf{v}$ and $\mathbf{w}$ with respect to S using $P_{S \leftarrow T}$.

(d) Find the coordinate vectors of $\mathbf{v}$ and $\mathbf{w}$ with respect to S directly.

(e) Find the transition matrix $Q_{T \leftarrow S}$ from the S- to the T-basis.

(f) Find the coordinate vectors of $\mathbf{v}$ and $\mathbf{w}$ with respect to T using $Q_{T \leftarrow S}$. Compare the answers with those of (a).

14. Let

$$S = \left\{ \begin{bmatrix} 1 \\ 0 \\ 1 \end{bmatrix}, \begin{bmatrix} -1 \\ 0 \\ 0 \end{bmatrix}, \begin{bmatrix} 0 \\ 1 \\ 2 \end{bmatrix} \right\}$$

and

$$T = \left\{ \begin{bmatrix} -1 \\ 1 \\ 0 \end{bmatrix}, \begin{bmatrix} 1 \\ 2 \\ -1 \end{bmatrix}, \begin{bmatrix} 0 \\ 1 \\ 0 \end{bmatrix} \right\}$$

be bases for R^3. Let

$$\mathbf{v} = \begin{bmatrix} 1 \\ 3 \\ 8 \end{bmatrix} \quad \text{and} \quad \mathbf{w} = \begin{bmatrix} -1 \\ 8 \\ -2 \end{bmatrix}.$$

Follow the directions for Exercise 13.

15. Let $S = \{t^2 + 1, t - 2, t + 3\}$ and $T = \{2t^2 + t, t^2 + 3, t\}$ be bases for P_2. Let $\mathbf{v} = 8t^2 - 4t + 6$ and $\mathbf{w} = 7t^2 - t + 9$. Follow the directions for Exercise 13.

16. Let $S = \{t^2 + t + 1, t^2 + 2t + 3, t^2 + 1\}$ and $T = \{t + 1, t^2, t^2 + 1\}$ be bases for P_2. Also let $\mathbf{v} = -t^2 + 4t + 5$ and $\mathbf{w} = 2t^2 - 6$. Follow the directions for Exercise 13.

17. Let

$$S = \left\{ \begin{bmatrix} 1 & 0 \\ 0 & 0 \end{bmatrix}, \begin{bmatrix} 0 & 1 \\ 1 & 0 \end{bmatrix}, \begin{bmatrix} 0 & 2 \\ 0 & 1 \end{bmatrix}, \begin{bmatrix} 0 & 0 \\ 1 & 1 \end{bmatrix} \right\}$$

and

$$T = \left\{ \begin{bmatrix} 1 & 1 \\ 0 & 0 \end{bmatrix}, \begin{bmatrix} 0 & 0 \\ 1 & 0 \end{bmatrix}, \begin{bmatrix} 0 & 0 \\ 0 & 1 \end{bmatrix}, \begin{bmatrix} 1 & 0 \\ 0 & 0 \end{bmatrix} \right\}$$

be bases for M_{22}. Let

$$\mathbf{v} = \begin{bmatrix} 1 & 1 \\ 1 & 1 \end{bmatrix} \quad \text{and} \quad \mathbf{w} = \begin{bmatrix} 1 & 2 \\ -2 & 1 \end{bmatrix}.$$

Follow the directions for Exercise 13.

18. Let

$$S = \left\{ \begin{bmatrix} -1 & -1 \\ 0 & 1 \end{bmatrix}, \begin{bmatrix} 1 & 0 \\ 0 & 1 \end{bmatrix}, \begin{bmatrix} 0 & -1 \\ 0 & 0 \end{bmatrix}, \begin{bmatrix} 1 & 0 \\ 1 & 0 \end{bmatrix} \right\}$$

and

$$T = \left\{ \begin{bmatrix} 1 & 1 \\ 0 & 0 \end{bmatrix}, \begin{bmatrix} 1 & 0 \\ 1 & -1 \end{bmatrix}, \begin{bmatrix} 1 & 0 \\ 0 & 0 \end{bmatrix}, \begin{bmatrix} 0 & 1 \\ 0 & 1 \end{bmatrix} \right\}$$

be bases for M_{22}. Let

$$\mathbf{v} = \begin{bmatrix} 0 & 0 \\ 3 & -1 \end{bmatrix} \quad \text{and} \quad \mathbf{w} = \begin{bmatrix} -2 & 3 \\ -1 & 3 \end{bmatrix}.$$

Follow the directions for Exercise 13.

19. Let $S = \{(1,-1),(2,1)\}$ and $T = \{(3,0),(4,-1)\}$ be bases for R^2. If $\mathbf{v}$ is in R^2 and

$$\left[\mathbf{v}\right]_T = \begin{bmatrix} 1 \\ 2 \end{bmatrix},$$

determine $\left[\mathbf{v}\right]_S$.

20. Let $S = \{t+1, t-2\}$ and $T = \{t-5, t-2\}$ be bases for P_1. If $\mathbf{v}$ is in P_1 and

$$\left[\mathbf{v}\right]_T = \begin{bmatrix} -1 \\ 3 \end{bmatrix},$$

determine $\left[\mathbf{v}\right]_S$.

21. Let $S = \{(-1,2,1),(0,1,1),(-2,2,1)\}$ and $T = \{(-1,1,0),(0,1,0),(0,1,1)\}$ be bases for R^3. If $\mathbf{v}$ is in R^3 and

$$\left[\mathbf{v}\right]_S = \begin{bmatrix} 2 \\ 0 \\ 1 \end{bmatrix},$$

determine $\left[\mathbf{v}\right]_T$.

22. Let $S = \{t^2, t-1, 1\}$ and $T = \{t^2+t+1, t+1, 1\}$ be bases for P_2. If $\mathbf{v}$ is in P_2 and

$$\left[\mathbf{v}\right]_S = \begin{bmatrix} 1 \\ 2 \\ 3 \end{bmatrix},$$

determine $\left[\mathbf{v}\right]_T$.

23. Let $S = \{\mathbf{v}_1, \mathbf{v}_2, \mathbf{v}_3\}$ and $T = \{\mathbf{w}_1, \mathbf{w}_2, \mathbf{w}_3\}$ be bases for R^3, where

$$\mathbf{v}_1 = (1,0,1), \quad \mathbf{v}_2 = (1,1,0), \quad \mathbf{v}_3 = (0,0,1).$$

If the transition matrix from T to S is

$$\begin{bmatrix} 1 & 1 & 2 \\ 2 & 1 & 1 \\ -1 & -1 & 1 \end{bmatrix},$$

determine T.

24. Let $S = \{\mathbf{v}_1, \mathbf{v}_2\}$ and $T = \{\mathbf{w}_1, \mathbf{w}_2\}$ be bases for P_1, where

$$\mathbf{w}_1 = t, \qquad \mathbf{w}_2 = t - 1.$$

If the transition matrix from S to T is

$$\begin{bmatrix} 2 & 3 \\ -1 & 2 \end{bmatrix},$$

determine S.

25. Let $S = \{\mathbf{v}_1, \mathbf{v}_2\}$ and $T = \{\mathbf{w}_1, \mathbf{w}_2\}$ be bases for R^2, where

$$\mathbf{v}_1 = (1,2), \qquad \mathbf{v}_2 = (0,1).$$

If the transition matrix from S to T is

$$\begin{bmatrix} 2 & 1 \\ 1 & 1 \end{bmatrix},$$

determine T.

26. Let $S = \{\mathbf{v}_1, \mathbf{v}_2\}$ and $T = \{\mathbf{w}_1, \mathbf{w}_2\}$ be bases for P_1, where

$$\mathbf{w}_1 = t - 1, \qquad \mathbf{w}_2 = t + 1.$$

If the transition matrix from T to S is

$$\begin{bmatrix} 1 & 2 \\ 2 & 3 \end{bmatrix},$$

determine S.

THEORETICAL EXERCISES

T.1. Let $S = \{\mathbf{v}_1, \mathbf{v}_2, \ldots, \mathbf{v}_n\}$ be a basis for the n-dimensional vector space V, and let $\mathbf{v}$ and $\mathbf{w}$ be two vectors in V. Show that $\mathbf{v} = \mathbf{w}$ if and only if $\left[\mathbf{v}\right]_S = \left[\mathbf{w}\right]_S$.

T.2. Show that if S is a basis for an n-dimensional vector space V, $\mathbf{v}$ and $\mathbf{w}$ are vectors in V, and k is a scalar, then

$$\left[\mathbf{v}+\mathbf{w}\right]_S = \left[\mathbf{v}\right]_S + \left[\mathbf{w}\right]_S$$

and

$$\left[k\mathbf{v}\right]_S = k\left[\mathbf{v}\right]_S.$$

T.3. Let S be a basis for an n-dimensional vector space V. Show that if $\{\mathbf{w}_1, \mathbf{w}_2, \ldots, \mathbf{w}_k\}$ is a linearly independent set of vectors in V, then

$$\left\{\left[\mathbf{w}_1\right]_S, \left[\mathbf{w}_2\right]_S, \ldots, \left[\mathbf{w}_k\right]_S\right\}$$

is a linearly independent set of vectors in R^n.

T.4. Let $S = \{\mathbf{v}_1, \mathbf{v}_2, \ldots, \mathbf{v}_n\}$ be a basis for an n-dimensional vector space V. Show that

$$\left\{\left[\mathbf{v}_1\right]_S, \left[\mathbf{v}_2\right]_S, \ldots, \left[\mathbf{v}_n\right]_S\right\}$$

is a basis for R^n.

In Exercises T.5 through T.7, let $S = \{\mathbf{v}_1, \mathbf{v}_2, \ldots, \mathbf{v}_n\}$ and $T = \{\mathbf{w}_1, \mathbf{w}_2, \ldots, \mathbf{w}_n\}$ be bases for the vector space R^n.

T.5. Let M_S be the $n \times n$ matrix whose jth column is $\mathbf{v}_j$ and let M_T be the $n \times n$ matrix whose jth column is $\mathbf{w}_j$. Show that M_S and M_T are nonsingular. (*Hint:* Consider the homogeneous systems $M_S\mathbf{x} = \mathbf{0}$ and $M_T\mathbf{x} = \mathbf{0}$.)

T.6. If $\mathbf{v}$ is a vector in V, show that
$$\mathbf{v} = M_S \begin{bmatrix} \mathbf{v} \end{bmatrix}_S \quad \text{and} \quad \mathbf{v} = M_T \begin{bmatrix} \mathbf{v} \end{bmatrix}_T.$$

T.7. (a) Use Equation (5) and Exercises T.5 and T.6 to show that
$$P_{S \leftarrow T} = M_S^{-1} M_T.$$

(b) Show that $P_{S \leftarrow T}$ is nonsingular.

(c) Verify the result in part (a) for Example 4.

MATLAB EXERCISES ■

Finding the coordinates of a vector with respect to a basis is a linear combination problem. Hence, once the corresponding linear system is constructed, we can use MATLAB routine **reduce** or **rref** to find its solution. The solution gives us the desired coordinates. (The discussion in Section 10.7 will be helpful as an aid for constructing the necessary linear system.)

ML.1. Let $V = R^3$ and
$$S = \left\{ \begin{bmatrix} 1 \\ 2 \\ 1 \end{bmatrix}, \begin{bmatrix} 2 \\ 1 \\ 0 \end{bmatrix}, \begin{bmatrix} 1 \\ 0 \\ 2 \end{bmatrix} \right\}.$$

Show that S is a basis for V and find $\begin{bmatrix} \mathbf{v} \end{bmatrix}_S$ for each of the following vectors.

(a) $\mathbf{v} = \begin{bmatrix} 8 \\ 4 \\ 7 \end{bmatrix}.$ (b) $\mathbf{v} = \begin{bmatrix} 2 \\ 0 \\ -3 \end{bmatrix}.$

(c) $\mathbf{v} = \begin{bmatrix} 4 \\ 3 \\ 3 \end{bmatrix}.$

ML.2. Let $V = R^4$ and $S = \{(1,0,1,1),(1,2,1,3),(0,2,1,1), (0,1,0,0)\}$. Show that S is a basis for V and find $\begin{bmatrix} \mathbf{v} \end{bmatrix}_S$ for each of the following vectors.

(a) $\mathbf{v} = (4,12,8,14).$ (b) $\mathbf{v} = \left(\tfrac{1}{2},0,0,0\right).$

(c) $\mathbf{v} = \left(1,1,1,\tfrac{7}{3}\right).$

ML.3. Let V be the vector space of all 2×2 matrices and
$$S = \left\{ \begin{bmatrix} 1 & 2 \\ 1 & 2 \end{bmatrix}, \begin{bmatrix} 0 & 2 \\ 1 & 0 \end{bmatrix}, \begin{bmatrix} 3 & 1 \\ -1 & 0 \end{bmatrix}, \begin{bmatrix} -1 & 0 \\ 0 & 0 \end{bmatrix} \right\}.$$

Show that S is a basis for V and find $\begin{bmatrix} \mathbf{v} \end{bmatrix}_S$ for each of the following vectors.

(a) $\mathbf{v} = \begin{bmatrix} 1 & 0 \\ 0 & 1 \end{bmatrix}.$ (b) $\mathbf{v} = \begin{bmatrix} 2 & \tfrac{10}{3} \\ \tfrac{7}{6} & 2 \end{bmatrix}.$

(c) $\mathbf{v} = \begin{bmatrix} 1 & 1 \\ 1 & 1 \end{bmatrix}.$

Finding the transition matrix $P_{S \leftarrow T}$ from the T-basis to the S-basis is also a linear combination problem. $P_{S \leftarrow T}$ is the matrix whose columns are the coordinates of the vectors in T with respect to the S-basis. Following the ideas developed in Example 4, we can find matrix $P_{S \leftarrow T}$ using routine **reduce** or **rref**. The idea is to construct a matrix A whose columns correspond to the vectors in S (see Section 10.7) and a matrix B whose columns correspond to the vectors in T. Then MATLAB command **rref([A B])** gives $[\mathbf{I}\ \mathbf{P_{S \leftarrow T}}]$.

In Exercises ML.4 through ML.6, use the MATLAB techniques described above to find the transition matrix $P_{S \leftarrow T}$ from the T-basis to the S-basis.

ML.4. $V = R^3$,
$$S = \left\{ \begin{bmatrix} 1 \\ 1 \\ 0 \end{bmatrix}, \begin{bmatrix} 0 \\ 1 \\ 1 \end{bmatrix}, \begin{bmatrix} 1 \\ 0 \\ 1 \end{bmatrix} \right\},$$
$$T = \left\{ \begin{bmatrix} 2 \\ 1 \\ 1 \end{bmatrix}, \begin{bmatrix} 1 \\ 2 \\ 1 \end{bmatrix}, \begin{bmatrix} 1 \\ 1 \\ 2 \end{bmatrix} \right\}.$$

ML.5. $V = P_3$, $S = \{t-1, t+1, t^2+t, t^3-t\}$, $T = \{t^2, 1-t, 2-t^2, t^3+t^2\}$.

ML.6. $V = R^4$, $S = \{(1,2,3,0),(0,1,2,3),(3,0,1,2), (2,3,0,1)\}$, $T = $ natural basis.

ML.7. Let $V = R^3$ and suppose that we have bases
$$S = \left\{ \begin{bmatrix} 1 \\ 1 \\ 1 \end{bmatrix}, \begin{bmatrix} 1 \\ 2 \\ 1 \end{bmatrix}, \begin{bmatrix} 0 \\ 1 \\ 1 \end{bmatrix} \right\},$$
$$T = \left\{ \begin{bmatrix} 1 \\ 0 \\ 1 \end{bmatrix}, \begin{bmatrix} 1 \\ 1 \\ 0 \end{bmatrix}, \begin{bmatrix} 0 \\ 1 \\ 2 \end{bmatrix} \right\},$$

and

$$U = \left\{ \begin{bmatrix} 2 \\ 1 \\ 1 \end{bmatrix}, \begin{bmatrix} -1 \\ 2 \\ 1 \end{bmatrix}, \begin{bmatrix} 1 \\ -2 \\ 1 \end{bmatrix} \right\}.$$

(a) Find the transition matrix P from U to T.

(b) Find the transition matrix Q from T to S.

(c) Find the transition matrix Z from U to S.

(d) Does $Z = PQ$ or QP?

4.8 ▼ Orthonormal Bases in R^n

From our work with the natural bases for R^2, R^3, and, in general, for R^n, we know that when these bases are present, the computations are kept to a minimum. A subspace W of R^n need not contain any of the natural basis vectors, but in this section we want to show that it has a basis with the same properties. That is, we want to show that W contains a basis S such that every vector in S is of unit length and every two vectors in S are orthogonal. The method used to obtain such a basis is the Gram–Schmidt process, which is presented below.

DEFINITION

A set $S = \{\mathbf{u}_1, \mathbf{u}_2, \dots, \mathbf{u}_k\}$ in R^n is called **orthogonal** if any two distinct vectors in S are orthogonal, that is, if $\mathbf{u}_i \cdot \mathbf{u}_j = 0$ for $i \neq j$. An **orthonormal** set of vectors is an orthogonal set of unit vectors. That is, $S = \{\mathbf{u}_1, \mathbf{u}_2, \dots, \mathbf{u}_k\}$ is orthonormal if $\mathbf{u}_i \cdot \mathbf{u}_j = 0$ for $i \neq j$, and $\mathbf{u}_i \cdot \mathbf{u}_i = 1$ for $i = 1, 2, \dots, k$.

EXAMPLE 1   ■   If $\mathbf{x}_1 = (1, 0, 2)$, $\mathbf{x}_2 = (-2, 0, 1)$, and $\mathbf{x}_3 = (0, 1, 0)$, then $\{\mathbf{x}_1, \mathbf{x}_2, \mathbf{x}_3\}$ is an orthogonal set in R^3. The vectors

$$\mathbf{u}_1 = \left(\frac{1}{\sqrt{5}}, 0, \frac{2}{\sqrt{5}} \right) \quad \text{and} \quad \mathbf{u}_2 = \left(-\frac{2}{\sqrt{5}}, 0, \frac{1}{\sqrt{5}} \right)$$

are unit vectors in the directions of $\mathbf{x}_1$ and $\mathbf{x}_2$, respectively. Since $\mathbf{x}_3$ is also of unit length, it follows that $\{\mathbf{u}_1, \mathbf{u}_2, \mathbf{x}_3\}$ is an orthonormal set. Also, span $\{\mathbf{x}_1, \mathbf{x}_2, \mathbf{x}_3\}$ is the same as span $\{\mathbf{u}_1, \mathbf{u}_2, \mathbf{x}_3\}$.   ■

EXAMPLE 2   ■   The natural basis

$$\{(1, 0, 0), (0, 1, 0), (0, 0, 1)\}$$

is an orthonormal set in R^3. More generally, the natural basis in R^n is an orthonormal set.   ■

An important result about orthogonal sets is the following theorem.

THEOREM 4.16   ■   *Let $S = \{\mathbf{u}_1, \mathbf{u}_2, \dots, \mathbf{u}_k\}$ be an orthogonal set of nonzero vectors in R^n. Then S is linearly independent.*

Proof

Consider the equation

$$c_1 \mathbf{u}_1 + c_2 \mathbf{u}_2 + \cdots + c_k \mathbf{u}_k = \mathbf{0}. \tag{1}$$

Taking the dot product of both sides of (1) with $\mathbf{u}_i$, $1 \leq i \leq k$, we have

$$(c_1 \mathbf{u}_1 + c_2 \mathbf{u}_2 + \cdots + c_k \mathbf{u}_k) \cdot \mathbf{u}_i = \mathbf{0} \cdot \mathbf{u}_i. \tag{2}$$

By properties (c) and (d) of Theorem 3.3, Section 3.2, the left side of (2) is

$$c_1(\mathbf{u}_1 \cdot \mathbf{u}_i) + c_2(\mathbf{u}_2 \cdot \mathbf{u}_i) + \cdots + c_k(\mathbf{u}_k \cdot \mathbf{u}_i),$$

and the right side is 0. Since $\mathbf{u}_j \cdot \mathbf{u}_i = 0$ if $i \neq j$, (2) becomes

$$0 = c_i(\mathbf{u}_i \cdot \mathbf{u}_i) = c_i\|\mathbf{u}_i\|^2. \tag{3}$$

By (a) of Theorem 3.3, Section 3.2, $\|\mathbf{u}_i\| \neq 0$, since $\mathbf{u}_i \neq \mathbf{0}$. Hence (3) implies that $c_i = 0$, $1 \leq i \leq k$, and S is linearly independent. ∎

COROLLARY 4.6 ■ *An orthonormal set of vectors in R^n is linearly independent.*

Proof   Exercise T.2. ∎

DEFINITION   From Theorem 4.9 of Section 4.4 and Corollary 4.6, it follows that an orthonormal set of n vectors in R^n is a basis for R^n (Exercise T.3). An **orthogonal (orthonormal) basis** for a vector space is a basis that is an orthogonal (orthonormal) set.

We have already seen that the computational effort required to solve a given problem is often reduced when we are dealing with the natural basis for R^n. This reduction in computational effort is due to the fact that we are dealing with an orthonormal basis. For example, if $S = \{\mathbf{u}_1, \mathbf{u}_2, \ldots, \mathbf{u}_n\}$ is a basis for R^n, then if $\mathbf{v}$ is any vector in V, we can write $\mathbf{v}$ as

$$\mathbf{v} = c_1\mathbf{u}_1 + c_2\mathbf{u}_2 + \cdots + c_n\mathbf{u}_n.$$

The coefficients c_1, c_2, ..., c_n are obtained by solving a linear system of n equations in n unknowns. (See Section 4.3.)

However, if S is orthonormal, we can obtain the same result with much less work. This is the content of the following theorem.

THEOREM 4.17 ■ *Let $S = \{\mathbf{u}_1, \mathbf{u}_2, \ldots, \mathbf{u}_n\}$ be an orthonormal basis for R^n and $\mathbf{v}$ any vector in R^n. Then*

$$\mathbf{v} = c_1\mathbf{u}_1 + c_2\mathbf{u}_2 + \cdots + c_n\mathbf{u}_n,$$

where

$$c_i = \mathbf{v} \cdot \mathbf{u}_i \qquad (1 \leq i \leq n).$$

Proof   Exercise T.4. ■

EXAMPLE 3 ■ Let $S = \{\mathbf{u}_1, \mathbf{u}_2, \mathbf{u}_3\}$ be an orthonormal basis for R^3, where

$$\mathbf{u}_1 = \left(\tfrac{2}{3}, -\tfrac{2}{3}, \tfrac{1}{3}\right), \quad \mathbf{u}_2 = \left(\tfrac{2}{3}, \tfrac{1}{3}, -\tfrac{2}{3}\right), \quad \text{and} \quad \mathbf{u}_3 = \left(\tfrac{1}{3}, \tfrac{2}{3}, \tfrac{2}{3}\right).$$

Write the vector $\mathbf{v} = (3, 4, 5)$ as a linear combination of the vectors in S.

Solution   We have

$$\mathbf{v} = c_1 \mathbf{u}_1 + c_2 \mathbf{u}_2 + c_3 \mathbf{u}_3.$$

Theorem 4.17 shows that c_1, c_2, and c_3 can be obtained without having to solve a linear system of three equations in three unknowns. Thus

$$c_1 = \mathbf{v} \cdot \mathbf{u}_1 = 1, \qquad c_2 = \mathbf{v} \cdot \mathbf{u}_2 = 0, \qquad c_3 = \mathbf{v} \cdot \mathbf{u}_3 = 7,$$

and $\mathbf{v} = \mathbf{u}_1 + 7\mathbf{u}_3$. ∎

THEOREM 4.18 ■
(*Gram*[*]–*Schmidt*[**]
Process)

Let W be a nonzero subspace of R^n with basis $S = \{\mathbf{u}_1, \mathbf{u}_2, \ldots, \mathbf{u}_m\}$. Then there exists an orthonormal basis $T = \{\mathbf{w}_1, \mathbf{w}_2, \ldots, \mathbf{w}_m\}$ for W.

Proof   The proof is constructive; that is, we develop the desired basis T. However, we first find an orthogonal basis $T^* = \{\mathbf{v}_1, \mathbf{v}_2, \ldots, \mathbf{v}_m\}$ for W.

First, we pick any one of the vectors in S, say $\mathbf{u}_1$, and call it $\mathbf{v}_1$. Thus $\mathbf{v}_1 = \mathbf{u}_1$. We now look for a vector $\mathbf{v}_2$ in the subspace W_1 of W spanned by $\{\mathbf{u}_1, \mathbf{u}_2\}$ that is orthogonal to $\mathbf{v}_1$. Since $\mathbf{v}_1 = \mathbf{u}_1$, W_1 is also the subspace spanned by $\{\mathbf{v}_1, \mathbf{u}_2\}$. Thus

$$\mathbf{v}_2 = c_1 \mathbf{v}_1 + c_2 \mathbf{u}_2.$$

We try to determine c_1 and c_2 so that $\mathbf{v}_1 \cdot \mathbf{v}_2 = 0$. Now

$$0 = \mathbf{v}_2 \cdot \mathbf{v}_1 = (c_1 \mathbf{v}_1 + c_2 \mathbf{u}_2) \cdot \mathbf{v}_1 = c_1 (\mathbf{v}_1 \cdot \mathbf{v}_1) + c_2 (\mathbf{u}_2 \cdot \mathbf{v}_1). \tag{4}$$

Since $\mathbf{v}_1 \neq \mathbf{0}$ (why?), $\mathbf{v}_1 \cdot \mathbf{v}_1 \neq 0$, and solving for c_1 and c_2 in (4), we have

$$c_1 = -c_2 \frac{\mathbf{u}_2 \cdot \mathbf{v}_1}{\mathbf{v}_1 \cdot \mathbf{v}_1}.$$

We may assign an arbitrary nonzero value to c_2. Thus, letting $c_2 = 1$, we obtain

$$c_1 = -\frac{\mathbf{u}_2 \cdot \mathbf{v}_1}{\mathbf{v}_1 \cdot \mathbf{v}_1}.$$

Hence

$$\mathbf{v}_2 = c_1 \mathbf{v}_1 + c_2 \mathbf{u}_2 = \mathbf{u}_2 - \left(\frac{\mathbf{u}_2 \cdot \mathbf{v}_1}{\mathbf{v}_1 \cdot \mathbf{v}_1} \right) \mathbf{v}_1.$$

Notice that at this point we have an orthogonal subset $\{\mathbf{v}_1, \mathbf{v}_2\}$ of W (see Figure 4.7).

Next, we look for a vector $\mathbf{v}_3$ in the subspace W_2 of W spanned by $\{\mathbf{u}_1, \mathbf{u}_2, \mathbf{u}_3\}$ that is orthogonal to both $\mathbf{v}_1$ and $\mathbf{v}_2$. Of course, W_2 is also the subspace spanned by $\{\mathbf{v}_1, \mathbf{v}_2, \mathbf{u}_3\}$ (why?). Thus

$$\mathbf{v}_3 = d_1 \mathbf{v}_1 + d_2 \mathbf{v}_2 + d_3 \mathbf{u}_3.$$

---

[*]Jörgen Pederson Gram (1850–1916) was a Danish actuary.

[**]Erhard Schmidt (1876–1959) taught at several leading German Universities and was a student of both Hermann Amandus Schwarz and David Hilbert. He made important contributions to the study of integral equations and partial differential equations and, as part of this study, he introduced the method for finding an orthonormal basis in 1907. In 1908 he wrote a paper on infinitely many linear equations in infinitely many unknowns, in which he founded the theory of Hilbert spaces and in which he again used his method.

FIGURE 4.7

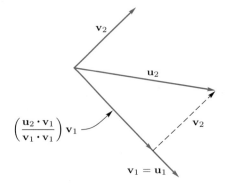

We now try to find d_1 and d_2 so that

$$\mathbf{v}_3 \cdot \mathbf{v}_1 = 0 \quad \text{and} \quad \mathbf{v}_3 \cdot \mathbf{v}_2 = 0.$$

Now

$$0 = \mathbf{v}_3 \cdot \mathbf{v}_1 = (d_1\mathbf{v}_1 + d_2\mathbf{v}_2 + d_3\mathbf{u}_3) \cdot \mathbf{v}_1 = d_1(\mathbf{v}_1 \cdot \mathbf{v}_1) + d_3\mathbf{u}_3 \cdot \mathbf{v}_1, \quad (5)$$
$$0 = \mathbf{v}_3 \cdot \mathbf{v}_2 = (d_1\mathbf{v}_1 + d_2\mathbf{v}_2 + d_3\mathbf{u}_3) \cdot \mathbf{v}_2 = d_2(\mathbf{v}_2 \cdot \mathbf{v}_2) + d_3\mathbf{u}_3 \cdot \mathbf{v}_2. \quad (6)$$

In obtaining the right sides of (5) and (6), we have used the fact that $\mathbf{v}_1 \cdot \mathbf{v}_2 = 0$. Observe that $\mathbf{v}_2 \neq \mathbf{0}$ (why?). Solving (5) and (6) for d_1 and d_2, respectively, we obtain

$$d_1 = -d_3 \frac{\mathbf{u}_3 \cdot \mathbf{v}_1}{\mathbf{v}_1 \cdot \mathbf{v}_1} \quad \text{and} \quad d_2 = -d_3 \frac{\mathbf{u}_3 \cdot \mathbf{v}_2}{\mathbf{v}_2 \cdot \mathbf{v}_2}.$$

We may assign an arbitrary nonzero value to d_3. Thus, letting $d_3 = 1$, we have

$$d_1 = -\frac{\mathbf{u}_3 \cdot \mathbf{v}_1}{\mathbf{v}_1 \cdot \mathbf{v}_1} \quad \text{and} \quad d_2 = -\frac{\mathbf{u}_3 \cdot \mathbf{v}_2}{\mathbf{v}_2 \cdot \mathbf{v}_2}.$$

Hence

$$\mathbf{v}_3 = \mathbf{u}_3 - \left(\frac{\mathbf{u}_3 \cdot \mathbf{v}_1}{\mathbf{v}_1 \cdot \mathbf{v}_1}\right)\mathbf{v}_1 - \left(\frac{\mathbf{u}_3 \cdot \mathbf{v}_2}{\mathbf{v}_2 \cdot \mathbf{v}_2}\right)\mathbf{v}_2.$$

Notice that at this point we have an orthogonal subset $\{\mathbf{v}_1, \mathbf{v}_2, \mathbf{v}_3\}$ of W (see Figure 4.8).

FIGURE 4.8

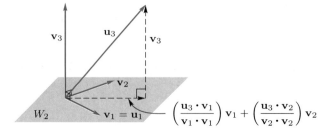

We next seek a vector $\mathbf{v}_4$ in the subspace W_3 of W spanned by the set $\{\mathbf{u}_1, \mathbf{u}_2, \mathbf{u}_3, \mathbf{u}_4\}$, and thus by $\{\mathbf{v}_1, \mathbf{v}_2, \mathbf{v}_3, \mathbf{u}_4\}$, that is orthogonal to $\mathbf{v}_1, \mathbf{v}_2,$ and $\mathbf{v}_3$.

We can then write

$$\mathbf{v}_4 = \mathbf{u}_4 - \left(\frac{\mathbf{u}_4 \cdot \mathbf{v}_1}{\mathbf{v}_1 \cdot \mathbf{v}_1}\right)\mathbf{v}_1 - \left(\frac{\mathbf{u}_4 \cdot \mathbf{v}_2}{\mathbf{v}_2 \cdot \mathbf{v}_2}\right)\mathbf{v}_2 - \left(\frac{\mathbf{u}_4 \cdot \mathbf{v}_3}{\mathbf{v}_3 \cdot \mathbf{v}_3}\right)\mathbf{v}_3.$$

We continue in this manner until we have an orthogonal set $T^* = \{\mathbf{v}_1, \mathbf{v}_2, \ldots, \mathbf{v}_m\}$ of m vectors. It then follows that T^* is a basis for W. If we now normalize the $\mathbf{v}_i$, that is, let

$$\mathbf{w}_i = \frac{1}{\|\mathbf{v}_i\|}\mathbf{v}_i \qquad (1 \le i \le m),$$

then $T = \{\mathbf{w}_1, \mathbf{w}_2, \ldots, \mathbf{w}_m\}$ is an orthonormal basis for W. ■

We now summarize the Gram–Schmidt process.

The Gram–Schmidt process for computing an orthonormal basis $T = \{\mathbf{w}_1, \mathbf{w}_2, \ldots, \mathbf{w}_m\}$ for a nonzero subspace W of R^n with basis $S = \{\mathbf{u}_1, \mathbf{u}_2, \ldots, \mathbf{u}_m\}$ is as follows.

Step 1. Let $\mathbf{v}_1 = \mathbf{u}_1$.

Step 2. Compute the vectors $\mathbf{v}_2, \mathbf{v}_3, \ldots, \mathbf{v}_m$, successively, one at a time, by the formula

$$\mathbf{v}_i = \mathbf{u}_i - \left(\frac{\mathbf{u}_i \cdot \mathbf{v}_1}{\mathbf{v}_1 \cdot \mathbf{v}_1}\right)\mathbf{v}_1 - \left(\frac{\mathbf{u}_i \cdot \mathbf{v}_2}{\mathbf{v}_2 \cdot \mathbf{v}_2}\right)\mathbf{v}_2 - \cdots - \left(\frac{\mathbf{u}_i \cdot \mathbf{v}_{i-1}}{\mathbf{v}_{i-1} \cdot \mathbf{v}_{i-1}}\right)\mathbf{v}_{i-1}.$$

The set of vectors $T^* = \{\mathbf{v}_1, \mathbf{v}_2, \ldots, \mathbf{v}_m\}$ is an orthogonal set.

Step 3. Let

$$\mathbf{w}_i = \frac{1}{\|\mathbf{v}_i\|}\mathbf{v}_i \qquad (1 \le i \le m).$$

Then $T = \{\mathbf{w}_1, \mathbf{w}_2, \ldots, \mathbf{w}_m\}$ is an orthonormal basis for W.

REMARK   It is not difficult to show that if $\mathbf{u}$ and $\mathbf{v}$ are vectors in R^n such that $\mathbf{u} \cdot \mathbf{v} = 0$, then $\mathbf{u} \cdot (c\mathbf{v}) = 0$ for any scalar c (Exercise T.7). This result can often be used to simplify hand computations in the Gram–Schmidt process. As soon as a vector $\mathbf{v}_i$ is computed in Step 2, multiply it by a proper scalar to clear any fractions that may be present. We shall use this approach in our computational work with the Gram–Schmidt process.

EXAMPLE 4 ■   Consider the basis $S = \{\mathbf{u}_1, \mathbf{u}_2, \mathbf{u}_3\}$ for R^3, where

$$\mathbf{u}_1 = (1, 1, 1), \quad \mathbf{u}_2 = (-1, 0, -1), \quad \text{and} \quad \mathbf{u}_3 = (-1, 2, 3).$$

Use the Gram–Schmidt process to transform S to an orthonormal basis for R^3.

Solution   **Step 1.** Let $\mathbf{v}_1 = \mathbf{u}_1 = (1, 1, 1)$.

Step 2. We now compute $\mathbf{v}_2$ and $\mathbf{v}_3$:

$$\mathbf{v}_2 = \mathbf{u}_2 - \left(\frac{\mathbf{u}_2 \cdot \mathbf{v}_1}{\mathbf{v}_1 \cdot \mathbf{v}_1}\right)\mathbf{v}_1 = (-1, 0, -1) - \left(\tfrac{-2}{3}\right)(1, 1, 1) = \left(-\tfrac{1}{3}, \tfrac{2}{3}, -\tfrac{1}{3}\right).$$

Multiplying $\mathbf{v}_2$ by 3 to clear fractions, we obtain $(-1, 2, -1)$, which we now use as $\mathbf{v}_2$. Then

$$\mathbf{v}_3 = \mathbf{u}_3 - \left(\frac{\mathbf{u}_3 \cdot \mathbf{v}_1}{\mathbf{v}_1 \cdot \mathbf{v}_1}\right)\mathbf{v}_1 - \left(\frac{\mathbf{u}_3 \cdot \mathbf{v}_2}{\mathbf{v}_2 \cdot \mathbf{v}_2}\right)\mathbf{v}_2$$

$$= (-1, 2, 3) - \tfrac{4}{3}(1, 1, 1) - \tfrac{2}{6}(-1, 2, -1) = (-2, 0, 2).$$

Thus

$$T^* = \{\mathbf{v}_1, \mathbf{v}_2, \mathbf{v}_3\} = \{(1, 1, 1), (-1, 2, -1), (-2, 0, 2)\}$$

is an orthogonal basis for R^3.

Step 3. Let

$$\mathbf{w}_1 = \frac{1}{\|\mathbf{v}_1\|}\,\mathbf{v}_1 = \frac{1}{\sqrt{3}}\,(1, 1, 1)$$

$$\mathbf{w}_2 = \frac{1}{\|\mathbf{v}_2\|}\,\mathbf{v}_2 = \frac{1}{\sqrt{6}}\,(-1, 2, -1)$$

$$\mathbf{w}_3 = \frac{1}{\|\mathbf{v}_3\|}\,\mathbf{v}_3 = \frac{1}{\sqrt{8}}\,(-2, 0, 2) = \left(-\frac{1}{\sqrt{2}}, 0, \frac{1}{\sqrt{2}}\right).$$

Then

$$T = \{\mathbf{w}_1, \mathbf{w}_2, \mathbf{w}_3\}$$

$$= \left\{\left(\frac{1}{\sqrt{3}}, \frac{1}{\sqrt{3}}, \frac{1}{\sqrt{3}}\right), \left(-\frac{1}{\sqrt{6}}, \frac{2}{\sqrt{6}}, -\frac{1}{\sqrt{6}}\right), \left(-\frac{1}{\sqrt{2}}, 0, \frac{1}{\sqrt{2}}\right)\right\}$$

is an orthonormal basis for R^3. ∎

EXAMPLE 5 ■ Let W be the subspace of R^4 with basis $S = \{\mathbf{u}_1, \mathbf{u}_2, \mathbf{u}_3\}$, where

$$\mathbf{u}_1 = (1, -2, 0, 1), \quad \mathbf{u}_2 = (-1, 0, 0, -1), \quad \text{and} \quad \mathbf{u}_3 = (1, 1, 0, 0).$$

Use the Gram–Schmidt process to transform S to an orthonormal basis for W.

Solution **Step 1.** Let $\mathbf{v}_1 = \mathbf{u}_1 = (1, -2, 0, 1)$.

Step 2. We now compute $\mathbf{v}_2$ and $\mathbf{v}_3$:

$$\mathbf{v}_2 = \mathbf{u}_2 - \left(\frac{\mathbf{u}_2 \cdot \mathbf{v}_1}{\mathbf{v}_1 \cdot \mathbf{v}_1}\right)\mathbf{v}_1 = (-1, 0, 0, -1) - \left(\tfrac{-2}{6}\right)(1, -2, 0, 1)$$

$$= \left(-\tfrac{2}{3}, -\tfrac{2}{3}, 0, -\tfrac{2}{3}\right).$$

Multiplying $\mathbf{v}_2$ by 3 to clear fractions, we obtain $(-2, -2, 0, -2)$, which we now use as $\mathbf{v}_2$. Then

$$\mathbf{v}_3 = \mathbf{u}_3 - \left(\frac{\mathbf{u}_3 \cdot \mathbf{v}_1}{\mathbf{v}_1 \cdot \mathbf{v}_1}\right)\mathbf{v}_1 - \left(\frac{\mathbf{u}_3 \cdot \mathbf{v}_2}{\mathbf{v}_2 \cdot \mathbf{v}_2}\right)\mathbf{v}_2$$

$$= (1, 1, 0, 0) - \left(\tfrac{-1}{6}\right)(1, -2, 0, 1) - \left(\tfrac{-4}{12}\right)(-2, -2, 0, -2)$$

$$= \left(\tfrac{1}{2}, 0, 0, -\tfrac{1}{2}\right).$$

Multiplying $\mathbf{v}_3$ by 2 to clear fractions, we obtain $(1, 0, 0, -1)$, which we now use as $\mathbf{v}_3$. Thus

$$T^* = \{(1, -2, 0, 1), (-2, -2, 0, -2), (1, 0, 0, -1)\}$$

is an orthogonal basis for W.

Step 3. Let

$$\mathbf{w}_1 = \frac{1}{\|\mathbf{v}_1\|} \mathbf{v}_1 = \frac{1}{\sqrt{6}} (1, -2, 0, 1)$$

$$\mathbf{w}_2 = \frac{1}{\|\mathbf{v}_2\|} \mathbf{v}_2 = \frac{1}{\sqrt{12}} (-2, -2, 0, -2) = \frac{1}{\sqrt{3}} (-1, -1, 0, -1)$$

$$\mathbf{w}_3 = \frac{1}{\|\mathbf{v}_3\|} \mathbf{v}_3 = \frac{1}{\sqrt{2}} (1, 0, 0, -1).$$

Then

$$T = \{\mathbf{w}_1, \mathbf{w}_2, \mathbf{w}_3\}$$

$$= \left\{ \left(\frac{1}{\sqrt{6}}, -\frac{2}{\sqrt{6}}, 0, \frac{1}{\sqrt{6}} \right), \left(-\frac{1}{\sqrt{3}}, -\frac{1}{\sqrt{3}}, 0, -\frac{1}{\sqrt{3}} \right), \left(\frac{1}{\sqrt{2}}, 0, 0, -\frac{1}{\sqrt{2}} \right) \right\}$$

is an orthonormal basis for W.   ■

REMARK   In solving Example 5, as soon as a vector is computed we multiplied it by an appropriate scalar to eliminate any fractions that may be present. This optional step results in simpler computations when working by hand. If this approach is taken, the resulting basis, while orthonormal, may differ from the orthonormal basis obtained by not clearing fractions. Most computer implementations of the Gram–Schmidt process, including those developed with MATLAB, do not clear fractions.

REMARK   We make one final observation with regard to the Gram–Schmidt process. In our proof of Theorem 4.18 we first obtained an orthogonal basis T^* and then normalized all the vectors in T^* to obtain the orthonormal basis T. Of course, an alternative course of action is to normalize each vector as soon as we produce it.

4.8   EXERCISES

1. Which of the following are orthogonal sets of vectors?
   (a) $\{(1, -1, 2), (0, 2, -1), (-1, 1, 1)\}$.
   (b) $\{(1, 2, -1, 1), (0, -1, -2, 0), (1, 0, 0, -1)\}$.
   (c) $\{(0, 1, 0, -1), (1, 0, 1, 1), (-1, 1, -1, 2)\}$.

2. Which of the following are orthonormal sets of vectors?
   (a) $\left\{ \left(\frac{1}{3}, \frac{2}{3}, \frac{2}{3} \right), \left(\frac{2}{3}, \frac{1}{3}, -\frac{2}{3} \right), \left(\frac{2}{3}, -\frac{2}{3}, \frac{1}{3} \right) \right\}$.
   (b) $\left\{ \left(\frac{1}{\sqrt{2}}, 0, -\frac{1}{\sqrt{2}} \right), \left(\frac{1}{\sqrt{3}}, \frac{1}{\sqrt{3}}, \frac{1}{\sqrt{3}} \right), (0, 1, 0) \right\}$.
   (c) $\{(0, 2, 2, 1), (1, 1, -2, 2), (0, -2, 1, 2)\}$.

In Exercises 3 and 4, let $V = R^3$.

3. Let $\mathbf{u} = (1, 1, -2)$ and $\mathbf{v} = (a, -1, 2)$. For what values of a are $\mathbf{u}$ and $\mathbf{v}$ orthogonal?

4. Let $\mathbf{u} = \left(\frac{1}{\sqrt{2}}, 0, \frac{1}{\sqrt{2}} \right)$, and $\mathbf{v} = \left(a, \frac{1}{\sqrt{2}}, -b \right)$. For what values of a and b is $\{\mathbf{u}, \mathbf{v}\}$ an orthonormal set?

5. Use the Gram–Schmidt process to find an orthonormal basis for the subspace of R^3 with basis $\{(1, -1, 0), (2, 0, 1)\}$.

6. Use the Gram–Schmidt process to find an orthonormal basis for the subspace of R^3 with basis $\{(1, 0, 2), (-1, 1, 0)\}$.

7. Use the Gram–Schmidt process to find an orthonormal basis for the subspace of R^4 with basis $\{(1, -1, 0, 1), (2, 0, 0, -1), (0, 0, 1, 0)\}$.

8. Use the Gram–Schmidt process to find an orthonormal basis for the subspace of R^4 with basis $\{(1, 1, -1, 0), (0, 2, 0, 1), (-1, 0, 0, 1)\}$.

9. Use the Gram–Schmidt process to transform the basis $\{(1, 2), (-3, 4)\}$ for R^2 into (a) an orthogonal basis; (b) an orthonormal basis.

10. (a) Use the Gram–Schmidt process to transform the basis $\{(1, 1, 1), (0, 1, 1), (1, 2, 3)\}$ for R^3 into an orthonormal basis for R^3.

    (b) Use Theorem 4.17 to write $(2, 3, 1)$ as a linear combination of the basis obtained in part (a).

11. Find an orthonormal basis for R^3 containing the vectors $\left(\frac{2}{3}, -\frac{2}{3}, \frac{1}{3}\right)$ and $\left(\frac{2}{3}, \frac{1}{3}, -\frac{2}{3}\right)$.

12. Use the Gram–Schmidt process to construct an orthonormal basis for the subspace W of R^3 *spanned* by $\{(1, 1, 1), (2, 2, 2), (0, 0, 1), (1, 2, 3)\}$.

13. Use the Gram–Schmidt process to construct an orthonormal basis for the subspace W of R^4 *spanned* by $\{(1, 1, 0, 0), (2, -1, 0, 1), (3, -3, 0, -2), (1, -2, 0, -3)\}$.

14. Find an orthonormal basis for the subspace of R^3 consisting of all vectors of the form $(a, a + b, b)$.

15. Find an orthonormal basis for the subspace of R^4 consisting of all vectors of the form
$$(a, a + b, c, b + c).$$

16. Find an orthonormal basis for the subspace of R^3 consisting of all vectors (a, b, c) such that
$$a + b + c = 0.$$

17. Find an orthonormal basis for the subspace of R^4 consisting of all vectors (a, b, c, d) such that
$$a - b - 2c + d = 0.$$

18. Find an orthonormal basis for the solution space of the homogeneous system
$$x_1 + x_2 - x_3 = 0$$
$$2x_1 + x_2 + 2x_3 = 0.$$

19. Find an orthonormal basis for the solution space of the homogeneous system
$$\begin{bmatrix} 1 & 1 & -1 \\ 2 & 1 & 3 \\ 1 & 2 & -6 \end{bmatrix} \begin{bmatrix} x_1 \\ x_2 \\ x_3 \end{bmatrix} = \begin{bmatrix} 0 \\ 0 \\ 0 \end{bmatrix}.$$

20. Consider the orthonormal basis
$$S = \left\{ \left(\frac{1}{\sqrt{2}}, \frac{1}{\sqrt{2}} \right), \left(-\frac{1}{\sqrt{2}}, \frac{1}{\sqrt{2}} \right) \right\}$$
for R^2. Using Theorem 4.17 write the vector $(2, 3)$ as a linear combination of the vectors in S.

21. Consider the orthonormal basis
$$S = \left\{ \left(\frac{1}{\sqrt{5}}, 0, \frac{2}{\sqrt{5}} \right), \left(-\frac{2}{\sqrt{5}}, 0, \frac{1}{\sqrt{5}} \right), (0, 1, 0) \right\}$$
for R^3. Using Theorem 4.17 write the vector $(2, -3, 1)$ as a linear combination of the vectors in S.

THEORETICAL EXERCISES

T.1. Verify that the natural basis for R^n is an orthonormal set.

T.2. Prove Corollary 4.6.

T.3. Show that an orthonormal set of n vectors in R^n is a basis for R^n.

T.4. Prove Theorem 4.17.

T.5. Let $\mathbf{u}, \mathbf{v}_1, \mathbf{v}_2, \ldots, \mathbf{v}_n$ be vectors in R^n. Show that if $\mathbf{u}$ is orthogonal to $\mathbf{v}_1, \mathbf{v}_2, \ldots, \mathbf{v}_n$, then $\mathbf{u}$ is orthogonal to every vector in span $\{\mathbf{v}_1, \mathbf{v}_2, \ldots, \mathbf{v}_n\}$.

T.6. Let $\mathbf{u}$ be a fixed vector in R^n. Show that the set of all vectors in R^n that are orthogonal to $\mathbf{u}$ is a subspace of R^n.

T.7. Let $\mathbf{u}$ and $\mathbf{v}$ be vectors in R^n. Show that if $\mathbf{u} \cdot \mathbf{v} = 0$, then $\mathbf{u} \cdot (c\mathbf{v}) = 0$ for any scalar c.

T.8. Suppose that $\{\mathbf{v}_1, \mathbf{v}_2, \ldots, \mathbf{v}_n\}$ is an orthonormal set in R^n. Let the matrix A be given by $A = \begin{bmatrix} \mathbf{v}_1 & \mathbf{v}_2 & \cdots & \mathbf{v}_n \end{bmatrix}$. Show that A is nonsingular and compute its inverse. Give three different examples of such a matrix in R^2 or R^3.

T.9. Suppose that $\{\mathbf{v}_1, \mathbf{v}_2, \ldots, \mathbf{v}_n\}$ is an orthogonal set in R^n. Let A be the matrix whose jth column is $\mathbf{v}_j$, $j = 1, 2, \ldots, n$. Prove or disprove: A is nonsingular.

T.10. Let $S = \{\mathbf{u}_1, \mathbf{u}_2, \ldots, \mathbf{u}_k\}$ be an orthonormal basis for a subspace W of R^n, where $n > k$.

Discuss how to construct an orthonormal basis for V that includes S.

T.11. Let $\{\mathbf{u}_1, \ldots, \mathbf{u}_k, \mathbf{u}_{k+1}, \ldots, \mathbf{u}_n\}$ be an orthonormal basis for R^n, $S = \text{span } \{\mathbf{u}_1, \ldots, \mathbf{u}_k\}$, and $T = \text{span } \{\mathbf{u}_{k+1}, \ldots, \mathbf{u}_n\}$. For any $\mathbf{x}$ in S and any $\mathbf{y}$ in T, show that $\mathbf{x} \cdot \mathbf{y} = 0$.

MATLAB EXERCISES ■

The Gram–Schmidt process takes a basis S for a subspace W of R^n and produces an orthonormal basis T for W. The algorithm to produce the orthonormal basis T that is given in this section is implemented in MATLAB in routine **gschmidt**. Type **help gschmidt** for directions.

ML.1. Use **gschmidt** to produce an orthonormal basis for R^3 from the basis

$$S = \left\{ \begin{bmatrix} 1 \\ 1 \\ 0 \end{bmatrix}, \begin{bmatrix} 1 \\ 0 \\ 0 \end{bmatrix}, \begin{bmatrix} 0 \\ 1 \\ 1 \end{bmatrix} \right\}.$$

Your answer will be in decimal form; rewrite it in terms of $\sqrt{2}$.

ML.2. Use **gschmidt** to produce an orthonormal basis for R^4 from the basis $S = \{(1, 0, 1, 1), (1, 2, 1, 3), (0, 2, 1, 1), (0, 1, 0, 0)\}$.

ML.3. In R^3, $S = \{(0, -1, 1), (0, 1, 1), (1, 1, 1)\}$ is a basis. Find an orthonormal basis T from S and then find $\begin{bmatrix} \mathbf{v} \end{bmatrix}_T$ for each of the following vectors.

  (a) $\mathbf{v} = (1, 2, 0)$.   (b) $\mathbf{v} = (1, 1, 1)$.

  (c) $\mathbf{v} = (-1, 0, 1)$.

ML.4. Find an orthonormal basis for the subspace of R^4 consisting of all vectors of the form

$$(a, 0, a + b, b + c),$$

where a, b, and c are any real numbers.

4.9 ▼ Orthogonal Complements

Let W_1 and W_2 be subspaces of a vector space V. Let $W_1 + W_2$ be the set of all vectors $\mathbf{v}$ in V such that $\mathbf{v} = \mathbf{w}_1 + \mathbf{w}_2$, where $\mathbf{w}_1$ is in W_1 and $\mathbf{w}_2$ is in W_2. In Exercise T.10 in Section 4.2, we asked you to show that $W_1 + W_2$ is a subspace of V. In Exercise T.11 in Section 4.2, we asked you to show that if $V = W_1 + W_2$ and $W_1 \cap W_2 = \{\mathbf{0}\}$, then V is the direct sum of W_1 and W_2 and we write $V = W_1 \oplus W_2$. Moreover, in this case every vector in V can be uniquely written as $\mathbf{w}_1 + \mathbf{w}_2$, where $\mathbf{w}_1$ is in W_1 and $\mathbf{w}_2$ is in W_2. In this section we show that if W is a subspace of R^n, then R^n can be written as a direct sum of W and another subspace of R^n. This subspace will be used to examine a basic relationship between four vector spaces associated with a matrix.

DEFINITION

Let W be a subspace of R^n. A vector $\mathbf{u}$ in R^n is said to be **orthogonal** to W if it is orthogonal to every vector in W. The set of all vectors in R^n that are orthogonal to all the vectors in W is called the **orthogonal complement** of W in R^n and is denoted by $W^\perp$ (read "W perp").

EXAMPLE 1 ■ Let W be the subspace of R^3 consisting of all multiples of the vector

$$\mathbf{w} = (2, -3, 4).$$

Thus $W = \text{span } \{\mathbf{w}\}$, so W is a one-dimensional subspace of W. Then a vector $\mathbf{u}$ in R^3 belongs to $W^\perp$ if and only if $\mathbf{u}$ is orthogonal to $c\mathbf{w}$, for any scalar c. It can be shown that $W^\perp$ is the plane with normal $\mathbf{w}$.   ■

Observe that if W is a subspace of R^n, then the zero vector always belongs to $W^\perp$ (Exercise T.1). Moreover, the orthogonal complement of R^n is the zero subspace and the orthogonal complement of the zero subspace is R^n itself (Exercise T.2).

THEOREM 4.19 ■ *Let W be a subspace of R^n. Then*

(a) $W^\perp$ *is a subspace of R^n.*

(b) $W \cap W^\perp = \{\mathbf{0}\}$.

Proof   (a) Let $\mathbf{u}_1$ and $\mathbf{u}_2$ be in $W^\perp$. Then $\mathbf{u}_1$ and $\mathbf{u}_2$ are orthogonal to each vector $\mathbf{w}$ in W. We now have

$$(\mathbf{u}_1 + \mathbf{u}_2) \cdot \mathbf{w} = \mathbf{u}_1 \cdot \mathbf{w} + \mathbf{u}_2 \cdot \mathbf{w} = 0 + 0 = 0,$$

so $\mathbf{u}_1 + \mathbf{u}_2$ is in $W^\perp$. Also, let $\mathbf{u}$ be in $W^\perp$ and c be a real scalar. Then for any vector $\mathbf{w}$ in W, we have

$$(c\mathbf{u}) \cdot \mathbf{w} = c(\mathbf{u} \cdot \mathbf{w}) = c0 = 0,$$

so $c\mathbf{u}$ is in W, which implies that $W^\perp$ is closed under vector addition and scalar multiplication and hence is a subspace of R^n.

(b) Let $\mathbf{u}$ be a vector in $W \cap W^\perp$. Then $\mathbf{u}$ is in both W and $W^\perp$, so $\mathbf{u} \cdot \mathbf{u} = 0$. From Theorem 3.3 in Section 3.2 it follows that $\mathbf{u} = \mathbf{0}$. ■

In Exercise T.3 we ask you to show that if W is a subspace of R^n that is spanned by a set of vectors S, then a vector $\mathbf{u}$ in R^n belongs to $W^\perp$ if and only if $\mathbf{u}$ is orthogonal to every vector in S. This result can be helpful in finding $W^\perp$, as shown in the next example.

EXAMPLE 2 ■ Let W be the subspace of R^4 with basis $\{\mathbf{w}_1, \mathbf{w}_2\}$, where

$$\mathbf{w}_1 = (1, 1, 0, 1) \quad \text{and} \quad \mathbf{w}_2 = (0, -1, 1, 1).$$

Find a basis for $W^\perp$.

Solution   Let $\mathbf{u} = (a, b, c, d)$ be a vector in $W^\perp$. Then $\mathbf{u} \cdot \mathbf{w}_1 = 0$ and $\mathbf{u} \cdot \mathbf{w}_2 = 0$. Thus we have

$$\mathbf{u} \cdot \mathbf{w}_1 = a + b + d = 0 \quad \text{and} \quad \mathbf{u} \cdot \mathbf{w}_2 = -b + c + d = 0.$$

Solving the homogeneous system

$$\begin{aligned} a + b \quad\ + d &= 0 \\ -b + c + d &= 0, \end{aligned}$$

we obtain (verify)

$$a = -r - s, \qquad b = r, \qquad c = -2s, \qquad d = s.$$

Then

$$\mathbf{u} = (-r - s, r, -2s, s) = r(-1, 1, 0, 0) + s(-1, 0, -2, 1).$$

Hence the vectors $(-1, 1, 0, 0)$ and $(-1, 0, -2, 1)$ span $W^\perp$. Since they are not multiples of each other, they are linearly independent and thus form a basis for $W^\perp$. ■

THEOREM 4.20 ■ *Let W be a subspace of R^n. Then*

$$R^n = W \oplus W^\perp.$$

Proof

Let $\dim W = m$. Then W has a basis consisting of m vectors. By the Gram–Schmidt process we can transform this basis to an orthonormal basis. Thus let $S = \{\mathbf{w}_1, \mathbf{w}_2, \ldots, \mathbf{w}_m\}$ be an orthonormal basis for W. If $\mathbf{v}$ is a vector in R^n, let

$$\mathbf{w} = (\mathbf{v} \cdot \mathbf{w}_1)\mathbf{w}_1 + (\mathbf{v} \cdot \mathbf{w}_2)\mathbf{w}_2 + \cdots + (\mathbf{v} \cdot \mathbf{w}_m)\mathbf{w}_m \tag{1}$$

and

$$\mathbf{u} = \mathbf{v} - \mathbf{w}. \tag{2}$$

Since $\mathbf{w}$ is a linear combination of vectors in S, $\mathbf{w}$ belongs to W. We next show that $\mathbf{u}$ lies in $W^\perp$ by showing that $\mathbf{u}$ is orthogonal to every vector in S, a basis for W. For each $\mathbf{w}_i$ in S, we have

$$\begin{aligned}
\mathbf{u} \cdot \mathbf{w}_i &= (\mathbf{v} - \mathbf{w}) \cdot \mathbf{w}_i = \mathbf{v} \cdot \mathbf{w}_i - \mathbf{w} \cdot \mathbf{w}_i \\
&= \mathbf{v} \cdot \mathbf{w}_i - [(\mathbf{v} \cdot \mathbf{w}_1)\mathbf{w}_1 + (\mathbf{v} \cdot \mathbf{w}_2)\mathbf{w}_2 + \cdots + (\mathbf{v} \cdot \mathbf{w}_m)\mathbf{w}_m] \cdot \mathbf{w}_i \\
&= \mathbf{v} \cdot \mathbf{w}_i - (\mathbf{v} \cdot \mathbf{w}_i)(\mathbf{w}_i \cdot \mathbf{w}_i) \\
&= 0
\end{aligned}$$

since $\mathbf{w}_i \cdot \mathbf{w}_j = 0$ for $i \neq j$ and $\mathbf{w}_i \cdot \mathbf{w}_i = 1$, $1 \leq i \leq m$. Thus $\mathbf{u}$ is orthogonal to every vector in W and so lies in $W^\perp$. Hence

$$\mathbf{v} = \mathbf{w} + \mathbf{u},$$

which means that $R^n = W + W^\perp$. From part (b) of Theorem 4.19, it follows that

$$R^n = W \oplus W^\perp. \qquad ■$$

REMARK   As pointed out at the beginning of this section, we also conclude that the vectors $\mathbf{w}$ and $\mathbf{u}$ defined by Equations (1) and (2) are unique.

THEOREM 4.21 ■ *If W is a subspace of R^n, then*

$$(W^\perp)^\perp = W.$$

Proof

First, if $\mathbf{w}$ is any vector in W, then $\mathbf{w}$ is orthogonal to every vector $\mathbf{u}$ in $W^\perp$, so $\mathbf{w}$ is in $(W^\perp)^\perp$. Hence W is a subspace of $(W^\perp)^\perp$. Conversely, let $\mathbf{v}$ be an arbitrary vector in $(W^\perp)^\perp$. Then, by Theorem 4.20, $\mathbf{v}$ can be written as

$$\mathbf{v} = \mathbf{w} + \mathbf{u},$$

where $\mathbf{w}$ is in W and $\mathbf{u}$ is in $W^\perp$. Since $\mathbf{u}$ is in $W^\perp$, it is orthogonal to $\mathbf{v}$ and $\mathbf{w}$. Thus

$$0 = \mathbf{u} \cdot \mathbf{v} = \mathbf{u} \cdot (\mathbf{w} + \mathbf{u}) = \mathbf{u} \cdot \mathbf{w} + \mathbf{u} \cdot \mathbf{u} = \mathbf{u} \cdot \mathbf{u}$$

or

$$\mathbf{u} \cdot \mathbf{u} = 0,$$

which implies that $\mathbf{u} = \mathbf{0}$. Then $\mathbf{v} = \mathbf{w}$, so $\mathbf{v}$ belongs to W. Hence $(W^\perp)^\perp = W$. ■

REMARK   Since W is the orthogonal complement of $W^\perp$ and $W^\perp$ is also the orthogonal complement of W, we say that W and $W^\perp$ are **orthogonal complements**.

Relations Among the Fundamental Vector Spaces Associated with a Matrix (Optional)

If A is a given $m \times n$ matrix, we associate the following four fundamental vector spaces with A: the null space of A, the row space of A, the null space of A^T, and the column space of A. The following theorem shows that pairs of these four vector spaces are orthogonal complements.

THEOREM 4.22 ■ *If A is a given $m \times n$ matrix, then*

(a) *The null space of A is the orthogonal complement of the row space of A.*

(b) *The null space of A^T is the orthogonal complement of the column space of A.*

Proof (a) Before proving the result, let us verify that the two vector spaces that we wish to show are the same have equal dimensions. If r is the rank of A, then the dimension of the null space of A is $n - r$ (Theorem 4.12 in Section 4.6). Since the dimension of the row space of A is r, then by Theorem 4.20, the dimension of its orthogonal complement is also $n - r$. Let the vector $\mathbf{x}$ in R^n be in the null space of A. Then $A\mathbf{x} = \mathbf{0}$. Let the vectors $\mathbf{v}_1, \mathbf{v}_2, \ldots, \mathbf{v}_m$ in R^n denote the rows of A. Then the entries in the $m \times 1$ matrix $A\mathbf{x}$ are $\mathbf{v}_1\mathbf{x}, \mathbf{v}_2\mathbf{x}, \ldots, \mathbf{v}_m\mathbf{x}$. Thus we have

$$\mathbf{v}_1\mathbf{x} = 0, \quad \mathbf{v}_2\mathbf{x} = 0, \ldots, \quad \mathbf{v}_m\mathbf{x} = 0. \tag{3}$$

Hence $\mathbf{x}$ is orthogonal to the vectors $\mathbf{v}_1, \mathbf{v}_2, \ldots, \mathbf{v}_m$, which span the row space of A. It then follows that $\mathbf{x}$ is orthogonal to every vector in the row space of A, so $\mathbf{x}$ lies in the orthogonal complement of the row space of A. Hence the null space of A is contained in the orthogonal complement of the row space of A. It then follows from Exercise T.7 in Section 4.4 that the null space of A equals the orthogonal complement of the row space of A.

(b) To establish the result, replace A by A^T in part (a) to conclude that the null space of A^T is the orthogonal complement of the row space of A^T. Since the row space of A^T is the column space of A, we have established part (b). ■

EXAMPLE 3 ■ Let

$$A = \begin{bmatrix} 1 & -2 & 1 & 0 & 2 \\ 1 & -1 & 4 & 1 & 3 \\ -1 & 3 & 2 & 1 & -1 \\ 2 & -3 & 5 & 1 & 5 \end{bmatrix}.$$

Compute the four fundamental vector spaces associated with A and verify Theorem 4.22.

Solution We first transform A to reduced row echelon form, obtaining (verify)

$$B = \begin{bmatrix} 1 & 0 & 7 & 2 & 4 \\ 0 & 1 & 3 & 1 & 1 \\ 0 & 0 & 0 & 0 & 0 \\ 0 & 0 & 0 & 0 & 0 \end{bmatrix}.$$

Solving the linear system $B\mathbf{x} = \mathbf{0}$, we find (verify) that

$$S = \left\{ \begin{bmatrix} -7 \\ -3 \\ 1 \\ 0 \\ 0 \end{bmatrix}, \begin{bmatrix} -2 \\ -1 \\ 0 \\ 1 \\ 0 \end{bmatrix}, \begin{bmatrix} -4 \\ -1 \\ 0 \\ 0 \\ 1 \end{bmatrix} \right\}$$

is a basis for the null space of A. Moreover,

$$T = \{(1, 0, 7, 2, 4), (0, 1, 3, 1, 1)\}$$

is a basis for the row space of A. Since the vectors in S and T are orthogonal, it follows that S is a basis for the orthogonal complement of the row space of A, where we take the vectors in S in horizontal form. Next,

$$A^T = \begin{bmatrix} 1 & 1 & -1 & 2 \\ -2 & -1 & 3 & -3 \\ 1 & 4 & 2 & 5 \\ 0 & 1 & 1 & 1 \\ 2 & 3 & -1 & 5 \end{bmatrix}.$$

Solving the linear system $A^T\mathbf{x} = \mathbf{0}$, we find (verify) that

$$S' = \left\{ \begin{bmatrix} 2 \\ -1 \\ 1 \\ 0 \end{bmatrix}, \begin{bmatrix} -1 \\ -1 \\ 0 \\ 1 \end{bmatrix} \right\}$$

is a basis for the null space of A^T. Transforming A^T to reduced row echelon form, we obtain (verify)

$$C = \begin{bmatrix} 1 & 0 & -2 & 1 \\ 0 & 1 & 1 & 1 \\ 0 & 0 & 0 & 0 \\ 0 & 0 & 0 & 0 \\ 0 & 0 & 0 & 0 \end{bmatrix}.$$

Then the nonzero rows of C read vertically yield the following basis for the column space of A:

$$T' = \left\{ \begin{bmatrix} 1 \\ 0 \\ -2 \\ 1 \end{bmatrix}, \begin{bmatrix} 0 \\ 1 \\ 1 \\ 1 \end{bmatrix} \right\}.$$

Since the vectors in S' and T' are orthogonal, it follows that S' is a basis for the orthogonal complement of the column space of A. ∎

EXAMPLE 4 ∎ Find a basis for the orthogonal complement of the subspace W of R^5 spanned by the vectors

$$\mathbf{w}_1 = (2, -1, 0, 1, 2), \qquad \mathbf{w}_2 = (1, 3, 1, -2, -4),$$
$$\mathbf{w}_3 = (3, 2, 1, -1, -2), \qquad \mathbf{w}_4 = (7, 7, 3, -4, -8), \qquad \mathbf{w}_5 = (1, -4, -1, -1, -2).$$

Solution 1

Let $\mathbf{u} = (a, b, c, d, e)$ be an arbitrary vector in $W^\perp$. Since $\mathbf{u}$ is orthogonal to each of the given vectors spanning W, we have a linear system of five equations in five unknowns, whose coefficient matrix is (verify)

$$A = \begin{bmatrix} 2 & -1 & 0 & 1 & 2 \\ 1 & 3 & 1 & -2 & -4 \\ 3 & 2 & 1 & -1 & -2 \\ 7 & 7 & 3 & -4 & -8 \\ 1 & -4 & -1 & -1 & -2 \end{bmatrix}.$$

Solving the homogeneous system $A\mathbf{x} = \mathbf{0}$, we obtain the following basis for the solution space (verify):

$$S = \left\{ \begin{bmatrix} -\frac{1}{7} \\ -\frac{2}{7} \\ 1 \\ 0 \\ 0 \end{bmatrix}, \begin{bmatrix} 0 \\ 0 \\ 0 \\ -2 \\ 1 \end{bmatrix} \right\}.$$

These vectors taken in horizontal form provide a basis for $W^\perp$.

Solution 2

Form the matrix whose rows are the given vectors. This matrix is A as shown in Solution 1 above, so the row space of A is W. By Theorem 4.22, $W^\perp$ is the null space of A. Thus we obtain the same basis for $W^\perp$ as in Solution 1. ■

Projections and Applications

In Theorem 4.20 and in the Remark following the theorem, we have shown that if W is a subspace of R^n with orthonormal basis $\{\mathbf{w}_1, \mathbf{w}_2, \ldots, \mathbf{w}_m\}$ and $\mathbf{v}$ is any vector in R^n, then there exist unique vectors $\mathbf{w}$ in W and $\mathbf{u}$ in $W^\perp$ such that

$$\mathbf{v} = \mathbf{w} + \mathbf{u}.$$

Moreover, as we saw in Equation (1),

$$\mathbf{w} = (\mathbf{v} \cdot \mathbf{w}_1)\mathbf{w}_1 + (\mathbf{v} \cdot \mathbf{w}_2)\mathbf{w}_2 + \cdots + (\mathbf{v} \cdot \mathbf{w}_m)\mathbf{w}_m,$$

which is called the **orthogonal projection** of $\mathbf{v}$ on W and is denoted by $\text{proj}_W \mathbf{v}$. In Figure 4.9, we illustrate Theorem 4.20 when W is a two-dimensional subspace of R^3 (a plane through the origin).

Often an orthonormal basis has many fractions, so it is helpful to also have a formula giving $\text{proj}_W \mathbf{v}$ when W has an *orthogonal* basis. In Exercise T.6, we ask you to show that if $\{\mathbf{w}_1, \mathbf{w}_2, \ldots, \mathbf{w}_m\}$ is an orthogonal basis for W, then

$$\text{proj}_W \mathbf{v} = \frac{\mathbf{v} \cdot \mathbf{w}_1}{\mathbf{w}_1 \cdot \mathbf{w}_1} \mathbf{w}_1 + \frac{\mathbf{v} \cdot \mathbf{w}_2}{\mathbf{w}_2 \cdot \mathbf{w}_2} \mathbf{w}_2 + \cdots + \frac{\mathbf{v} \cdot \mathbf{w}_m}{\mathbf{w}_m \cdot \mathbf{w}_m} \mathbf{w}_m.$$

FIGURE 4.9

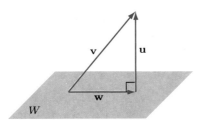

EXAMPLE 5 ■ Let W be the two-dimensional subspace of R^3 with orthonormal basis $\{\mathbf{w}_1, \mathbf{w}_2\}$, where

$$\mathbf{w}_1 = \left(\frac{2}{3}, -\frac{1}{3}, -\frac{2}{3}\right) \quad \text{and} \quad \mathbf{w}_2 = \left(\frac{1}{\sqrt{2}}, 0, \frac{1}{\sqrt{2}}\right).$$

Using the standard inner product on R^3, find the orthogonal projection of

$$\mathbf{v} = (2, 1, 3)$$

on W and the vector $\mathbf{u}$ that is orthogonal to every vector in W.

Solution   From Equation (1) we have

$$\mathbf{w} = \text{proj}_W \mathbf{v} = (\mathbf{v} \cdot \mathbf{w}_1)\mathbf{w}_1 + (\mathbf{v} \cdot \mathbf{w}_2)\mathbf{w}_2 = -\frac{1}{3}\mathbf{w}_1 + \frac{5}{\sqrt{2}}\mathbf{w}_2 = \left(\frac{41}{18}, -\frac{1}{9}, \frac{49}{18}\right)$$

and

$$\mathbf{u} = \mathbf{v} - \mathbf{w} = \left(-\frac{5}{18}, \frac{10}{9}, \frac{5}{18}\right).$$ ■

It is clear from Figure 4.9 that the distance from $\mathbf{v}$ to the plane W is given by the length of the vector $\mathbf{u} = \mathbf{v} - \mathbf{w}$, that is, by

$$\|\mathbf{v} - \text{proj}_W \mathbf{v}\|.$$

We prove this result in general in Theorem 4.23.

EXAMPLE 6 ■ Let W be the subspace of R^3 defined in Example 5 and let $\mathbf{v} = (1, 1, 0)$. Find the distance from $\mathbf{v}$ to W.

Solution   We first compute

$$\text{proj}_W \mathbf{v} = (\mathbf{v} \cdot \mathbf{w}_1)\mathbf{w}_1 + (\mathbf{v} \cdot \mathbf{w}_2)\mathbf{w}_2 = \mathbf{w}_1 + \frac{1}{\sqrt{2}}\mathbf{w}_2 = \left(\frac{7}{6}, \frac{1}{3}, -\frac{1}{6}\right).$$

Then

$$\mathbf{v} - \text{proj}_W \mathbf{v} = (1, 1, 0) - \left(\frac{7}{6}, \frac{1}{3}, -\frac{1}{6}\right) = \left(-\frac{1}{6}, \frac{2}{3}, \frac{1}{6}\right)$$

and

$$\|\mathbf{v} - \text{proj}_W \mathbf{v}\| = \sqrt{\frac{1}{36} + \frac{4}{9} + \frac{1}{36}} = \frac{\sqrt{2}}{2},$$

so the distance from $\mathbf{v}$ to W is $\dfrac{\sqrt{2}}{2}$.   ■

In Example 6, $\|\mathbf{v} - \text{proj}_W \mathbf{v}\|$ represented the distance in 3-space from $\mathbf{v}$ to the plane W. We can generalize this notion of distance from a vector in R^n to a subspace W of R^n. We can show that the vector in W that is closest to $\mathbf{v}$ is in fact $\text{proj}_W \mathbf{v}$, so $\|\mathbf{v} - \text{proj}_W \mathbf{v}\|$ represents the distance from $\mathbf{v}$ to W.

THEOREM 4.23 ■ *Let W be a subspace of R^n. Then for vector $\mathbf{v}$ belonging to R^n, the vector in W closest to $\mathbf{v}$ is $\text{proj}_W \mathbf{v}$. That is, $\|\mathbf{v} - \mathbf{w}\|$, for $\mathbf{w}$ belonging to W, is minimized when $\mathbf{w} = \text{proj}_W \mathbf{v}$.*

Proof   Let $\mathbf{w}$ be any vector in W. Then

$$\mathbf{v} - \mathbf{w} = (\mathbf{v} - \text{proj}_W \mathbf{v}) + (\text{proj}_W \mathbf{v} - \mathbf{w}).$$

Since $\mathbf{w}$ and $\text{proj}_W \mathbf{v}$ are both in W, $\text{proj}_W \mathbf{v} - \mathbf{w}$ is in W. By Theorem 4.20, $\mathbf{v} - \text{proj}_W \mathbf{v}$ is orthogonal to every vector in W, so

$$\begin{aligned}
\|\mathbf{v} - \mathbf{w}\|^2 &= (\mathbf{v} - \mathbf{w}) \cdot (\mathbf{v} - \mathbf{w}) \\
&= \big((\mathbf{v} - \text{proj}_W \mathbf{v}) + (\text{proj}_W \mathbf{v} - \mathbf{w})\big) \cdot \big((\mathbf{v} - \text{proj}_W \mathbf{v}) + (\text{proj}_W \mathbf{v} - \mathbf{w})\big) \\
&= \|\mathbf{v} - \text{proj}_W \mathbf{v}\|^2 + \|\text{proj}_W \mathbf{v} - \mathbf{w}\|^2.
\end{aligned}$$

If $\mathbf{w} \neq \text{proj}_W \mathbf{v}$, then $\|\text{proj}_W \mathbf{v} - \mathbf{w}\|^2$ is positive and

$$\|\mathbf{v} - \mathbf{w}\|^2 > \|\mathbf{v} - \text{proj}_W \mathbf{v}\|^2.$$

Thus it follows that $\text{proj}_W \mathbf{v}$ is the vector in W that minimizes $\|\mathbf{v} - \mathbf{w}\|^2$ and hence minimizes $\|\mathbf{v} - \mathbf{w}\|$.   ■

In Example 5,

$$\mathbf{w} = \text{proj}_W \mathbf{v} = \left(\tfrac{41}{18}, -\tfrac{1}{9}, \tfrac{49}{18}\right)$$

is the vector in $W = \text{span}\{\mathbf{w}_1, \mathbf{w}_2\}$ that is closest to $\mathbf{v} = (2, 1, 3)$.

4.9   EXERCISES

1. Let W be the subspace of R^3 spanned by the vector $\mathbf{w} = (2, -1, 3)$.

   (a) Find a basis for $W^\perp$.

   (b) Describe $W^\perp$ geometrically. (You may use a verbal or pictorial description.)

2. Let $W = \text{span}\{(1, 2, -1), (-1, 3, 2)\}$.

   (a) Find a basis for $W^\perp$.

   (b) Describe $W^\perp$ geometrically. (You may use a verbal or pictorial description.)

3. Let W be the subspace of R^5 spanned by the vectors $\mathbf{w}_1, \mathbf{w}_2, \mathbf{w}_3, \mathbf{w}_4, \mathbf{w}_5$, where

$$\mathbf{w}_1 = (2, -1, 1, 3, 0), \qquad \mathbf{w}_2 = (1, 2, 0, 1, -2),$$
$$\mathbf{w}_3 = (4, 3, 1, 5, -4), \qquad \mathbf{w}_4 = (3, 1, 2, -1, 1),$$
$$\mathbf{w}_5 = (2, -1, 2, -2, 3).$$

Find a basis for $W^\perp$.

In Exercises 4 and 5, compute the four fundamental vector spaces associated with A and verify Theorem 4.22.

4. $A = \begin{bmatrix} 1 & 5 & 3 & 7 \\ 2 & 0 & -4 & -6 \\ 4 & 7 & -1 & 2 \end{bmatrix}$.

5. $A = \begin{bmatrix} 2 & -1 & 3 & 4 \\ 0 & -3 & 7 & -2 \\ 1 & 1 & -2 & 3 \\ 1 & 4 & -9 & 5 \end{bmatrix}.$

In Exercises 6 and 7, find $\text{proj}_W \mathbf{v}$ for the given vector $\mathbf{v}$ and subspace W.

6. Let W be the subspace of R^3 with basis

$$\left(\frac{1}{\sqrt{5}}, 0, \frac{2}{\sqrt{5}}\right), \qquad \left(-\frac{2}{\sqrt{5}}, 0, \frac{1}{\sqrt{5}}\right).$$

(a) $\mathbf{v} = (3, 4, -1)$.   (b) $\mathbf{v} = (2, 1, 3)$.
(c) $\mathbf{v} = (-5, 0, 1)$.

7. Let W be the subspace of R^4 with basis $(1, 1, 0, 1), (0, 1, 1, 0), (-1, 0, 0, 1)$.

(a) $\mathbf{v} = (2, 1, 3, 0)$.   (b) $\mathbf{v} = (0, -1, 1, 0)$.
(c) $\mathbf{v} = (0, 2, 0, 3)$.

8. Let W be the subspace of R^3 with orthonormal basis $\{\mathbf{w}_1, \mathbf{w}_2\}$, where

$$\mathbf{w}_1 = (0, 1, 0) \quad \text{and} \quad \mathbf{w}_2 = \left(\frac{1}{\sqrt{5}}, 0, \frac{2}{\sqrt{5}}\right).$$

Write the vector $\mathbf{v} = (1, 2, -1)$ as $\mathbf{w} + \mathbf{u}$, with $\mathbf{w}$ in W and $\mathbf{u}$ in $W^\perp$.

9. Let W be the subspace of R^4 with orthonormal basis $\{\mathbf{w}_1, \mathbf{w}_2, \mathbf{w}_3\}$, where

$$\mathbf{w}_1 = \left(\frac{1}{\sqrt{2}}, 0, 0, -\frac{1}{\sqrt{2}}\right), \quad \mathbf{w}_2 = (0, 0, 1, 0),$$

$$\text{and} \quad \mathbf{w}_3 = \left(\frac{1}{\sqrt{2}}, 0, 0, \frac{1}{\sqrt{2}}\right).$$

Write the vector $\mathbf{v} = (1, 0, 2, 3)$ as $\mathbf{w} + \mathbf{u}$, with $\mathbf{w}$ in W and $\mathbf{u}$ in $W^\perp$.

10. Let W be the subspace of R^3 defined in Exercise 8, and let $\mathbf{v} = (-1, 0, 1)$. Find the distance from $\mathbf{v}$ to W.

11. Let W be the subspace of R^4 defined in Exercise 9, and let $\mathbf{v} = (1, 2, -1, 0)$. Find the distance from $\mathbf{v}$ to W.

12. Find the distance from the point $(2, 3, -1)$ to the plane $3x - 2y + z = 0$. (*Hint*: First find an orthonormal basis for the plane.)

THEORETICAL EXERCISES

T.1. Show that if W is a subspace of R^n, then the zero vector of R^n belongs to $W^\perp$.

T.2. Show that the orthogonal complement of R^n is the zero subspace and the orthogonal complement of the zero subspace is R^n itself.

T.3. Show that if W is a subspace of R^n that is spanned by a set of vectors S, then a vector $\mathbf{u}$ in R^n belongs to $W^\perp$ if and only if $\mathbf{u}$ is orthogonal to every vector in S.

T.4. Let A be an $m \times n$ matrix. Show that every vector $\mathbf{v}$ in R^n can be written uniquely as $\mathbf{w} + \mathbf{u}$, where $\mathbf{w}$ is in the null space of A and $\mathbf{u}$ is in the column space of A^T.

T.5. Let W be a subspace of R^n. Show that if $\mathbf{w}_1, \mathbf{w}_2, \ldots, \mathbf{w}_r$ is a basis for W and $\mathbf{u}_1, \mathbf{u}_2, \ldots, \mathbf{u}_s$ is a basis for $W^\perp$, then $\mathbf{w}_1, \mathbf{w}_2, \ldots, \mathbf{w}_r, \mathbf{u}_1, \mathbf{u}_2, \ldots, \mathbf{u}_s$ is a basis for R^n, and $n = \dim W + \dim W^\perp$.

T.6. Let W be a subspace of R^n and let $\{\mathbf{w}_1, \mathbf{w}_2, \ldots, \mathbf{w}_m\}$ be an orthogonal basis for W. Show that if $\mathbf{v}$ is any vector in R^n, then

$$\text{proj}_W \mathbf{v} = \frac{\mathbf{v} \cdot \mathbf{w}_1}{\mathbf{w}_1 \cdot \mathbf{w}_1} \mathbf{w}_1 + \frac{\mathbf{v} \cdot \mathbf{w}_2}{\mathbf{w}_2 \cdot \mathbf{w}_2} \mathbf{w}_2 + \cdots$$
$$+ \frac{\mathbf{v} \cdot \mathbf{w}_m}{\mathbf{w}_m \cdot \mathbf{w}_m} \mathbf{w}_m.$$

MATLAB EXERCISES

ML.1. Find the projection of $\mathbf{v}$ onto $\mathbf{w}$. (Recall that we have routines **dot** and **norm** available in MATLAB.)

(a) $\mathbf{v} = \begin{bmatrix} 1 \\ 5 \\ -1 \\ 2 \end{bmatrix}$, $\mathbf{w} = \begin{bmatrix} 0 \\ 1 \\ 2 \\ 1 \end{bmatrix}.$

(b) $\mathbf{v} = \begin{bmatrix} 1 \\ -2 \\ 3 \\ 0 \\ 1 \end{bmatrix}$, $\mathbf{w} = \begin{bmatrix} 1 \\ 1 \\ 1 \\ 1 \\ 1 \end{bmatrix}$.

ML.2. Let $S = \{\mathbf{w1}, \mathbf{w2}\}$, where

$$\mathbf{w1} = \begin{bmatrix} 1 \\ 0 \\ 1 \\ 1 \end{bmatrix} \quad \text{and} \quad \mathbf{w2} = \begin{bmatrix} 1 \\ 1 \\ -1 \\ 0 \end{bmatrix},$$

and let $W = \text{span } S$.

(a) Show that S is an orthogonal basis for W.

(b) Let

$$\mathbf{v} = \begin{bmatrix} 2 \\ 1 \\ 2 \\ 1 \end{bmatrix}.$$

Compute $\text{proj}_{\mathbf{w1}} \mathbf{v}$.

(c) For vector $\mathbf{v}$ in part (b), compute $\text{proj}_W \mathbf{v}$.

ML.3. Plane P in R^3 has orthogonal basis $\{\mathbf{w1}, \mathbf{w2}\}$, where

$$\mathbf{w1} = \begin{bmatrix} 1 \\ 2 \\ 3 \end{bmatrix} \quad \text{and} \quad \mathbf{w2} = \begin{bmatrix} 0 \\ -3 \\ 2 \end{bmatrix}.$$

(a) Find the projection of

$$\mathbf{v} = \begin{bmatrix} 2 \\ 4 \\ 8 \end{bmatrix}$$

onto P.

(b) Find the distance from $\mathbf{v}$ to P.

ML.4. Let W be the subspace of R^4 with basis

$$S = \left\{ \begin{bmatrix} 1 \\ 1 \\ 0 \\ 1 \end{bmatrix}, \begin{bmatrix} 2 \\ -1 \\ 0 \\ 0 \end{bmatrix}, \begin{bmatrix} 0 \\ 1 \\ 0 \\ 1 \end{bmatrix} \right\} \quad \text{and} \quad \mathbf{v} = \begin{bmatrix} 0 \\ 0 \\ 1 \\ 1 \end{bmatrix}.$$

(a) Find $\text{proj}_W \mathbf{v}$.

(b) Find the distance from $\mathbf{v}$ to W.

ML.5. Let

$$T = \begin{bmatrix} 1 & 0 \\ 0 & 1 \\ 1 & 1 \\ 1 & 0 \\ 1 & 0 \end{bmatrix} \quad \text{and} \quad \mathbf{b} = \begin{bmatrix} 1 \\ 1 \\ 1 \\ 1 \\ 1 \end{bmatrix}.$$

(a) Show that system $T\mathbf{x} = \mathbf{b}$ is inconsistent.

(b) Since $T\mathbf{x} = \mathbf{b}$ is inconsistent, $\mathbf{b}$ is not in the column space of T. One approach to find an approximate solution is to find a vector $\mathbf{y}$ in the column space of T so that $T\mathbf{y}$ is as close as possible to $\mathbf{b}$. We can do this by finding the projection $\mathbf{p}$ of $\mathbf{b}$ onto the column space of T. Find this projection $\mathbf{p}$ (which will be $T\mathbf{y}$).

KEY IDEAS FOR REVIEW

☐ **Vector space.** See page 197.

☐ **Theorem 4.2.** Let V be a vector space with operations $\oplus$ and $\odot$ and let W be a nonempty subset of V. Then W is a subspace of V if and only if
(α) If $\mathbf{u}$ and $\mathbf{v}$ are any vectors in W, then $\mathbf{u} \oplus \mathbf{v}$ is in W.
(β) If $\mathbf{u}$ is in W and c is a scalar, then $c \odot \mathbf{u}$ is in W.

☐ $S = \{\mathbf{v}_1, \mathbf{v}_2, \ldots, \mathbf{v}_k\}$ is linearly dependent if and only if one of the vectors in S is a linear combination of all the other vectors in S.

☐ **Theorem 4.5.** If $S = \{\mathbf{v}_1, \mathbf{v}_2, \ldots, \mathbf{v}_n\}$ is a basis for a vector space V, then every vector in V can be written in one and only one way as a linear combination of the vectors in S.

☐ **Theorem 4.7.** If $S = \{\mathbf{v}_1, \mathbf{v}_2, \ldots, \mathbf{v}_n\}$ is a basis for a vector space V and $T = \{\mathbf{w}_1, \mathbf{w}_2, \ldots, \mathbf{w}_r\}$ is a linearly independent set of vectors in V, then $r \leq n$.

☐ **Corollary 4.1.** All bases for a vector space V must have the same number of vectors.

☐ **Theorem 4.9.** Let V be an n-dimensional vector space, and let $S = \{\mathbf{v}_1, \mathbf{v}_2, \ldots, \mathbf{v}_n\}$ be a set of n vectors in V.

(a) If S is linearly independent, then it is a basis for V.

(b) If S spans V, then it is a basis for V.

☐ **Basis for solution space of** $A\mathbf{x} = \mathbf{0}$. See page 239.

☐ **Theorem 4.11.** Row rank A = column rank A.

☐ **Theorem 4.12.** If A is an $m \times n$ matrix, then rank A + nullity $A = n$.

☐ **Theorem 4.13.** An $n \times n$ matrix A is nonsingular if and only if rank $A = n$.

☐ **Corollary 4.2.** If A is an $n \times n$ matrix, then rank $A = n$ if and only if $\det(A) \neq 0$.

☐ **Corollary 4.5.** If A is $n \times n$, then $A\mathbf{x} = \mathbf{0}$ has a nontrivial solution if and only if rank $A < n$.

☐ **Theorem 4.14.** The linear system $A\mathbf{x} = \mathbf{b}$ has a solution if and only if rank A = rank $\begin{bmatrix} A & \vdots & \mathbf{b} \end{bmatrix}$.

☐ **List of Nonsingular Equivalences.** The following statements are equivalent for an $n \times n$ matrix A.

1. A is nonsingular.
2. $A\mathbf{x} = \mathbf{0}$ has only the trivial solution.
3. A is row equivalent to I_n.
4. The linear system $A\mathbf{x} = \mathbf{b}$ has a unique solution for every $n \times 1$ matrix $\mathbf{b}$.
5. $\det(A) \neq 0$.
6. A has rank n.
7. A has nullity 0.

8. The rows of A form a linearly independent set of n vectors in R^n.

9. The columns of A form a linearly independent set of n vectors in R^n.

☐ **Theorem 4.15.** Let $S = \{\mathbf{v}_1, \mathbf{v}_2, \ldots, \mathbf{v}_n\}$ and $T = \{\mathbf{w}_1, \mathbf{w}_2, \ldots, \mathbf{w}_n\}$ be bases for the n-dimensional vector space V. Let $P_{S \leftarrow T}$ be the transition matrix from T to S. Then $P_{S \leftarrow T}$ is nonsingular and $P_{S \leftarrow T}^{-1}$ is the transition matrix from S to T.

☐ **Theorem 4.16.** An orthogonal set of nonzero vectors in R^n is linearly independent.

☐ **Theorem 4.17.** If $\{\mathbf{u}_1, \mathbf{u}_2, \ldots, \mathbf{u}_n\}$ is an orthonormal basis for R^n and $\mathbf{v}$ is a vector in R^n, then $\mathbf{v} = (\mathbf{v} \cdot \mathbf{u}_1)\mathbf{u}_1 + (\mathbf{v} \cdot \mathbf{u}_2)\mathbf{u}_2 + \cdots + (\mathbf{v} \cdot \mathbf{u}_n)\mathbf{u}_n$.

☐ **Theorem 4.18 (Gram–Schmidt Process).** See page 271.

☐ **Theorem 4.20.** Let W be a subspace of R^n. Then $R^n = W \oplus W^\perp$.

☐ **Theorem 4.22.** If A is a given $m \times n$ matrix, then

(a) The null space of A is the orthogonal complement of the row space of A.

(b) The null space of A^T is the orthogonal complement of the column space of A.

SUPPLEMENTARY EXERCISES ▪

1. Determine whether the set of all ordered pairs of real numbers with the operations
$$(x, y) \oplus (x', y') = (x - x', y + y')$$
$$c \odot (x, y) = (cx, cy)$$
is a vector space.

2. Consider the set W of all vectors in R^4 of the form (a, b, c, d), where $a = b + c$ and $d = a + 1$. Is W a subspace of R^4?

3. Is the vector $(4, 2, 1)$ a linear combination of the vectors $(1, 2, 1)$, $(-1, 1, 2)$, and $(-3, -3, 0)$?

4. Do the vectors $(-1, 2, 1)$, $(2, 1, -1)$, and $(0, 5, -1)$ span R^3?

5. Is the set of vectors
$$\{t^2 + 2t + 2, 2t^2 + 3t + 1, -t - 3\}$$
linearly dependent or linearly independent? If it is linearly dependent, write one of the vectors as a linear combination of the other two.

6. Does the set of vectors
$$\{t - 1, t^2 + 2t + 1, t^2 + t - 2\}$$
form a basis for P_2?

7. Find a basis for the subspace of R^4 consisting of all vectors of the form
$$(a + b, b + c, a - b - 2c, b + c).$$
What is the dimension of this subspace?

8. Find the dimension of the subspace of P_2 consisting of all polynomials $a_0 t^2 + a_1 t + a_2$, where $a_2 = 0$.

9. Find a basis for the solution space of the homogeneous system
$$x_1 + 2x_2 - x_3 + x_4 + 2x_5 = 0$$
$$x_1 + x_2 + 2x_3 - 3x_4 + x_5 = 0.$$
What is the dimension of the solution space? Find a basis for $V = \text{span } S$.

10. Let
$$S = \{(1, 2, -1, 2), (0, -1, 3, -6),$$
$$(2, 3, 1, -2), (3, 2, 1, -4), (1, -1, 0, -2)\}.$$
Find a basis for $V = \mathrm{span}\ S$.

11. For what value(s) of λ is the set
$$\{t + 3, 2t + \lambda^2 + 2\}$$
linearly independent?

12. Let
$$A = \begin{bmatrix} \lambda^2 & 0 & 1 \\ -1 & \lambda & -2 \\ -1 & 0 & -1 \end{bmatrix}.$$
For what value(s) of λ does the homogeneous system $A\mathbf{x} = \mathbf{0}$ have only the trivial solution?

13. For what values of a is the vector $(a^2, a, 1)$ in span $\{(1, 2, 3), (1, 1, 1), (0, 1, 2)\}$?

14. For what values of a is the vector $(a^2, -3a, -2)$ in span $\{(1, 2, 3), (0, 1, 1), (1, 3, 4)\}$?

15. Let $\mathbf{v}_1$, $\mathbf{v}_2$, $\mathbf{v}_3$ be vectors in a vector space. Verify that
$$\mathrm{span}\ \{\mathbf{v}_1, \mathbf{v}_2, \mathbf{v}_3\}$$
$$= \mathrm{span}\ \{\mathbf{v}_1 + \mathbf{v}_2, \mathbf{v}_1 - \mathbf{v}_3, \mathbf{v}_1 + \mathbf{v}_3\}.$$

16. For what values of k will the set S form a basis for R^6?
$$S = \{(1, 2, -2, 1, 1, 1), (0, 2, 0, 0, 0, 0),$$
$$(0, 5, k, 1, 0, 0), (0, 2, 1, k, 0, 0),$$
$$(0, 1, -2, 4, -3, 1), (0, 3, 1, 1, 1, 0)\}.$$

17. Consider the vector space R^2.

(a) For what values of m and b will all vectors of the form $(x, mx + b)$ be a subspace of R^2?

(b) For what values of r will the set of all vectors of the form (x, rx^2) be a subspace of R^2?

18. Let A be a fixed $n \times n$ matrix and let the set of all $n \times n$ matrices B such that $AB = BA$ be denoted by $C(A)$. Is $C(A)$ a subspace of the vector space of all $n \times n$ matrices?

19. Show that a subspace W of R^3 coincides with R^3 if and only if it contains the vectors $(1, 0, 0)$, $(0, 1, 0)$, and $(0, 0, 1)$.

20. Let A be a fixed $m \times n$ matrix and define W to be the subset of all $m \times 1$ matrices $\mathbf{b}$ in R^m for which the linear system $A\mathbf{x} = \mathbf{b}$ has a solution.

(a) Is W a subspace of R^m?

(b) What is the relationship between W and the column space of A?

21. (**Calculus Required**). Let $C[a, b]$ denote the set of all real-valued continuous functions defined on $[a, b]$. If f and g are in $C[a, b]$, we define $f \oplus g$ by $(f \oplus g)(t) = f(t) + g(t)$, for t in $[a, b]$. If f is in $C[a, b]$ and c is a scalar, we define $c \odot f$ by $(c \odot f)(t) = cf(t)$, for t in $[a, b]$.

(a) Show that $C[a, b]$ is a real vector space.

(b) Let $W(k)$ be the set of all functions in $C[a, b]$ with $f(a) = k$. For what values of k will $W(k)$ be a subspace of $C[a, b]$?

(c) Let $t_1, t_2, \ldots, t_n$ be a fixed set of points in $[a, b]$. Show that the subset of all functions f in $C[a, b]$ that have roots at $t_1, t_2, \ldots, t_n$, that is, $f(t_i) = 0$ for $i = 1, 2, \ldots, n$, forms a subspace.

22. Let $V = \mathrm{span}\ \{\mathbf{v}_1, \mathbf{v}_2\}$, where
$$\mathbf{v}_1 = (1, 0, 2) \quad \text{and} \quad \mathbf{v}_2 = (1, 1, 1).$$
Find a basis S for R^3 that includes $\mathbf{v}_1$ and $\mathbf{v}_2$. (*Hint*: Use Theorem 4.8; see Example 9 in Section 4.4.)

23. Find the rank of the matrix
$$\begin{bmatrix} 1 & 2 & -1 & 3 & 1 \\ 0 & 1 & -3 & 2 & 3 \\ 2 & 3 & 1 & 4 & -1 \\ -1 & 2 & 2 & 2 & -5 \\ 3 & 1 & -1 & 2 & 4 \end{bmatrix}.$$

24. Let A be an 8×5 matrix. How large can the dimension of the row space be? Explain.

25. What can you say about the dimension of the solution space of a homogeneous system of 8 equations in 10 unknowns?

26. Is it possible that all nontrivial solutions of a homogeneous system of 5 equations in 7 unknowns be multiples of each other? Explain.

27. Let $S = \{(1, 0, -1), (0, 1, 1), (0, 0, 1)\}$ and $T = \{(1, 0, 0), (0, 1, -1), (1, -1, 2)\}$ be bases for R^3. Let $\mathbf{v} = (2, 3, 5)$.

(a) Find the coordinate vector of $\mathbf{v}$ with respect to T directly.

(b) Find the coordinate vector of $\mathbf{v}$ with respect to S directly.

(c) Find the transition matrix $P_{S \leftarrow T}$ from T to S.

(d) Find the coordinate vector of $\mathbf{v}$ with respect to S using $P_{S \leftarrow T}$ and the answer to part (a).

(e) Find the transition matrix $Q_{T \leftarrow S}$ from S to T.

(f) Find the coordinate vector of **v** with respect to T using $Q_{T \leftarrow S}$ and the answer to part (b).

28. Let $S = \{\mathbf{v}_1, \mathbf{v}_2\}$ and $T = \{\mathbf{w}_1, \mathbf{w}_2\}$ be bases for P_1, where

$$\mathbf{v}_1 = t + 2, \qquad \mathbf{v}_2 = 2t - 1.$$

If the transition matrix from T to S is

$$\begin{bmatrix} 3 & -1 \\ 2 & 1 \end{bmatrix},$$

determine T.

29. Let

$$\mathbf{u} = \left(\frac{1}{\sqrt{2}}, 0, -\frac{1}{\sqrt{2}} \right) \quad \text{and} \quad \mathbf{v} = (a, -1, -b).$$

For what values of a and b is $\{\mathbf{u}, \mathbf{v}\}$ an orthonormal set?

30. Exercise T.6 of Section 4.8 shows that the set of all vectors in R^n that are orthogonal to a fixed vector **u** is a subspace of R^n. For $\mathbf{u} = (1, -2, 1)$ find an orthogonal basis for the subspace of vectors in R^3 that are orthogonal to **u**. [*Hint*: Solve the linear system $\mathbf{u} \cdot \mathbf{v} = 0$, when $\mathbf{v} = (y_1, y_2, y_3)$.]

31. Find an orthonormal basis for the set of vectors **x** such that $A\mathbf{x} = \mathbf{0}$ when

(a) $A = \begin{bmatrix} 1 & 0 & 5 & -2 \\ 0 & 1 & -2 & -5 \end{bmatrix}$.

(b) $A = \begin{bmatrix} 1 & 0 & 5 & -2 \\ 0 & 1 & -2 & 4 \end{bmatrix}$.

32. Given the orthonormal basis

$$S = \left\{ \left(\frac{1}{\sqrt{2}}, 0, -\frac{1}{\sqrt{2}} \right), (0, 1, 0), \left(\frac{1}{\sqrt{2}}, 0, \frac{1}{\sqrt{2}} \right) \right\}$$

for R^3, write the vector $(1, 2, 3)$ as a linear combination of the vectors in S.

33. Use the Gram–Schmidt process to find an orthonormal basis for the subspace of R^4 with basis $\{(1, 0, 0, -1), (1, -1, 0, 0), (0, 1, 0, 1)\}$.

34. Let W be the subspace of R^4 spanned by the vectors $\mathbf{w}_1, \mathbf{w}_2, \mathbf{w}_3, \mathbf{w}_4$, where

$$\mathbf{w}_1 = (2, 0, -1, 3), \qquad \mathbf{w}_2 = (1, 2, 2, -5),$$

$$\mathbf{w}_3 = (3, 2, 1, -2), \qquad \mathbf{w}_4 = (7, 2, -1, 4).$$

Find a basis for $W^\perp$.

35. Let $W = \text{span} \{(1, 0, 1), (0, 1, 0)\}$ in R^3.

(a) Find a basis for $W^\perp$.

(b) Show that vectors $(1, 0, 1)$, $(0, 1, 0)$ and the basis for $W^\perp$ from part (a) form a basis for R^3.

(c) Write the vector **v** as $\mathbf{w} + \mathbf{u}$ with **w** in W and **u** in $W^\perp$.

   (i) $\mathbf{v} = (1, 0, 0)$.

   (ii) $\mathbf{v} = (1, 2, 3)$.

36. Let W be the subspace of R^3 defined in Exercise 35. Find the distance from $\mathbf{v} = (2, 1, -2)$ to W.

37. Compute the four fundamental vector spaces associated with

$$A = \begin{bmatrix} 2 & 3 & -1 & 2 \\ 1 & 1 & -2 & 3 \\ 2 & 1 & 4 & 2 \end{bmatrix}.$$

THEORETICAL EXERCISES

T.1. Let A be an $n \times n$ matrix. Show that A is nonsingular if and only if the dimension of the solution space of the homogeneous system $A\mathbf{x} = \mathbf{0}$ is zero.

T.2. Let $\{\mathbf{v}_1, \mathbf{v}_2, \ldots, \mathbf{v}_n\}$ be a basis for an n-dimensional vector space V. Show that if A is a nonsingular $n \times n$ matrix, then $\{A\mathbf{v}_1, A\mathbf{v}_2, \ldots, A\mathbf{v}_n\}$ is also a basis for V.

T.3. Let $S = \{\mathbf{v}_1, \mathbf{v}_2, \ldots, \mathbf{v}_n\}$ be a basis for R^n. Show that if **u** is a vector in R^n that is orthogonal to each vector in S, then $\mathbf{u} = \mathbf{0}$.

T.4. Show that rank $A = $ rank A^T, for any $m \times n$ matrix A.

T.5. Let A and B be $m \times n$ matrices that are row equivalent.

(a) Show that rank $A = $ rank B.

(b) Show that for **x** in R^n, $A\mathbf{x} = \mathbf{0}$ if and only if $B\mathbf{x} = \mathbf{0}$.

T.6. Let A be $m \times n$ and B be $n \times k$.

(a) Show that

$$\text{rank } (AB) \leq \min\{\text{rank } A, \text{rank } B\}.$$

(b) Find A and B such that

$$\text{rank } (AB) < \min\{\text{rank } A, \text{rank } B\}.$$

(c) If $k = n$ and B is nonsingular, show that
$$\text{rank } (AB) = \text{rank } A.$$

(d) If $m = n$ and A is nonsingular, show that
$$\text{rank } (AB) = \text{rank } B.$$

(e) For nonsingular matrices P and Q, what is rank (PAQ)?

T.7. Let $S = \{\mathbf{v}_1, \mathbf{v}_2, \ldots, \mathbf{v}_k\}$ be an orthonormal basis for R^n and let $\{a_1, \ldots, a_k\}$ be any set of scalars

none of which is zero. Show that $T = \{a_1\mathbf{v}_1, a_2\mathbf{v}_2, \ldots, a_k\mathbf{v}_k\}$ is an orthogonal basis for V. How should the scalars $a_1, \ldots, a_k$ be chosen so that T is an orthonormal basis for R^n?

T.8. Let B be an $m \times n$ matrix with orthonormal columns $\mathbf{b}_1, \mathbf{b}_2, \ldots, \mathbf{b}_n$.

(a) Show that $m \geq n$.

(b) Show that $B^T B = I_n$.

CHAPTER TEST ▪

1. Consider the set W of all vectors in R^3 of the form (a, b, c), where $a + b + c = 0$. Is W a subspace of R^3?

2. Find a basis for the solution space of the homogeneous system
$$x_1 + 3x_2 + 3x_3 - x_4 + 2x_5 = 0$$
$$x_1 + 2x_2 + 2x_3 - 2x_4 + 2x_5 = 0$$
$$x_1 + x_2 + x_3 - 3x_4 + 2x_5 = 0.$$

3. Does the set of vectors $\{(1, -1, 1), (1, -3, 1), (1, -2, 2)\}$ form a basis for R^3?

4. For what value(s) of λ is the set of vectors $\{(\lambda^2 - 5, 1, 0), (2, -2, 3), (2, 3, -3)\}$ linearly dependent?

5. Use the Gram–Schmidt process to find an orthonormal basis for the subspace of R^4 with basis $\{(1, 0, -1, 0), (1, -1, 0, 0), (3, 1, 0, 0)\}$.

6. Answer each of the following as true or false.

(a) All vectors of the form $(a, 0, -a)$ form a subspace of R^3.

(b) In R^n, $\|c\mathbf{x}\| = c\|\mathbf{x}\|$.

(c) Every set of vectors in R^3 containing two vectors is linearly independent.

(d) The solution space of the homogeneous system $A\mathbf{x} = \mathbf{0}$ is spanned by the columns of A.

(e) If the columns of an $n \times n$ matrix form a basis for R^n, so do the rows.

(f) If A is an 8×8 matrix such that the homogeneous system $A\mathbf{x} = \mathbf{0}$ has only the trivial solution, then rank $A < 8$.

(g) Every orthonormal set of five vectors in R^5 is a basis for R^5.

(h) Every linearly independent set of vectors in R^3 contains three vectors.

(i) If A is an $n \times n$ symmetric matrix, then rank $A = n$.

(j) Every set of vectors spanning R^3 contains at least three vectors.

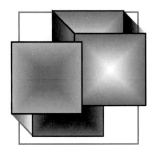

5

Eigenvalues and Eigenvectors

5.1 ▾ Diagonalization

In this chapter every matrix considered is a square matrix. Let A be an $n \times n$ matrix. Then, as we have seen in Section 3.3, the function $L: R^n \to R^n$ defined by $L(\mathbf{x}) = A\mathbf{x}$, for $\mathbf{x}$ in R^n, is a linear transformation. A question of considerable importance in a great many applied problems is the determination of vectors $\mathbf{x}$, if there are any, such that $\mathbf{x}$ and $A\mathbf{x}$ are parallel. Such questions arise in all applications involving vibrations; they arise in aerodynamics, elasticity, nuclear physics, mechanics, chemical engineering, biology, differential equations, and so on. In this section we shall formulate this problem precisely; we also define some pertinent terminology. In the next section we solve this problem for symmetric matrices and briefly discuss the situation in the general case.

DEFINITION   Let A be an $n \times n$ matrix. The real number λ is called an **eigenvalue** of A if there exists a nonzero vector $\mathbf{x}$ in R^n such that

$$A\mathbf{x} = \lambda\mathbf{x}. \tag{1}$$

Every nonzero vector $\mathbf{x}$ satisfying (1) is called an **eigenvector of A associated with the eigenvalue** λ. We might mention that the word "eigenvalue" is a hybrid one ("eigen" in German means "proper"). Eigenvalues are also called **proper values, characteristic values**, and **latent values**; and eigenvectors are also called **proper vectors**, and so on, accordingly.

Note that $\mathbf{x} = \mathbf{0}$ always satisfies (1), but $\mathbf{0}$ is not an eigenvector, since we insist that an eigenvector be a nonzero vector.

In some applications one encounters matrices with complex entries and vector spaces with scalars that are complex numbers (see Sections A.1 and A.2, respectively). In such a setting the preceding definition of eigenvalue is modified so that an eigenvalue can be a real *or* a complex number. An introduction to this approach, a treatment usually presented in more advanced books, is given in Section A.2. Throughout the rest of this book, unless stated otherwise, we require that an eigenvalue be a real number.

EXAMPLE 1 ■ If A is the identity matrix I_n, then the only eigenvalue is $\lambda = 1$; every nonzero vector in R^n is an eigenvector of A associated with the eigenvalue $\lambda = 1$:

$$I_n\mathbf{x} = 1\mathbf{x}.$$ ■

EXAMPLE 2 ■ Let

$$A = \begin{bmatrix} 0 & \frac{1}{2} \\ \frac{1}{2} & 0 \end{bmatrix}.$$

Then

$$A\begin{bmatrix} 1 \\ 1 \end{bmatrix} = \begin{bmatrix} 0 & \frac{1}{2} \\ \frac{1}{2} & 0 \end{bmatrix}\begin{bmatrix} 1 \\ 1 \end{bmatrix} = \begin{bmatrix} \frac{1}{2} \\ \frac{1}{2} \end{bmatrix} = \frac{1}{2}\begin{bmatrix} 1 \\ 1 \end{bmatrix}$$

so that

$$\mathbf{x}_1 = \begin{bmatrix} 1 \\ 1 \end{bmatrix}$$

is an eigenvector of A associated with the eigenvalue $\lambda_1 = \frac{1}{2}$. Also,

$$A\begin{bmatrix} 1 \\ -1 \end{bmatrix} = \begin{bmatrix} 0 & \frac{1}{2} \\ \frac{1}{2} & 0 \end{bmatrix}\begin{bmatrix} 1 \\ -1 \end{bmatrix} = \begin{bmatrix} -\frac{1}{2} \\ \frac{1}{2} \end{bmatrix} = -\frac{1}{2}\begin{bmatrix} 1 \\ 1 \end{bmatrix}$$

so that

$$\mathbf{x}_2 = \begin{bmatrix} 1 \\ -1 \end{bmatrix}$$

is an eigenvector of A associated with the eigenvalue $\lambda_2 = -\frac{1}{2}$. Figure 5.1 shows that $\mathbf{x}_1$ and $A\mathbf{x}_1$ are parallel, and $\mathbf{x}_2$ and $A\mathbf{x}_2$ are parallel also. This illustrates the fact that if $\mathbf{x}$ is an eigenvector of A, then $\mathbf{x}$ and $A\mathbf{x}$ are parallel. ■

FIGURE 5.1

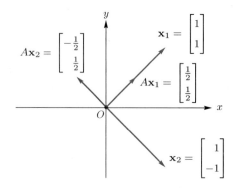

FIGURE 5.2

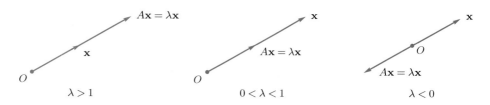

$Ax = \lambda x$

x

O

$\lambda > 1$

x

$Ax = \lambda x$

O

$0 < \lambda < 1$

x

O

$Ax = \lambda x$

$\lambda < 0$

Let λ be an eigenvalue of A with corresponding eigenvector $\mathbf{x}$. In Figure 5.2 we show $\mathbf{x}$ and $A\mathbf{x}$ for the cases $\lambda > 1$, $0 < \lambda < 1$, and $\lambda < 0$.

An eigenvalue λ of A can have associated with it many different eigenvectors. In fact, if $\mathbf{x}$ is an eigenvector of A associated with λ (i.e., $A\mathbf{x} = \lambda\mathbf{x}$) and r is any nonzero real number, then

$$A(r\mathbf{x}) = r(A\mathbf{x}) = r(\lambda\mathbf{x}) = \lambda(r\mathbf{x}).$$

Thus $r\mathbf{x}$ is also an eigenvector of A associated with λ.

EXAMPLE 3 ■ Let

$$A = \begin{bmatrix} 0 & 0 \\ 0 & 1 \end{bmatrix}.$$

Then

$$A\begin{bmatrix} 1 \\ 0 \end{bmatrix} = \begin{bmatrix} 0 & 0 \\ 0 & 1 \end{bmatrix} \begin{bmatrix} 1 \\ 0 \end{bmatrix} = \begin{bmatrix} 0 \\ 0 \end{bmatrix} = 0\begin{bmatrix} 1 \\ 0 \end{bmatrix}$$

so that $\mathbf{x}_1 = \begin{bmatrix} 1 \\ 0 \end{bmatrix}$ is an eigenvector of A associated with the eigenvalue $\lambda_1 = 0$. Also,

$$\mathbf{x}_2 = \begin{bmatrix} 0 \\ 1 \end{bmatrix}$$

is an eigenvector of A associated with the eigenvalue $\lambda_2 = 1$ (verify).   ■

Example 3 points out the fact that although the zero vector, by definition, cannot be an eigenvector, the number zero can be an eigenvalue.

EXAMPLE 4 ■ Let

$$A = \begin{bmatrix} 1 & 1 \\ -2 & 4 \end{bmatrix}.$$

We wish to find the eigenvalues of A and their associated eigenvectors. Thus we wish to find all real numbers λ and all nonzero vectors

$$\mathbf{x} = \begin{bmatrix} x_1 \\ x_2 \end{bmatrix}$$

satisfying (1), that is,

$$\begin{bmatrix} 1 & 1 \\ -2 & 4 \end{bmatrix} \begin{bmatrix} x_1 \\ x_2 \end{bmatrix} = \lambda \begin{bmatrix} x_1 \\ x_2 \end{bmatrix}. \tag{2}$$

Equation (2) becomes

$$x_1 + \ x_2 = \lambda x_1$$
$$-2x_1 + 4x_2 = \lambda x_2,$$

or

$$(\lambda - 1)x_1 - \ x_2 = 0$$
$$2x_1 + (\lambda - 4)x_2 = 0. \tag{3}$$

Equation (3) is a homogeneous system of two equations in two unknowns. From Corollary 2.3 of Section 2.2, it follows that the homogeneous system in (3) has a nontrivial solution if and only if the determinant of its coefficient matrix is zero; thus if and only if

$$\begin{bmatrix} \lambda - 1 & -1 \\ 2 & \lambda - 4 \end{bmatrix} = 0.$$

This means that

$$(\lambda - 1)(\lambda - 4) + 2 = 0,$$

or

$$\lambda^2 - 5\lambda + 6 = 0 = (\lambda - 3)(\lambda - 2).$$

Hence

$$\lambda_1 = 2 \quad \text{and} \quad \lambda_2 = 3$$

are the eigenvalues of A. To find all eigenvectors of A associated with $\lambda_1 = 2$, we form the linear system

$$A\mathbf{x} = 2\mathbf{x},$$

or

$$\begin{bmatrix} 1 & 1 \\ -2 & 4 \end{bmatrix} \begin{bmatrix} x_1 \\ x_2 \end{bmatrix} = 2 \begin{bmatrix} x_1 \\ x_2 \end{bmatrix}.$$

This gives

$$x_1 + \ x_2 = 2x_1$$
$$-2x_1 + 4x_2 = 2x_2$$

or

$$(2 - 1)x_1 - \ x_2 = 0$$
$$2x_1 + (2 - 4)x_2 = 0$$

or

$$x_1 - \ x_2 = 0$$
$$2x_1 - 2x_2 = 0.$$

Note that we could have obtained this last homogeneous system by merely substituting $\lambda = 2$ in (3). All solutions to this last system are given by

$$x_1 = x_2$$
$$x_2 = \text{any real number } r.$$

Hence all eigenvectors associated with the eigenvalue $\lambda_1 = 2$ are given by $\begin{bmatrix} r \\ r \end{bmatrix}$,

r any nonzero real number. In particular, $\mathbf{x}_1 = \begin{bmatrix} 1 \\ 1 \end{bmatrix}$ is an eigenvector associated with $\lambda_1 = 2$. Similarly, for $\lambda_2 = 3$ we obtain, from (3),

$$(3-1)x_1 - \qquad x_2 = 0$$
$$2x_1 + (3-4)x_2 = 0$$

or

$$2x_1 - x_2 = 0$$
$$2x_1 - x_2 = 0.$$

All solutions to this last homogeneous system are given by

$$x_1 = \tfrac{1}{2}x_2$$
$$x_2 = \text{any real number } r.$$

Hence all eigenvectors associated with the eigenvalue $\lambda_2 = 3$ are given by $\begin{bmatrix} \frac{1}{2}r \\ r \end{bmatrix}$,

r any nonzero real number. In particular, $\mathbf{x}_2 = \begin{bmatrix} 1 \\ 2 \end{bmatrix}$ is an eigenvector associated with the eigenvalue $\lambda_2 = 3$. ∎

In Examples 1, 2, and 3 we found eigenvalues and eigenvectors by inspection, whereas in Example 4 we proceeded in a more systematic fashion. We use the procedure of Example 4 as our standard method, as follows.

DEFINITION Let $A = \begin{bmatrix} a_{ij} \end{bmatrix}$ be an $n \times n$ matrix. The determinant

$$f(\lambda) = \det(\lambda I_n - A) = \begin{vmatrix} \lambda - a_{11} & -a_{12} & \cdots & -a_{1n} \\ -a_{21} & \lambda - a_{22} & \cdots & -a_{2n} \\ \vdots & \vdots & & \vdots \\ -a_{n1} & -a_{n2} & \cdots & \lambda - a_{nn} \end{vmatrix} \qquad (4)$$

is called the **characteristic polynomial of** A. The equation

$$f(\lambda) = \det(\lambda I_n - A) = 0$$

is called the **characteristic equation of** A.

EXAMPLE 5 ∎ Let

$$A = \begin{bmatrix} 1 & 2 & -1 \\ 1 & 0 & 1 \\ 4 & -4 & 5 \end{bmatrix}.$$

The characteristic polynomial of A is (verify)

$$f(\lambda) = \det(\lambda I_3 - A) = \begin{vmatrix} \lambda - 1 & -2 & 1 \\ -1 & \lambda - 0 & -1 \\ -4 & 4 & \lambda - 5 \end{vmatrix} = \lambda^3 - 6\lambda^2 + 11\lambda - 6.$$

∎

Recall from Chapter 2 that each term in the expansion of the determinant of an $n \times n$ matrix is a product of n elements of the matrix, containing exactly one element from each row and exactly one element from each column. Thus, if we expand $f(\lambda) = \det(\lambda I_n - A)$, we obtain a polynomial of degree n. A polynomial of degree n with real coefficients has n roots, some of which may be complex numbers. The expression involving λ^n in the characteristic polynomial of A comes from the product

$$(\lambda - a_{11})(\lambda - a_{22}) \cdots (\lambda - a_{nn}),$$

so the coefficient of λ^n is 1. We can then write

$$f(\lambda) = \det(\lambda I_n - A) = \lambda^n + c_1 \lambda^{n-1} + c_2 \lambda^{n-2} + \cdots + c_{n-1}\lambda + c_n.$$

If we let $\lambda = 0$ in $\det(\lambda I_n - A)$ as well as in the expression on the right, then we get $\det(-A) = c_n$, which shows that the constant term c_n is $(-1)^n \det(A)$. This result can be used to establish the following theorem.

THEOREM 5.1 ■ *An $n \times n$ matrix A is singular if and only if 0 is an eigenvalue of A.*

Proof   Exercise T.8(b).   ■

We now extend our List of Nonsingular Equivalences.

List of Nonsingular Equivalences

The following statements are equivalent for an $n \times n$ matrix A.

1. A is nonsingular.
2. $A\mathbf{x} = \mathbf{0}$ has only the trivial solution.
3. A is row equivalent to I_n.
4. The linear system $A\mathbf{x} = \mathbf{b}$ has a unique solution for every $n \times 1$ matrix $\mathbf{b}$.
5. $\det(A) \neq 0$.
6. A has rank n.
7. A has nullity 0.
8. The rows of A form a linearly independent set of n vectors in R^n.
9. The columns of A form a linearly independent set of n vectors in R^n.
10. Zero is *not* an eigenvalue of A.

We now connect the characteristic polynomial of a matrix with its eigenvalues in the following theorem.

THEOREM 5.2 ■ *The eigenvalues of A are the real roots of the characteristic polynomial of A.*

Proof   Let λ be an eigenvalue of A with associated eigenvector $\mathbf{x}$. Then $A\mathbf{x} = \lambda\mathbf{x}$, which can be rewritten as

$$A\mathbf{x} = (\lambda I_n)\mathbf{x}$$

or

$$(\lambda I_n - A)\mathbf{x} = \mathbf{0}, \tag{5}$$

a homogeneous system of n equations in n unknowns. This system has a non-trivial solution if and only if the determinant of its coefficient matrix vanishes (Corollary 2.3, Section 2.2), that is, if and only if $\det(\lambda I_n - A) = 0$.

Conversely, if λ is a real root of the characteristic polynomial of A, then $\det(\lambda I_n - A) = 0$, so the homogeneous system (5) has a nontrivial solution $\mathbf{x}$. Hence λ is an eigenvalue of A. ■

Thus, to find the eigenvalues of a given matrix A, we must find the real roots of its characteristic polynomial $f(\lambda)$. There are many methods for finding approximations to the roots of a polynomial, some of them more effective than others; indeed, many computer programs are available to find the roots of a polynomial. Two results that are sometimes useful in this connection are (1) the product of all the roots of the polynomial

$$f(\lambda) = \lambda^n + a_1\lambda^{n-1} + \cdots + a_{n-1}\lambda + a_n$$

is $(-1)^n a_n$, and (2) if $a_1, a_2, \ldots, a_n$ are integers, then $f(\lambda)$ cannot have a rational root that is not already an integer. Thus, as possible rational roots of $f(\lambda)$, one need only try the integer factors of a_n. Of course, $f(\lambda)$ might well have irrational roots. However, to minimize the computational effort and as a convenience to the reader, all the characteristic polynomials to be considered in the rest of this chapter have only integer roots, and each of these roots is a factor of the constant term of the characteristic polynomial of A. The corresponding eigenvectors are obtained by substituting the value of λ in Equation (5) and solving the resulting homogeneous system. The solution to this type of problem has already been studied in Section 4.5.

EXAMPLE 6 ■ Consider the matrix of Example 5. The characteristic polynomial is

$$f(\lambda) = \lambda^3 - 6\lambda^2 + 11\lambda - 6.$$

The possible integer roots of $f(\lambda)$ are ± 1, ± 2, ± 3, and ± 6. By substituting these values in $f(\lambda)$, we find that $f(1) = 0$, so $\lambda = 1$ is a root of $f(\lambda)$. Hence $(\lambda - 1)$ is a factor of $f(\lambda)$. Dividing $f(\lambda)$ by $(\lambda - 1)$, we obtain (verify)

$$f(\lambda) = (\lambda - 1)(\lambda^2 - 5\lambda + 6).$$

Factoring $\lambda^2 - 5\lambda + 6$, we have

$$f(\lambda) = (\lambda - 1)(\lambda - 2)(\lambda - 3).$$

The eigenvalues of A are then

$$\lambda_1 = 1, \qquad \lambda_2 = 2, \qquad \lambda_3 = 3.$$

To find an eigenvector $\mathbf{x}_1$ associated with $\lambda_1 = 1$, we form the system

$$(1I_3 - A)\mathbf{x} = \mathbf{0},$$

$$\begin{bmatrix} 1-1 & -2 & 1 \\ -1 & 1 & -1 \\ -4 & 4 & 1-5 \end{bmatrix} \begin{bmatrix} x_1 \\ x_2 \\ x_3 \end{bmatrix} = \begin{bmatrix} 0 \\ 0 \\ 0 \end{bmatrix}$$

or

$$\begin{bmatrix} 0 & -2 & 1 \\ -1 & 1 & -1 \\ -4 & 4 & -4 \end{bmatrix} \begin{bmatrix} x_1 \\ x_2 \\ x_3 \end{bmatrix} = \begin{bmatrix} 0 \\ 0 \\ 0 \end{bmatrix}.$$

A solution is

$$\begin{bmatrix} -\frac{1}{2}r \\ \frac{1}{2}r \\ r \end{bmatrix}$$

for any real number r. Thus, for $r = 2$,

$$\mathbf{x}_1 = \begin{bmatrix} -1 \\ 1 \\ 2 \end{bmatrix}$$

is an eigenvector of A associated with $\lambda_1 = 1$.

To find an eigenvector $\mathbf{x}_2$ associated with $\lambda_2 = 2$, we form the system

$$(2I_3 - A)\mathbf{x} = \mathbf{0},$$

that is,

$$\begin{bmatrix} 2-1 & -2 & 1 \\ -1 & 2 & -1 \\ -4 & 4 & 2-5 \end{bmatrix} \begin{bmatrix} x_1 \\ x_2 \\ x_3 \end{bmatrix} = \begin{bmatrix} 0 \\ 0 \\ 0 \end{bmatrix}$$

or

$$\begin{bmatrix} 1 & -2 & 1 \\ -1 & 2 & -1 \\ -4 & 4 & -3 \end{bmatrix} \begin{bmatrix} x_1 \\ x_2 \\ x_3 \end{bmatrix} = \begin{bmatrix} 0 \\ 0 \\ 0 \end{bmatrix}.$$

A solution is

$$\begin{bmatrix} -\frac{1}{2}r \\ \frac{1}{4}r \\ r \end{bmatrix}$$

for any real number r. Thus, for $r = 4$,

$$\mathbf{x}_2 = \begin{bmatrix} -2 \\ 1 \\ 4 \end{bmatrix}$$

is an eigenvector of A associated with $\lambda_2 = 2$.

To find an eigenvector $\mathbf{x}_3$ associated with $\lambda_3 = 3$, we form the system

$$(3I_3 - A)\mathbf{x} = \mathbf{0},$$

and find that a solution is (verify)

$$\begin{bmatrix} -\frac{1}{4}r \\ \frac{1}{4}r \\ r \end{bmatrix}$$

for any real number r. Thus, for $r = 4$,

$$\mathbf{x}_3 = \begin{bmatrix} -1 \\ 1 \\ 4 \end{bmatrix}$$

is an eigenvector of A associated with $\lambda_3 = 3$.   ■

Of course, the characteristic polynomial of a given matrix may have imaginary roots and it may even have no real roots. However, for the matrices that we are most interested in, symmetric matrices, all the roots of the characteristic polynomial are real. We shall state this result in Section 5.2.

EXAMPLE 7 ■ Let $A = \begin{bmatrix} 0 & 1 \\ -1 & 0 \end{bmatrix}$. Then the characteristic polynomial of A is

$$f(\lambda) = \lambda^2 + 1,$$

which has no real roots. Thus we say that A has no eigenvalues.   ■

Similar Matrices

We shall now develop an equivalent formulation of the eigenvalue–eigenvector problem. Throughout this chapter this equivalent formulation merely provides another way of describing the eigenvalue–eigenvector problem. In Section 6.3, this equivalent approach will shed much light on the eigenvalue–eigenvector problem.

DEFINITION

A matrix B is said to be **similar** to a matrix A if there is a nonsingular matrix P such that

$$B = P^{-1}AP.$$

EXAMPLE 8 ■ Let

$$A = \begin{bmatrix} 1 & 1 \\ -2 & 4 \end{bmatrix}$$

be the matrix of Example 4. Let

$$P = \begin{bmatrix} 1 & 1 \\ 1 & 2 \end{bmatrix}.$$

Then

$$P^{-1} = \begin{bmatrix} 2 & -1 \\ -1 & 1 \end{bmatrix}$$

and

$$B = P^{-1}AP = \begin{bmatrix} 2 & -1 \\ -1 & 1 \end{bmatrix} \begin{bmatrix} 1 & 1 \\ -2 & 4 \end{bmatrix} \begin{bmatrix} 1 & 1 \\ 1 & 2 \end{bmatrix} = \begin{bmatrix} 2 & 0 \\ 0 & 3 \end{bmatrix}.$$

Thus B is similar to A.   ■

We shall let the reader (Exercise T.1) show that the following elementary properties hold for similarity:

    1. A is similar to A.

    2. If B is similar to A, then A is similar to B.

    3. If A is similar to B and B is similar to C, then A is similar to C.

By property 2 we replace the statements "A is similar to B" and "B is similar to A" by "A and B are similar."

In Section 6.3, we discuss the meaning of similarity of matrices.

DEFINITION   We shall say that the matrix A is **diagonalizable** if it is similar to a diagonal matrix. In this case we also say that A **can be diagonalized**.

EXAMPLE 9 ■ If A and B are as in Example 8, then A is diagonalizable, since it is similar to B. ■

THEOREM 5.3 ■ *An $n \times n$ matrix A is diagonalizable if and only if it has n linearly independent eigenvectors. In this case A is similar to a diagonal matrix D, with $P^{-1}AP = D$, whose diagonal elements are the eigenvalues of A, while P is a matrix whose columns are respectively the n linearly independent eigenvectors of A.*

Proof   Suppose that A is similar to D. Then

$$P^{-1}AP = D,$$

so that

$$AP = PD. \qquad (6)$$

Let

$$D = \begin{bmatrix} \lambda_1 & 0 & \cdots & 0 \\ 0 & \lambda_2 & \cdots & 0 \\ \vdots & & & \vdots \\ 0 & \cdots & 0 & \lambda_n \end{bmatrix},$$

and let $\mathbf{x}_j$, $j = 1, 2, \ldots, n$ be the jth column of P. From Exercise T.9 in Section 1.3, it follows that the jth column of the matrix AP is $A\mathbf{x}_j$, and the jth column of PD is $\lambda_j \mathbf{x}_j$.

Thus from (6) we have

$$A\mathbf{x}_j = \lambda_j \mathbf{x}_j. \qquad (7)$$

Since P is a nonsingular matrix, by Theorem 4.13 in Section 4.6 its columns are linearly independent and so are all nonzero. Hence λ_j is an eigenvalue of A and $\mathbf{x}_j$ is a corresponding eigenvector.

Conversely, suppose that $\lambda_1, \lambda_2, \ldots, \lambda_n$ are n eigenvalues of A and that the corresponding eigenvectors $\mathbf{x}_1, \mathbf{x}_2, \ldots, \mathbf{x}_n$ are linearly independent. Let $P = \begin{bmatrix} \mathbf{x}_1 & \mathbf{x}_2 & \cdots & \mathbf{x}_n \end{bmatrix}$ be the matrix whose jth column is $\mathbf{x}_j$. Since the columns of P are linearly independent, it follows from Theorem 4.13 in Section 4.6 that P is nonsingular. From (7) we obtain (6), which implies that A is diagonalizable. This completes the proof. ■

Observe that in Theorem 5.3 the order of the columns of P determines the order of the diagonal entries in D.

EXAMPLE 10 ■ Let A be as in Example 4. The eigenvalues are $\lambda_1 = 2$ and $\lambda_2 = 3$. The corresponding eigenvectors

$$\mathbf{x}_1 = \begin{bmatrix} 1 \\ 1 \end{bmatrix} \quad \text{and} \quad \mathbf{x}_2 = \begin{bmatrix} 1 \\ 2 \end{bmatrix}$$

are linearly independent. Hence A is diagonalizable. Here

$$P = \begin{bmatrix} 1 & 1 \\ 1 & 2 \end{bmatrix} \quad \text{and} \quad P^{-1} = \begin{bmatrix} 2 & -1 \\ -1 & 1 \end{bmatrix}.$$

Thus, as in Example 8,

$$P^{-1}AP = \begin{bmatrix} 2 & -1 \\ -1 & 1 \end{bmatrix} \begin{bmatrix} 1 & 1 \\ -2 & 4 \end{bmatrix} \begin{bmatrix} 1 & 1 \\ 1 & 2 \end{bmatrix} = \begin{bmatrix} 2 & 0 \\ 0 & 3 \end{bmatrix}.$$

On the other hand, if we let $\lambda_1 = 3$ and $\lambda_2 = 2$, then

$$\mathbf{x}_1 = \begin{bmatrix} 1 \\ 2 \end{bmatrix} \quad \text{and} \quad \mathbf{x}_2 = \begin{bmatrix} 1 \\ 1 \end{bmatrix}.$$

Then

$$P = \begin{bmatrix} 1 & 1 \\ 2 & 1 \end{bmatrix} \quad \text{and} \quad P^{-1} = \begin{bmatrix} -1 & 1 \\ 2 & -1 \end{bmatrix}.$$

Hence

$$P^{-1}AP = \begin{bmatrix} -1 & 1 \\ 2 & -1 \end{bmatrix} \begin{bmatrix} 1 & 1 \\ -2 & 4 \end{bmatrix} \begin{bmatrix} 1 & 1 \\ 2 & 1 \end{bmatrix} = \begin{bmatrix} 3 & 0 \\ 0 & 2 \end{bmatrix}. \qquad ■$$

EXAMPLE 11 ■ Let

$$A = \begin{bmatrix} 1 & 1 \\ 0 & 1 \end{bmatrix}.$$

The eigenvalues of A are $\lambda_1 = 1$ and $\lambda_2 = 1$. Eigenvectors associated with λ_1 and λ_2 are vectors of the form

$$\begin{bmatrix} r \\ 0 \end{bmatrix},$$

where r is any nonzero real number. Since A does not have two linearly independent eigenvectors, we conclude that A is not diagonalizable. ■

The following is a useful theorem because it identifies a large class of matrices that can be diagonalized.

THEOREM 5.4 ■ *A matrix A is diagonalizable if all the roots of its characteristic polynomial are real and distinct.*

Proof   Let $\lambda_1, \lambda_2, \ldots, \lambda_n$ be the distinct eigenvalues of A and let $S = \{\mathbf{x}_1, \mathbf{x}_2, \ldots, \mathbf{x}_n\}$ be a set of associated eigenvectors. We wish to show that S is linearly independent.

Suppose that S is linearly dependent. Then Theorem 4.4 of Section 4.3 implies that some vector $\mathbf{x}_j$ is a linear combination of the preceding vectors in S. We can assume that $S_1 = \{\mathbf{x}_1, \mathbf{x}_2, \ldots, \mathbf{x}_{j-1}\}$ is linearly independent, for otherwise one of the vectors in S_1 is a linear combination of the preceding ones, and we can choose a new set S_2, and so on. We thus have that S_1 is linearly independent and that

$$\mathbf{x}_j = c_1\mathbf{x}_1 + c_2\mathbf{x}_2 + \cdots + c_{j-1}\mathbf{x}_{j-1}, \tag{8}$$

where $c_1, c_2, \ldots, c_{j-1}$ are real numbers. Premultiplying (multiplying on the left) both sides of Equation (8) by A, we obtain

$$\begin{aligned} A\mathbf{x}_j &= A(c_1\mathbf{x}_1 + c_2\mathbf{x}_2 + \cdots + c_{j-1}\mathbf{x}_{j-1}) \\ &= c_1 A\mathbf{x}_1 + c_2 A\mathbf{x}_2 + \cdots + c_{j-1}A\mathbf{x}_{j-1}. \end{aligned} \tag{9}$$

Since $\lambda_1, \lambda_2, \ldots, \lambda_j$ are eigenvalues of A and $\mathbf{x}_1, \mathbf{x}_2, \ldots, \mathbf{x}_j$, its associated eigenvectors, we know that $A\mathbf{x}_i = \lambda_i\mathbf{x}_i$ for $i = 1, 2, \ldots, j$. Substituting in (9), we have

$$\lambda_j\mathbf{x}_j = c_1\lambda_1\mathbf{x}_1 + c_2\lambda_2\mathbf{x}_2 + \cdots + c_{j-1}\lambda_{j-1}\mathbf{x}_{j-1}. \tag{10}$$

Multiplying (8) by λ_j, we obtain

$$\lambda_j\mathbf{x}_j = \lambda_j c_1\mathbf{x}_1 + \lambda_j c_2\mathbf{x}_2 + \cdots + \lambda_j c_{j-1}\mathbf{x}_{j-1}. \tag{11}$$

Subtracting (11) from (10), we have

$$\begin{aligned} 0 &= \lambda_j\mathbf{x}_j - \lambda_j\mathbf{x}_j \\ &= c_1(\lambda_1 - \lambda_j)\mathbf{x}_1 + c_2(\lambda_2 - \lambda_j)\mathbf{x}_2 + \cdots + c_{j-1}(\lambda_{j-1} - \lambda_j)\mathbf{x}_{j-1}. \end{aligned}$$

Since S_1 is linearly independent, we must have

$$c_1(\lambda_1 - \lambda_j) = 0, \quad c_2(\lambda_2 - \lambda_j) = 0, \ldots, \quad c_{j-1}(\lambda_{j-1} - \lambda_j) = 0.$$

Now

$$\lambda_1 - \lambda_j \neq 0, \quad \lambda_2 - \lambda_j \neq 0, \ldots, \quad \lambda_{j-1} - \lambda_j \neq 0$$

(because the λ's are distinct), which implies that

$$c_1 = c_2 = \cdots = c_{j-1} = 0.$$

From (8) we conclude that $\mathbf{x}_j = \mathbf{0}$, which is impossible if $\mathbf{x}_j$ is an eigenvector. Hence S is linearly independent, and from Theorem 5.3 it follows that A is diagonalizable.   ■

REMARK   In the proof of Theorem 5.4, we have actually established the following somewhat stronger result: Let A be an $n \times n$ matrix and let $\lambda_1, \lambda_2, \ldots, \lambda_k$ be k distinct eigenvalues of A with associated eigenvectors $\mathbf{x}_1, \mathbf{x}_2, \ldots, \mathbf{x}_k$. Then $\mathbf{x}_1, \mathbf{x}_2, \ldots, \mathbf{x}_k$ are linearly independent (Exercise T.15).

If all the roots of the characteristic polynomial of A are real and not all distinct, then A may or may not be diagonalizable. The characteristic polynomial of A can be written as the product of n factors, each of the form $\lambda - \lambda_j$, where λ_j is a root of the characteristic polynomial. Now the eigenvalues of A are the real roots of the characteristic polynomial of A. Thus the characteristic polynomial can be written as

$$(\lambda - \lambda_1)^{k_1}(\lambda - \lambda_2)^{k_2} \cdots (\lambda - \lambda_r)^{k_r},$$

where $\lambda_1, \lambda_2, \ldots, \lambda_r$ are the distinct eigenvalues of A, and $k_1, k_2, \ldots, k_r$ are integers whose sum is n. The integer k_i is called the **multiplicity** of λ_i. Thus in Example 11, $\lambda = 1$ is an eigenvalue of

$$A = \begin{bmatrix} 1 & 1 \\ 0 & 1 \end{bmatrix}$$

of multiplicity 2. It can be shown that if the roots of the characteristic polynomial of A are all real, then A can be diagonalized if and only if for each eigenvalue λ_j of multiplicity k_j we can find k_j linearly independent eigenvectors. This means that the solution space of the linear system $(\lambda_j I_n - A)\mathbf{x} = \mathbf{0}$ has dimension k_j. It can also be shown that if λ_j is an eigenvalue of A of multiplicity k_j, then we can never find more than k_j linearly independent eigenvectors associated with λ_j. We consider the following examples.

EXAMPLE 12 ■ Let

$$A = \begin{bmatrix} 0 & 0 & 1 \\ 0 & 1 & 2 \\ 0 & 0 & 1 \end{bmatrix}.$$

The characteristic polynomial of A is $f(\lambda) = \lambda(\lambda - 1)^2$, so the eigenvalues of A are $\lambda_1 = 0$, $\lambda_2 = 1$, and $\lambda_3 = 1$. Thus $\lambda_2 = 1$ is an eigenvalue of multiplicity 2. We now consider the eigenvectors associated with the eigenvalues $\lambda_2 = \lambda_3 = 1$. They are obtained by solving the linear system $(1I_3 - A)\mathbf{x} = \mathbf{0}$:

$$\begin{bmatrix} 1 & 0 & -1 \\ 0 & 0 & -2 \\ 0 & 0 & 0 \end{bmatrix} \begin{bmatrix} x_1 \\ x_2 \\ x_3 \end{bmatrix} = \begin{bmatrix} 0 \\ 0 \\ 0 \end{bmatrix}.$$

A solution is any vector of the form

$$\begin{bmatrix} 0 \\ r \\ 0 \end{bmatrix},$$

where r is any real number, so the dimension of the solution space of the linear system $(1I_3 - A)\mathbf{x} = \mathbf{0}$ is 1. There do not exist two linearly independent eigenvectors associated with $\lambda_2 = 1$. Thus A cannot be diagonalized.   ■

EXAMPLE 13 ■ Let

$$A = \begin{bmatrix} 0 & 0 & 0 \\ 0 & 1 & 0 \\ 1 & 0 & 1 \end{bmatrix}.$$

The characteristic polynomial of A is $f(\lambda) = \lambda(\lambda - 1)^2$, so the eigenvalues of A are $\lambda_1 = 0$, $\lambda_2 = 1$, $\lambda_3 = 1$; $\lambda_2 = 1$ is again an eigenvalue of multiplicity 2. Now we consider the solution space of $(1I_3 - A)\mathbf{x} = \mathbf{0}$, that is, of

$$\begin{bmatrix} 1 & 0 & 0 \\ 0 & 0 & 0 \\ -1 & 0 & 0 \end{bmatrix} \begin{bmatrix} x_1 \\ x_2 \\ x_3 \end{bmatrix} = \begin{bmatrix} 0 \\ 0 \\ 0 \end{bmatrix}.$$

A solution is any vector of the form

$$\begin{bmatrix} 0 \\ r \\ s \end{bmatrix}$$

for any real numbers r and s. Thus we can take as eigenvectors $\mathbf{x}_2$ and $\mathbf{x}_3$ the vectors

$$\mathbf{x}_2 = \begin{bmatrix} 0 \\ 1 \\ 0 \end{bmatrix} \quad \text{and} \quad \mathbf{x}_3 = \begin{bmatrix} 0 \\ 0 \\ 1 \end{bmatrix}.$$

Now we look for an eigenvector associated with $\lambda_1 = 0$. We have to solve $(0I_3 - A)\mathbf{x} = \mathbf{0}$, or

$$\begin{bmatrix} 0 & 0 & 0 \\ 0 & -1 & 0 \\ -1 & 0 & -1 \end{bmatrix} \begin{bmatrix} x_1 \\ x_2 \\ x_3 \end{bmatrix} = \begin{bmatrix} 0 \\ 0 \\ 0 \end{bmatrix}.$$

A solution is any vector of the form

$$\begin{bmatrix} t \\ 0 \\ -t \end{bmatrix}$$

for any real number t. Thus

$$\mathbf{x}_1 = \begin{bmatrix} 1 \\ 0 \\ -1 \end{bmatrix}$$

is an eigenvector associated with $\lambda_1 = 0$. Since $\mathbf{x}_1$, $\mathbf{x}_2$, and $\mathbf{x}_3$ are linearly independent, A can be diagonalized. ∎

Thus an $n \times n$ matrix may fail to be diagonalizable either because not all the roots of its characteristic polynomial are real numbers, or because it does not have n linearly independent eigenvectors.

Eigenvalues and eigenvectors satisfy many important properties. For example, if A is an upper (lower) triangular matrix, then the eigenvalues of A are the elements on the main diagonal of A. Also, let λ_j be a particular eigenvalue of A. The set S consisting of all eigenvectors of A associated with λ_j as well as the zero vector is a subspace of R^n (Exercise T.2) called the **eigenspace associated with** λ_j. Other properties are developed in the exercises of this section.

> The procedure for diagonalizing a matrix A is as follows.
>
> ***Step 1.*** Form the characteristic polynomial $f(\lambda) = \det(\lambda I_n - A)$ of A.
>
> ***Step 2.*** Find the roots of the characteristic polynomial of A. If the roots are not all real, then A cannot be diagonalized.
>
> ***Step 3.*** For each eigenvalue λ_j of A of multiplicity k_j, find a basis for the solution space of $(\lambda_j I_n - A)\mathbf{x} = \mathbf{0}$ (the eigenspace associated with λ_j). If the dimension of the eigenspace is less than k_j, then A is not diagonalizable. We thus determine n linearly independent eigenvectors of A. In Section 4.5 we solved the problem of finding a basis for the solution space of a homogeneous system.
>
> ***Step 4.*** Let P be the matrix whose columns are the n linearly independent eigenvectors determined in Step 3. Then $P^{-1}AP = D$, a diagonal matrix whose diagonal elements are the eigenvalues of A that correspond to the columns of P.

It must be pointed out that the method for finding the eigenvalues of a matrix by obtaining the roots of the characteristic polynomial is not a practical one for $n > 4$, owing to the need for evaluating a determinant. An efficient numerical method for finding eigenvalues will be presented in Section 9.5.

Application (Population Growth)

Consider a population of animals that can live to a maximum age of n years (or any other time unit). Suppose that the number of males in the population is always a fixed percentage of the female population. Thus, in studying the growth of the entire population, we can ignore the male population and concentrate our attention on the female population. We divide the female population into $n + 1$ age groups as follows:

$$x_i^{(k)} = \text{number of females of age } i \text{ who are alive at time } k,\ 0 \le k \le n;$$
$$f_i = \text{fraction of females of age } i \text{ who will be alive a year later;}$$
$$b_i = \text{average number of females born to a female of age } i.$$

Let

$$\mathbf{x}^{(k)} = \begin{bmatrix} x_0^{(k)} \\ x_1^{(k)} \\ \vdots \\ x_n^{(k)} \end{bmatrix} \qquad (0 \le k \le n)$$

denote the age distribution vector at time k.

The number of females in the first age group (age zero) at time $k + 1$ is merely the total number of females born from time k to time $k + 1$. There are $x_0^{(k)}$ females in the first age group at time k and each of these females, on the average, produces b_0 female offspring, so the first age group produces a total of $b_0 x_0^{(k)}$ females. Similarly, the $x_1^{(k)}$ females in the second age group (age 1) produces a total of $b_1 x_1^{(k)}$ females. Thus

$$x_0^{(k+1)} = b_0 x_0^{(k)} + b_1 x_1^{(k)} + \cdots + b_n x_n^{(k)}. \tag{12}$$

The number $x_1^{(k+1)}$ of females in the second age group at time $k+1$ is the number of females from the first age group at time k who are alive a year later. Thus

$$x_1^{(k+1)} = \left(\begin{array}{c} \text{fraction of females in} \\ \text{first age group who are} \\ \text{alive a year later} \end{array} \right) \times \left(\begin{array}{c} \text{number of females in} \\ \text{first age group} \end{array} \right)$$

or

$$x_1^{(k+1)} = f_0 x_0^{(k)},$$

and in general

$$x_j^{(k+1)} = f_{j-1} x_{j-1}^{(k)} \qquad (1 \le j \le n). \tag{13}$$

We can write (12) and (13) using matrix notation as

$$\mathbf{x}^{(k+1)} = A\mathbf{x}^{(k)} \qquad (1 \le k \le n), \tag{14}$$

where

$$A = \begin{bmatrix} b_0 & b_1 & b_2 & \cdots & b_{n-1} & b_n \\ f_0 & 0 & 0 & \cdots & 0 & 0 \\ 0 & f_1 & 0 & \cdots & 0 & 0 \\ \vdots & \vdots & \vdots & & \vdots & \vdots \\ 0 & 0 & 0 & \cdots & f_{n-1} & 0 \end{bmatrix}.$$

We can use Equation (14) to try to determine a distribution of the population by age groups at time $k+1$ so that the number of females in each age group at time $k+1$ will be a fixed multiple of the number in the corresponding age group at time k. That is, if λ is the multiplier, we want

$$\mathbf{x}^{(k+1)} = \lambda \mathbf{x}^{(k)}$$

or

$$A\mathbf{x}^{(k)} = \lambda \mathbf{x}^{(k)}.$$

Thus λ is an eigenvalue of A and $\mathbf{x}^{(k)}$ is a corresponding eigenvector. If $\lambda = 1$, the number of females in each age group will be the same year after year. If we can find an eigenvector $\mathbf{x}^{(k)}$ corresponding to the eigenvalue $\lambda = 1$, we say that we have a **stable age distribution**.

EXAMPLE 14 ■
(*Population Growth*)

Consider a beetle that can live to a maximum age of 2 years and whose population dynamics are represented by the matrix

$$A = \begin{bmatrix} 0 & 0 & 6 \\ \frac{1}{2} & 0 & 0 \\ 0 & \frac{1}{3} & 0 \end{bmatrix}.$$

We find that $\lambda = 1$ is an eigenvalue of A with corresponding eigenvector

$$\begin{bmatrix} 6 \\ 3 \\ 1 \end{bmatrix}.$$

Thus, if the numbers of females in the three groups are proportional to 6:3:1, we have a stable age distribution. That is, if we have 600 females in the first age group, 300 in the second, and 100 in the third, then year after year the number of females in each age group will remain the same.   ∎

Population growth problems of this type have applications to animal harvesting. For further discussion of elementary applications of eigenvalues and eigenvectors, see D. R. Hill, *Experiments in Computational Matrix Algebra*, New York: Random House, 1988 or D. R. Hill and D. E. Zitarelli, *Linear Algebra Labs with MATLAB*, 2nd ed., Upper Saddle River, N.J.: Prentice Hall, 1996.

The theoretical exercises in this section contain many useful properties of eigenvalues. The reader is encouraged to develop a list of facts about eigenvalues, eigenvectors, and diagonalizable matrices.

Section 8.6, Differential Equations, and Section 8.7, The Fibonacci Sequence, which can be covered at this time, use material from this section.

▼ *Preview of an Application*

DIFFERENTIAL EQUATIONS (SECTION 8.6) (CALCULUS REQUIRED)

A differential equation is an equation that involves an unknown function and its derivatives. Differential equations occur in a wide variety of applications.

As an example, suppose that we have a system consisting of two interconnected tanks each containing a brine solution.

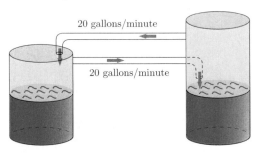

20 gallons/minute

20 gallons/minute

Tank A   Tank B

Tank A contains $x(t)$ pounds of salt in 200 gallons of brine and tank B contains $y(t)$ pounds of salt in 300 gallons of brine. The mixture in each tank is kept uniform by constant stirring. When $t = 0$, brine is pumped from tank A to tank B at 20 gallons/minute and from tank B to tank A at 20 gallons/minute.

When we set up a mathematical model for this problem, we find that $x(t)$ and $y(t)$ must satisfy the following system of differential equations:

$$x'(t) = -\tfrac{1}{10}x(t) + \tfrac{2}{30}y(t)$$
$$y'(t) = \tfrac{1}{10}x(t) - \tfrac{2}{30}y(t),$$

which can be written in matrix form as

$$\begin{bmatrix} x'(t) \\ y'(t) \end{bmatrix} = \begin{bmatrix} -\tfrac{1}{10} & \tfrac{2}{30} \\ \tfrac{1}{10} & -\tfrac{2}{30} \end{bmatrix} \begin{bmatrix} x(t) \\ y(t) \end{bmatrix}.$$

Section 8.6 gives an introduction to the solution of homogeneous linear systems of differential equations.

▼ *Preview of an Application*

THE FIBONACCI SEQUENCE (SECTION 8.7)

The sequence of numbers

$$1, 1, 2, 3, 5, 8, 13, 21, 34, 55, 89, \ldots$$

is called the *Fibonacci sequence*. The first two numbers in this sequence are 1, 1 and the next number is obtained by adding the two preceding ones. Thus, in general, if $u_0 = 1$ and $u_1 = 1$, then for $n \geq 2$

$$u_n = u_{n-1} + u_{n-2}. \qquad (*)$$

The Fibonacci sequence occurs in a wide variety of applications such as the distribution of leaves on some trees, the arrangement of seeds on sunflowers, search techniques in numerical methods, generating random numbers in statistics, and others.

The equation given above can be used to *successively* compute values of u_n for any desired value of n, which can be tedious if n is large. In addition to $(*)$ we write

$$u_{n-1} = u_{n-1} \qquad (**)$$

and define

$$\mathbf{w}_k = \begin{bmatrix} u_{k+1} \\ u_k \end{bmatrix} \quad \text{and} \quad A = \begin{bmatrix} 1 & 1 \\ 1 & 0 \end{bmatrix}, \qquad 0 \leq k \leq n-1.$$

Then $(*)$ and $(**)$ can be written in matrix form as

$$\mathbf{w}_{n-1} = A\mathbf{w}_{n-2}.$$

By diagonalizing the matrix A, we obtain the following formula for calculating u_n directly:

$$u_n = \frac{1}{\sqrt{5}} \left[\left(\frac{1 + \sqrt{5}}{2} \right)^{n+1} - \left(\frac{-1 - \sqrt{5}}{2} \right)^{n+1} \right].$$

Section 8.7 provides a brief discussion of the Fibonacci sequence of numbers.

5.1   EXERCISES

In Exercises 1 through 3, find the characteristic polynomial of each matrix.

1. $\begin{bmatrix} 1 & 2 & 1 \\ 0 & 1 & 2 \\ -1 & 3 & 2 \end{bmatrix}$.   **2.** $\begin{bmatrix} 2 & 1 \\ -1 & 3 \end{bmatrix}$.

3. $\begin{bmatrix} 4 & -1 & 3 \\ 0 & 2 & 1 \\ 0 & 0 & 3 \end{bmatrix}$.

In Exercises 4 through 11, find the characteristic polynomial, eigenvalues, and eigenvectors of each matrix.

4. $\begin{bmatrix} 0 & 1 & 2 \\ 0 & 0 & 3 \\ 0 & 0 & 0 \end{bmatrix}$.   **5.** $\begin{bmatrix} 1 & 0 & 0 \\ -1 & 3 & 0 \\ 3 & 2 & -2 \end{bmatrix}$.

6. $\begin{bmatrix} 1 & 1 \\ 1 & 1 \end{bmatrix}$.   **7.** $\begin{bmatrix} 1 & -1 \\ 2 & 4 \end{bmatrix}$.

8. $\begin{bmatrix} 2 & -2 & 3 \\ 0 & 3 & -2 \\ 0 & -1 & 2 \end{bmatrix}$.   **9.** $\begin{bmatrix} 2 & 2 & 3 \\ 1 & 2 & 1 \\ 2 & -2 & 1 \end{bmatrix}$.

10. $\begin{bmatrix} 2 & 0 & 0 \\ 3 & -1 & 0 \\ 0 & 4 & 3 \end{bmatrix}$.   **11.** $\begin{bmatrix} 1 & 2 & 3 & 4 \\ 0 & -1 & 3 & 2 \\ 0 & 0 & 3 & 3 \\ 0 & 0 & 0 & 2 \end{bmatrix}$.

In Exercises 12 through 16, find which of the matrices are diagonalizable.

12. $\begin{bmatrix} 1 & 4 \\ 1 & -2 \end{bmatrix}$.   **13.** $\begin{bmatrix} 1 & 0 \\ -2 & 1 \end{bmatrix}$.

14. $\begin{bmatrix} 1 & 1 & -2 \\ 4 & 0 & 4 \\ 1 & -1 & -4 \end{bmatrix}$.   **15.** $\begin{bmatrix} 1 & 2 & 3 \\ 0 & -1 & 2 \\ 0 & 0 & 2 \end{bmatrix}$.

16. $\begin{bmatrix} 3 & 1 & 0 \\ 0 & 3 & 1 \\ 0 & 0 & 3 \end{bmatrix}$.

In Exercises 17 through 21, find for each matrix A, if possible, a nonsingular matrix P such that $P^{-1}AP$ is diagonal.

17. $\begin{bmatrix} 4 & 2 & 3 \\ 2 & 1 & 2 \\ -1 & -2 & 0 \end{bmatrix}$.   **18.** $\begin{bmatrix} 1 & 1 & 2 \\ 0 & 1 & 0 \\ 0 & 1 & 3 \end{bmatrix}$.

19. $\begin{bmatrix} 1 & 2 & 3 \\ 0 & 1 & 0 \\ 2 & 1 & 2 \end{bmatrix}$.   **20.** $\begin{bmatrix} 0 & -1 \\ 2 & 3 \end{bmatrix}$.

21. $\begin{bmatrix} 3 & -2 & 1 \\ 0 & 2 & 0 \\ 0 & 0 & 0 \end{bmatrix}$.

In Exercises 22 and 23, find bases for the eigenspaces associated with each eigenvalue.

22. $\begin{bmatrix} 2 & 3 & 0 \\ 0 & 1 & 0 \\ 0 & 0 & 2 \end{bmatrix}$.   **23.** $\begin{bmatrix} 2 & 2 & 3 & 4 \\ 0 & 2 & 3 & 2 \\ 0 & 0 & 1 & 1 \\ 0 & 0 & 0 & 1 \end{bmatrix}$.

24. Let $D = \begin{bmatrix} 2 & 0 \\ 0 & -2 \end{bmatrix}$. Compute D^9.

25. Let $A = \begin{bmatrix} 3 & -5 \\ 1 & -3 \end{bmatrix}$. Compute A^9.

(*Hint*: Find a matrix P such that $P^{-1}AP$ is a diagonal matrix D and show that $A^9 = PD^9P^{-1}$).

26. Consider a living organism that can live to a maximum age of 2 years and whose matrix is

$$A = \begin{bmatrix} 0 & 0 & 8 \\ \frac{1}{4} & 0 & 0 \\ 0 & \frac{1}{2} & 0 \end{bmatrix}.$$

Find a stable age distribution.

THEORETICAL EXERCISES ▪

T.1. Prove:

 (a) A is similar to A.

 (b) If B is similar to A, then A is similar to B.

 (c) If A is similar to B and B is similar to C, then A is similar to C.

T.2. Let λ_j be a particular eigenvalue of A. Show that the set S consisting of all eigenvectors of A associated with λ_j, as well as the zero vector, is a subspace of R^n. This subspace is called the **eigenspace associated with λ_j**.

T.3. Show that if A and B are similar matrices, they have the same characteristic polynomials and hence the same eigenspaces.

T.4. Show that if A is an upper (lower) triangular matrix, then the eigenvalues of A are the elements on the main diagonal of A.

T.5. Show that A and A^T have the same eigenvalues. What, if anything, can we say about the associated eigenvectors of A and A^T?

T.6. If λ is an eigenvalue of A with associated eigenvector $\mathbf{x}$, show that λ^k is an eigenvalue of $A^k = A \cdot A \cdots A$ (k factors) with associated eigenvector $\mathbf{x}$, where k is a positive integer.

T.7. An $n \times n$ matrix A is called **nilpotent** if $A^k = O$ for some positive integer k. Show that if A is nilpotent, then the only eigenvalue of A is 0. (*Hint*: Use Exercise T.6.)

T.8. Let A be an $n \times n$ matrix.
  (a) Show that $\det(A)$ is the product of all the roots of the characteristic polynomial of A.
  (b) Show that A is singular if and only if 0 is an eigenvalue of A.

T.9. Show that if A is nonsingular and diagonalizable, then A^{-1} is diagonalizable.

T.10. Let λ be an eigenvalue of the nonsingular matrix A with associated eigenvector $\mathbf{x}$. Show that $1/\lambda$ is an eigenvalue of A^{-1} with associated eigenvector $\mathbf{x}$.

T.11. Let

$$A = \begin{bmatrix} a & b \\ c & d \end{bmatrix}.$$

Find necessary and sufficient conditions for A to be diagonalizable.

T.12. If A and B are nonsingular, show that AB and BA are similar.

T.13. Show that if A is diagonalizable, then
  (a) A^T is diagonalizable.
  (b) A^k is diagonalizable, where k is a positive integer.

T.14. Let A be an $n \times n$ matrix and let $B = P^{-1}AP$ be similar to A. Show that if $\mathbf{x}$ is an eigenvector of A associated with the eigenvalue λ of A, then $P^{-1}\mathbf{x}$ is an eigenvector of B associated with the eigenvalue λ of B.

T.15. Let $\lambda_1, \lambda_2, \ldots, \lambda_k$ be distinct eigenvalues of a matrix A with associated eigenvectors $\mathbf{x}_1, \mathbf{x}_2, \ldots, \mathbf{x}_k$. Show that $\mathbf{x}_1, \mathbf{x}_2, \ldots, \mathbf{x}_k$ are linearly independent. (*Hint*: See the proof of Theorem 5.4.)

T.16. Let A and B be $n \times n$ matrices such that $A\mathbf{x} = \lambda\mathbf{x}$ and $B\mathbf{x} = \mu\mathbf{x}$. Show that:
  (a) $(A + B)\mathbf{x} = (\lambda + \mu)\mathbf{x}$.
  (b) $(AB)\mathbf{x} = (\lambda\mu)\mathbf{x}$.

T.17. Show that if A is a matrix all of whose columns add up to 1, then $\lambda = 1$ is an eigenvalue of A. (*Hint*: Consider the product $A^T\mathbf{x}$, where $\mathbf{x}$ is a vector all of whose entries are 1 and use Exercise T.5.)

MATLAB EXERCISES

MATLAB has a pair of commands that can be used to find the characteristic polynomial and eigenvalues of a matrix. Command **poly(A)** gives the coefficients of the characteristic polynomial of matrix A, starting with the highest-degree term. If we set $\mathbf{v} = \mathbf{poly(A)}$ and then use command **roots(v)**, we obtain the roots of the characteristic polynomial of A. This process can also find complex eigenvalues, which are discussed in Appendix A.2.

  Once we have an eigenvalue λ of A, we can use **rref** or **homsoln** to find a corresponding eigenvector from the linear system $(\lambda I - A)\mathbf{x} = \mathbf{0}$.

ML.1. Find the characteristic polynomial of each of the following matrices using MATLAB.

  (a) $A = \begin{bmatrix} 1 & 2 \\ 2 & -1 \end{bmatrix}.$

  (b) $A = \begin{bmatrix} 2 & 4 & 0 \\ 1 & 2 & 1 \\ 0 & 4 & 2 \end{bmatrix}.$

  (c) $A = \begin{bmatrix} 1 & 0 & 0 & 0 \\ 2 & -2 & 0 & 0 \\ 0 & 0 & 2 & -1 \\ 0 & 0 & -1 & 2 \end{bmatrix}.$

ML.2. Use the **poly** and **roots** commands in MATLAB to find the eigenvalues of the following matrices:

(a) $A = \begin{bmatrix} 1 & -3 \\ 3 & -5 \end{bmatrix}$.

(b) $A = \begin{bmatrix} 3 & -1 & 4 \\ -1 & 0 & 1 \\ 4 & 1 & 2 \end{bmatrix}$.

(c) $A = \begin{bmatrix} 2 & -2 & 0 \\ 1 & -1 & 0 \\ 1 & -1 & 0 \end{bmatrix}$.

(d) $A = \begin{bmatrix} 2 & 4 \\ 3 & 6 \end{bmatrix}$.

ML.3. In each of the following cases, λ is an eigenvalue of A. Use MATLAB to find a corresponding eigenvector.

(a) $\lambda = 3$, $A = \begin{bmatrix} 1 & 2 \\ -1 & 4 \end{bmatrix}$.

(b) $\lambda = -1$, $A = \begin{bmatrix} 4 & 0 & 0 \\ 1 & 3 & 0 \\ 2 & 1 & -1 \end{bmatrix}$.

(c) $\lambda = 2$, $A = \begin{bmatrix} 2 & 1 & 2 \\ 2 & 2 & -2 \\ 3 & 1 & 1 \end{bmatrix}$.

ML.4. Use MATLAB to determine if A is diagonalizable. If it is, find a nonsingular matrix P so that $P^{-1}AP$ is diagonal.

(a) $A = \begin{bmatrix} 0 & 2 \\ -1 & 3 \end{bmatrix}$.

(b) $A = \begin{bmatrix} 1 & -3 \\ 3 & -5 \end{bmatrix}$.

(c) $A = \begin{bmatrix} 0 & 0 & 4 \\ 5 & 3 & 6 \\ 6 & 0 & 5 \end{bmatrix}$.

ML.5. Use MATLAB and the hint in Exercise 25 to compute A^{30}, where

$$A = \begin{bmatrix} -1 & 1 & -1 \\ -2 & 2 & -1 \\ -2 & 2 & -1 \end{bmatrix}.$$

ML.6. Repeat Exercise ML.5 for

$$A = \begin{bmatrix} -1 & 1.5 & -1.5 \\ -2 & 2.5 & -1.5 \\ -2 & 2.0 & -1.0 \end{bmatrix}.$$

Display your answer in both **format short** and **format long**.

ML.7. Use MATLAB to investigate the sequences

$$A, A^3, A^5, \ldots \quad \text{and} \quad A^2, A^4, A^6, \ldots$$

for matrix A in Exercise ML.5. Write a brief description of the behavior of these sequences. Describe $\lim_{n \to \infty} A^n$.

5.2 ▼ Diagonalization of Symmetric Matrices

In this section we consider the diagonalization of symmetric matrices $(A = A^T)$. We restrict our attention to this case because it is easier to handle than that of general matrices and also because symmetric matrices arise in many applied problems.

As an example of such a problem, consider the task of identifying the conic represented by the equation

$$2x^2 + 2xy + 2y^2 = 9,$$

which can be written in matrix form as

$$\begin{bmatrix} x & y \end{bmatrix} \begin{bmatrix} 2 & 1 \\ 1 & 2 \end{bmatrix} \begin{bmatrix} x \\ y \end{bmatrix} = 9.$$

Observe that the matrix used here is a symmetric matrix. This problem is discussed in detail in Section 8.9. We shall merely remark here that the solution calls for the determination of the eigenvalues and eigenvectors of the matrix

$$\begin{bmatrix} 2 & 1 \\ 1 & 2 \end{bmatrix}.$$

The x- and y-axes are then rotated to a new set of axes, which lie along the eigenvectors of the matrix. In the new set of axes, the given conic can be readily identified.

In Example 7 of Section 5.1 we saw a matrix with the property that the roots of its characteristic polynomial were imaginary numbers. This cannot happen for a symmetric matrix, as the following theorem asserts. We omit the proof (see D. R. Hill, *Experiments in Computational Matrix Algebra*, New York: Random House, 1988).

THEOREM 5.5 ■ *All the roots of the characteristic polynomial of a symmetric matrix are real numbers.* ■

COROLLARY 5.1 ■ *If A is a symmetric matrix all of whose eigenvalues are distinct, then A is diagonalizable.*

Proof   Since A is symmetric, Theorem 5.5 tells us that all its eigenvalues are real. Then from Theorem 5.4 it follows that A can be diagonalized. ■

THEOREM 5.6 ■ *If A is a symmetric matrix, then eigenvectors that belong to distinct eigenvalues of A are orthogonal.*

Proof   First, we shall let the reader verify (Exercise T.1) the property that if $\mathbf{x}$ and $\mathbf{y}$ are vectors in R^n, then

$$(A\mathbf{x}) \cdot \mathbf{y} = \mathbf{x} \cdot (A^T\mathbf{y}). \tag{1}$$

Now let $\mathbf{x}_1$ and $\mathbf{x}_2$ be eigenvectors of A associated with the distinct eigenvalues λ_1 and λ_2 of A. We then have

$$A\mathbf{x}_1 = \lambda_1\mathbf{x}_1 \quad \text{and} \quad A\mathbf{x}_2 = \lambda_2\mathbf{x}_2.$$

Now

$$\begin{aligned} \lambda_1(\mathbf{x}_1 \cdot \mathbf{x}_2) &= (\lambda_1\mathbf{x}_1) \cdot \mathbf{x}_2 = (A\mathbf{x}_1) \cdot \mathbf{x}_2 \\ &= \mathbf{x}_1 \cdot (A^T\mathbf{x}_2) = \mathbf{x}_1 \cdot (A\mathbf{x}_2) \\ &= \mathbf{x}_1 \cdot (\lambda_2\mathbf{x}_2) = \lambda_2(\mathbf{x}_1 \cdot \mathbf{x}_2), \end{aligned} \tag{2}$$

where we have used the fact that $A = A^T$. Thus

$$\lambda_1(\mathbf{x}_1 \cdot \mathbf{x}_2) = \lambda_2(\mathbf{x}_1 \cdot \mathbf{x}_2)$$

and subtracting, we obtain

$$\begin{aligned} 0 &= \lambda_1(\mathbf{x}_1 \cdot \mathbf{x}_2) - \lambda_2(\mathbf{x}_1 \cdot \mathbf{x}_2) \\ &= (\lambda_1 - \lambda_2)(\mathbf{x}_1 \cdot \mathbf{x}_2). \end{aligned} \tag{3}$$

Since $\lambda_1 \neq \lambda_2$, we conclude that $\mathbf{x}_1 \cdot \mathbf{x}_2 = 0$, so $\mathbf{x}_1$ and $\mathbf{x}_2$ are orthogonal. ■

EXAMPLE 1 ■ Let

$$A = \begin{bmatrix} 0 & 0 & -2 \\ 0 & -2 & 0 \\ -2 & 0 & 3 \end{bmatrix}.$$

We find that the characteristic polynomial of A is (verify)

$$f(\lambda) = (\lambda + 2)(\lambda - 4)(\lambda + 1),$$

so the eigenvalues of A are

$$\lambda_1 = -2, \quad \lambda_2 = 4, \quad \text{and} \quad \lambda_3 = -1.$$

Then we can find the associated eigenvectors by solving the linear system $(\lambda I_3 - A)\mathbf{x} = \mathbf{0}$; and obtain the respective eigenvectors (verify)

$$\mathbf{x}_1 = \begin{bmatrix} 0 \\ 1 \\ 0 \end{bmatrix}, \quad \mathbf{x}_2 = \begin{bmatrix} -1 \\ 0 \\ 2 \end{bmatrix}, \quad \text{and} \quad \mathbf{x}_3 = \begin{bmatrix} 2 \\ 0 \\ 1 \end{bmatrix}.$$

It is easy to check that $\{\mathbf{x}_1, \mathbf{x}_2, \mathbf{x}_3\}$ is an orthogonal set of vectors in R^3 (and is thus linearly independent by Theorem 4.16, Section 4.8). Thus A is diagonalizable and is similar to

$$D = \begin{bmatrix} -2 & 0 & 0 \\ 0 & 4 & 0 \\ 0 & 0 & -1 \end{bmatrix}.$$

■

We recall that if A can be diagonalized, then there exists a nonsingular matrix P such that $P^{-1}AP$ is diagonal. Moreover, the columns of P are eigenvectors of A. Now, if the eigenvectors of A form an orthogonal set S, as happens when A is symmetric and the eigenvalues of A are distinct, then since any scalar multiple of an eigenvector of A is also an eigenvector of A, we can normalize S to obtain an orthonormal set

$$T = \{\mathbf{x}_1, \mathbf{x}_2, \dots, \mathbf{x}_n\}$$

of eigenvectors of A. The jth column of P is the eigenvector $\mathbf{x}_j$ associated with λ_j, and we now examine what type of matrix P must be. We can write P as

$$P = \begin{bmatrix} \mathbf{x}_1 & \mathbf{x}_2 & \cdots & \mathbf{x}_n \end{bmatrix}.$$

Then

$$P^T = \begin{bmatrix} \mathbf{x}_1^T \\ \mathbf{x}_2^T \\ \vdots \\ \mathbf{x}_n^T \end{bmatrix},$$

where $\mathbf{x}_i^T$, $1 \leq i \leq n$, is the transpose of the $n \times 1$ matrix (or vector) $\mathbf{x}_i$. We find that the i, jth entry in $P^T P$ is $\mathbf{x}_i \cdot \mathbf{x}_j$ (verify). Since

$$\mathbf{x}_i \cdot \mathbf{x}_j = \begin{cases} 1 & \text{if } i = j \\ 0 & \text{if } i \neq j, \end{cases}$$

then $P^T P = I_n$. Thus $P^T = P^{-1}$. Such matrices are important enough to have a special name.

DEFINITION   A nonsingular matrix A is called an **orthogonal matrix** if

$$A^{-1} = A^T.$$

We can also say that a nonsingular matrix A is orthogonal if $A^T A = I_n$.

EXAMPLE 2   ■   Let

$$A = \begin{bmatrix} \frac{2}{3} & -\frac{2}{3} & \frac{1}{3} \\ \frac{2}{3} & \frac{1}{3} & -\frac{2}{3} \\ \frac{1}{3} & \frac{2}{3} & \frac{2}{3} \end{bmatrix}.$$

It is easy to check that $A^T A = I_n$. Hence A is an orthogonal matrix.   ■

EXAMPLE 3   ■   Let A be the matrix of Example 1. We already know that the set of eigenvectors

$$\left\{ \begin{bmatrix} 0 \\ 1 \\ 0 \end{bmatrix}, \begin{bmatrix} -1 \\ 0 \\ 2 \end{bmatrix}, \begin{bmatrix} 2 \\ 0 \\ 1 \end{bmatrix} \right\}$$

is orthogonal. If we normalize these vectors, we find that

$$T = \left\{ \begin{bmatrix} 0 \\ 1 \\ 0 \end{bmatrix}, \begin{bmatrix} -\frac{1}{\sqrt{5}} \\ 0 \\ \frac{2}{\sqrt{5}} \end{bmatrix}, \begin{bmatrix} \frac{2}{\sqrt{5}} \\ 0 \\ \frac{1}{\sqrt{5}} \end{bmatrix} \right\}$$

is an orthonormal set of vectors. The matrix P such that $P^{-1}AP$ is diagonal is the matrix whose columns are the vectors in T. Thus

$$P = \begin{bmatrix} 0 & -\frac{1}{\sqrt{5}} & \frac{2}{\sqrt{5}} \\ 1 & 0 & 0 \\ 0 & \frac{2}{\sqrt{5}} & \frac{1}{\sqrt{5}} \end{bmatrix}.$$

We leave it to the reader to verify (Exercise 4) that P is an orthogonal matrix and that

$$P^{-1}AP = P^T AP = \begin{bmatrix} -2 & 0 & 0 \\ 0 & 4 & 0 \\ 0 & 0 & -1 \end{bmatrix}.$$ ■

The following theorem is not difficult to show, and we leave its proof to the reader (Exercise T.3).

THEOREM 5.7 ■ *The $n \times n$ matrix A is orthogonal if and only if the columns (and rows) of A form an orthonormal set of vectors in R^n.*   ■

If A is an orthogonal matrix, then it is easy to show that $\det(A) = \pm 1$ (Exercise T.4).

We now turn to the general situation for a symmetric matrix; even if A has eigenvalues whose multiplicities are greater than one, it turns out that we can still diagonalize A. We omit the proof of the theorem. For a proof, see J. M. Ortega, *Matrix Theory, a Second Course*, New York: Plenum Press, 1987.

THEOREM 5.8 ■ *If A is a symmetric $n \times n$ matrix, then there exists an orthogonal matrix P such that $P^{-1}AP = D$, a diagonal matrix. The eigenvalues of A lie on the main diagonal of D.*   ■

Thus, not only is a symmetric matrix always diagonalizable, but it is diagonalizable by means of an orthogonal matrix. In such a case, we say that A is **orthogonally diagonalizable**.

It can be shown that if a symmetric matrix A has an eigenvalue λ_j of multiplicity k_j, then the solution space of the linear system $(\lambda_j I_n - A)\mathbf{x} = \mathbf{0}$ (the eigenspace of λ_j) has dimension k_j. This means that there exist k_j linearly independent eigenvectors of A associated with the eigenvalue λ_j. By the Gram–Schmidt process, we can construct an orthonormal basis for the solution space. Thus we obtain a set of k_j orthonormal eigenvectors associated with the eigenvalue λ_j. Since eigenvectors associated with distinct eigenvalues are orthogonal, if we form the set of all eigenvectors, we get an orthonormal set. Hence the matrix P whose columns are the eigenvectors is orthogonal.

The procedure for diagonalizing a symmetric matrix A by an orthogonal matrix P is as follows.

Step 1. Form the characteristic polynomial $f(\lambda) = \det(\lambda I_n - A)$.

Step 2. Find the roots of the characteristic polynomial of A. These will all be real.

Step 3. For each eigenvalue λ_j of A of multiplicity k_j, find a basis of k_j eigenvectors for the solution space of $(\lambda_j I_n - A)\mathbf{x} = \mathbf{0}$ (the eigenspace of λ_j).

Step 4. For each eigenspace, transform the basis obtained in Step 3 to an orthonormal basis by the Gram–Schmidt process. The totality of all these orthonormal bases determines an orthonormal set of n linearly independent eigenvectors of A.

Step 5. Let P be the matrix whose columns are the n linearly independent eigenvectors determined in Step 4. Then P is an orthogonal matrix and $P^{-1}AP = P^{T}AP = D$, a diagonal matrix whose diagonal elements are the eigenvalues of A that correspond to the columns of P.

EXAMPLE 4 ∎ Let

$$A = \begin{bmatrix} 0 & 2 & 2 \\ 2 & 0 & 2 \\ 2 & 2 & 0 \end{bmatrix}.$$

The characteristic polynomial of A is (verify)

$$f(\lambda) = (\lambda + 2)^2(\lambda - 4),$$

so the eigenvalues are

$$\lambda_1 = -2, \quad \lambda_2 = -2, \quad \text{and} \quad \lambda_3 = 4.$$

That is, -2 is an eigenvalue whose multiplicity is 2. Next, to find the eigenvectors associated with λ_1 and λ_2, we solve the homogeneous linear system $(-2I_3 - A)\mathbf{x} = \mathbf{0}$:

$$\begin{bmatrix} -2 & -2 & -2 \\ -2 & -2 & -2 \\ -2 & -2 & -2 \end{bmatrix} \begin{bmatrix} x_1 \\ x_2 \\ x_3 \end{bmatrix} = \begin{bmatrix} 0 \\ 0 \\ 0 \end{bmatrix}. \tag{4}$$

A basis for the solution space of (4) consists of the eigenvectors (verify)

$$\mathbf{x}_1 = \begin{bmatrix} -1 \\ 1 \\ 0 \end{bmatrix} \quad \text{and} \quad \mathbf{x}_2 = \begin{bmatrix} -1 \\ 0 \\ 1 \end{bmatrix}.$$

Now $\mathbf{x}_1$ and $\mathbf{x}_2$ are not orthogonal, since $\mathbf{x}_1 \cdot \mathbf{x}_2 \neq 0$. We can use the Gram–Schmidt process to obtain an orthonormal basis for the solution space of (4) (the eigenspace of $\lambda_1 = -2$) as follows. Let

$$\mathbf{y}_1 = \mathbf{x}_1 = \begin{bmatrix} -1 \\ 1 \\ 0 \end{bmatrix}$$

and

$$\mathbf{y}_2 = \mathbf{x}_2 - \left(\frac{\mathbf{x}_2 \cdot \mathbf{y}_1}{\mathbf{y}_1 \cdot \mathbf{y}_1} \right) \mathbf{y}_1 = \begin{bmatrix} -\frac{1}{2} \\ -\frac{1}{2} \\ 1 \end{bmatrix}.$$

Let

$$\mathbf{y}_2^* = 2\mathbf{y}_2 = \begin{bmatrix} -1 \\ -1 \\ 2 \end{bmatrix}.$$

The set $\{\mathbf{y}_1, \mathbf{y}_2^*\}$ is an orthogonal set of eigenvectors. Normalizing these eigenvectors, we obtain

$$\mathbf{z}_1 = \frac{1}{\|\mathbf{y}_1\|} \mathbf{y}_1 = \frac{1}{\sqrt{2}} \begin{bmatrix} -1 \\ 1 \\ 0 \end{bmatrix} \quad \text{and} \quad \mathbf{z}_2 = \frac{1}{\|\mathbf{y}_2^*\|} \mathbf{y}_2^* = \frac{1}{\sqrt{6}} \begin{bmatrix} -1 \\ -1 \\ 2 \end{bmatrix}.$$

The set $\{\mathbf{z}_1, \mathbf{z}_2\}$ is an orthonormal basis of eigenvectors of A for the solution space of (4). Now we find a basis for the solution space of $(4I_3 - A)\mathbf{x} = \mathbf{0}$,

$$
\begin{bmatrix} 4 & -2 & -2 \\ -2 & 4 & -2 \\ -2 & -2 & 4 \end{bmatrix} \begin{bmatrix} x_1 \\ x_2 \\ x_3 \end{bmatrix} = \begin{bmatrix} 0 \\ 0 \\ 0 \end{bmatrix}, \tag{5}
$$

to consist of (verify)

$$
\mathbf{x}_3 = \begin{bmatrix} 1 \\ 1 \\ 1 \end{bmatrix}.
$$

Normalizing this vector, we have the eigenvector

$$
\mathbf{z}_3 = \frac{1}{\sqrt{3}} \begin{bmatrix} 1 \\ 1 \\ 1 \end{bmatrix}
$$

as an orthonormal basis for the solution space of (5). Since eigenvectors associated with distinct eigenvalues are orthogonal, we observe that $\mathbf{z}_3$ is orthogonal to both $\mathbf{z}_1$ and $\mathbf{z}_2$. Thus the set $\{\mathbf{z}_1, \mathbf{z}_2, \mathbf{z}_3\}$ is an orthonormal basis for R^3 consisting of eigenvectors of A. The matrix P is the matrix whose jth column is $\mathbf{z}_j$:

$$
P = \begin{bmatrix} -\dfrac{1}{\sqrt{2}} & -\dfrac{1}{\sqrt{6}} & \dfrac{1}{\sqrt{3}} \\ \dfrac{1}{\sqrt{2}} & -\dfrac{1}{\sqrt{6}} & \dfrac{1}{\sqrt{3}} \\ 0 & \dfrac{2}{\sqrt{6}} & \dfrac{1}{\sqrt{3}} \end{bmatrix}.
$$

We leave it to the reader to verify that

$$
P^{-1}AP = P^{T}AP = \begin{bmatrix} -2 & 0 & 0 \\ 0 & -2 & 0 \\ 0 & 0 & 4 \end{bmatrix}.
$$
∎

EXAMPLE 5 ∎ Let

$$
A = \begin{bmatrix} 1 & 2 & 0 & 0 \\ 2 & 1 & 0 & 0 \\ 0 & 0 & 1 & 2 \\ 0 & 0 & 2 & 1 \end{bmatrix}.
$$

The characteristic polynomial of A is (verify)

$$
f(\lambda) = (\lambda + 1)^2 (\lambda - 3)^2,
$$

so the eigenvalues of A are

$$
\lambda_1 = -1, \quad \lambda_2 = -1, \quad \lambda_3 = 3, \quad \text{and} \quad \lambda_4 = 3.
$$

We find (verify) that a basis for the solution space of

$$
(-1I_3 - A)\mathbf{x} = \mathbf{0} \tag{6}
$$

consists of the eigenvectors

$$\mathbf{x}_1 = \begin{bmatrix} 1 \\ -1 \\ 0 \\ 0 \end{bmatrix} \quad \text{and} \quad \mathbf{x}_2 = \begin{bmatrix} 0 \\ 0 \\ 1 \\ -1 \end{bmatrix},$$

which are orthogonal. Normalizing these eigenvectors, we obtain

$$\mathbf{z}_1 = \begin{bmatrix} \frac{1}{\sqrt{2}} \\ -\frac{1}{\sqrt{2}} \\ 0 \\ 0 \end{bmatrix} \quad \text{and} \quad \mathbf{z}_2 = \begin{bmatrix} 0 \\ 0 \\ \frac{1}{\sqrt{2}} \\ -\frac{1}{\sqrt{2}} \end{bmatrix}$$

as an orthonormal basis of eigenvectors for the solution space of (6). We also find (verify) that a basis for the solution space of

$$(3I_3 - A)\mathbf{x} = \mathbf{0} \tag{7}$$

consists of the eigenvectors

$$\mathbf{x}_3 = \begin{bmatrix} 1 \\ 1 \\ 0 \\ 0 \end{bmatrix} \quad \text{and} \quad \mathbf{x}_4 = \begin{bmatrix} 0 \\ 0 \\ 1 \\ 1 \end{bmatrix},$$

which are orthogonal. Normalizing these eigenvectors, we obtain

$$\mathbf{z}_3 = \begin{bmatrix} \frac{1}{\sqrt{2}} \\ \frac{1}{\sqrt{2}} \\ 0 \\ 0 \end{bmatrix} \quad \text{and} \quad \mathbf{z}_4 = \begin{bmatrix} 0 \\ 0 \\ \frac{1}{\sqrt{2}} \\ \frac{1}{\sqrt{2}} \end{bmatrix}$$

as an orthonormal basis of eigenvectors for the solution space of (7). Since eigenvectors associated with distinct eigenvalues are orthogonal, we conclude that

$$\{\mathbf{z}_1, \mathbf{z}_2, \mathbf{z}_3, \mathbf{z}_4\}$$

is an orthonormal basis for R^4 consisting of eigenvectors of A. The matrix P is the matrix whose jth column is $\mathbf{z}_j$:

$$P = \begin{bmatrix} \frac{1}{\sqrt{2}} & 0 & \frac{1}{\sqrt{2}} & 0 \\ -\frac{1}{\sqrt{2}} & 0 & \frac{1}{\sqrt{2}} & 0 \\ 0 & \frac{1}{\sqrt{2}} & 0 & \frac{1}{\sqrt{2}} \\ 0 & -\frac{1}{\sqrt{2}} & 0 & \frac{1}{\sqrt{2}} \end{bmatrix}.$$

■

Suppose now that A is an $n \times n$ matrix that is orthogonally diagonalizable. Thus we have an orthogonal matrix P such that $P^{-1}AP$ is a diagonal matrix D. Then $P^{-1}AP = D$, or $A = PDP^{-1}$. Since $P^{-1} = P^T$, we can write $A = PDP^T$. Then

$$A^T = (PDP^T)^T = (P^T)^T D^T P^T = PDP^T = A$$

($D = D^T$ since D is a diagonal matrix). Thus A is symmetric. This result, together with Theorem 5.8, yields the following theorem.

THEOREM 5.9 ■ *An $n \times n$ matrix A is orthogonally diagonalizable if and only if A is symmetric.*

■

Some remarks about nonsymmetric matrices are in order at this point. Theorem 5.4 assures us that A is diagonalizable if all the roots of its characteristic polynomial are real and distinct. We also studied examples, in Section 5.1, of nonsymmetric matrices with repeated eigenvalues that were diagonalizable and others that were not diagonalizable. There are some striking differences between the symmetric and nonsymmetric cases, which we now summarize. If A is nonsymmetric, then the roots of its characteristic polynomial need not all be real numbers; if an eigenvalue λ_j has multiplicity k_j, then the solution space of $(\lambda_j I_n - A)\mathbf{x} = \mathbf{0}$ may have dimension $< k_j$; if the roots of the characteristic polynomial of A are all real, it is still possible for A not to have n linearly independent eigenvectors (which means that A cannot be diagonalized); eigenvectors associated with distinct eigenvalues need not be orthogonal. Thus, in Example 13 of Section 5.1, the eigenvectors $\mathbf{x}_1$ and $\mathbf{x}_3$ associated with the eigenvalues $\lambda_1 = 0$ and $\lambda_3 = 1$, respectively, are not orthogonal. If a matrix A cannot be diagonalized, then we can often find a matrix B similar to A that is "nearly diagonal." The matrix B is said to be in **Jordan canonical form**; its treatment lies beyond the scope of this book but is studied in advanced books on linear algebra.*

It should be noted that in many applications, we need only find a diagonal matrix D that is similar to the given matrix A; that is, we do not need the orthogonal matrix P such that $P^{-1}AP = D$. Many of the matrices to be diagonalized in applied problems are either symmetric, or all the roots of their characteristic polynomial are real. Of course, the methods for finding eigenvalues that have been presented in this chapter are not recommended for matrices with more than four rows because of the need to evaluate determinants. In Section 9.5 we shall consider an efficient numerical method for diagonalizing a symmetric matrix.

Sections 8.8, 8.9, and 8.10 on Quadratic Forms, Conic Sections, and Quadric Surfaces, respectively, which can be taken up at this time, use material from this section.

---

*See, for example, K. Hoffman and R. Kunze, *Linear Algebra*, 2nd ed., Upper Saddle River, N.J.: Prentice Hall, Inc., 1971, or J. M. Ortega, *Matrix Theory, a Second Course*, New York: Plenum Press, 1987.

▼ *Preview of an Application*

QUADRATIC FORMS (SECTION 8.8)
CONIC SECTIONS (SECTION 8.9)
QUADRIC SURFACES (SECTION 8.10)

The conic sections are the curves obtained by intersecting a right circular cone with a plane. Depending on the position of the plane relative to the cone, the result can be an ellipse, circle, parabola, or hyperbola, or degenerate forms of these such as a point, a line, a pair of lines, or the empty set. These curves occur in a wide variety of applications in everyday life, ranging from the orbit of the planets to navigation devices and the design of headlights in automobiles.

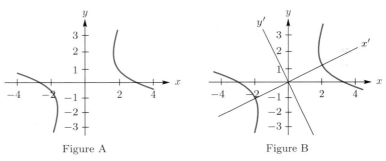

Figure A   Figure B

Conic sections are in standard form when they are centered at the origin and their foci lie on a coordinate axis. In this case, they are easily identified by their simple equations. For example, the graph of the equation

$$x^2 + 4xy - 2y^2 = 8$$

shown in Figure A is a hyperbola, which is not in standard form. This equation can be written in matrix form as

$$\begin{bmatrix} x & y \end{bmatrix} \begin{bmatrix} 1 & 2 \\ 2 & -2 \end{bmatrix} \begin{bmatrix} x \\ y \end{bmatrix} = 8.$$

The left side of this equation is called a quadratic form and these are discussed in Section 8.8. By diagonalizing the matrix

$$\begin{bmatrix} 1 & 2 \\ 2 & -2 \end{bmatrix}$$

that occurs in the quadratic form, we can transform the given equation to

$$\frac{x'^2}{4} - \frac{y'^2}{\frac{8}{3}} = 1$$

whose graph is the hyperbola in standard form shown in Figure B. As you can see, the x- and y-axes have been rotated to the x'- and y'-axes. Section 8.9 discusses the rotation of axes to identify conic sections.

Section 8.10 discusses the analogous problem of identifying quadric surfaces in R^3.

5.2 EXERCISES

1. Verify that

$$P = \begin{bmatrix} \frac{2}{3} & -\frac{2}{3} & \frac{1}{3} \\ \frac{2}{3} & \frac{1}{3} & -\frac{2}{3} \\ \frac{1}{3} & \frac{2}{3} & \frac{2}{3} \end{bmatrix}$$

is an orthogonal matrix.

2. Find the inverse of each of the following orthogonal matrices.

(a) $A = \begin{bmatrix} 1 & 0 & 0 \\ 0 & \cos\theta & \sin\theta \\ 0 & -\sin\theta & \cos\theta \end{bmatrix}$.

(b) $B = \begin{bmatrix} 1 & 0 & 0 \\ 0 & \frac{1}{\sqrt{2}} & -\frac{1}{\sqrt{2}} \\ 0 & -\frac{1}{\sqrt{2}} & -\frac{1}{\sqrt{2}} \end{bmatrix}$.

3. Verify Theorem 5.7 for the matrices in Exercise 2.

4. Verify that the matrix P in Example 3 is an orthogonal matrix.

In Exercises 5 through 10, orthogonally diagonalize each given matrix A, giving the diagonal matrix D and the diagonalizing orthogonal matrix P.

5. $\begin{bmatrix} 2 & 2 \\ 2 & 2 \end{bmatrix}$.

6. $\begin{bmatrix} 0 & 0 & 1 \\ 0 & 0 & 0 \\ 1 & 0 & 0 \end{bmatrix}$.

7. $\begin{bmatrix} 0 & 0 & 0 \\ 0 & 2 & 2 \\ 0 & 2 & 2 \end{bmatrix}$.

8. $\begin{bmatrix} 0 & 0 & 0 & 0 \\ 0 & 0 & 0 & 0 \\ 0 & 0 & 0 & 1 \\ 0 & 0 & 1 & 0 \end{bmatrix}$.

9. $\begin{bmatrix} 0 & -1 & -1 \\ -1 & 0 & -1 \\ -1 & -1 & 0 \end{bmatrix}$.

10. $\begin{bmatrix} -1 & 2 & 2 \\ 2 & -1 & 2 \\ 2 & 2 & -1 \end{bmatrix}$.

In Exercises 11 through 18, orthogonally diagonalize each given matrix.

11. $\begin{bmatrix} 2 & 1 \\ 1 & 2 \end{bmatrix}$.

12. $\begin{bmatrix} 2 & 2 & 0 & 0 \\ 2 & 2 & 0 & 0 \\ 0 & 0 & 2 & 2 \\ 0 & 0 & 2 & 2 \end{bmatrix}$.

13. $\begin{bmatrix} 1 & 1 & 0 \\ 1 & 1 & 0 \\ 0 & 0 & 1 \end{bmatrix}$.

14. $\begin{bmatrix} 1 & 0 & 0 \\ 0 & 3 & -2 \\ 0 & -2 & 3 \end{bmatrix}$.

15. $\begin{bmatrix} 1 & 0 & 0 \\ 0 & 1 & 1 \\ 0 & 1 & 1 \end{bmatrix}$.

16. $\begin{bmatrix} 0 & 0 & 0 & 1 \\ 0 & 0 & 0 & 0 \\ 0 & 0 & 0 & 0 \\ 1 & 0 & 0 & 0 \end{bmatrix}$.

17. $\begin{bmatrix} 2 & 1 & 1 \\ 1 & 2 & 1 \\ 1 & 1 & 2 \end{bmatrix}$.

18. $\begin{bmatrix} -3 & 0 & -1 \\ 0 & -2 & 0 \\ -1 & 0 & -3 \end{bmatrix}$.

THEORETICAL EXERCISES ■

T.1. Show that if $\mathbf{x}$ and $\mathbf{y}$ are vectors in R^n, then $(A\mathbf{x}) \cdot \mathbf{y} = \mathbf{x} \cdot (A^T\mathbf{y})$.

T.2. Show that if A is an $n \times n$ orthogonal matrix and $\mathbf{x}$ and $\mathbf{y}$ are vectors in R^n, then $(A\mathbf{x}) \cdot (A\mathbf{y}) = \mathbf{x} \cdot \mathbf{y}$.

T.3. Prove Theorem 5.7.

T.4. Show that if A is an orthogonal matrix, then $\det(A) = \pm 1$.

T.5. Prove Theorem 5.8 for the 2×2 case by studying the possible roots of the characteristic polynomial of A.

T.6. Show that if A and B are orthogonal matrices, then AB is an orthogonal matrix.

T.7. Show that if A is an orthogonal matrix, then A^{-1} is also orthogonal.

T.8. (a) Verify that the matrix

$$\begin{bmatrix} \cos\theta & -\sin\theta \\ \sin\theta & \cos\theta \end{bmatrix}$$

is orthogonal.

(b) Show that if A is an orthogonal 2×2 matrix, then there exists a real number θ such that either

$$A = \begin{bmatrix} \cos\theta & -\sin\theta \\ \sin\theta & \cos\theta \end{bmatrix}$$

or

$$A = \begin{bmatrix} \cos\theta & \sin\theta \\ \sin\theta & -\cos\theta \end{bmatrix}.$$

T.9. Show that if $A^T A \mathbf{y} = \mathbf{y}$ for any $\mathbf{y}$ in R^n, then $A^T A = I_n$.

T.10. Show that if A is nonsingular and orthogonally diagonalizable, then A^{-1} is orthogonally diagonalizable.

MATLAB EXERCISES

The MATLAB command **eig** will produce the eigenvalues and a set of orthonormal eigenvectors for a symmetric matrix A. Use the command in the form

$$[\mathbf{V}, \mathbf{D}] = \mathbf{eig}(\mathbf{A}).$$

The matrix V will contain the orthonormal eigenvectors, and matrix D will be diagonal containing the corresponding eigenvalues.

ML.1. Use **eig** to find the eigenvalues of A and an orthogonal matrix P so that $P^T A P$ is diagonal.

(a) $A = \begin{bmatrix} 6 & 6 \\ 6 & 6 \end{bmatrix}$.

(b) $A = \begin{bmatrix} 1 & 2 & 2 \\ 2 & 1 & 2 \\ 2 & 2 & 1 \end{bmatrix}$.

(c) $A = \begin{bmatrix} 4 & 1 & 0 \\ 1 & 4 & 1 \\ 0 & 1 & 4 \end{bmatrix}$.

ML.2. Command **eig** can be applied to any matrix, but the matrix V of eigenvectors need not be orthogonal. For each of the following, use **eig** to determine which matrices A are such that V is orthogonal. If V is not orthogonal, then discuss briefly whether it can or cannot be replaced by an orthogonal matrix of eigenvectors.

(a) $A = \begin{bmatrix} 1 & 2 \\ -1 & 4 \end{bmatrix}$.

(b) $A = \begin{bmatrix} 2 & 1 & 2 \\ 2 & 2 & -2 \\ 3 & 1 & 1 \end{bmatrix}$.

(c) $A = \begin{bmatrix} 1 & -3 \\ 3 & -5 \end{bmatrix}$.

(d) $A = \begin{bmatrix} 1 & 0 & 0 \\ 0 & 1 & 1 \\ 0 & 1 & 1 \end{bmatrix}$.

KEY IDEAS FOR REVIEW

☐ **Theorem 5.2.** The eigenvalues of A are the real roots of the characteristic polynomial of A.

☐ **Theorem 5.3.** An $n \times n$ matrix A is diagonalizable if and only if it has n linearly independent eigenvectors. In this case A is similar to a diagonal matrix D, with $D = P^{-1}AP$, whose diagonal elements are the eigenvalues of A, and P is a matrix whose columns are n linearly independent eigenvectors of A.

☐ **Theorem 5.5.** All the roots of the characteristic polynomial of a symmetric matrix are real numbers.

☐ **Theorem 5.6.** If A is a symmetric matrix, then eigenvectors that belong to distinct eigenvalues of A are orthogonal.

☐ **Theorem 5.7.** The $n \times n$ matrix A is orthogonal if and only if the columns of A form an orthonormal set of vectors in R^n.

☐ **Theorem 5.8.** If A is a symmetric $n \times n$ matrix, then there exists an orthogonal matrix P

$(P^{-1} = P^T)$ such that $P^{-1}AP = D$, a diagonal matrix. The eigenvalues of A lie on the main diagonal of D.

☐ **List of Nonsingular Equivalences.** The following statements are equivalent for an $n \times n$ matrix A.

1. A is nonsingular.
2. $A\mathbf{x} = \mathbf{0}$ has only the trivial solution.
3. A is row equivalent to I_n.
4. The linear system $A\mathbf{x} = \mathbf{b}$ has a unique solution for every $n \times 1$ matrix $\mathbf{b}$.
5. $\det(A) \neq 0$.
6. A has rank n.
7. A has nullity 0.
8. The rows of A form a linearly independent set of n vectors in R^n.
9. The columns of A form a linearly independent set of n vectors in R^n.
10. Zero is *not* an eigenvalue of A.

SUPPLEMENTARY EXERCISES ■

1. Find the characteristic polynomial, eigenvalues, and eigenvectors of the matrix
$$\begin{bmatrix} -2 & 0 & 0 \\ 3 & 2 & 3 \\ 4 & -1 & 6 \end{bmatrix}.$$

In Exercises 2 and 3, determine whether the given matrix is diagonalizable.

2. $A = \begin{bmatrix} -1 & 0 & 0 \\ 3 & 2 & 0 \\ 4 & -1 & 2 \end{bmatrix}.$

3. $A = \begin{bmatrix} 2 & 2 & 0 \\ 5 & -1 & 3 \\ 0 & 0 & 0 \end{bmatrix}.$

In Exercises 4 and 5, find, if possible, a nonsingular matrix P and a diagonal matrix D so that A is similar to D.

4. $A = \begin{bmatrix} 1 & 0 & 0 \\ 2 & -1 & -6 \\ 2 & 0 & -2 \end{bmatrix}.$

5. $A = \begin{bmatrix} 1 & 0 & 1 \\ 0 & 1 & 0 \\ -1 & 0 & 1 \end{bmatrix}.$

6. If possible, find a diagonal matrix D so that
$$A = \begin{bmatrix} 0 & 0 & 0 \\ 0 & 1 & 0 \\ 2 & 2 & 1 \end{bmatrix}$$
is similar to D.

7. Find bases for the eigenspaces associated with each eigenvalue of the matrix
$$A = \begin{bmatrix} 0 & 0 & 1 \\ 0 & 2 & 0 \\ 0 & 0 & 2 \end{bmatrix}.$$

8. Is the matrix
$$\begin{bmatrix} \frac{2}{3} & \frac{1}{\sqrt{5}} & 1 \\ \frac{2}{3} & 0 & 0 \\ \frac{1}{3} & -\frac{2}{\sqrt{5}} & 0 \end{bmatrix}$$
orthogonal?

In Exercises 9 and 10, orthogonally diagonalize the given matrix A, giving the orthogonal matrix P and the diagonal matrix D.

9. $A = \begin{bmatrix} 1 & 1 & 1 \\ 1 & 1 & 1 \\ 1 & 1 & 1 \end{bmatrix}.$

10. $A = \begin{bmatrix} -3 & 0 & -4 \\ 0 & 5 & 0 \\ -4 & 0 & 3 \end{bmatrix}.$

11. If $A^2 = A$, what are possible values for the eigenvalues of A? Justify your answer.

12. Let $p_1(\lambda)$ be the characteristic polynomial of A_{11} and $p_2(\lambda)$ the characteristic polynomial of A_{22}. What is the characteristic polynomial of each of the following partitioned matrices?

(a) $A = \begin{bmatrix} A_{11} & O \\ O & A_{22} \end{bmatrix}.$ (b) $A = \begin{bmatrix} A_{11} & O \\ A_{12} & A_{22} \end{bmatrix}.$

(*Hint*: See Supplementary Exercises T.7 and T.8 in Chapter 2.)

THEORETICAL EXERCISES ■

T.1. Show that if a matrix A is similar to a diagonal matrix D, then $\text{Tr}(A) = \text{Tr}(D)$, where $\text{Tr}(A)$ is the trace of A. [*Hint*: See Supplementary Exercise T.1 in Chapter 1, where part (c) establishes $\text{Tr}(AB) = \text{Tr}(BA)$.]

T.2. Show that $\text{Tr}(A)$ is the sum of the eigenvalues of A (see Supplementary Exercise T.1 in Chapter 1).

T.3. Let A and B be similar matrices. Show that:
(a) A^T and B^T are similar.

(b) $\text{rank } A = \text{rank } B$.
(c) A is nonsingular if and only if B is nonsingular.
(d) If A and B are nonsingular, then A^{-1} and B^{-1} are similar.
(e) $\text{Tr}(A) = \text{Tr}(B)$.

T.4. Show that if A is an orthogonal matrix, then A^T is also orthogonal.

T.5. Let A be an orthogonal matrix. Show that cA is orthogonal if and only if $c = \pm 1$.

T.6. Let

$$A = \begin{bmatrix} a & b \\ c & d \end{bmatrix}.$$

Show that the characteristic polynomial $f(\lambda)$ of A is given by

$$f(\lambda) = \lambda^2 - \mathrm{Tr}(A)\lambda + \det(A),$$

where $\mathrm{Tr}(A)$ denotes the trace of A (see Supplementary Exercise T.1 in Chapter 1).

T.7. The **Cayley–Hamilton theorem** states that a matrix satisfies its characteristic equation; that is, if A is an $n \times n$ matrix with characteristic polynomial

$$f(\lambda) = \lambda^n + a_1\lambda^{n-1} + \cdots + a_{n-1}\lambda + a_n,$$

then

$$A^n + a_1 A^{n-1} + \cdots + a_{n-1}A + a_n I_n = O.$$

The proof and applications of this result, unfortunately, lie beyond the scope of this book.

Verify the Cayley–Hamilton theorem for the following matrices.

(a) $\begin{bmatrix} 1 & 2 & 3 \\ 2 & -1 & 5 \\ 3 & 2 & 1 \end{bmatrix}$. (b) $\begin{bmatrix} 1 & 2 & 3 \\ 0 & 2 & 2 \\ 0 & 0 & -3 \end{bmatrix}$.

(c) $\begin{bmatrix} 3 & 3 \\ 2 & 4 \end{bmatrix}$.

T.8. Let A be an $n \times n$ matrix whose characteristic polynomial is

$$f(\lambda) = \lambda^n + a_1\lambda^{n-1} + \cdots + a_{n-1}\lambda + a_n.$$

If A is nonsingular, show that

$$A^{-1}$$
$$= -\frac{1}{a_n}(A^{n-1} + a_1 A^{n-2} + \cdots + a_{n-2}A + a_{n-1}I_n).$$

[*Hint*: Use the Cayley–Hamilton theorem (Exercise T.7).]

CHAPTER TEST ■

1. If possible, find a nonsingular matrix P and a diagonal matrix D so that A is similar to D, where

$$A = \begin{bmatrix} 1 & 0 & 0 \\ 5 & 2 & 0 \\ 4 & 3 & 2 \end{bmatrix}.$$

2. Verify that the following matrix is orthogonal.

$$\begin{bmatrix} -\frac{1}{\sqrt{2}} & \frac{1}{\sqrt{3}} & \frac{1}{\sqrt{6}} \\ 0 & \frac{1}{\sqrt{3}} & -\frac{2}{\sqrt{6}} \\ \frac{1}{\sqrt{2}} & \frac{1}{\sqrt{3}} & \frac{1}{\sqrt{6}} \end{bmatrix}$$

3. Orthogonally diagonalize A, giving the orthogonal matrix P and the diagonal matrix D.

$$A = \begin{bmatrix} -1 & -4 & -8 \\ -4 & -7 & 4 \\ -8 & 4 & -1 \end{bmatrix}.$$

4. Answer each of the following as true or false.

(a) If A is an $n \times n$ orthogonal matrix, then $\mathrm{rank}\, A < n$.

(b) If A is diagonalizable, then each of its eigenvalues has multiplicity one.

(c) If none of the eigenvalues of A are zero, then $\det(A) \neq 0$.

(d) If A and B are similar, then $\det(A) = \det(B)$.

(e) If $\mathbf{x}$ and $\mathbf{y}$ are eigenvectors of A associated with the distinct eigenvalues λ_1 and λ_2, respectively, then $\mathbf{x} + \mathbf{y}$ is an eigenvector of A associated with the eigenvalue $\lambda_1 + \lambda_2$.

6

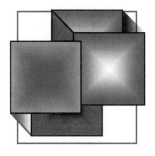

Linear Transformations and Matrices

In Section 3.3 we gave the definition, basic properties, and some examples of linear transformations mapping R^n into R^m. In this chapter we consider linear transformations mapping a vector space V into a vector space W.

6.1 ▾ Definition and Examples

DEFINITION

Let V and W be vector spaces. A **linear transformation** L of V into W is a function assigning a unique vector $L(\mathbf{u})$ in W to each $\mathbf{u}$ in V such that:

(a) $L(\mathbf{u} + \mathbf{v}) = L(\mathbf{u}) + L(\mathbf{v})$, for every $\mathbf{u}$ and $\mathbf{v}$ in V.

(b) $L(k\mathbf{u}) = kL(\mathbf{u})$, for every $\mathbf{u}$ in V and every scalar k.

In the definition above, observe that in (a) the $+$ in $\mathbf{u} + \mathbf{v}$ on the left side of the equation refers to the addition operation in V, whereas the $+$ in $L(\mathbf{u}) + L(\mathbf{v})$ on the right side of the equation refers to the addition operation in W. Similarly, in (b) the scalar product $k\mathbf{u}$ is in V, while the scalar product $kL(\mathbf{u})$ is in W.

As in Section 3.3, we shall write the fact that L maps V into W, even if it is not a linear transformation, as

$$L \colon V \to W.$$

If $V = W$, the linear transformation $L \colon V \to V$ is also called a **linear operator** on V.

In Section 3.3 we gave a number of examples of linear transformations mapping R^n into R^m. Thus, the following are linear transformations that we have already discussed:

Projection: $L: R^3 \to R^2$ defined by $L(x, y, z) = (x, y)$.

Dilation: $L: R^3 \to R^3$ defined by $L(\mathbf{u}) = r\mathbf{u}, \ r > 1$.

Contraction: $L: R^3 \to R^3$ defined by $L(\mathbf{u}) = r\mathbf{u}, \ 0 < r < 1$.

Reflection: $L: R^2 \to R^2$ defined by $L(x, y) = (x, -y)$.

Rotation: $L: R^2 \to R^2$ defined by $L(\mathbf{u}) = \begin{bmatrix} \cos\phi & -\sin\phi \\ \sin\phi & \cos\phi \end{bmatrix} \mathbf{u}$.

Recall that P_1 is the vector space of all polynomials of degree ≤ 1; in general, P_n is the vector space of all polynomials of degree $\leq n$, and M_{nn} is the vector space of all $n \times n$ matrices.

As in Section 3.3, to verify that a given function is a linear transformation, we have to check that conditions (a) and (b) in the definition above are satisfied.

EXAMPLE 1 ■ Let $L: P_1 \to P_2$ be defined by

$$L(at + b) = t(at + b).$$

Show that L is a linear transformation.

Solution   Let $at + b$ and $ct + d$ be vectors in P_1 and let k be a scalar. Then

$$L[(at + b) + (ct + d)] = t[(at + b) + (ct + d)]$$
$$= t(at + b) + t(ct + d) = L(at + b) + L(ct + d)$$

and

$$L[k(at + b)] = t[k(at + b)] = k[t(at + b)] = kL(at + b).$$

Hence L is a linear transformation.   ■

EXAMPLE 2 ■ Let $L: P_1 \to P_2$ be defined by

$$L[p(t)] = tp(t) + t^2.$$

Is L a linear transformation?

Solution   Let $p(t)$ and $q(t)$ be vectors in P_1 and let k be a scalar. Then

$$L[p(t) + q(t)] = t[p(t) + q(t)] + t^2 = tp(t) + tq(t) + t^2,$$

and

$$L[p(t)] + L[q(t)] = [tp(t) + t^2] + [tq(t) + t^2] = t[p(t) + q(t)] + 2t^2.$$

Since $L[p(t) + q(t)] \neq L[p(t)] + L[q(t)]$, we conclude that L is not a linear transformation.   ■

EXAMPLE 3 ■ Let $L: M_{mn} \to M_{nm}$ be defined by

$$L(A) = A^T$$

for A in M_{mn}. Is L a linear transformation?

Solution   Let A and B be in M_{mn}. Then by Theorem 1.4 in Section 1.4, we have

$$L(A + B) = (A + B)^T = A^T + B^T = L(A) + L(B),$$

and, if k is a scalar,

$$L(kA) = (kA)^T = kA^T = kL(A).$$

Hence L is a linear transformation.   ■

EXAMPLE 4 ■ Let W be the vector space of all real-valued functions and let V be the subspace
(*Calculus Required*)   of all differentiable functions. Let $L: V \to W$ be defined by

$$L(f) = f',$$

where f' is the derivative of f. It is easy to show (Exercise 13), from the properties of differentiation, that L is a linear transformation.   ■

EXAMPLE 5 ■ Let $V = C[0, 1]$ denote the vector space of all real-valued continuous functions
(*Calculus Required*)   defined on $[0, 1]$. Let $W = R^1$. Define $L: V \to W$ by

$$L(f) = \int_0^1 f(x)\, dx.$$

It is easy to show (Exercise 14), from the properties of integration, that L is a linear transformation.   ■

EXAMPLE 6 ■ Let V be an n-dimensional vector space and $S = \{\mathbf{v}_1, \mathbf{v}_2, \ldots, \mathbf{v}_n\}$ a basis for V. If $\mathbf{v}$ is a vector in V, then

$$\mathbf{v} = c_1\mathbf{v}_1 + c_2\mathbf{v}_2 + \cdots + c_n\mathbf{v}_n,$$

where $c_1, c_2, \ldots, c_n$ are the coordinates of $\mathbf{v}$ with respect to S (see Section 4.7). We define $L: V \to R^n$ by

$$L(\mathbf{v}) = \begin{bmatrix} \mathbf{v} \end{bmatrix}_S.$$

It is easy to show (Exercise 15) that L is a linear transformation.   ■

EXAMPLE 7 ■ Let A be an $m \times n$ matrix. In Example 7 of Section 3.3 we have observed that if $L: R^n \to R^m$ is defined by

$$L(\mathbf{x}) = A\mathbf{x}$$

for $\mathbf{x}$ in R^n, then L is a linear transformation. Specific cases of this type of linear transformation have been seen in Examples 6, 8, 9, and 10 in Section 3.3.   ■

The following two theorems give some basic properties of linear transformations.

THEOREM 6.1 ■ *If $L: V \to W$ is a linear transformation, then*

$$L(c_1\mathbf{v}_1 + c_2\mathbf{v}_2 + \cdots + c_k\mathbf{v}_k) = c_1L(\mathbf{v}_1) + c_2L(\mathbf{v}_2) + \cdots + c_kL(\mathbf{v}_k)$$

for any vectors $\mathbf{v}_1, \mathbf{v}_2, \ldots, \mathbf{v}_k$ in V and any scalars $c_1, c_2, \ldots, c_k$.

Proof   Exercise T.1.   ■

THEOREM 6.2 ■ *Let $L: V \to W$ be a linear transformation. Then:*

(a) *$L(\mathbf{0}_V) = \mathbf{0}_W$, where $\mathbf{0}_V$ and $\mathbf{0}_W$ are the zero vectors in V and W, respectively.*

(b) *$L(\mathbf{u} - \mathbf{v}) = L(\mathbf{u}) - L(\mathbf{v})$.*

Proof   (a) We have

$$\mathbf{0}_V = \mathbf{0}_V + \mathbf{0}_V.$$

Then

$$L(\mathbf{0}_V) = L(\mathbf{0}_V + \mathbf{0}_V) = L(\mathbf{0}_V) + L(\mathbf{0}_V). \tag{1}$$

Adding $-L(\mathbf{0}_V)$ to both sides of Equation (1), we obtain

$$L(\mathbf{0}_V) = \mathbf{0}_W.$$

(b) Exercise T.2.   ■

A function f mapping a set V into a set W can be specified by a formula that assigns to every member of V a unique element of W. On the other hand, we can also specify a function by listing next to each member of V its assigned element of W. An example of this would be provided by listing the names of all charge account customers of a department store along with their charge account number. At first glance, it appears impossible to describe a linear transformation $L: V \to W$ of a vector space $V \neq \{\mathbf{0}\}$ into a vector space W in this latter manner, since V has infinitely many members in it. However, the following very useful theorem tells us that once we know what L does to a basis for V, then we have completely determined L. Thus, for a finite-dimensional vector space V, it is possible to describe L by giving only the images of a finite number of vectors in V.

THEOREM 6.3 ■ *Let $L: V \to W$ be a linear transformation of an n-dimensional vector space V into a vector space W. Also, let $S = \{\mathbf{v}_1, \mathbf{v}_2, \ldots, \mathbf{v}_n\}$ be a basis for V. If $\mathbf{u}$ is any vector in V, then $L(\mathbf{u})$ is completely determined by $\{L(\mathbf{v}_1), L(\mathbf{v}_2), \ldots, L(\mathbf{v}_n)\}$.*

Proof   Since $\mathbf{u}$ is in V, we can write

$$\mathbf{u} = c_1 \mathbf{v}_1 + c_2 \mathbf{v}_2 + \cdots + c_n \mathbf{v}_n, \tag{2}$$

where $c_1, c_2, \ldots, c_n$ are uniquely determined real numbers. Then

$$L(\mathbf{u}) = L(c_1 \mathbf{v}_1 + c_2 \mathbf{v}_2 + \cdots + c_n \mathbf{v}_n) = c_1 L(\mathbf{v}_1) + c_2 L(\mathbf{v}_2) + \cdots + c_n L(\mathbf{v}_n),$$

by Theorem 6.1. Thus $L(\mathbf{u})$ has been completely determined by the elements $L(\mathbf{v}_1), L(\mathbf{v}_2), \ldots, L(\mathbf{v}_n)$.   ■

It might be noted that in the proof of Theorem 6.3, the real numbers c_i determined in Equation (2) depend on the basis vectors in S. Thus, if we change S, then we may change the c_i's.

EXAMPLE 8 ■ Let $L\colon P_1 \to P_2$ be a linear transformation for which we know that

$$L(t+1) = t^2 - 1 \quad \text{and} \quad L(t-1) = t^2 + t.$$

(a) What is $L(7t+3)$?

(b) What is $L(at+b)$?

Solution

(a) First, note that $\{t+1, t-1\}$ is a basis for P_1 (verify). Next, we find that (verify)

$$7t + 3 = 5(t+1) + 2(t-1).$$

Then

$$L(7t+3) = L(5(t+1) + 2(t-1))$$

$$= 5L(t+1) + 2L(t-1)$$

$$= 5(t^2 - 1) + 2(t^2 + t) = 7t^2 + 2t - 5.$$

(b) Writing $at+b$ as a linear combination of the given basis vectors, we find that (verify)

$$at + b = \left(\frac{a+b}{2}\right)(t+1) + \left(\frac{a-b}{2}\right)(t-1).$$

Then

$$L(at+b) = L\left(\left(\frac{a+b}{2}\right)(t+1) + \left(\frac{a-b}{2}\right)(t-1)\right)$$

$$= \left(\frac{a+b}{2}\right)L(t+1) + \left(\frac{a-b}{2}\right)L(t-1)$$

$$= \left(\frac{a+b}{2}\right)(t^2 - 1) + \left(\frac{a-b}{2}\right)(t^2 + t)$$

$$= at^2 + \left(\frac{a-b}{2}\right)t - \left(\frac{a+b}{2}\right).$$

■

6.1   EXERCISES

1. Which of the following are linear transformations?

(a) $L(x, y) = (x + y, x - y)$.

(b) $L\left(\begin{bmatrix} x \\ y \\ z \end{bmatrix}\right) = \begin{bmatrix} x + 1 \\ y - z \end{bmatrix}$.

(c) $L\left(\begin{bmatrix} x \\ y \\ z \end{bmatrix}\right) = \begin{bmatrix} 1 & 2 & 3 \\ -1 & 2 & 4 \end{bmatrix} \begin{bmatrix} x \\ y \\ z \end{bmatrix}$.

2. Which of the following are linear transformations?

(a) $L(x, y, z) = (0, 0)$.

(b) $L(x, y, z) = (1, 2, -1)$.

(c) $L(x, y, z) = (x^2 + y, y - z)$.

3. Let $L: P_1 \to P_2$ be defined as indicated. Is L a linear transformation? Justify your answer.

(a) $L[p(t)] = tp(t) + p(0)$.

(b) $L[p(t)] = tp(t) + t^2$.

(c) $L(at + b) = at^2 + (a - b)t$.

4. Let $L: P_2 \to P_1$ be defined as indicated. Is L a linear transformation? Justify your answer.

(a) $L(at^2 + bt + c) = at + b + 1$.

(b) $L(at^2 + bt + c) = 2at - b$.

(c) $L(at^2 + bt + c) = (a + 2)t + (b - a)$.

5. Let $L: P_2 \to P_2$ be defined as indicated. Is L a linear transformation? Justify your answer.

(a) $L(at^2 + bt + c) = (a + 1)t^2 + (b - c)t + (a + c)$.

(b) $L(at^2 + bt + c) = at^2 + (b - c)t + (a - b)$.

(c) $L(at^2 + bt + c) = 0$.

6. Let C be a fixed $n \times n$ matrix and let $L: M_{nn} \to M_{nn}$ be defined by $L(A) = CA$. Show that L is a linear transformation.

7. Let $L: M_{22} \to M_{22}$ be defined by
$$L\left(\begin{bmatrix} a & b \\ c & d \end{bmatrix}\right) = \begin{bmatrix} b & c - d \\ c + d & 2a \end{bmatrix}.$$
Is L a linear transformation?

8. Let $L: M_{22} \to M_{22}$ be defined by
$$L\left(\begin{bmatrix} a & b \\ c & d \end{bmatrix}\right) = \begin{bmatrix} a - 1 & b + 1 \\ 2c & 3d \end{bmatrix}.$$
Is L a linear transformation?

9. Let $L: M_{22} \to R^1$ be defined by
$$L\left(\begin{bmatrix} a & b \\ c & d \end{bmatrix}\right) = a + d.$$
Is L a linear transformation?

10. Let $L: M_{22} \to R^1$ be defined by
$$L\left(\begin{bmatrix} a & b \\ c & d \end{bmatrix}\right) = a + b - c - d + 1.$$
Is L a linear transformation?

11. Consider the function $L: M_{34} \to M_{24}$ defined by
$$L(A) = \begin{bmatrix} 2 & 3 & 1 \\ 1 & 2 & -3 \end{bmatrix} A$$
for A in M_{34}.

(a) Find $L\left(\begin{bmatrix} 1 & 2 & 0 & -1 \\ 3 & 0 & 2 & 3 \\ 4 & 1 & -2 & 1 \end{bmatrix}\right)$.

(b) Show that L is a linear transformation.

12. Let $L: M_{nn} \to R^1$ be defined by $L(A) = a_{11}a_{22} \cdots a_{nn}$, for an $n \times n$ matrix $A = \begin{bmatrix} a_{ij} \end{bmatrix}$. Is L a linear transformation?

13. (**Calculus Required**). Verify that the function in Example 4 is a linear transformation.

14. (**Calculus Required**). Verify that the function in Example 5 is a linear transformation.

15. Verify that the function in Example 6 is a linear transformation.

16. Verify that the function in Example 7 is a linear transformation.

17. Let $L: R^2 \to R^2$ be a linear transformation for which we know that $L(1, 1) = (1, -2)$, $L(-1, 1) = (2, 3)$.

(a) What is $L(-1, 5)$?

(b) What is $L(a_1, a_2)$?

18. Let $L: P_2 \to P_3$ be a linear transformation for which we know that $L(1) = 1$, $L(t) = t^2$, and $L(t^2) = t^3 + t$.

(a) Find $L(2t^2 - 5t + 3)$.

(b) Find $L(at^2 + bt + c)$.

19. Let $L: P_1 \to P_1$ be a linear transformation for which we know that $L(t + 1) = 2t + 3$ and $L(t - 1) = 3t - 2$.

(a) Find $L(6t - 4)$.

(b) Find $L(at + b)$.

THEORETICAL EXERCISES

T.1. Prove Theorem 6.1.

T.2. Prove (b) of Theorem 6.2.

T.3. Show that $L: V \rightarrow W$ is a linear transformation if and only if

$$L(a\mathbf{u} + b\mathbf{v}) = aL(\mathbf{u}) + bL(\mathbf{v}),$$

for any scalars a and b and any vectors $\mathbf{u}$ and $\mathbf{v}$ in V.

T.4. Consider the function $\text{Tr}: M_{nn} \rightarrow R^1$ (the **trace**): If $A = \begin{bmatrix} a_{ij} \end{bmatrix}$ is in V, then $\text{Tr}(A) = a_{11} + a_{22} + \cdots + a_{nn}$. Show that Tr is a linear transformation (see Supplementary Exercise T.1 in Chapter 1).

T.5. Let $L: M_{nn} \rightarrow M_{nn}$ be the function defined by

$$L(A) = \begin{cases} A^{-1} & \text{if } A \text{ is nonsingular} \\ O & \text{if } A \text{ is singular} \end{cases}$$

for A in M_{nn}. Is L a linear transformation?

T.6. Let V and W be vector spaces. Show that the function $O: V \rightarrow W$ defined by $O(\mathbf{v}) = \mathbf{0}_W$ is a linear transformation, which is called the **zero linear transformation**.

T.7. Let $I: V \rightarrow V$ be defined by $I(\mathbf{v}) = \mathbf{v}$, for $\mathbf{v}$ in V. Show that I is a linear transformation, which is called the **identity operator** on V.

T.8. Let $L: V \rightarrow W$ be a linear transformation from a vector space V into a vector space W. The **image** of a subspace V_1 of V is defined as

$$L(V_1) = \{\mathbf{w} \text{ in } W \mid \mathbf{w} = L(\mathbf{v}) \text{ for some } \mathbf{v} \text{ in } V\}.$$

Show that $L(V_1)$ is a subspace of V.

T.9. Let L_1 and L_2 be linear transformations from a vector space V into a vector space W. Let $\{\mathbf{v}_1, \mathbf{v}_2, \ldots, \mathbf{v}_n\}$ be basis for V. Show that if $L_1(\mathbf{v}_i) = L_2(\mathbf{v}_i)$ for $i = 1, 2, \ldots, n$, then $L_1(\mathbf{v}) = L_2(\mathbf{v})$ for any $\mathbf{v}$ in V.

MATLAB EXERCISES

MATLAB cannot be used to show that a function between vector spaces is a linear transformation. However, MATLAB can be used to construct an example that shows that a function is not a linear transformation. The following exercises illustrate this point.

ML.1. Let $L: M_{nn} \rightarrow R^1$ be defined by $L(A) = \det(A)$.

(a) Find a pair of 2×2 matrices A and B such that

$$L(A + B) \neq L(A) + L(B).$$

Use MATLAB to do the computations. It follows that L is not a linear transformation.

(b) Find a pair of 3×3 matrices A and B such that

$$L(A + B) \neq L(A) + L(B).$$

It follows that L is not a linear transformation. Use MATLAB to do the computations.

ML.2. Let $L: M_{nn} \rightarrow R^1$ be defined by $L(A) = \text{rank } A$.

(a) Find a pair of 2×2 matrices A and B such that

$$L(A + B) \neq L(A) + L(B).$$

It follows that L is not a linear transformation. Use MATLAB to do the computations.

(b) Find a pair of 3×3 matrices A and B such that

$$L(A + B) \neq L(A) + L(B).$$

It follows that L is not a linear transformation. Use MATLAB to do the computations.

6.2 ▾ The Kernel and Range of a Linear Transformation

In this section we study special types of linear transformations; we formulate the notions of one-to-one linear transformations and onto linear transformations. We also develop methods for determining when a linear transformation is one-to-one or onto.

DEFINITION   A linear transformation $L\colon V \to W$ is said to be **one-to-one** if for all $\mathbf{v}_1$, $\mathbf{v}_2$ in V, $\mathbf{v}_1 \neq \mathbf{v}_2$ implies that $L(\mathbf{v}_1) \neq L(\mathbf{v}_2)$. An equivalent statement is that L is one-to-one if for all $\mathbf{v}_1$, $\mathbf{v}_2$ in V, $L(\mathbf{v}_1) = L(\mathbf{v}_2)$ implies that $\mathbf{v}_1 = \mathbf{v}_2$.

This definition says that L is one-to-one if $L(\mathbf{v}_1)$ and $L(\mathbf{v}_2)$ are distinct whenever $\mathbf{v}_1$ and $\mathbf{v}_2$ are distinct (Figure 6.1).

FIGURE 6.1

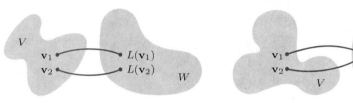

(a) L is one-to-one.   (b) L is not one-to-one.

EXAMPLE 1 ■ Let $L\colon R^2 \to R^2$ be defined by

$$L(x, y) = (x + y, x - y).$$

To determine whether L is one-to-one, we let

$$\mathbf{v}_1 = (a_1, a_2) \quad \text{and} \quad \mathbf{v}_2 = (b_1, b_2).$$

Then if

$$L(\mathbf{v}_1) = L(\mathbf{v}_2),$$

we have

$$a_1 + a_2 = b_1 + b_2$$
$$a_1 - a_2 = b_1 - b_2.$$

Adding these equations, we obtain $2a_1 = 2b_1$, or $a_1 = b_1$, which implies that $a_2 = b_2$. Hence, $\mathbf{v}_1 = \mathbf{v}_2$ and L is one-to-one.   ■

EXAMPLE 2 ■ Let $L\colon R^3 \to R^2$ be the linear transformation defined in Example 1 of Section 3.3 (the projection function) by

$$L(x, y, z) = (x, y).$$

Since $(1, 3, 3) \neq (1, 3, -2)$ but

$$L(1, 3, 3) = L(1, 3, -2) = (1, 3),$$

we conclude that L is not one-to-one.   ■

We shall now develop some more efficient ways of determining whether or not a linear transformation is one-to-one.

DEFINITION
Let $L\colon V \to W$ be a linear transformation. The **kernel** of L, ker L, is the subset of V consisting of all vectors $\mathbf{v}$ such that $L(\mathbf{v}) = \mathbf{0}_W$.

We observe that property (a) of Theorem 6.2 in Section 6.1 assures us that ker L is never an empty set, since $\mathbf{0}_V$ is in ker L.

EXAMPLE 3 ■ Let $L\colon R^3 \to R^2$ be as defined in Example 2. The vector $(0, 0, 2)$ is in ker L, since $L(0, 0, 2) = (0, 0)$. However, the vector $(2, -3, 4)$ is not in ker L, since $L(2, -3, 4) = (2, -3)$. To find ker L, we must determine all $\mathbf{x}$ in R^3 so that $L(\mathbf{x}) = \mathbf{0}$. That is, we seek $\mathbf{x} = (x_1, x_2, x_3)$ so that

$$L(\mathbf{x}) = L(x_1, x_2, x_3) = \mathbf{0} = (0, 0).$$

However, $L(\mathbf{x}) = (x_1, x_2)$. Thus $(x_1, x_2) = (0, 0)$, so $x_1 = 0$, $x_2 = 0$, and x_3 can be any real number. Hence, ker L consists of all vectors in R^3 of the form $(0, 0, r)$, where r is any real number. It is clear that ker L consists of the z-axis in three-dimensional space R^3. ■

EXAMPLE 4 ■ If L is as defined in Example 1, then ker L consists of all vectors $\mathbf{x}$ in R^2 such that $L(\mathbf{x}) = \mathbf{0}$. Thus we must solve the linear system

$$x + y = 0$$
$$x - y = 0$$

for x and y. The only solution is $\mathbf{x} = \mathbf{0}$, so ker $L = \{\mathbf{0}\}$. ■

EXAMPLE 5 ■ If $L\colon R^4 \to R^2$ is defined by

$$L\left(\begin{bmatrix} x \\ y \\ z \\ w \end{bmatrix}\right) = \begin{bmatrix} x + y \\ z + w \end{bmatrix},$$

then ker L consists of all vectors $\mathbf{u}$ in R^4 such that $L(\mathbf{u}) = \mathbf{0}$. This leads to the linear system

$$x + y \qquad\; = 0$$
$$z + w = 0.$$

Thus ker L consists of all vectors of the form

$$\begin{bmatrix} r \\ -r \\ s \\ -s \end{bmatrix},$$

where r and s are any real numbers. ■

THEOREM 6.4 ■ *If $L\colon V \to W$ is a linear transformation, then ker L is a subspace of V.*

Proof First, observe that $\ker L$ is not an empty set, since $\mathbf{0}_V$ is in $\ker L$. Also, let $\mathbf{u}$ and $\mathbf{v}$ be in $\ker L$. Then since L is a linear transformation,

$$L(\mathbf{u} + \mathbf{v}) = L(\mathbf{u}) + L(\mathbf{v}) = \mathbf{0}_W + \mathbf{0}_W = \mathbf{0}_W,$$

so $\mathbf{u} + \mathbf{v}$ is in $\ker L$. Also, if c is a scalar, then since L is a linear transformation,

$$L(c\mathbf{u}) = cL(\mathbf{u}) = c\mathbf{0}_W = \mathbf{0}_W,$$

so $c\mathbf{u}$ is in $\ker L$. Hence $\ker L$ is a subspace of V. ∎

EXAMPLE 6 ■ If L is as in Example 1, then $\ker L$ is the subspace $\{\mathbf{0}\}$; its dimension is zero.
∎

EXAMPLE 7 ■ If L is as in Example 2, then a basis for $\ker L$ is

$$\{(0, 0, 1)\}$$

and $\dim(\ker L) = 1$. Thus $\ker L$ consists of the z-axis in three-dimensional space R^3. ∎

EXAMPLE 8 ■ If L is as in Example 5, then a basis for $\ker L$ consists of the vectors

$$\begin{bmatrix} 1 \\ -1 \\ 0 \\ 0 \end{bmatrix} \quad \text{and} \quad \begin{bmatrix} 0 \\ 0 \\ 1 \\ -1 \end{bmatrix};$$

thus $\dim(\ker L) = 2$. ∎

If $L\colon R^n \to R^m$ is a linear transformation defined by $L(\mathbf{x}) = A\mathbf{x}$, where A is an $m \times n$ matrix, then the kernel of L is the solution space of the homogeneous system $A\mathbf{x} = \mathbf{0}$.

An examination of the elements in $\ker L$ allows us to decide whether L is or is not one-to-one.

THEOREM 6.5 ■ *A linear transformation* $L\colon V \to W$ *is one-to-one if and only if* $\ker L = \{\mathbf{0}_V\}$.

Proof Let L be one-to-one. We show that $\ker L = \{\mathbf{0}_V\}$. Let $\mathbf{x}$ be in $\ker L$. Then $L(\mathbf{x}) = \mathbf{0}_W$. Also, we already know that $L(\mathbf{0}_V) = \mathbf{0}_W$. Thus $L(\mathbf{x}) = L(\mathbf{0}_V)$. Since L is one-to-one, we conclude that $\mathbf{x} = \mathbf{0}_V$. Hence $\ker L = \{\mathbf{0}_V\}$.

Conversely, suppose that $\ker L = \{\mathbf{0}_V\}$. We wish to show that L is one-to-one. Assume that $L(\mathbf{u}) = L(\mathbf{v})$, for $\mathbf{u}$ and $\mathbf{v}$ in V. Then

$$L(\mathbf{u}) - L(\mathbf{v}) = \mathbf{0}_W,$$

so by Theorem 6.2, $L(\mathbf{u} - \mathbf{v}) = \mathbf{0}_W$, which means that $\mathbf{u} - \mathbf{v}$ is in $\ker L$. Therefore, $\mathbf{u} - \mathbf{v} = \mathbf{0}_V$, so $\mathbf{u} = \mathbf{v}$. Thus L is one-to-one. ∎

Note that we can also state Theorem 6.5 as: *L is one-to-one if and only if* $\dim(\ker L) = 0$.

The proof of Theorem 6.5 has also established the following result, which we state as Corollary 6.1.

COROLLARY 6.1 ■ *Any two solutions to* $L(\mathbf{x}) = \mathbf{b}$ *differ by an element of the kernel of L.*

Proof   Exercise T.1.   ■

EXAMPLE 9 ■ The linear transformation in Example 1 is one-to-one; the one in Example 2 is not. ■

In Section 6.3 we shall prove that for every linear transformation $L \colon R^n \to R^m$, we can find a unique $m \times n$ matrix A so that if $\mathbf{x}$ is in R^n, then $L(\mathbf{x}) = A\mathbf{x}$. It then follows that to find $\ker L$, we need to find the solution space of the homogeneous system $A\mathbf{x} = \mathbf{0}$. Hence to find $\ker L$ we need only use techniques with which we are already familiar.

DEFINITION   If $L \colon V \to W$ is a linear transformation, then the **range** of L, denoted by range L, is the set of all vectors in W that are images, under L, of vectors in V. Thus a vector $\mathbf{w}$ is in range L if we can find some vector $\mathbf{v}$ in V such that $L(\mathbf{v}) = \mathbf{w}$. If range $L = W$, we say that L is **onto**.

THEOREM 6.6 ■ *If* $L \colon V \to W$ *is a linear transformation, then* range L *is a subspace of* W.

Proof   First, observe that range L is not an empty set, since $\mathbf{0}_W = L(\mathbf{0}_V)$, so $\mathbf{0}_W$ is in range L. Let $\mathbf{w}_1$ and $\mathbf{w}_2$ be in range L. Then $\mathbf{w}_1 = L(\mathbf{v}_1)$ and $\mathbf{w}_2 = L(\mathbf{v}_2)$ for some $\mathbf{v}_1$ and $\mathbf{v}_2$ in V. Now

$$\mathbf{w}_1 + \mathbf{w}_2 = L(\mathbf{v}_1) + L(\mathbf{v}_2) = L(\mathbf{v}_1 + \mathbf{v}_2),$$

which implies that $\mathbf{w}_1 + \mathbf{w}_2$ is in range L. Also, if c is a scalar, then $c\mathbf{w}_1 = cL(\mathbf{v}_1) = L(c\mathbf{v}_1)$, so $c\mathbf{w}_1$ is in range L. Hence range L is a subspace of W.   ■

EXAMPLE 10 ■ Let L be the linear transformation defined in Example 2. To find out whether L is onto, we choose any vector $\mathbf{y} = (y_1, y_2)$ in R^2 and seek a vector $\mathbf{x} = (x_1, x_2, x_3)$ in R^3 such that $L(\mathbf{x}) = \mathbf{y}$. Since $L(\mathbf{x}) = (x_1, x_2)$, we find that if $x_1 = y_1$ and $x_2 = y_2$, then $L(\mathbf{x}) = \mathbf{y}$. Therefore, L is onto and the dimension of range L is 2. ■

EXAMPLE 11 ■ Let $L \colon R^3 \to R^3$ be defined by

$$L\left(\begin{bmatrix} a_1 \\ a_2 \\ a_3 \end{bmatrix}\right) = \begin{bmatrix} 1 & 0 & 1 \\ 1 & 1 & 2 \\ 2 & 1 & 3 \end{bmatrix} \begin{bmatrix} a_1 \\ a_2 \\ a_3 \end{bmatrix}.$$

(a) Is L onto?

(b) Find a basis for range L.

(c) Find ker L.

(d) Is L one-to-one?

Solution (a) Given any

$$\mathbf{w} = \begin{bmatrix} a \\ b \\ c \end{bmatrix}$$

in R^3, where a, b, and c are any real numbers, can we find

$$\mathbf{v} = \begin{bmatrix} a_1 \\ a_2 \\ a_3 \end{bmatrix}$$

so that $L(\mathbf{v}) = \mathbf{w}$? We seek a solution to the linear system

$$\begin{bmatrix} 1 & 0 & 1 \\ 1 & 1 & 2 \\ 2 & 1 & 3 \end{bmatrix} \begin{bmatrix} a_1 \\ a_2 \\ a_3 \end{bmatrix} = \begin{bmatrix} a \\ b \\ c \end{bmatrix}$$

and we find the reduced row echelon form of the augmented matrix to be (verify)

$$\begin{bmatrix} 1 & 0 & 1 & \vdots & a \\ 0 & 1 & 1 & \vdots & b - a \\ 0 & 0 & 0 & \vdots & c - b - a \end{bmatrix}.$$

Thus a solution exists only for $c - b - a = 0$, so L is not onto.

(b) To find a basis for range L, we note that

$$L\left(\begin{bmatrix} a_1 \\ a_2 \\ a_3 \end{bmatrix}\right) = \begin{bmatrix} 1 & 0 & 1 \\ 1 & 1 & 2 \\ 2 & 1 & 3 \end{bmatrix} \begin{bmatrix} a_1 \\ a_2 \\ a_3 \end{bmatrix} = \begin{bmatrix} a_1 + a_3 \\ a_1 + a_2 + 2a_3 \\ 2a_1 + a_2 + 3a_3 \end{bmatrix}$$

$$= a_1 \begin{bmatrix} 1 \\ 1 \\ 2 \end{bmatrix} + a_2 \begin{bmatrix} 0 \\ 1 \\ 1 \end{bmatrix} + a_3 \begin{bmatrix} 1 \\ 2 \\ 3 \end{bmatrix}.$$

This means that

$$\left\{ \begin{bmatrix} 1 \\ 1 \\ 2 \end{bmatrix}, \begin{bmatrix} 0 \\ 1 \\ 1 \end{bmatrix}, \begin{bmatrix} 1 \\ 2 \\ 3 \end{bmatrix} \right\}$$

spans range L. That is, range L is the subspace of R^3 spanned by the columns of the matrix defining L.

The first two vectors in this set are linearly independent, since they are not constant multiples of each other. The third vector is the sum of the first two. Therefore, the first two vectors form a basis for range L, and $\dim(\text{range } L) = 2$.

(c) To find ker L, we wish to find all $\mathbf{v}$ in R^3 so that $L(\mathbf{v}) = \mathbf{0}_{R^3}$. Solving the resulting homogeneous system, we find (verify) that $a_1 = -a_3$ and $a_2 = -a_3$. Thus ker L consists of all vectors of the form

$$\begin{bmatrix} -a \\ -a \\ a \end{bmatrix} = a \begin{bmatrix} -1 \\ -1 \\ 1 \end{bmatrix},$$

where a is any real number. Moreover, $\dim(\ker L) = 1$.

(d) Since ker $L \neq \{\mathbf{0}_{R^3}\}$, it follows from Theorem 6.5 that L is not one-to-one.

■

The problem of finding a basis for ker L always reduces to the problem of finding a basis for the solution space of a homogeneous system; this latter problem has been solved in Example 1 of Section 4.5.

If range L is a subspace of R^m, then a basis for range L can be obtained by the method discussed in the alternate constructive proof of Theorem 4.6 or by the procedure given in Section 4.6. Both approaches are illustrated in the next example.

EXAMPLE 12 ■ Let $L\colon R^4 \to R^3$ be defined by

$$L(a_1, a_2, a_3, a_4) = (a_1 + a_2, a_3 + a_4, a_1 + a_3).$$

Find a basis for range L.

Solution We have

$$L(a_1, a_2, a_3, a_4) = a_1(1, 0, 1) + a_2(1, 0, 0) + a_3(0, 1, 1) + a_4(0, 1, 0).$$

Thus

$$S = \{(1, 0, 1), (1, 0, 0), (0, 1, 1), (0, 1, 0)\}$$

spans range L. To find a subset of S that is a basis for range L, we proceed as in Theorem 4.6 by first writing

$$a_1(1, 0, 1) + a_2(1, 0, 0) + a_3(0, 1, 1) + a_4(0, 1, 0) = (0, 0, 0).$$

The reduced row echelon form of the augmented matrix of this homogeneous system is (verify)

$$\begin{bmatrix} 1 & 0 & 0 & -1 & \vdots & 0 \\ 0 & 1 & 0 & 1 & \vdots & 0 \\ 0 & 0 & 1 & 1 & \vdots & 0 \end{bmatrix}.$$

Since the leading 1's appear in columns 1, 2, and 3, we conclude that the first three vectors in S form a basis for range L. Thus

$$\{(1, 0, 1), (1, 0, 0), (0, 1, 1)\}$$

is a basis for range L.

Alternatively, we may proceed as in Section 4.6 to form the matrix whose rows are the given vectors

$$\begin{bmatrix} 1 & 0 & 1 \\ 1 & 0 & 0 \\ 0 & 1 & 1 \\ 0 & 1 & 0 \end{bmatrix}.$$

Transforming this matrix to reduced row echelon form, we obtain (verify)

$$\begin{bmatrix} 1 & 0 & 0 \\ 0 & 1 & 0 \\ 0 & 0 & 1 \\ 0 & 0 & 0 \end{bmatrix}.$$

Hence $\{(1,0,0),(0,1,0),(0,0,1)\}$ is a basis for range L. ∎

To determine if a linear transformation is one-to-one or onto, we must solve a linear system. This is one further demonstration of the frequency with which linear systems must be solved to answer many questions in linear algebra. Finally, from Example 11 we saw that

$$\dim(\ker L) + \dim(\text{range } L) = \dim(\text{domain } L).$$

This very important result is always true and we now prove it in the following theorem.

THEOREM 6.7 ∎ *If $L\colon V \to W$ is a linear transformation of an n-dimensional vector space V into a vector space W, then*

$$\dim(\ker L) + \dim(\text{range } L) = \dim V. \tag{1}$$

Proof Let $k = \dim(\ker L)$. If $k = n$, then $\ker L = V$ (Exercise T.7, Section 4.4), which implies that $L(\mathbf{v}) = \mathbf{0}_W$ for every $\mathbf{v}$ in V. Hence range $L = \{\mathbf{0}_W\}$, $\dim(\text{range } L) = 0$, and the conclusion holds. Next, suppose that $1 \le k < n$. We shall prove that $\dim(\text{range } L) = n - k$. Let $\{\mathbf{v}_1, \mathbf{v}_2, \ldots, \mathbf{v}_k\}$ be a basis for $\ker L$. By Theorem 4.8 we can extend this basis to a basis

$$S = \{\mathbf{v}_1, \mathbf{v}_2, \ldots, \mathbf{v}_k, \mathbf{v}_{k+1}, \ldots, \mathbf{v}_n\}$$

for V. We prove that the set

$$T = \{L(\mathbf{v}_{k+1}), L(\mathbf{v}_{k+2}), \ldots, L(\mathbf{v}_n)\}$$

is a basis for range L.

First, we show that T spans range L. Let $\mathbf{w}$ be any vector in range L. Then $\mathbf{w} = L(\mathbf{v})$ for some $\mathbf{v}$ in V. Since S is a basis for V, we can find a unique set of real numbers $a_1, a_2, \ldots, a_n$ such that $\mathbf{v} = a_1\mathbf{v}_1 + a_2\mathbf{v}_2 + \cdots + a_n\mathbf{v}_n$. Then

$$
\begin{aligned}
\mathbf{w} &= L(\mathbf{v}) \\
&= L(a_1\mathbf{v}_1 + a_2\mathbf{v}_2 + \cdots + a_k\mathbf{v}_k + a_{k+1}\mathbf{v}_{k+1} + \cdots + a_n\mathbf{v}_n) \\
&= a_1 L(\mathbf{v}_1) + a_2 L(\mathbf{v}_2) + \cdots + a_k L(\mathbf{v}_k) + a_{k+1} L(\mathbf{v}_{k+1}) + \cdots + a_n L(\mathbf{v}_n) \\
&= a_{k+1} L(\mathbf{v}_{k+1}) + \cdots + a_n L(\mathbf{v}_n)
\end{aligned}
$$

because $\mathbf{v}_1, \mathbf{v}_2, \ldots, \mathbf{v}_k$ are in ker L. Hence T spans range L.

Now we show that T is linearly independent. Suppose that

$$a_{k+1}L(\mathbf{v}_{k+1}) + a_{k+2}L(\mathbf{v}_{k+2}) + \cdots + a_n L(\mathbf{v}_n) = \mathbf{0}_W.$$

Then by (b) of Theorem 6.2

$$L(a_{k+1}\mathbf{v}_{k+1} + a_{k+2}\mathbf{v}_{k+2} + \cdots + a_n\mathbf{v}_n) = \mathbf{0}_W.$$

Hence the vector $a_{k+1}\mathbf{v}_{k+1} + a_{k+2}\mathbf{v}_{k+2} + \cdots + a_n\mathbf{v}_n$ is in ker L, and we can write

$$a_{k+1}\mathbf{v}_{k+1} + a_{k+2}\mathbf{v}_{k+2} + \cdots + a_n\mathbf{v}_n = b_1\mathbf{v}_1 + b_2\mathbf{v}_2 + \cdots + b_k\mathbf{v}_k,$$

where $b_1, b_2, \ldots, b_k$ are uniquely determined real numbers. We then have

$$b_1\mathbf{v}_1 + b_2\mathbf{v}_2 + \cdots + b_k\mathbf{v}_k - a_{k+1}\mathbf{v}_{k+1} - a_{k+2}\mathbf{v}_{k+2} - \cdots - a_n\mathbf{v}_n = \mathbf{0}_V.$$

Since S is linearly independent, we find that

$$b_1 = b_2 = \cdots = b_k = a_{k+1} = a_{k+2} = \cdots = a_n = 0.$$

Hence T is linearly independent and forms a basis for range L.

If $k = 0$, then ker L has no basis; we let $\{\mathbf{v}_1, \mathbf{v}_2, \ldots, \mathbf{v}_n\}$ be a basis for V. The proof now proceeds as above.   ∎

The dimension of ker L is also called the **nullity** of L, and the dimension of range L is called the **rank** of L. With this terminology the conclusion of Theorem 6.7 is very similar to that of Theorem 4.12. This is not a coincidence, since in the next section we shall show how to attach a unique $m \times n$ matrix to L, whose properties reflect those of L.

We have seen that a linear transformation may be one-to-one and not onto or onto and not one-to-one. However, the following corollary shows that each of these properties implies the other if the vector spaces V and W have the same dimensions.

COROLLARY 6.2 ∎ *Let $L\colon V \to W$ be a linear transformation and let $\dim V = \dim W$.*

(a) *If L is one-to-one, then it is onto.*

(b) *If L is onto, then it is one-to-one.*

Proof   Exercise T.2.   ∎

EXAMPLE 13 ∎ Let $L\colon P_2 \to P_2$ be the linear transformation defined by

$$L(at^2 + bt + c) = (a + 2b)t + (b + c).$$

(a) Is $-4t^2 + 2t - 2$ in ker L?

(b) Is $t^2 + 2t + 1$ in range L?

(c) Find a basis for ker L.

(d) Is L one-to-one?

(e) Find a basis for range L.

(f) Is L onto?

(g) Verify Theorem 6.7.

Solution

(a) Since
$$L(-4t^2 + 2t - 2) = (-4 + 2 \cdot 2)t + (-2 + 2) = 0,$$
we conclude that $-4t^2 + 2t - 2$ is in ker L.

(b) The vector $t^2 + 2t + 1$ is in range L if we can find a vector $at^2 + bt + c$ in P_2 such that
$$L(at^2 + bt + c) = t^2 + 2t + 1.$$
Since $L(at^2 + bt + c) = (a + 2b)t + (b + c)$, we have
$$(a + 2b)t + (b + c) = t^2 + 2t + 1.$$
The left side of this equation can also be written as $0t^2 + (a + 2b)t + (b + c)$. Thus
$$0t^2 + (a + 2b)t + (b + c) = t^2 + 2t + 1.$$
We must then have
$$0 = 1$$
$$a + 2b = 2$$
$$b + c = 1.$$

Since this linear system has no solution, the given vector is not in range L.

(c) The vector $at^2 + bt + c$ is in ker L if
$$L(at^2 + bt + c) = \mathbf{0},$$
that is, if
$$(a + 2b)t + (b + c) = 0.$$
Then
$$a + 2b \quad\ = 0$$
$$b + c = 0.$$

Transforming the augmented matrix of this linear system to reduced row echelon form, we find (verify) that a basis for the solution space is
$$\left\{ \begin{bmatrix} 2 \\ -1 \\ 1 \end{bmatrix} \right\},$$
so a basis for ker L is $\{2t^2 - t + 1\}$.

(d) Since ker L does not consist only of the zero vector, L is not one-to-one.

(e) Every vector in range L has the form
$$(a + 2b)t + (b + c),$$
so the vectors t and 1 span range L. Since these vectors are also linearly independent, they form a basis for range L.

(f) The dimension of P_2 is 3, while range L is a subspace of P_2 of dimension 2, so range $L \neq P_2$. Hence, L is not onto.

(g) From (c), $\dim(\ker L) = 1$, and from (e), $\dim(\operatorname{range} L) = 2$, so

$$3 = \dim P_2 = \dim(\ker L) + \dim(\operatorname{range} L). \qquad \blacksquare$$

If $L \colon R^n \to R^n$ is a linear transformation defined by $L(\mathbf{x}) = A\mathbf{x}$, where A is an $n \times n$ matrix, then using Theorem 6.7, Equation (1), and Corollary 4.2, we can show (Exercise T.4) that L is one-to-one if and only if $\det(A) \neq 0$.

We now make one final remark for a linear system $A\mathbf{x} = \mathbf{b}$, where A is $n \times n$. We again consider the linear transformation $L \colon R^n \to R^n$ defined by $L(\mathbf{x}) = A\mathbf{x}$, for $\mathbf{x}$ in R^n. If A is a nonsingular matrix, then $\dim(\operatorname{range} L) = \operatorname{rank} A = n$, so $\dim(\ker L) = 0$. Thus L is one-to-one and hence onto. This means that the given linear system has a unique solution (of course, we already knew this result from other considerations). However, if A is singular, then $\operatorname{rank} A < n$. This means that $\dim(\ker L) = n - \operatorname{rank} A > 0$, so L is not one-to-one and not onto. Therefore, there exists a vector $\mathbf{b}$ in R^n, for which the system $A\mathbf{x} = \mathbf{b}$ has no solution. Moreover, since A is singular, $A\mathbf{x} = \mathbf{0}$ has a nontrivial solution $\mathbf{x}_0$. If $A\mathbf{x} = \mathbf{b}$ has a solution $\mathbf{y}$, then $\mathbf{x}_0 + \mathbf{y}$ is a solution to $A\mathbf{x} = \mathbf{b}$ (verify). Thus, for A singular, if a solution to $A\mathbf{x} = \mathbf{b}$ exists, then it is not unique.

6.2 EXERCISES

1. Let $L \colon R^2 \to R^2$ be the linear transformation defined by $L(a_1, a_2) = (a_1, 0)$.

   (a) Is $(0, 2)$ in $\ker L$?

   (b) Is $(2, 2)$ in $\ker L$?

   (c) Is $(3, 0)$ in range L?

   (d) Is $(3, 2)$ in range L?

   (e) Find $\ker L$.

   (f) Find range L.

2. Let $L \colon R^2 \to R^2$ be the linear transformation defined by

   $$L\left(\begin{bmatrix} a_1 \\ a_2 \end{bmatrix}\right) = \begin{bmatrix} 1 & 2 \\ 2 & 4 \end{bmatrix} \begin{bmatrix} a_1 \\ a_2 \end{bmatrix}.$$

   (a) Is $\begin{bmatrix} 1 \\ 2 \end{bmatrix}$ in $\ker L$?

   (b) Is $\begin{bmatrix} 2 \\ -1 \end{bmatrix}$ in $\ker L$?

   (c) Is $\begin{bmatrix} 3 \\ 6 \end{bmatrix}$ in range L?

   (d) Is $\begin{bmatrix} 2 \\ 3 \end{bmatrix}$ in range L?

   (e) Find $\ker L$.

   (f) Find a set of vectors spanning range L.

3. Let $L \colon R^2 \to R^3$ be defined by

   $$L(x, y) = (x, x + y, y).$$

   (a) Find $\ker L$.

   (b) Is L one-to-one?

   (c) Is L onto?

4. Let $L \colon R^4 \to R^3$ be defined by

   $$L(x, y, z, w) = (x + y, z + w, x + z).$$

   (a) Find a basis for $\ker L$.

   (b) Find a basis for range L.

   (c) Verify Theorem 6.7.

5. Let $L \colon R^5 \to R^4$ be defined by

   $$L\left(\begin{bmatrix} x_1 \\ x_2 \\ x_3 \\ x_4 \\ x_5 \end{bmatrix}\right) = \begin{bmatrix} 1 & 0 & -1 & 3 & -1 \\ 1 & 0 & 0 & 2 & -1 \\ 2 & 0 & -1 & 5 & -1 \\ 0 & 0 & -1 & 1 & 0 \end{bmatrix} \begin{bmatrix} x_1 \\ x_2 \\ x_3 \\ x_4 \\ x_5 \end{bmatrix}.$$

   (a) Find a basis for $\ker L$.

   (b) Find a basis for range L.

   (c) Verify Theorem 6.7.

6. Let $L: R^3 \to R^3$ be defined by

$$L\left(\begin{bmatrix} x \\ y \\ z \end{bmatrix}\right) = \begin{bmatrix} 4 & 2 & 2 \\ 2 & 3 & -1 \\ -1 & 1 & -2 \end{bmatrix} \begin{bmatrix} x \\ y \\ z \end{bmatrix}.$$

(a) Is L one-to-one?

(b) Find the dimension of range L.

7. Let $L: R^4 \to R^3$ be defined by

$$L\left(\begin{bmatrix} x \\ y \\ z \\ w \end{bmatrix}\right) = \begin{bmatrix} x+y \\ y-z \\ z-w \end{bmatrix}.$$

(a) Is L onto?

(b) Find the dimension of $\ker L$.

(c) Verify Theorem 6.7.

8. Let $L: R^3 \to R^3$ be defined by

$$L(x, y, z) = (x - y, x + 2y, z).$$

(a) Find a basis for $\ker L$.

(b) Find a basis for range L.

(c) Verify Theorem 6.7.

9. Verify Theorem 6.7 for the following linear transformations.

(a) $L(x, y) = (x + y, y)$.

(b) $L\left(\begin{bmatrix} x \\ y \\ z \end{bmatrix}\right) = \begin{bmatrix} 4 & -1 & -1 \\ 2 & 2 & 3 \\ 2 & -3 & -4 \end{bmatrix} \begin{bmatrix} x \\ y \\ z \end{bmatrix}.$

(c) $L(x, y, z) = (x + y - z, x + y, y + z)$.

10. Let $L: R^4 \to R^4$ be defined by

$$L\left(\begin{bmatrix} x \\ y \\ z \\ w \end{bmatrix}\right) = \begin{bmatrix} 1 & 2 & 1 & 3 \\ 2 & 1 & -1 & 2 \\ 1 & 0 & 0 & -1 \\ 4 & 1 & -1 & 0 \end{bmatrix} \begin{bmatrix} x \\ y \\ z \\ w \end{bmatrix}.$$

(a) Find a basis for $\ker L$.

(b) Find a basis for range L.

(c) Verify Theorem 6.7.

11. Let $L: P_2 \to P_2$ be the linear transformation defined by $L(at^2 + bt + c) = (a + c)t^2 + (b + c)t$.

(a) Is $t^2 - t - 1$ in $\ker L$?

(b) Is $t^2 + t - 1$ in $\ker L$?

(c) Is $2t^2 - t$ in range L?

(d) Is $t^2 - t + 2$ in range L?

(e) Find a basis for $\ker L$.

(f) Find a basis for range L.

12. Let $L: P_3 \to P_3$ be the linear transformation defined by

$$L(at^3 + bt^2 + ct + d) = (a - b)t^3 + (c - d)t.$$

(a) Is $t^3 + t^2 + t - 1$ in $\ker L$?

(b) Is $t^3 - t^2 + t - 1$ in $\ker L$?

(c) Is $3t^3 + t$ in range L?

(d) Is $3t^3 - t^2$ in range L?

(e) Find a basis for $\ker L$.

(f) Find a basis for range L.

13. Let $L: M_{22} \to M_{22}$ be the linear transformation defined by

$$L\left(\begin{bmatrix} a & b \\ c & d \end{bmatrix}\right) = \begin{bmatrix} a+b & b+c \\ a+d & b+d \end{bmatrix}.$$

(a) Find a basis for $\ker L$.

(b) Find a basis for range L.

14. Let $L: P_2 \to R^2$ be the linear transformation defined by $L(at^2 + bt + c) = (a, b)$.

(a) Find a basis for $\ker L$.

(b) Find a basis for range L.

15. Let $L: M_{22} \to M_{22}$ be the linear transformation defined by

$$L(\mathbf{v}) = \begin{bmatrix} 1 & 2 \\ 1 & 1 \end{bmatrix} \mathbf{v} - \mathbf{v} \begin{bmatrix} 1 & 2 \\ 1 & 1 \end{bmatrix}.$$

(a) Find a basis for $\ker L$.

(b) Find a basis for range L.

16. Let $L: M_{22} \to M_{22}$ be the linear transformation defined by $L(A) = A^T$.

(a) Find a basis for $\ker L$.

(b) Find a basis for range L.

17. (**Calculus Required**). Let $L: P_2 \to P_1$ be the linear transformation defined by

$$L[p(t)] = p'(t).$$

(a) Find a basis for $\ker L$.

(b) Find a basis for range L.

18. (**Calculus Required**). Let $L: P_2 \to R^1$ be the linear transformation defined by

$$L[p(t)] = \int_0^1 p(t)\, dt.$$

(a) Find a basis for $\ker L$.

(b) Find a basis for range L.

19. Let $L: R^4 \to R^6$ be a linear transformation.

(a) If $\dim(\ker L) = 2$, what is $\dim(\text{range } L)$?

(b) If $\dim(\text{range } L) = 3$, what is $\dim(\ker L)$?

20. Let $L: V \to R^5$ be a linear transformation.

(a) If L is onto and $\dim(\ker L) = 2$, what is $\dim V$?

(b) If L is one-to-one and onto, what is $\dim V$?

THEORETICAL EXERCISES ■

T.1. Prove Corollary 6.1.

T.2. Prove Corollary 6.2.

T.3. Let A be an $m \times n$ matrix and let $L: R^n \to R^m$ be defined by $L(\mathbf{x}) = A\mathbf{x}$ for $\mathbf{x}$ in R^n. Show that the column space of A is the range of L.

T.4. Let $L: R^n \to R^n$ be a linear transformation defined by $L(\mathbf{x}) = A\mathbf{x}$, where A is an $n \times n$ matrix. Show that L is one-to-one if and only if $\det(A) \neq 0$. [*Hint*: Use Theorem 6.7, Equation (1), and Corollary 4.2.]

T.5. Let $L: V \to W$ be a linear transformation. If $\{\mathbf{v}_1, \mathbf{v}_2, \ldots, \mathbf{v}_k\}$ spans V, show that $\{L(\mathbf{v}_1), L(\mathbf{v}_2), \ldots, L(\mathbf{v}_k)\}$ spans range L.

T.6. Let $L: V \to W$ be a linear transformation.
   (a) Show that $\dim(\operatorname{range} L) \leq \dim V$.
   (b) Show that if L is onto, then $\dim W \leq \dim V$.

T.7. Let $L: V \to W$ be a linear transformation, and let $S = \{\mathbf{v}_1, \mathbf{v}_2, \ldots, \mathbf{v}_n\}$ be a set of vectors in V. Show that if $T = \{L(\mathbf{v}_1), L(\mathbf{v}_2), \ldots, L(\mathbf{v}_n)\}$ is linearly independent, then so is S. (*Hint*: Assume that S is linearly dependent. What can we say about T?)

T.8. Let $L: V \to W$ be a linear transformation. Show that L is one-to-one if and only if $\dim(\operatorname{range} L) = \dim V$.

T.9. Let $L: V \to W$ be a linear transformation. Show that L is one-to-one if and only if the image of every linearly independent set of vectors in V is a linearly independent set of vectors in W.

T.10. Let $L: V \to W$ be a linear transformation, and let $\dim V = \dim W$. Show that L is one-to-one if and only if the image under L of a basis for V is a basis for W.

T.11. Let V be an n-dimensional vector space and $S = \{\mathbf{v}_1, \mathbf{v}_2, \ldots, \mathbf{v}_n\}$ be a basis for V. Let $L: V \to R^n$ be defined by $L(\mathbf{v}) = \begin{bmatrix} \mathbf{v} \end{bmatrix}_S$. Show that
   (a) L is a linear transformation.
   (b) L is one-to-one.
   (c) L is onto.

MATLAB EXERCISES ■

In order to use MATLAB in this section, you should first read Section 10.8. Find a basis for the kernel and range of the linear transformation $L(\mathbf{x}) = A\mathbf{x}$ for each of the following matrices A.

ML.1. $A = \begin{bmatrix} 1 & 2 & 5 & 5 \\ -2 & -3 & -8 & -7 \end{bmatrix}$.

ML.2. $A = \begin{bmatrix} -3 & 2 & -7 \\ 2 & -1 & 4 \\ 2 & -2 & 6 \end{bmatrix}$.

ML.3. $A = \begin{bmatrix} 3 & 3 & -3 & 1 & 11 \\ -4 & -4 & 7 & -2 & -19 \\ 2 & 2 & -3 & 1 & 9 \end{bmatrix}$.

6.3 ▼ The Matrix of a Linear Transformation

We have shown in Theorem 3.8 that if $L: R^n \to R^m$ is a linear transformation, then there is a unique $m \times n$ matrix A so that $L(\mathbf{x}) = A\mathbf{x}$ for $\mathbf{x}$ in R^n. In this section we generalize this result for a linear transformation $L: V \to W$ of a finite-dimensional vector space V into a finite-dimensional vector space W.

The Matrix of a Linear Transformation

THEOREM 6.8 ■ *Let $L: V \to W$ be a linear transformation of an n-dimensional vector space V into an m-dimensional vector space W $(n \neq 0$ and $m \neq 0)$ and let $S = \{\mathbf{v}_1, \mathbf{v}_2, \ldots, \mathbf{v}_n\}$ and $T = \{\mathbf{w}_1, \mathbf{w}_2, \ldots, \mathbf{w}_m\}$ be bases for V and W, respectively. Then the $m \times n$ matrix A, whose jth column is the coordinate vector $\left[L(\mathbf{v}_j)\right]_T$ of $L(\mathbf{v}_j)$ with respect to T, is associated with L and has the following property: If $\mathbf{x}$ is in V, then*

$$\left[L(\mathbf{x})\right]_T = A\left[\mathbf{x}\right]_S, \tag{1}$$

where $\left[\mathbf{x}\right]_S$ and $\left[L(\mathbf{x})\right]_T$ are the coordinate vectors of $\mathbf{x}$ and $L(\mathbf{x})$ with respect to the respective bases S and T. Moreover, A is the only matrix with this property.

Proof The proof is a constructive one; that is, we show how to construct the matrix A. It is more complicated than the proof of Theorem 3.8. Consider the vector $\mathbf{v}_j$ in V for $j = 1, 2, \ldots, n$. Then $L(\mathbf{v}_j)$ is a vector in W, and since T is a basis for W, we can express this vector as a linear combination of the vectors in T in a unique manner. Thus

$$L(\mathbf{v}_j) = c_{1j}\mathbf{w}_1 + c_{2j}\mathbf{w}_2 + \cdots + c_{mj}\mathbf{w}_m \quad (1 \leq j \leq n). \tag{2}$$

This means that the coordinate vector of $L(\mathbf{v}_j)$ with respect to T is

$$\left[L(\mathbf{v}_j)\right]_T = \begin{bmatrix} c_{1j} \\ c_{2j} \\ \vdots \\ c_{mj} \end{bmatrix}.$$

We now define the $m \times n$ matrix A by choosing $\left[L(\mathbf{v}_j)\right]_T$ as the jth column of A and show that this matrix satisfies the properties stated in the theorem. We leave the rest of the proof as Exercise T.1 and amply illustrate it in the examples below. ■

DEFINITION The matrix A of Theorem 6.8 is called the **matrix representing L with respect to the bases S and T**, or the **matrix of L with respect to S and T**.

We now summarize the procedure given in Theorem 6.8.

The procedure for computing the matrix of a linear transformation $L: V \to W$ with respect to the bases $S = \{\mathbf{v}_1, \mathbf{v}_2, \ldots, \mathbf{v}_n\}$ and $T = \{\mathbf{w}_1, \mathbf{w}_2, \ldots, \mathbf{w}_m\}$ for V and W, respectively, is as follows.

Step 1. Compute $L(\mathbf{v}_j)$ for $j = 1, 2, \ldots, n$.

Step 2. Find the coordinate vector $\left[L(\mathbf{v}_j)\right]_T$ of $L(\mathbf{v}_j)$ with respect to the basis T. This means that we have to express $L(\mathbf{v}_j)$ as a linear combination of the vectors in T [see Equation (2)].

Step 3. The matrix A of L with respect to S and T is formed by choosing $\left[L(\mathbf{v}_j)\right]_T$ as the jth column of A.

FIGURE 6.2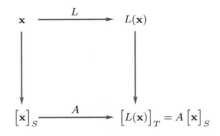

Figure 6.2 gives a graphical interpretation of Equation (1), that is, of Theorem 6.8. The top horizontal arrow represents the linear transformation L from the n-dimensional vector space V into the m-dimensional vector space W and takes the vector $\mathbf{x}$ in V to the vector $L(\mathbf{x})$ in W. The bottom horizontal line represents the matrix A. Then $\left[L(\mathbf{x})\right]_T$, a coordinate vector in R^m, is obtained simply by multiplying $\left[\mathbf{x}\right]_S$, a coordinate vector in R^n, by the matrix A. We can thus always work with matrices rather than with linear transformations.

Physicists and others who deal at great length with linear transformations perform most of their computations with the matrices of the linear transformations.

EXAMPLE 1 ■ Let $L\colon R^3 \to R^2$ be defined by

$$L\left(\begin{bmatrix} x \\ y \\ z \end{bmatrix}\right) = \begin{bmatrix} x+y \\ y-z \end{bmatrix}. \tag{3}$$

Let

$$S = \{\mathbf{v}_1, \mathbf{v}_2, \mathbf{v}_3\} \quad \text{and} \quad T = \{\mathbf{w}_1, \mathbf{w}_2\}$$

be bases for R^3 and R^2, respectively, where

$$\mathbf{v}_1 = \begin{bmatrix} 1 \\ 0 \\ 0 \end{bmatrix}, \qquad \mathbf{v}_2 = \begin{bmatrix} 0 \\ 1 \\ 0 \end{bmatrix}, \qquad \mathbf{v}_3 = \begin{bmatrix} 0 \\ 0 \\ 1 \end{bmatrix},$$

$$\mathbf{w}_1 = \begin{bmatrix} 1 \\ 0 \end{bmatrix}, \quad \text{and} \quad \mathbf{w}_2 = \begin{bmatrix} 0 \\ 1 \end{bmatrix}.$$

We now find the matrix A of L with respect to S and T. We have

$$L(\mathbf{v}_1) = \begin{bmatrix} 1+0 \\ 0-0 \end{bmatrix} = \begin{bmatrix} 1 \\ 0 \end{bmatrix},$$

$$L(\mathbf{v}_2) = \begin{bmatrix} 0+1 \\ 1-0 \end{bmatrix} = \begin{bmatrix} 1 \\ 1 \end{bmatrix},$$

$$L(\mathbf{v}_3) = \begin{bmatrix} 0+0 \\ 0-1 \end{bmatrix} = \begin{bmatrix} 0 \\ -1 \end{bmatrix}.$$

Since T is the natural basis for R^2, the coordinate vectors of $L(\mathbf{v}_1)$, $L(\mathbf{v}_2)$, and $L(\mathbf{v}_3)$ with respect to T are the same as $L(\mathbf{v}_1)$, $L(\mathbf{v}_2)$, and $L(\mathbf{v}_3)$, respectively. That is,

$$\left[L(\mathbf{v}_1)\right]_T = \begin{bmatrix} 1 \\ 0 \end{bmatrix}, \qquad \left[L(\mathbf{v}_2)\right]_T = \begin{bmatrix} 1 \\ 1 \end{bmatrix}, \qquad \left[L(\mathbf{v}_3)\right]_T = \begin{bmatrix} 0 \\ -1 \end{bmatrix}.$$

Hence

$$A = \begin{bmatrix} 1 & 1 & 0 \\ 0 & 1 & -1 \end{bmatrix}.$$

$\blacksquare$

EXAMPLE 2 $\blacksquare$ Let $L\colon R^3 \to R^2$ be defined as in Example 1. Now let

$$S = \{\mathbf{v}_1, \mathbf{v}_2, \mathbf{v}_3\} \quad \text{and} \quad T = \{\mathbf{w}_1, \mathbf{w}_2\}$$

be bases for R^3 and R^2, respectively, where

$$\mathbf{v}_1 = \begin{bmatrix} 1 \\ 0 \\ 1 \end{bmatrix}, \qquad \mathbf{v}_2 = \begin{bmatrix} 0 \\ 1 \\ 1 \end{bmatrix}, \qquad \mathbf{v}_3 = \begin{bmatrix} 1 \\ 1 \\ 1 \end{bmatrix},$$

$$\mathbf{w}_1 = \begin{bmatrix} 1 \\ 2 \end{bmatrix}, \quad \text{and} \quad \mathbf{w}_2 = \begin{bmatrix} -1 \\ 1 \end{bmatrix}.$$

Find the matrix of L with respect to S and T.

Solution   We have

$$L(\mathbf{v}_1) = \begin{bmatrix} 1 \\ -1 \end{bmatrix}, \qquad L(\mathbf{v}_2) = \begin{bmatrix} 1 \\ 0 \end{bmatrix}, \qquad L(\mathbf{v}_3) = \begin{bmatrix} 2 \\ 0 \end{bmatrix}.$$

To find the coordinate vectors $\left[L(\mathbf{v}_1)\right]_T$, $\left[L(\mathbf{v}_2)\right]_T$, and $\left[L(\mathbf{v}_3)\right]_T$, we write

$$L(\mathbf{v}_1) = \begin{bmatrix} 1 \\ -1 \end{bmatrix} = a_1 \mathbf{w}_1 + a_2 \mathbf{w}_2 = a_1 \begin{bmatrix} 1 \\ 2 \end{bmatrix} + a_2 \begin{bmatrix} -1 \\ 1 \end{bmatrix},$$

$$L(\mathbf{v}_2) = \begin{bmatrix} 1 \\ 0 \end{bmatrix} = b_1 \mathbf{w}_1 + b_2 \mathbf{w}_2 = b_1 \begin{bmatrix} 1 \\ 2 \end{bmatrix} + b_2 \begin{bmatrix} -1 \\ 1 \end{bmatrix},$$

$$L(\mathbf{v}_3) = \begin{bmatrix} 2 \\ 0 \end{bmatrix} = c_1 \mathbf{w}_1 + c_2 \mathbf{w}_2 = c_1 \begin{bmatrix} 1 \\ 2 \end{bmatrix} + c_2 \begin{bmatrix} -1 \\ 1 \end{bmatrix}.$$

That is, we must solve three linear systems, each of two equations in two unknowns. Since their coefficient matrix is the same, we solve them all at once, as in Example 4 of Section 4.7. Thus we form the matrix

$$\begin{bmatrix} 1 & -1 & \vdots & 1 & \vdots & 1 & \vdots & 2 \\ 2 & 1 & \vdots & -1 & \vdots & 0 & \vdots & 0 \end{bmatrix},$$

which we transform to reduced row echelon form, obtaining (verify)

$$\begin{bmatrix} 1 & 0 & \vdots & 0 & \vdots & \frac{1}{3} & \vdots & \frac{2}{3} \\ 0 & 1 & \vdots & -1 & \vdots & -\frac{2}{3} & \vdots & -\frac{4}{3} \end{bmatrix}.$$

Hence the matrix A of L with respect to S and T is

$$A = \begin{bmatrix} 0 & \frac{1}{3} & \frac{2}{3} \\ -1 & -\frac{2}{3} & -\frac{4}{3} \end{bmatrix}.$$

Equation (1) is then

$$\left[L(\mathbf{x})\right]_T = \begin{bmatrix} 0 & \frac{1}{3} & \frac{2}{3} \\ -1 & -\frac{2}{3} & -\frac{4}{3} \end{bmatrix} \left[\mathbf{x}\right]_S. \tag{4}$$

To illustrate Equation (4), let

$$\mathbf{x} = \begin{bmatrix} 1 \\ 6 \\ 3 \end{bmatrix}.$$

Then from the definition of L, Equation (3), we have

$$L(\mathbf{x}) = \begin{bmatrix} 1+6 \\ 6-3 \end{bmatrix} = \begin{bmatrix} 7 \\ 3 \end{bmatrix}.$$

Now (verify)

$$\left[\mathbf{x}\right]_S = \begin{bmatrix} -3 \\ 2 \\ 4 \end{bmatrix}.$$

Then from (4),

$$\left[L(\mathbf{x})\right]_T = A\left[\mathbf{x}\right]_S = \begin{bmatrix} \frac{10}{3} \\ -\frac{11}{3} \end{bmatrix}.$$

Hence

$$L(\mathbf{x}) = \frac{10}{3}\begin{bmatrix} 1 \\ 2 \end{bmatrix} - \frac{11}{3}\begin{bmatrix} -1 \\ 1 \end{bmatrix} = \begin{bmatrix} 7 \\ 3 \end{bmatrix},$$

which agrees with the previous value for $L(\mathbf{x})$. ■

Notice that the matrices obtained in Examples 1 and 2 are different even though L is the same in both cases. It can be shown that there is a relationship between these two matrices. The study of this relationship is beyond the scope of this book.

The procedure used in Example 2 can be used to find the matrix representing a linear transformation $L: R^n \to R^m$ with respect to given bases S and T for R^n and R^m, respectively.

The procedure for computing the matrix of a linear transformation $L\colon R^n \to R^m$ with respect to the bases $S = \{\mathbf{v}_1, \mathbf{v}_2, \ldots, \mathbf{v}_n\}$ and $T = \{\mathbf{w}_1, \mathbf{w}_2, \ldots, \mathbf{w}_m\}$ for R^n and R^m, respectively, is as follows.

Step 1. Compute $L(\mathbf{x}_j)$ for $j = 1, 2, \ldots, n$.

Step 2. Form the matrix

$$\begin{bmatrix} \mathbf{w}_1 & \mathbf{w}_2 & \cdots & \mathbf{w}_m \vdots L(\mathbf{v}_1) \vdots L(\mathbf{v}_2) \vdots \cdots \vdots L(\mathbf{v}_n) \end{bmatrix},$$

which we transform to reduced row echelon form, obtaining the matrix

$$\begin{bmatrix} I_n \vdots A \end{bmatrix}.$$

Step 3. The matrix A is the matrix representing L with respect to bases S and T.

EXAMPLE 3 ■ Let $L\colon R^3 \to R^2$ be as defined in Example 1. Now let

$$S = \{\mathbf{v}_3, \mathbf{v}_2, \mathbf{v}_1\} \quad \text{and} \quad T = \{\mathbf{w}_1, \mathbf{w}_2\},$$

where $\mathbf{v}_1$, $\mathbf{v}_2$, $\mathbf{v}_3$, $\mathbf{w}_1$, and $\mathbf{w}_2$ are as in Example 2. Then the matrix of L with respect to S and T is

$$A = \begin{bmatrix} \frac{2}{3} & \frac{1}{3} & 0 \\ -\frac{4}{3} & -\frac{2}{3} & -1 \end{bmatrix}.$$

Note that if we change the order of the vectors in the bases S and T, then the matrix A of L may change. ■

EXAMPLE 4 ■ Let $L\colon R^3 \to R^2$ be defined by

$$L\left(\begin{bmatrix} x \\ y \\ z \end{bmatrix} \right) = \begin{bmatrix} 1 & 1 & 1 \\ 1 & 2 & 3 \end{bmatrix} \begin{bmatrix} x \\ y \\ z \end{bmatrix}.$$

Let

$$S = \{\mathbf{v}_1, \mathbf{v}_2, \mathbf{v}_3\} \quad \text{and} \quad T = \{\mathbf{w}_1, \mathbf{w}_2\}$$

be the natural bases for R^3 and R^2, respectively. Find the matrix of L with respect to S and T.

Solution   We have

$$L(\mathbf{v}_1) = \begin{bmatrix} 1 & 1 & 1 \\ 1 & 2 & 3 \end{bmatrix} \begin{bmatrix} 1 \\ 0 \\ 0 \end{bmatrix} = \begin{bmatrix} 1 \\ 1 \end{bmatrix} = 1\mathbf{w}_1 + 1\mathbf{w}_2, \quad \text{so} \quad \left[L(\mathbf{v}_1)\right]_T = \begin{bmatrix} 1 \\ 1 \end{bmatrix};$$

$$L(\mathbf{v}_2) = \begin{bmatrix} 1 & 1 & 1 \\ 1 & 2 & 3 \end{bmatrix} \begin{bmatrix} 0 \\ 1 \\ 0 \end{bmatrix} = \begin{bmatrix} 1 \\ 2 \end{bmatrix} = 1\mathbf{w}_1 + 2\mathbf{w}_2, \quad \text{so} \quad \left[L(\mathbf{v}_2)\right]_T = \begin{bmatrix} 1 \\ 2 \end{bmatrix}.$$

Also,

$$\left[L(\mathbf{v}_3)\right]_T = \begin{bmatrix} 1 \\ 3 \end{bmatrix} \quad \text{(verify)}.$$

Then the matrix of L with respect to S and T is

$$A = \begin{bmatrix} 1 & 1 & 1 \\ 1 & 2 & 3 \end{bmatrix}.$$

Of course, the reason that A is the same matrix as the one involved in the definition of L is that the natural bases are being used for R^3 and R^2. ■

EXAMPLE 5 ■ Let $L: R^3 \to R^2$ be defined as in Example 4. Now let

$$S = \{\mathbf{v}_1, \mathbf{v}_2, \mathbf{v}_3\} \quad \text{and} \quad T = \{\mathbf{w}_1, \mathbf{w}_2\},$$

where

$$\mathbf{v}_1 = \begin{bmatrix} 1 \\ 1 \\ 0 \end{bmatrix}, \qquad \mathbf{v}_2 = \begin{bmatrix} 0 \\ 1 \\ 1 \end{bmatrix}, \qquad \mathbf{v}_3 = \begin{bmatrix} 0 \\ 0 \\ 1 \end{bmatrix},$$

$$\mathbf{w}_1 = \begin{bmatrix} 1 \\ 2 \end{bmatrix}, \quad \text{and} \quad \mathbf{w}_2 = \begin{bmatrix} 1 \\ 3 \end{bmatrix}.$$

Find the matrix of L with respect to S and T.

Solution   We have

$$L(\mathbf{v}_1) = \begin{bmatrix} 2 \\ 3 \end{bmatrix}, \quad L(\mathbf{v}_2) = \begin{bmatrix} 1 \\ 2 \end{bmatrix}, \quad \text{and} \quad L(\mathbf{v}_3) = \begin{bmatrix} 1 \\ 3 \end{bmatrix}.$$

We now form (verify)

$$\begin{bmatrix} \mathbf{w}_1 & \mathbf{w}_2 & \vdots & L(\mathbf{v}_1) & \vdots & L(\mathbf{v}_2) & \vdots & L(\mathbf{v}_3) \end{bmatrix} = \begin{bmatrix} 1 & 1 & \vdots & 2 & \vdots & 1 & \vdots & 1 \\ 2 & 3 & \vdots & 3 & \vdots & 2 & \vdots & 3 \end{bmatrix}.$$

Transforming this matrix to reduced row echelon form, we obtain (verify)

$$\begin{bmatrix} 1 & 0 & \vdots & 3 & \vdots & 1 & \vdots & 0 \\ 0 & 1 & \vdots & -1 & \vdots & 1 & \vdots & 1 \end{bmatrix},$$

so the matrix of L with respect to S and T is

$$A = \begin{bmatrix} 3 & 1 & 0 \\ -1 & 1 & 1 \end{bmatrix}.$$

This matrix is, of course, far different from the one that defined L. Thus, although a matrix A may be involved in the definition of a linear transformation L, we cannot conclude that it is necessarily the matrix representing L with respect to two given bases S and T. ■

EXAMPLE 6 ■ Let $L: P_1 \to P_2$ be defined by $L[p(t)] = tp(t)$.

(a) Find the matrix of L with respect to the bases $S = \{t, 1\}$ and $T = \{t^2, t, 1\}$ for P_1 and P_2, respectively.

(b) If $p(t) = 3t - 2$, compute $L[p(t)]$ directly and using the matrix obtained in (a).

Solution   (a) We have

$$L(t) = t \cdot t = t^2 = 1(t^2) + 0(t) + 0(1), \quad \text{so} \quad \left[L(t)\right]_T = \begin{bmatrix} 1 \\ 0 \\ 0 \end{bmatrix};$$

$$L(1) = t \cdot 1 = t = 0(t^2) + 1(t) + 0(1), \quad \text{so} \quad \left[L(1)\right]_T = \begin{bmatrix} 0 \\ 1 \\ 0 \end{bmatrix}.$$

Hence the matrix of L with respect to S and T is

$$A = \begin{bmatrix} 1 & 0 \\ 0 & 1 \\ 0 & 0 \end{bmatrix}.$$

(b) Computing $L[p(t)]$ directly, we have

$$L[p(t)] = tp(t) = t(3t - 2) = 3t^2 - 2t.$$

To compute $L[p(t)]$ using A, we first write

$$p(t) = 3 \cdot t + (-2)1, \quad \text{so} \quad \left[p(t)\right]_S = \begin{bmatrix} 3 \\ -2 \end{bmatrix}.$$

Then

$$\left[L[p(t)]\right]_T = A\left[p(t)\right]_S = \begin{bmatrix} 1 & 0 \\ 0 & 1 \\ 0 & 0 \end{bmatrix} \begin{bmatrix} 3 \\ -2 \end{bmatrix} = \begin{bmatrix} 3 \\ -2 \\ 0 \end{bmatrix}.$$

Hence

$$L[p(t)] = 3t^2 + (-2)t + 0(1) = 3t^2 - 2t. \qquad \blacksquare$$

EXAMPLE 7 ■ Let $L: P_1 \to P_2$ be as defined in Example 6.

(a) Find the matrix of L with respect to the bases $S = \{t, 1\}$ and $T = \{t^2, t - 1, t + 1\}$ for P_1 and P_2, respectively.

(b) If $p(t) = 3t - 2$, compute $L[p(t)]$ using the matrix obtained in (a).

Solution   (a) We have (verify)

$$L(t) = t^2 = 1(t^2) + 0(t - 1) + 0(t + 1), \quad \text{so} \quad \left[L(t)\right]_T = \begin{bmatrix} 1 \\ 0 \\ 0 \end{bmatrix};$$

$$L(1) = t = 0(t^2) + \tfrac{1}{2}(t-1) + \tfrac{1}{2}(t+1), \quad \text{so} \quad \big[L(1)\big]_T = \begin{bmatrix} 0 \\ \tfrac{1}{2} \\ \tfrac{1}{2} \end{bmatrix}.$$

Then the matrix of L with respect to S and T is

$$A = \begin{bmatrix} 1 & 0 \\ 0 & \tfrac{1}{2} \\ 0 & \tfrac{1}{2} \end{bmatrix}.$$

(b) We have

$$\big[L[p(t)]\big]_T = A\,\big[p(t)\big]_S = \begin{bmatrix} 1 & 0 \\ 0 & \tfrac{1}{2} \\ 0 & \tfrac{1}{2} \end{bmatrix} \begin{bmatrix} 3 \\ -2 \end{bmatrix} = \begin{bmatrix} 3 \\ -1 \\ -1 \end{bmatrix}.$$

Hence

$$L[p(t)] = 3t^2 + (-1)(t-1) + (-1)(t+1) = 3t^2 - 2t. \qquad \blacksquare$$

Suppose that $L\colon V \to W$ is a linear transformation and that A is the matrix of L with respect to bases for V and W. Then the problem of finding $\ker L$ reduces to the problem of finding the solution space of $A\mathbf{x} = \mathbf{0}$. Moreover, the problem of finding range L reduces to the problem of finding the column space of A.

If $L\colon V \to V$ is a linear operator (a linear transformation from V to W, where $V = W$) and V is an n-dimensional vector space, then to obtain a matrix representing L, we fix bases S and T for V and obtain the matrix of L with respect to S and T. However, it is often convenient in this case to choose $S = T$. To avoid redundancy in this case, we refer to A as the **matrix of L with respect to S**.

Let $I\colon V \to V$ be the identity linear operator on an n-dimensional vector space defined by $I(\mathbf{v}) = \mathbf{v}$ for every $\mathbf{v}$ in V. If S is a basis for V, then the matrix of I with respect to S is I_n (Exercise T.2). Let T be another basis for V. Then the matrix of I with respect to S and T is the transition matrix (see Section 4.7) from the S-basis to the T-basis (Exercise T.5).

If $L\colon R^n \to R^n$ is a linear operator defined by $L(\mathbf{x}) = A\mathbf{x}$, for $\mathbf{x}$ in R^n, then we can show that L is one-to-one and onto if and only if A is nonsingular.

We can now extend our List of Nonsingular Equivalences.

> ### List of Nonsingular Equivalences
> The following statements are equivalent for an $n \times n$ matrix A.
>
> 1. A is nonsingular.
> 2. $A\mathbf{x} = \mathbf{0}$ has only the trivial solution.
> 3. A is row equivalent to I_n.
> 4. The linear system $A\mathbf{x} = \mathbf{b}$ has a unique solution for every $n \times 1$ matrix $\mathbf{b}$.
> 5. $\det(A) \neq 0$.
> 6. A has rank n.
> 7. A has nullity 0.
> 8. The rows of A form a linearly independent set of n vectors in R^n.
> 9. The columns of A form a linearly independent set of n vectors in R^n.
> 10. Zero is *not* an eigenvalue of A.
> 11. The linear operator $L \colon R^n \to R^n$ defined by $L(\mathbf{x}) = A\mathbf{x}$, for $\mathbf{x}$ in R^n, is one-to-one and onto.

Change of Basis Leads to a New Matrix of a Linear Operator

If $L \colon V \to V$ is a linear operator and S is a basis for V, then the matrix A representing L with respect to S will change if we use the basis T for V instead of S. The following theorem tells us how to find the matrix of L with respect to T using the matrix A.

THEOREM 6.9 ■ *Let $L \colon V \to V$ be a linear operator, where V is an n-dimensional vector space. Let $S = \{\mathbf{v}_1, \mathbf{v}_2, \ldots, \mathbf{v}_n\}$ and $T = \{\mathbf{w}_1, \mathbf{w}_2, \ldots, \mathbf{w}_n\}$ be bases for V and let P be the transition matrix from T to S.* If A is the matrix representing L with respect to S, then $P^{-1}AP$ is the matrix representing L with respect to T.*

Proof If P is the transition matrix from T to S and $\mathbf{x}$ is a vector in V, then by Equation (5) in Section 4.7 we have

$$[\mathbf{x}]_S = P\,[\mathbf{x}]_T, \tag{5}$$

where the jth column of P is the coordinate vector $[\mathbf{w}_j]_S$ of $\mathbf{w}_j$ with respect to S. From Theorem 4.15 (Section 4.7) we know that P^{-1} is the transition matrix from S to T, where the jth column of P^{-1} is the coordinate vector $[\mathbf{v}_j]_T$ of $\mathbf{v}_j$ with respect to T. By Equation (5) in Section 4.7 we have

$$[\mathbf{y}]_T = P^{-1}\,[\mathbf{y}]_S \tag{6}$$

for $\mathbf{y}$ in V. If A is the matrix representing L with respect to S, then

$$[L(\mathbf{x})]_S = A\,[\mathbf{x}]_S \tag{7}$$

---

*In Section 4.7, P was denoted by $P_{S \leftarrow T}$. In this section we denote it by P to simplify the notation.

for $\mathbf{x}$ in V. Substituting $\mathbf{y} = L(\mathbf{x})$ in (6), we have

$$\left[L(\mathbf{x})\right]_T = P^{-1}\left[L(\mathbf{x})\right]_S.$$

Using first (7) and then (5) in this last equation, we obtain

$$\left[L(\mathbf{x})\right]_T = P^{-1}\left[L(\mathbf{x})\right]_S = P^{-1}A\left[\mathbf{x}\right]_S = P^{-1}AP\left[\mathbf{x}\right]_T.$$

The equation

$$\left[L(\mathbf{x})\right]_T = P^{-1}AP\left[\mathbf{x}\right]_T$$

means that $B = P^{-1}AP$ is the matrix representing L with respect to T.   ■

Theorem 6.9 can be illustrated by the diagram shown in Figure 6.3. This figure shows that there are two ways of going from $\mathbf{x}$ in V to $L(\mathbf{x})$: directly, using the matrix B; or indirectly, using the matrices P, A, and P^{-1}.

FIGURE 6.3

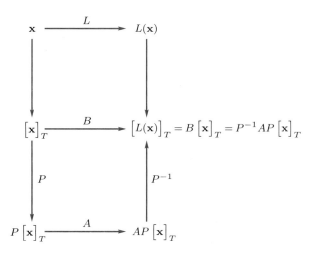

EXAMPLE 8   ■   Let $L\colon R^2 \to R^2$ be defined by

$$L\left(\begin{bmatrix} a_1 \\ a_2 \end{bmatrix}\right) = \begin{bmatrix} a_1 + a_2 \\ a_1 - 2a_2 \end{bmatrix}.$$

Let

$$S = \left\{ \begin{bmatrix} 1 \\ 0 \end{bmatrix}, \begin{bmatrix} 0 \\ 1 \end{bmatrix} \right\} \quad \text{and} \quad T = \left\{ \begin{bmatrix} 1 \\ -1 \end{bmatrix}, \begin{bmatrix} 2 \\ 1 \end{bmatrix} \right\}$$

be bases for R^2. We can easily show (verify) that

$$A = \begin{bmatrix} 1 & 1 \\ 1 & -2 \end{bmatrix}$$

is the matrix representing L with respect to S.

The transition matrix P from T to S is the matrix whose jth column is the coordinate vector of the jth vector in the basis T with respect to S. Thus

$$P = \begin{bmatrix} 1 & 2 \\ -1 & 1 \end{bmatrix}.$$

The transition matrix from S to T is (verify)

$$P^{-1} = \tfrac{1}{3} \begin{bmatrix} 1 & -2 \\ 1 & 1 \end{bmatrix}.$$

Then the matrix representing L with respect to T is (verify)

$$P^{-1}AP = \begin{bmatrix} -2 & 1 \\ 1 & 1 \end{bmatrix}.$$

On the other hand, we can compute the matrix of L with respect to T directly. We have

$$L\left(\begin{bmatrix} 1 \\ -1 \end{bmatrix}\right) = \begin{bmatrix} 0 \\ 3 \end{bmatrix}, \qquad L\left(\begin{bmatrix} 2 \\ 1 \end{bmatrix}\right) = \begin{bmatrix} 3 \\ 0 \end{bmatrix}.$$

We now form the matrix

$$\begin{bmatrix} 1 & 2 & \vdots & 0 & \vdots & 3 \\ -1 & 1 & \vdots & 3 & \vdots & 0 \end{bmatrix},$$

which we transform to reduced row echelon form, obtaining

$$\begin{bmatrix} 1 & 0 & \vdots & -2 & \vdots & 1 \\ 0 & 1 & \vdots & 1 & \vdots & 1 \end{bmatrix}.$$

Hence the matrix representing L with respect to T is

$$\begin{bmatrix} -2 & 1 \\ 1 & 1 \end{bmatrix},$$

as before. ■

Diagonalizability and Similarity Revisited

Recall from Section 5.1 that an $n \times n$ matrix B is similar to an $n \times n$ matrix A if there is a nonsingular matrix P such that $B = P^{-1}AP$. It then follows from Theorem 6.9 that any two matrices representing the same linear operator $L\colon V \to V$ with respect to different bases are similar. Conversely, it can be shown (we omit the proof) that if A and B are similar matrices, then they represent the same linear transformation $L\colon V \to V$ with respect to bases for V.

We can now prove the following theorem, which follows from Theorem 5.3.

THEOREM 6.10 ■ *Consider the linear operator $L\colon R^n \to R^n$ defined by $L(\mathbf{x}) = A\mathbf{x}$ for $\mathbf{x}$ in R^n. Then A is diagonalizable with n linearly independent eigenvectors $\mathbf{x}_1, \mathbf{x}_2, \ldots, \mathbf{x}_n$ if and only if the matrix of L with respect to $S = \{\mathbf{x}_1, \mathbf{x}_2, \ldots, \mathbf{x}_n\}$ is diagonal.*

Proof   Suppose that A is diagonalizable. Then by Theorem 5.3 it has n linearly independent eigenvectors $\mathbf{x}_1, \mathbf{x}_2, \ldots, \mathbf{x}_n$ with corresponding eigenvalues $\lambda_1, \lambda_2, \ldots, \lambda_n$. Since n linearly independent vectors in R^n form a basis (Theorem 4.9 in Section 4.4), we conclude that $S = \{\mathbf{x}_1, \mathbf{x}_2, \ldots, \mathbf{x}_n\}$ is a basis for R^n. Now

$$L(\mathbf{x}_j) = A\mathbf{x}_j = \lambda_j \mathbf{x}_j = 0\mathbf{x}_1 + \cdots + 0\mathbf{x}_{j-1} + \lambda_j \mathbf{x}_j + 0\mathbf{x}_{j+1} + \cdots + 0\mathbf{x}_n,$$

so the coordinate vector $\left[L(\mathbf{x}_j)\right]_S$ of $L(\mathbf{x}_j)$ with respect to S is

$$\begin{bmatrix} 0 \\ \vdots \\ 0 \\ \lambda_j \\ 0 \\ \vdots \\ 0 \end{bmatrix} \quad \longleftarrow j\text{th row.} \tag{8}$$

Hence the matrix of L with respect to S is

$$\begin{bmatrix} \lambda_1 & 0 & \cdots & 0 \\ 0 & \lambda_2 & \cdots & 0 \\ \vdots & & & \vdots \\ 0 & \cdots & 0 & \lambda_n \end{bmatrix}. \tag{9}$$

Conversely, suppose that there is a basis $S = \{\mathbf{x}_1, \mathbf{x}_2, \ldots, \mathbf{x}_n\}$ for R^n with respect to which the matrix of L is diagonal, say, of the form in (9). Then the coordinate vector of $L(\mathbf{x}_j)$ with respect to S is (8), so

$$L(\mathbf{x}_j) = 0\mathbf{x}_1 + \cdots + 0\mathbf{x}_{j-1} + \lambda_j \mathbf{x}_j + 0\mathbf{x}_{j+1} + \cdots + 0\mathbf{x}_n = \lambda_j \mathbf{x}_j.$$

Since $L(\mathbf{x}_j) = A\mathbf{x}_j$, we have

$$A\mathbf{x}_j = \lambda_j \mathbf{x}_j,$$

which means that $\mathbf{x}_1, \mathbf{x}_2, \ldots, \mathbf{x}_n$ are eigenvectors of A. Since they form a basis for R^n, they are linearly independent, and by Theorem 5.3 we conclude that A is diagonalizable.   ∎

REMARK   Let $L: R^n \to R^n$ be a linear transformation defined by $L(\mathbf{x}) = A\mathbf{x}$. If A is diagonalizable, then, by Theorem 5.3, R^n has a basis S consisting of eigenvectors of A. Moreover, by Theorem 6.10, the matrix of L with respect to S is the diagonal matrix whose entries on the main diagonal are the eigenvalues of A. Thus the problem of diagonalizing A becomes the problem of finding a basis S for R^n such that the matrix of L with respect to S will be diagonal.

Orthogonal Matrices Revisited

We now look at the geometric implications of orthogonal matrices. Let A be an orthogonal $n \times n$ matrix and consider the linear transformation $L: R^n \to R^n$ defined by $L(\mathbf{x}) = A\mathbf{x}$, for $\mathbf{x}$ in R^n. We first compute $L(\mathbf{x}) \cdot L(\mathbf{y})$ for any vectors

$\mathbf{x}$ and $\mathbf{y}$ in R^n. We have, using Exercise T.14 in Section 1.3 and the fact that $A^T A = I_n$,

$$\begin{aligned} L(\mathbf{x}) \cdot L(\mathbf{y}) = (A\mathbf{x}) \cdot (A\mathbf{y}) &= (A\mathbf{x})^T (A\mathbf{y}) \\ &= \mathbf{x}^T (A^T A)\mathbf{y} \qquad\qquad (10) \\ &= \mathbf{x} \cdot (A^T A\mathbf{y}) = \mathbf{x} \cdot (I_n \mathbf{y}) = \mathbf{x} \cdot \mathbf{y}. \end{aligned}$$

This means that L preserves the inner product of two vectors, and consequently L preserves length. It is, of course, clear that if θ is the angle between vectors $\mathbf{x}$ and $\mathbf{y}$ in R^n, then the angle between $L(\mathbf{x})$ and $L(\mathbf{y})$ is also θ (Exercise T.8). A linear transformation satisfying Equation (10) is called an **isometry**. Conversely, let $L\colon R^n \to R^n$ be an isometry, so that

$$L(\mathbf{x}) \cdot L(\mathbf{y}) = \mathbf{x} \cdot \mathbf{y},$$

for any $\mathbf{x}$ and $\mathbf{y}$ in R^n. Let A be the matrix of L with respect to the natural basis for R^n. Then

$$L(\mathbf{x}) = A\mathbf{x}.$$

If $\mathbf{x}$ and $\mathbf{y}$ are any vectors in R^n, we have, using Equation (1) in Section 5.2,

$$\mathbf{x} \cdot \mathbf{y} = L(\mathbf{x}) \cdot L(\mathbf{y}) = (A\mathbf{x}) \cdot (A\mathbf{y}) = \mathbf{x} \cdot (A^T A\mathbf{y}).$$

Since this holds for all $\mathbf{x}$ in R^n, we conclude by Exercise T.8 in Section 3.2 that

$$A^T A\mathbf{y} = \mathbf{y}$$

for any $\mathbf{y}$ in R^n. It then follows from Exercise T.20 in Section 1.4 that $A^T A = I_n$, so A is an orthogonal matrix.

6.3 EXERCISES

1. Let $L\colon R^2 \to R^2$ be defined by
$$L(x, y) = (x - 2y, x + 2y).$$
Let $S = \{(1, -1), (0, 1)\}$ be a basis for R^2 and let T be the natural basis for R^2. Find the matrix representing L with respect to
(a) S. (b) S and T. (c) T and S. (d) T.
(e) Compute $L(2, -1)$ using the definition of L and also using the matrices obtained in (a), (b), (c), and (d).

2. Let $L\colon R^2 \to R^2$ be defined by
$$L\left(\begin{bmatrix} x \\ y \end{bmatrix}\right) = \begin{bmatrix} x + 2y \\ 2x - y \end{bmatrix}.$$
Let S be the natural basis for R^2 and let
$$T = \left\{ \begin{bmatrix} -1 \\ 2 \end{bmatrix}, \begin{bmatrix} 2 \\ 0 \end{bmatrix} \right\}$$

be another basis for R^2. Find the matrix representing L with respect to
(a) S. (b) S and T. (c) T and S. (d) T.
(e) Compute
$$L\left(\begin{bmatrix} 1 \\ 2 \end{bmatrix}\right)$$
using the definition of L and also using the matrices obtained in (a), (b), (c), and (d).

3. Let $L\colon R^2 \to R^3$ be defined by
$$L\left(\begin{bmatrix} x \\ y \end{bmatrix}\right) = \begin{bmatrix} x - 2y \\ 2x + y \\ x + y \end{bmatrix}.$$

Let S and T be the natural bases for R^2 and R^3, respectively. Also, let

$$S' = \left\{ \begin{bmatrix} 1 \\ -1 \end{bmatrix}, \begin{bmatrix} 0 \\ 1 \end{bmatrix} \right\}$$

and

$$T' = \left\{ \begin{bmatrix} 1 \\ 1 \\ 0 \end{bmatrix}, \begin{bmatrix} 0 \\ 1 \\ 1 \end{bmatrix}, \begin{bmatrix} 1 \\ -1 \\ 1 \end{bmatrix} \right\}$$

be bases for R^2 and R^3, respectively. Find the matrix representing L with respect to
(a) S and T.   (b) S' and T'.
(c) Compute

$$L\left(\begin{bmatrix} 1 \\ 2 \end{bmatrix} \right)$$

using the definition of L and also using the matrices obtained in (a) and (b).

4. Let $L: R^3 \to R^3$ be defined by
$$L(x, y, z) = (x + 2y + z, 2x - y, 2y + z).$$

Let S be the natural basis for R^3 and let $T = \{(1,0,1), (0,1,1), (0,0,1)\}$ be another basis for R^3. Find the matrix of L with respect to
(a) S.   (b) S and T.   (c) T and S.   (d) T.
(e) Compute $L(1, 1, -2)$ using the definition of L and also using the matrices obtained in (a), (b), (c), and (d).

5. Let $L: R^3 \to R^2$ be defined by

$$L\left(\begin{bmatrix} x \\ y \\ z \end{bmatrix} \right) = \begin{bmatrix} x + y \\ y - z \end{bmatrix}.$$

Let S and T be the natural bases for R^3 and R^2, respectively. Also, let

$$S' = \left\{ \begin{bmatrix} 1 \\ 1 \\ 0 \end{bmatrix}, \begin{bmatrix} 0 \\ 1 \\ 0 \end{bmatrix}, \begin{bmatrix} -1 \\ 1 \\ 1 \end{bmatrix} \right\}$$

and

$$T' = \left\{ \begin{bmatrix} -1 \\ 1 \end{bmatrix}, \begin{bmatrix} 1 \\ 2 \end{bmatrix} \right\}$$

be bases for R^3 and R^2, respectively. Find the matrix of L with respect to
(a) S and T.   (b) S' and T'.
(c) Compute

$$L\left(\begin{bmatrix} 1 \\ 2 \\ 3 \end{bmatrix} \right)$$

using the definition of L and also using the matrices obtained in (a) and (b).

6. Let $L: R^2 \to R^3$ be defined by

$$L\left(\begin{bmatrix} x \\ y \end{bmatrix} \right) = \begin{bmatrix} 1 & 1 \\ 1 & -1 \\ 1 & 2 \end{bmatrix} \begin{bmatrix} x \\ y \end{bmatrix}.$$

(a) Find the matrix of L with respect to the natural bases for R^2 and R^3.
(b) Find the matrix of L with respect to the bases S' and T' of Exercise 3.
(c) Compute

$$L\left(\begin{bmatrix} 2 \\ -3 \end{bmatrix} \right)$$

using the definition of L and also using the matrices obtained in (a) and (b).

7. Let $L: P_1 \to P_3$ be defined by $L[p(t)] = t^2 p(t)$. Let
$$S = \{t, 1\} \quad \text{and} \quad S' = \{t, t + 1\}$$
be bases for P_1. Let
$$T = \{t^3, t^2, t, 1\} \quad \text{and} \quad T' = \{t^3, t^2 - 1, t, t + 1\}$$
be bases for P_3. Find the matrix of L with respect to
(a) S and T.   (b) S' and T'.

8. Let $L: P_1 \to P_2$ be defined by $L[p(t)] = tp(t) + p(0)$. Let
$$S = \{t, 1\} \quad \text{and} \quad S' = \{t + 1, t - 1\}$$
be bases for P_1. Let
$$T = \{t^2, t, 1\} \quad \text{and} \quad T' = \{t^2 + 1, t - 1, t + 1\}$$
be bases for P_2. Find the matrix of L with respect to
(a) S and T.   (b) S' and T'.
(c) Find $L(-3t + 3)$ using the definition of L and also using the matrices obtained in (a) and (b).

9. Let $L: M_{22} \to M_{22}$ be defined by $L(A) = A^T$. Let

$$S = \left\{ \begin{bmatrix} 1 & 0 \\ 0 & 0 \end{bmatrix}, \begin{bmatrix} 0 & 1 \\ 0 & 0 \end{bmatrix}, \begin{bmatrix} 0 & 0 \\ 1 & 0 \end{bmatrix}, \begin{bmatrix} 0 & 0 \\ 0 & 1 \end{bmatrix} \right\}$$

and

$$T = \left\{ \begin{bmatrix} 1 & 1 \\ 0 & 0 \end{bmatrix}, \begin{bmatrix} 0 & 1 \\ 0 & 0 \end{bmatrix}, \begin{bmatrix} 0 & 0 \\ 1 & 1 \end{bmatrix}, \begin{bmatrix} 1 & 0 \\ 0 & 1 \end{bmatrix} \right\}$$

be bases for M_{22}. Find the matrix of L with respect to
(a) S.   (b) S and T.   (c) T and S.   (d) T.

10. Let
$$C = \begin{bmatrix} 1 & 2 \\ 2 & 3 \end{bmatrix}.$$

Let $L: M_{22} \to M_{22}$ be defined by $L(A) = CA$. Let S and T be the bases for M_{22} defined in Exercise 9. Find the matrix of L with respect to

(a) S. (b) S and T. (c) T and S. (d) T.

11. Let $L: R^3 \to R^3$ be the linear transformation whose matrix with respect to the natural basis for R^3 is
$$\begin{bmatrix} 1 & 3 & 1 \\ 1 & 2 & 0 \\ 0 & 1 & 1 \end{bmatrix}.$$

Find

(a) $L\left(\begin{bmatrix} 1 \\ 2 \\ 3 \end{bmatrix} \right)$. (b) $L\left(\begin{bmatrix} 0 \\ 1 \\ 1 \end{bmatrix} \right)$.

12. Let
$$C = \begin{bmatrix} 1 & 2 \\ 3 & 4 \end{bmatrix},$$

and let $L: M_{22} \to M_{22}$ be the linear transformation defined by $L(A) = AC - CA$ for A in M_{22}. Let S and T be the ordered bases for M_{22} defined in Exercise 9. Find the matrix of L with respect to

(a) S. (b) T. (c) S and T. (d) T and S.

13. Let $L: R^2 \to R^2$ be a linear transformation. Suppose that the matrix of L with respect to the basis
$$S = \{\mathbf{v}_1, \mathbf{v}_2\}$$
is
$$A = \begin{bmatrix} 2 & -3 \\ -1 & 4 \end{bmatrix},$$
where
$$\mathbf{v}_1 = \begin{bmatrix} 1 \\ 2 \end{bmatrix} \quad \text{and} \quad \mathbf{v}_2 = \begin{bmatrix} 1 \\ -1 \end{bmatrix}.$$

(a) Compute $\left[L(\mathbf{v}_1) \right]_S$ and $\left[L(\mathbf{v}_2) \right]_S$.

(b) Compute $L(\mathbf{v}_1)$ and $L(\mathbf{v}_2)$.

(c) Compute
$$L\left(\begin{bmatrix} -2 \\ 3 \end{bmatrix} \right).$$

14. Let the matrix of $L: R^3 \to R^2$ with respect to the bases
$$S = \{\mathbf{v}_1, \mathbf{v}_2, \mathbf{v}_3\} \quad \text{and} \quad T = \{\mathbf{w}_1, \mathbf{w}_2\}$$

be
$$A = \begin{bmatrix} 1 & 2 & 1 \\ -1 & 1 & 0 \end{bmatrix},$$
where
$$\mathbf{v}_1 = \begin{bmatrix} -1 \\ 1 \\ 0 \end{bmatrix}, \quad \mathbf{v}_2 = \begin{bmatrix} 0 \\ 1 \\ 1 \end{bmatrix}, \quad \text{and} \quad \mathbf{v}_3 = \begin{bmatrix} 1 \\ 0 \\ 0 \end{bmatrix}$$

and
$$\mathbf{w}_1 = \begin{bmatrix} 1 \\ 2 \end{bmatrix}, \quad \mathbf{w}_2 = \begin{bmatrix} 1 \\ -1 \end{bmatrix}.$$

(a) Compute $\left[L(\mathbf{v}_1) \right]_T$, $\left[L(\mathbf{v}_2) \right]_T$, and $\left[L(\mathbf{v}_3) \right]_T$.

(b) Compute $L(\mathbf{v}_1)$, $L(\mathbf{v}_2)$, and $L(\mathbf{v}_3)$.

(c) Compute
$$L\left(\begin{bmatrix} 2 \\ 1 \\ -1 \end{bmatrix} \right).$$

(d) Compute
$$L\left(\begin{bmatrix} a \\ b \\ c \end{bmatrix} \right).$$

15. Let $L: P_1 \to P_2$ be a linear transformation. Suppose that the matrix of L with respect to the basis $S = \{\mathbf{v}_1, \mathbf{v}_2\}$ and $T = \{\mathbf{w}_1, \mathbf{w}_2, \mathbf{w}_3\}$ is
$$A = \begin{bmatrix} 1 & 0 \\ 2 & 1 \\ -1 & -2 \end{bmatrix},$$
where
$$\mathbf{v}_1 = t + 1 \quad \text{and} \quad \mathbf{v}_2 = t - 1;$$
$$\mathbf{w}_1 = t^2 + 1, \quad \mathbf{w}_2 = t, \quad \text{and} \quad \mathbf{w}_3 = t - 1.$$

(a) Compute $\left[L(\mathbf{v}_1) \right]_T$ and $\left[L(\mathbf{v}_2) \right]_T$.

(b) Compute $L(\mathbf{v}_1)$ and $L(\mathbf{v}_2)$.

(c) Compute $L(2t + 1)$.

(d) Compute $L(at + b)$.

16. Let $L: R^3 \to R^3$ be defined by
$$L\left(\begin{bmatrix} 1 \\ 0 \\ 0 \end{bmatrix} \right) = \begin{bmatrix} 1 \\ 1 \\ 0 \end{bmatrix},$$
$$L\left(\begin{bmatrix} 0 \\ 1 \\ 0 \end{bmatrix} \right) = \begin{bmatrix} 2 \\ 0 \\ 1 \end{bmatrix},$$
$$L\left(\begin{bmatrix} 0 \\ 0 \\ 1 \end{bmatrix} \right) = \begin{bmatrix} 1 \\ 0 \\ 1 \end{bmatrix}.$$

(a) Find the matrix of L with respect to the natural basis S for R^3.

(b) Find
$$L\left(\begin{bmatrix}1\\2\\3\end{bmatrix}\right)$$
using the definition of L and also using the matrix obtained in (a).

(c) Find $L\left(\begin{bmatrix}a\\b\\c\end{bmatrix}\right)$.

17. Let $L: P_1 \to P_1$ be defined by
$$L(t+1) = t-1,$$
$$L(t-1) = 2t+1.$$

(a) Find the matrix of L with respect to the basis $S = \{t+1, t-1\}$ for P_1.

(b) Find $L(2t+3)$ using the definition of L and also using the matrix obtained in (a).

(c) Find $L(at+b)$.

18. Let the matrix of $L: R^2 \to R^2$ with respect to the basis
$$S = \left\{\begin{bmatrix}1\\-1\end{bmatrix}, \begin{bmatrix}0\\1\end{bmatrix}\right\}$$
be
$$\begin{bmatrix}1 & 2\\-2 & 3\end{bmatrix}.$$
Find the matrix of L with respect to the natural basis for R^3.

19. Let the matrix of $L: P_1 \to P_1$ with respect to the basis $S = \{t+1, t-1\}$ be
$$\begin{bmatrix}2 & 3\\-1 & -2\end{bmatrix}.$$
Find the matrix of L with respect to the basis $\{t, 1\}$ for P_1.

20. Let $L: R^3 \to R^3$ be defined by
$$L\left(\begin{bmatrix}a_1\\a_2\\a_3\end{bmatrix}\right) = \begin{bmatrix}a_1 - a_2 + a_3\\a_1 + a_2\\a_2 - a_3\end{bmatrix}.$$

Let S be the natural basis for R^3 and let
$$T = \left\{\begin{bmatrix}1\\0\\1\end{bmatrix}, \begin{bmatrix}0\\1\\-1\end{bmatrix}, \begin{bmatrix}0\\0\\1\end{bmatrix}\right\}$$
be another basis for R^3.

(a) Compute the matrix of L with respect to S.

(b) Compute the matrix of L with respect to T directly.

(c) Compute the matrix of L using Theorem 6.9.

21. (**Calculus Required**). Let $L: P_3 \to P_3$ be defined by
$$L[p(t)] = p''(t) + p(0).$$
Let
$$S = \{1, t, t^2, t^3\} \quad \text{and} \quad T = \{t^3, t^2 - 1, t, 1\}$$
be bases for P_3.

(a) Compute the matrix of L with respect to S.

(b) Compute the matrix of L with respect to T directly.

(c) Compute the matrix of L with respect to T using Theorem 6.9.

22. (**Calculus Required**). Let V be the vector space with basis $S = \{\sin t, \cos t\}$, and let
$$T = \{\sin t - \cos t, \sin t + \cos t\}$$
be another basis for V. Find the matrix of the linear operator $L: V \to V$ defined by $L(f) = f'$ with respect to

(a) S. (b) T. (c) S and T. (d) T and S.

23. Let V be the vector space with basis $S = \{e^t, e^{-t}\}$. Find the matrix of the linear operator $L: V \to V$ defined by $L(f) = f'$ with respect to S.

24. For the orthogonal matrix
$$A = \begin{bmatrix}\frac{1}{\sqrt{2}} & -\frac{1}{\sqrt{2}}\\-\frac{1}{\sqrt{2}} & -\frac{1}{\sqrt{2}}\end{bmatrix}$$
verify that $(A\mathbf{x}) \cdot (A\mathbf{y}) = \mathbf{x} \cdot \mathbf{y}$ for any $\mathbf{x}$ and $\mathbf{y}$ in R^2.

25. Let $L: R^2 \to R^2$ be the linear transformation performing a counterclockwise rotation through $45°$; and let A be the matrix of L with respect to the natural basis for R^2. Show that A is orthogonal.

26. Let $L: R^2 \to R^2$ be defined by
$$L\left(\begin{bmatrix}x\\y\end{bmatrix}\right) = \begin{bmatrix}\frac{1}{\sqrt{2}} & \frac{1}{\sqrt{2}}\\\frac{1}{\sqrt{2}} & -\frac{1}{\sqrt{2}}\end{bmatrix}\begin{bmatrix}x\\y\end{bmatrix}.$$
Show that L is an isometry of R^2.

THEORETICAL EXERCISES ■

T.1. Complete the proof of Theorem 6.8.

T.2. Let $I: V \to V$ be the identity linear operator on an n-dimensional vector space V defined by $I(\mathbf{v}) = \mathbf{v}$ for every $\mathbf{v}$ in V. Show that the matrix of I with respect to a basis S for V is I_n.

T.3. Let $O: V \to W$ be the zero linear transformation, defined by $O(\mathbf{x}) = \mathbf{0}_W$, for any $\mathbf{x}$ in V. Show that the matrix of O with respect to any bases for V and W is the $m \times n$ zero matrix (where $n = \dim V$, $m = \dim W$).

T.4. Let $L: V \to V$ be a linear operator defined by $L(\mathbf{v}) = c\mathbf{v}$, where c is a fixed constant. Show that the matrix of L with respect to any basis for V is a scalar matrix (see Section 1.2).

T.5. Let $I: V \to V$ be the identity operator on an n-dimensional vector space V defined by $I(\mathbf{v}) = \mathbf{v}$ for every $\mathbf{v}$ in V. Show that the matrix of I with respect to the bases S and T for V is the transition matrix from the S-basis to the T-basis.

T.6. Let $L: V \to V$ be a linear operator. A nonempty subspace U of V is called **invariant** under L if $L(U)$ is contained in U. Let L be a linear operator with invariant subspace U. Show that if $\dim U = m$, and $\dim V = n$, then the matrix of L

with respect to a basis S for V is of the form

$$\begin{bmatrix} A & B \\ O & C \end{bmatrix},$$

where A is $m \times m$, B is $m \times (n-m)$, O is the zero $(n-m) \times m$ matrix, and C is $(n-m) \times (n-m)$.

T.7. Let $L: R^n \to R^n$ be a linear operator defined by $L(\mathbf{x}) = A\mathbf{x}$, for $\mathbf{x}$ in R^n. Show that L is one-to-one and onto if and only if A is nonsingular.

T.8. Let A be an $n \times n$ orthogonal matrix and let $L: R^n \to R^n$ be the linear transformation defined by $L(\mathbf{x}) = A\mathbf{x}$, for $\mathbf{x}$ in R^n. Let θ be the angle between the vectors $\mathbf{x}$ and $\mathbf{y}$ in R^n. Show that if A is orthogonal, then the angle between $L(\mathbf{x})$ and $L(\mathbf{y})$ is also θ.

T.9. Let $L: R^n \to R^n$ be a linear operator and $S = \{\mathbf{v}_1, \mathbf{v}_2, \dots, \mathbf{v}_n\}$ an orthonormal basis for R^n. Show that L is an isometry if and only if $T = \{L(\mathbf{v}_1), L(\mathbf{v}_2), \dots, L(\mathbf{v}_n)\}$ is an orthonormal basis for R^n.

T.10. Show that if a linear transformation $L: R^n \to R^n$ preserves length ($\|L(\mathbf{x})\| = \|\mathbf{x}\|$ for all $\mathbf{x}$ in R^n), then it also preserves the inner product; that is, $L(\mathbf{x}) \cdot L(\mathbf{y}) = \mathbf{x} \cdot \mathbf{y}$ for all $\mathbf{x}$ and $\mathbf{y}$ in R^n. [*Hint:* See Equation (10).]

MATLAB EXERCISES ■

In MATLAB, follow the steps given in this section to find the matrix of $L: R^n \to R^m$. The solution technique used in the MATLAB exercises of Section 4.7 will be helpful here.

ML.1. Let $L: R^3 \to R^2$ be given by

$$L\left(\begin{bmatrix} x \\ y \\ z \end{bmatrix}\right) = \begin{bmatrix} 2x - y \\ x + y - 3z \end{bmatrix}.$$

Find the matrix A representing L with respect to the bases

$$S = \{\mathbf{v}_1, \mathbf{v}_2, \mathbf{v}_3\} = \left\{ \begin{bmatrix} 1 \\ 1 \\ 1 \end{bmatrix}, \begin{bmatrix} 1 \\ 2 \\ 1 \end{bmatrix}, \begin{bmatrix} 0 \\ 1 \\ -1 \end{bmatrix} \right\}$$

and

$$T = \{\mathbf{w}_1, \mathbf{w}_2\} = \left\{ \begin{bmatrix} 1 \\ 2 \end{bmatrix}, \begin{bmatrix} 2 \\ 1 \end{bmatrix} \right\}.$$

ML.2. Let $L: R^3 \to R^4$ be given by $L(\mathbf{v}) = C\mathbf{v}$, where

$$C = \begin{bmatrix} 1 & 2 & 0 \\ 2 & 1 & -1 \\ 3 & 1 & 0 \\ -1 & 0 & 2 \end{bmatrix}.$$

Find the matrix A representing L with respect to the bases

$$S = \{\mathbf{v}_1, \mathbf{v}_2, \mathbf{v}_3\} = \left\{ \begin{bmatrix} 1 \\ 0 \\ 1 \end{bmatrix}, \begin{bmatrix} 2 \\ 0 \\ 1 \end{bmatrix}, \begin{bmatrix} 0 \\ 1 \\ 2 \end{bmatrix} \right\}$$

and

$$T = \{\mathbf{w}_1, \mathbf{w}_2, \mathbf{w}_3\}$$

$$= \left\{ \begin{bmatrix} 1 \\ 1 \\ 1 \\ 2 \end{bmatrix}, \begin{bmatrix} 1 \\ 1 \\ 1 \\ 0 \end{bmatrix}, \begin{bmatrix} 0 \\ 1 \\ 1 \\ -1 \end{bmatrix}, \begin{bmatrix} 0 \\ 0 \\ 1 \\ 0 \end{bmatrix} \right\}.$$

ML.3. Let $L: R^2 \to R^2$ be defined by

$$L\left(\begin{bmatrix} x \\ y \end{bmatrix} \right) = \begin{bmatrix} -x + 2y \\ 3x - y \end{bmatrix}$$

and let

$$S = \{\mathbf{v}_1, \mathbf{v}_2\} = \left\{ \begin{bmatrix} 1 \\ 2 \end{bmatrix}, \begin{bmatrix} -1 \\ 1 \end{bmatrix} \right\}$$

and

$$T = \{\mathbf{w}_1, \mathbf{w}_2\} = \left\{ \begin{bmatrix} -2 \\ 1 \end{bmatrix}, \begin{bmatrix} 1 \\ 1 \end{bmatrix} \right\}$$

be bases for R^2.

(a) Find the matrix A representing L with respect to S.

(b) Find the matrix B representing L with respect to T.

(c) Find the transition matrix P from T to S.

(d) Verify that $B = P^{-1}AP$.

KEY IDEAS FOR REVIEW

☐ **Theorem 6.1.** If $L: V \to W$ is a linear transformation, then

$$L(c_1\mathbf{v}_1 + c_2\mathbf{v}_2 + \cdots + c_k\mathbf{v}_k)$$
$$= c_1 L(\mathbf{v}_1) + c_2 L(\mathbf{v}_2) + \cdots + c_k L(\mathbf{v}_k).$$

☐ **Theorem 6.3.** Let $L: V \to W$ be a linear transformation and let $S = \{\mathbf{v}_1, \mathbf{v}_2, \ldots, \mathbf{v}_n\}$ be a basis for V. If $\mathbf{u}$ is any vector in V, then $L(\mathbf{u})$ is completely determined by $\{L(\mathbf{v}_1), L(\mathbf{v}_2), \ldots, L(\mathbf{v}_n)\}$.

☐ **Theorem 6.4.** If $L: V \to W$ is a linear transformation, then ker L is a subspace of V.

☐ **Theorem 6.5.** A linear transformation $L: V \to W$ is one-to-one if and only if ker $L = \{\mathbf{0}_V\}$.

☐ **Theorem 6.6.** If $L: V \to W$ is a linear transformation, then range L is a subspace of W.

☐ **Theorem 6.7.** If $L: V \to W$ is a linear transformation, then

$$\dim(\ker L) + \dim(\text{range } L) = \dim V.$$

☐ Matrix of a linear transformation $L: V \to W$. See Theorem 6.8.

☐ **Theorem 6.9.** Let $L: V \to V$ be a linear transformation where V is an n-dimensional vector space. Let $S = \{\mathbf{v}_1, \mathbf{v}_2, \ldots, \mathbf{v}_n\}$ and $T = \{\mathbf{w}_1, \mathbf{w}_2, \ldots, \mathbf{w}_n\}$ be bases for V and P be the transition matrix from T to S. If A is the matrix representing L with respect to S, then $P^{-1}AP$ is the matrix representing L with respect to T.

☐ **Theorem 6.10.** Consider the linear operator $L: R^n \to R^n$ defined by $L(\mathbf{x}) = A\mathbf{x}$ for $\mathbf{x}$ in R^n. Then A is diagonalizable with n linearly independent eigenvectors $\mathbf{x}_1, \mathbf{x}_2, \ldots, \mathbf{x}_n$ if and only if the matrix of L with respect to $S = \{\mathbf{x}_1, \mathbf{x}_2, \ldots, \mathbf{x}_n\}$ is diagonal.

☐ **List of Nonsingular Equivalences.** The following statements are equivalent for an $n \times n$ matrix A.

1. A is nonsingular.
2. $A\mathbf{x} = \mathbf{0}$ has only the trivial solution.
3. A is row equivalent to I_n.
4. The linear system $A\mathbf{x} = \mathbf{b}$ has a unique solution for every $n \times 1$ matrix $\mathbf{b}$.
5. $\det(A) \neq 0$.
6. A has rank n.
7. A has nullity 0.
8. The rows of A form a linearly independent set of n vectors in R^n.
9. The columns of A form a linearly independent set of n vectors in R^n.
10. Zero is *not* an eigenvalue of A.
11. The linear operator $L: R^n \to R^n$ defined by $L(\mathbf{x}) = A\mathbf{x}$, for $\mathbf{x}$ in R^n, is one-to-one and onto.

SUPPLEMENTARY EXERCISES ▪

1. Is $L: R^2 \to R^2$ defined by
$$L(x, y) = (x + y, 2x - y)$$
a linear transformation?

2. Is $L: P_2 \to P_2$ defined by
$$L(at^2 + bt + c) = (a - 1)t^2 + bt - c$$
a linear transformation?

3. Let $L: P_1 \to P_1$ be the linear transformation defined by
$$L(t - 1) = t + 2, \quad L(t + 1) = 2t + 1.$$
Find $L(5t + 1)$. [*Hint*: $\{t - 1, t + 1\}$ is a basis for P_1.]

4. Let $L: P_2 \to P_2$ be defined by
$$L(at^2 + bt + c) = (a - 2c)t^2 + (b - c)t.$$
   (a) Find a basis for ker L.
   (b) Is L one-to-one?

5. Let $L: R^2 \to R^3$ be defined by
$$L(x, y) = (x + y, x - y, x + 2y).$$
   (a) Find a basis for range L.
   (b) Is L onto?

6. Find the dimension of the solution space of $A\mathbf{x} = \mathbf{0}$, where
$$A = \begin{bmatrix} 1 & -2 & -1 & 2 & -1 \\ 4 & -7 & 1 & 4 & 0 \\ 1 & -3 & -6 & 4 & -4 \\ 2 & -3 & 3 & 2 & 1 \end{bmatrix}.$$

7. Let $S = \{t^2 - 1, t + 2, t - 1\}$ be a basis for P_2.
   (a) Find the coordinate vector of $2t^2 - 2t + 6$ with respect to S.
   (b) If the coordinate vector of $p(t)$ with respect to S is
$$\begin{bmatrix} 2 \\ -1 \\ 3 \end{bmatrix},$$
   find $p(t)$.

8. Let $L: P_2 \to P_2$ be the linear transformation that is defined by
$$L(at^2 + bt + c) = (a + 2c)t^2 + (b - c)t + (a - c).$$
   Let $S = \{t^2, t, 1\}$ and $T = \{t^2 - 1, t, t - 1\}$ be bases for P_2.
   (a) Find the matrix of L with respect to S and T.
   (b) If $p(t) = 2t^2 - 3t + 1$, compute $L[p(t)]$ using the matrix obtained in part (a).

9. Let $L: P_1 \to P_1$ be a linear transformation. Suppose that the matrix of L with respect to the basis $S = \{p_1(t), p_2(t)\}$ is
$$A = \begin{bmatrix} 2 & -3 \\ 1 & 2 \end{bmatrix},$$
   where
$$p_1(t) = t - 2 \quad \text{and} \quad p_2(t) = t + 1.$$
   (a) Compute $\left[L[p_1(t)]\right]_S$ and $\left[L[p_2(t)]\right]_S$.
   (b) Compute $L[p_1(t)]$ and $L[p_2(t)]$.
   (c) Compute $L(t + 2)$.

10. Let $L: P_3 \to P_3$ be defined by
$$L(at^3 + bt^2 + ct + d) = 3at^2 + 2bt + c.$$
   Find the matrix of L with respect to the basis $S = \{t^3, t^2, t, 1\}$ for P_3.

11. Let $L: R^3 \to R^3$ be the linear transformation defined by
$$L\left(\begin{bmatrix} x \\ y \\ z \end{bmatrix}\right) = \begin{bmatrix} y + z \\ x + z \\ x + y \end{bmatrix}.$$
   Let S be the natural basis for R^3 and let
$$T = \left\{ \begin{bmatrix} 0 \\ 1 \\ 1 \end{bmatrix}, \begin{bmatrix} 1 \\ 0 \\ 1 \end{bmatrix}, \begin{bmatrix} 1 \\ 1 \\ 0 \end{bmatrix} \right\}$$
   be another basis for R^3. Find the matrix of L with respect to S and T.

12. Let $L: M_{nn} \to R^1$ be defined by $L(A) = \det(A)$, for A in M_{nn}. Is L a linear transformation? Justify your answer.

13. Let A be a fixed $n \times n$ matrix. Define $L: M_{nn} \to M_{nn}$ by $L(B) = AB - BA$ for B in M_{nn}. Is L a linear transformation? Justify your answer.

14. Let P be a fixed nonsingular $n \times n$ matrix. Define $L: M_{nn} \to M_{nn}$ by $L(A) = P^{-1}AP$ for A in M_{nn}. Is L a linear transformation? Justify your answer.

15. (**Calculus Required**).   Let $V = C[0, 1]$, the vector space of all real-valued continuous functions that are defined on $[0, 1]$ and let $L: V \to R^1$ be given by $L(f) = f(0)$, for f in V.
   (a) Show that L is a linear transformation.
   (b) Describe the kernel of L and give examples of polynomials, quotients of polynomials, and trigonometric functions that belong to ker L.
   (c) If we redefine L by $L(f) = f\left(\frac{1}{2}\right)$, is it still a linear transformation? Explain.

THEORETICAL EXERCISES

T.1. Let V be an n-dimensional vector space with basis $S = \{\mathbf{v}_1, \mathbf{v}_2, \ldots, \mathbf{v}_n\}$. Show that
$$\left\{ \left[\mathbf{v}_1\right]_S, \left[\mathbf{v}_2\right]_S, \ldots, \left[\mathbf{v}_n\right]_S \right\}$$
is the natural basis for R^n.

T.2. Let V be an n-dimensional vector space with basis $S = \{\mathbf{v}_1, \mathbf{v}_2, \ldots, \mathbf{v}_n\}$. If $\mathbf{v}$ and $\mathbf{w}$ are vectors in V and c is a scalar, show that
$$\left[\mathbf{v} + \mathbf{w}\right]_S = \left[\mathbf{v}\right]_S + \left[\mathbf{w}\right]_S$$
$$\left[c\mathbf{v}\right]_S = c\left[\mathbf{v}\right]_S.$$

T.3. Let V and W be two vector spaces of dimensions n and m, respectively. If $L_1: V \to W$ and $L_2: V \to W$ are linear transformations, we define
$$L_1 \boxplus L_2: V \to W$$
by
$$(L_1 \boxplus L_2)(\mathbf{v}) = L_1(\mathbf{v}) + L_2(\mathbf{v}),$$
for $\mathbf{v}$ in V. Also, if $L: V \to W$ is a linear transformation and c is a scalar, we define
$$c \boxdot L: V \to W$$
by
$$(c \boxdot L)(\mathbf{v}) = cL(\mathbf{v}),$$
for $\mathbf{v}$ in V.
(a) Show that $L_1 \boxplus L_2$ is a linear transformation.
(b) Show that $c \boxdot L$ is a linear transformation.

(c) Let $V = R^3$, $W = R^2$, $L_1: V \to W$, and $L_2: V \to W$ be defined by
$$L_1(\mathbf{v}) = L_1(v_1, v_2, v_3) = (v_1 + v_2, v_2 + v_3)$$
$$L_2(\mathbf{v}) = L_2(v_1, v_2, v_3) = (v_1 + v_3, v_2).$$
Compute $(L_1 \boxplus L_2)(\mathbf{v})$ and $(-2 \boxdot L_1)(\mathbf{v})$.

T.4. Show that the set U of all linear transformations of an n-dimensional vector space into an m-dimensional vector space W is a vector space under the operations $\boxplus$ and $\boxdot$ defined in Supplementary Exercise T.3.

T.5. Let $L: R^n \to R^m$ be a linear transformation defined by $L(\mathbf{x}) = A\mathbf{x}$, $\mathbf{x}$ in R^n, where A is an $m \times n$ matrix.
(a) Show that L is one-to-one if and only if rank $A = n$.
(b) Show that L is onto if and only if rank $A = m$.

T.6. Let V be an n-dimensional vector space and $S = \{\mathbf{v}_1, \mathbf{v}_2, \ldots, \mathbf{v}_n\}$ a basis for V. Define $L: R^n \to V$ as follows: If $\mathbf{v} = (a_1, a_2, \ldots, a_n)$ is a vector in R^n, let
$L(\mathbf{v}) = a_1\mathbf{v}_1 + a_2\mathbf{v}_2 + \cdots + a_n\mathbf{v}_n$. Show that
(a) L is a linear transformation.
(b) L is one-to-one.
(c) L is onto.

CHAPTER TEST

1. Let $L: R^2 \to R^3$ be a linear transformation for which we know that
$$L\left(\begin{bmatrix} 1 \\ -1 \end{bmatrix}\right) = \begin{bmatrix} 1 \\ 2 \\ -1 \end{bmatrix}$$
and
$$L\left(\begin{bmatrix} 2 \\ -1 \end{bmatrix}\right) = \begin{bmatrix} 0 \\ 1 \\ 2 \end{bmatrix}.$$
Find $L\left(\begin{bmatrix} 8 \\ -5 \end{bmatrix}\right)$.

2. Let $L: R^3 \to R^3$ be defined by
$$L(x, y, z) = (x + 2y + z, x + y, 2y + z).$$

(a) Find a basis for ker L.
(b) Is L one-to-one?

3. Let $L: R^2 \to R^3$ be defined by
$$L(x, y) = (x + y, x - y, 2x + y).$$
(a) Find a basis for range L.
(b) Is L onto?

4. Find the dimension of the solution space of $A\mathbf{x} = \mathbf{0}$, where
$$A = \begin{bmatrix} 1 & 0 & -1 & 2 & 3 \\ 2 & 1 & -2 & 2 & 1 \\ 0 & 1 & -6 & 8 & 8 \\ 1 & -1 & -1 & 4 & 8 \end{bmatrix}.$$

5. Let $L: R^2 \to R^2$ be the linear transformation defined by

$$L\left(\begin{bmatrix} x \\ y \end{bmatrix}\right) = \begin{bmatrix} x + 2y \\ x - y \end{bmatrix}.$$

Let

$$S = \left\{ \begin{bmatrix} 1 \\ -1 \end{bmatrix}, \begin{bmatrix} 0 \\ 2 \end{bmatrix} \right\}$$

and

$$T = \left\{ \begin{bmatrix} 1 \\ 2 \end{bmatrix}, \begin{bmatrix} -1 \\ 2 \end{bmatrix} \right\}$$

be bases for R^2. Find the matrix of L with respect to S and T.

6. Answer each of the following as true or false.

(a) If $L: P_2 \to P_1$ is the linear transformation defined by

$$L(at^2 + bt + c) = (a - c)t + (b + c),$$

then $t^2 + 2t + 1$ is in ker L.

(b) If $L: R^2 \to R^2$ is the linear transformation defined by

$$L(x, y) = (x - y, x + y),$$

then $(2, 3)$ is in range L.

(c) If A is a 4×7 matrix whose rank is 4, then the homogeneous system $A\mathbf{x} = \mathbf{0}$ has a nontrivial solution.

(d) There are linear transformations $L: R^3 \to R^5$ that are onto.

(e) If $L: R^7 \to R^5$ is a linear transformation such that $\dim(\ker L) = 3$, then $\dim(\text{range } L) = 2$.

CUMULATIVE REVIEW OF PART I

Answer each of the following as true (T) or false (F). Justify your answer.

1. If an $n \times n$ matrix A is singular, then A has either a row or a column of zeros.

2. A diagonal matrix is nonsingular if and only if none of the entries on its main diagonal are zero.

3. Two vectors in R^3 always span a two-dimensional subspace.

4. If $|AB| = 12$ and $|A| = 4$, then $|B| = 48$.

5. Let $L: R^6 \to R^{10}$ be a linear transformation defined by $L(\mathbf{x}) = A\mathbf{x}$ for $\mathbf{x}$ in R^6. If $\dim(\text{range } L) = 3$, then $\dim(\ker L) = 7$.

6. If AB is singular, then A is singular or B is singular.

7. Let $L: R^n \to R^n$ be a linear transformation defined by $L(\mathbf{x}) = A\mathbf{x}$. Then L is onto if and only if $\det(A) \neq 0$.

8. The columns of a 5×8 matrix whose rank is 5 form a linearly dependent set.

9. If A is an $m \times n$ matrix with $m < n$, then the linear system $A\mathbf{x} = \mathbf{b}$ has a solution for every $m \times 1$ matrix $\mathbf{b}$.

10. Let $L: R^n \to R^n$ be the linear transformation defined by $L(\mathbf{x}) = A\mathbf{x}$, for $\mathbf{x}$ in R^n. Then L is onto if and only if A is nonsingular.

11. If A is an $n \times n$ matrix, then rank $A < n$ if and only if some eigenvalue of A is zero.

12. Let A be an $n \times n$ matrix. If $A\mathbf{x} = A\mathbf{y}$, then $\mathbf{x} = \mathbf{y}$.

13. Span $\{(1, 1, 0), (0, 1, -1), (1, 0, 1)\} = R^3$.

14. If no row in an $n \times n$ matrix is a multiple of another row of A, then $\det(A) \neq 0$.

15. If $L: V \to W$ is a linear transformation, then $\ker L = V$ if and only if range $L = \{\mathbf{0}_W\}$.

16. The inverse of a nonsingular diagonal matrix is a diagonal matrix.

17. If A and B are $n \times n$ matrices, then $(A + B)^2 = A^2 + 2AB + B^2$.

18. If $\mathbf{u}$ and $\mathbf{v}$ are vectors in R^n, then $\|\mathbf{u} - \mathbf{v}\|^2 = \|\mathbf{u}\|^2 - \|\mathbf{v}\|^2$.

19. There are real vector spaces containing exactly seven vectors.

20. If $\mathbf{u}_1, \mathbf{u}_2, \ldots, \mathbf{u}_k$ are orthogonal to $\mathbf{v}$, then every vector in span $\{\mathbf{u}_1, \mathbf{u}_2, \ldots, \mathbf{u}_k\}$ is orthogonal to $\mathbf{v}$.

21. $\det(A) = 0$ if and only if some eigenvalue of A is zero.

22. If λ is an eigenvalue of A of multiplicity k, then the dimension of the eigenspace associated with λ is k.

23. If A and B are $n \times n$ matrices, then $(A^T B^T)^T = BA$.

24. Let A be an $n \times n$ singular matrix. If the linear system $A\mathbf{x} = \mathbf{b}$ has a solution for $\mathbf{b} \neq 0$, then it has infinitely many solutions.

25. Let $L: R^n \to R^n$ be a linear transformation defined by $L(\mathbf{x}) = A\mathbf{x}$, for $\mathbf{x}$ in R^n. Then A is singular if and only if $\ker L = \{\mathbf{0}_V\}$.

26. The product of two diagonal matrices is always a diagonal matrix.

27. If A is an $n \times n$ matrix that is row equivalent to I_n, then A is singular.

28. If
$$A = \begin{bmatrix} a-3 & 2 & 1 \\ -1 & a & 1 \\ 0 & 0 & 1 \end{bmatrix},$$
then the only value of a for which the linear system $A\mathbf{x} = \mathbf{0}$ has a nontrivial solution is $a = 2$.

29. If A is a 3×3 matrix and $|A| = 3$, then $\left| \frac{1}{2} A^{-1} \right| = \frac{8}{3}$.

30. The linear system $A\mathbf{x} = \mathbf{b}$ has a solution if and only if $\mathbf{b}$ is in the column space of A.

31. Let $L: R^n \to R^n$ be a linear transformation defined by $L(\mathbf{x}) = A\mathbf{x}$, for $\mathbf{x}$ in R^n. Then $\dim(\text{range } L) = n$ if and only if rank $A = n$.

32. If W is a subspace of a finite-dimensional vector space V such that $\dim W = \dim V$, then $W = V$.

33. If A is similar to B, then rank $A = $ rank B.

34. The set of all vectors of the form $(a, b, -a)$ is a subspace of R^3.

35. If the columns of an $n \times n$ matrix span R^n, then the rows are linearly independent.

36. If A is an $n \times n$ matrix, then rank $A = n$.

37. Let λ_1 and λ_2 be eigenvalues of A with associated eigenvectors $\mathbf{x}_1$ and $\mathbf{x}_2$. If $\lambda_1 = \lambda_2$, then $\mathbf{x}_1$ and $\mathbf{x}_2$ are linearly independent.

38. Every orthogonal set of n vectors in R^n is a basis for R^n.

39. The set of all solutions to the linear system $A\mathbf{x} = \mathbf{b}$, where A is $m \times n$ and $\mathbf{b} \neq \mathbf{0}$, is a subspace of R^n.

40. If $\mathbf{u}$ is a vector in R^n such that $\mathbf{u} \cdot \mathbf{v} = 0$ for all $\mathbf{v}$ in R^n, then $\mathbf{u} = \mathbf{0}$.

41. Every set of linearly independent vectors in R^3 contains three vectors.

42. If $\det(A) = 0$, then the linear system $A\mathbf{x} = \mathbf{b}$, $\mathbf{b} \neq \mathbf{0}$, has no solution.

43. Let V be an n-dimensional vector space. If a set of m vectors spans V, then $m = n$.

44. Every set of five orthonormal vectors is a basis for R^5.

45. It is possible for a vector space V to have more than one orthonormal basis.

46. If the set of vectors S spans a vector space V, then every subset of S also spans V.

47. If an $n \times n$ matrix A is diagonalizable, then A^3 is diagonalizable.

48. If $\mathbf{x}$ and $\mathbf{y}$ are eigenvectors of A associated with the eigenvalue λ, then $\mathbf{x} + \mathbf{y}$ is also an eigenvector of A associated with λ.

49. If $\mathbf{x}_0$ is a nontrivial solution to the homogeneous system $A\mathbf{x} = \mathbf{0}$, where A is $n \times n$, then $\mathbf{x}_0$ is an eigenvector of A associated with the eigenvalue 0.

50. If A is similar to B, then A^n is similar to B^n for each positive integer n.

51. If A is an $m \times n$ matrix, then

$\dim(\text{null space of } A)$
$\qquad + \dim(\text{column space of } A) = n.$

52. The set of solutions to any linear system is always a subspace.

53. The number of linearly independent eigenvectors of a matrix is always greater than or equal to the number of distinct eigenvalues.

54. Every finite set of vectors in a vector space that contains the zero vector is linearly dependent.

55. If the rows of a 4×6 matrix are linearly independent, then the column rank $= 4$.

56. If $\mathbf{v}_1$, $\mathbf{v}_2$, $\mathbf{v}_3$, $\mathbf{v}_4$, and $\mathbf{v}_5$ are linearly dependent vectors in a vector space, then $\mathbf{v}_1$, $\mathbf{v}_2$, and $\mathbf{v}_3$ are linearly dependent.

57. $\det(ABC) = \det(BAC)$.

58. The set of all $n \times n$ symmetric matrices form a subspace of M_{nn}.

59. Every linear system $A\mathbf{x} = \mathbf{0}$, where A is an $m \times n$ matrix, has a solution $\mathbf{x} \neq \mathbf{0}$ if $m < n$.

60. If V is a subspace of R^5, then $1 \leq \dim V < 5$.

61. A diagonalizable $n \times n$ matrix must always have n distinct eigenvalues.

62. Every symmetric matrix is diagonalizable.

63. If c and d are scalars and $\mathbf{u}$ is a vector in R^n such that $c\mathbf{u} = d\mathbf{u}$, then $c = d$.

64. The linear transformation $L\colon P_2 \to P_2$ defined by
$$L(at^2 + bt + c) = 2at + b$$
is one-to-one.

65. If $\mathbf{u}$ and $\mathbf{v}$ are any vectors in R^n, then
$-1 \le \mathbf{u} \cdot \mathbf{v} \le 1$.

66. If A and B are $n \times n$ matrices, then
$(A + B)^{-1} = A^{-1} + B^{-1}$.

67. The rows of an $n \times n$ matrix A always form a basis for the row space of A.

68. If A and B are $n \times n$ matrices with $\det(AB) \ne 0$, then A and B are both nonsingular.

69. If A and B are $n \times n$ orthogonal matrices, then AB and BA are orthogonal matrices.

70. If A is a nonsingular matrix and $A\mathbf{u} = A\mathbf{v}$, then $\mathbf{u} = \mathbf{v}$.

71. If A and B are $n \times n$ nonsingular matrices, then $(3AB)^{-1} = \frac{1}{3}B^{-1}A^{-1}$.

72. If A is a 3×3 matrix and $\det(A) = 2$, then $\det(3A) = 6$.

73. If A is a singular matrix, then A^2 is singular.

74. If λ is an eigenvalue of an $n \times n$ matrix A, then $\lambda I_n - A$ is a singular matrix.

75. If A is an $n \times n$ matrix such that $A^2 = O$, then $A = O$.

PART II

APPLICATIONS

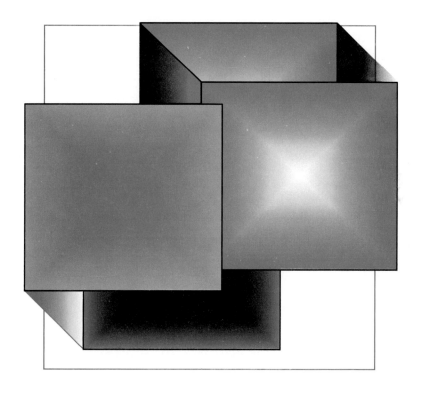

7

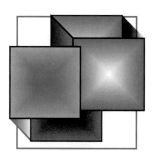

Linear Programming

PREREQUISITE: Section 1.5, Solutions of Equations.

In this chapter we provide an introduction to the basic ideas and techniques of linear programming. Linear programming is a recent area of applied mathematics, developed in the late 1940s, to solve a number of problems for the federal government. Since then, it has been applied to an amazing number of problems in many areas. As a vital tool in management science and operations research, it has resulted in enormous savings of money. In the first section we give some examples of linear programming problems, formulate their mathematical models, and describe a geometric solution method. In the second section we present an algebraic method for solving linear programming problems. In the third section, dealing with duality, we discuss several applied interpretations of two related linear programming problems.

7.1 ▼ The Linear Programming Problem; Geometric Solution

In many problems in business and industry we are interested in making decisions that will maximize or minimize some quantity. For example, a plant manager may want to determine the most economical way of shipping goods from the factory to the markets, a hospital may want to design a diet satisfying certain nutritional requirements at a minimum cost, an investor may want to select investments that will maximize profits, or a manufacturer may wish to blend ingredients, subject to given specifications, to maximize profit. In this section we give several examples of linear programming problems and show how mathematical models for them can be formulated. Their geometric solution is also considered in this section.

EXAMPLE 1 ■
(*A Production Problem*)

A small manufacturer of photographic products prepares two types of film developers each day, Fine and Extra Fine, using as the raw material solutions A and B. Suppose that each quart of Fine contains 2 ounces of solution A and 1 ounce of solution B, while each quart of Extra Fine contains 1 ounce of solution A and 2 ounces of solution B. Suppose also that the profit on each quart of Fine is 8 cents and that it is 10 cents on each quart of Extra Fine. If the firm has 50 ounces of solution A and 70 ounces of solution B available each day, how many quarts of Fine and how many quarts of Extra Fine should be made each day to maximize profit (assuming that the shop can sell all that is made)?

Mathematical Formulation

Let x be the number of quarts of Fine to be made and let y be the number of quarts of Extra Fine to be made. Since each quart of Fine contains 2 ounces of solution A and each quart of Extra Fine contains 1 ounce of solution A, the total amount of solution A required is

$$2x + y.$$

Similarly, since each quart of Fine contains one ounce of solution B and each quart of Extra Fine contains 2 ounces of solution B, the total amount of solution B required is

$$x + 2y.$$

Since we only have 50 ounces of solution A and 70 ounces of solution B available, we must have

$$2x + \ y \le 50$$
$$x + 2y \le 70.$$

Of course, x and y cannot be negative, so we must also have

$$x \ge 0 \quad \text{and} \quad y \ge 0.$$

Since the profit on each quart of Fine is 8 cents and the profit on each quart of Extra Fine is 10 cents, the total profit (in cents) is

$$z = 8x + 10y.$$

Our problem can be stated in mathematical form as: Find values of x and y that will maximize

$$z = 8x + 10y$$

subject to the following restrictions that must be satisfied by x and y:

$$2x + \ y \le 50$$
$$x + 2y \le 70$$
$$x \ge \ 0$$
$$y \ge \ 0.$$
∎

EXAMPLE 2 ∎ *(Pollution)*

A manufacturer of a certain chemical product has two plants where the product is made. Plant X can make at most 30 tons per week and plant Y can make at most 40 tons per week. The manufacturer wants to make a total of at least 50 tons per week. The amount of particulate matter found weekly in the atmosphere over a nearby town is measured and found to be 20 pounds for each ton of the product made by plant X and 30 pounds for each ton of the product made at plant Y. How many tons should be made weekly at each plant to minimize the total amount of particulate matter in the atmosphere?

Mathematical Formulation

Let x and y denote the number of tons of the product made at plants X and Y, respectively, per week. Then the total amount manufactured per week is

$$x + y.$$

Since we want to make at least 50 tons per week, we must have

$$x + y \geq 50.$$

Since plant X can make at most 30 tons, we must have

$$x \leq 30.$$

Similarly, since plant Y can make at most 40 tons, we must have

$$y \leq 40.$$

Of course, x and y cannot be negative, so we require

$$x \geq 0$$
$$y \geq 0.$$

The total amount of particulate matter (in pounds) is

$$z = 20x + 30y,$$

which we want to minimize. Thus a mathematical statement of our problem is: Find values of x and y that will minimize

$$z = 20x + 30y$$

subject to the following restrictions that must be satisfied by x and y:

$$x + y \geq 50$$
$$x \leq 30$$
$$y \leq 40$$
$$x \geq 0$$
$$y \geq 0.$$ ∎

EXAMPLE 3 ∎ (*The Diet Problem*)

A nutritionist is planning a menu that includes foods A and B as its main staples. Suppose that each ounce of food A contains 2 units of protein, 1 unit of iron, and 1 unit of thiamine; each ounce of food B contains 1 unit of protein, 1 unit of iron, and 3 units of thiamine. Suppose that each ounce of A costs 30 cents, while each ounce of B costs 40 cents. The nutritionist wants the meal to provide at least 12 units of protein, at least 9 units of iron, and at least 15 units of thiamine. How many ounces of each of the foods should be used to minimize the cost of the meal?

Mathematical Formulation

Let x denote the number of ounces of food A to be used and let y be the number of ounces of food B. The number of units of protein supplied by the meal is

$$2x + y,$$

so we must have
$$2x + y \geq 12.$$
The number of units of iron supplied by the meal is
$$x + y,$$
so we must have
$$x + y \geq 9.$$
Since the number of units of thiamine supplied by the meal is
$$x + 3y,$$
we must have
$$x + 3y \geq 15.$$
Of course, we also require
$$x \geq 0, \quad y \geq 0.$$
The cost of the meal (in cents) is
$$z = 30x + 40y,$$
which we want to minimize. Thus a mathematical formulation of our problem is: Find values of x and y that will minimize
$$z = 30x + 40y$$
subject to the restrictions
$$
\begin{aligned}
2x + \ y &\geq 12 \\
x + \ y &\geq \ 9 \\
x + 3y &\geq 15 \\
x &\geq \ 0 \\
y &\geq \ 0.
\end{aligned}
$$
■

These examples are typical of linear programming problems, the general form being: Find values of $x_1, x_2, \ldots, x_n$ that will minimize or maximize
$$z = c_1 x_1 + c_2 x_2 + \cdots + c_n x_n \tag{1}$$
subject to

$$
\left.
\begin{aligned}
a_{11} x_1 + a_{12} x_2 + \cdots + a_{1n} x_n \ (\leq)(\geq)(=) \ b_1 \\
a_{21} x_1 + a_{22} x_2 + \cdots + a_{2n} x_n \ (\leq)(\geq)(=) \ b_2 \\
\vdots \qquad \vdots \qquad\qquad \vdots \qquad\qquad \vdots \\
a_{m1} x_1 + a_{m2} x_2 + \cdots + a_{mn} x_n \ (\leq)(\geq)(=) \ b_m
\end{aligned}
\right\} \tag{2}
$$

$$x_j \geq 0 \qquad (1 \leq j \leq n), \tag{3}$$

where in (2) one and only one of the symbols $\leq$, $\geq$, $=$, occurs in each statement. The linear function in (1) is called the **objective function**. The equalities or inequalities in (2) and (3) are called **constraints**. The term **linear** in linear programming means that the objective function (1) and each of the constraints in (2) are linear functions of the variables $x_1, x_2, \ldots, x_n$. The word "programming," *not* to be confused with its use in computer programming, refers to the applications in planning or allocation problems.

Geometric Solution

We now develop a geometric method for solving linear programming problems in two variables. This will enable us to solve the problems posed earlier. Since linear programming deals with systems of linear inequalities, we first consider these from a geometric point of view.

Consider Example 1. Find the set of points that maximize

$$z = 8x + 10y \tag{4}$$

subject to the constraints

$$\begin{aligned} 2x + \ y &\le 50 \\ x + 2y &\le 70 \\ x &\ge \ 0 \\ y &\ge \ 0. \end{aligned} \tag{5}$$

The set of points satisfying the system of four inequalities (5) consists of those points satisfying each of the inequalities. In Figure 7.1(a) we show the set of points satisfying the inequality $x \ge 0$, in Figure 7.1(b) the set of points satisfying the inequality $y \ge 0$. The set of points satisfying both inequalities $x \ge 0$ and $y \ge 0$ is the intersection of the regions in Figure 7.1(a) and (b). This set of points, the set of all points in the first quadrant, is shown in Figure 7.1(c).

FIGURE 7.1

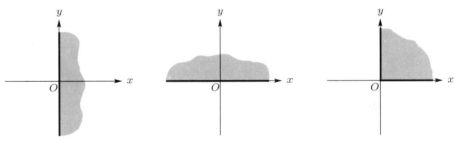

(a) Set of points satisfying $x \ge 0$.

(b) Set of points satisfying $y \ge 0$.

(c) Set of points satisfying $x \ge 0$ and $y \ge 0$.

We must now consider the set of points satisfying the inequality

$$2x + y \le 50. \tag{6}$$

First, consider the set of points satisfying the strict inequality

$$2x + y < 50. \tag{7}$$

The straight line

$$2x + y = 50 \tag{8}$$

divides the set of points not on the line into two regions (Figure 7.2); one region contains all points satisfying (7) and the other region contains all points not satisfying (7). The line itself, drawn in dashed form, belongs to neither region.

FIGURE 7.2

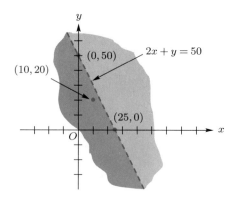

To determine the region determined by inequality (7), we choose any point not on the line as a test point and see on which side of the line it lies. The point $(10, 20)$ does not lie on the line (8) and can thus serve as a test point. Since its coordinates satisfy inequality (7), the correct region determined by (7) is shown in Figure 7.3(a). Another possible test point is $(20, 20)$, since it does not lie on the line. Its coordinates do not satisfy inequality (7), so the test point does not lie in the correct region. Thus the correct region would again be as shown in Figure 7.3(a).

FIGURE 7.3

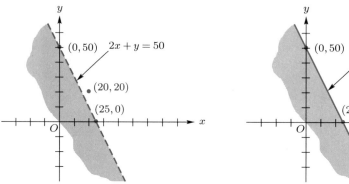

(a) Set of points satisfying $2x + y < 50$. (b) Set of points satisfying $2x + y \leq 50$.

The set of points satisfying inequality (6) consists of the set of points satisfying (7) as well as the points on the line (8). Thus the region, shown in Figure 7.3(b), includes the straight line, which is drawn solidly.

Next, the set of points satisfying the inequality

$$x + 2y \leq 70 \tag{9}$$

is shown in Figure 7.4.

The set of points satisfying inequalities (6) and (9) is the intersection of the regions in Figures 7.3(b) and 7.4; it is the shaded region shown in Figure 7.5.

FIGURE 7.4

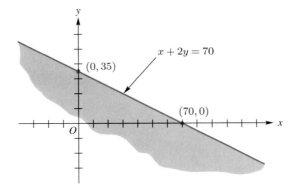

Set of points satisfying $x + 2y \leq 70$.

Finally, the set of points satisfying the four inequalities (5) is the intersection of the regions in Figures 7.1(c) and 7.5. That is, it is the set of points in Figure 7.5 that lie in the first quadrant. This set of points is shown in Figure 7.6.

FIGURE 7.5

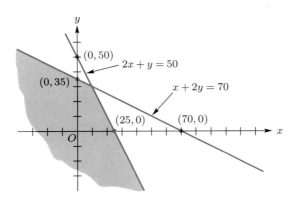

Set of points satisfying $2x + y \leq 50$ and $x + 2y \leq 70$.

Thus the set of points satisfying a system of inequalities is the intersection of the sets of points satisfying each of the inequalities.

A point satisfying the constraints of a linear programming problem is called a **feasible solution**; the set of all such points is the **feasible region**. It should be noted that not every linear programming problem has a feasible region, as the next example shows.

EXAMPLE 4 ■ Find the set of points that maximize

$$z = 8x + 10y \tag{10}$$

FIGURE 7.6

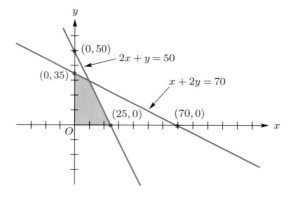

subject to the constraints

$$2x + 3y \leq 6$$
$$x + 2y \geq 6$$
$$x \geq 0$$
$$y \geq 0.$$

(11)

In Figure 7.7 we have labeled as Region I the set of points satisfying the first, third, and fourth inequalities; that is, Region I is the set of points in the first quadrant satisfying the first inequality in (11). Similarly, Region II is the set of points satisfying the second, third, and fourth inequalities. It is clear that there are no points satisfying all four inequalities. ■

FIGURE 7.7

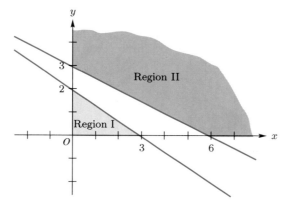

To solve the linear programming problem in Example 1, we must find a feasible solution that makes the objective function (1) as large as possible. Such a solution is called an **optimal solution**. Since there are infinitely many feasible points, it appears, at first glance, quite difficult to tell whether a feasible solution is optimal or not. In Figure 7.8 we have reproduced Figure 7.6 by showing the feasible region for this linear programming problem along with a number of indicated feasible points: O, A, B, C, D, and E. Points F and G are not in the feasible region.

FIGURE 7.8

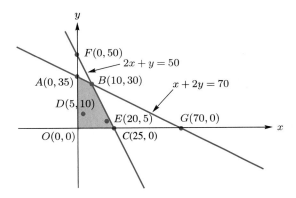

In Table 7.1 we have tabulated the value of the objective function (in cents) for each of the points O, A, B, C, D, and E. Among the feasible solutions O, A, B, C, D, and E we see that the value of the objective function is largest for the point $B(10, 30)$. However, we do not know if there is another feasible solution at which the value of the objective function will be larger than it is at B. Since there are infinitely many feasible solutions in the region, we cannot examine the value of the objective function at every one of these. However, we shall be able to find optimal solutions without examining all feasible solutions, but we first consider some auxiliary notions.

Table 7.1

Point	Value of $z = 8x + 10y$ (cents)
$O(0,0)$	0
$A(0,35)$	350
$B(10,30)$	380
$C(25,0)$	200
$D(5,10)$	140
$E(20,5)$	210

DEFINITION   The **line segment** joining the distinct points $\mathbf{x}_1$ and $\mathbf{x}_2$ of R^n is the set of all points in R^n of the form $\lambda\mathbf{x}_1 + (1 - \lambda)\mathbf{x}_2$, $0 \le \lambda \le 1$. Observe that if $\lambda = 0$, we get $\mathbf{x}_2$, and if $\lambda = 1$, we get $\mathbf{x}_1$.

DEFINITION   A nonempty set S of points in R^n is called **convex** if the line segment joining any two points in S also lies entirely in S.

FIGURE 7.9
Convex sets in R^2

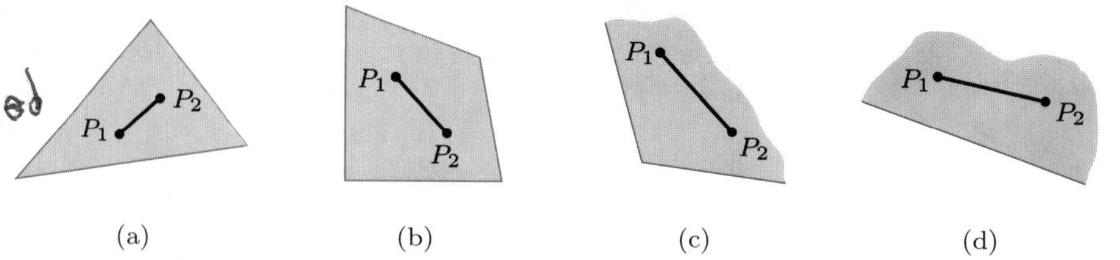

Bounded

(a)   (b)   (c)   (d)

FIGURE 7.10
Sets in R^2 that
are not convex

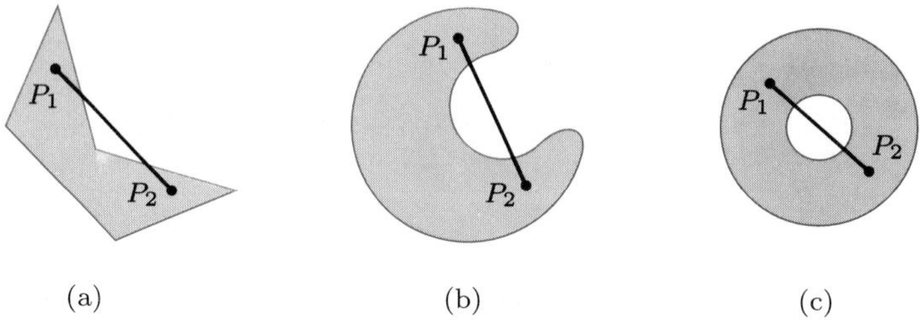

unBound

(a)   (b)   (c)

EXAMPLE 5 ■ The shaded sets in Figure 7.9(a), (b), (c), and (d) are convex sets in R^2. The shaded sets in Figure 7.10(a), (b), and (c) are *not* convex sets in R^2, since the line joining the indicated points does not lie entirely in the set. ■

EXAMPLE 6 ■ It is not too difficult to show that the feasible region of a linear programming problem is a convex set. The proof of this result for a broad class of linear programming problems is outlined in Exercise T.1 of Section 7.2. ■

The reader may also check that the feasible regions for the linear programming problems in Examples 2 and 3 are convex.

We shall now limit our further discussion of convex sets to such sets in R^2, although the ideas presented here can be generalized to convex sets in R^n.

Convex sets are either **bounded** or **unbounded**. A bounded convex set is one that can be enclosed by a sufficiently large rectangle; an unbounded convex set is one that cannot be so enclosed. The convex sets in Figure 7.9(a) and (b) are bounded; the convex sets in Figure 7.9(c) and (d) are unbounded.

DEFINITION   An **extreme point** in a convex set S is a point *in* S that is not an interior point of any line segment in S.

EXAMPLE 7 ■ In Figure 7.11 we have marked each extreme point of the convex sets in Figure 7.9(a), (b), and (c) with a solid dot. The convex set in Figure 7.9(d) has no extreme points. ■

FIGURE 7.11

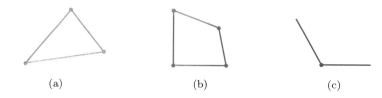

(a)   (b)   (c)

The basic result connecting convex sets, extreme points, and linear programming problems is the following theorem, whose proof we omit.

THEOREM 7.1 ■ *Let S be the feasible region of a linear programming problem.*

(a) *If S is bounded, then the objective function*

$$z = ax + by$$

assumes both a maximum and a minimum value on S; these values occur at extreme points of S.

(b) *If S is unbounded, then there may or may not be a maximum or minimum value on S. If a maximum or minimum value does exist on S, it occurs at an extreme point.* ■

Thus, when the feasible region S of a linear programming problem is bounded, a method for solving the problem consists of finding the extreme points of S and evaluating the objective function $z = ax + by$ at each extreme point. An optimal solution is an extreme point at which the value of z is a maximum or a minimum.

The procedure for solving a linear programming problem in two variables geometrically is as follows.

Step 1. Sketch the feasible region S.

Step 2. Determine all the extreme points of S.

Step 3. Evaluate the objective function at each extreme point.

Step 4. Choose an extreme point at which the objective function is largest (smallest) for a maximization (minimization) problem.

EXAMPLE 8 ■ Consider Example 1 again. The feasible region shown in Figure 7.8 is bounded and its extreme points are $O(0,0)$, $A(0,35)$, $B(10,30)$, and $C(25,0)$. From Table 7.1 we see that the value of z is a maximum at the extreme point $B(10,30)$. Thus the optimal solution is

$$x = 10, \quad y = 30.$$

This means that the manufacturer should make 10 quarts of Fine and 30 quarts of Extra Fine to maximize its profit. If the firm follows this course of action, its maximum profit will be $3.80 each day. In this problem we chose small numbers to reduce the computational effort. In a real problem of this type the numbers would be much larger. ■

Referring to Figure 7.8, we note that the points $(70, 0)$ and $(0, 50)$ are points of intersection of boundary lines. However, they are not extreme points in the feasible region, since they are *not* feasible solutions; that is, they do not lie in the feasible region.

EXAMPLE 9 ■ Solve Example 2 geometrically.

Solution The feasible region S of this linear programming problem is shown in Figure 7.12 (verify). Since S is bounded, we can find the minimum value of z by evaluating z at each of the extreme points. In Table 7.2 we have tabulated the value of the objective function for each of the points $A(10, 40)$, $B(30, 40)$, and $C(30, 20)$. The value of z is a minimum at the extreme point $C(30, 20)$. Thus the optimal solution is

$$x = 30, \quad y = 20,$$

which means that the manufacturer should make 30 tons of the product at plant X and 20 tons at plant Y. If this course of action is followed, the total amount of particulate matter over the town will be 1200 pounds weekly. ■

FIGURE 7.12

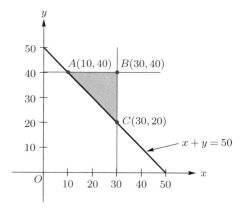

Table 7.2

Point	Value of $z = 20x + 30y$ (pounds)
$A(10, 40)$	1400
$B(30, 40)$	1800
$C(30, 20)$	1200

The feasible region of Example 3 is not bounded (verify). We shall not deal further with such problems in this book.

In general, a linear programming problem may have

1. No feasible solution; that is, there are no points satisfying all the constraints in Equations (2) and (3).

2. A unique optimal solution.

3. More than one optimal solution (see Exercise 4).

4. No maximum (or minimum) value in the feasible region; that is, it may be possible to choose a point in the feasible region to make the objective function as large (or as small) as we please.

Standard Linear Programming Problems

We shall now focus our attention on a special class of linear programming problems and will show that every linear programming problem can be transformed into a problem of this special type.

DEFINITION   We shall refer to the following linear programming problem as a **standard linear programming problem**. Find values of $x_1, x_2, \ldots, x_n$ that will maximize

$$z = c_1 x_1 + c_2 x_2 + \cdots + c_n x_n \tag{12}$$

subject to the constraints

$$\left. \begin{aligned} a_{11} x_1 + a_{12} x_2 + \cdots + a_{1n} x_n &\leq b_1 \\ a_{21} x_1 + a_{22} x_2 + \cdots + a_{2n} x_n &\leq b_2 \\ \vdots \qquad \vdots \qquad\qquad \vdots \qquad \vdots \\ a_{m1} x_1 + a_{m2} x_2 + \cdots + a_{mn} x_n &\leq b_m \end{aligned} \right\} \tag{13}$$

$$x_j \geq 0 \qquad (1 \leq j \leq n). \tag{14}$$

EXAMPLE 10 ■ Example 1 is a standard linear programming problem. ■

EXAMPLE 11 ■ The linear programming problem:

$$\text{Minimize} \quad z = 3x - 4y$$

subject to

$$2x - 3y \leq 6$$
$$x + y \leq 8$$
$$x \geq 0$$
$$y \geq 0$$

is *not* a standard linear programming problem, since the objective function is to be minimized and not maximized. ■

EXAMPLE 12 ■ The linear programming problem:

$$\text{Maximize} \quad z = 12x - 15y$$

subject to

$$3x - y \geq 4$$
$$2x + 3y \leq 6$$
$$x \geq 0$$
$$y \geq 0$$

is *not* a standard linear programming problem, since one of the inequalities is of the form $\geq$; in a standard linear programming problem every inequality of (13) must be of the form $\leq$. ■

EXAMPLE 13 ■ The linear programming problem:

$$\text{Maximize} \quad z = 8x + 10y$$

subject to

$$3x + y = 4$$
$$2x - 3y \leq 5$$
$$x \geq 0$$
$$y \geq 0$$

is *not* a standard linear programming problem, since the first constraint is an equation and not an inequality of the form $\leq$. ■

Every linear programming problem can be transformed into a standard linear programming problem.

Minimization Problem As a Maximization Problem

Every maximization problem can be viewed as a minimization problem, and conversely. This follows from the observation that

$$\text{minimum of } c_1 x_1 + c_2 x_2 + \cdots + c_n x_n$$
$$= - \text{ maximum of } \{-(c_1 x_1 + c_2 x_2 + \cdots + c_n x_n)\}. \tag{15}$$

Reversing an Inequality

Consider the inequality

$$d_1 x_1 + d_2 x_2 + \cdots + d_n x_n \geq -b.$$

Multiplying both sides of this inequality by -1 reverses the inequality, yielding

$$- d_1 x_1 - d_2 x_2 - \cdots - d_n x_n \leq b.$$

EXAMPLE 14 ■ Consider the linear programming problem:

$$\text{Minimize} \quad w = 5x - 2y$$

subject to

$$2x - 3y \geq -5$$
$$3x + 2y \leq 12$$
$$x \geq 0$$
$$y \geq 0. \tag{16}$$

Using (15) and multiplying the first inequality in (16) by (-1), we then obtain the following standard linear programming problem:

$$\text{Maximize} \quad z = -(5x - 2y) = -5x + 2y$$

subject to

$$-2x + 3y \leq 5$$
$$3x + 2y \leq 12$$
$$x \geq 0$$
$$y \geq 0.$$

Once this new maximization problem has been solved, we take the negative of the maximum value of z to obtain the minimum value of w. ■

Slack Variables

It is not too difficult to change a number of equalities into inequalities of the form $\leq$. Thus Example 13 can be transformed into a standard linear programming problem. In this book we shall not encounter linear programming problems of the type in Example 13, so we shall not continue to examine such problems.

It is not easy to handle systems of linear inequalities algebraically. However, as we have seen in Chapter 1, it is not difficult at all to deal with systems of linear equations. Accordingly, we shall change our given standard linear programming problem into a problem where we must find nonnegative variables maximizing a linear objective function and satisfying a system of linear equations. Every solution to the given problem yields a solution to the new problem, and conversely, every solution to the new problem yields a solution to the given problem.

Consider the constraint

$$d_1 x_1 + d_2 x_2 + \cdots + d_n x_n \leq b. \tag{17}$$

Since the left side of (17) is not larger than the right side, we can make (17) into an equation by adding the unknown nonnegative quantity u to its left side, to obtain

$$d_1 x_1 + d_2 x_2 + \cdots + d_n x_n + u = b. \tag{18}$$

The quantity u in (18) is called a **slack variable**, since it takes up the slack between the two sides of the inequality.

We now change each of the constraints in (2) (which we assume represent a standard linear programming problem with only the $\leq$-type inequalities) into an equation by introducing a nonnegative slack variable. Thus the ith inequality

$$a_{i1} x_1 + a_{i2} x_2 + \cdots + a_{in} x_n \leq b_i \qquad (1 \leq i \leq m) \tag{19}$$

is converted into the equation

$$a_{i1} x_1 + a_{i2} x_2 + \cdots + a_{in} x_n + x_{n+i} = b_i \qquad (1 \leq i \leq m),$$

by introducing the nonnegative slack variable x_{n+i}. Our new problem can now be stated as follows.

New Problem

Find values of $x_1, x_2, \ldots, x_n, x_{n+1}, \ldots, x_{n+m}$ that will maximize

$$z = c_1 x_1 + c_2 x_2 + \cdots + c_n x_n \tag{20}$$

subject to

$$\left. \begin{aligned}
a_{11} x_1 + a_{12} x_2 + \cdots + a_{1n} x_n + x_{n+1} &= b_1 \\
a_{21} x_1 + a_{22} x_2 + \cdots + a_{2n} x_n \qquad + x_{n+2} &= b_2 \\
\vdots \qquad \vdots \qquad \vdots \qquad \ddots \qquad \vdots \\
a_{m1} x_1 + a_{m2} x_2 + \cdots + a_{mn} x_n \qquad + x_{n+m} &= b_m
\end{aligned} \right\} \tag{21}$$

$$x_1 \geq 0, \ldots, x_n \geq 0, x_{n+1} \geq 0, x_{n+2} \geq 0, \ldots, x_{n+m} \geq 0. \tag{22}$$

Thus the new problem has m equations in $m + n$ unknowns. Solving the original problem is equivalent to solving the new problem in the following sense. If $x_1, x_2, \ldots, x_n$ is a feasible solution to the given problem as defined by (1), (2), and (3), then

$$x_1 \geq 0, \quad x_2 \geq 0, \ldots, \quad x_n \geq 0.$$

Also, $x_1, x_2, \ldots, x_n$ satisfy each of the constraints in (2). Let x_{n+i}, $1 \leq i \leq m$, be defined by

$$x_{n+i} = b_i - a_{i1}x_1 - a_{i2}x_2 - \cdots - a_{in}x_n.$$

That is, x_{n+i} is the difference between the right side of inequality (19) and its left side. Then

$$x_{n+1} \geq 0, \quad x_{n+2} \geq 0, \ldots, \quad x_{n+m} \geq 0$$

so that $x_1, x_2, \ldots, x_n, x_{n+1}, \ldots, x_{n+m}$ satisfy (21) and (22).

Conversely, suppose that $x_1, x_2, \ldots, x_n, x_{n+1}, \ldots, x_{n+m}$ satisfy (21) and (22). It is then clear that $x_1, x_2, \ldots, x_n$ satisfy (2) and (3).

EXAMPLE 15 ■ Consider the problem of Example 1. Introducing the slack variables u and v, we formulate our new problem as: Find values of x, y, u, and v that will maximize

$$z = 8x + 10y$$

subject to

$$
\begin{aligned}
2x + y + u \phantom{{}+v} &= 50 \\
x + 2y \phantom{{}+u} + v &= 70
\end{aligned}
$$

$$x \geq 0, \quad y \geq 0, \quad u \geq 0, \quad v \geq 0.$$

The slack variable u is the difference between the total amount of solution A available, 50 ounces, and the amount $2x + y$ of solution A actually used. The slack variable v is the difference between the total amount of solution B available, 70 ounces, and the amount $x + 2y$ of solution B actually used.

Consider the feasible solution to the given problem

$$x = 5, \qquad y = 10,$$

which represents point D in Figure 7.8. We then obtain the slack variables

$$u = 50 - 2(5) - 10 = 50 - 10 - 10 = 30$$

and

$$v = 70 - 5 - 2(10) = 70 - 5 - 20 = 45$$

so that

$$x = 5, \qquad y = 10, \qquad u = 30, \qquad v = 45$$

is a feasible solution to the new problem. Of course, the solution $x = 5$, $y = 10$ is not an optimal solution, since $z = 8(5) + 10(10) = 140$, and we recall that the maximum value of z is attained for

$$x = 10, \qquad y = 30.$$

In this case, the corresponding optimal solution to the new problem is

$$x = 10, \qquad y = 30, \qquad u = 0, \qquad v = 0. \qquad ■$$

7.1   EXERCISES

In Exercises 1 through 9, formulate mathematically each linear programming problem.

1. A steel producer makes two types of steel: regular and special. A ton of regular steel requires 2 hours in the open-hearth furnace and 5 hours in the soaking pit; a ton of special steel requires 2 hours in the open-hearth furnace and 3 hours in the soaking pit. The open-hearth furnace is available 8 hours per day and the soaking pit is available 15 hours per day. The profit on a ton of regular steel is $120 and it is $100 on a ton of special steel. Determine how many tons of each type of steel should be made to maximize the profit.

2. A trust fund is planning to invest up to $6000 in two types of bonds: A and B. Bond A is safer than bond B and carries a dividend of 8 percent, and bond B carries a dividend of 10 percent. Suppose that the fund's rules state that no more than $4000 may be invested in bond B, while at least $1500 must be invested in bond A. How much should be invested in each type of bond to maximize the fund's return?

3. Solve Exercise 2 if the fund has the following additional rule: "The amount invested in bond B cannot exceed one half the amount invested in bond A."

4. A trash-removal company carries industrial waste in sealed containers in its fleet of trucks. Suppose that each container from the Smith Corporation weighs 6 pounds and is 3 cubic feet in volume, while each container from the Johnson Corporation weighs 12 pounds and is 1 cubic foot in volume. The company charges the Smith Corporation 30 cents for each container carried on a trip, and 60 cents for each container from the Johnson Corporation. If a truck cannot carry more than 18,000 pounds or more than 1800 cubic feet in volume, how many containers from each customer should the company carry in a truck on each trip to maximize the revenue per truckload?

5. A television producer designs a program based on a comedian and time for commercials. The advertiser insists on at least 2 minutes of advertising time, the station insists on no more than 4 minutes of advertising time, and the comedian insists on at least 24 minutes of the comedy program. Also, the total time allotted for the advertising and comedy portions of the program cannot exceed 30 minutes. If it has been determined that each minute of advertising (very creative) attracts 40,000 viewers and each minute of the comedy program attracts 45,000 viewers, how should the time be divided between advertising and programming to maximize the number of viewers per minute?

6. A small generator burns two types of fuel: low sulfur (L) and high sulfur (H) to produce electricity. For each hour of use, each gallon of L emits 3 units of sulfur dioxide, generates 4 kilowatts, and costs 60 cents, while each gallon of H emits 5 units of sulfur dioxide, generates 4 kilowatts, and costs 50 cents. The environmental protection agency insists that the maximum amount of sulfur dioxide that may be emitted per hour is 15 units. Suppose that at least 16 kilowatts must be generated per hour. How many gallons of L and how many gallons of H should be used hourly to minimize the cost of the fuel used?

7. The Protein Diet Club serves a luncheon consisting of two dishes, A and B. Suppose that each unit of A has 1 gram of fat, 1 gram of carbohydrate, and 4 grams of protein, whereas each unit of B has 2 grams of fat, 1 gram of carbohydrate, and 6 grams of protein. If the dietician planning the luncheon wants to provide no more than 10 grams of fat or more than 7 grams of carbohydrate, how many units of A and how many units of B should be served to maximize the amount of protein consumed?

8. In designing a new airline route, a company is considering two types of planes, types A and B. Each type A plane can carry 40 passengers and requires 2 mechanics for servicing; each type B plane can carry 60 passengers and requires 3 mechanics for servicing. Suppose that the company must transport at least 300 people daily and that insurance rules for the size of the hangar allow no more than 12 mechanics on the payroll. If each type A plane costs $10,000,000 and each type B plane costs $15,000,000, how many planes of each type should be bought to minimize the cost?

9. An animal feed producer makes two types of grain: A and B. Each unit of grain A contains 2 grams of fat, 1 gram of protein, and 80 calories. Each unit of grain B contains 3 grams of fat, 3 grams of protein, and 60 calories. Suppose that the producer wants each unit of the final product to yield at least 18 grams of fat, at least 12 grams of protein, and at least 480 calories. If each unit of A costs 10 cents and each unit of B costs 12 cents, how many units of each type of grain should the producer use to minimize the cost?

In Exercises 10 through 13, sketch the set of points satisfying the given system of inequalities.

10.
$$x \leq 4$$
$$x \geq 2$$
$$y \leq 4$$
$$y \geq 1$$
$$x + y \leq 6.$$

11. $2x - y \leq 6$
$$2x + y \leq 10$$
$$x \geq 0$$
$$y \geq 0.$$

12. $x + y \leq 3$
$$5x + 4y \geq 20$$
$$x \geq 0$$
$$y \geq 0.$$

13. $x + y \geq 4$
$$x + 4y \geq 8$$
$$x \geq 0$$
$$y \geq 0.$$

In Exercises 14 and 15, solve the given linear programming problem geometrically.

14. Maximize $z = 3x + 2y$
subject to
$$2x - 3y \leq 6$$
$$x + y \leq 4$$
$$x \geq 0$$
$$y \geq 0.$$

15. Minimize $z = 3x - y$
subject to
$$-3x + 2y \leq 6$$
$$5x + 4y \geq 20$$
$$8x + 3y \leq 24$$
$$x \geq 0$$
$$y \geq 0.$$

16. Solve the problem in Exercise 1 geometrically.

17. Solve the problem in Exercise 2 geometrically.

18. Solve the problem in Exercise 3 geometrically.

19. Solve the problem in Exercise 4 geometrically.

20. Solve the problem in Exercise 5 geometrically.

21. Solve the problem in Exercise 6 geometrically.

22. Solve the problem in Exercise 7 geometrically.

23. Solve the problem in Exercise 8 geometrically.

24. Which of the following are standard linear programming problems?

(a) Maximize $z = 2x - 3y$
subject to
$$2x - 3y \leq 4$$
$$3x + 2y \geq 5.$$

(b) Minimize $z = 2x + 3y$
subject to
$$2x + 3y \leq 4$$
$$3x + 2y \leq 5$$
$$x \geq 0$$
$$y \geq 0.$$

(c) Minimize $z = 2x_1 - 3x_2 + x_3$
subject to
$$2x_1 + 3x_2 + 2x_3 \leq 6$$
$$3x_1 \quad\quad - 2x_3 \leq 4$$
$$x_1 \leq 0$$
$$x_2 \geq 0$$
$$x_3 \geq 0.$$

(d) Maximize $z = 2x + 2y$
subject to
$$2x + 3y \leq 4$$
$$3x \leq 5$$
$$x \geq 0.$$

25. Which of the following are standard linear programming problems?

(a) Maximize $z = 3x_1 + 2x_2 + x_3$
subject to
$$2x_1 + 3x_2 + x_3 \leq 4 \quad \text{Standard}$$
$$3x_1 - 2x_2 \quad\quad \leq 5$$
$$x_1 \geq 0$$
$$x_2 \geq 0$$
$$x_3 \geq 0.$$

(b) Minimize $z = 2x + 3y$
subject to

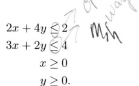

$$2x + 4y \leq 2$$
$$3x + 2y \leq 4$$
$$x \geq 0$$
$$y \geq 0.$$

(c) Maximize $z = 3x_1 + 4x_2 + x_3$
  subject to

$$2x_1 + 4x_2 + x_3 \leq 2$$
$$3x_1 - 2x_2 + x_3 \leq 4$$
$$2x_1 \qquad + x_3 \leq 8 \quad \text{NO}$$
$$x_1 \geq 0$$
$$x_2 \geq 0.$$

X_3 - is missing

(d) Maximize $z = 2x_1 + 3x_2 + x_3$
  subject to

$$2x_1 + 3x_2 + 5x_3 \leq 8$$
$$3x_1 - 2x_2 + 2x_3 = 4 \quad \text{no}$$
$$2x_1 + x_2 + 3x_3 \leq 6$$
$$x_1 \geq 0$$
$$x_2 \geq 0$$
$$x_3 \geq 0.$$

In Exercises 26 and 27, formulate each problem as a standard linear programming problem.

26. Minimize $z = -2x_1 + 3x_2 + 2x_3$
  subject to

$$2x_1 + x_2 + 2x_3 \leq 12$$
$$x_1 + x_2 - 3x_3 \leq 8$$
$$x_1 \geq 0$$
$$x_2 \geq 0$$
$$x_3 \geq 0.$$

27. Maximize $z = 3x_1 - x_2 + 6x_3$
  subject to

$$2x_1 + 4x_2 + x_3 \leq 4$$
$$-3x_1 + 2x_2 - 3x_3 \geq -4$$
$$2x_1 + x_2 - x_3 \leq 8$$
$$x_1 \geq 0$$
$$x_2 \geq 0$$
$$x_3 \geq 0.$$

In Exercises 28 and 29, formulate the given linear programming problem as a new problem with slack variables.

28. Maximize $z = 2x + 8y$
  subject to

$$2x + 3y \leq 18$$
$$3x - 2y \leq 6$$
$$x \geq 0$$
$$y \geq 0.$$

29. Maximize $z = 2x_1 + 3x_2 + 7x_3$
  subject to

$$3x_1 + x_2 - 4x_3 \leq 3$$
$$x_1 - 2x_2 + 6x_3 \leq 21$$
$$x_1 - x_2 - x_3 \leq 9$$
$$x_1 \geq 0$$
$$x_2 \geq 0$$
$$x_3 \geq 0.$$

7.2 ▼ The Simplex Method

The simplex method for solving linear programming problems was developed by George B. Dantzig* in connection with his work on planning problems for the federal government. In this section we present the essential features of the method, illustrating them with examples. A number of proofs will be omitted, and for further details the interested reader may consult the references given at

---

*George B. Dantzig (1914–) was born in Portland, Oregon. He received a bachelor's degree from the University of Maryland, master's degree from the University of Michigan, and a Ph.D. degree from the University of California at Berkeley. He is presently Professor of Operations Research and computer Science at Stanford University. His simplex method for solving linear programming problems, developed in 1947, was a major breakthrough in the newly developed area of operations research. The rapid growth of computing power coupled with its decreasing cost has produced a great many computer implementations of the simplex method and has resulted in savings of billions of dollars for industry and government. Professor Dantzig has held important positions in government, industry, and universities, and is the recipient of numerous honors and awards.

the end of this chapter. It is convenient to introduce matrix terminology into our further discussion of linear programming.

Matrix Notation

We again restrict our attention to the standard linear programming problem: Maximize

$$c_1 x_1 + c_2 x_2 + \cdots + c_n x_n \tag{1}$$

subject to

$$\left.\begin{array}{l} a_{11}x_1 + a_{12}x_2 + \cdots + a_{1n}x_n \le b_1 \\ a_{21}x_1 + a_{22}x_2 + \cdots + a_{2n}x_n \le b_2 \\ \quad\vdots \qquad\quad \vdots \qquad\qquad\quad \vdots \qquad \vdots \\ a_{m1}x_1 + a_{m2}x_2 + \cdots + a_{mn}x_n \le b_m \end{array}\right\} \tag{2}$$

$$x_j \ge 0 \quad (1 \le j \le n). \tag{3}$$

If we let

$$A = \begin{bmatrix} a_{11} & a_{12} & \cdots & a_{1n} \\ a_{21} & a_{22} & \cdots & a_{2n} \\ \vdots & \vdots & & \vdots \\ a_{m1} & a_{m2} & \cdots & a_{mn} \end{bmatrix}, \qquad \mathbf{x} = \begin{bmatrix} x_1 \\ x_2 \\ \vdots \\ x_n \end{bmatrix}, \qquad \mathbf{b} = \begin{bmatrix} b_1 \\ b_2 \\ \vdots \\ b_m \end{bmatrix},$$

and

$$\mathbf{c} = \begin{bmatrix} c_1 \\ c_2 \\ \vdots \\ c_n \end{bmatrix},$$

then the given problem can be stated as follows: Find a vector $\mathbf{x}$ in R^n that will maximize the objective function

$$z = \mathbf{c}^T \mathbf{x} \tag{4}$$

subject to

$$A\mathbf{x} \le \mathbf{b} \tag{5}$$

$$\mathbf{x} \ge \mathbf{0}, \tag{6}$$

where $\mathbf{x} \ge \mathbf{0}$ means that each entry of $\mathbf{x}$ is nonnegative and $A\mathbf{x} \le \mathbf{b}$ means that each entry of $A\mathbf{x}$ is less than or equal to the corresponding entry in $\mathbf{b}$.

A vector $\mathbf{x}$ in R^n satisfying (5) and (6) is called a **feasible solution** to the given problem, and a feasible solution maximizing the objective function (4) is called an **optimal solution**.

EXAMPLE 1 ■ We can write the problem in Example 1 of Section 7.1 in matrix form as follows: Find a vector $\mathbf{x} = \begin{bmatrix} x \\ y \end{bmatrix}$ in R^2 that will maximize

$$z = \begin{bmatrix} 8 & 10 \end{bmatrix} \begin{bmatrix} x \\ y \end{bmatrix}$$

subject to

$$\begin{bmatrix} 2 & 1 \\ 1 & 2 \end{bmatrix} \begin{bmatrix} x \\ y \end{bmatrix} \le \begin{bmatrix} 50 \\ 70 \end{bmatrix}$$

$$\begin{bmatrix} x \\ y \end{bmatrix} \ge \begin{bmatrix} 0 \\ 0 \end{bmatrix}.$$

Feasible solutions include the vectors

$$\begin{bmatrix} 0 \\ 0 \end{bmatrix}, \quad \begin{bmatrix} 0 \\ 35 \end{bmatrix}, \quad \begin{bmatrix} 10 \\ 30 \end{bmatrix}, \quad \begin{bmatrix} 25 \\ 0 \end{bmatrix}, \quad \begin{bmatrix} 5 \\ 10 \end{bmatrix}, \quad \text{and} \quad \begin{bmatrix} 20 \\ 5 \end{bmatrix}.$$

An optimal solution is the vector

$$\begin{bmatrix} 10 \\ 30 \end{bmatrix}. \qquad \blacksquare$$

The new problem with slack variables can also be written in matrix form as follows: Find a vector $\mathbf{x}$ that will maximize

$$z = \mathbf{c}^T \mathbf{x} \tag{7}$$

subject to

$$A\mathbf{x} = \mathbf{b} \tag{8}$$

$$\mathbf{x} \ge \mathbf{0}, \tag{9}$$

where now

$$A = \begin{bmatrix} a_{11} & a_{12} & \cdots & a_{1n} & 1 & 0 & \cdots & 0 \\ a_{21} & a_{22} & \cdots & a_{2n} & 0 & 1 & \cdots & 0 \\ \vdots & \vdots & & \vdots & \vdots & \vdots & & \vdots \\ a_{m1} & a_{m2} & \cdots & a_{mn} & 0 & 0 & \cdots & 1 \end{bmatrix}, \qquad \mathbf{x} = \begin{bmatrix} x_1 \\ x_2 \\ \vdots \\ x_n \\ x_{n+1} \\ \vdots \\ x_{n+m} \end{bmatrix},$$

$$\mathbf{b} = \begin{bmatrix} b_1 \\ b_2 \\ \vdots \\ b_m \end{bmatrix}, \quad \text{and} \quad \mathbf{c} = \begin{bmatrix} c_1 \\ c_2 \\ \vdots \\ c_n \\ 0 \\ \vdots \\ 0 \end{bmatrix}.$$

A vector $\mathbf{x}$ satisfying (8) and (9) is called a **feasible solution** to the new problem, and a feasible solution maximizing the objective function (7) is called an **optimal solution**. Throughout this chapter, we now make the additional assumption that in all standard linear programming problems,

$$b_1 \ge 0, \quad b_2 \ge 0, \dots, \quad b_m \ge 0.$$

We shall use Example 1 of Section 7.1 as our principal illustrative example in this section.

Illustrative Problem

Find values of x and y that will maximize

$$z = 8x + 10y \tag{10}$$

subject to

$$2x + y \leq 50$$
$$x + 2y \leq 70 \tag{11}$$

$$x \geq 0, \quad y \geq 0. \tag{12}$$

The new problem with slack variables u and v is: Find values of x, y, u, and v that will maximize

$$z = 8x + 10y \tag{13}$$

subject to

$$2x + y + u \quad = 50$$
$$x + 2y \quad + v = 70 \tag{14}$$

$$x \geq 0, \quad y \geq 0, \quad u \geq 0, \quad v \geq 0. \tag{15}$$

DEFINITION   The vector $\mathbf{x}$ in R^{n+m} is called a **basic solution** to the new problem if it is obtained by setting n of the variables in (8) equal to zero and solving for the remaining m variables. The m variables that we solve for are called **basic variables**, and the n variables set equal to zero are called **nonbasic variables**. The vector $\mathbf{x}$ is called a **basic feasible solution** if it is a basic solution that also satisfies (9).

Basic feasible solutions are important because the following theorem can be established.

THEOREM 7.2 ∎ *If a linear programming problem has an optimal solution, then it has a basic feasible solution that is optimal.*   ∎

Thus to solve a linear programming problem we need only search for basic feasible solutions. In our illustrative example we can select two of the four variables x, y, u, and v as nonbasic variables by setting them equal to zero and solving for the remaining two variables; that is, we solve for the basic variables. Thus, if

$$x = y = 0,$$

then

$$u = 50, \qquad v = 70.$$

The vector

$$\mathbf{x}_1 = \begin{bmatrix} 0 \\ 0 \\ 50 \\ 70 \end{bmatrix}$$

is a basic solution, which gives rise to the feasible solution

$$\begin{bmatrix} 0 \\ 0 \end{bmatrix}$$

to the original problem specified by (10), (11), and (12). The variables x and y are nonbasic, and the variables u and v are basic. The convex region of solutions to the original problem has been sketched in Figure 7.8. Thus the vector $\mathbf{x}_1$ corresponds to the extreme point O.

If we let the variables x and u be nonbasic ($x = u = 0$), then $y = 50$ and $v = -30$. The vector

$$\mathbf{x}_2 = \begin{bmatrix} 0 \\ 50 \\ 0 \\ -30 \end{bmatrix}$$

is a basic solution that is not feasible, since v is negative. It corresponds to the point F in Figure 7.8, which is not a feasible solution to the original problem.

In Table 7.3 we have tabulated all the possible choices for basic solutions. The nonbasic variables are shaded and the corresponding point from Figure 7.8 is indicated in the table. It can be seen in this example, and proved in general, that every basic feasible solution determines an extreme point, and conversely, each extreme point determines a basic feasible solution.

Table 7.3

x	y	u	v	Type of Solution	Corresponding Point in Figure 7.8
0	0	50	70	Basic feasible solution	O
0	50	0	−30	Not a basic feasible solution	F
0	35	15	0	Basic feasible solution	A
25	0	0	45	Basic feasible solution	C
70	0	−90	0	Not a basic feasible solution	G
10	30	0	0	Basic feasible solution	B

One method of solving our linear programming problem would be to obtain all the basic solutions, discard those which are not feasible, and evaluate the objective function at each basic feasible solution, selecting that one, or ones, for which we get a maximum value of the objective function. The number of possible basic solutions is

$$\binom{n+m}{n} = \frac{(n+m)!}{m!\,n!}.$$

That is, it is the number of ways of selecting n objects out of $m + n$ given objects.

The simplex method is a procedure that enables us to go from a given extreme point (basic feasible solution) to an adjacent extreme point in such a way that the value of the objective function increases as we move from extreme point to extreme point until we either obtain an optimal solution or find that the given problem has no finite optimal solution. The simplex method thus consists of two steps: (1) a way of checking whether a given basic feasible solution is an optimal solution, and (2) a way of obtaining another basic feasible solution with a larger value of the objective function. In most practical problems the simplex method does not consider every basic feasible solution; rather, it works with only a small number of these. However, examples have been given where a large number of basic feasible solutions have been examined by the simplex method. We shall now turn to a detailed discussion of this powerful method, using our illustrative example as a guide.

Selecting an Initial Basic Feasible Solution

We can take all the original (nonslack) variables as our nonbasic variables and set them equal to zero. We then solve for the slack variables, our basic variables. In our example, we set

$$x = y = 0$$

and solve for u and v:

$$u = 50, \qquad v = 70.$$

Thus the initial basic feasible solution is the vector

$$\begin{bmatrix} 0 \\ 0 \\ 50 \\ 70 \end{bmatrix},$$

which yields the origin as an extreme point.

It is convenient to develop a tabular method for displaying the given problem and the initial basic feasible solution. First, we write (13) as the equation

$$-8x - 10y + z = 0, \tag{13'}$$

with z being considered as another variable. We now form the **initial tableau** (Tableau 1). The variables x, y, u, v, and z are written in the top row as labels on the corresponding columns. Constraints (14) are entered in the top rows followed by Equation (13′) in the bottom row. The bottom row of the tableau is called the **objective row**. Along the left-hand side of the tableau we indicate which variable is a basic variable in the corresponding equation. Thus u is a basic variable in the first equation and v is a basic variable in the second equation.

Tableau 1

	x	y	u	v	z	
u	2	1	1	0	0	50
v	1	2	0	1	0	70
	-8	-10	0	0	1	0

A basic variable can also be described as a variable, other than z, which is present in exactly one equation, and there it appears with a coefficient of $+1$. In the tableau, the value of the basic variable is explicitly given in the rightmost column.

The initial tableau (Tableau 1) shows the values of the basic variables u and v, and therefore the nonbasic variables have values

$$x = 0, \qquad y = 0.$$

The value of the objective function for this initial basic feasible solution is

$$c_1 x + c_2 y + 0(u) + 0(v) = 8(0) + 10(0) + 0(u) + 0(v) = 0,$$

which is the entry in the objective row and rightmost column. It is clear that this solution is not optimal, for using the bottom row of the initial tableau, we can write

$$z = 0 + 8x + 10y - 0u - 0v. \tag{16}$$

Now the value of z can be increased by increasing either x or y, since both of these variables appear in (16) with positive coefficients. Since (16) contains terms with positive coefficients if and only if the objective row of our initial tableau has negative entries under the columns labeled with variables, we see that we can increase z by increasing any variable with a negative entry in the objective row. Thus we obtain the following optimality criterion for determining whether the feasible solution indicated in a tableau is an optimal solution yielding a maximum value for the objective function z.

In general, for the problem given by (7), (8), and (9), the initial tableau is Tableau 2.

Tableau 2

	x_1	x_2	$\cdots$	x_n	x_{n+1}	x_{n+2}	$\cdots$	x_{n+m}	z	
x_{n+1}	a_{11}	a_{12}	$\cdots$	a_{1n}	1	0	$\cdots$	0	0	b_1
x_{n+2}	a_{21}	a_{22}	$\cdots$	a_{2n}	0	1	$\cdots$	0	0	b_2
$\vdots$	$\vdots$						$\vdots$			$\vdots$
x_{n+m}	a_{m1}	a_{m2}	$\cdots$	a_{mn}	0	0	$\cdots$	1	0	b_m
	$-c_1$	$-c_2$	$\cdots$	$-c_n$	0	0	$\cdots$	0	1	0

Optimality Criterion

If the objective row of a tableau has no negative entries in the columns labeled with variables, then the indicated solution is optimal and we can stop our computation.

Selecting the Entering Variable

If the objective row of a tableau has negative entries in the columns labeled with the variables, then the indicated solution is not optimal and we must continue our search for an optimal solution.

The simplex method moves from one extreme point to an adjacent extreme point in such a way that the value of the objective function increases. This is done by increasing *one* variable at a time. The largest increase in z per unit increase in a variable occurs for the variable with the most negative entry in the objective row. In Tableau 1, the most negative entry in the objective row is -10, and since it occurs in the y-column, this is the variable to be increased. The variable to be increased is called the **entering variable**, because in the next iteration it will become a basic variable, thereby *entering* the set of basic variables. If there are several candidates for entering variables, choose one. An increase in y must be accompanied by a decrease in some of the other variables. This can be seen if we solve the equations in (14) for u and v:

$$u = 50 - 2x - y$$
$$v = 70 - x - 2y.$$

Since we only increase y, we keep $x = 0$, and obtain

$$u = 50 - y$$
$$v = 70 - 2y \qquad (17)$$

so that as y increases, both u and v decrease. Equations (17) also show by how much we can increase y. That is, since u and v must be nonnegative, we must have

$$y \le \tfrac{50}{1} = 50$$
$$y \le \tfrac{70}{2} = 35.$$

We now see that the allowable increase in y can be no larger than the smaller of the ratios $\frac{50}{1}$ and $\frac{70}{2}$. Taking y as 35, we obtain the new basic feasible solution.

$$x = 0, \qquad y = 35, \qquad u = 15, \qquad v = 0.$$

The basic variables are y and u; the variables x and v are nonbasic. The objective function for this solution now has the value

$$z = 8(0) + 10(35) + 0(15) + 0(0) = 350,$$

which is much better than the earlier value of 0. This solution yields the extreme point A in Figure 7.8, which is adjacent to O.

Choosing the Departing Variable

Since the variable $v = 0$, it is not basic and is called the **departing variable**, for it has *departed* from the set of basic variables. The column of the entering variable is called the **pivotal column**; the row that is labeled by the departing variable is called the **pivotal row**.

Let us now look more closely at the selection of the departing variable. The choice of this variable was closely related to the determination of how far we could increase the entering variable (y in our example). To obtain this number, we formed the ratios (called **θ-ratios**) of the entries above the objective row in the rightmost column of the tableau by the corresponding entries in the pivotal column. The smallest of these ratios tells how far the entering variable can be increased. The basic variable labeling the row for which this smallest ratio occurs (the pivotal row) is then the departing variable. In our example, the θ-ratios, formed by using the rightmost column and the y-column, are $\frac{50}{1}$ and $\frac{70}{2}$. The smaller of these ratios, 35, occurs for the second row, which means that the second row is the pivotal row and the basic variable, v, labeling it becomes the departing variable and is no longer basic. If the smallest of the ratios is not selected, then one of the variables in the new solution becomes negative and the new solution is no longer feasible (Exercise T.2). What happens if there are entries in the pivotal column that are either zero or negative? If any entry in the pivotal column is negative, then the corresponding ratio is also negative; in this case the equation associated with the negative entry imposes no restriction on how far the entering variable can be increased. Suppose, for example, that the y-column in our initial tableau is

$$\begin{bmatrix} -3 \\ 2 \end{bmatrix} \quad \text{instead of} \quad \begin{bmatrix} 1 \\ 2 \end{bmatrix}.$$

Then instead of (17) we have

$$u = 50 + 3y$$
$$v = 70 - 2y,$$

and since u must be nonnegative, we have

$$y \geq -\tfrac{50}{3},$$

which puts no limitation at all on how far y can be increased. If an entry in the pivotal column is zero, then the corresponding ratio cannot be formed (we cannot divide by zero), and again the associated equation puts no limitation on how far the entering variable can be increased. Thus, in forming the ratios, we only use the positive entries above the objective row in the pivotal column.

If all the entries above the objective row in the pivotal column are either zero or negative, then the entering variable can be made as large as we please. This means that the problem has no finite optimal solution.

Obtaining a New Tableau

We must now obtain a new tableau indicating the new basic variables and the new basic feasible solution. Solving the second equation in (14) for y, we obtain

$$y = 35 - \tfrac{1}{2}x - \tfrac{1}{2}v, \tag{18}$$

and substituting this expression for y in the first equation in (14), we have

$$\tfrac{3}{2}x + u - \tfrac{1}{2}v = 15. \tag{19}$$

The second equation in (14) can be written, upon dividing by 2 (the coefficient of y), as

$$\tfrac{1}{2}x + y + \tfrac{1}{2}v = 35. \tag{20}$$

Substituting (18) for y in (13′), we have

$$-3x + 5v + z = 350. \tag{21}$$

Since $x = 0$, $v = 0$, we obtain the value of z for the current basic feasible solution as

$$z = 350.$$

Equations (19), (20), and (21) yield our new tableau (Tableau 3). Observe in this tableau that we have labeled the basic variables in each row.

Tableau 3

	x	y	u	v	z	
u	$\frac{3}{2}$	0	1	$-\frac{1}{2}$	0	15
y	$\frac{1}{2}$	1	0	$\frac{1}{2}$	0	35
	-3	0	0	5	1	350

Comparing Tableau 1 with Tableau 3, we observe that we can transform the former to the latter by elementary row operations as follows.

Step 1. Locate and circle the entry in the pivotal row and pivotal column. This entry is called the **pivot**. Mark the pivotal column by placing an arrow ↓ above the entering variable and mark the pivotal row by placing an arrow ← to the left of the departing variable.

Step 2. If the pivot is k, multiply the pivotal row by $1/k$, making the entry that was the pivot a 1.

Step 3. Add the appropriate multiples of the pivotal row to all other rows (including the objective row) so that all elements in the pivotal column except for the 1 where the pivot was located become zero.

Step 4. In the new tableau replace the label on the pivotal row by the entering variable.

These four steps form a process called **pivotal elimination**. It is one of the iterations of the procedure described in Section 1.5 for transforming a matrix to reduced row echelon form.

We now repeat Tableau 1, with the arrows placed next to the entering and departing variables and with the pivot circled.

Performing pivotal elimination on Tableau 1 yields Tableau 3. We now repeat the entire procedure, using Tableau 3 as our initial tableau. Since the most negative entry in the objective row of Tableau 3, -3, occurs in the first column, x is the entering variable and the first column is the pivotal column.

Tableau 1

$\downarrow$

	x	y	u	v	z	
u	2	1	1	0	0	50
$\leftarrow v$	1	②	0	1	0	70
	-8	-10	0	0	1	0

To determine the departing variable, we form the ratios of the entries in the rightmost column (except for the objective row) by the corresponding entries of the pivotal column for those entries in the pivotal column which are positive and select the smallest of these ratios. Since both entries in the pivotal column are positive, the ratios are $15/\frac{3}{2}$ and $35/\frac{1}{2} = 10$. The smaller of these, $15/\frac{3}{2} = 10$, occurs for the first row, so the departing variable is u and the pivotal row is the first row. Tableau 3, with the pivotal column, pivotal row, and circled pivot, is shown again here.

Tableau 3

$\downarrow$

	x	y	u	v	z	
$\leftarrow u$	③⁄②	0	1	$-\frac{1}{2}$	0	15
y	$\frac{1}{2}$	1	0	$\frac{1}{2}$	0	35
	-3	0	0	5	1	350

Performing pivotal elimination on Tableau 3 yields Tableau 4.

Tableau 4

	x	y	u	v	z	
x	1	0	$\frac{2}{3}$	$-\frac{1}{3}$	0	10
y	0	1	$-\frac{1}{3}$	$\frac{2}{3}$	0	30
	0	0	2	4	1	380

Since the objective row of Tableau 4 has no negative entries in the columns labeled with variables, we conclude, by the optimality criterion, that we are finished and that the indicated solution is optimal. Thus the optimal solution is

$$x = 10, \qquad y = 30, \qquad u = 0, \qquad v = 0,$$

which corresponds to the extreme point $B(10, 30)$. Thus the simplex method started from the extreme point $O(0,0)$ moved to the adjacent extreme point $A(0, 35)$ and then to the extreme point $B(10, 30)$, which is adjacent to A. The value of the objective function increased from 0 to 350 to 380, respectively.

We now summarize the simplex method.

The procedure for carrying out the simplex method is as follows.

Step 1. Set up the initial tableau.

Step 2. Apply the optimality test. If the objective row has no negative entries in the columns labeled with variables, then the indicated solution is optimal; we stop our computations.

Step 3. Choose a pivotal column by determining the column with the most negative entry in the objective row. If there are several candidates for a pivotal column, choose any one.

Step 4. Choose a pivotal row. Form the ratios of the entries above the objective row in the rightmost column by the corresponding entries of the pivotal column for those entries in the pivotal column which are positive. The pivotal row is the row for which the smallest of these ratios occurs. If there is a tie, so that the smallest ratio occurs at more than one row, choose any one of the qualifying rows. If none of the entries in the pivotal column above the objective row is positive, the problem has no finite optimum. We stop our computation.

Step 5.   Perform pivotal elimination to construct a new tableau and return to Step 2.

Figure 7.13 gives a flow chart for the simplex method.

We have restricted *our* discussion of the simplex method to standard linear programming problems in which all the right-hand entries are nonnegative. It should be noted that the method applies to the general linear programming problem. For additional details we refer the reader to the references at the end of this chapter.

EXAMPLE 2 ■ Maximize

$$z = 4x_1 + 8x_2 + 5x_3$$

subject to

$$x_1 + 2x_2 + 3x_3 \leq 18$$
$$x_1 + 4x_2 + x_3 \leq 6$$
$$2x_1 + 6x_2 + 4x_3 \leq 15$$
$$x_1 \geq 0, \quad x_2 \geq 0, \quad x_3 \geq 0.$$

The new problem with slack variables is

$$\text{Maximize} \quad z = 4x_1 + 8x_2 + 5x_3$$

FIGURE 7.13
The simplex method

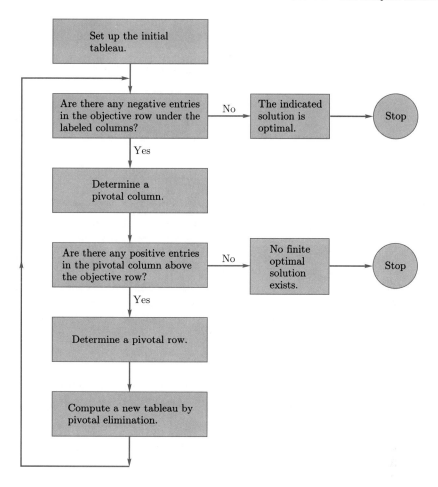

subject to

$$x_1 + 2x_2 + 3x_3 + x_4 \qquad\qquad = 18$$
$$x_1 + 4x_2 + x_3 \qquad + x_5 \qquad = 6$$
$$2x_1 + 6x_2 + 4x_3 \qquad\qquad + x_6 = 15$$

$$x_1 \geq 0, \quad x_2 \geq 0, \quad x_3 \geq 0, \quad x_4 \geq 0, \quad x_5 \geq 0, \quad x_6 \geq 0.$$

The initial tableau and the succeeding tableaux are

	x_1	x_2	x_3	x_4	x_5	x_6	z	
x_4	1	2	3	1	0	0	0	18
$\leftarrow x_5$	1	④	1	0	1	0	0	6
x_6	2	6	4	0	0	1	0	15
	-4	-8	-5	0	0	0	1	0

↓

	x_1	x_2	x_3	x_4	x_5	x_6	z	
x_4	$\frac{1}{2}$	0	$\frac{5}{2}$	1	$-\frac{1}{2}$	0	0	15
x_2	$\frac{1}{4}$	1	$\frac{1}{4}$	0	$\frac{1}{4}$	0	0	$\frac{3}{2}$
$\leftarrow x_6$	$\frac{1}{2}$	0	$\boxed{\frac{5}{2}}$	0	$-\frac{3}{2}$	1	0	6
	-2	0	-3	0	2	0	1	12

↓

	x_1	x_2	x_3	x_4	x_5	x_6	z	
x_4	0	0	0	1	1	-1	0	9
$\leftarrow x_2$	$\boxed{\frac{1}{5}}$	1	0	0	$\frac{2}{5}$	$-\frac{1}{10}$	0	$\frac{9}{10}$
x_3	$\frac{1}{5}$	0	1	0	$-\frac{3}{5}$	$\frac{2}{5}$	0	$\frac{12}{5}$
	$-\frac{7}{5}$	0	0	0	$\frac{1}{5}$	$\frac{6}{5}$	1	$\frac{96}{5}$

	x_1	x_2	x_3	x_4	x_5	x_6	z	
x_4	0	0	0	1	1	-1	0	9
x_1	1	5	0	0	2	$-\frac{1}{2}$	0	$\frac{9}{2}$
x_3	0	-1	1	0	-1	$\frac{1}{2}$	0	$\frac{3}{2}$
	0	7	0	0	3	$\frac{1}{2}$	1	$\frac{51}{2}$

Hence an optimal solution is

$$x_1 = \tfrac{9}{2}, \qquad x_2 = 0, \qquad x_3 = \tfrac{3}{2}.$$

The slack variables are

$$x_4 = 9, \qquad x_5 = 0, \qquad x_6 = 0,$$

and the optimal value of z is $\frac{51}{2}$. ■

A number of difficulties can arise in using the simplex method. We shall briefly describe one of these, and we again refer the reader to the references at the end of this chapter for an extended discussion of computational considerations.

Degeneracy

Suppose that a basic variable becomes zero in one of the tableaux in the simplex method. Then one of the ratios used to determine the next pivotal row may be zero, in which case the pivotal row is the one labeled with the basic variable that is zero (in this case, zero is the smallest of the ratios). Recall that the entering variable is increased from zero to the smallest ratio. Since the smallest ratio is zero, the entering variable remains at the value zero. In the new tableau all the old basic variables have the same values that they had in the old tableau; the value of the objective function has not been increased. The new tableau looks like the old one, except that a basic variable with value zero has become nonbasic, and its place has been taken by a nonbasic variable also entering with value zero. A basic feasible solution in which one or more of the basic variables are zero is called a **degenerate basic feasible solution**. It can be shown that when a tie occurs for the smallest ratio in determining the pivotal row, a degenerate solution will arise and there are several optimal solutions. When no degenerate solution occurs, the value of the objective function increases as we move from one basic feasible solution to another. The procedure stops after a finite number of steps, since the number of basic feasible solutions is finite. However, if we have a degenerate basic feasible solution, we may return to a basic feasible solution that had already been found at an earlier iteration and thus enter an infinite cycle. Fortunately, cycling is encountered only occasionally in practical linear programming problems, although several examples have been carefully constructed to show that cycling can occur. In actual practice, when degeneracy occurs, it is handled by merely ignoring it. The value of the objective function may remain constant for a few iterations and will then start to increase. Moreover, a technique called **perturbation** has been developed for handling degeneracy. This technique calls for making slight changes in the rightmost column of the tableau so that the troublesome ties in the ratios will no longer exist.

Karmarkar's Method

In 1984 N. Karmarkar developed a new method for solving linear programming problems that is claimed to be much faster than the simplex method for very large problems. Much of the current work in linear programming is concentrating on a better understanding of Karmarkar's method and of its applications to practical problems. A description of the method may be found in Strang's book mentioned in the Further Readings at the end of this chapter.

Computer Implementations

There are many computer programs implementing the simplex method and other algorithms in the area of mathematical programming (which includes integer programming and nonlinear programming). Some of these programs can do extensive data manipulation to prepare the input for the problem, solve the problem, and then prepare elaborate reports that can be used by management in decision making. Moreover, some programs can handle large problems having as many as 8200 inequalities and 100,000 unknowns.

The M-files available to users of this book (see the Preface for how to obtain these M-files) contain two routines for solving linear programming problems. The first routine, **lpstep**, provides a step-by-step solution procedure. The second routine, **linprog**, solves a linear programming problem directly. Both routines use the techniques described in this section.

The MATLAB optimization toolbox, designed to run with MATLAB will solve sizable linear programming problems presented in mathematical form.

LINDO Systems [1415 North Dayton Avenue, Chicago, Illinois, 60622; telephone, (800) 441-2378] makes available three programs that are inexpensive, are easy to use, and will run on personal computers. They are especially suitable for solving problems encountered in a course covering linear programming and related topics. The programs are: LINGO, a program that formulates a mathematical model of the problem and then proceeds to solve it; and WHAT'S BEST!, a spreadsheet-based program that solves a mathematically presented problem. This program works with a spreadsheet produced by one of the four or five most widely used spreadsheet programs that it supports. The third program, LINDO, solves linear programming problems presented in mathematical form.

These programs solve linear programming and related problems and are available for a variety of platforms, including PC's, Macintoshes, and various workstations. They are available in a number of different versions, whose capabilities differ according to the platform.

▼ *Preview of an Application*

THE THEORY OF GAMES (SECTION 8.11)

Suppose that a new specialized type of modem, with a potential market of 1,000,000 customers, is produced by two firms: Brown Inc. and Green Inc. Each firm can advertise the product in one of two monthly computer magazines: *World Access* and *Connections*. A marketing company determines that if both firms advertise only in *World Access*, then Brown Inc. gets 400,000 customers (and Green Inc. gets 600,000 customers). If both manufacturers advertise in *Connections*, then each gets 500,000 customers. If Brown Inc. advertises only in *World Access* and Green Inc. only in *Connections*, then Brown gets 600,000 customers (and Green gets 400,000 customers). If Brown advertises only in *Connections* and Green only in *World Access*, they each get 500,000 customers. The problem is that of determining the best advertising course of action for each firm.

This problem can be viewed as a game between Brown and Green. The following matrix describes this situation, where the entries in the matrix indicate the number of customers secured by Brown Inc.

		Green	
		World Access	*Connections*
Brown	*World Access*	400,000	600,000
	Connections	500,000	500,000

The theory of games, developed in the 1920s, is an area of mathematics that deals with these types of problems. A brief introduction to the theory of games is presented in Section 8.11.

7.2   EXERCISES

In Exercises 1 through 4, write the initial simplex tableau for each given linear programming problem.

1. Maximize $z = 3x + 7y$
   subject to
$$3x - 2y \le 7$$
$$2x + 5y \le 6$$
$$2x + 3y \le 8$$
$$x \ge 0, \quad y \ge 0.$$

2. Maximize $z = 2x_1 + 3x_2 - 4x_3$
   subject to
$$3x_1 - 2x_2 + x_3 \le 4$$
$$2x_1 + 4x_2 + 5x_3 \le 6$$
$$x_1 \ge 0, \quad x_2 \ge 0, \quad x_3 \ge 0.$$

3. Maximize $z = 2x_1 + 2x_2 + 3x_3 + x_4$
   subject to
$$3x_1 - 2x_2 + x_3 + x_4 \le 6$$
$$x_1 + x_2 + x_3 + x_4 \le 8$$
$$2x_1 - 3x_2 - x_3 + 2x_4 \le 10$$
$$x_1 \ge 0, \quad x_2 \ge 0, \quad x_3 \ge 0, \quad x_4 \ge 0.$$

4. Maximize $z = 2x_1 - 3x_2 + x_3$
   subject to
$$x_1 - 2x_2 + 4x_3 \le 5$$
$$2x_1 + 2x_2 + 4x_3 \le 5$$
$$3x_1 + x_2 - x_3 \le 7$$
$$x_1 \ge 0, \quad x_2 \ge 0, \quad x_3 \ge 0.$$

In Exercises 5 through 11, solve each linear programming problem by the simplex method. Some of these problems may have no finite optimal solution.

5. Maximize $z = 2x + 3y$
   subject to
$$3x + 5y \le 6$$
$$2x + 3y \le 7$$
$$x \ge 0, \quad y \ge 0.$$

6. Maximize $z = 2x + 5y$
   subject to
$$3x + 7y \le 6$$
$$2x + 6y \le 7$$
$$3x + 2y \le 5$$
$$x \ge 0, \quad y \ge 0.$$

7. Maximize $z = 2x + 5y$
   subject to
$$2x - 3y \le 4$$
$$x - 2y \le 6$$
$$x \ge 0, \quad y \ge 0.$$

8. Maximize $z = 3x_1 + 2x_2 + 4x_3$
   subject to
$$x_1 - x_2 - x_3 \le 6$$
$$-2x_1 + x_2 - 2x_3 \le 7$$
$$3x_1 + x_2 - 4x_3 \le 8$$
$$x_1 \ge 0, \quad x_2 \ge 0, \quad x_3 \ge 0.$$

9. Maximize $z = 2x_1 - 4x_2 + 5x_3$
   subject to
$$3x_1 + 2x_2 + x_3 \le 6$$
$$3x_1 - 6x_2 + 7x_3 \le 9$$
$$x_1 \ge 0, \quad x_2 \ge 0, \quad x_3 \ge 0.$$

10. Maximize $z = 2x_1 + 4x_2 - 3x_3$
   subject to
$$5x_1 + 2x_2 + x_3 \le 5$$
$$3x_1 - 2x_2 + 3x_3 \le 10$$
$$4x_1 + 5x_2 - x_3 \le 20$$
$$x_1 \ge 0, \quad x_2 \ge 0, \quad x_3 \ge 0.$$

11. Maximize $z = x_1 + 2x_2 - x_3 + 5x_4$
   subject to
$$2x_1 + 3x_2 + x_3 - x_4 \le 8$$
$$3x_1 + x_2 - 4x_3 + 5x_4 \le 9$$
$$x_1 \ge 0, \quad x_2 \ge 0, \quad x_3 \ge 0, \quad x_4 \ge 0.$$

12. Solve Exercise 1 in Section 7.1 by the simplex method.

13. Solve Exercise 4 in Section 7.1 by the simplex method.

14. Solve Exercise 7 in Section 7.1 by the simplex method.

15. A power plant burns coal, oil, and gas to generate electricity. Suppose that each ton of coal generates 600 kilowatt hours, emits 20 units of sulfur dioxide and 15 units of particulate matter, and costs \$200; each ton of oil generates 550 kilowatt hours, emits 18 units of sulfur dioxide and 12 units of particulate matter, and costs \$220; each ton of gas generates 500 kilowatt hours, emits 15 units of sulfur dioxide and 10 units of particulate matter, and costs \$250. The environmental protection agency restricts the daily emission of sulfur dioxide to no more than 60 units and no more than 75 units of particulate matter. If the power plant wants to spend no more than \$2000 per day on fuel, how much fuel of each type should be bought to maximize the amount of energy generated?

THEORETICAL EXERCISES

T.1. Consider the standard linear programming problem:

Maximize $z = \mathbf{c}^T \mathbf{x}$

subject to

$$A\mathbf{x} \leq \mathbf{b}$$
$$\mathbf{x} \geq \mathbf{0},$$

where A is an $m \times n$ matrix. Show that the feasible region (the set of all feasible solutions) of this problem is a convex set. [*Hint*: If $\mathbf{u}$ and $\mathbf{v}$ are in R^n, the line segment joining them is the set of points $\lambda\mathbf{u} + (1 - \lambda)\mathbf{v}$, where $0 \leq \lambda \leq 1$.]

T.2. Suppose that in selecting the departing variable the minimum θ-ratio is not chosen. Show that the resulting solution is not feasible.

MATLAB EXERCISES

Routine **lpstep** in MATLAB provides a step-by-step procedure for solving linear programming problems using techniques described in this section. Before using **lpstep**, you must formulate the initial tableau into a matrix for entry into MATLAB. The matrix representing the tableau has the same form as discussed in this section except that there are no variable names for rows and columns. For more information, use **help lpstep**.

ML.1. For the illustrative problem given in Equations (10) through (12), the initial tableau is displayed in the text as Tableau 1. To use **lpstep**, use the following MATLAB commands and answer the questions posed. You can use the choices for this problem given in the discussion following the steps for pivotal elimination.

$$\mathbf{A} = [2\ 1\ 1\ 0\ 0\ 50; 1\ 2\ 0\ 1\ 0\ 70;$$
$$-8\ -10\ 0\ 0\ 1\ 0]$$
$$\mathbf{lpstep(A)}$$

ML.2. Solve Exercise 5 using **lpstep**.

ML.3. Solve Exercise 6 using **lpstep**.

ML.4. Solve Exercise 8 using **lpstep**.

ML.5. Maximize $z = 8x_1 + 9x_2 + 5x_3$

subject to

$$x_1 + x_2 + 2x_3 \leq 2$$
$$2x_1 + 3x_2 + 4x_3 \leq 3$$
$$6x_1 + 6x_2 + 2x_3 \leq 8$$
$$x_i \geq 0, \quad i = 1, 2, 3.$$

ML.6. Maximize $z = x_1 + 2x_2 + x_3 + x_4$

subject to

$$2x_1 + x_2 + 3x_3 + x_4 \leq 8$$
$$2x_1 + 3x_2 + 4x_4 \leq 12$$
$$3x_1 + 2x_2 + 2x_3 \leq 18$$
$$x_i \geq 0, \quad i = 1, 2, 3, 4.$$

ML.7. Routine **linprog** solves linear programming problems of the form described in this section directly. That is, the steps are performed automatically. Check your solutions to Exercises 10 and 12.

7.3 ▼ Duality

In this section we show how to associate a minimization problem with each linear programming problem in standard form. There are some interesting interpretations of the associated problem which we also discuss.

Consider the following pair of linear programming problems:

$$\text{Maximize} \quad z = \mathbf{c}^T \mathbf{x}$$

subject to

$$\begin{aligned} A\mathbf{x} &\le \mathbf{b} \\ \mathbf{x} &\ge \mathbf{0} \end{aligned} \tag{1}$$

and

$$\text{Minimize} \quad z' = \mathbf{b}^T \mathbf{y}$$

subject to

$$\begin{aligned} A^T \mathbf{y} &\ge \mathbf{c} \\ \mathbf{y} &\ge \mathbf{0}, \end{aligned} \tag{2}$$

where A is $m \times n$, $\mathbf{b}$ is $m \times 1$, $\mathbf{c}$ is $n \times 1$, $\mathbf{x}$ is $n \times 1$, and $\mathbf{y}$ is $m \times 1$.

These problems are called **dual problems**. The problem given by (1) is called the **primal problem**; the problem given by (2) is called the **dual problem**.

EXAMPLE 1 ■ If the primal problem is

$$\text{Maximize} \quad z = \begin{bmatrix} 3 & 4 \end{bmatrix} \begin{bmatrix} x_1 \\ x_2 \end{bmatrix}$$

subject to

$$\begin{bmatrix} 2 & 3 \\ 3 & -1 \\ 5 & 4 \end{bmatrix} \begin{bmatrix} x_1 \\ x_2 \end{bmatrix} \le \begin{bmatrix} 3 \\ 4 \\ 2 \end{bmatrix}$$

$$x_1 \ge 0, \quad x_2 \ge 0,$$

then the dual problem is

$$\text{Minimize} \quad z' = \begin{bmatrix} 3 & 4 & 2 \end{bmatrix} \begin{bmatrix} y_1 \\ y_2 \\ y_3 \end{bmatrix}$$

subject to

$$\begin{bmatrix} 2 & 3 & 5 \\ 3 & -1 & 4 \end{bmatrix} \begin{bmatrix} y_1 \\ y_2 \\ y_3 \end{bmatrix} \ge \begin{bmatrix} 3 \\ 4 \end{bmatrix}$$

$$y_1 \ge 0, \quad y_2 \ge 0, \quad y_3 \ge 0.$$

■

Observe that in formulating the dual problem, the coefficients of the ith constraint of the primal problem become the coefficients of the variable y_i in the constraints of the dual problem. Conversely, the coefficients of x_j in the primal problem become the coefficients of the jth constraint in the dual problem. Moreover, the coefficients of the objective function of the primal problem become the right-hand sides of the constraints of the dual problem, and conversely.

Since problem (2) can be rewritten as a standard linear programming problem, we can ask for the dual of the dual problem. The answer is provided by the following theorem.

THEOREM 7.3 ■ *Given a primal problem, as in* (1), *the dual of its dual problem is the primal problem.*

Proof   Consider the dual problem (2), which we rewrite in standard form as

$$\text{Maximize} \quad z' = -\mathbf{b}^T \mathbf{y}$$

subject to

$$-A^T \mathbf{y} \le -\mathbf{c}$$
$$\mathbf{y} \ge \ \ \mathbf{0}. \tag{3}$$

Now the dual of (3) is

$$\text{Minimize} \quad z'' = -\mathbf{c}^T \mathbf{w}$$

subject to

$$(-A^T)^T \mathbf{w} \ge (-\mathbf{b}^T)^T$$
$$\mathbf{w} \ge \mathbf{0}$$

or

$$\text{Maximize} \quad z'' = \mathbf{c}^T \mathbf{w}$$

subject to

$$A\mathbf{w} \le \mathbf{b}$$
$$\mathbf{w} \ge \mathbf{0}. \tag{4}$$

Writing $\mathbf{w} = \mathbf{x}$, we see that problem (4) is the primal problem.   ■

EXAMPLE 2 ■ Find the dual problem of the linear programming problem

$$\text{Minimize} \quad z' = \begin{bmatrix} 2 & 3 \end{bmatrix} \begin{bmatrix} y_1 \\ y_2 \end{bmatrix}$$

subject to

$$\begin{bmatrix} 3 & 4 \\ 1 & 2 \\ 5 & 3 \end{bmatrix} \begin{bmatrix} y_1 \\ y_2 \end{bmatrix} \ge \begin{bmatrix} 5 \\ 2 \\ 7 \end{bmatrix}$$
$$y_1 \ge 0, \quad y_2 \ge 0.$$

Solution   The given problem is the dual of the primal problem that we seek to formulate. This primal problem is obtained by taking the dual of the given problem, obtaining

$$\text{Maximize} \quad z = 5x_1 + 2x_2 + 7x_3$$

subject to

$$3x_1 + x_2 + 5x_3 \leq 2$$
$$4x_1 + 2x_2 + 3x_3 \leq 3$$
$$x_1 \geq 0, \quad x_2 \geq 0, \quad x_3 \geq 0.$$ ∎

The following theorem, whose proof we omit, gives the relations between the optimal solutions to the primal and dual problems.

THEOREM 7.4 ∎ *If either the primal problem or dual problem has an optimal solution with finite*
(*Duality Theorem*) *objective value, then the other problem also has an optimal solution. Moreover, the objective values of the two problems are equal.* ∎

It can also be shown that when the primal problem is solved by the simplex method, the final tableau contains the optimal solution to the dual problem in the objective row under the columns of the slack variables. That is, y_1, the first dual variable, is found in the objective row under the first slack variable; y_2 is found under the second slack variable, and so on. We can use these facts to solve the diet problem, Example 3 in Section 7.1, as follows.

EXAMPLE 3 ∎ Consider the diet problem of Example 3, Section 7.1 (using z' instead of z and y_1, y_2 instead of x, y):

$$\text{Minimize} \quad z' = 30y_1 + 40y_2$$

subject to

$$2y_1 + y_2 \geq 12$$
$$y_1 + y_2 \geq 9 \tag{5}$$
$$y_1 + 3y_2 \geq 15$$
$$y_1 \geq 0, \quad y_2 \geq 0.$$

The dual of this problem is

$$\text{Maximize} \quad z = 12x_1 + 9x_2 + 15x_3$$

subject to

$$2x_1 + x_2 + x_3 \leq 30$$
$$x_1 + x_2 + 3x_3 \leq 40 \tag{6}$$
$$x_1 \geq 0, \quad x_2 \geq 0, \quad x_3 \geq 0.$$

This is a standard linear programming problem. Introducing the slack variables x_4 and x_5, we obtain

$$\text{Maximize} \quad z = 12x_1 + 9x_2 + 15x_3$$

subject to

$$2x_1 + x_2 + x_3 + x_4 \qquad = 30$$
$$x_1 + x_2 + 3x_3 \qquad + x_5 = 40$$

$$x_1 \geq 0, \quad x_2 \geq 0, \quad x_3 \geq 0, \quad x_4 \geq 0, \quad x_5 \geq 0.$$

We now apply the simplex method, obtaining the following tableaux.

$\downarrow$

	x_1	x_2	x_3	x_4	x_5	z	
x_4	2	1	1	1	0	0	30
$\leftarrow x_5$	1	1	③	0	1	0	40
	-12	-9	-15	0	0	1	0

$\downarrow$

	x_1	x_2	x_3	x_4	x_5	z	
$\leftarrow x_4$	$\boxed{\frac{5}{3}}$	$\frac{2}{3}$	0	1	$-\frac{1}{3}$	0	$\frac{50}{3}$
x_3	$\frac{1}{3}$	$\frac{1}{3}$	1	0	$\frac{1}{3}$	0	$\frac{40}{3}$
	-7	-4	0	0	5	1	200

$\downarrow$

	x_1	x_2	x_3	x_4	x_5	z	
$\leftarrow x_1$	1	$\boxed{\frac{2}{5}}$	0	$\frac{3}{5}$	$-\frac{1}{5}$	0	10
x_3	0	$\frac{1}{5}$	1	$-\frac{1}{5}$	$\frac{2}{5}$	0	10
	0	$-\frac{6}{5}$	0	$\frac{21}{5}$	$\frac{18}{5}$	1	270

	x_1	x_2	x_3	x_4	x_5	z	
x_2	$\frac{5}{2}$	1	0	$\frac{3}{2}$	$-\frac{1}{2}$	0	25
x_3	$-\frac{1}{2}$	0	1	$-\frac{1}{2}$	$\frac{1}{2}$	0	5
	3	0	0	6	3	1	300

The optimal solution to problem (6) is

$$x_1 = 0, \qquad x_2 = 25, \qquad x_3 = 5,$$

and the value of z is 300.

The optimal solution to the given problem (5), the dual of (6), is found in the objective row under the x_4 and x_5 columns:

$$y_1 = 6, \qquad y_2 = 3.$$

The value of $z' = 30(6) + 40(3) = 300$, as we expect from Theorem 7.4 (the Duality Theorem). ∎

Economic Interpretation of the Dual Problem

We now give two economic interpretations of the dual linear programming problem. The primal problem (1) can be interpreted as follows. We have m resources, inputs or raw materials, and n activities, each of which produces a product. Let

b_i = amount of resource i that is available $(1 \leq i \leq m)$;

a_{ij} = amount of resource i that is used by 1 unit of activity j;

c_j = contribution to the total profit z from 1 unit of activity j $(1 \leq j \leq n)$;

x_j = level of activity j.

We illustrate these interpretations with Example 1 of Section 7.1, the photo shop problem, which we restate for the convenience of the reader (we use x_1 and x_2 instead of x and y as in Section 7.1).

$$\text{Maximize} \quad z = 8x_1 + 10x_2$$

subject to

$$2x_1 + x_2 \leq 50$$
$$x_1 + 2x_2 \leq 70$$
$$x_1 \geq 0, \quad x_2 \geq 0.$$

Here $b_1 = 50$ and $b_2 = 70$ are the amounts of solution A and B, respectively, available each day; activity 1 is the preparation of 1 quart of Fine developer, activity 2 is the preparation of 1 quart of Extra Fine developer; $a_{12} = 1$ is the amount of solution A used in making 1 quart of Extra Fine and so on; $c_1 = 8$ cents is the profit on 1 quart of Fine, $c_2 = 10$ cents is the profit on 1 quart of Extra Fine; x_1 and x_2 are the number of quarts of Fine and Extra Fine, respectively, to be made. The dual of the photo shop problem is

$$\text{Minimize} \quad z' = 50y_1 + 70y_2$$

subject to

$$2y_1 + y_2 \geq 8$$
$$y_1 + 2y_2 \geq 10$$
$$y_1 \geq 0, \quad y_2 \geq 0.$$

Consider the jth constraint of the dual general problem (2):

$$a_{1j}y_1 + a_{2j}y_2 + \cdots + a_{mj}y_m \geq c_j. \tag{7}$$

Since a_{ij} represents amount of resource i per unit of output j, and c_j is the value per unit of output j, we now show from Equation (7) that the dual variable y_j represents "value per unit of resource i" as follows: The dimensions of c_j/a_{ij} are

$$\frac{\text{value/} \cancel{\text{unit of output } j}}{\text{amount of resource } i / \cancel{\text{unit of output } j}} = \frac{\text{value}}{\text{amount of resource } i}.$$

Thus the dual variable y_i acts as a price or cost of one unit of resource i. The dual variables are called **shadow prices**, **fictitious prices**, or **accounting prices**.

First Interpretation

From (7) we see that the left side of this equation is the contribution to the total profit z of the resources that are used in making 1 unit of the jth product. Since the profit of 1 unit of the jth product is c_j, Equation (7) merely says that the contribution of the resources to the total profit has to be at least c_j, for otherwise, we should be using the resources in some better way. In our example, the optimal solution to the dual problem obtained from Tableau 4 in Section 7.2 is

$$y_1 = 2, \qquad y_2 = 4.$$

This means that a shadow price of 2 cents has been attached to each ounce of solution A, and a shadow price of 4 cents has been attached to each unit of solution B. Thus the contribution of the resources used in making 1 quart of Fine to the profit is (in cents)

$$2y_1 + y_2 = 2(2) + 4 = 8,$$

so that the first constraint of the dual has been satisfied. Of course, the constraints

$$y_1 \geq 0, \quad y_2 \geq 0$$

in the dual merely state that the contribution of each resource to the total profit is nonnegative; if some y_1 were negative, we would be better off not using this resource. Finally, the objective function

$$z' = b_1 y_1 + b_2 y_2 + \cdots + b_m y_m$$

is to be minimized; this can be interpreted to be minimizing the total shadow value of the resources that are being used to make the products.

Second Interpretation

If $\mathbf{x}_0$ and $\mathbf{y}_0$ are optimal solutions to the primal and dual problems, respectively, then, by the Duality Theorem (Theorem 7.4), the maximum profit z_0 satisfies the equation

$$z_0 = \mathbf{b}^T \mathbf{y}_0 = b_1 y_1^0 + b_2 y_2^0 + \cdots + b_m y_m^0, \tag{8}$$

where

$$\mathbf{y}_0 = \begin{bmatrix} y_1^0 \\ \vdots \\ y_m^0 \end{bmatrix}.$$

From Equation (8) we see that the manufacturer can increase the profit by increasing the available amount of at least one of the resources. If b_i is increased by 1 unit, the profit will increase by y_i^0. Thus y_i^0 represents the **marginal value** of the ith resource. Similarly, y_i^0 represents the loss incurred if 1 unit of the ith resource is not used. Thus it can be considered as a replacement value of the ith resource for insurance purposes. Then the objective function of the dual problem

$$z' = b_1 y_1 + b_2 y_2 + \cdots + b_m y_m$$

is the total replacement value. Thus an insurance company would want to minimize this objective function, since it would want to pay out as little as possible to settle a claim.

Applications

In the Theory of Games, Section 8.11, the strategies for two competing players are obtained by solving two dual linear programming problems.

Further Readings ▶

CALVERT, JAMES E., and WILLIAM L. VOXMAN. *Linear Programming.* Philadelphia: Harcourt Brace Jovanovich, 1989.

GASS, SAUL I. *Linear Programming,* 5th ed. New York: McGraw-Hill Book Company, 1985.

KOLMAN, BERNARD, and ROBERT E. BECK. *Elementary Linear Programming with Applications,* 2nd ed. Boston: Academic Press, Inc., 1995.

KUESTER, JAMES L., and JOE H. MIZE. *Optimization Techniques with FORTRAN.* New York: McGraw-Hill Book Company, 1974.

MURTY, KATTA G. *Linear and Combinatorial Programming.* New York: John Wiley & Sons, Inc., 1989.

NERING, EVAR D., and ALBERT W. TUCKER. *Linear Programs and Related Problems,* Boston: Academic Press, Inc., 1993.

SCHRAGE, LINUS. *Linear, Quadratic, and Integer Programming with LINDO,* 5th ed. South San Francisco, Calif.: Scientific Press, 1991.

STRANG, GILBERT. *Introduction to Applied Mathematics.* Wellesley, Mass.: Wellesley-Cambridge Press, 1985.

SULTAN, ALAN. *Linear Programming.* Boston: Academic Press, Inc., 1993.

7.3 EXERCISES

In Exercises 1 through 4, state the dual of the given linear programming problem.

1. Maximize $z = 3x_1 + 2x_2$
subject to

$$4x_1 + 3x_2 \leq 7$$
$$5x_1 - 2x_2 \leq 6$$
$$6x_1 + 8x_2 \leq 9$$
$$x_1 \geq 0, \quad x_2 \geq 0.$$

2. Maximize $z = 10x_1 + 12x_2 + 15x_3$

subject to

$$x_1 + 3x_2 + 4x_3 \leq 5$$
$$2x_1 + 4x_2 - 5x_3 \leq 6$$
$$x_1 \geq 0, \quad x_2 \geq 0, \quad x_3 \geq 0.$$

3. Minimize $z = 3x_1 + 5x_2$

subject to

$$2x_1 + 3x_2 \geq 7$$
$$8x_1 - 9x_2 \geq 12$$
$$10x_1 + 15x_2 \geq 18$$
$$x_1 \geq 0, \quad x_2 \geq 0.$$

4. Minimize $z' = 14x_1 + 12x_2 + 18x_3$
   subject to

$$3x_1 + 5x_2 - 4x_3 \geq 9$$
$$5x_1 + 2x_2 + 7x_3 \geq 12$$
$$x_1 \geq 0, \quad x_2 \geq 0, \quad x_3 \geq 0.$$

5. Verify that the dual of the dual of the linear programming problem

   Maximize $z = 3x_1 + 6x_2 + 9x_3$

   subject to

$$3x_1 + 2x_2 - 3x_3 \leq 12$$
$$5x_1 + 4x_2 + 7x_3 \leq 18$$
$$x_1 \geq 0, \quad x_2 \geq 0, \quad x_3 \geq 0,$$

   is the given problem.

In Exercises 6 through 9, solve the dual of the indicated problem by the method of Example 3.

6. Exercise 5 in Section 7.2.

7. Exercise 6 in Section 7.2.

8. Exercise 9 in Section 7.2.

9. Exercise 10 in Section 7.2.

10. A natural cereal consists of dates, nuts, and raisins. Suppose that each ounce of dates contains 24 units of protein, 6 units of iron, and costs 15 cents; each ounce of nuts contains 2 units of protein, 6 units of iron, and costs 18 cents; and each ounce of raisins contains 4 units of protein, 2 units of iron, and costs 12 cents. If each box of cereal is to contain at least 24 units of protein and at least 36 units of iron, how many ounces of each ingredient should be used to minimize the cost of a box of cereal? (*Hint*: Solve the dual problem.)

KEY IDEAS FOR REVIEW ▪

☐ **Linear programming problem.** See page 378.

☐ **Theorem 7.1.** Let S be the feasible region of a linear programming problem.

   (a) If S is bounded, then the objective function $z = ax + by$ assumes both a maximum and a minimum value on S; these values occur at extreme points of S.

   (b) If S is unbounded, then there may or may not be a maximum or minimum value on S. If a maximum or minimum value does exist on S, it occurs at an extreme point.

☐ **Theorem 7.2.** If a linear programming problem has an optimal solution, then it has a basic feasible solution that is optimal.

☐ **Simplex method.** See page 393.

☐ **Primal and dual problems.** See page 412.

☐ **Theorem 7.4 (Duality Theorem).** If either the primal problem or dual problem has an optimal solution with finite objective value, then the other problem also has an optimal solution. Moreover, the objective values of the two problems are equal.

SUPPLEMENTARY EXERCISES ▪

1. Solve the following linear programming problem geometrically.

   Maximize $z = 2x + 3y$

   subject to

$$3x + y \leq 6$$
$$x + y \leq 4$$
$$x + 2y \leq 6$$
$$x \geq 0, \quad y \geq 0.$$

2. Solve the following linear programming problem geometrically.

A microprocessor manufacturer makes two types of microprocessors, model A and model B. The size of the work force limits the total daily production to at most 600 microprocessors. On the other hand, the suppliers of components limit production to at most 400 model A units and 500 model B units. If the net profit on each model A unit is $80 and it is $100 on each model B unit, how many microprocessors of each type should the manufacturer produce daily to maximize the profit?

3. Solve the following linear programming problem by the simplex method.

$$\text{Maximize} \quad z = 50x + 100y$$

subject to

$$x + 2y \le 16$$
$$3x + 2y \le 24$$
$$2x + 2y \le 18$$
$$x \ge 0, \quad y \ge 0.$$

4. Determine the dual of the following linear programming problem.

$$\text{Minimize} \quad z = 6x_1 + 5x_2$$

subject to

$$2x_1 + 3x_2 \ge 6$$
$$5x_1 + 2x_2 \ge 10$$
$$x_1 \ge 0, \quad x_2 \ge 0.$$

5. Solve the problem in Exercise 4 by solving its dual.

CHAPTER TEST ▪

1. Solve the following linear programming problem geometrically.

A farmer who has a 120-acre farm plants corn and wheat. The expenses are $12 for each acre of corn planted and $24 for each acre of wheat planted. Each acre of corn requires 32 bushels of storage and yields a profit of $40; each acre of wheat requires 8 bushels of storage and yields a profit of $50. If the total amount of storage available is 160 bushels and the farmer has $1200 of capital, how many acres of corn and how many acres of wheat should be planted to maximize profit?

2. Solve the following linear programming problem by the simplex method.

$$\text{Maximize} \quad z = 8x_1 + 9x_2 + 5x_3$$

subject to

$$x_1 + x_2 + 2x_3 \le 2$$
$$2x_1 + 3x_2 + 4x_3 \le 3$$
$$3x_1 + 3x_2 + x_3 \le 4$$
$$x_1 \ge 0, \quad x_2 \ge 0, \quad x_3 \ge 0.$$

3. Determine the dual of the following linear programming problem.

$$\text{Minimize} \quad z = 3x_1 + 4x_2$$

subject to

$$x_1 + 4x_2 \ge 8$$
$$2x_1 + 3x_2 \ge 12$$
$$2x_1 + x_2 \ge 6$$
$$x_1 \ge 0, \quad x_2 \ge 0.$$

8

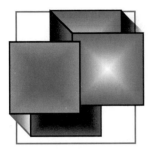

Applications

8.1 ▾ Graph Theory

Prerequisite. Section 1.4, Properties of Matrix Operations.

Graph theory is a new area of mathematics that is being widely used in formulating models in many problems in business, the social sciences, and the physical sciences. These applications include communications problems and the study of organizations and social structures. In this section we present a very brief introduction to the subject as it relates to matrices and show how these elementary and simple notions can be used in formulating models of some important problems.

Graphs

DEFINITION

A **graph** G is a finite set of points, called **vertices** or **nodes**, together with a finite set of **edges**, each of which joins a pair of vertices. An edge joining a vertex to itself is called a **loop**.

The vertices of a graph are represented by dots and the edges by straight or curved line segments.

EXAMPLE 1 ■ Figure 8.1 shows examples of graphs. The graph in Figure 8.1(a) has four vertices, P_1, P_2, P_3, and P_4, and six edges, P_1P_2, P_1P_3, P_1P_4, P_2P_3, P_2P_4, and P_3P_4. In this representation the edges P_1P_4 and P_2P_3 intersect at a point other than a vertex, so they are drawn as shown. The graph in Figure 8.1(b) has four vertices, P_1, P_2, P_3, and P_4, and two edges, P_1P_2 and P_1P_3. Vertices P_2 and P_3

FIGURE 8.1

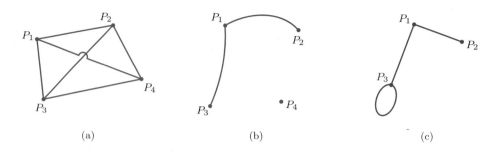

(a)   (b)   (c)

are not joined by an edge. Moreover, vertex P_4 is not joined to any vertex. The graph in Figure 8.1(c) has three vertices, P_1, P_2, and P_3, and three edges, P_1P_2, P_1P_3, and P_3P_3. The edge P_3P_3 is a loop. ∎

A graph with no loops is called **complete** if every pair of vertices is joined by exactly one edge. The complete graph with n vertices is denoted by K_n. Thus every vertex in K_n is connected to every other vertex. The graph in Figure 8.1(a) is the complete graph K_4.

EXAMPLE 2 ∎ A bowling league consists of seven teams: T_1, T_2, T_3, T_4, T_5, T_6, and T_7. Suppose that after a number of games have been played we have the following situation:

T_1 has played T_2, T_3, and T_5;

T_2 has played T_1, T_3, and T_5;

T_3 has played T_1, T_2, and T_4;

T_4 has played T_3 and T_7;

T_5 has played T_1 and T_2;

T_6 has not played anyone;

T_7 has played T_4.

This situation can be described by a graph in which the teams are represented by the vertices, and an edge joining two vertices means that the corresponding teams have played each other. We thus obtain the graph in Figure 8.2. ∎

FIGURE 8.2

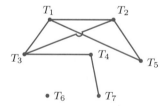

EXAMPLE 3 ∎ A pair of vertices in a graph may be joined by more than one edge, as shown in Figure 8.3. In this case we say that we have a **multiple edge**. Thus, in a communications network there might be two lines between P_1 and P_2 and one line between P_2 and P_3. ∎

FIGURE 8.3

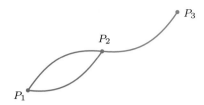

A graph is determined by the set of vertices and by the set of edges joining the vertices, not by the particular appearance of the configuration. Thus two graphs G and G' are said to be **equal** or **isomorphic** if they have the same number of vertices, the same number of edges, and if the vertices (respectively, edges) of G may be put into one-to-one correspondence with the vertices (respectively, edges) of G' in such a way that if edge e of G corresponds to edge e' of G', and the end points of e are P_i and P_j, then the end points of e' are the vertices corresponding to P_i and P_j.

EXAMPLE 4 ■ The graph shown in Figure 8.4 is equal to the graph shown in Figure 8.1(a) (verify). ■

FIGURE 8.4

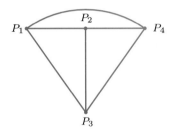

Matrices provide a convenient way of describing graphs, and since matrices lend themselves well to computer use, they make it possible to use the computer for extensive computational work in graph theory.

If we are given a graph G with n vertices, we can associate a symmetric $n \times n$ matrix $M(G)$ with it by letting the i, jth element be the number of edges from the vertex P_i to vertex P_j. The matrix $M(G)$ is called the **matrix representing** G.

EXAMPLE 5 ■ The matrix representing the graph in Figure 8.5 is

$$
M(G) = \begin{array}{c} \\ P_1 \\ P_2 \\ P_3 \\ P_4 \\ P_5 \\ P_6 \end{array}
\begin{array}{c}
\begin{array}{cccccc} P_1 & P_2 & P_3 & P_4 & P_5 & P_6 \end{array} \\
\left[
\begin{array}{cccccc}
1 & 1 & 0 & 0 & 0 & 0 \\
1 & 0 & 0 & 1 & 0 & 0 \\
0 & 0 & 0 & 2 & 1 & 0 \\
0 & 1 & 2 & 0 & 0 & 1 \\
0 & 0 & 1 & 0 & 0 & 2 \\
0 & 0 & 0 & 1 & 2 & 0
\end{array}
\right]
\end{array}.
$$

■

FIGURE 8.5

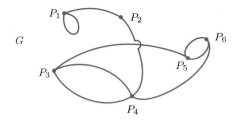

G

Conversely, a given $n \times n$ symmetric matrix with nonnegative integer entries gives rise to a graph G with the given matrix representing G.

EXAMPLE 6 ■ The matrix

$$\begin{bmatrix} 0 & 2 & 1 & 0 & 0 \\ 2 & 0 & 0 & 1 & 0 \\ 1 & 0 & 0 & 1 & 1 \\ 0 & 1 & 1 & 0 & 2 \\ 0 & 0 & 1 & 2 & 0 \end{bmatrix}$$

determines the graph G, shown in Figure 8.6. ■

FIGURE 8.6

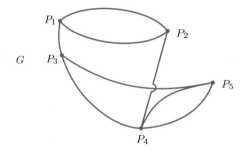

G

Of course, it is easy to see that graphs G and G' are equal if and only if there is a labeling of the vertices for which the matrices representing them are equal.

There is another matrix associated with a graph, which we now define.

DEFINITION   If G is a graph that has n vertices, then the matrix $A(G)$, whose i, jth element is 1 if there is at least one edge between P_i and P_j and zero otherwise, is called the **adjacency matrix** of G. Note that $A(G)$ is a symmetric matrix.

EXAMPLE 7 ■ The adjacency matrix of the graph in Figure 8.5 is

$$A(G) = \begin{bmatrix} 1 & 1 & 0 & 0 & 0 & 0 \\ 1 & 0 & 0 & 1 & 0 & 0 \\ 0 & 0 & 0 & 1 & 1 & 0 \\ 0 & 1 & 1 & 0 & 0 & 1 \\ 0 & 0 & 1 & 0 & 0 & 1 \\ 0 & 0 & 0 & 1 & 1 & 0 \end{bmatrix}.$$

■

FIGURE 8.7

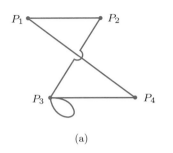

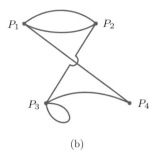

(a)   (b)

Of course, from a symmetric matrix whose entries are zeros and ones we can obtain a graph whose adjacency matrix is the given matrix. However, this graph need not be unique.

EXAMPLE 8 ■ The matrix

$$\begin{bmatrix} 0 & 1 & 0 & 1 \\ 1 & 0 & 1 & 0 \\ 0 & 1 & 1 & 1 \\ 1 & 0 & 1 & 0 \end{bmatrix}$$

is the adjacency matrix of the graphs in Figure 8.7.   ■

Sometimes the adjacency matrix of G may equal the matrix representing G. This will occur if G has no multiple edges.

Figure 8.2 indicated the teams that played each other. However, if we wanted to indicate the winner in each competition, we need to develop the notion of a directed graph.

Digraphs

A **directed graph**, or a **digraph**, is a finite set of points, called **vertices** or **nodes**, together with a finite set of **directed edges**, each of which joins an ordered pair of distinct vertices. Thus a digraph contains no loops. Let us also assume that there are no multiple edges. Moreover, the directed edge P_iP_j is now different from the directed edge P_jP_i. The matrix $A(G)$, whose i, jth element is 1 if there is a directed edge from P_i to P_j and zero otherwise, is called the **adjacency matrix** of G. The adjacency matrix of a digraph need not be symmetric.

EXAMPLE 9 ■ In Figure 8.8 four examples of digraphs are shown. The digraph in Figure 8.8(a) has vertices P_1, P_2, and P_3 and directed edges P_1P_2 and P_2P_3; the digraph in Figure 8.8(b) has vertices P_1, P_2, P_3, and P_4 and directed edges P_1P_2 and P_1P_3; the digraph in Figure 8.8(c) has vertices P_1, P_2, and P_3 and directed edges P_1P_2, P_1P_3, and P_3P_1; the digraph in Figure 8.8(d) has vertices P_1, P_2, and P_3 and directed edges P_2P_1, P_2P_3, P_1P_3, and P_3P_1. The pair of directed edges P_1P_3 and P_3P_1 are indicated by a curved or straight segment with a double arrow, $P_1 \longleftrightarrow P_3$.   ■

FIGURE 8.8

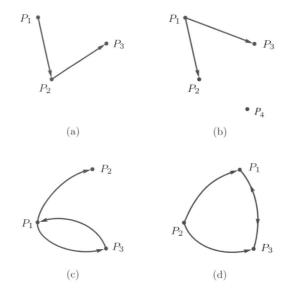

EXAMPLE 10 ■ Consider Example 2 again. Suppose we now know that

T_1 has defeated T_2 and T_5 and lost to T_3;

T_2 has defeated T_5 and lost to T_1 and T_3;

T_3 has defeated T_1 and T_2 and lost to T_4;

T_4 has defeated T_3 and lost to T_7;

T_5 has lost to T_1 and T_2;

T_6 has not played anyone;

T_7 has defeated T_4.

We now obtain the directed graph in Figure 8.9, where $T_i \to T_j$ means that T_i defeated T_j. ■

FIGURE 8.9

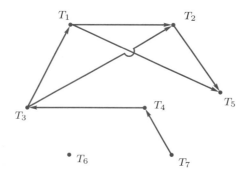

Of course, digraphs can be used in a great many situations including communications problems, family relationships, social structures, street maps, flow charts, transportation problems, electrical circuits, and ecological chains. We shall deal with some of these below and in the exercises.

Models in Sociology and in Communications

Suppose that we have n individuals $P_1, P_2, \ldots, P_n$, some of whom are associated with each other. We assume that no one is associated with himself. Examples of such relations are

1. P_i has access to P_j. In this case it may or may not be the case that if P_i has access to P_j, then P_j has access to P_i. For example, many emergency telephones on turnpikes allow a distressed traveler to contact a nearby emergency station but make no provision for the station to contact the traveler. This model can thus be represented by a digraph G as follows. Let $P_1, P_2, \ldots, P_n$ be the vertices of G and draw a directed edge from P_i to P_j if P_i has access to P_j. It is important to observe that this relation need not be transitive. That is, P_i may have access to P_j and P_j may have access to P_k, but P_i need not have access to P_k.

2. P_i influences P_j. This situation is identical to that in 1: if P_i influences P_j, then it may or may not happen that P_j influences P_i.

3. For every pair of individuals, P_i, P_j, either P_i dominates P_j or P_j dominates P_i, but not both. The graph representing this situation is the complete directed graph with n vertices. Such graphs are often called **dominance digraphs**.

EXAMPLE 11 ■ Suppose that six individuals have been meeting in group therapy for a long time and their leader, who is not part of the group, has drawn the digraph G in Figure 8.10 to describe the influence relations among the various individuals.

FIGURE 8.10

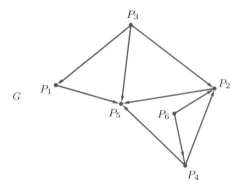

The adjacency matrix of G is

$$
A(G) = \begin{array}{c c} & \begin{array}{cccccc} P_1 & P_2 & P_3 & P_4 & P_5 & P_6 \end{array} \\ \begin{array}{c} P_1 \\ P_2 \\ P_3 \\ P_4 \\ P_5 \\ P_6 \end{array} & \left[\begin{array}{cccccc} 0 & 0 & 0 & 0 & 1 & 0 \\ 0 & 0 & 0 & 0 & 1 & 0 \\ 1 & 1 & 0 & 0 & 1 & 0 \\ 0 & 1 & 0 & 0 & 1 & 0 \\ 0 & 0 & 0 & 0 & 0 & 0 \\ 0 & 1 & 0 & 1 & 0 & 0 \end{array} \right] \end{array}.
$$

Looking at the rows of $A(G)$, we see that P_3 has three 1's in its row so that P_3 influences three people — more than any other individual. Thus P_3 would be declared the leader of the group. On the other hand, P_5 influences no one.   ■

EXAMPLE 12   ■   Consider a communication network whose digraph G is shown in Figure 8.11. The adjacency matrix of G is

$$
A(G) = \begin{array}{c} \\ P_1 \\ P_2 \\ P_3 \\ P_4 \\ P_5 \\ P_6 \end{array} \begin{array}{c} \begin{array}{cccccc} P_1 & P_2 & P_3 & P_4 & P_5 & P_6 \end{array} \\ \left[\begin{array}{cccccc} 0 & 0 & 0 & 0 & 1 & 0 \\ 0 & 0 & 0 & 0 & 1 & 1 \\ 1 & 1 & 0 & 0 & 1 & 0 \\ 0 & 1 & 0 & 0 & 1 & 1 \\ 0 & 1 & 0 & 0 & 0 & 1 \\ 1 & 0 & 1 & 0 & 0 & 0 \end{array} \right] \end{array}.
$$

■

FIGURE 8.11

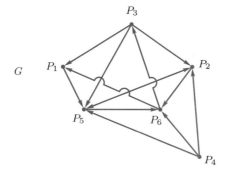

G

Although the relation "P_i has access to P_j" need not be transitive, we can speak of two-stage access. We say that P_i has **two-stage access** to P_k if we can find an individual P_j such that P_i has access to P_j and P_j has access to P_k. Similarly, P_i has r-stage access to P_k if we can find $r-1$ individuals $P_{j_1}, \ldots, P_{j_{r-1}}$, such that P_i has access to P_{j_1}, P_{j_1} has access to $P_{j_2}, \ldots, P_{j_{r-2}}$ has access to $P_{j_{r-1}}$, and $P_{j_{r-1}}$ has access to P_k. Some of the $r+1$ individuals $P_i, P_{j_1}, \ldots, P_{j_{r-1}}, P_k$ may be the same.

The following theorem, whose proof we omit, can be established.

THEOREM 8.1   ■   *Let $A(G)$ be the adjacency matrix of a digraph G and let the rth power of $A(G)$ be B_r:*

$$
[A(G)]^r = B_r = [\, b_{ij}^{(r)} \,].
$$

Then the i, jth element in B_r, $b_{ij}^{(r)}$, is the number of ways in which P_i has access to P_j in r stages.   ■

The sum of the elements in the jth column of $[A(G)]^r$ gives the number of ways in which P_j is reached by all the other individuals in r stages.

If we let

$$A(G) + [A(G)]^2 + \cdots + [A(G)]^r = C = \begin{bmatrix} c_{ij} \end{bmatrix}, \tag{1}$$

then c_{ij} is the number of ways in which P_i has access to P_j in one, two, ..., or r stages.

Similarly, we speak of r-stage dominance, r-stage influence, and so forth. We can also use this model to study the spread of a rumor. Thus c_{ij} in (1) is the number of ways in which P_i has spread the rumor to P_j in one, two, ..., or r stages. In the influence relation, r-stage influence shows the effect of indirect influence.

EXAMPLE 13 ■ If G is the digraph in Figure 8.11, then we find that

$$[A(G)]^2 = \begin{array}{c} \\ P_1 \\ P_2 \\ P_3 \\ P_4 \\ P_5 \\ P_6 \end{array} \begin{array}{cccccc} P_1 & P_2 & P_3 & P_4 & P_5 & P_6 \\ \begin{bmatrix} 0 & 1 & 0 & 0 & 0 & 1 \\ 1 & 1 & 1 & 0 & 0 & 1 \\ 0 & 1 & 0 & 0 & 2 & 2 \\ 1 & 1 & 1 & 0 & 1 & 2 \\ 1 & 0 & 1 & 0 & 1 & 1 \\ 1 & 1 & 0 & 0 & 2 & 0 \end{bmatrix} \end{array}$$

and

$$A(G) + [A(G)]^2 = C = \begin{array}{c} \\ P_1 \\ P_2 \\ P_3 \\ P_4 \\ P_5 \\ P_6 \end{array} \begin{array}{cccccc} P_1 & P_2 & P_3 & P_4 & P_5 & P_6 \\ \begin{bmatrix} 0 & 1 & 0 & 0 & 1 & 1 \\ 1 & 1 & 1 & 0 & 1 & 2 \\ 1 & 2 & 0 & 0 & 3 & 2 \\ 1 & 2 & 1 & 0 & 2 & 3 \\ 1 & 1 & 1 & 0 & 1 & 2 \\ 2 & 1 & 1 & 0 & 2 & 0 \end{bmatrix} \end{array}.$$

Since $c_{35} = 3$, there are three ways in which P_3 has access to P_5 in one or two stages: $P_3 \rightarrow P_5$, $P_3 \rightarrow P_2 \rightarrow P_5$, and $P_3 \rightarrow P_1 \rightarrow P_5$.   ■

In studying organizational structures, we often find subsets of people in which any pair of individuals is related. This is an example of a clique, which we now define.

DEFINITION ■ A **clique** in a digraph is a subset S of the vertices satisfying the following properties:

(a) S contains three or more vertices.

(b) If P_i and P_j are in S, then there is a directed edge from P_i to P_j and a directed edge from P_j to P_i.

(c) There is no larger subset T of the vertices that satisfies property (b) and contains S [that is, S is a maximal subset satisfying (b)].

FIGURE 8.12

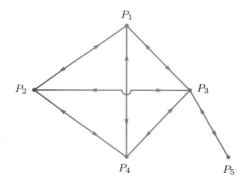

EXAMPLE 14 ■ Consider the digraph in Figure 8.12. The set $\{P_1, P_2, P_3\}$ satisfies conditions (a) and (b) for a clique, but it is not a clique, since it fails to satisfy condition (c). That is, $\{P_1, P_2, P_3\}$ is contained in $\{P_1, P_2, P_3, P_4\}$, which satisfies conditions (a), (b), and (c). Thus the only clique in this digraph is $\{P_1, P_2, P_3, P_4\}$.   ■

EXAMPLE 15 ■ Consider the digraph in Figure 8.13. In this case we have two cliques:

$$\{P_1, P_2, P_3, P_4\} \quad \text{and} \quad \{P_4, P_5, P_6\}.$$

Moreover, P_4 belongs to both cliques.   ■

FIGURE 8.13

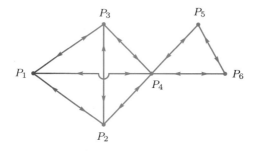

For large digraphs it is difficult to determine cliques. The following approach provides a useful method for detecting cliques that can easily be implemented on a computer. If $A(G) = [a_{ij}]$ is the given adjacency matrix of a digraph, form a new matrix $S = [s_{ij}]$:

$$s_{ij} = s_{ji} = 1 \quad \text{if} \quad a_{ij} = a_{ji} = 1;$$

otherwise, let $s_{ij} = s_{ji} = 0$. Thus $s_{ij} = 1$ if P_i and P_j have access to each other; otherwise, $s_{ij} = 0$. It should be noted that S is a symmetric matrix ($S = S^T$).

EXAMPLE 16 ■ Consider a digraph with adjacency matrix

$$
A(G) = \begin{array}{c} \\ P_1 \\ P_2 \\ P_3 \\ P_4 \\ P_5 \\ P_6 \end{array}
\begin{array}{c} \begin{array}{cccccc} P_1 & P_2 & P_3 & P_4 & P_5 & P_6 \end{array} \\
\left[\begin{array}{cccccc}
0 & 0 & 1 & 1 & 1 & 0 \\
1 & 0 & 1 & 1 & 1 & 1 \\
0 & 1 & 0 & 1 & 1 & 1 \\
1 & 0 & 1 & 0 & 0 & 1 \\
1 & 1 & 0 & 1 & 0 & 1 \\
0 & 1 & 1 & 1 & 1 & 0
\end{array} \right] \end{array}.
$$

Then

$$
S = \begin{array}{c} \\ P_1 \\ P_2 \\ P_3 \\ P_4 \\ P_5 \\ P_6 \end{array}
\begin{array}{c} \begin{array}{cccccc} P_1 & P_2 & P_3 & P_4 & P_5 & P_6 \end{array} \\
\left[\begin{array}{cccccc}
0 & 0 & 0 & 1 & 1 & 0 \\
0 & 0 & 1 & 0 & 1 & 1 \\
0 & 1 & 0 & 1 & 0 & 1 \\
1 & 0 & 1 & 0 & 0 & 1 \\
1 & 1 & 0 & 0 & 0 & 1 \\
0 & 1 & 1 & 1 & 1 & 0
\end{array} \right] \end{array}.
$$

The following theorem can be proved.

THEOREM 8.2 ■ *Let $A(G)$ be the adjacency matrix of a digraph and $S = [s_{ij}]$ be the symmetric matrix defined above, with $S^3 = [s_{ij}^{(3)}]$, where $s_{ij}^{(3)}$ is the i, jth element in S^3. Then P_i belongs to a clique if and only if the diagonal entry $s_{ii}^{(3)}$ is positive.* ■

Let us briefly consider why we examine the diagonal entries of S^3 in Theorem 8.2. First, note that the diagonal entry $s_{ii}^{(3)}$ of S^3 gives the number of ways in which P_i has access to himself in three stages. If $s_{ii}^{(3)} > 0$, then there is at least one way in which P_i has access to himself. Since a digraph has no loops, this access must occur through two individuals: $P_i \rightarrow P_j \rightarrow P_k \rightarrow P_i$. Thus $s_{ij} \neq 0$. But $s_{ij} \neq 0$ implies that $s_{ji} \neq 0$; so $P_j \rightarrow P_i$. Similarly, any two of the individuals in $\{P_i, P_j, P_k\}$ have access to each other. This means that P_i, P_j, and P_k all belong to the same clique. The opposite direction (if P_i is in a clique, then $s_{ii}^{(3)} > 0$) is left as an exercise.

The procedure for determining a clique in a digraph is as follows.

Step 1. If $A = [a_{ij}]$ is the adjacency matrix of the given digraph, compute the symmetric matrix $S = [s_{ij}]$, where

$$
s_{ij} = s_{ji} = 1 \quad \text{if} \quad a_{ij} = a_{ji} = 1;
$$

otherwise, $s_{ij} = 0$.

Step 2. Compute $S^3 = [s_{ij}^{(3)}]$.

Step 3. P_i belongs to a clique if and only if $s_{ii}^{(3)}$ is positive.

FIGURE 8.14

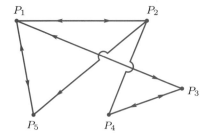

EXAMPLE 17 ■ Consider the digraph in Figure 8.14, whose adjacency matrix is

$$
A(G) = \begin{array}{c} \\ P_1 \\ P_2 \\ P_3 \\ P_4 \\ P_5 \end{array}
\begin{array}{c}
\begin{array}{ccccc} P_1 & P_2 & P_3 & P_4 & P_5 \end{array} \\
\left[\begin{array}{ccccc}
0 & 1 & 1 & 0 & 1 \\
1 & 0 & 0 & 0 & 1 \\
1 & 0 & 0 & 1 & 0 \\
0 & 1 & 1 & 0 & 0 \\
1 & 0 & 0 & 0 & 0
\end{array} \right].
\end{array}
$$

Then (verify)

$$
S = \begin{array}{c} \\ P_1 \\ P_2 \\ P_3 \\ P_4 \\ P_5 \end{array}
\begin{array}{c}
\begin{array}{ccccc} P_1 & P_2 & P_3 & P_4 & P_5 \end{array} \\
\left[\begin{array}{ccccc}
0 & 1 & 1 & 0 & 1 \\
1 & 0 & 0 & 0 & 0 \\
1 & 0 & 0 & 1 & 0 \\
0 & 0 & 1 & 0 & 0 \\
1 & 0 & 0 & 0 & 0
\end{array} \right]
\end{array}
$$

and

$$
S^3 = \begin{array}{c} \\ P_1 \\ P_2 \\ P_3 \\ P_4 \\ P_5 \end{array}
\begin{array}{c}
\begin{array}{ccccc} P_1 & P_2 & P_3 & P_4 & P_5 \end{array} \\
\left[\begin{array}{ccccc}
0 & 3 & 4 & 0 & 3 \\
3 & 0 & 0 & 1 & 0 \\
4 & 0 & 0 & 2 & 0 \\
0 & 1 & 2 & 0 & 1 \\
3 & 0 & 0 & 1 & 0
\end{array} \right].
\end{array}
$$

Since every diagonal entry in S^3 is zero, we conclude that there are no cliques.
■

EXAMPLE 18 ■ Consider the digraph in Figure 8.15, whose adjacency matrix is

$$
A(G) = \begin{array}{c} \\ P_1 \\ P_2 \\ P_3 \\ P_4 \\ P_5 \end{array}
\begin{array}{c}
\begin{array}{ccccc} P_1 & P_2 & P_3 & P_4 & P_5 \end{array} \\
\left[\begin{array}{ccccc}
0 & 0 & 1 & 1 & 1 \\
0 & 0 & 1 & 0 & 1 \\
1 & 0 & 0 & 0 & 1 \\
0 & 0 & 1 & 0 & 1 \\
1 & 1 & 1 & 0 & 0
\end{array} \right].
\end{array}
$$

FIGURE 8.15

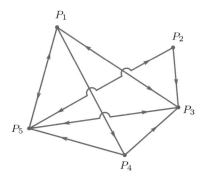

Then (verify)

$$S = \begin{array}{c} \\ P_1 \\ P_2 \\ P_3 \\ P_4 \\ P_5 \end{array} \begin{array}{ccccc} P_1 & P_2 & P_3 & P_4 & P_5 \\ \left[\begin{array}{ccccc} 0 & 0 & 1 & 0 & 1 \\ 0 & 0 & 0 & 0 & 1 \\ 1 & 0 & 0 & 0 & 1 \\ 0 & 0 & 0 & 0 & 0 \\ 1 & 1 & 1 & 0 & 0 \end{array}\right] \end{array}$$

and

$$S^3 = \begin{array}{c} \\ P_1 \\ P_2 \\ P_3 \\ P_4 \\ P_5 \end{array} \begin{array}{ccccc} P_1 & P_2 & P_3 & P_4 & P_5 \\ \left[\begin{array}{ccccc} 2 & 1 & 3 & 0 & 4 \\ 1 & 0 & 1 & 0 & 3 \\ 3 & 1 & 2 & 0 & 4 \\ 0 & 0 & 0 & 0 & 0 \\ 4 & 3 & 4 & 0 & 2 \end{array}\right] \end{array}.$$

Since s_{11}, s_{33}, and s_{55} are positive, we conclude that P_1, P_3, and P_5 belong to cliques and in fact they form the only clique in this digraph. ■

We now consider the notion of a strongly connected digraph.

DEFINITION   A **path** joining two individuals P_i and P_k in a digraph is a sequence of distinct vertices $P_i, P_a, P_b, P_c, \ldots, P_r, P_k$ and directed edges $P_i P_a, P_a P_b, \ldots, P_r P_k$.

EXAMPLE 19   ■   Consider the digraph in Figure 8.16. The sequence

$$P_2 \to P_3 \to P_4 \to P_5$$

is a path. The sequence

$$P_2 \to P_3 \to P_4 \to P_2 \to P_5$$

is not a path, since the vertex P_2 is repeated. ■

DEFINITION   The digraph G is said to be **strongly connected** if for every two distinct vertices P_i and P_j there is a path from P_i to P_j and a path from P_j to P_i. Otherwise, G is said to be **not strongly connected**.

FIGURE 8.16

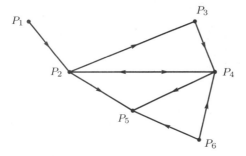

An example of a strongly connected digraph is provided by the streets of a city.

For many digraphs it is a tedious task to determine whether or not they are strongly connected. First, observe that if our digraph has n vertices, then the number of edges in a path from P_i to P_j cannot exceed $n - 1$, since all the vertices of a path are distinct and we cannot go through more than the n vertices in the digraph. If $[A(G)]^r = [b_{ij}^{(r)}]$, then $b_{ij}^{(r)}$ is the number of ways of getting from P_i to P_j in r stages. A way of getting from P_i to P_j in r stages need not be a path, since it may contain repeated vertices. If these repeated vertices and all edges between repeated vertices are deleted, we do obtain a path between P_i and P_j with at most r edges. For example, if we have $P_1 \to P_2 \to P_4 \to P_3 \to P_2 \to P_5$, we can eliminate the second P_2 and all edges between the P_2's and obtain the path $P_1 \to P_2 \to P_5$. Hence, if the i, jth element in

$$[A(G)] + [A(G)]^2 + \cdots + [A(G)]^{n-1}$$

is zero, then there is no path from P_i to P_j. Thus the following theorem, half of whose proof we have sketched here, provides a test for strongly connected digraphs.

THEOREM 8.3 ■ *A digraph with n vertices is strongly connected if and only if its adjacency matrix $A(G)$ has the property that*

$$[A(G)] + [A(G)]^2 + \cdots + [A(G)]^{n-1} = E$$

has no zero entries.   ■

The procedure for determining if a digraph G with n vertices is strongly connected is as follows.

Step 1. If $A(G)$ is the adjacency matrix of the digraph, compute

$$[A(G)] + [A(G)]^2 + \cdots + [A(G)]^{n-1} = E.$$

Step 2. G is strongly connected if and only if E has no zero entries.

FIGURE 8.17

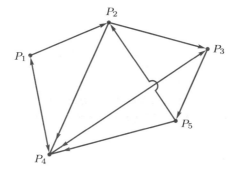

EXAMPLE 20 ■ Consider the digraph in Figure 8.17. The adjacency matrix is

$$A(G) = \begin{array}{c} \\ P_1 \\ P_2 \\ P_3 \\ P_4 \\ P_5 \end{array} \begin{array}{ccccc} P_1 & P_2 & P_3 & P_4 & P_5 \\ \left[\begin{array}{ccccc} 0 & 1 & 0 & 1 & 0 \\ 0 & 0 & 1 & 1 & 0 \\ 0 & 0 & 0 & 1 & 1 \\ 1 & 0 & 1 & 0 & 0 \\ 0 & 1 & 0 & 1 & 0 \end{array}\right] \end{array}.$$

Then (verify)

$$A(G) + [A(G)]^2 + [A(G)]^3 + [A(G)]^4 = E = \begin{array}{c} \\ P_1 \\ P_2 \\ P_3 \\ P_4 \\ P_5 \end{array} \begin{array}{ccccc} P_1 & P_2 & P_3 & P_4 & P_5 \\ \left[\begin{array}{ccccc} 5 & 5 & 7 & 10 & 3 \\ 5 & 4 & 8 & 10 & 3 \\ 5 & 4 & 7 & 10 & 4 \\ 5 & 4 & 7 & 10 & 4 \\ 5 & 5 & 7 & 10 & 3 \end{array}\right] \end{array}.$$

Since all the entries in E are positive, the given digraph is strongly connected.   ■

This approach can be used to trace the spread of a contaminant in a group of individuals; if there is a path from P_i to P_j, then the contaminant can spread from P_i to P_j.

Further Readings ▶

BERGE, C. *The Theory of Graphs and Its Applications.* London: Methuen & Company Ltd., 1962.

BUSACKER, R. G., and T. L. SAATY. *Finite Graphs and Networks*: *An Introduction with Applications.* New York: McGraw-Hill Book Company, 1965.

CHARTRAND, GARY. *Introduction to Graph Theory.* New York: Dover, 1985.

JOHNSTON, J. B., G. PRICE, and F. S. VAN VLECK. *Linear Equations and Matrices.* Reading, Mass.: Addison-Wesley Publishing Company, Inc., 1966.

KOLMAN, B., R. C. BUSBY, and S. ROSS. *Discrete Mathematical Structures*, 3rd ed. Upper Saddle River, N.J.: Prentice Hall, Inc., 1996.

ORE, O. *Graphs and Their Uses.* New York: Random House, Inc., 1963.

TRUDEAU, R. J. *Introduction to Graph Theory.* New York: Dover, 1994.

TUCKER, ALAN. *Applied Combinatorics*, 3rd ed. New York: Wiley, 1994.

8.1 EXERCISES

1. Draw the complete graph on five vertices.

2. Write the matrices $M(G)$ representing digraphs (a) and (b), as in Example 5.

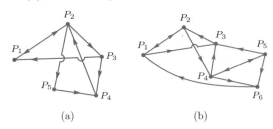

(a)   (b)

3. Which of the digraphs (a), (b), and (c) are equal? (*Hint*: If two digraphs are equal, the number of directed edges emanating from corresponding vertices must be the same.)

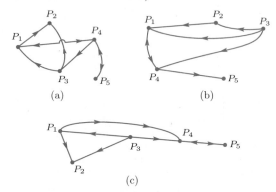

(a)   (b)

(c)

4. Draw a digraph determined by the given matrix.

(a) $\begin{bmatrix} 0 & 1 & 0 & 0 & 0 \\ 0 & 0 & 1 & 1 & 0 \\ 1 & 0 & 0 & 1 & 1 \\ 1 & 1 & 0 & 0 & 1 \\ 1 & 0 & 1 & 1 & 0 \end{bmatrix}.$

(b) $\begin{bmatrix} 0 & 1 & 1 & 1 \\ 1 & 0 & 0 & 1 \\ 1 & 1 & 0 & 0 \\ 0 & 1 & 0 & 0 \end{bmatrix}.$

5. Write the adjacency matrices of the digraphs in Exercise 2.

6. Consider a group of five people, P_1, P_2, P_3, P_4, and P_5, who have been stationed on a remote island to operate a weather station. The following social interactions have been observed:

P_1 gets along with P_2, P_3, and P_4.
P_2 gets along with P_1 and P_5.
P_3 gets along with P_1, P_2, and P_4.
P_4 gets along with P_2, P_3, and P_5.
P_5 gets along with P_4.

Draw a digraph G describing the situation. Write the adjacency matrix representing G.

7. Which of the following matrices can be adjacency matrices for a dominance digraph?

(a) $\begin{bmatrix} 0 & 1 & 1 & 0 \\ 0 & 0 & 1 & 1 \\ 0 & 0 & 0 & 0 \\ 0 & 0 & 1 & 0 \end{bmatrix}.$

(b) $\begin{bmatrix} 0 & 0 & 1 & 0 & 0 \\ 1 & 0 & 1 & 0 & 1 \\ 0 & 0 & 0 & 0 & 1 \\ 1 & 1 & 1 & 0 & 1 \\ 1 & 0 & 0 & 0 & 0 \end{bmatrix}.$

8. The following data have been obtained by studying a group of six individuals in the course of a sociological study:

Carter influences Smith and Gordon.
Gordon influences Jones.
Smith is influenced by Peters.
Russell is influenced by Carter, Smith, and Gordon.
Peters is influenced by Russell.
Jones influences Carter and Russell.
Smith influences Jones.
Carter is influenced by Peters.
Peters influences Jones and Gordon.

(a) Who influences the most people?
(b) Who is influenced by the most people?

9. Consider a communications network among five individuals with adjacency matrix

$$\begin{array}{c} \\ P_1 \\ P_2 \\ P_3 \\ P_4 \\ P_5 \end{array} \begin{array}{ccccc} P_1 & P_2 & P_3 & P_4 & P_5 \\ \begin{bmatrix} 0 & 1 & 1 & 1 & 1 \\ 0 & 0 & 0 & 1 & 1 \\ 0 & 1 & 0 & 1 & 0 \\ 0 & 0 & 0 & 0 & 1 \\ 1 & 0 & 1 & 0 & 0 \end{bmatrix} \end{array}.$$

(a) In how many ways does P_2 have access to P_1 through one individual?

(b) What is the smallest number of individuals through which P_2 has access to himself?

10. Consider the following influence relation among five individuals.

P_1 influences P_2, P_4, and P_5.

P_2 influences P_3 and P_4.

P_3 influences P_1 and P_4.

P_4 influences P_5.

P_5 influences P_2 and P_3.

(a) Can P_4 influence P_1 in at most two stages?

(b) In how many ways does P_1 influence P_4 in exactly three stages?

(c) In how many ways does P_1 influence P_4 in one, two, or three stages?

In Exercises 11 through 13, determine a clique, if there is one, for the digraph with given adjacency matrix.

11. $\begin{bmatrix} 0 & 0 & 0 & 0 & 0 \\ 1 & 0 & 1 & 1 & 1 \\ 0 & 1 & 0 & 1 & 0 \\ 1 & 1 & 1 & 0 & 0 \\ 0 & 0 & 1 & 1 & 0 \end{bmatrix}.$

12. $\begin{bmatrix} 0 & 1 & 1 & 0 & 1 & 1 \\ 1 & 0 & 0 & 0 & 0 & 0 \\ 0 & 0 & 0 & 1 & 0 & 0 \\ 1 & 0 & 0 & 0 & 1 & 1 \\ 1 & 0 & 1 & 1 & 0 & 1 \\ 1 & 0 & 1 & 1 & 1 & 0 \end{bmatrix}.$

13. $\begin{bmatrix} 0 & 1 & 1 & 0 & 1 \\ 1 & 0 & 1 & 1 & 1 \\ 0 & 0 & 0 & 1 & 0 \\ 1 & 0 & 0 & 0 & 1 \\ 0 & 1 & 0 & 1 & 0 \end{bmatrix}.$

14. Consider a communication network among five individuals with adjacency matrix

$$\begin{bmatrix} 0 & 1 & 0 & 0 & 0 \\ 0 & 0 & 1 & 0 & 1 \\ 0 & 0 & 0 & 1 & 0 \\ 1 & 1 & 0 & 0 & 0 \\ 0 & 0 & 1 & 1 & 0 \end{bmatrix}.$$

(a) Can P_3 get a message to P_5 in at most two stages?

(b) What is the minimum number of stages that will guarantee that every person can get a message to any other (different) person?

(c) What is the minimum number of stages that will guarantee that every person can get a message to any person (including himself)?

15. Determine whether the digraph with given adjacency matrix is strongly connected.

(a) $\begin{bmatrix} 0 & 1 & 1 & 1 \\ 0 & 0 & 1 & 1 \\ 1 & 0 & 0 & 1 \\ 0 & 0 & 1 & 0 \end{bmatrix}.$

(b) $\begin{bmatrix} 0 & 1 & 0 & 0 & 0 \\ 0 & 0 & 1 & 0 & 0 \\ 1 & 0 & 0 & 0 & 0 \\ 0 & 0 & 0 & 0 & 1 \\ 0 & 0 & 1 & 1 & 0 \end{bmatrix}.$

16. Determine whether the digraph with given adjacency matrix is strongly connected.

(a) $\begin{bmatrix} 0 & 0 & 1 & 1 & 1 \\ 1 & 0 & 1 & 1 & 0 \\ 0 & 1 & 0 & 0 & 0 \\ 0 & 1 & 0 & 0 & 1 \\ 1 & 1 & 0 & 0 & 0 \end{bmatrix}.$

(b) $\begin{bmatrix} 0 & 0 & 0 & 0 & 1 \\ 0 & 0 & 1 & 1 & 0 \\ 0 & 1 & 0 & 0 & 1 \\ 1 & 0 & 0 & 0 & 0 \\ 0 & 0 & 0 & 1 & 0 \end{bmatrix}.$

17. A group of five acrobats performs a pyramiding act in which there must be a supporting path between any two different persons or the pyramid collapses. In the following adjacency matrix, $a_{ij} = 1$ means that P_i supports P_j.

$$\begin{bmatrix} 0 & 0 & 1 & 1 & 0 \\ 1 & 0 & 0 & 0 & 1 \\ 1 & 0 & 0 & 1 & 0 \\ 1 & 0 & 0 & 0 & 1 \\ 0 & 1 & 1 & 0 & 0 \end{bmatrix}.$$

Who can be left out of the pyramid without having a collapse occur?

THEORETICAL EXERCISES ▮

T.1. Show that a digraph which has the directed edges P_iP_j and P_jP_i cannot be a dominance graph.

T.2. Prove Theorem 8.1.

T.3. Prove Theorem 8.2.

MATLAB EXERCISES ▮

MATLAB operators $+$ and $^\wedge$ can be used to compute sums and powers of a matrix. Hence the computations in Theorems 8.2 and 8.3 are easily performed in MATLAB.

ML.1. Solve Exercise 11 using MATLAB.

ML.2. Determine a clique, if there is one, for the digraph with the following adjacency matrix.

$$\begin{bmatrix} 0 & 1 & 1 & 0 & 1 \\ 1 & 0 & 0 & 1 & 0 \\ 0 & 1 & 0 & 0 & 1 \\ 0 & 1 & 1 & 0 & 1 \\ 1 & 0 & 0 & 1 & 0 \end{bmatrix}$$

ML.3. Solve Exercise 16 using MATLAB.

8.2 ▾ Electrical Circuits

Prerequisite. Section 1.5, Solutions of Linear Systems of Equations.

In this section we introduce the basic laws of electrical-circuit analysis and then use these to analyze electrical circuits consisting of batteries, resistors, and wires.

A **battery** is a source of direct current (or voltage) in the circuit, a **resistor** is a device, such as a lightbulb, that reduces the current in a circuit by converting electrical energy into thermal energy, and a **wire** is a conductor that allows a free flow of electrical current. A simple electrical circuit is a closed connection of resistors, batteries, and wires. When circuits are represented by diagrams, batteries, resistors, and wires are depicted as follows:

Batteries   Resistors   Wires

Figure 8.18 shows a simple electrical circuit consisting of three batteries and four resistors connected by wires.

FIGURE 8.18

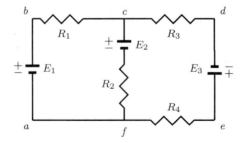

The physical quantities used when discussing electrical circuits are current, resistance, and electrical potential difference across a battery. Electrical potential difference is denoted by E and is measured in volts (V). Current is denoted by I and is measured in amperes (A). Resistance is denoted by R and is measured in ohms (Ω). These units are all related by the equation

$$\text{One Volt} = (\text{One Ampere}) \times (\text{One Ohm}).$$

The electrical potential difference of a battery is taken as positive when measured from the negative terminal ($-$) to the positive terminal ($+$), and negative when measured from the positive terminal ($+$) to the negative terminal ($-$). The electrical potential difference across a resistor (denoted by V) depends on the current flowing through the resistor and its resistance, and is given by Ohm's law:

$$V = \pm IR.$$

The negative ($-$) sign is used when the difference is measured across the resistor in the direction of the current flow, while the positive ($+$) sign is used when the difference is measured across the resistor in the direction opposite the current flow.

All electrical circuits consist of voltage loops and current nodes. A **voltage loop** is a closed connection within the circuit. For example, Figure 8.18 contains the three voltage loops

$$a \rightarrow b \rightarrow c \rightarrow f \rightarrow a,$$
$$c \rightarrow d \rightarrow e \rightarrow f \rightarrow c,$$

and

$$a \rightarrow b \rightarrow c \rightarrow d \rightarrow e \rightarrow f \rightarrow a.$$

A **current node** is a point where three or more segments of wire meet. For example, Figure 8.18 contains two current nodes at points

$$c \quad \text{and} \quad f.$$

Points a, b, d, and e are not current nodes since only two wire segments meet at these points.

The physical laws that govern the flow of current in an electrical circuit are conservation of energy and conservation of charge.

- The *conservation of energy* appears in what is known as **Kirchhoff's voltage law**: Around any voltage loop, the total electrical potential difference is zero.

- The *conservation of charge* appears in what is known as **Kirchhoff's current law**: At any current node, the flow of all currents into the node equals the flow of all currents out of the node. This guarantees that no charge builds up or is depleted at a node, so the current flow is steady through the node.

We are now ready to apply these ideas and the methods for solving linear systems to solving problems involving electrical circuits, which have the following general format. In a circuit containing batteries, resistors, and wires, determine all *unknown* values of electrical potential difference across the batteries, resistance in the resistors, and currents flowing through the wires, given enough *known* values of these same quantities. The following example illustrates this for the standard situation when the unknowns are the currents.

EXAMPLE 1 ■ Figure 8.19 shows the circuit from Figure 8.18 with the batteries having the indicated electrical potentials, measured from the negative terminal to the positive one, and the resistors having the indicated resistances. The problem is to determine the currents that flow through each segment of the circuit.

FIGURE 8.19

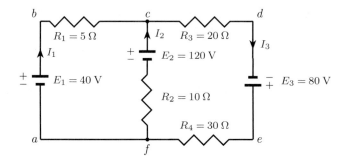

We begin by assigning currents to each segment of the circuit that begins at some node point and ends at some other node point (with no other node points in between). For example, in Figure 8.19, we assign I_1 to the segment $f \to a \to b \to c$, I_2 to the segment $f \to c$, and I_3 to the segment $c \to d \to e \to f$. In addition, we arbitrarily assign directions to these currents as indicated by the arrows in Figure 8.19. If the assigned direction is correct, the computed value of the current will be positive, while if the assigned direction is incorrect, the computed value of the current will be negative. The latter situation then indicates that the actual direction of current flow is just opposite to the assigned one. Using Kirchhoff's current law (the sum of the incoming currents = the sum of the outgoing currents) at points c and f we have

$$I_1 + I_2 = I_3$$

and

$$I_3 = I_1 + I_2, \tag{1}$$

respectively. Since these two equations contain the same information, only one of them is needed. In general, if a circuit has n nodes, Kirchhoff's current law will yield $n - 1$ useful equations and one equation that is a linear combination of the other $n - 1$ equations.

Next we use Kirchhoff's voltage law. Starting at point a and moving through the battery from $(-)$ to $(+)$ to point b, the potential is $+40$ V. Moving from point b to point c through the 5-Ω resistor results in a potential difference of $-5I_1$. Moving from point c to point f through the 120-V battery and 10-Ω resistor results in a potential difference of -120 V (across the battery) and a potential difference of $+10I_2$ (across the resistor). Finally, moving along the wire from point f to point a results in no potential difference. To summarize, applying Kirchhoff's voltage law around the closed loop $a \to b \to c \to f \to a$ leads to

$$(+E_1) + (-R_1 I_1) + (-E_2) + (+R_2 I_2) = 0$$

or

$$(+40) + (-5I_1) + (-120) + (10I_2) = 0,$$

which simplifies to

$$I_1 - 2I_2 = -16. \tag{2}$$

In a similar way, applying Kirchhoff's voltage law around the closed loop $c \to d \to e \to f \to c$ leads to

$$(-R_3 I_3) + (+E_3) + (-R_4 I_3) + (-R_2 I_2) + (+E_2) = 0$$

or

$$(-20I_3) + (+80) + (-30I_3) + (-10I_2) + (+120) = 0.$$

This simplifies to $10I_2 + 50I_3 = 200$, or

$$I_2 + 5I_3 = 20. \tag{3}$$

Note that the equation resulting from the voltage loop, $a \to b \to c \to d \to e \to f \to a$ becomes

$$(+E_1) + (-R_1 I_1) + (-R_3 I_3) + (+E_3) + (-R_4 I_3) = 0$$

or

$$(+40) + (-5I_1) + (-20I_3) + (+80) + (-30I_3) = 0,$$

which simplifies to

$$I_1 + 10I_3 = 24.$$

But this is just the linear combination Equation $(2) + 2$ Equation (3), and it therefore provides no new information. Thus this equation is redundant and can be omitted. In general, a larger outer loop like $a \to b \to c \to d \to e \to f \to a$ provides no new information if all its inner loops, like $a \to b \to c \to f \to a$ and $c \to d \to e \to f \to c$, have already been included.

Equations (1), (2), and (3) lead to the linear system

$$\begin{bmatrix} 1 & 1 & -1 \\ 1 & -2 & 0 \\ 0 & 1 & 5 \end{bmatrix} \begin{bmatrix} I_1 \\ I_2 \\ I_3 \end{bmatrix} = \begin{bmatrix} 0 \\ -16 \\ 20 \end{bmatrix}.$$

Solving this for I_1, I_2, and I_3 yields (verify)

$$I_1 = -3.5 \, \text{A}, \quad I_2 = 6.25 \, \text{A}, \quad \text{and} \quad I_3 = 2.75 \, \text{A}.$$

The *negative* value of I_1 indicates that its true direction is opposite that assigned in Figure 8.19. ■

In general, for an electrical circuit consisting of batteries, resistors, and wires and having n different current assignments, Kirchhoff's voltage and current laws will always lead to n linear equations that have a unique solution.

8.2 EXERCISES

In Exercises 1 through 4, determine the unknown currents in the given circuit.

In Exercises 5 through 8, determine the unknowns in the given circuit.

1.

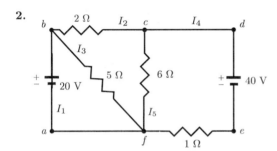

5.

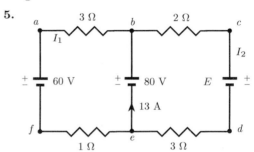

2.

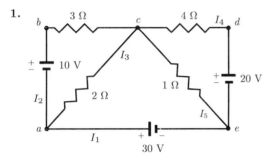

6.

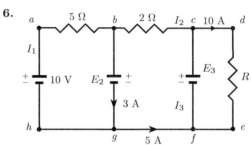

3.

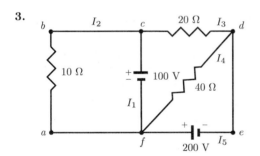

7.

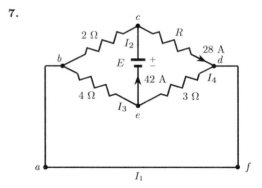

4.

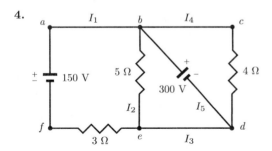

8.

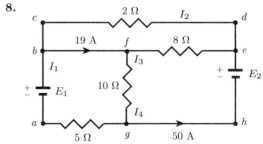

THEORETICAL EXERCISES ■

T.1. For the following circuit, show that

$$I_1 = \left(\frac{R_2}{R_1 + R_2}\right) I = \left(\frac{R}{R_1}\right) I$$

and

$$I_2 = \left(\frac{R_1}{R_1 + R_2}\right) I = \left(\frac{R}{R_2}\right) I,$$

where

$$\frac{1}{R} = \frac{1}{R_1} + \frac{1}{R_2}.$$

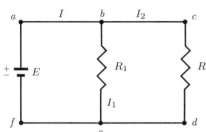

T.2. For the following circuit, show that

$$I_1 = \left(\frac{R}{R_1}\right) I, \qquad I_2 = \left(\frac{R}{R_2}\right) I,$$

and

$$I_3 = \left(\frac{R}{R_3}\right) I,$$

where

$$\frac{1}{R} = \frac{1}{R_1} + \frac{1}{R_2} + \frac{1}{R_3}.$$

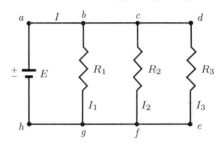

8.3 ▼ Markov* Chains

Prerequisites. Section 1.5, Solutions of Linear Systems of Equations. Basic Ideas of Probability. The Notion of a Limit.

Consider a system that is, at any one time, in one and only one of a finite number of states. For example, the weather in a certain area is either rainy or dry; a person is either a smoker or a nonsmoker; we either go or do not go to college; we live in an urban, suburban, or rural area; we are in the lower-, middle-, or upper-income bracket; we buy a Chevrolet, Ford, or other make of car. As time goes on, the system may move from one state to another, and we assume that the state of the system is observed at fixed time periods (for example, each day, each hour, and so on). In many applications we know the present state of the system and wish to predict the state at the next observation period, or at some future observation period. We often can predict the probability of the system being in a particular state at a future observation period from its past history. The applications mentioned above are of this type.

---

*Andrei Andreevitch Markov (1856–1922) lived most of his life in St. Petersburg, where his father was an official of the Russian forestry department. He was first a student and then a professor at St. Petersburg University. A political liberal, he participated in the protests against the Czarist rule in the first decade of the twentieth century. Although he was interested in many aspects of mathematical analysis, his most important work was in helping to lay the foundations of modern probability theory. His ideas on Markov processes were motivated by a desire to give a strictly rigorous proof of the law of large numbers and to extend the applicability of this law; these ideas appeared in a series of papers between 1906 and 1912.

DEFINITION

A **Markov chain** or **Markov process** is a process in which the probability of the system being in a particular state at a given observation period depends only on its state at the immediately preceding observation period.

Suppose that the system has n possible states. For each $i = 1, 2, \ldots, n$, $j = 1, 2, \ldots, n$ let t_{ij} be the probability that if the system is in state j at a certain observation period, it will be in state i at the next observation period; t_{ij} is called a **transition probability**. Moreover, t_{ij} applies to every time period; that is, it does not change with time.

Since t_{ij} is a probability, we must have

$$0 \le t_{ij} \le 1 \qquad (1 \le i, j \le n).$$

Also, if the system is in state j at a certain observation period, then it must be in one of the n states (it may remain in state j) at the next observation period. Thus we have

$$t_{1j} + t_{2j} + \cdots + t_{nj} = 1. \tag{1}$$

It is convenient to arrange the transition probabilities as the $n \times n$ matrix $T = \begin{bmatrix} t_{ij} \end{bmatrix}$, which is called the **transition matrix** of the Markov chain. Other names for a transition matrix are **Markov matrix**, **stochastic matrix**, and **probability matrix**. We see that the entries in each column of T are nonnegative and, from Equation (1), add up to 1.

EXAMPLE 1 ■ Suppose that the weather in a certain city is either rainy or dry. As a result of extensive record keeping it has been determined that the probability of a rainy day following a dry day is $\frac{1}{3}$, and the probability of a rainy day following a rainy day is $\frac{1}{2}$. Let state D be a dry day and state R be a rainy day. Then the transition matrix of this Markov chain is

$$T = \begin{matrix} & \begin{matrix} \text{D} & \;\; \text{R} \end{matrix} \\ \begin{bmatrix} \frac{2}{3} & \frac{1}{2} \\ \frac{1}{3} & \frac{1}{2} \end{bmatrix} & \begin{matrix} \text{D} \\ \text{R} \end{matrix} \end{matrix}.$$

■

EXAMPLE 2 ■ A market research organization is studying a large group of coffee buyers who buy a can of coffee each week. It is found that 50 percent of those presently using brand A will again buy brand A next week, 25 percent will switch to brand B, and 25 percent will switch to another brand. Of those using brand B now, 30 percent will again buy brand B next week, 60 percent will switch to brand A, and 10 percent will switch to another brand. Of those using another brand now, 30 percent will again buy another brand next week, 40 percent will switch to brand A, and 30 percent will switch to brand B. Let states A, B, and D denote brand A, brand B, and another brand, respectively. The probability that a person presently using brand A will switch to brand B is 0.25, the probability that a person presently using brand B will again buy brand B is 0.3, and so on. Thus the transition matrix of this Markov chain is

$$T = \begin{matrix} \text{A} & \text{B} & \text{D} \\ \begin{bmatrix} 0.50 & 0.60 & 0.40 \\ 0.25 & 0.30 & 0.30 \\ 0.25 & 0.10 & 0.30 \end{bmatrix} & \begin{matrix} \text{A} \\ \text{B} \\ \text{D} \end{matrix} \end{matrix} .$$

■

We shall now use the transition matrix of the Markov process to determine the probability of the system being in any of the n states at future times.

Let

$$\mathbf{x}^{(k)} = \begin{bmatrix} p_1^{(k)} \\ p_2^{(k)} \\ \vdots \\ p_n^{(k)} \end{bmatrix} \qquad (k \geq 0)$$

denote the **state vector** of the Markov process at the observation period k, where $p_j^{(k)}$ is the probability that the system is in state j at the observation period k. The state vector $\mathbf{x}^{(0)}$, at the observation period 0, is called the **initial state vector**.

The following theorem, whose proof we omit, is proved by using the basic ideas of probability theory.

THEOREM 8.4 ■ *If T is the transition matrix of a Markov process, then the state vector $\mathbf{x}^{(k+1)}$, at the $(k+1)$th observation period, can be determined from the state vector $\mathbf{x}^{(k)}$, at the kth observation period, as*

$$\mathbf{x}^{(k+1)} = T\mathbf{x}^{(k)}. \qquad (2) \quad ■$$

From (2) we have

$$\mathbf{x}^{(1)} = T\mathbf{x}^{(0)}$$
$$\mathbf{x}^{(2)} = T\mathbf{x}^{(1)} = T(T\mathbf{x}^{(0)}) = T^2\mathbf{x}^{(0)}$$
$$\mathbf{x}^{(3)} = T\mathbf{x}^{(2)} = T(T^2\mathbf{x}^{(0)}) = T^3\mathbf{x}^{(0)},$$

and, in general,

$$\mathbf{x}^{(n)} = T^n\mathbf{x}^{(0)}.$$

Thus the transition matrix and the initial state vector completely determine every other state vector.

EXAMPLE 3 ■ Consider Example 1 again. Suppose that when we begin our observations (day 0), it is dry, so the initial state vector is

$$\mathbf{x}^{(0)} = \begin{bmatrix} 1 \\ 0 \end{bmatrix}.$$

Then the state vector on day 1 (the day after we begin our observation) is

$$\mathbf{x}^{(1)} = T\mathbf{x}^{(0)} = \begin{bmatrix} 0.67 & 0.5 \\ 0.33 & 0.5 \end{bmatrix} \begin{bmatrix} 1 \\ 0 \end{bmatrix} = \begin{bmatrix} 0.67 \\ 0.33 \end{bmatrix}.$$

Thus the probability of no rain on day 1 is 0.67 and the probability of rain on that day is 0.33. Similarly,

$$\mathbf{x}^{(2)} = T\mathbf{x}^{(1)} = \begin{bmatrix} 0.67 & 0.5 \\ 0.33 & 0.5 \end{bmatrix} \begin{bmatrix} 0.67 \\ 0.33 \end{bmatrix} = \begin{bmatrix} 0.614 \\ 0.386 \end{bmatrix}$$

$$\mathbf{x}^{(3)} = T\mathbf{x}^{(2)} = \begin{bmatrix} 0.67 & 0.5 \\ 0.33 & 0.5 \end{bmatrix} \begin{bmatrix} 0.614 \\ 0.386 \end{bmatrix} = \begin{bmatrix} 0.604 \\ 0.396 \end{bmatrix}$$

$$\mathbf{x}^{(4)} = T\mathbf{x}^{(3)} = \begin{bmatrix} 0.67 & 0.5 \\ 0.33 & 0.5 \end{bmatrix} \begin{bmatrix} 0.604 \\ 0.396 \end{bmatrix} = \begin{bmatrix} 0.603 \\ 0.397 \end{bmatrix}$$

$$\mathbf{x}^{(5)} = T\mathbf{x}^{(4)} = \begin{bmatrix} 0.67 & 0.5 \\ 0.33 & 0.5 \end{bmatrix} \begin{bmatrix} 0.603 \\ 0.397 \end{bmatrix} = \begin{bmatrix} 0.603 \\ 0.397 \end{bmatrix}.$$

From the fourth day on, the state vector of the system is always the same,

$$\begin{bmatrix} 0.603 \\ 0.397 \end{bmatrix}.$$

This means that from the fourth day on, it is dry about 60 percent of the time, and it rains about 40 percent of the time. ■

EXAMPLE 4 ■ Consider Example 2 again. Suppose that when our survey begins, we find that brand A has 20 percent of the market, brand B has 20 percent of the market, and the other brands have 60 percent of the market. Thus

$$\mathbf{x}^{(0)} = \begin{bmatrix} 0.2 \\ 0.2 \\ 0.6 \end{bmatrix}.$$

The state vector after the first week is

$$\mathbf{x}^{(1)} = T\mathbf{x}^{(0)} = \begin{bmatrix} 0.50 & 0.60 & 0.40 \\ 0.25 & 0.30 & 0.30 \\ 0.25 & 0.10 & 0.30 \end{bmatrix} \begin{bmatrix} 0.2 \\ 0.2 \\ 0.6 \end{bmatrix} = \begin{bmatrix} 0.4600 \\ 0.2900 \\ 0.2500 \end{bmatrix}.$$

Similarly,

$$\mathbf{x}^{(2)} = T\mathbf{x}^{(1)} = \begin{bmatrix} 0.50 & 0.60 & 0.40 \\ 0.25 & 0.30 & 0.30 \\ 0.25 & 0.10 & 0.30 \end{bmatrix} \begin{bmatrix} 0.4600 \\ 0.2900 \\ 0.2500 \end{bmatrix} = \begin{bmatrix} 0.5040 \\ 0.2770 \\ 0.2190 \end{bmatrix}$$

$$\mathbf{x}^{(3)} = T\mathbf{x}^{(2)} = \begin{bmatrix} 0.50 & 0.60 & 0.40 \\ 0.25 & 0.30 & 0.30 \\ 0.25 & 0.10 & 0.30 \end{bmatrix} \begin{bmatrix} 0.5040 \\ 0.2770 \\ 0.2190 \end{bmatrix} = \begin{bmatrix} 0.5058 \\ 0.2748 \\ 0.2194 \end{bmatrix}$$

$$\mathbf{x}^{(4)} = T\mathbf{x}^{(3)} = \begin{bmatrix} 0.50 & 0.60 & 0.40 \\ 0.25 & 0.30 & 0.30 \\ 0.25 & 0.10 & 0.30 \end{bmatrix} \begin{bmatrix} 0.5058 \\ 0.2748 \\ 0.2194 \end{bmatrix} = \begin{bmatrix} 0.5055 \\ 0.2747 \\ 0.2198 \end{bmatrix}$$

$$\mathbf{x}^{(5)} = T\mathbf{x}^{(4)} = \begin{bmatrix} 0.50 & 0.60 & 0.40 \\ 0.25 & 0.30 & 0.30 \\ 0.25 & 0.10 & 0.30 \end{bmatrix} \begin{bmatrix} 0.5055 \\ 0.2747 \\ 0.2198 \end{bmatrix} = \begin{bmatrix} 0.5055 \\ 0.2747 \\ 0.2198 \end{bmatrix}.$$

Thus as n increases, the state vectors approach the fixed vector

$$\begin{bmatrix} 0.5055 \\ 0.2747 \\ 0.2198 \end{bmatrix}.$$

This means that in the long run, brand A will command about 51 percent of the market, brand B will retain about 27 percent, and the other brands will command about 22 percent. ■

In the last two examples we have seen that as the number of observation periods increases, the state vectors converge to a fixed vector. In this case we say that the Markov process has reached equilibrium. The fixed vector is called the **steady-state vector**. Markov processes are generally used to determine the behavior of a system in the long run; for example, the share of the market that a certain manufacturer can expect to retain on a somewhat permanent basis. Thus the question of whether or not a Markov process reaches equilibrium is of paramount importance. The following example shows that not every Markov process reaches equilibrium.

EXAMPLE 5 ■ Let

$$T = \begin{bmatrix} 0 & 1 \\ 1 & 0 \end{bmatrix} \quad \text{and} \quad \mathbf{x}^{(0)} = \begin{bmatrix} \frac{1}{3} \\ \frac{2}{3} \end{bmatrix}.$$

Then

$$\mathbf{x}^{(1)} = \begin{bmatrix} \frac{2}{3} \\ \frac{1}{3} \end{bmatrix}, \quad \mathbf{x}^{(2)} = \begin{bmatrix} \frac{1}{3} \\ \frac{2}{3} \end{bmatrix}, \quad \mathbf{x}^{(3)} = \begin{bmatrix} \frac{2}{3} \\ \frac{1}{3} \end{bmatrix}, \quad \mathbf{x}^{(4)} = \begin{bmatrix} \frac{1}{3} \\ \frac{2}{3} \end{bmatrix}, \dots.$$

Thus the state vectors oscillate between the vectors

$$\begin{bmatrix} \frac{2}{3} \\ \frac{1}{3} \end{bmatrix} \quad \text{and} \quad \begin{bmatrix} \frac{1}{3} \\ \frac{2}{3} \end{bmatrix}$$

and do not converge to a fixed vector. ■

However, if we demand that the transition matrix of a Markov process satisfy a rather reasonable property, we obtain a large class of Markov processes, many of which arise in practical applications, that *do* reach equilibrium. We now proceed to formulate these ideas.

DEFINITION   The vector

$$\mathbf{u} = \begin{bmatrix} u_1 \\ u_2 \\ \vdots \\ u_n \end{bmatrix}$$

is called a **probability vector** if $u_i \geq 0 \; (1 \leq i \leq n)$ and

$$u_1 + u_2 + \cdots + u_n = 1.$$

EXAMPLE 6 ■ The vectors

$$\begin{bmatrix} \frac{1}{2} \\ \frac{1}{4} \\ \frac{1}{4} \end{bmatrix} \quad \text{and} \quad \begin{bmatrix} \frac{1}{3} \\ \frac{2}{3} \\ 0 \end{bmatrix}$$

are probability vectors; the vectors

$$\begin{bmatrix} \frac{1}{5} \\ \frac{1}{5} \\ \frac{2}{5} \end{bmatrix} \quad \text{and} \quad \begin{bmatrix} \frac{1}{3} \\ \frac{1}{2} \\ \frac{1}{2} \end{bmatrix}$$

are not probability vectors. (Why not?)   ■

DEFINITION   A transition matrix T of a Markov process is called **regular** if all the entries in some power of T are positive. A Markov process is called **regular** if its transition matrix is regular.

The Markov processes in Examples 1 and 2 are regular, since all the entries in the transition matrices themselves are positive.

EXAMPLE 7 ■ The transition matrix

$$T = \begin{bmatrix} 0.2 & 1 \\ 0.8 & 0 \end{bmatrix}$$

is regular, since

$$T^2 = \begin{bmatrix} 0.84 & 0.2 \\ 0.16 & 0.8 \end{bmatrix}.$$

■

We now state the following fundamental theorem of regular Markov processes; the proof, which we omit, can be found in the book by Kemeny and Snell given in Further Readings at the end of the section.

THEOREM 8.5 ■ *If T is the transition matrix of a regular Markov process, then:*

(a) *As $n \to \infty$, T^n approaches a matrix*

$$A = \begin{bmatrix} u_1 & u_1 & \cdots & u_1 \\ u_2 & u_2 & \cdots & u_2 \\ \vdots & \vdots & & \vdots \\ u_n & u_n & \cdots & u_n \end{bmatrix},$$

all of whose columns are identical.

(b) *Every column*

$$\mathbf{u} = \begin{bmatrix} u_1 \\ u_2 \\ \vdots \\ u_n \end{bmatrix}$$

of A is a probability vector all of whose components are positive. That is,
$u_i > 0$ $(1 \le i \le n)$ *and*

$$u_1 + u_2 + \cdots + u_n = 1.$$ ∎

We next establish the following result.

THEOREM 8.6 ∎ *If T is a regular transition matrix and A and $\mathbf{u}$ are as in Theorem 8.5, then:*

(a) *For any probability vector $\mathbf{x}$, $T^n\mathbf{x} \to \mathbf{u}$ as $n \to \infty$, so that $\mathbf{u}$ is a steady-state vector.*

(b) *The steady-state vector $\mathbf{u}$ is the unique probability vector satisfying the matrix equation $T\mathbf{u} = \mathbf{u}$.*

Proof (a) Let

$$\mathbf{x} = \begin{bmatrix} x_1 \\ x_2 \\ \vdots \\ x_n \end{bmatrix}$$

be a probability vector. Since $T^n \to A$ as $n \to \infty$, we have

$$T^n\mathbf{x} \to A\mathbf{x}.$$

Now

$$A\mathbf{x} = \begin{bmatrix} u_1 & u_1 & \cdots & u_1 \\ u_2 & u_2 & \cdots & u_2 \\ \vdots & \vdots & & \vdots \\ u_n & u_n & \cdots & u_n \end{bmatrix} \begin{bmatrix} x_1 \\ x_2 \\ \vdots \\ x_n \end{bmatrix} = \begin{bmatrix} u_1x_1 + u_1x_2 + \cdots + u_1x_n \\ u_2x_1 + u_2x_2 + \cdots + u_2x_n \\ \vdots \\ u_nx_1 + u_nx_2 + \cdots + u_nx_n \end{bmatrix}$$

$$= \begin{bmatrix} u_1(x_1 + x_2 + \cdots + x_n) \\ u_2(x_1 + x_2 + \cdots + x_n) \\ \vdots \\ u_n(x_1 + x_2 + \cdots + x_n) \end{bmatrix} = \begin{bmatrix} u_1 \\ u_2 \\ \vdots \\ u_n \end{bmatrix},$$

since $x_1 + x_2 + \cdots + x_n = 1$. Hence $T^n\mathbf{x} \to \mathbf{u}$.

(b) Since $T^n \to A$, we also have $T^{n+1} \to A$. However,

$$T^{n+1} = TT^n,$$

so $T^{n+1} \to TA$. Hence $TA = A$. Equating corresponding columns of this equation (using Exercise T.9 of Section 1.3), we have $T\mathbf{u} = \mathbf{u}$. To show that $\mathbf{u}$ is unique, we let $\mathbf{v}$ be another probability vector such that $T\mathbf{v} = \mathbf{v}$. From (a), $T^n\mathbf{v} \to \mathbf{u}$, and since $T\mathbf{v} = \mathbf{v}$, we have $T^n\mathbf{v} = \mathbf{v}$ for all n. Hence $\mathbf{v} = \mathbf{u}$.

∎

REMARK (For those who have read Section 5.1.) Since the columns of A add up to 1, it follows from Exercise T.17 in Section 5.1 that $\lambda = 1$ is an eigenvalue of A.

Moreover, a steady-state vector $\mathbf{u}$ of a regular matrix T is that eigenvector for $\lambda = 1$ which is also a probability vector.

In Examples 3 and 4 we calculated steady-state vectors by computing the powers $T^n\mathbf{x}$. An alternative way of computing the steady-state vector of a regular transition matrix is as follows. From (b) of Theorem 8.6, we write

$$T\mathbf{u} = \mathbf{u}$$

as

$$T\mathbf{u} = I_n\mathbf{u}$$

or

$$(I_n - T)\mathbf{u} = \mathbf{0}. \tag{3}$$

We have shown that the homogeneous system (3) has a unique solution $\mathbf{u}$ that is a probability vector, so that

$$u_1 + u_2 + \cdots + u_n = 1. \tag{4}$$

The first procedure for computing the steady-state vector $\mathbf{u}$ of a regular transition matrix T is as follows.

Step 1. Compute the powers $T^n\mathbf{x}$, where $\mathbf{x}$ is any probability vector.

Step 2. $\mathbf{u}$ is the limit of the powers $T^n\mathbf{x}$.

The second procedure for computing the steady-state vector $\mathbf{u}$ of a regular transition matrix T is as follows.

Step 1. Solve the homogeneous system

$$(I_n - T)\mathbf{u} = \mathbf{0}.$$

Step 2. From the infinitely many solutions obtained in Step 1, determine a unique solution $\mathbf{u}$ by requiring that its components satisfy Equation (4).

EXAMPLE 8 ■ Consider the matrix of Example 2. The homogeneous system (3) is (verify)

$$\begin{bmatrix} 0.50 & -0.60 & -0.40 \\ -0.25 & 0.70 & -0.30 \\ -0.25 & -0.10 & 0.70 \end{bmatrix} \begin{bmatrix} u_1 \\ u_2 \\ u_3 \end{bmatrix} = \begin{bmatrix} 0 \\ 0 \\ 0 \end{bmatrix}.$$

The reduced row echelon form of the augmented matrix is (verify)

$$\begin{bmatrix} 1 & 0 & -2.30 & \vdots & 0 \\ 0 & 1 & -1.25 & \vdots & 0 \\ 0 & 0 & 0.00 & \vdots & 0 \end{bmatrix}.$$

Hence a solution is

$$u_1 = 2.3r$$
$$u_2 = 1.25r$$
$$u_3 = r,$$

where r is any real number. From (4), we have

$$2.3r + 1.25r + r = 1,$$

or

$$r = \frac{1}{4.55} \approx 0.2198.$$

Hence

$$u_1 = 0.5055$$
$$u_2 = 0.2747$$
$$u_3 = 0.2198.$$

These results agree with those obtained in Example 4.   ∎

Further Readings ▶

KEMENY, JOHN G., and J. LAURIE SNELL. *Finite Markov Chains*. New York: Springer-Verlag, 1976.

MAKI, D. P., and M. THOMPSON. *Mathematical Models and Applications: With Emphasis on the Social, Life, and Management Sciences*. Upper Saddle River, N.J.: Prentice Hall, Inc., 1973.

ROBERTS, FRED S. *Discrete Mathematical Models with Applications to Social, Biological, and Environmental Problems*. Upper Saddle River, N.J.: Prentice Hall, Inc., 1976.

8.3   EXERCISES

1. Which of the following can be transition matrices of a Markov process?

(a) $\begin{bmatrix} 0.3 & 0.7 \\ 0.4 & 0.6 \end{bmatrix}$.   (b) $\begin{bmatrix} 0.2 & 0.3 & 0.1 \\ 0.8 & 0.5 & 0.7 \\ 0.0 & 0.2 & 0.2 \end{bmatrix}$.

(c) $\begin{bmatrix} 0.55 & 0.33 \\ 0.45 & 0.67 \end{bmatrix}$.   (d) $\begin{bmatrix} 0.3 & 0.4 & 0.2 \\ 0.2 & 0.0 & 0.8 \\ 0.1 & 0.3 & 0.6 \end{bmatrix}$.

2. Which of the following are probability vectors?

(a) $\begin{bmatrix} \frac{1}{2} \\ \frac{1}{3} \\ \frac{2}{3} \end{bmatrix}$ (b) $\begin{bmatrix} 0 \\ 1 \\ 0 \end{bmatrix}$.

(c) $\begin{bmatrix} \frac{1}{4} \\ \frac{1}{6} \\ \frac{1}{3} \\ \frac{1}{4} \end{bmatrix}$ (d) $\begin{bmatrix} \frac{1}{5} \\ \frac{2}{5} \\ \frac{1}{10} \\ \frac{2}{10} \end{bmatrix}$.

3. Consider the transition matrix

$$T = \begin{bmatrix} 0.7 & 0.4 \\ 0.3 & 0.6 \end{bmatrix}.$$

(a) If $\mathbf{x}^{(0)} = \begin{bmatrix} 1 \\ 0 \end{bmatrix}$, compute $\mathbf{x}^{(1)}$, $\mathbf{x}^{(2)}$, and $\mathbf{x}^{(3)}$ to three decimal places.

(b) Show that T is regular and find its steady-state vector.

4. Consider the transition matrix

$$T = \begin{bmatrix} 0 & 0.2 & 0.0 \\ 0 & 0.3 & 0.3 \\ 1 & 0.5 & 0.7 \end{bmatrix}.$$

(a) If

$$\mathbf{x}^{(0)} = \begin{bmatrix} 0 \\ 1 \\ 0 \end{bmatrix},$$

compute $\mathbf{x}^{(1)}$, $\mathbf{x}^{(2)}$, $\mathbf{x}^{(3)}$, and $\mathbf{x}^{(4)}$ to three decimal places.

(b) Show that T is regular and find its steady-state vector.

5. Which of the following transition matrices are regular?

(a) $\begin{bmatrix} 0 & \frac{1}{2} \\ 1 & \frac{1}{2} \end{bmatrix}.$ (b) $\begin{bmatrix} \frac{1}{2} & 0 & 0 \\ 0 & 1 & \frac{1}{2} \\ \frac{1}{2} & 0 & \frac{1}{2} \end{bmatrix}.$

(c) $\begin{bmatrix} 1 & \frac{1}{3} & 0 \\ 0 & \frac{1}{3} & 1 \\ 0 & \frac{1}{3} & 0 \end{bmatrix}.$ (d) $\begin{bmatrix} \frac{1}{4} & \frac{3}{5} & \frac{1}{2} \\ \frac{1}{2} & 0 & 0 \\ \frac{1}{4} & \frac{2}{5} & \frac{1}{2} \end{bmatrix}.$

6. Show that each of the following transition matrices reaches a state of equilibrium.

(a) $\begin{bmatrix} \frac{1}{2} & 1 \\ \frac{1}{2} & 0 \end{bmatrix}.$ (b) $\begin{bmatrix} 0.4 & 0.2 \\ 0.6 & 0.8 \end{bmatrix}.$

(c) $\begin{bmatrix} \frac{1}{3} & 1 & \frac{1}{2} \\ \frac{1}{3} & 0 & \frac{1}{4} \\ \frac{1}{3} & 0 & \frac{1}{4} \end{bmatrix}.$ (d) $\begin{bmatrix} 0.3 & 0.1 & 0.4 \\ 0.2 & 0.4 & 0.0 \\ 0.5 & 0.5 & 0.6 \end{bmatrix}.$

7. Let

$$T = \begin{bmatrix} \frac{1}{2} & 0 \\ \frac{1}{2} & 1 \end{bmatrix}.$$

(a) Show that T is not regular.

(b) Show that $T^n\mathbf{x} \to \begin{bmatrix} 0 \\ 1 \end{bmatrix}$ for any probability vector $\mathbf{x}$. Thus a Markov chain may have a unique steady-state vector even though its transition matrix is not regular.

8. Find the steady-state vector of each of the following regular matrices.

(a) $\begin{bmatrix} \frac{1}{3} & \frac{1}{2} \\ \frac{2}{3} & \frac{1}{2} \end{bmatrix}.$ (b) $\begin{bmatrix} 0.3 & 0.1 \\ 0.7 & 0.9 \end{bmatrix}.$

(c) $\begin{bmatrix} \frac{1}{4} & \frac{1}{2} & \frac{1}{3} \\ 0 & \frac{1}{2} & \frac{2}{3} \\ \frac{3}{4} & 0 & 0 \end{bmatrix}.$ (d) $\begin{bmatrix} 0.4 & 0.0 & 0.1 \\ 0.2 & 0.5 & 0.3 \\ 0.4 & 0.5 & 0.6 \end{bmatrix}.$

9. (**Psychology**) A behavioral psychologist places a rat each day in a cage with two doors, A and B. The rat can go through door A, where it receives an electric shock, or through door B, where it receives some food. A record is made of the door through which the rat passes. At the start of the experiment, on a Monday, the rat is equally likely to go through door A as through door B. After going through door A, and receiving a shock, the probability of going through the same door on the next day is 0.3. After going through door B, and receiving food, the probability of going through the same door on the next day is 0.6.

(a) Write the transition matrix for the Markov process.

(b) What is the probability of the rat going through door A on Thursday (the third day after starting the experiment)?

(c) What is the steady-state vector?

10. (**Business**) The subscription department of a magazine sends out a letter to a large mailing list inviting subscriptions for the magazine. Some of the people receiving this letter already subscribe to the magazine, while others do not. From this mailing list, 60 percent of those who already subscribe will subscribe again, while 25 percent of those who do not now subscribe will subscribe.

(a) Write the transition matrix for this Markov process.

(b) On the last letter it was found that 40 percent of those receiving it, ordered a subscription. What percentage of those receiving the current letter can be expected to order a subscription?

11. (Sociology) A study has determined that the occupation of a boy, as an adult, depends upon the occupation of his father and is given by the following transition matrix where P = professional, F = farmer and L = laborer.

Father's occupation

		P	F	L
Son's occupation	P	0.8	0.3	0.2
	F	0.1	0.5	0.2
	L	0.1	0.2	0.6

Thus the probability that the son of a professional will also be a professional is 0.8, and so on.

(a) What is the probability that the grandchild of a professional will also be a professional?

(b) In the long run, what proportion of the population will be farmers?

12. (Genetics) Consider a plant that can have red flowers (R), pink flowers (P), or white flowers (W), depending upon the genotypes RR, RW, and WW. When we cross each of these genotypes with a genotype RW, we obtain the transition matrix

Flowers of parent plant

		R	P	W
Flowers of offspring plant	R	0.5	0.25	0.0
	P	0.5	0.50	0.5
	W	0.0	0.25	0.5

Suppose that each successive generation is produced by crossing only with plants of RW genotype. When the process reaches equilibrium, what percentage of the plants will have red, pink, or white flowers?

13. (Mass Transit) A new mass transit system has just gone into operation. The transit authority has made studies that predict the percentage of commuters who will change to mass transit (M) or continue driving their automobile (A). The following transition matrix has been obtained:

This year

		M	A
Next year	M	0.7	0.2
	A	0.3	0.8

Suppose that the population of the area remains constant, and that initially 30 percent of the commuters use mass transit and 70 percent use their automobiles.

(a) What percentage of the commuters will be using the mass transit system after 1 year? After 2 years?

(b) What percentage of the commuters will be using the mass transit system in the long run?

MATLAB EXERCISES

The computation of the sequence of vectors $\mathbf{x}^{(1)}, \mathbf{x}^{(2)}, \ldots$ as in Examples 3 and 4 can easily be done using MATLAB commands. Once the transition matrix T and initial state vector $\mathbf{x}^{(0)}$ are entered into MATLAB, the state vector for the kth observation period is obtained from the MATLAB command

$$\mathbf{T}^{\wedge}\mathbf{k} * \mathbf{x}$$

ML.1. Use MATLAB to verify the computations of state vectors in Example 3 for periods 1 through 5.

ML.2. In Example 4, if the initial state is changed to

$$\begin{bmatrix} 0.1 \\ 0.3 \\ 0.6 \end{bmatrix},$$

determine $\mathbf{x}^{(5)}$.

ML.3. In MATLAB, enter **help sum** and determine the action of command **sum** on an $m \times n$ matrix. Use command **sum** to determine which of the following are Markov matrices.

(a) $\begin{bmatrix} \frac{2}{3} & \frac{1}{3} & \frac{1}{2} \\ \frac{1}{3} & \frac{1}{3} & \frac{1}{4} \\ 0 & \frac{1}{3} & \frac{1}{4} \end{bmatrix}.$

(b) $\begin{bmatrix} 0.5 & 0.6 & 0.7 \\ 0.3 & 0.2 & 0.3 \\ 0.1 & 0.2 & 0.0 \end{bmatrix}.$

(d) $\begin{bmatrix} 0.66 & 0.25 & 0.125 \\ 0.33 & 0.25 & 0.625 \\ 0.00 & 0.50 & 0.250 \end{bmatrix}.$

8.4 ▾ Least Squares

Prerequisites. Section 1.5, Solutions of Linear Systems of Equations. Section 1.6, The Inverse of a Matrix.

From Chapter 1 we recall that an $m \times n$ linear system $A\mathbf{x} = \mathbf{b}$ is inconsistent if it has no solution. In the proof of Theorem 4.14 in Section 4.6 we show that $A\mathbf{x} = \mathbf{b}$ is consistent if and only if $\mathbf{b}$ belongs to the column space of A. Equivalently, $A\mathbf{x} = \mathbf{b}$ is inconsistent if and only if $\mathbf{b}$ is *not* in the column space of A. Inconsistent systems do indeed arise in many situations and we must determine how to deal with them. Our approach is to change the problem so that we do not require that the matrix equation $A\mathbf{x} = \mathbf{b}$ be satisfied. Instead, we seek a vector $\widehat{\mathbf{x}}$ in R^n such that $A\widehat{\mathbf{x}}$ is as close to $\mathbf{b}$ as possible. If W is the column space of A, then from Theorem 4.23 in Section 4.9, it follows that the vector in W that is closest to $\mathbf{b}$ is $\mathrm{proj}_W \mathbf{b}$. That is, $\|\mathbf{b} - \mathbf{w}\|$, for $\mathbf{w}$ in W, is minimized when $\mathbf{w} = \mathrm{proj}_W \mathbf{b}$. Thus, if we find $\widehat{\mathbf{x}}$ such that $A\widehat{\mathbf{x}} = \mathrm{proj}_W \mathbf{b}$, then we are assured that $\|\mathbf{b} - A\widehat{\mathbf{x}}\|$ will be as small as possible. As shown in the proof of Theorem 4.23, $\mathbf{b} - \mathrm{proj}_W \mathbf{b} = \mathbf{b} - A\widehat{\mathbf{x}}$ is orthogonal to every vector in W. It then follows that $\mathbf{b} - A\widehat{\mathbf{x}}$ is orthogonal to each column of A. In terms of a matrix equation, we have

$$A^T(A\widehat{\mathbf{x}} - \mathbf{b}) = \mathbf{0}$$

or, equivalently,

$$A^T A\widehat{\mathbf{x}} = A^T \mathbf{b}. \tag{1}$$

Any solution to (1) is called a **least squares solution** to the linear system $A\mathbf{x} = \mathbf{b}$. (**Warning:** In general, $A\widehat{\mathbf{x}} \neq \mathbf{b}$.) Equation (1) is called the **normal system** of equations associated with $A\mathbf{x} = \mathbf{b}$, or just the normal system. Observe that if A is nonsingular, a least squares solution to $A\mathbf{x} = \mathbf{b}$ is just the usual solution $\mathbf{x} = A^{-1}\mathbf{b}$ (see Exercise T.2).

To compute a least squares solution $\widehat{\mathbf{x}}$ to the linear system $A\mathbf{x} = \mathbf{b}$, we can proceed as follows. Let $\{\mathbf{w}_1, \mathbf{w}_2, \ldots, \mathbf{w}_m\}$ be an orthonormal basis for the column space W of A. Then Equation (1) of Section 4.9 yields

$$\mathrm{proj}_W \mathbf{b} = (\mathbf{b} \cdot \mathbf{w}_1)\mathbf{w}_1 + (\mathbf{b} \cdot \mathbf{w}_2)\mathbf{w}_2 + \cdots + (\mathbf{b} \cdot \mathbf{w}_m)\mathbf{w}_m.$$

Recall that to find an orthonormal basis for W we first find a basis for W by transforming A^T to reduced row echelon form and then taking the transposes of the nonzero rows as a basis for W. Next apply the Gram–Schmidt process to this basis to find an orthonormal basis for W. The procedure just outlined is theoretically valid when we assume that exact arithmetic is used. However, even small numerical errors, due to, say, roundoff, may adversely affect the results. Thus more sophisticated algorithms are required for numerical applications. [See D. Hill, *Experiments in Computational Matrix Algebra* (New York: Random House, 1988), distributed by McGraw-Hill.] We shall not pursue the general case here, but turn our attention to an important special case.

REMARK An alternate method for finding $\mathrm{proj}_W \mathbf{b}$ is as follows. Solve Equation (1) for $\widehat{\mathbf{x}}$, the least squares solution to the linear system $A\mathbf{x} = \mathbf{b}$. Then $A\widehat{\mathbf{x}}$ will be $\mathrm{proj}_W \mathbf{b}$.

THEOREM 8.7 ■ *If A is an $m \times n$ matrix with* rank $A = n$, *then $A^T A$ is nonsingular and the linear system $A\mathbf{x} = \mathbf{b}$ has a unique least squares solution given by $\widehat{\mathbf{x}} = (A^T A)^{-1} A^T \mathbf{b}$. That is, the normal system of equations has a unique solution.*

Proof   If A has rank n, then the columns of A are linearly independent. The matrix $A^T A$ is nonsingular provided the linear system $A^T A \mathbf{x} = \mathbf{0}$ has only the zero solution. Multiplying both sides of $A^T A \mathbf{x} = \mathbf{0}$ by $\mathbf{x}^T$ gives

$$\mathbf{0} = \mathbf{x}^T A^T A \mathbf{x} = (A\mathbf{x})^T (A\mathbf{x}) = (A\mathbf{x}) \cdot (A\mathbf{x}).$$

It then follows from Theorem 3.3 in Section 3.2 that $A\mathbf{x} = \mathbf{0}$. But this implies that we have a linear combination of the linearly independent columns of A that is zero, hence $\mathbf{x} = \mathbf{0}$. Thus $A^T A$ is nonsingular and Equation (1) has the unique solution $\widehat{\mathbf{x}} = (A^T A)^{-1} A^T \mathbf{b}$.   ■

The procedure for finding the least squares solution $\widehat{\mathbf{x}}$ to $A\mathbf{x} = \mathbf{b}$ is as follows.

Step 1. Form $A^T A$ and $A^T \mathbf{b}$.

Step 2. Solve the normal system

$$A^T A \mathbf{x} = A^T \mathbf{b}$$

for $\mathbf{x}$ by Gauss–Jordan reduction.

EXAMPLE 1 ■ Determine a least squares solution to $A\mathbf{x} = \mathbf{b}$, where

$$A = \begin{bmatrix} 1 & 2 & -1 & 3 \\ 2 & 1 & 1 & 2 \\ -2 & 3 & 4 & 1 \\ 4 & 2 & 1 & 0 \\ 0 & 2 & 1 & 3 \\ 1 & -1 & 2 & 0 \end{bmatrix}, \qquad \mathbf{b} = \begin{bmatrix} 1 \\ 5 \\ -2 \\ 1 \\ 3 \\ 5 \end{bmatrix}.$$

Solution   Using row reduction, we can show that rank $A = 4$ (verify). We now form the normal system $A^T A \mathbf{x} = A^T \mathbf{b}$ (verify),

$$\begin{bmatrix} 26 & 5 & -1 & 5 \\ 5 & 23 & 13 & 17 \\ -1 & 13 & 24 & 6 \\ 5 & 17 & 6 & 23 \end{bmatrix} \mathbf{x} = \begin{bmatrix} 24 \\ 4 \\ 10 \\ 20 \end{bmatrix}.$$

By Theorem 8.7, the normal system has a unique solution. Applying Gauss–Jordan reduction, we have the unique least squares solution (verify)

$$\widehat{\mathbf{x}} \approx \begin{bmatrix} 0.9990 \\ -2.0643 \\ 1.1039 \\ 1.8902 \end{bmatrix}.$$

If W is the column space of A, then (verify)

$$\operatorname{proj}_W \mathbf{b} = A\widehat{\mathbf{x}} \approx \begin{bmatrix} 1.4371 \\ 4.8181 \\ -1.8852 \\ 0.9713 \\ 2.6459 \\ 5.2712 \end{bmatrix},$$

which is the vector in W such that $\|\mathbf{b} - \mathbf{w}\|$, $\mathbf{w}$ in W, is minimized. That is,

$$\min_{\mathbf{w} \text{ in } W} \|\mathbf{b} - \mathbf{w}\| = \|\mathbf{b} - A\widehat{\mathbf{x}}\|. \qquad \blacksquare$$

When A is an $m \times n$ matrix whose rank is n, it is computationally more efficient to solve Equation (1) by Gauss–Jordan reduction than to determine $(A^T A)^{-1}$ and then form the product $(A^T A)^{-1} A^T \mathbf{b}$. An even better approach is to use the QR-factorization of A, which is discussed in Section 9.4. Thus we revisit least squares in Section 9.4.

Least Squares Line Fit

The problem of gathering and analyzing data is one that is present in many facets of human activity. Frequently, we measure a value of y for a given value of x and then plot the points (x, y) on graph paper. From the resulting graph we try to develop a relationship between the variables x and y that can then be used to predict new values of y for given values of x.

EXAMPLE 2 ■ In the manufacture of product XXX, the amount of the compound beta present in the product is controlled by the amount of the ingredient alpha used in the process. In manufacturing a gallon of XXX, the amount of alpha used and the amount of beta present are recorded. The following data were obtained:

Alpha Used (ounces/gallon)	3	4	5	6	7	8	9	10	11	12
Beta Present (ounces/gallon)	4.5	5.5	5.7	6.6	7.0	7.7	8.5	8.7	9.5	9.7

The points in this table are plotted in Figure 8.20. ■

Suppose it is assumed that the relationship between the amount of alpha used and the amount of beta present is given by a linear equation so that the graph is a straight line. It would thus not be reasonable to connect the plotted points by drawing a curve that passes through every point. Moreover, the data are of a *probabilistic* nature. That is, they are not *deterministic* in the sense that if we repeated the experiment, we would expect slightly different values of beta for the same values of alpha, since all measurements are subject to experimental

FIGURE 8.20

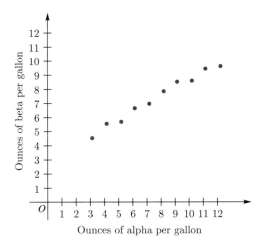

error. Thus the plotted points do not lie exactly on a straight line. We shall now use the method of least squares to obtain the straight line that "best fits" the given data. This line is called the **least squares line**.

Suppose that we are given n points $(x_1, y_1), (x_2, y_2), \ldots, (x_n, y_n)$, where at least two of the x_i are distinct. We are interested in finding the least squares line

$$y = b_1 x + b_0 \tag{2}$$

that "best fits the data." If the points $(x_1, y_1), (x_2, y_2), \ldots, (x_n, y_n)$ were exactly on the least squares line, we would have

$$y_i = b_1 x_i + b_0. \tag{3}$$

Since some of these points may not lie exactly on the line, we have

$$y_i = b_1 x_i + b_0 + d_i, \qquad i = 1, 2, \ldots, n, \tag{4}$$

FIGURE 8.21

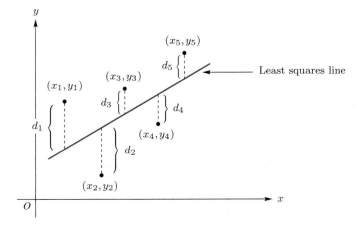

where d_i is the vertical deviation from the point (x_i, y_i) to the (desired) least squares line. The quantity d_i can be positive, negative, or zero. In Figure 8.21 we show five data points (x_1, y_1), (x_2, y_2), (x_3, y_3), (x_4, y_4), (x_5, y_5) and their corresponding deviations $d_1, d_2, \ldots, d_5$ from the least squares line.

If we let

$$
\mathbf{b} = \begin{bmatrix} y_1 \\ y_2 \\ \vdots \\ y_n \end{bmatrix}, \quad A = \begin{bmatrix} x_1 & 1 \\ x_2 & 1 \\ \vdots & \vdots \\ x_n & 1 \end{bmatrix}, \quad \mathbf{x} = \begin{bmatrix} b_1 \\ b_0 \end{bmatrix}, \quad \text{and} \quad \mathbf{d} = \begin{bmatrix} d_1 \\ d_2 \\ \vdots \\ d_n \end{bmatrix},
$$

then we may write the n equations in (4) as a single matrix equation

$$
\mathbf{b} = A\mathbf{x} + \mathbf{d}.
$$

Since the linear system $A\mathbf{x} = \mathbf{b}$ is usually inconsistent, we find a least squares solution $\widehat{\mathbf{x}}$ to $A\mathbf{x} = \mathbf{b}$. Recall that at least two of the x_i are distinct, so rank $A = 2$. Theorem 8.7 then implies that $A\mathbf{x} = \mathbf{b}$ has a unique least squares solution given by $\widehat{\mathbf{x}} = (A^T A)^{-1} A^T \mathbf{b}$. We are then assured that $\|\mathbf{b} - A\widehat{\mathbf{x}}\|$ will be minimized. Since $\mathbf{d} = \mathbf{b} - A\widehat{\mathbf{x}}$, it follows that $\|\mathbf{d}\|$ is minimized. Equivalently,

$$
\|\mathbf{d}\|^2 = d_1^2 + d_2^2 + \cdots + d_n^2
$$

will be minimized. Let $\widehat{\mathbf{x}} = (\widehat{x}_1, \widehat{x}_2, \ldots, \widehat{x}_n)$ and $\widehat{y}_i$ be the corresponding value obtained by using $\widehat{x}_i$ in Equation (3). That is, $(\widehat{x}_i, \widehat{y}_i)$, $i = 1, 2, \ldots, n$, are the points on the least squares line. Then the least squares solution $\widehat{\mathbf{x}}$ minimizes the sum of the squares of the deviations between the given values y_i and the values $\widehat{y}_i$ predicted by the least squares line.

The procedure for finding the least squares line $y = b_1 x + b_0$ for the data $(x_1, y_1), (x_2, y_2), \ldots, (x_n, y_n)$, where at least two of the x_i are different, is as follows.

Step 1. Let

$$
\mathbf{b} = \begin{bmatrix} y_1 \\ y_2 \\ \vdots \\ y_n \end{bmatrix}, \quad A = \begin{bmatrix} x_1 & 1 \\ x_2 & 1 \\ \vdots & \vdots \\ x_n & 1 \end{bmatrix}, \quad \text{and} \quad \mathbf{x} = \begin{bmatrix} b_1 \\ b_0 \end{bmatrix}.
$$

Step 2. Solve the normal system

$$
A^T A \mathbf{x} = A^T \mathbf{b}
$$

for $\mathbf{x}$ by Gauss–Jordan reduction.

EXAMPLE 3 ■ (a) Find an equation of the least squares line for the data in Example 2.

(b) Use the equation obtained in part (a) to predict the number of ounces of beta present in a gallon of product XXX if 30 ounces of alpha are used per gallon.

Solution   (a) We have

$$
\mathbf{b} = \begin{bmatrix} 4.5 \\ 5.5 \\ 5.7 \\ 6.6 \\ 7.0 \\ 7.7 \\ 8.5 \\ 8.7 \\ 9.5 \\ 9.7 \end{bmatrix}, \quad A = \begin{bmatrix} 3 & 1 \\ 4 & 1 \\ 5 & 1 \\ 6 & 1 \\ 7 & 1 \\ 8 & 1 \\ 9 & 1 \\ 10 & 1 \\ 11 & 1 \\ 12 & 1 \end{bmatrix}, \quad \mathbf{x} = \begin{bmatrix} b_1 \\ b_0 \end{bmatrix}.
$$

Then

$$
A^T A = \begin{bmatrix} 645 & 75 \\ 75 & 10 \end{bmatrix} \quad \text{and} \quad A^T \mathbf{b} = \begin{bmatrix} 598.6 \\ 73.4 \end{bmatrix}.
$$

Solving the normal system $A^T A \mathbf{x} = A^T \mathbf{b}$ by Gauss–Jordan reduction we obtain (verify)

$$
\mathbf{x} = \begin{bmatrix} b_1 \\ b_0 \end{bmatrix} = \begin{bmatrix} 0.583 \\ 2.967 \end{bmatrix}.
$$

Then

$$
b_1 = 0.583 \quad \text{and} \quad b_0 = 2.967.
$$

Hence, an equation for the least squares line, shown in Figure 8.22, is

$$
y = 0.583x + 2.967, \tag{5}
$$

where y is the amount of beta present and x is the amount of alpha used.

(b) If $x = 30$, then substituting in (5), we obtain

$$
y = 20.457.
$$

Thus there would be 20.457 ounces of beta present in a gallon of XXX. ■

FIGURE 8.22

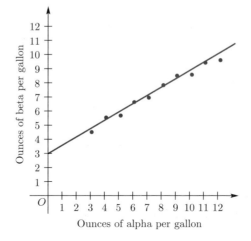

Ounces of alpha per gallon

Least Squares Polynomial Fit

The method presented for obtaining the least squares line fit for a given set of points can be easily generalized to the problem of finding a polynomial of prescribed degree that "best fits" the given data.

Thus, suppose we are given n data points $(x_1, y_1), (x_2, y_2), \ldots, (x_n, y_n)$, where at least $m + 1$ of the x_i are distinct, and wish to construct a mathematical model of the form

$$y = a_m x^m + a_{m-1} x^{m-1} + \cdots + a_1 x + a_0, \qquad m \le n - 1,$$

that "best fits" these data. As in the least squares line fit, since some of the n data points do not lie exactly on the graph of the least squares polynomial, we have

$$y_i = a_m x_i^m + a_{m-1} x_i^{m-1} + \cdots + a_1 x_i + a_0 + d_i, \qquad i = 1, 2, \ldots, n. \qquad (6)$$

Letting

$$\mathbf{b} = \begin{bmatrix} y_1 \\ y_2 \\ \vdots \\ y_n \end{bmatrix}, \qquad A = \begin{bmatrix} x_1^m & x_1^{m-1} & \cdots & x_1^2 & x_1 & 1 \\ x_2^m & x_2^{m-1} & \cdots & x_2^2 & x_2 & 1 \\ \vdots & \vdots & & \vdots & \vdots & \vdots \\ x_n^m & x_n^{m-1} & \cdots & x_n^2 & x_n & 1 \end{bmatrix}, \qquad \mathbf{x} = \begin{bmatrix} a_m \\ a_{m-1} \\ \vdots \\ a_1 \\ a_0 \end{bmatrix},$$

and

$$\mathbf{d} = \begin{bmatrix} d_1 \\ d_2 \\ \vdots \\ d_n \end{bmatrix},$$

we may write the n equations in (6) as the matrix equation

$$\mathbf{b} = A\mathbf{x} + \mathbf{d}.$$

As in the case of the least squares line fit, a solution $\widehat{\mathbf{x}}$ to the normal system

$$A^T A \mathbf{x} = A^T \mathbf{b}$$

is a least squares solution to $A\mathbf{x} = \mathbf{b}$. With this solution we will be assured that $\|\mathbf{d}\| = \|\mathbf{b} - A\widehat{\mathbf{x}}\|$ is minimized.

The procedure for finding the least squares polynomial

$$y = a_m x^m + a_{m-1} x^{m-1} + \cdots + a_1 x + a_0$$

that best fits the data $(x_1, y_1), (x_2, y_2), \ldots, (x_n, y_n)$, where $m \leq n - 1$ and at least $m + 1$ of the x_i are distinct, is as follows.

Step 1. Form

$$
\mathbf{b} = \begin{bmatrix} y_1 \\ y_2 \\ \vdots \\ y_n \end{bmatrix}, \quad
A = \begin{bmatrix} x_1^m & x_1^{m-1} & \cdots & x_1^2 & x_1 & 1 \\ x_2^m & x_2^{m-1} & \cdots & x_2^2 & x_2 & 1 \\ \vdots & \vdots & & \vdots & \vdots & \vdots \\ x_n^m & x_n^{m-1} & \cdots & x_n^2 & x_n & 1 \end{bmatrix}, \quad \text{and} \quad
\mathbf{x} = \begin{bmatrix} a_m \\ a_{m-1} \\ \vdots \\ a_1 \\ a_0 \end{bmatrix}.
$$

Step 2. Solve the normal system

$$A^T A \mathbf{x} = A^T \mathbf{b}$$

for $\mathbf{x}$ by Gauss–Jordan reduction.

EXAMPLE 4 ■ The following data show atmospheric pollutants y_i (relative to an EPA standard) at half-hour intervals t_i.

t_i	1	1.5	2	2.5	3	3.5	4	4.5	5
y_i	−0.15	0.24	0.68	1.04	1.21	1.15	0.86	0.41	−0.08

A plot of these data points as shown in Figure 8.23, suggests that a quadratic polynomial

$$y = a_2 t^2 + a_1 t + a_0$$

may produce a good model for these data.

FIGURE 8.23

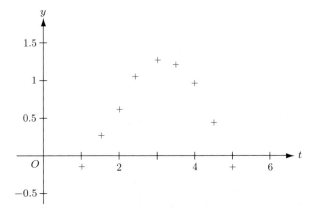

We now have

$$A = \begin{bmatrix} 1 & 1 & 1 \\ 2.25 & 1.5 & 1 \\ 4 & 2 & 1 \\ 6.25 & 2.5 & 1 \\ 9 & 3 & 1 \\ 12.25 & 3.5 & 1 \\ 16 & 4 & 1 \\ 20.25 & 4.5 & 1 \\ 25 & 5 & 1 \end{bmatrix}, \quad \mathbf{x} = \begin{bmatrix} a_2 \\ a_1 \\ a_0 \end{bmatrix}, \quad \mathbf{b} = \begin{bmatrix} -0.15 \\ 0.24 \\ 0.68 \\ 1.04 \\ 1.21 \\ 1.15 \\ 0.86 \\ 0.41 \\ -0.08 \end{bmatrix}.$$

The rank of A is 3 (verify), and the normal system is

$$\begin{bmatrix} 1583.25 & 378 & 96 \\ 378 & 96 & 27 \\ 96 & 27 & 9 \end{bmatrix} \begin{bmatrix} a_2 \\ a_1 \\ a_0 \end{bmatrix} = \begin{bmatrix} 54.65 \\ 16.71 \\ 5.36 \end{bmatrix}.$$

Applying Gauss–Jordan reduction we obtain (verify)

$$\widehat{\mathbf{x}} \approx \begin{bmatrix} -0.3274 \\ 2.0067 \\ -1.9317 \end{bmatrix},$$

so we obtain the quadratic polynomial model

$$y = -0.3274t^2 + 2.0067t - 1.9317.$$

Figure 8.24 shows the data set indicated with $+$ and a graph of y. We see that y is close to each data point but is not required to go through the data. ■

FIGURE 8.24

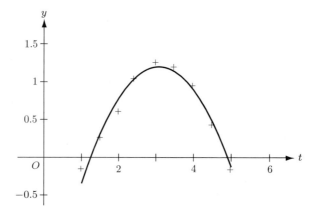

8.4   EXERCISES

In Exercises 1 through 4, find a least squares solution to $A\mathbf{x} = \mathbf{b}$.

1. $A = \begin{bmatrix} 2 & 1 \\ 1 & 0 \\ 0 & -1 \\ -1 & 1 \end{bmatrix}$ and $\mathbf{b} = \begin{bmatrix} 3 \\ 1 \\ 2 \\ -1 \end{bmatrix}$.

2. $A = \begin{bmatrix} 3 & -2 \\ 2 & -3 \\ 1 & -1 \\ 2 & 3 \\ 3 & 4 \end{bmatrix}$ and $\mathbf{b} = \begin{bmatrix} 2 \\ -1 \\ 0 \\ 1 \\ 0 \end{bmatrix}$.

3. $A = \begin{bmatrix} 1 & 2 & 1 \\ 1 & 3 & 2 \\ 2 & 5 & 3 \\ 2 & 0 & 1 \\ 3 & 1 & 1 \end{bmatrix}$ and $\mathbf{b} = \begin{bmatrix} -1 \\ 2 \\ 0 \\ 1 \\ -2 \end{bmatrix}$.

4. $A = \begin{bmatrix} -1 & 0 & 0 & 4 \\ 4 & -2 & 0 & 0 \\ 3 & 0 & 1 & 0 \\ 0 & 0 & -1 & 2 \\ 0 & 2 & -2 & 0 \\ 2 & 0 & 0 & 1 \end{bmatrix}$ and $\mathbf{b} = \begin{bmatrix} -1 \\ 1 \\ 2 \\ 0 \\ 0 \\ 1 \end{bmatrix}$.

In Exercises 5 through 8, find the least squares line for the given data points.

5. $(2, 1)$, $(3, 2)$, $(4, 3)$, $(5, 2)$.

6. $(3, 2)$, $(4, 3)$, $(5, 2)$, $(6, 4)$, $(7, 3)$.

7. $(2, 3)$, $(3, 4)$, $(4, 3)$, $(5, 4)$, $(6, 3)$, $(7, 4)$.

8. $(3, 3)$, $(4, 5)$, $(5, 4)$, $(6, 5)$, $(7, 5)$, $(8, 6)$, $(9, 5)$, $(10, 6)$.

In Exercises 9 and 10, find a quadratic least squares polynomial for the given data.

9. $(0, 3.2)$, $(0.5, 1.6)$, $(1, 2)$, $(2, -0.4)$, $(2.5, -0.8)$, $(3, -1.6)$, $(4, 0.3)$, $(5, 2.2)$.

10. $(0.5, -1.6)$, $(1, 0.4)$, $(1.5, 0.7)$, $(2, 1.8)$, $(2.5, 1.6)$, $(3, 2.2)$, $(3.5, 1.7)$, $(4, 2.2)$, $(4.5, 1.6)$, $(5, 1.5)$.

11. In an experiment designed to determine the extent of a person's natural orientation, a subject is put in a special room and kept there for a certain length of time. He is then asked to find a way out of a maze and record is made of the time it takes the subject to accomplish this task. The following data are obtained:

Time in Room (hours)	1	2	3	4	5	6
Time to Find Way Out of Maze (minutes)	0.8	2.1	2.6	2.0	3.1	3.3

Let x denote the number of hours in the room and let y denote the number of minutes that it takes the subject to find his way out.

(a) Find the least squares line relating x and y.

(b) Use the equation obtained in (a) to estimate the time it will take the subject to find his way out of the maze after 10 hours in the room.

12. A steel producer gathers the following data.

Year	1997	1998	1999	2000	2001	2002
Annual Sales (millions of dollars)	1.2	2.3	3.2	3.6	3.8	5.1

Represent the years $1999, \ldots, 2002$ as 0, 1, 2, 3, 4, 5, respectively, and let x denote the year. Let y denote the annual sales (in millions of dollars).

(a) Find the least squares line relating x and y.

(b) Use the equation obtained in (a) to estimate the annual sales for the year 2006.

13. A sales organization obtains the following data relating the number of salespersons to annual sales.

Number of Salespersons	5	6	7	8	9	10
Annual Sales (millions of dollars)	2.3	3.2	4.1	5.0	6.1	7.2

Let x denote the number of salespersons and let y denote the annual sales (in millions of dollars).

(a) Find the least squares line relating x and y.

(b) Use the equation obtained in (a) to estimate the annual sales when there are 14 salespersons.

14. The distributor of a new car has obtained the following data.

Number of Weeks After Introduction of Car	Gross Receipts per Week (millions of dollars)
1	0.8
2	0.5
3	3.2
4	4.3
5	4
6	5.1
7	4.3
8	3.8
9	1.2
10	0.8

Let x denote the gross receipts per week (in millions of dollars) t weeks after the introduction of the car.

(a) Find a least squares quadratic polynomial for the given data.

(b) Use the equation in part (a) to estimate the gross receipts 12 weeks after the introduction of the car.

15. Given $A\mathbf{x} = \mathbf{b}$, where

$$A = \begin{bmatrix} 1 & 3 & -3 \\ 2 & 4 & -2 \\ 0 & -1 & 2 \\ 1 & 2 & -1 \end{bmatrix} \quad \text{and} \quad \mathbf{b} = \begin{bmatrix} 1 \\ 0 \\ 0 \\ 1 \end{bmatrix}.$$

(a) Show that rank $A = 2$.

(b) Since rank $A \neq$ number of columns, Theorem 8.7 cannot be used to determine a least squares solution $\widehat{\mathbf{x}}$. Follow the general procedure as discussed prior to Theorem 8.7 to find a least squares solution. Is the solution unique?

THEORETICAL EXERCISES ▦

T.1. Suppose that we wish to find the least squares line for the n data points $(x_1, y_1), (x_2, y_2), \ldots, (x_n, y_n)$, so that $m = 1$ in (6). Show that if at least two x-coordinates are unequal, then the matrix $A^T A$ is nonsingular, where A is the matrix resulting from the n equations in (6).

T.2. Let A be $n \times n$ and nonsingular. From the normal system of equations in (1), show that the least squares solution to $A\mathbf{x} = \mathbf{b}$ is $\widehat{\mathbf{x}} = A^{-1}\mathbf{b}$.

MATLAB EXERCISES ▦

Routine **lsqline** in MATLAB will compute the least squares line for data you supply and graph both the line and the data points. To use **lsqline**, put the x-coordinates of your data into a vector $\mathbf{x}$ and the corresponding y-coordinates into a vector $\mathbf{y}$ and then type **lsqline$(\mathbf{x}, \mathbf{y})$**. For more information, use **help lsqline**.

ML.1. Solve Exercise 7 in MATLAB using **lsqline**.

ML.2. Use **lsqline** to determine the solution to Exercise 11. Then estimate the time it will take the subject to find his way out of the maze after 7 hours; 8 hours; 9 hours.

ML.3. An experiment was conducted on the temperatures of a fluid in a newly designed container. The following data were obtained.

Time (minutes)	0	2	3	5	9
Temperature (°F)	185	170	166	152	110

(a) Determine the least squares line.

(b) Estimate the temperature at $x = 1, 6, 8$ minutes.

(c) Estimate the time at which the temperature of the fluid was 160°F.

ML.4. At time $t = 0$ an object is dropped from a height of 1 meter above a fluid. A recording device registers the height of the object above the surface of the fluid at $\frac{1}{2}$ second intervals, with a negative value indicating the object is below the surface of the fluid. The following table of data is the result.

Time (seconds)	Depth (meters)
0	1
0.5	0.88
1	0.54
1.5	0.07
2	−0.42
2.5	−0.80
3	−0.99
3.5	−0.94
4	−0.65
4.5	−0.21

(a) Determine the least squares quadratic polynomial.

(b) Estimate the depth at $t = 5$ and $t = 6$ seconds.

(c) Estimate the time the object breaks through the surface of the fluid the second time.

ML.5. Determine the least squares quadratic polynomial for the table of data given below. Use this data model to predict the value of y when $x = 7$.

x	y
−3	0.5
−2.5	0
−2	−1.125
−1.5	−1.875
−1	−1
0	0.9375
0.5	2.8750
1	4.75
1.5	8.25
2	11.5

8.5 ▾ Linear Economic Models

Prerequisite. Chapter 1, Linear Equations and Matrices.

As society has grown more complex, increasing attention has been paid to the analysis of economic behavior. Problems of economic behavior are more difficult to handle than problems in the physical sciences for many reasons. For example, we may not know all the factors or variables that must be considered, what data need to be gathered, when we have enough data, or the resulting mathematical problem may be too difficult to solve.

In the 1930s Wassily Leontief, a Harvard University economics professor, developed a pioneering approach to the mathematical analysis of economic behavior. He was awarded the 1973 Nobel Prize in Economics for his work. In this section we give a very brief introduction to the application of linear algebra to economics.

Our approach leans heavily on the presentations in the books by Gale and by Johnston, Price, and van Vleck given in Further Readings, which may be consulted by the reader for a more extensive treatment of this material.

The Leontief Closed Model

EXAMPLE 1* ■ Consider a simple society consisting of a farmer who only produces all the food, a carpenter who builds only all the homes, and a tailor who only makes all the clothes. For convenience, we may select our units so that each individual produces one unit of each commodity during the year. Suppose that during the year the portion of each commodity that is consumed by each individual is given in Table 8.1.

Table 8.1

Goods Consumed by:	Goods Produced by:		
	Farmer	Carpenter	Tailor
Farmer	$\frac{7}{16}$	$\frac{1}{2}$	$\frac{3}{16}$
Carpenter	$\frac{5}{16}$	$\frac{1}{6}$	$\frac{5}{16}$
Tailor	$\frac{1}{4}$	$\frac{1}{3}$	$\frac{1}{2}$

Thus the farmer consumes $\frac{7}{16}$ of his own produce while the carpenter consumes $\frac{5}{16}$ of the farmer's produce, the carpenter consumes $\frac{5}{16}$ of the clothes made by the tailor, and so on. Let p_1 be the price per unit of food, p_2 the price per unit of housing, and p_3 the price per unit of clothes. We assume that everyone pays the same price for a commodity. Thus the farmer pays the same price for his food as the tailor and carpenter, although he grew the food himself. We are interested in determining the prices p_1, p_2, and p_3 so that we have a state of equilibrium, which is defined as: *No one makes money or loses money.*

The farmer's expenditures are

$$\tfrac{7}{16}\,p_1 + \tfrac{1}{2}\,p_2 + \tfrac{3}{16}\,p_3,$$

while his income is p_1 because he produces one unit of food. Since expenditures must equal income, we have

$$\tfrac{7}{16}\,p_1 + \tfrac{1}{2}\,p_2 + \tfrac{3}{16}\,p_3 = p_1. \tag{1}$$

Similarly, for the carpenter we obtain

$$\tfrac{5}{16}\,p_1 + \tfrac{1}{6}\,p_2 + \tfrac{5}{16}\,p_3 = p_2, \tag{2}$$

and for the tailor we have

$$\tfrac{1}{4}\,p_1 + \tfrac{1}{3}\,p_2 + \tfrac{1}{2}\,p_3 = p_3. \tag{3}$$

Equations (1), (2), and (3) can be written in matrix notation as

$$A\mathbf{p} = \mathbf{p}, \tag{4}$$

---

*This example is due to Johnston, Price, and van Vleck and is described in their book given in Further Readings. The general model for this example is also presented by Gale.

where

$$A = \begin{bmatrix} \frac{7}{16} & \frac{1}{2} & \frac{3}{16} \\ \frac{5}{16} & \frac{1}{6} & \frac{5}{16} \\ \frac{1}{4} & \frac{1}{3} & \frac{1}{2} \end{bmatrix}, \qquad \mathbf{p} = \begin{bmatrix} p_1 \\ p_2 \\ p_3 \end{bmatrix}.$$

Equation (4) can be rewritten as

$$(I_n - A)\mathbf{p} = \mathbf{0}, \tag{5}$$

which is a homogeneous system.

Our problem is that of finding a solution $\mathbf{p}$ to (5) whose components p_i will be nonnegative with at least one positive p_i, since $\mathbf{p} = \mathbf{0}$ means that all the prices are zero, which makes no sense.

Solving (5), we obtain (verify)

$$\mathbf{p} = r \begin{bmatrix} 4 \\ 3 \\ 4 \end{bmatrix},$$

where r is any real number. Letting r be a positive number, we determine the *relative* prices of the commodities. For example, letting $r = 1000$, we find that food costs $4000 per unit, housing costs $3000 per unit, and clothes cost $4000 per unit.   ∎

EXAMPLE 2
(*The Exchange Model*)

Consider now the general problem where we have n manufacturers $M_1, M_2, \ldots, M_n$, and n goods $G_1, G_2, \ldots, G_n$, with M_i making only G_i. Consider a fixed interval of time, say one year, and suppose that M_i only makes one unit of G_i during the year.

In producing good G_i, manufacturer M_i may consume amounts of goods $G_1, G_2, \ldots, G_i, \ldots, G_n$. For example, iron, along with many other ingredients, is used in the manufacture of iron. Let a_{ij} be the amount of good G_j consumed by manufacturer M_i. Then

$$0 \le a_{ij} \le 1.$$

Suppose that the model is **closed**; that is, no goods leave or enter the system. This means that the total consumption of each good must equal its total production. Since the total production of G_j is 1, we have

$$a_{1j} + a_{2j} + \cdots + a_{nj} = 1 \qquad (1 \le j \le n).$$

If the price per unit of G_k is p_k, then manufacturer M_i pays

$$a_{i1}p_1 + a_{i2}p_2 + \cdots + a_{in}p_n \tag{6}$$

for the goods he uses.

Our problem is that of determining the prices $p_1, p_2, \ldots, p_n$ so that no manufacturer makes money or loses money. That is, so that each manufacturer's income will equal his expenses. Since M_i manufactures only one unit, his income is p_i. Thus from (6) we have

$$a_{11}p_1 + a_{12}p_2 + \cdots + a_{1n}p_n = p_1$$
$$a_{21}p_1 + a_{22}p_2 + \cdots + a_{2n}p_n = p_2$$
$$\vdots \qquad \vdots \qquad \qquad \vdots \qquad \vdots$$
$$a_{n1}p_1 + a_{n2}p_2 + \cdots + a_{nn}p_n = p_n,$$

which can be written in matrix form as

$$A\mathbf{p} = \mathbf{p}, \tag{7}$$

where

$$A = \begin{bmatrix} a_{ij} \end{bmatrix} \quad \text{and} \quad \mathbf{p} = \begin{bmatrix} p_1 \\ p_2 \\ \vdots \\ p_n \end{bmatrix}.$$

Equation (7) can be rewritten as

$$(I_n - A)\mathbf{p} = \mathbf{0}. \tag{8}$$

Thus our problem is that of finding a vector

$$\mathbf{p} \geq \mathbf{0},$$

with at least one component that is positive and satisfies Equation (8). ∎

DEFINITION   An $n \times n$ matrix $A = \begin{bmatrix} a_{ij} \end{bmatrix}$ is called an **exchange matrix** if it satisfies the following two properties:

(a) $a_{ij} \geq 0$ (each entry is nonnegative).

(b) $a_{1j} + a_{2j} + \cdots + a_{nj} = 1$, for $j = 1, 2, \ldots, n$ (the entries in each column add up to 1).

EXAMPLE 3 ∎ Matrix A of Example 1 is an exchange matrix, as is the matrix A of Example 2. ∎

Our general problem can now be stated as: Given an exchange matrix A, find a vector $\mathbf{p} \geq \mathbf{0}$ with at least one positive component, satisfying Equation (8). It can be shown that this problem always has a solution (see the book by Gale, p. 264, given in Further Readings).

In our general problem we required that each manufacturer's income equal his expenses. Instead, we could have required that each manufacturer's expenses not exceed his income. This would have led to

$$A\mathbf{p} \leq \mathbf{p} \tag{9}$$

instead of $A\mathbf{p} = \mathbf{p}$. However, it can be shown (Exercise T.1) that if Equation (9) holds, then Equation (7) will hold. Thus, if no manufacturer spends more than he earns, then everyone's income equals his expenses. An economic interpretation of this statement is that in the Leontief closed model if some manufacturer is making a profit, then at least one manufacturer is taking a loss.

REMARK   (For those who have read Section 5.1.) Let A be an exchange matrix for a Leontief closed model. It follows from Exercise T.17 in Section 5.1 that A has an eigenvalue $\lambda = 1$. Moreover, a price vector $\mathbf{p}$ is an eigenvector of A associated with $\lambda = 1$ that has nonnegative components.

An International Trade Model

EXAMPLE 4 ■   Suppose that n countries $C_1, C_2, \ldots, C_n$ are engaged in trading with each other and that a common currency is in use. We assume that prices are fixed throughout the discussion and that C_j's income y_j comes entirely from selling its goods either internally or to other countries. We also assume that the fraction of C_j's income that is spent on imports from C_i is a fixed number a_{ij} which does not depend on C_j's income y_j. Since the a_{ij}'s are fractions of y_j, we have

$$a_{ij} \geq 0$$
$$a_{1j} + a_{2j} + \cdots + a_{nj} = 1,$$

so that $A = \begin{bmatrix} a_{ij} \end{bmatrix}$ is an exchange matrix. We now wish to determine the total income y_i for each country C_i. Since the value of C_i's exports to C_j is $a_{ij}y_j$, the total income of C_i is

$$a_{i1}y_1 + a_{i2}y_2 + \cdots + a_{in}y_n.$$

Hence, we must have

$$a_{i1}y_1 + a_{i2}y_2 + \cdots + a_{in}y_n = y_i.$$

In matrix notation we must find

$$\mathbf{p} = \begin{bmatrix} y_1 \\ y_2 \\ \vdots \\ y_n \end{bmatrix} \geq \mathbf{0},$$

with at least one $y_i > 0$, so that

$$A\mathbf{p} = \mathbf{p},$$

which is our earlier problem.   ■

The Leontief Open Model

Suppose that we have n goods $G_1, G_2, \ldots, G_n$ and n activities $M_1, M_2, \ldots, M_n$. Assume that each activity M_i produces only one good G_i and that G_i is produced only by M_i. Let $c_{ij} \geq 0$ be the dollar value of G_i that has to be consumed to produce 1 dollar's worth of G_j. The matrix $C = \begin{bmatrix} c_{ij} \end{bmatrix}$ is called the **consumption matrix**. Observe that c_{ii} may be positive, which means that we may require some amount of G_i to make 1 dollar's worth of G_i.

Let x_i be the dollar value of G_i produced in a fixed period of time, say, 1 year. The vector

$$\mathbf{x} = \begin{bmatrix} x_1 \\ x_2 \\ \vdots \\ x_n \end{bmatrix} \qquad (x_i \geq 0) \tag{10}$$

is called the **production vector**. The expression

$$c_{i1}x_1 + c_{i2}x_2 + \cdots + c_{in}x_n$$

is the total value of the product G_i that is consumed, as determined by the production vector to make x_1 dollar's worth of G_1, x_2 dollar's worth of G_2, and so on. Observe that the expression given by Equation (10) is the ith entry of the matrix product $C\mathbf{x}$. The difference between the dollar value of G_i that is produced and the total dollar value of G_i that is consumed,

$$x_i - (c_{i1}x_1 + c_{i2}x_2 + \cdots + c_{in}x_n), \tag{11}$$

is called the **net production**.

Observe that the expression in Equation (11) is the ith entry in

$$\mathbf{x} - C\mathbf{x} = (I_n - C)\mathbf{x}.$$

Now let d_i represent the dollar value of the outside demand for G_i and let

$$\mathbf{d} = \begin{bmatrix} d_1 \\ d_2 \\ \vdots \\ d_n \end{bmatrix} \qquad (d_i \geq 0)$$

be the **demand vector**.

Our problem can now be stated: Given a demand vector $\mathbf{d} \geq \mathbf{0}$, can we find a production vector $\mathbf{x}$ such that the outside demand $\mathbf{d}$ is met without any surplus? That is, can we find a vector $\mathbf{x} \geq \mathbf{0}$ so that the following equation is satisfied?

$$(I_n - C)\mathbf{x} = \mathbf{d}. \tag{12}$$

EXAMPLE 5 ■ Let

$$C = \begin{bmatrix} \frac{1}{4} & \frac{1}{2} \\ \frac{2}{3} & \frac{1}{3} \end{bmatrix}$$

be a consumption matrix. Then

$$I_2 - C = \begin{bmatrix} 1 & 0 \\ 0 & 1 \end{bmatrix} - \begin{bmatrix} \frac{1}{4} & \frac{1}{2} \\ \frac{2}{3} & \frac{1}{3} \end{bmatrix} = \begin{bmatrix} \frac{3}{4} & -\frac{1}{2} \\ -\frac{2}{3} & \frac{2}{3} \end{bmatrix}.$$

Equation (12) becomes

$$\begin{bmatrix} \frac{3}{4} & -\frac{1}{2} \\ -\frac{2}{3} & \frac{2}{3} \end{bmatrix} \begin{bmatrix} x_1 \\ x_2 \end{bmatrix} = \begin{bmatrix} d_1 \\ d_2 \end{bmatrix},$$

so

$$\begin{bmatrix} x_1 \\ x_2 \end{bmatrix} = \begin{bmatrix} \frac{3}{4} & -\frac{1}{2} \\ -\frac{2}{3} & \frac{2}{3} \end{bmatrix}^{-1} \begin{bmatrix} d_1 \\ d_2 \end{bmatrix} = \begin{bmatrix} 4 & 3 \\ 4 & \frac{9}{2} \end{bmatrix} \begin{bmatrix} d_1 \\ d_2 \end{bmatrix} \geq \mathbf{0},$$

since $d_1 \geq 0$ and $d_2 \geq 0$. Thus we can obtain a production vector for any given demand vector. ■

In general, if $(I_n - C)^{-1}$ exists and is $\geq \mathbf{0}$, then $\mathbf{x} = (I_n - C)^{-1}\mathbf{d} \geq \mathbf{0}$ is a production vector for any given demand vector. However, for a given consumption matrix there may be no solution to Equation (12).

EXAMPLE 6 ■ Consider the consumption matrix

$$C = \begin{bmatrix} \frac{1}{2} & \frac{1}{2} \\ \frac{1}{2} & \frac{3}{4} \end{bmatrix}.$$

Then

$$I_2 - C = \begin{bmatrix} \frac{1}{2} & -\frac{1}{2} \\ -\frac{1}{2} & \frac{1}{4} \end{bmatrix}$$

and

$$(I_2 - C)^{-1} = \begin{bmatrix} -2 & -4 \\ -4 & -4 \end{bmatrix},$$

so that

$$\mathbf{x} = (I_2 - C)^{-1}\mathbf{d}$$

is not a production vector for $\mathbf{d} \neq \mathbf{0}$, since all of its components are negative. Thus the problem has no solution. If $\mathbf{d} = \mathbf{0}$, we do have a solution, namely, $\mathbf{x} = \mathbf{0}$, which means that if there is no outside demand, then nothing is produced. ■

DEFINITION   An $n \times n$ consumption matrix C is called **productive** if $(I_n - C)^{-1}$ exists and $(I_n - C)^{-1} \geq \mathbf{0}$. That is, C is productive if $(I_n - C)$ is nonsingular and every entry of $(I_n - C)^{-1}$ is nonnegative. In this case, the model is also called **productive**.

It follows that if C is productive, then for any demand vector $\mathbf{d} \geq \mathbf{0}$, the equation

$$(I_n - C)\mathbf{x} = \mathbf{d}$$

has a unique solution $\mathbf{x} \geq \mathbf{0}$.

EXAMPLE 7 ■ Consider the consumption matrix

$$C = \begin{bmatrix} \frac{1}{2} & \frac{1}{3} \\ \frac{1}{4} & \frac{1}{3} \end{bmatrix}.$$

Then

$$(I_2 - C) = \begin{bmatrix} \frac{1}{2} & -\frac{1}{3} \\ -\frac{1}{4} & \frac{2}{3} \end{bmatrix},$$

and

$$(I_2 - C)^{-1} = 4 \begin{bmatrix} \frac{2}{3} & \frac{1}{3} \\ \frac{1}{4} & \frac{1}{2} \end{bmatrix}.$$

Thus C is productive. If $\mathbf{d} \geq \mathbf{0}$ is a demand vector, then the equation $(I_n - C)\mathbf{x} = \mathbf{d}$ has the unique solution $\mathbf{x} = (I_n - C)^{-1}\mathbf{d} \geq \mathbf{0}$. ■

Results are given in more advanced books (see the book by Johnston, p. 251, given in Further Readings) for telling when a given consumption matrix is productive.

Further Readings ▶

GALE, DAVID. *The Theory of Linear Economic Models.* New York: McGraw-Hill Book Company, 1960.

JOHNSTON, B., G. PRICE, and F. S. VAN VLECK. *Linear Equations and Matrices.* Reading, Mass.: Addison-Wesley Publishing Co., Inc., 1966.

8.5   EXERCISES

1. Which of the following matrices are exchange matrices?

(a) $\begin{bmatrix} \frac{1}{3} & 0 & 1 \\ \frac{2}{3} & 1 & 0 \\ \frac{1}{2} & -\frac{1}{2} & 0 \end{bmatrix}.$ (b) $\begin{bmatrix} \frac{1}{2} & \frac{1}{3} & \frac{3}{4} \\ \frac{1}{2} & \frac{1}{3} & \frac{1}{4} \\ 0 & \frac{1}{3} & 0 \end{bmatrix}.$

(c) $\begin{bmatrix} \frac{1}{3} & -\frac{2}{3} & \frac{1}{2} \\ \frac{2}{3} & \frac{2}{3} & \frac{1}{2} \\ 0 & 1 & 0 \end{bmatrix}.$ (d) $\begin{bmatrix} 1 & \frac{1}{4} & \frac{5}{6} \\ 0 & \frac{1}{4} & \frac{1}{6} \\ 0 & \frac{1}{2} & 0 \end{bmatrix}.$

In Exercises 2 through 4, find a vector $\mathbf{p} \geq \mathbf{0}$, with at least one positive component, satisfying Equation (8) for the given exchange matrix.

2. $\begin{bmatrix} \frac{1}{3} & \frac{2}{3} & 0 \\ \frac{1}{3} & 0 & \frac{1}{4} \\ \frac{1}{3} & \frac{1}{3} & \frac{3}{4} \end{bmatrix}.$ 3. $\begin{bmatrix} \frac{1}{2} & 1 & \frac{2}{3} \\ 0 & 0 & 0 \\ \frac{1}{2} & 0 & \frac{1}{3} \end{bmatrix}.$

4. $\begin{bmatrix} 0 & \frac{1}{3} & 1 \\ \frac{1}{6} & \frac{1}{6} & 0 \\ \frac{5}{6} & \frac{1}{2} & 0 \end{bmatrix}.$

5. Consider the simple economy of Example 1. Suppose that the farmer consumes $\frac{2}{5}$ of the food, $\frac{1}{3}$ of the shelter, and $\frac{1}{2}$ of the clothes; that the carpenter consumes $\frac{2}{5}$ of the food, $\frac{1}{3}$ of the shelter, and $\frac{1}{2}$ of the clothes; that the tailor consumes $\frac{1}{5}$ of

the food, $\frac{1}{3}$ of the shelter, and none of the clothes. Find the exchange matrix A for this problem and a vector $\mathbf{p} \geq \mathbf{0}$ with at least one positive component satisfying Equation (8).

6. Consider the international trade model consisting of three countries, C_1, C_2, and C_3. Suppose that the fraction of C_1's income spent on imports from C_1 is $\frac{1}{4}$, from C_2 is $\frac{1}{2}$, and from C_3 is $\frac{1}{4}$; that the fraction of C_2's income spent on imports from C_1 is $\frac{2}{5}$, from C_2 is $\frac{1}{5}$, and from C_3 is a $\frac{2}{5}$; that the fraction of C_3's income spent on imports from C_1 is $\frac{1}{2}$, from C_2 is $\frac{1}{2}$, and from C_3 is 0. Find the income of each country.

In Exercises 7 through 10, determine which matrices are productive.

7. $\begin{bmatrix} \frac{1}{2} & \frac{1}{3} & 0 \\ 0 & \frac{2}{3} & 0 \\ 1 & 0 & 2 \end{bmatrix}$.

8. $\begin{bmatrix} 0 & \frac{2}{3} & 0 \\ \frac{1}{2} & 0 & 0 \\ 0 & \frac{1}{4} & 0 \end{bmatrix}$.

9. $\begin{bmatrix} 0 & \frac{1}{3} & \frac{1}{2} \\ \frac{1}{2} & 0 & \frac{1}{4} \\ \frac{1}{4} & \frac{1}{3} & 0 \end{bmatrix}$.

10. $\begin{bmatrix} 0 & \frac{1}{3} & \frac{1}{3} \\ \frac{1}{4} & 0 & \frac{1}{6} \\ \frac{1}{3} & \frac{2}{3} & 0 \end{bmatrix}$.

11. Suppose that the consumption matrix for the linear production model is

$$\begin{bmatrix} \frac{1}{2} & \frac{1}{2} \\ \frac{1}{2} & \frac{1}{4} \end{bmatrix}.$$

(a) Find the production vector for the demand vector $\begin{bmatrix} 1 \\ 3 \end{bmatrix}$.

(b) Find the production vector for the demand vector $\begin{bmatrix} 2 \\ 0 \end{bmatrix}$.

12. A small town has three primary industries: a copper mine, a railroad, and an electric utility. To produce $1 of copper, the copper mine uses $0.20 of copper, $0.10 of transportation, and $0.20 of electric power. To provide $1 of transportation, the railroad uses $0.10 of copper, $0.10 of transportation, and $0.40 of electric power. To provide $1 of electric power, the electric utility uses $0.20 of copper, $0.20 of transportation, and $0.30 of electric power. Suppose that during the year there is an outside demand of 1.2 million dollars for copper, 0.8 million dollars for transportation, and 1.5 million dollars for electric power. How much should each industry produce to satisfy the demands?

THEORETICAL EXERCISE ▦

T.1. In the exchange model (Example 3), show that $A\mathbf{p} \leq \mathbf{p}$ implies that $A\mathbf{p} = \mathbf{p}$.

8.6 ▼ Differential Equations

Prerequisite. Section 5.1, Diagonalization. Calculus required.

A **differential equation** is an equation that involves an unknown function and its derivatives. An important, simple example of a differential equation is

$$\frac{dy}{dx} = ry,$$

where r is a constant. The idea here is to find a function f that will satisfy the given differential equation; that is, $f' = rf$. This differential equation is discussed further below.

Differential equations occur often in all branches of science and engineering; linear algebra is helpful in the formulation and solution of differential equations. In this section we provide only a brief survey of the approach; books on differential equations deal with the subject in much greater detail, and several suggestions for further reading are given at the end of this section.

Homogeneous Linear Systems

We consider the **homogeneous linear system** of differential equations

$$
\begin{aligned}
x_1'(t) &= a_{11}x_1(t) + a_{12}x_2(t) + \cdots + a_{1n}x_n(t) \\
x_2'(t) &= a_{21}x_1(t) + a_{22}x_2(t) + \cdots + a_{2n}x_n(t) \\
&\vdots \\
x_n'(t) &= a_{n1}x_1(t) + a_{n2}x_2(t) + \cdots + a_{nn}x_n(t),
\end{aligned}
\tag{1}
$$

where the a_{ij} are known constants. We seek functions $x_1(t), x_2(t), \ldots, x_n(t)$ defined and differentiable on the real line satisfying (1).

We can write (1) in matrix form by letting

$$
\mathbf{x}(t) = \begin{bmatrix} x_1(t) \\ x_2(t) \\ \vdots \\ x_n(t) \end{bmatrix}, \qquad
A = \begin{bmatrix} a_{11} & a_{12} & \cdots & a_{1n} \\ a_{21} & a_{22} & \cdots & a_{2n} \\ \vdots & \vdots & & \vdots \\ a_{n1} & a_{n2} & \cdots & a_{nn} \end{bmatrix},
$$

and defining

$$
\mathbf{x}'(t) = \begin{bmatrix} x_1'(t) \\ x_2'(t) \\ \vdots \\ x_n'(t) \end{bmatrix}.
$$

Then (1) can be written as

$$
\mathbf{x}'(t) = A\mathbf{x}(t).
\tag{2}
$$

We shall often write (2) more briefly as

$$
\mathbf{x}' = A\mathbf{x}.
$$

With this notation, a vector function

$$
\mathbf{x}(t) = \begin{bmatrix} x_1(t) \\ x_2(t) \\ \vdots \\ x_n(t) \end{bmatrix}
$$

satisfying (2) is called a **solution** to the given system. It can be shown (Exercise T.1) that the set of all solutions to the homogeneous linear system of differential equations (1) is a subspace of the vector space of differentiable real-valued functions.

We leave it to the reader to verify that if $\mathbf{x}^{(1)}(t), \mathbf{x}^{(2)}(t), \ldots, \mathbf{x}^{(n)}(t)$ are all solutions to (2), then any linear combination

$$
\mathbf{x}(t) = b_1\mathbf{x}^{(1)}(t) + b_2\mathbf{x}^{(2)}(t) + \cdots + b_n\mathbf{x}^{(n)}(t)
\tag{3}
$$

is also a solution to (2).

A set of vector functions $\{\mathbf{x}^{(1)}(t), \mathbf{x}^{(2)}(t), \ldots, \mathbf{x}^{(n)}(t)\}$ is said to be a **fundamental system** for (1) if every solution to (1) can be written in the form (3). In this case, the right side of (3), where $b_1, b_2, \ldots, b_n$ are arbitrary constants, is said to be the **general solution** to (2).

It can be shown (see the book by Boyce and DiPrima or the book by Cullen cited in Further Readings) that any system of the form (2) has a fundamental system (in fact, infinitely many).

In general, differential equations arise in the course of solving physical problems. Typically, once a general solution to the differential equation has been obtained, the physical constraints of the problem impose certain definite values on the arbitrary constants in the general solution, giving rise to a **particular solution**. An important particular solution is obtained by finding a solution $\mathbf{x}(t)$ to Equation (2) such that $\mathbf{x}(0) = \mathbf{x}_0$, an **initial condition**, where $\mathbf{x}_0$ is a given vector. This problem is called an **initial value problem**. If the general solution (3) is known, then the initial value problem can be solved by setting $t = 0$ in (3) and determining the constants $b_1, b_2, \ldots, b_n$ so that

$$\mathbf{x}_0 = b_1 \mathbf{x}^{(1)}(0) + b_2 \mathbf{x}^{(2)}(0) + \cdots + b_n \mathbf{x}^{(n)}(0).$$

It is easily seen that this is actually an $n \times n$ linear system with unknowns $b_1, b_2, \ldots, b_n$. This linear system can also be written as

$$C\mathbf{b} = \mathbf{x}_0, \tag{4}$$

where

$$\mathbf{b} = \begin{bmatrix} b_1 \\ b_2 \\ \vdots \\ b_n \end{bmatrix}$$

and C is the $n \times n$ matrix whose columns are $\mathbf{x}^{(1)}(0), \mathbf{x}^{(2)}(0), \ldots, \mathbf{x}^{(n)}(0)$, respectively. It can be shown (see the book by Boyce and DiPrima or the book by Cullen cited in Further Readings) that if $\mathbf{x}^{(1)}(t), \mathbf{x}^{(2)}(t), \ldots, \mathbf{x}^{(n)}(t)$ form a fundamental system for (1), then C is nonsingular, so (4) always has a unique solution.

EXAMPLE 1 ■ The simplest system of the form (1) is the single equation

$$\frac{dx}{dt} = ax, \tag{5}$$

where a is a constant. From calculus, the solutions to this equation are of the form

$$x = be^{at}; \tag{6}$$

that is, this is the general solution to (5). To solve the initial value problem

$$\frac{dx}{dt} = ax, \qquad x(0) = x_0,$$

we set $t = 0$ in (6) and obtain $b = x_0$. Thus the solution to the initial value problem is

$$x = x_0 e^{at}.$$

$\blacksquare$

The system (2) is said to be **diagonal** if the matrix A is diagonal. Then (1) can be rewritten as

$$
\begin{aligned}
x_1'(t) &= a_{11}x_1(t) \\
x_2'(t) &= \qquad\quad a_{22}x_2(t) \\
&\;\;\vdots \\
x_n'(t) &= \qquad\qquad\qquad a_{nn}x_n(t).
\end{aligned}
\tag{7}
$$

This system is easy to solve, since the equations can be solved separately. Applying the results of Example 1 to each equation in (7), we obtain

$$
\begin{aligned}
x_1(t) &= b_1 e^{a_{11}t} \\
x_2(t) &= b_2 e^{a_{22}t} \\
&\;\;\vdots \qquad \vdots \\
x_n(t) &= b_n e^{a_{nn}t},
\end{aligned}
\tag{8}
$$

where $b_1, b_2, \ldots, b_n$ are arbitrary constants. Writing (8) in vector form yields

$$
\mathbf{x}(t) =
\begin{bmatrix} b_1 e^{a_{11}t} \\ b_2 e^{a_{22}t} \\ \vdots \\ b_n e^{a_{nn}t} \end{bmatrix}
= b_1 \begin{bmatrix} 1 \\ 0 \\ 0 \\ \vdots \\ 0 \end{bmatrix} e^{a_{11}t}
+ b_2 \begin{bmatrix} 0 \\ 1 \\ 0 \\ \vdots \\ 0 \end{bmatrix} e^{a_{22}t}
+ \cdots + b_n \begin{bmatrix} 0 \\ 0 \\ \vdots \\ 0 \\ 1 \end{bmatrix} e^{a_{nn}t}.
$$

This implies that the vector functions

$$
\mathbf{x}^{(1)}(t) = \begin{bmatrix} 1 \\ 0 \\ 0 \\ \vdots \\ 0 \end{bmatrix} e^{a_{11}t}, \quad
\mathbf{x}^{(2)}(t) = \begin{bmatrix} 0 \\ 1 \\ 0 \\ \vdots \\ 0 \end{bmatrix} e^{a_{22}t}, \ldots, \quad
\mathbf{x}^{(n)}(t) = \begin{bmatrix} 0 \\ 0 \\ \vdots \\ 0 \\ 1 \end{bmatrix} e^{a_{nn}t}
$$

form a fundamental system for the diagonal system (7).

EXAMPLE 2 $\blacksquare$ The diagonal system

$$
\begin{bmatrix} x_1' \\ x_2' \\ x_3' \end{bmatrix}
= \begin{bmatrix} 3 & 0 & 0 \\ 0 & -2 & 0 \\ 0 & 0 & 4 \end{bmatrix}
\begin{bmatrix} x_1 \\ x_2 \\ x_3 \end{bmatrix}
\tag{9}
$$

can be written as three equations:

$$
\begin{aligned}
x_1' &= \;\;\;3x_1 \\
x_2' &= -2x_2 \\
x_3' &= \;\;\;4x_3.
\end{aligned}
$$

Solving these equations, we obtain

$$x_1 = b_1 e^{3t}, \qquad x_2 = b_2 e^{-2t}, \qquad x_3 = b_3 e^{4t},$$

where b_1, b_2, and b_3 are arbitrary constants. Thus

$$\mathbf{x}(t) = \begin{bmatrix} b_1 e^{3t} \\ b_2 e^{-2t} \\ b_3 e^{4t} \end{bmatrix} = b_1 \begin{bmatrix} 1 \\ 0 \\ 0 \end{bmatrix} e^{3t} + b_2 \begin{bmatrix} 0 \\ 1 \\ 0 \end{bmatrix} e^{-2t} + b_3 \begin{bmatrix} 0 \\ 0 \\ 1 \end{bmatrix} e^{4t}$$

is the general solution to (9) and the functions

$$\mathbf{x}^{(1)}(t) = \begin{bmatrix} 1 \\ 0 \\ 0 \end{bmatrix} e^{3t}, \qquad \mathbf{x}^{(2)}(t) = \begin{bmatrix} 0 \\ 1 \\ 0 \end{bmatrix} e^{-2t}, \qquad \mathbf{x}^{(3)}(t) = \begin{bmatrix} 0 \\ 0 \\ 1 \end{bmatrix} e^{4t}$$

form a fundamental system for (9).   ■

If the system (2) is not diagonal, then it cannot be solved as simply as the system in the preceding example. However, there is an extension of this method that yields the general solution in the case where A is diagonalizable. Suppose that A is diagonalizable and P is a nonsingular matrix such that

$$P^{-1}AP = D, \tag{10}$$

where D is diagonal. Then multiplying the given system

$$\mathbf{x}' = A\mathbf{x}$$

on the left by P^{-1}, we obtain

$$P^{-1}\mathbf{x}' = P^{-1}A\mathbf{x}.$$

Since $P^{-1}P = I_n$, we can rewrite the last equation as

$$P^{-1}\mathbf{x}' = (P^{-1}AP)(P^{-1}\mathbf{x}). \tag{11}$$

Temporarily, let

$$\mathbf{u} = P^{-1}\mathbf{x}. \tag{12}$$

Since P^{-1} is a constant matrix,

$$\mathbf{u}' = P^{-1}\mathbf{x}'. \tag{13}$$

Therefore, substituting (10), (12), and (13) into (11), we obtain

$$\mathbf{u}' = D\mathbf{u}. \tag{14}$$

Equation (14) is a diagonal system and can be solved by the methods just discussed. Before proceeding, however, let us recall from Theorem 5.3 in Section 5.1 that

$$D = \begin{bmatrix} \lambda_1 & 0 & \cdots & 0 \\ 0 & \lambda_2 & \cdots & 0 \\ \vdots & \vdots & & \vdots \\ 0 & 0 & \cdots & \lambda_n \end{bmatrix},$$

where $\lambda_1, \lambda_2, \ldots, \lambda_n$ are the eigenvalues of A, and that the columns of P are linearly independent eigenvectors of A associated, respectively, with $\lambda_1, \lambda_2, \ldots, \lambda_n$. From the discussion just given for diagonal systems, the general solution to (14) is

$$\mathbf{u}(t) = b_1 \mathbf{u}^{(1)}(t) + b_2 \mathbf{u}^{(2)}(t) + \cdots + b_n \mathbf{u}^{(n)}(t) = \begin{bmatrix} b_1 e^{\lambda_1 t} \\ b_2 e^{\lambda_2 t} \\ \vdots \\ b_n e^{\lambda_n t} \end{bmatrix},$$

where

$$\mathbf{u}^{(1)}(t) = \begin{bmatrix} 1 \\ 0 \\ 0 \\ \vdots \\ 0 \end{bmatrix} e^{\lambda_1 t}, \quad \mathbf{u}^{(2)}(t) = \begin{bmatrix} 0 \\ 1 \\ 0 \\ \vdots \\ 0 \end{bmatrix} e^{\lambda_2 t}, \ldots, \quad \mathbf{u}^{(n)}(t) = \begin{bmatrix} 0 \\ 0 \\ \vdots \\ 0 \\ 1 \end{bmatrix} e^{\lambda_n t} \quad (15)$$

and $b_1, b_2, \ldots, b_n$ are arbitrary constants. From Equation (12), $\mathbf{x} = P\mathbf{u}$, so the general solution to the given system $\mathbf{x}' = A\mathbf{x}$ is

$$\mathbf{x}(t) = P\mathbf{u}(t) = b_1 P\mathbf{u}^{(1)}(t) + b_2 P\mathbf{u}^{(2)}(t) + \cdots + b_n P\mathbf{u}^{(n)}(t). \quad (16)$$

However, since the constant vectors in (15) are the columns of the identity matrix and $PI_n = P$, (16) can be rewritten as

$$\mathbf{x}(t) = b_1 \mathbf{p}_1 e^{\lambda_1 t} + b_2 \mathbf{p}_2 e^{\lambda_2 t} + \cdots + b_n \mathbf{p}_n e^{\lambda_n t}, \quad (17)$$

where $\mathbf{p}_1, \mathbf{p}_2, \ldots, \mathbf{p}_n$ are the columns of P, and therefore eigenvectors of A associated with $\lambda_1, \lambda_2, \ldots, \lambda_n$, respectively.

We summarize the discussion above in the following theorem.

THEOREM 8.8 ■ *If the $n \times n$ matrix A has n linearly independent eigenvectors $\mathbf{p}_1, \mathbf{p}_2, \ldots, \mathbf{p}_n$ associated with the eigenvalues $\lambda_1, \lambda_2, \ldots, \lambda_n$, respectively, then the general solution to the homogeneous linear system of differential equations*

$$\mathbf{x}' = A\mathbf{x}$$

is given by (17). ■

The procedure for obtaining the general solution to the homogeneous linear system $\mathbf{x}'(t) = A\mathbf{x}(t)$, where A is diagonalizable, is as follows.

Step 1. Compute the eigenvalues $\lambda_1, \lambda_2, \ldots, \lambda_n$ of A.

Step 2. Compute eigenvectors $\mathbf{p}_1, \mathbf{p}_2, \ldots, \mathbf{p}_n$ of A associated, respectively, with $\lambda_1, \lambda_2, \ldots, \lambda_n$.

Step 3. The general solution is given by Equation (17).

EXAMPLE 3 ■ For the system

$$\mathbf{x}' = \begin{bmatrix} 1 & -1 \\ 2 & 4 \end{bmatrix} \mathbf{x},$$

the matrix

$$A = \begin{bmatrix} 1 & -1 \\ 2 & 4 \end{bmatrix}$$

has eigenvalues $\lambda_1 = 2$ and $\lambda_2 = 3$ with associated eigenvectors (verify)

$$\mathbf{p}_1 = \begin{bmatrix} 1 \\ -1 \end{bmatrix} \quad \text{and} \quad \mathbf{p}_2 = \begin{bmatrix} 1 \\ -2 \end{bmatrix}.$$

These eigenvectors are automatically linearly independent, since they are associated with distinct eigenvalues (proof of Theorem 5.4 in Section 5.1). Hence the general solution to the given system is

$$\mathbf{x}(t) = b_1 \begin{bmatrix} 1 \\ -1 \end{bmatrix} e^{2t} + b_2 \begin{bmatrix} 1 \\ -2 \end{bmatrix} e^{3t}.$$

In terms of components, this can be written as

$$x_1(t) = \quad b_1 e^{2t} + \quad b_2 e^{3t}$$
$$x_2(t) = -b_1 e^{2t} - 2b_2 e^{3t}.$$ ■

EXAMPLE 4 ■ Consider the following homogeneous linear system of differential equations:

$$\mathbf{x}' = \begin{bmatrix} x_1' \\ x_2' \\ x_3' \end{bmatrix} = \begin{bmatrix} 0 & 1 & 0 \\ 0 & 0 & 1 \\ 8 & -14 & 7 \end{bmatrix} \begin{bmatrix} x_1 \\ x_2 \\ x_3 \end{bmatrix}.$$

The characteristic polynomial of A is (verify)

$$f(\lambda) = \lambda^3 - 7\lambda^2 + 14\lambda - 8$$

or

$$f(\lambda) = (\lambda - 1)(\lambda - 2)(\lambda - 4),$$

so the eigenvalues of A are $\lambda_1 = 1$, $\lambda_2 = 2$, and $\lambda_3 = 4$. Associated eigenvectors are (verify)

$$\begin{bmatrix} 1 \\ 1 \\ 1 \end{bmatrix}, \quad \begin{bmatrix} 1 \\ 2 \\ 4 \end{bmatrix}, \quad \begin{bmatrix} 1 \\ 4 \\ 16 \end{bmatrix},$$

respectively. The general solution is then given by

$$\mathbf{x}(t) = b_1 \begin{bmatrix} 1 \\ 1 \\ 1 \end{bmatrix} e^t + b_2 \begin{bmatrix} 1 \\ 2 \\ 4 \end{bmatrix} e^{2t} + b_3 \begin{bmatrix} 1 \\ 4 \\ 16 \end{bmatrix} e^{4t},$$

where b_1, b_2, and b_3 are arbitrary constants. ■

EXAMPLE 5 ■ For the system of Example 4 solve the initial value problem determined by the **initial conditions** $x_1(0) = 4$, $x_2(0) = 6$, and $x_3(0) = 8$.

Solution   We write our general solution in the form $\mathbf{x} = P\mathbf{u}$ as

$$\mathbf{x}(t) = \begin{bmatrix} 1 & 1 & 1 \\ 1 & 2 & 4 \\ 1 & 4 & 16 \end{bmatrix} \begin{bmatrix} b_1 e^t \\ b_2 e^{2t} \\ b_3 e^{4t} \end{bmatrix}.$$

Now

$$\mathbf{x}(0) = \begin{bmatrix} 4 \\ 6 \\ 8 \end{bmatrix} = \begin{bmatrix} 1 & 1 & 1 \\ 1 & 2 & 4 \\ 1 & 4 & 16 \end{bmatrix} \begin{bmatrix} b_1 e^0 \\ b_2 e^0 \\ b_3 e^0 \end{bmatrix}$$

or

$$\begin{bmatrix} 1 & 1 & 1 \\ 1 & 2 & 4 \\ 1 & 4 & 16 \end{bmatrix} \begin{bmatrix} b_1 \\ b_2 \\ b_3 \end{bmatrix} = \begin{bmatrix} 4 \\ 6 \\ 8 \end{bmatrix}. \tag{18}$$

Solving (18) by Gauss–Jordan reduction, we obtain (verify)

$$b_1 = \tfrac{4}{3}, \qquad b_2 = 3, \qquad b_3 = -\tfrac{1}{3}.$$

Therefore, the solution to the initial value problem is

$$\mathbf{x}(t) = \tfrac{4}{3} \begin{bmatrix} 1 \\ 1 \\ 1 \end{bmatrix} e^t + 3 \begin{bmatrix} 1 \\ 2 \\ 4 \end{bmatrix} e^{2t} - \tfrac{1}{3} \begin{bmatrix} 1 \\ 4 \\ 16 \end{bmatrix} e^{4t}. \qquad\blacksquare$$

We now recall several facts from Chapter 5. If A does not have distinct eigenvalues, then we may or may not be able to diagonalize A. Let λ be an eigenvalue of A of multiplicity k. Then A can be diagonalized if and only if the dimension of the eigenspace associated with λ is k, that is, if and only if the rank of the matrix $(\lambda I_n - A)$ is $n - k$ (verify). If the rank of $(\lambda I_n - A)$ is $n - k$, then we can find k linearly independent eigenvectors of A associated with λ.

EXAMPLE 6 ■ Consider the linear system

$$\mathbf{x}' = A\mathbf{x} = \begin{bmatrix} 1 & 0 & 0 \\ 0 & 3 & -2 \\ 0 & -2 & 3 \end{bmatrix} \mathbf{x}.$$

The eigenvalues of A are $\lambda_1 = \lambda_2 = 1$ and $\lambda_3 = 5$ (verify). The rank of the matrix

$$(1I_3 - A) = \begin{bmatrix} 0 & 0 & 0 \\ 0 & -2 & 2 \\ 0 & 2 & -2 \end{bmatrix}$$

is 1 and the linearly independent eigenvectors

$$\begin{bmatrix} 1 \\ 0 \\ 0 \end{bmatrix} \quad \text{and} \quad \begin{bmatrix} 0 \\ 1 \\ 1 \end{bmatrix}$$

are associated with the eigenvalue 1 (verify). The eigenvector

$$\begin{bmatrix} 0 \\ 1 \\ -1 \end{bmatrix}$$

is associated with the eigenvalue 5 (verify). The general solution to the given system is then

$$\mathbf{x}(t) = b_1 \begin{bmatrix} 1 \\ 0 \\ 0 \end{bmatrix} e^t + b_2 \begin{bmatrix} 0 \\ 1 \\ 1 \end{bmatrix} e^t + b_3 \begin{bmatrix} 0 \\ 1 \\ -1 \end{bmatrix} e^{5t},$$

where b_1, b_2, and b_3 are arbitrary constants. ■

If we cannot diagonalize A as in the examples, we have a considerably more difficult situation. Methods for dealing with such problems are discussed in more advanced books (see Further Readings).

Application—A Diffusion Process

The following example is a modification of an example presented by Derrick and Grossman in *Elementary Differential Equations with Applications* (see Further Readings).

EXAMPLE 7 ■ Consider two adjoining cells separated by a permeable membrane and suppose that a fluid flows from the first cell to the second one at a rate (in milliliters per minute) that is numerically equal to three times the volume (in milliliters) of the fluid in the first cell. It then flows out of the second cell at a rate (in milliliters per minute) that is numerically equal to twice the volume in the second cell. Let $x_1(t)$ and $x_2(t)$ denote the volumes of the fluid in the first and second cells at time t, respectively. Assume that initially the first cell has 40 milliliters of fluid, while the second one has 5 milliliters of fluid. Find the volume of fluid in each cell at time t.

Solution The change in volume of the fluid in each cell is the difference between the amount flowing in and the amount flowing out. Since no fluid flows into the first cell, we have

$$\frac{dx_1(t)}{dt} = -3x_1(t),$$

where the minus sign indicates that the fluid is flowing out of the cell. The flow $3x_1(t)$ from the first cell flows into the second cell. The flow out of the second cell is $2x_2(t)$. Thus the change in volume of the fluid in the second cell is given by

$$\frac{dx_2(t)}{dt} = 3x_1(t) - 2x_2(t).$$

We have then obtained the linear system

$$\frac{dx_1(t)}{dt} = -3x_1(t)$$

$$\frac{dx_2(t)}{dt} = 3x_1(t) - 2x_2(t),$$

which can be written in matrix form as

$$\begin{bmatrix} x_1'(t) \\ x_2'(t) \end{bmatrix} = \begin{bmatrix} -3 & 0 \\ 3 & -2 \end{bmatrix} \begin{bmatrix} x_1(t) \\ x_2(t) \end{bmatrix}.$$

The eigenvalues of the matrix

$$A = \begin{bmatrix} -3 & 0 \\ 3 & -2 \end{bmatrix}$$

are (verify)

$$\lambda_1 = -3, \qquad \lambda_2 = -2$$

and corresponding associated eigenvectors are (verify)

$$\begin{bmatrix} 1 \\ -3 \end{bmatrix}, \qquad \begin{bmatrix} 0 \\ 1 \end{bmatrix}.$$

Hence the general solution is given by

$$\mathbf{x}(t) = \begin{bmatrix} x_1(t) \\ x_2(t) \end{bmatrix} = b_1 \begin{bmatrix} 1 \\ -3 \end{bmatrix} e^{-3t} + b_2 \begin{bmatrix} 0 \\ 1 \end{bmatrix} e^{-2t}.$$

Using the initial conditions, we find that (verify)

$$b_1 = 40, \qquad b_2 = 125.$$

Thus the volume of fluid in each cell at time t is given by

$$x_1(t) = 40e^{-3t}$$

$$x_2(t) = -120e^{-3t} + 125e^{-2t}.$$
■

It should also be pointed out that many differential equations cannot be solved in the sense that we can write a formula for the solution. Numerical methods, some of which are studied in numerical analysis, exist for obtaining numerical solutions to differential equations; computer codes for some of these methods are widely available.

Further Readings ▶

BOYCE, W. E., and R. C. DiPRIMA. *Elementary Differential Equations*, 5th ed. New York: John Wiley & Sons, Inc., 1992.

CULLEN, C. G. *Linear Algebra and Differential Equations*, 2nd ed. Boston: PWS-Kent, 1991.

DERRICK, W. R., and S. I. GROSSMAN. *Elementary Differential Equations with Applications*, 2nd ed. Reading, Mass.: Addison-Wesley, 1981.

DETTMAN, J. H. *Introduction to Linear Algebra and Differential Equations.* New York: McGraw-Hill, 1974.

GOODE, S. W. *Differential Equations and Linear Algebra.* Upper Saddle River, N.J.: Prentice Hall, Inc., 1991.

RABENSTEIN, A. L. *Elementary Differential Equations with Linear Algebra*, 4th ed. Philadelphia: W. B. Saunders, Harcourt Brace Jovanovich, 1992.

8.6   EXERCISES

1. Consider the homogeneous linear system of differential equations

$$\begin{bmatrix} x_1' \\ x_2' \\ x_3' \end{bmatrix} = \begin{bmatrix} -3 & 0 & 0 \\ 0 & 4 & 0 \\ 0 & 0 & 2 \end{bmatrix} \begin{bmatrix} x_1 \\ x_2 \\ x_3 \end{bmatrix}.$$

(a) Find the general solution.

(b) Find the solution to the initial value problem determined by the initial conditions $x_1(0) = 3$, $x_2(0) = 4$, $x_3(0) = 5$.

2. Consider the homogeneous linear system of differential equations

$$\begin{bmatrix} x_1' \\ x_2' \\ x_3' \end{bmatrix} = \begin{bmatrix} 1 & 0 & 0 \\ 0 & -2 & 1 \\ 0 & 0 & 3 \end{bmatrix} \begin{bmatrix} x_1 \\ x_2 \\ x_3 \end{bmatrix}.$$

(a) Find the general solution.

(b) Find the solution to the initial value problem determined by the initial conditions $x_1(0) = 2$, $x_2(0) = 7$, $x_3(0) = 20$.

3. Find the general solution to the homogeneous linear system of differential equations

$$\begin{bmatrix} x_1' \\ x_2' \\ x_3' \end{bmatrix} = \begin{bmatrix} 4 & 0 & 0 \\ 3 & -5 & 0 \\ 2 & 1 & 2 \end{bmatrix} \begin{bmatrix} x_1 \\ x_2 \\ x_3 \end{bmatrix}.$$

4. Find the general solution to the homogeneous linear system of differential equations

$$\begin{bmatrix} x_1' \\ x_2' \\ x_3' \end{bmatrix} = \begin{bmatrix} 2 & 3 & 0 \\ 0 & 1 & 0 \\ 0 & 0 & 2 \end{bmatrix} \begin{bmatrix} x_1 \\ x_2 \\ x_3 \end{bmatrix}.$$

5. Find the general solution to the homogeneous linear system of differential equations

$$\begin{bmatrix} x_1' \\ x_2' \\ x_3' \end{bmatrix} = \begin{bmatrix} 5 & 0 & 0 \\ 0 & -4 & 3 \\ 0 & 3 & 4 \end{bmatrix} \begin{bmatrix} x_1 \\ x_2 \\ x_3 \end{bmatrix}.$$

6. Find the general solution to the homogeneous linear system of differential equations

$$\begin{bmatrix} x_1' \\ x_2' \end{bmatrix} = \begin{bmatrix} 3 & -2 \\ -2 & 3 \end{bmatrix} \begin{bmatrix} x_1 \\ x_2 \end{bmatrix}.$$

7. Find the general solution to the homogeneous linear system of differential equations

$$\begin{bmatrix} x_1' \\ x_2' \\ x_3' \end{bmatrix} = \begin{bmatrix} 1 & 2 & 3 \\ 0 & 1 & 0 \\ 2 & 1 & 2 \end{bmatrix} \begin{bmatrix} x_1 \\ x_2 \\ x_3 \end{bmatrix}.$$

8. Find the general solution to the homogeneous linear system of differential equations

$$\begin{bmatrix} x_1' \\ x_2' \\ x_3' \end{bmatrix} = \begin{bmatrix} 1 & 1 & 2 \\ 0 & 1 & 0 \\ 0 & 1 & 3 \end{bmatrix} \begin{bmatrix} x_1 \\ x_2 \\ x_3 \end{bmatrix}.$$

9. Consider two competing species that live in the same forest and let $x_1(t)$ and $x_2(t)$ denote the respective populations of the species at time t. Suppose that the initial populations are $x_1(0) = 500$ and $x_2(0) = 200$. If the growth rates of the species are given by

$$\begin{aligned} x_1'(t) &= -3x_1(t) + 6x_2(t) \\ x_2'(t) &= x_1(t) - 2x_2(t), \end{aligned}$$

what is the population of each species at time t?

THEORETICAL EXERCISE ▪

T.1. Show that the set of all solutions to the homogeneous linear system of differential equations $\mathbf{x}' = A\mathbf{x}$ is a subspace of the vector space of all differentiable real-valued functions. This subspace is called the **solution space** of the given linear system.

MATLAB EXERCISES ▪

For the linear system $\mathbf{x}' = A\mathbf{x}$, this section shows how to construct the general solution $\mathbf{x}(t)$ provided that A is diagonalizable. In MATLAB we can determine the eigenvalues and eigenvectors of A as in Section 5.1, or we can use the command **eig**, which is discussed in Section 5.2. The **eig** command may give eigenvectors different from those found by your hand computations, but recall that eigenvectors are not unique.

ML.1. Solve Example 4 using the **eig** command in MATLAB.

ML.2. Solve Example 6 using the **eig** command in MATLAB.

ML.3. Solve Example 7 using the **eig** command in MATLAB.

8.7 ▾ The Fibonacci Sequence

Prerequisite. Section 5.1, Diagonalization.

In 1202, Leonardo of Pisa, also called Fibonacci,* wrote a book on mathematics in which he posed the following problem. A pair of newborn rabbits begins to breed at the age of 1 month, and thereafter produces one pair of offspring per month. Suppose that we start with a pair of newly born rabbits and that none of the rabbits produced from this pair die. How many pairs of rabbits will there be at the beginning of each month?

The breeding pattern of the rabbits is shown in Figure 8.25, where an arrow indicates an offspring pair. At the beginning of month 0, we have the newly born pair of rabbits P_1. At the beginning of month 1 we still have only the original pair of rabbits P_1, which have not yet produced any offspring. At the beginning of month 2 we have the original pair P_1 and its first pair of offspring, P_2. At the beginning of month 3 we have the original pair P_1, its first pair of offspring P_2 born at the beginning of month 2, and its second pair of offspring, P_3. At the beginning of month 4 we have P_1, P_2, and P_3; P_4, the offspring of P_1; and P_5, the offspring of P_2. Let u_n denote the number of pairs of rabbits at the beginning of month n. We see that

$$u_0 = 1, \qquad u_1 = 1, \qquad u_2 = 2, \qquad u_3 = 3, \qquad u_4 = 5, \qquad u_5 = 8.$$

---

*Leonardo Fibonacci of Pisa (about 1170–1250) was born and lived most of his life in Pisa, Italy. When he was about 20, his father was appointed director of Pisan commercial interests in northern Africa, now a part of Algeria. Leonardo accompanied his father to Africa and for several years traveled extensively throughout the Mediterranean area on behalf of his father. During these travels he learned the Hindu–Arabic method of numeration and calculation and decided to promote its use in Italy. This was one purpose of his most famous book, *Liber Abaci*, which appeared in 1202 and contained the rabbit problem stated here.

FIGURE 8.25

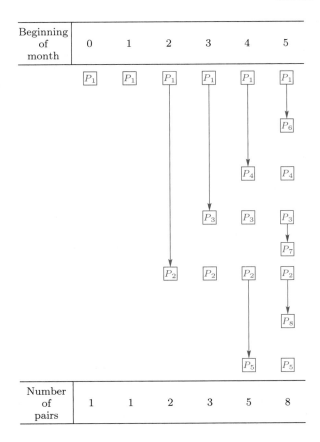

Beginning of month	0	1	2	3	4	5

| Number of pairs | 1 | 1 | 2 | 3 | 5 | 8 |

To obtain a formula for u_n, we proceed as follows. The number of pairs of rabbits that are alive at the beginning of month n is u_{n-1}, the number of pairs who were alive the previous month, plus the number of pairs newly born at the beginning of month n. The latter number is u_{n-2}, since a pair of rabbits produces a pair of offspring starting with its second month of life. Thus

$$u_n = u_{n-1} + u_{n-2}. \qquad (1)$$

That is, each number is the sum of its two predecessors. The resulting sequence of numbers, called a **Fibonacci sequence**, occurs in a remarkably varied number of applications, such as the distribution of leaves on certain trees, the arrangements of seeds on sunflowers, search techniques in numerical analysis, generating random numbers in statistics, and others.

To compute u_n by using the **recursion relation** (or difference equation) (1), we have to compute $u_0, u_1, \ldots, u_{n-2}, u_{n-1}$. This can be rather tedious for large n. We now develop a formula that will enable us to calculate u_n directly.

In addition to Equation (1) we write

$$u_{n-1} = u_{n-1},$$

so we now have

$$u_n = u_{n-1} + u_{n-2}$$
$$u_{n-1} = u_{n-1},$$

which can be written in matrix form as

$$\begin{bmatrix} u_n \\ u_{n-1} \end{bmatrix} = \begin{bmatrix} 1 & 1 \\ 1 & 0 \end{bmatrix} \begin{bmatrix} u_{n-1} \\ u_{n-2} \end{bmatrix}. \tag{2}$$

We now define in general

$$\mathbf{w}_k = \begin{bmatrix} u_{k+1} \\ u_k \end{bmatrix} \quad \text{and} \quad A = \begin{bmatrix} 1 & 1 \\ 1 & 0 \end{bmatrix} \qquad (0 \le k \le n-1)$$

so that

$$\mathbf{w}_0 = \begin{bmatrix} u_1 \\ u_0 \end{bmatrix} = \begin{bmatrix} 1 \\ 1 \end{bmatrix},$$

$$\mathbf{w}_1 = \begin{bmatrix} u_2 \\ u_1 \end{bmatrix} = \begin{bmatrix} 2 \\ 1 \end{bmatrix}, \dots, \quad \mathbf{w}_{n-2} = \begin{bmatrix} u_{n-1} \\ u_{n-2} \end{bmatrix}, \quad \text{and} \quad \mathbf{w}_{n-1} = \begin{bmatrix} u_n \\ u_{n-1} \end{bmatrix}.$$

Then (2) can be written as

$$\mathbf{w}_{n-1} = A\mathbf{w}_{n-2}.$$

Thus

$$\mathbf{w}_1 = A\mathbf{w}_0$$
$$\mathbf{w}_2 = A\mathbf{w}_1 = A(A\mathbf{w}_0) = A^2\mathbf{w}_0$$
$$\mathbf{w}_3 = A\mathbf{w}_2 = A(A^2\mathbf{w}_0) = A^3\mathbf{w}_0 \tag{3}$$
$$\vdots$$
$$\mathbf{w}_{n-1} = A^{n-1}\mathbf{w}_0.$$

Hence, to find u_n, we merely have to calculate A^{n-1}, which is still rather tedious if n is large. To avoid this difficulty, we find a diagonal matrix B that is similar to A. The characteristic equation of A is

$$\begin{vmatrix} \lambda - 1 & -1 \\ -1 & \lambda \end{vmatrix} = \lambda^2 - \lambda - 1 = 0.$$

The eigenvalues of A are (verify)

$$\lambda_1 = \frac{1 + \sqrt{5}}{2} \quad \text{and} \quad \lambda_2 = \frac{1 - \sqrt{5}}{2}$$

so that

$$D = \begin{bmatrix} \frac{1+\sqrt{5}}{2} & 0 \\ 0 & \frac{1-\sqrt{5}}{2} \end{bmatrix}.$$

Corresponding eigenvectors are (verify)

$$\mathbf{x}_1 = \begin{bmatrix} \frac{1+\sqrt{5}}{2} \\ 1 \end{bmatrix} \quad \text{and} \quad \mathbf{x}_2 = \begin{bmatrix} \frac{1-\sqrt{5}}{2} \\ 1 \end{bmatrix}.$$

Thus

$$P = \begin{bmatrix} \frac{1+\sqrt{5}}{2} & \frac{1-\sqrt{5}}{2} \\ 1 & 1 \end{bmatrix}, \qquad P^{-1} = \begin{bmatrix} \frac{1}{\sqrt{5}} & -\frac{1-\sqrt{5}}{2\sqrt{5}} \\ -\frac{1}{\sqrt{5}} & \frac{1+\sqrt{5}}{2\sqrt{5}} \end{bmatrix},$$

and

$$A = PDP^{-1}.$$

Hence (verify) for any nonnegative integer k,

$$A^k = PD^kP^{-1}.$$

Since D is diagonal, D^k is easy to calculate; its entries are the diagonal entries of D raised to the kth power. From (3) we have

$$\mathbf{w}_{n-1} = A^{n-1}\mathbf{w}_0 = PD^{n-1}P^{-1}\mathbf{w}_0$$

$$= \begin{bmatrix} \frac{1+\sqrt{5}}{2} & \frac{1-\sqrt{5}}{2} \\ 1 & 1 \end{bmatrix} \begin{bmatrix} \left(\frac{1+\sqrt{5}}{2}\right)^{n-1} & 0 \\ 0 & \left(\frac{1-\sqrt{5}}{2}\right)^{n-1} \end{bmatrix} \begin{bmatrix} \frac{1}{\sqrt{5}} & -\frac{1-\sqrt{5}}{2\sqrt{5}} \\ -\frac{1}{\sqrt{5}} & \frac{1+\sqrt{5}}{2\sqrt{5}} \end{bmatrix} \begin{bmatrix} 1 \\ 1 \end{bmatrix}.$$

This equation gives the formula (verify)

$$u_n = \frac{1}{\sqrt{5}}\left[\left(\frac{1+\sqrt{5}}{2}\right)^{n+1} - \left(\frac{1-\sqrt{5}}{2}\right)^{n+1}\right]$$

for calculating u_n directly.

Using a hand held calculator, we find that for $n = 50$, u_{50} is approximately 20.365 billion.

Further Readings ▶

A Primer for the Fibonacci Numbers. San Jose State University, San Jose, Calif., 1972.

VOROBYOV, N. N., *The Fibonacci Numbers.* Boston: D. C. Heath and Company, 1963.

8.7   EXERCISES

1. Compute the eigenvalues and eigenvectors of

$$A = \begin{bmatrix} 1 & 1 \\ 1 & 0 \end{bmatrix}$$

and verify that they are as given in the text.

2. Verify that if $A = PBP^{-1}$ and k is a positive integer, then $A^k = PB^kP^{-1}$.

3. Using a hand calculator or MATLAB, compute

(a) u_8.   (b) u_{12}.   (c) u_{20}.

4. Consider the rabbit-breeding problem but suppose that each pair of rabbits now produces two pairs of rabbits starting with its second month of life and continuing every month thereafter.

(a) Formulate a recursion relation for the number u_n of rabbits at the beginning of month n.

(b) Develop a formula for calculating u_n directly.

THEORETICAL EXERCISE ■

T.1. Let $A = \begin{bmatrix} 1 & 1 \\ 1 & 0 \end{bmatrix}$. It can then be shown that

$$A^{n+1} = \begin{bmatrix} u_{n+1} & u_n \\ u_n & u_{n-1} \end{bmatrix}.$$

Use this result to obtain the formula

$$u_{n+1}u_{n-1} - u_n^2 = (-1)^{n+1}.$$

8.8 ▼ Quadratic Forms

Prerequisite. Section 5.2, Diagonalization of Symmetric Matrices.

In your precalculus and calculus courses you have seen that the graph of the equation

$$ax^2 + 2bxy + cy^2 = d, \tag{1}$$

where a, b, c, and d are real numbers, is a **conic section** centered at the origin of a rectangular Cartesian coordinate system in two-dimensional space. Similarly, the graph of the equation

$$ax^2 + 2dxy + 2exz + by^2 + 2fyz + cz^2 = g, \tag{2}$$

where a, b, c, d, e, f, and g are real numbers, is a **quadric surface** centered at the origin of a rectangular Cartesian coordinate system in three-dimensional space. If a conic section or quadric surface is not centered at the origin, its equations are more complicated than those given in (1) and (2).

The identification of the conic section or quadric surface that is the graph of a given equation often requires the rotation and translation of the coordinate axes. These methods can best be understood as an application of eigenvalues and eigenvectors of matrices and will be discussed in Sections 8.9 and 8.10.

The expressions on the left sides of Equations (1) and (2) are examples of quadratic forms. Quadratic forms arise in statistics, mechanics, and in other problems in physics; in quadratic programming; in the study of maxima and minima of functions of several variables; and in other applied problems. In this section we use our results on eigenvalues and eigenvectors of matrices to give a brief treatment of real quadratic forms in n variables. In Section 8.9 we apply these results to the classification of the conic sections, and in Section 8.10 to the classification of the quadric surfaces.

DEFINITION   If A is a symmetric matrix, then the function $g\colon R^n \to R^1$ (a real-valued function on R^n) defined by

$$g(\mathbf{x}) = \mathbf{x}^T A\mathbf{x},$$

where

$$\mathbf{x} = \begin{bmatrix} x_1 \\ x_2 \\ \vdots \\ x_n \end{bmatrix},$$

is called a **real quadratic form in the variables** $x_1, x_2, \ldots, x_n$. The matrix A is called the **matrix of the quadratic form** g. We shall also denote the quadratic form by $g(\mathbf{x})$.

EXAMPLE 1 ■ Write the left side of (1) as the quadratic form in the variables x and y.

Solution   Let

$$\mathbf{x} = \begin{bmatrix} x \\ y \end{bmatrix} \quad \text{and} \quad A = \begin{bmatrix} a & b \\ b & c \end{bmatrix}.$$

Then the left side of (1) is the quadratic form

$$g(\mathbf{x}) = \mathbf{x}^T A\mathbf{x}.$$ ■

EXAMPLE 2 ■ Write the left side of (2) as the quadratic form.

Solution   Let

$$\mathbf{x} = \begin{bmatrix} x \\ y \\ z \end{bmatrix} \quad \text{and} \quad A = \begin{bmatrix} a & d & e \\ d & b & f \\ e & f & c \end{bmatrix}.$$

Then the left side of (2) is the quadratic form

$$g(\mathbf{x}) = \mathbf{x}^T A\mathbf{x}.$$ ■

EXAMPLE 3 ■ The following expressions are quadratic forms:

(a) $3x^2 - 5xy - 7y^2 = \begin{bmatrix} x & y \end{bmatrix} \begin{bmatrix} 3 & -\frac{5}{2} \\ -\frac{5}{2} & -7 \end{bmatrix} \begin{bmatrix} x \\ y \end{bmatrix}.$

(b) $3x^2 - 7xy + 5xz + 4y^2 - 4yz - 3z^2 = \begin{bmatrix} x & y & z \end{bmatrix} \begin{bmatrix} 3 & -\frac{7}{2} & \frac{5}{2} \\ -\frac{7}{2} & 4 & -2 \\ \frac{5}{2} & -2 & -3 \end{bmatrix} \begin{bmatrix} x \\ y \\ z \end{bmatrix}.$ ■

Suppose now that $g(\mathbf{x}) = \mathbf{x}^T A \mathbf{x}$ is a quadratic form. To simplify the quadratic form, we change from the variables $x_1, x_2, \ldots, x_n$ to the variables $y_1, y_2, \ldots, y_n$, where we assume that the old variables are related to the new variables by $\mathbf{x} = P\mathbf{y}$ for some orthogonal matrix P. Then

$$g(\mathbf{x}) = \mathbf{x}^T A \mathbf{x} = (P\mathbf{y})^T A (P\mathbf{y}) = \mathbf{y}^T (P^T A P) \mathbf{y} = \mathbf{y}^T B \mathbf{y},$$

where $B = P^T A P$. We shall let you verify that if A is a symmetric matrix, then $P^T A P$ is also symmetric (Exercise T.1). Thus

$$h(\mathbf{y}) = \mathbf{y}^T B \mathbf{y}$$

is another quadratic form and $g(\mathbf{x}) = h(\mathbf{y})$.

This situation is important enough to formulate the following definitions.

DEFINITION   If A and B are $n \times n$ matrices, we say that B is **congruent** to A if $B = P^T A P$ for a nonsingular matrix P.

In light of Exercise T.2 we do not distinguish between the statements "A is congruent to B," and "B is congruent to A." Each of these statements can be replaced by "A and B are congruent."

DEFINITION   Two quadratic forms g and h with matrices A and B, respectively, are said to be **equivalent** if A and B are congruent.

The congruence of matrices and equivalence of forms are more general concepts than similarity of symmetric matrices by an orthogonal matrix P, since P is required only to be nonsingular. We shall consider here the more restrictive situation with P orthogonal.

EXAMPLE 4 ■   Consider the quadratic form in the variables x and y defined by

$$g(\mathbf{x}) = 2x^2 + 2xy + 2y^2 = \begin{bmatrix} x & y \end{bmatrix} \begin{bmatrix} 2 & 1 \\ 1 & 2 \end{bmatrix} \begin{bmatrix} x \\ y \end{bmatrix}. \tag{3}$$

We now change from the variables x and y to the variables x' and y'. Suppose that the old variables are related to the new variables by the equations

$$x = \frac{1}{\sqrt{2}} x' - \frac{1}{\sqrt{2}} y' \quad \text{and} \quad y = \frac{1}{\sqrt{2}} x' + \frac{1}{\sqrt{2}} y', \tag{4}$$

which can be written in matrix form as

$$\mathbf{x} = \begin{bmatrix} x \\ y \end{bmatrix} = \begin{bmatrix} \frac{1}{\sqrt{2}} & -\frac{1}{\sqrt{2}} \\ \frac{1}{\sqrt{2}} & \frac{1}{\sqrt{2}} \end{bmatrix} \begin{bmatrix} x' \\ y' \end{bmatrix} = P\mathbf{y},$$

where the orthogonal (hence nonsingular) matrix

$$P = \begin{bmatrix} \frac{1}{\sqrt{2}} & -\frac{1}{\sqrt{2}} \\ \frac{1}{\sqrt{2}} & \frac{1}{\sqrt{2}} \end{bmatrix} \quad \text{and} \quad \mathbf{y} = \begin{bmatrix} x' \\ y' \end{bmatrix}.$$

We shall soon see why and how this particular matrix P was selected. Substituting in (3), we obtain

$$g(\mathbf{x}) = \mathbf{x}^T A \mathbf{x} = (P\mathbf{y})^T A (P\mathbf{y}) = \mathbf{y}^T P^T A P \mathbf{y}$$

$$= \begin{bmatrix} x' & y' \end{bmatrix} \begin{bmatrix} \frac{1}{\sqrt{2}} & -\frac{1}{\sqrt{2}} \\ \frac{1}{\sqrt{2}} & \frac{1}{\sqrt{2}} \end{bmatrix}^T \begin{bmatrix} 2 & 1 \\ 1 & 2 \end{bmatrix} \begin{bmatrix} \frac{1}{\sqrt{2}} & -\frac{1}{\sqrt{2}} \\ \frac{1}{\sqrt{2}} & \frac{1}{\sqrt{2}} \end{bmatrix} \begin{bmatrix} x' \\ y' \end{bmatrix}$$

$$= \begin{bmatrix} x' & y' \end{bmatrix} \begin{bmatrix} 3 & 0 \\ 0 & 1 \end{bmatrix} \begin{bmatrix} x' \\ y' \end{bmatrix} = h(\mathbf{y})$$

$$= 3x'^2 + y'^2$$

Thus the matrices

$$\begin{bmatrix} 2 & 1 \\ 1 & 2 \end{bmatrix} \quad \text{and} \quad \begin{bmatrix} 3 & 0 \\ 0 & 1 \end{bmatrix}$$

are congruent and the quadratic forms g and h are equivalent.   ■

We now turn to the question of how to select the matrix P.

THEOREM 8.9 ■ *Any quadratic form in n variables $g(\mathbf{x}) = \mathbf{x}^T A \mathbf{x}$ is equivalent by means of an*
(*Principal Axes*   *orthogonal matrix P to a quadratic form, $h(\mathbf{y}) = \lambda_1 y_1^2 + \lambda_2 y_2^2 + \cdots + \lambda_n y_n^2$, where*
Theorem)

$$\mathbf{y} = \begin{bmatrix} y_1 \\ y_2 \\ \vdots \\ y_n \end{bmatrix}$$

and $\lambda_1, \lambda_2, \ldots, \lambda_n$ are the eigenvalues of the matrix A of g.

Proof   If A is the matrix of g, then, since A is symmetric, we know, by Theorem 5.8 in Section 5.2, that A can be diagonalized by an orthogonal matrix. This means that there exists an orthogonal matrix P such that $D = P^{-1} A P$ is a diagonal matrix. Since P is orthogonal, $P^{-1} = P^T$, so $D = P^T A P$. Moreover, the elements on the main diagonal of D are the eigenvalues, $\lambda_1, \lambda_2, \ldots, \lambda_n$ of A, which are real numbers. The quadratic form h with matrix D is given by

$$h(\mathbf{y}) = \lambda_1 y_1^2 + \lambda_2 y_2^2 + \cdots + \lambda_n y_n^2;$$

g and h are equivalent.   ■

EXAMPLE 5 ■ Consider the quadratic form g in the variables x, y, and z defined by

$$g(\mathbf{x}) = 2x^2 + 4y^2 + 6yz - 4z^2.$$

Determine a quadratic form h of the form in Theorem 8.9 to which g is equivalent.

Solution The matrix of g is

$$A = \begin{bmatrix} 2 & 0 & 0 \\ 0 & 4 & 3 \\ 0 & 3 & -4 \end{bmatrix},$$

and the eigenvalues of A are

$$\lambda_1 = 2, \quad \lambda_2 = 5, \quad \text{and} \quad \lambda_3 = -5 \quad \text{(verify)}.$$

Let h be the quadratic form in the variables x', y', and z' defined by

$$h(\mathbf{y}) = 2x'^2 + 5y'^2 - 5z'^2.$$

Then g and h are equivalent by means of some orthogonal matrix. Note that $\widehat{h}(\mathbf{y}) = -5x'^2 + 2y'^2 + 5z'^2$ is also equivalent to g. ∎

Observe that to apply Theorem 8.9 to diagonalize a given quadratic form, as shown in Example 5, we do not need to know the eigenvectors of A (nor the matrix P); we only require the eigenvalues of A.

To understand the significance of Theorem 8.9, we consider quadratic forms in two and three variables. As we have already observed at the beginning of this section, the graph of the equation

$$g(\mathbf{x}) = \mathbf{x}^T A \mathbf{x} = 1,$$

where $\mathbf{x}$ is a vector in R^2 and A is a symmetric 2×2 matrix, is a conic section centered at the origin of the xy-plane. From Theorem 8.9 it follows that there is a Cartesian coordinate system in the xy-plane with respect to which the equation of this conic section is

$$ax'^2 + by'^2 = 1,$$

where a and b are real numbers. Similarly, the graph of the equation

$$g(\mathbf{x}) = \mathbf{x}^T A \mathbf{x} = 1,$$

where $\mathbf{x}$ is a vector in R^3 and A is a symmetric 3×3 matrix, is a quadric surface centered at the origin of the xyz Cartesian coordinate system. From Theorem 8.9 it follows that there is a Cartesian coordinate system in 3-space with respect to which the equation of the quadric surface is

$$ax'^2 + by'^2 + cz'^2 = 1,$$

where a, b, and c are real numbers. The principal axes of the conic or surface lie along the new coordinate axes, and this is the reason for calling Theorem 8.9 the **principal axes theorem**.

EXAMPLE 6 ■ Consider the conic section whose equation is

$$g(\mathbf{x}) = 2x^2 + 2xy + 2y^2 = 9.$$

From Example 4 it follows that this conic section can also be described by the equation

$$h(\mathbf{y}) = 3x'^2 + y'^2 = 9,$$

which can be rewritten as

$$\frac{x'^2}{3} + \frac{y'^2}{9} = 1.$$

The graph of this equation is an ellipse (Figure 8.26) whose axis is along the y'-axis. The major axis is of length 6; the minor axis is of length $2\sqrt{3}$. We now note that there is a very close connection between the eigenvectors of the matrix of (3) and the location of the x'- and y'-axes.

FIGURE 8.26

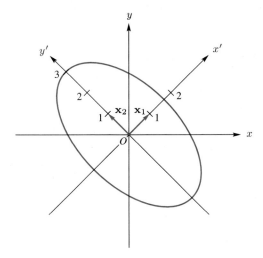

Since $\mathbf{x} = P\mathbf{y}$, we have $\mathbf{y} = P^{-1}\mathbf{x} = P^T\mathbf{x} = P\mathbf{x}$ (P is orthogonal and, in this example, also symmetric). Thus

$$x' = \frac{1}{\sqrt{2}}x + \frac{1}{\sqrt{2}}y \quad \text{and} \quad y' = -\frac{1}{\sqrt{2}}x + \frac{1}{\sqrt{2}}y.$$

This means that, in terms of the x- and y-axes, the x'-axis lies along the vector

$$\mathbf{x}_1 = \begin{bmatrix} \frac{1}{\sqrt{2}} \\ \frac{1}{\sqrt{2}} \end{bmatrix}$$

and the y'-axis lies along the vector

$$\mathbf{x}_2 = \begin{bmatrix} -\frac{1}{\sqrt{2}} \\ \frac{1}{\sqrt{2}} \end{bmatrix}.$$

Now $\mathbf{x}_1$ and $\mathbf{x}_2$ are the columns of the matrix

$$P = \begin{bmatrix} \frac{1}{\sqrt{2}} & -\frac{1}{\sqrt{2}} \\ \frac{1}{\sqrt{2}} & \frac{1}{\sqrt{2}} \end{bmatrix},$$

which in turn are eigenvectors of the matrix of (3). Thus the x'- and y'-axes lie along the eigenvectors of the matrix of (3) (see Figure 8.26).   ∎

The situation described in Example 6 is true in general. That is, the principal axes of a conic section or quadric surface lie along the eigenvectors of the matrix of the quadratic form.

Let $g(\mathbf{x}) = \mathbf{x}^T A \mathbf{x}$ be a quadratic form in n variables. Then we know that g is equivalent to the quadratic form

$$h(\mathbf{y}) = \lambda_1 y_1^2 + \lambda_2 y_2^2 + \cdots + \lambda_n y_n^2,$$

where $\lambda_1, \lambda_2, \ldots, \lambda_n$ are eigenvalues of the symmetric matrix A of g, and hence are all real. We can label the eigenvalues so that all the positive eigenvalues of A, if any, are listed first, followed by all the negative eigenvalues, if any, followed by the zero eigenvalues, if any. Thus let $\lambda_1, \lambda_2, \ldots, \lambda_p$ be positive, $\lambda_{p+1}, \lambda_{p+2}, \ldots, \lambda_r$ be negative, and $\lambda_{r+1}, \lambda_{r+2}, \ldots, \lambda_n$ be zero. We now define the diagonal matrix H whose entries on the main diagonal are

$$\frac{1}{\sqrt{\lambda_1}}, \frac{1}{\sqrt{\lambda_2}}, \ldots, \frac{1}{\sqrt{\lambda_p}}, \frac{1}{\sqrt{-\lambda_{p+1}}}, \frac{1}{\sqrt{-\lambda_{p+2}}}, \ldots, \frac{1}{\sqrt{-\lambda_r}}, 1, 1, \ldots, 1,$$

with $n - r$ ones. Let D be the diagonal matrix whose entries on the main diagonal are $\lambda_1, \lambda_2, \ldots, \lambda_p, \lambda_{p+1}, \ldots, \lambda_r, \lambda_{r+1}, \ldots, \lambda_n$; A and D are congruent. Let $D_1 = H^T D H$ be the matrix whose diagonal elements are $1, 1, \ldots, 1, -1, \ldots, -1$, $0, 0, \ldots, 0$ (p ones, $r - p$ negative ones, and $n - r$ zeros); D and D_1 are then congruent. From Exercise T.2 it follows that A and D_1 are congruent. In terms of quadratic forms, we have established Theorem 8.10.

THEOREM 8.10 ■ *A quadratic form $g(\mathbf{x}) = \mathbf{x}^T A \mathbf{x}$ in n variables is equivalent to a quadratic form*

$$h(\mathbf{y}) = y_1^2 + y_2^2 + \cdots + y_p^2 - y_{p+1}^2 - y_{p+2}^2 - \cdots - y_r^2.$$ ■

It is clear that the rank of the matrix D_1 is r, the number of nonzero entries on its main diagonal. Now it can be shown that congruent matrices have equal ranks. Since the rank of D_1 is r, the rank of A is also r. We also refer to r as the **rank** of the quadratic form g whose matrix is A. It can be shown that the number p of positive terms in the quadratic form h of Theorem 8.10 is unique; that is, no matter how we simplify the given quadratic form g to obtain an equivalent quadratic form, the latter will always have p positive terms. Hence the quadratic form h in Theorem 8.10 is unique; it is often called the **canonical form** of a quadratic form in n variables. The difference between the number of positive eigenvalues and the number of negative eigenvalues is $s = p - (r - p) = 2p - r$ and is called the **signature** of the quadratic form. Thus, if g and h are equivalent quadratic forms, then they have equal ranks and signatures. However, it can also be shown that if g and h have equal ranks and signatures, then they are equivalent.

EXAMPLE 7 ■ Consider the quadratic form in x_1, x_2, x_3, given by

$$g(\mathbf{x}) = 3x_2^2 + 8x_2 x_3 - 3x_3^2 = \mathbf{x}^T A \mathbf{x} = \begin{bmatrix} x_1 & x_2 & x_3 \end{bmatrix} \begin{bmatrix} 0 & 0 & 0 \\ 0 & 3 & 4 \\ 0 & 4 & -3 \end{bmatrix} \begin{bmatrix} x_1 \\ x_2 \\ x_3 \end{bmatrix}.$$

The eigenvalues of A are (verify)

$$\lambda_1 = 5, \quad \lambda_2 = -5, \quad \text{and} \quad \lambda_3 = 0.$$

In this case A is congruent to

$$D = \begin{bmatrix} 5 & 0 & 0 \\ 0 & -5 & 0 \\ 0 & 0 & 0 \end{bmatrix}.$$

If we let

$$H = \begin{bmatrix} \frac{1}{\sqrt{5}} & 0 & 0 \\ 0 & \frac{1}{\sqrt{5}} & 0 \\ 0 & 0 & 1 \end{bmatrix},$$

then

$$D_1 = H^T D H = \begin{bmatrix} 1 & 0 & 0 \\ 0 & -1 & 0 \\ 0 & 0 & 0 \end{bmatrix}$$

and A are congruent, and the given quadratic form is equivalent to the canonical form

$$h(\mathbf{y}) = y_1^2 - y_2^2.$$

The rank of g is 2, and since $p = 1$, the signature $s = 2p - r = 0.$ ■

As a final application of quadratic forms we consider positive definite symmetric matrices.

DEFINITION   A symmetric $n \times n$ matrix A is called **positive definite** if $\mathbf{x}^T A \mathbf{x} > 0$ for every nonzero vector $\mathbf{x}$ in R^n.

If A is a symmetric matrix, then $\mathbf{x}^T A \mathbf{x}$ is a quadratic form $g(\mathbf{x}) = \mathbf{x}^T A \mathbf{x}$ and, by Theorem 8.9, g is equivalent to h, where

$$h(\mathbf{y}) = \lambda_1 y_1^2 + \lambda_2 y_2^2 + \cdots + \lambda_p y_p^2 + \lambda_{p+1} y_{p+1}^2 + \lambda_{p+2} y_{p+2}^2 + \cdots + \lambda_r y_r^2.$$

Now A is positive definite if and only if $h(\mathbf{y}) > 0$ for each $\mathbf{y} \neq \mathbf{0}$. However, this can happen if and only if all summands in $h(\mathbf{y})$ are positive and $r = n$. These remarks have established the following theorem.

THEOREM 8.11 ■ *A symmetric matrix A is positive definite if and only if all the eigenvalues of A are positive.*   ■

A quadratic form is then called **positive definite** if its matrix is positive definite.

8.8   EXERCISES

In Exercises 1 and 2, write each quadratic form as $\mathbf{x}^T A\mathbf{x}$, where A is a symmetric matrix.

1. (a) $-3x^2 + 5xy - 2y^2$.
   (b) $2x_1^2 + 3x_1x_2 - 5x_1x_3 + 7x_2x_3$.
   (c) $3x_1^2 + x_2^2 - 2x_3^2 + x_1x_2 - x_1x_3 - 4x_2x_3$.

2. (a) $x_1^2 - 3x_2^2 + 4x_3^2 - 4x_1x_2 + 6x_2x_3$.
   (b) $4x^2 - 6xy - 2y^2$.
   (c) $-2x_1x_2 + 4x_1x_3 + 6x_2x_3$.

In Exercises 3 and 4, for each given symmetric matrix A find a diagonal matrix D that is congruent to A.

3. (a) $A = \begin{bmatrix} -1 & 0 & 0 \\ 0 & 1 & 1 \\ 0 & 1 & 1 \end{bmatrix}$.

   (b) $A = \begin{bmatrix} 1 & 1 & 1 \\ 1 & 1 & 1 \\ 1 & 1 & 1 \end{bmatrix}$.

   (c) $A = \begin{bmatrix} 0 & 2 & 2 \\ 2 & 0 & 2 \\ 2 & 2 & 0 \end{bmatrix}$.

4. (a) $A = \begin{bmatrix} 3 & 4 & 0 \\ 4 & -3 & 0 \\ 0 & 0 & 5 \end{bmatrix}$.

   (b) $A = \begin{bmatrix} 2 & 1 & 1 \\ 1 & 2 & 1 \\ 1 & 1 & 2 \end{bmatrix}$.

   (c) $A = \begin{bmatrix} 0 & 0 & 1 \\ 0 & 1 & 0 \\ 1 & 0 & 0 \end{bmatrix}$.

In Exercises 5 through 10, find a quadratic form of the type in Theorem 8.9 that is equivalent to the given quadratic form.

5. $2x^2 - 4xy - y^2$.

6. $x_1^2 + x_2^2 + x_3^2 + 2x_2x_3$.

7. $2x_1x_3$.

8. $2x_2^2 + 2x_3^2 + 4x_2x_3$.

9. $-2x_1^2 - 4x_2^2 + 4x_3^2 - 6x_2x_3$.

10. $6x_1x_2 + 8x_2x_3$.

In Exercises 10 through 16, find a quadratic form of the type in Theorem 8.10 that is equivalent to the given quadratic form.

11. $2x^2 + 4xy + 2y^2$.

12. $x_1^2 + x_2^2 + x_3^2 + 2x_1x_2$.

13. $2x_1^2 + 4x_2^2 + 4x_3^2 + 10x_2x_3$.

14. $2x_1^2 + 3x_2^2 + 3x_3^2 + 4x_2x_3$.

15. $-3x_1^2 + 2x_2^2 + 2x_3^2 + 4x_2x_3$.

16. $-3x_1^2 + 5x_2^2 + 3x_3^2 - 8x_1x_3$.

17. Let $g(\mathbf{x}) = 4x_2^2 + 4x_3^2 - 10x_2x_3$ be a quadratic form in three variables. Find a quadratic form of the type in Theorem 8.10 that is equivalent to g. What is the rank of g? What is the signature of g?

18. Let $g(\mathbf{x}) = 3x_1^2 - 3x_2^2 - 3x_3^2 + 4x_2x_3$ be a quadratic form in three variables. Find a quadratic form of the type in Theorem 8.10 that is equivalent to g. What is the rank of g? What is the signature of g?

19. Find all quadratic forms $g(\mathbf{x}) = \mathbf{x}^T A\mathbf{x}$ in two variables of the type described in Theorem 8.10. What conics do the equations $\mathbf{x}^T A\mathbf{x} = 1$ represent?

20. Find all quadratic forms $g(\mathbf{x}) = \mathbf{x}^T A\mathbf{x}$ in two variables of rank 1 of the type described in Theorem 8.10. What conics do the equations $\mathbf{x}^T A\mathbf{x} = 1$ represent?

In Exercises 21 and 22, which of the given quadratic forms in three variables are equivalent?

21. $g_1(\mathbf{x}) = x_1^2 + x_2^2 + x_3^2 + 2x_1x_2$.
    $g_2(\mathbf{x}) = 2x_2^2 + 2x_3^2 + 2x_2x_3$.
    $g_3(\mathbf{x}) = 3x_2^2 - 3x_3^2 + 8x_2x_3$.
    $g_4(\mathbf{x}) = 3x_2^2 + 3x_3^2 - 4x_2x_3$.

22. $g_1(\mathbf{x}) = x_2^2 + 2x_1x_3$.
    $g_2(\mathbf{x}) = 2x_1^2 + 2x_2^2 + x_3^2 + 2x_1x_2 + 2x_1x_3 + 2x_2x_3$.
    $g_3(\mathbf{x}) = 2x_1x_2 + 2x_1x_3 + 2x_2x_3$.
    $g_4(\mathbf{x}) = 4x_1^2 + 3x_2^2 + 4x_3^2 + 10x_1x_3$.

In Exercises 23 and 24, which of the given matrices are positive definite?

23. (a) $\begin{bmatrix} 2 & -1 \\ -1 & 2 \end{bmatrix}$.   (b) $\begin{bmatrix} 2 & 1 \\ 1 & 2 \end{bmatrix}$.

    (c) $\begin{bmatrix} 3 & 1 & 0 \\ 1 & 3 & 0 \\ 0 & 0 & 3 \end{bmatrix}$.   (d) $\begin{bmatrix} 1 & 0 & 0 \\ 0 & 2 & 0 \\ 0 & 0 & -3 \end{bmatrix}$.

    (e) $\begin{bmatrix} 2 & 2 \\ 2 & 2 \end{bmatrix}$.

24. (a) $\begin{bmatrix} 0 & -1 \\ -1 & 0 \end{bmatrix}$.   (b) $\begin{bmatrix} 1 & 1 \\ 1 & 1 \end{bmatrix}$.

(c) $\begin{bmatrix} 0 & 0 & 0 \\ 0 & 1 & 2 \\ 0 & 2 & 1 \end{bmatrix}$.

(d) $\begin{bmatrix} 7 & 4 & 4 \\ 4 & 7 & 4 \\ 4 & 4 & 7 \end{bmatrix}$.

(e) $\begin{bmatrix} 2 & 0 & 0 & 0 \\ 0 & 1 & 0 & 0 \\ 0 & 0 & 3 & 4 \\ 0 & 0 & 4 & -3 \end{bmatrix}$.

THEORETICAL EXERCISES ▪

T.1. Show that if A is a symmetric matrix, then $P^T A P$ is also symmetric.

T.2. If A, B, and C are $n \times n$ symmetric matrices, show the following.

(a) A and A are congruent.

(b) If A and B are congruent, then B and A are congruent.

(c) If A and B are congruent and if B and C are congruent, then A and C are congruent.

T.3. Show that if A is symmetric, then A is congruent to a diagonal matrix D.

T.4. Let

$$A = \begin{bmatrix} a & b \\ b & d \end{bmatrix}$$

be a 2×2 symmetric matrix. Show that A is positive definite if and only if $\det(A) > 0$ and $a > 0$.

T.5. Show that a symmetric matrix A is positive definite if and only if $A = P^T P$ for a nonsingular matrix P.

MATLAB EXERCISES ▪

In MATLAB, eigenvalues of a matrix A can be determined using the command **eig(A)**. (See the MATLAB Exercises in Section 5.2.) Hence the rank and signature of a quadratic form are easily determined.

ML.1. Determine the rank and signature of the quadratic form with matrix A.

(a) $A = \begin{bmatrix} -1 & 0 & 0 \\ 0 & 1 & 1 \\ 0 & 1 & 1 \end{bmatrix}$.

(b) $A = \begin{bmatrix} 1 & 1 & 1 \\ 1 & 1 & 1 \\ 1 & 1 & 1 \end{bmatrix}$.

(c) $A = \begin{bmatrix} 2 & 1 & 0 & -2 \\ 1 & -1 & 1 & 3 \\ 0 & 1 & 2 & -1 \\ -2 & 3 & -1 & 0 \end{bmatrix}$.

(d) $A = \begin{bmatrix} 2 & -1 & 0 & 0 \\ -1 & 2 & -1 & 0 \\ 0 & -1 & 2 & -1 \\ 0 & 0 & -1 & 2 \end{bmatrix}$.

ML.2. Use **eig** to determine which of the matrices in Exercise ML.1 are positive definite.

8.9 ▾ Conic Sections

Prerequisite. Section 8.8. Quadratic Forms.

In this section we discuss the classification of the conic sections in the plane. A **quadratic equation** in the variables x and y has the form

$$ax^2 + 2bxy + cy^2 + dx + ey + f = 0, \tag{1}$$

where a, b, c, d, e, and f are real numbers. The graph of Equation (1) is a **conic section**, a curve so named because it is obtained by intersecting a plane with a right circular cone that has two nappes. In Figure 8.27 we show that a plane cuts the cone in a circle, ellipse, parabola, or hyperbola. Degenerate cases of the conic sections are a point, a line, a pair of lines, or the empty set.

FIGURE 8.27
The nondegenerate
conic sections

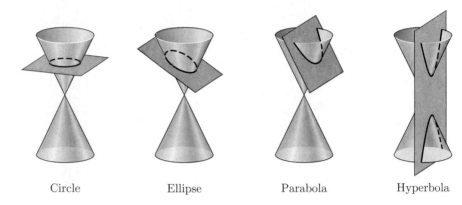

Circle   Ellipse   Parabola   Hyperbola

The nondegenerate conics are said to be in **standard position** if their graphs and equations are as given in Figure 8.28. The equation is said to be in **standard form**.

EXAMPLE 1 ■ Identify the graph of the given equation.

(a) $4x^2 + 25y^2 - 100 = 0$.

(b) $9y^2 - 4x^2 = -36$.

(c) $x^2 + 4y = 0$.

(d) $y^2 = 0$.

(e) $x^2 + 9y^2 + 9 = 0$.

(f) $x^2 + y^2 = 0$.

Solution   (a) We rewrite the given equation as

$$\frac{4}{100}x^2 + \frac{25}{100}y^2 = \frac{100}{100}$$

or

$$\frac{x^2}{25} + \frac{y^2}{4} = 1,$$

whose graph is an ellipse in standard position with $a = 5$ and $b = 2$. Thus the x-intercepts are $(5, 0)$ and $(-5, 0)$ and the y-intercepts are $(0, 2)$ and $(0, -2)$.

(b) Rewriting the given equation as

$$\frac{x^2}{9} - \frac{y^2}{4} = 1,$$

we see that its graph is a hyperbola in standard position with $a = 3$ and $b = 2$. The x-intercepts are $(3, 0)$ and $(-3, 0)$.

FIGURE 8.28
The conic sections
in standard position

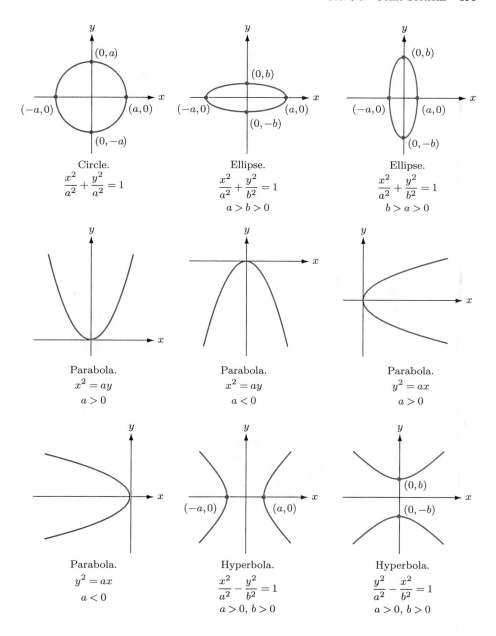

Circle.
$$\frac{x^2}{a^2} + \frac{y^2}{a^2} = 1$$

Ellipse.
$$\frac{x^2}{a^2} + \frac{y^2}{b^2} = 1$$
$$a > b > 0$$

Ellipse.
$$\frac{x^2}{a^2} + \frac{y^2}{b^2} = 1$$
$$b > a > 0$$

Parabola.
$$x^2 = ay$$
$$a > 0$$

Parabola.
$$x^2 = ay$$
$$a < 0$$

Parabola.
$$y^2 = ax$$
$$a > 0$$

Parabola.
$$y^2 = ax$$
$$a < 0$$

Hyperbola.
$$\frac{x^2}{a^2} - \frac{y^2}{b^2} = 1$$
$$a > 0,\ b > 0$$

Hyperbola.
$$\frac{y^2}{a^2} - \frac{x^2}{b^2} = 1$$
$$a > 0,\ b > 0$$

(c) Rewriting the given equation as

$$x^2 = -4y,$$

we see that its graph is a parabola in standard position with $a = -4$, so it opens downward.

(d) Every point satisfying the given equation must have a y-coordinate equal to zero. Thus the graph of this equation consists of all the points on the x-axis.

(e) Rewriting the given equation as

$$x^2 + 9y^2 = -9,$$

we conclude that there are no points in the plane whose coordinates satisfy the given equation.

(f) The only point satisfying the equation is the origin $(0,0)$, so the graph of this equation is the single point consisting of the origin. ■

We next turn to the study of conic sections whose graphs are not in standard position. First, notice that the equations of the conic sections whose graphs are in standard position do not contain an xy-term (called a **cross-product term**). If a cross-product term appears in the equation, the graph is a conic section that has been rotated from its standard position [see Figure 8.29(a)]. Also notice that none of the equations in Figure 8.29(b) contain an x^2-term and an x-term or a y^2-term and a y-term. If either of these cases occurs and there is no xy-term in the equation, the graph is a conic section that has been translated from its standard position [see Figure 8.29(b)]. On the other hand, if an xy-term is present, the graph is a conic section that has been rotated and possibly also translated [see Figure 8.29(c)].

FIGURE 8.29

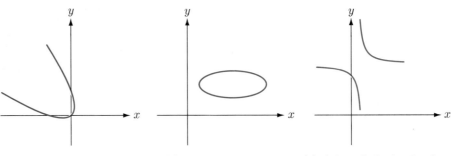

(a) A parabola that has been rotated.

(b) An ellipse that has been translated.

(c) A hyperbola that has been rotated and translated.

To identify a nondegenerate conic section whose graph is not in standard position, we proceed as follows:

Step 1. If a cross-product term is present in the given equation, rotate the xy-coordinate axes by means of an orthogonal linear transformation so that in the resulting equation the xy-term no longer appears.

Step 2. If an xy-term is not present in the given equation, but an x^2-term and an x-term, or a y^2-term and a y-term appear, translate the xy-coordinate axes by completing the square so that the graph of the resulting equation will be in standard position with respect to the origin of the new coordinate system.

Thus, if an xy-term appears in a given equation, we first rotate the xy-coordinate axes and then, if necessary, translate the rotated axes. In the next example, we deal with the case requiring only a translation of axes.

EXAMPLE 2 ■ Identify and sketch the graph of the equation

$$x^2 - 4y^2 + 6x + 16y - 23 = 0. \tag{2}$$

Also write its equation in standard form.

Solution   Since there is no cross-product term, we only need to translate axes. Completing the squares in the x- and y-terms, we have

$$x^2 + 6x + 9 - 4(y^2 - 4y + 4) - 23 = 9 - 16$$
$$(x+3)^2 - 4(y-2)^2 = 23 + 9 - 16 = 16. \tag{3}$$

Letting

$$x' = x + 3 \quad \text{and} \quad y' = y - 2,$$

we can rewrite Equation (3) as

$$x'^2 - 4y'^2 = 16$$

or in standard form as

$$\frac{x'^2}{16} - \frac{y'^2}{4} = 1. \tag{4}$$

If we translate the xy-coordinate system to the $x'y'$-coordinate system, whose origin is at $(-3, 2)$, then the graph of Equation (4) is a hyperbola in standard position with respect to the $x'y'$-coordinate system (see Figure 8.30). ■

FIGURE 8.30

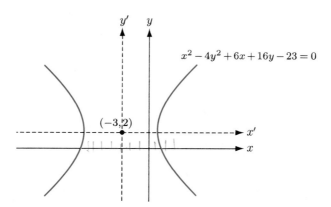

We now turn to the problem of identifying the graph of Equation (1), where we assume that $b \neq 0$, that is, a cross-product term is present. This equation can be written in matrix form as

$$\mathbf{x}^T A \mathbf{x} + B \mathbf{x} + f = 0, \tag{5}$$

where

$$\mathbf{x} = \begin{bmatrix} x \\ y \end{bmatrix}, \quad A = \begin{bmatrix} a & b \\ b & c \end{bmatrix}, \quad \text{and} \quad B = \begin{bmatrix} d & e \end{bmatrix}.$$

Since A is a symmetric matrix, we know from Section 5.2 that it can be diagonalized by an orthogonal matrix P. Thus

$$P^T A P = \begin{bmatrix} \lambda_1 & 0 \\ 0 & \lambda_2 \end{bmatrix},$$

where λ_1 and λ_2 are the eigenvalues of A and the columns of P are $\mathbf{x}_1$ and $\mathbf{x}_2$, orthonormal eigenvectors of A associated with λ_1 and λ_2, respectively.

Letting

$$\mathbf{x} = P\mathbf{y}, \quad \text{where} \quad \mathbf{y} = \begin{bmatrix} x' \\ y' \end{bmatrix},$$

we can rewrite Equation (5) as

$$(P\mathbf{y})^T A (P\mathbf{y}) + B(P\mathbf{y}) + f = 0$$
$$\mathbf{y}^T (P^T A P)\mathbf{y} + BP\mathbf{y} + f = 0$$

or

$$\begin{bmatrix} x' & y' \end{bmatrix} \begin{bmatrix} \lambda_1 & 0 \\ 0 & \lambda_2 \end{bmatrix} \begin{bmatrix} x' \\ y' \end{bmatrix} + B(P\mathbf{y}) + f = 0 \tag{6}$$

or

$$\lambda_1 x'^2 + \lambda_2 y'^2 + d'x' + e'y' + f = 0. \tag{7}$$

Equation (7) is the resulting equation for the given conic section and it has no cross-product term.

As discussed in Section 8.8, the x' and y' coordinate axes lie along the eigenvectors $\mathbf{x}_1$ and $\mathbf{x}_2$, respectively. Since P is an orthogonal matrix, $\det(P) = \pm 1$ and, if necessary, we can interchange the columns of P (the eigenvectors $\mathbf{x}_1$ and $\mathbf{x}_2$ of A) or multiply a column of P by -1, so that $\det(P) = 1$. As noted in Section 8.8, it then follows that P is the matrix of a counterclockwise rotation of R^2 through an angle θ that can be determined as follows. First, it is not difficult to show that if $b \neq 0$, then $x_{11} \neq 0$. Since θ is the angle between the directed line segment from the origin to the point (x_{11}, x_{21}), it follows that

$$\theta = \tan^{-1} \left(\frac{x_{21}}{x_{11}} \right).$$

EXAMPLE 3 ■ Identify and sketch the graph of the equation

$$5x^2 - 6xy + 5y^2 - 24\sqrt{2}\, x + 8\sqrt{2}\, y + 56 = 0. \tag{8}$$

Write the equation in standard form.

Solution   Rewriting the given equation in matrix form, we obtain

$$\begin{bmatrix} x & y \end{bmatrix} \begin{bmatrix} 5 & -3 \\ -3 & 5 \end{bmatrix} \begin{bmatrix} x \\ y \end{bmatrix} + \begin{bmatrix} -24\sqrt{2} & 8\sqrt{2} \end{bmatrix} \begin{bmatrix} x \\ y \end{bmatrix} + 56 = 0.$$

We now find the eigenvalues of the matrix

$$A = \begin{bmatrix} 5 & -3 \\ -3 & 5 \end{bmatrix}.$$

Thus

$$|\lambda I_2 - A| = \begin{vmatrix} \lambda - 5 & 3 \\ 3 & \lambda - 5 \end{vmatrix}$$
$$= (\lambda - 5)(\lambda - 5) - 9 = \lambda^2 - 10\lambda + 16$$
$$= (\lambda - 2)(\lambda - 8),$$

so the eigenvalues of A are

$$\lambda_1 = 2, \quad \lambda_2 = 8.$$

Associated eigenvectors are obtained by solving the homogeneous system

$$(\lambda I_2 - A)\mathbf{x} = \mathbf{0}.$$

Thus, for $\lambda_1 = 2$, we have

$$\begin{bmatrix} -3 & 3 \\ 3 & -3 \end{bmatrix} \mathbf{x} = \mathbf{0},$$

so an eigenvalue of A associated with $\lambda_1 = 2$ is

$$\begin{bmatrix} 1 \\ 1 \end{bmatrix}.$$

For $\lambda_2 = 8$ we have

$$\begin{bmatrix} 3 & 3 \\ 3 & 3 \end{bmatrix} \mathbf{x} = \mathbf{0},$$

so an eigenvector of A associated with $\lambda_2 = 8$ is

$$\begin{bmatrix} -1 \\ 1 \end{bmatrix}.$$

Normalizing these eigenvectors, we obtain the orthogonal matrix

$$P = \begin{bmatrix} \frac{1}{\sqrt{2}} & -\frac{1}{\sqrt{2}} \\ \frac{1}{\sqrt{2}} & \frac{1}{\sqrt{2}} \end{bmatrix}.$$

Then

$$P^T A P = \begin{bmatrix} 2 & 0 \\ 0 & 8 \end{bmatrix}.$$

Letting $\mathbf{x} = P\mathbf{y}$, we write the transformed equation for the given conic section, Equation (7), as

$$2x'^2 + 8y'^2 - 16x' + 32y' + 56 = 0$$

or

$$x'^2 + 4y'^2 - 8x' + 16y' + 28 = 0.$$

To identify the graph of this equation, we need to translate axes, so we complete the squares, obtaining

$$(x' - 4)^2 + 4(y' + 2)^2 + 28 = 16 + 16$$

$$(x' - 4)^2 + 4(y' + 2)^2 = 4$$

$$\frac{(x' - 4)^2}{4} + \frac{(y' + 2)^2}{1} = 1. \tag{9}$$

Letting

$$x'' = x' - 4 \quad \text{and} \quad y'' = y' + 2,$$

we find that Equation (9) becomes

$$\frac{x''^2}{4} + \frac{y''^2}{1} = 1, \tag{10}$$

whose graph is an ellipse in standard position with respect to the $x''y''$-coordinate axes, as shown in Figure 8.31, where the origin of the $x''y''$-coordinate system is at $(4, -2)$. Equation (10) is the standard form of the equation of the ellipse. Since

$$\mathbf{x}_1 = \begin{bmatrix} \frac{1}{\sqrt{2}} \\ \frac{1}{\sqrt{2}} \end{bmatrix},$$

the xy-coordinate axes have been rotated through the angle θ, where

$$\theta = \tan^{-1}\left(\frac{\frac{1}{\sqrt{2}}}{\frac{1}{\sqrt{2}}} \right) = \tan^{-1} 1,$$

so $\theta = 45°$. ■

FIGURE 8.31

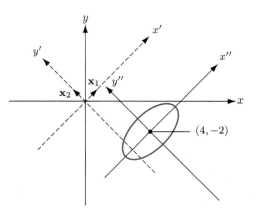

$$5x^2 - 6xy + 5y^2 - 24\sqrt{2}\,x + 8\sqrt{2}\,y + 56 = 0$$

The graph of a given quadratic equation in x and y can be identified from the equation that is obtained after rotating axes, that is, from Equation (6) or (7). The identification of the conic section given by these equations is shown in Table 8.2.

Table 8.2: Identification of the Conic Sections

λ_1, λ_2 both nonzero		Exactly one of λ_1, λ_2 is zero
$\lambda_1\lambda_2 > 0$	$\lambda_1\lambda_2 < 0$	
Ellipse	Hyperbola	Parabola

8.9   EXERCISES

In Exercises 1 through 10, identify the graph of the equation.

1. $x^2 + 9y^2 - 9 = 0$.

2. $x^2 = 2y$.

3. $25y^2 - 4x^2 = 100$.

4. $y^2 - 16 = 0$.

5. $3x^2 - y^2 = 0$.

6. $y = 0$.

7. $4x^2 + 4y^2 - 9 = 0$.

8. $-25x^2 + 9y^2 + 225 = 0$.

9. $4x^2 + y^2 = 0$.

10. $9x^2 + 4y^2 + 36 = 0$.

In Exercises 11 through 18, translate axes to identify the graph of the equation and write the equation in standard form.

11. $x^2 + 2y^2 - 4x - 4y + 4 = 0$.

12. $x^2 - y^2 + 4x - 6y - 9 = 0$.

13. $x^2 + y^2 - 8x - 6y = 0$.

14. $x^2 - 4x + 4y + 4 = 0$.

15. $y^2 - 4y = 0$.

16. $4x^2 + 5y^2 - 30y + 25 = 0$.

17. $x^2 + y^2 - 2x - 6y + 10 = 0$.

18. $2x^2 + y^2 - 12x - 4y + 24 = 0$.

In Exercises 19 through 24, rotate axes to identify the graph of the equation and write the equation in standard form.

19. $x^2 + xy + y^2 = 6$.

20. $xy = 1$.

21. $9x^2 + y^2 + 6xy = 4$.

22. $x^2 + y^2 + 4xy = 9$.

23. $4x^2 + 4y^2 - 10xy = 0$.

24. $9x^2 + 6y^2 + 4xy - 5 = 0$.

In Exercises 25 through 30, identify the graph of the equation and write the equation in standard form.

25. $9x^2 + y^2 + 6xy - 10\sqrt{10}\,x + 10\sqrt{10}\,y + 90 = 0$.

26. $5x^2 + 5y^2 - 6xy - 30\sqrt{2}\,x + 18\sqrt{2}\,y + 82 = 0$.

27. $5x^2 + 12xy - 12\sqrt{13}\,x = 36$.

28. $6x^2 + 9y^2 - 4xy - 4\sqrt{5}\,x - 18\sqrt{5}\,y = 5$.

29. $x^2 - y^2 + 2\sqrt{3}\,xy + 6x = 0$.

30. $8x^2 + 8y^2 - 16xy + 33\sqrt{2}\,x - 31\sqrt{2}\,y + 70 = 0$.

8.10 ▼ Quadric Surfaces

Prerequisite. Section 8.9, Conic Sections.

In Section 8.9 conic sections were used to provide geometric models for quadratic forms in two variables. In this section we investigate quadratic forms in three variables and use particular surfaces called quadric surfaces as geometric models. Quadric surfaces are often studied and sketched in analytic geometry and calculus. Here we use Theorems 8.9 and 8.10 to develop a classification scheme for quadric surfaces.

A **second-degree polynomial equation** in three variables x, y, and z has the form

$$ax^2 + by^2 + cz^2 + 2dxy + 2exz + 2fyz + gx + hy + iz = j, \qquad (1)$$

where coefficients a through j are real numbers with $a, b, \ldots, f$ not all zero. Equation (1) can be written in matrix form as

$$\mathbf{x}^T A \mathbf{x} + B \mathbf{x} = j, \tag{2}$$

where

$$A = \begin{bmatrix} a & d & e \\ d & b & f \\ e & f & c \end{bmatrix}, \quad B = \begin{bmatrix} g & h & i \end{bmatrix}, \quad \text{and} \quad \mathbf{x} = \begin{bmatrix} x \\ y \\ z \end{bmatrix}.$$

We call $\mathbf{x}^T A \mathbf{x}$ the **quadratic form (in three variables) associated with the second-degree polynomial** in (1). As in Section 8.8, the symmetric matrix A is called the matrix of the quadratic form.

The graph of (1) in R^3 is called a **quadric surface**. As in the case of the classification of conic sections in Section 8.9, the classification of (1) as to the type of surface represented depends on the matrix A. Using the ideas in Section 8.9, we have the following strategies to determine a simpler equation for a quadric surface.

---

1. If A is not diagonal, then a rotation of axes is used to eliminate any cross-product terms xy, xz, or yz.

2. If $B = \begin{bmatrix} g & h & i \end{bmatrix} \neq \mathbf{0}$, then a translation of axes is used to eliminate any first-degree terms.

The resulting equation will have the standard form

$$\lambda_1 x''^2 + \lambda_2 y''^2 + \lambda_3 z''^2 = k$$

or, in matrix form,

$$\mathbf{y}^T C \mathbf{y} = k, \tag{3}$$

where

$$\mathbf{y} = \begin{bmatrix} x'' \\ y'' \\ z'' \end{bmatrix},$$

k is some real constant, and C is a diagonal matrix with diagonal entries λ_1, λ_2, λ_3, which are the eigenvalues of A.

We now turn to the classification of quadric surfaces.

DEFINITION Let A be an $n \times n$ symmetric matrix. The **inertia** of A, denoted $\text{In}(A)$, is an ordered triple of numbers

$$(\text{pos}, \text{neg}, \text{zer}),$$

where pos, neg, and zer are the number of positive, negative, and zero eigenvalues of A, respectively.

EXAMPLE 1 ■ Find the inertia of each of the following matrices:

$$A_1 = \begin{bmatrix} 2 & 2 \\ 2 & 2 \end{bmatrix}, \quad A_2 = \begin{bmatrix} 2 & 1 \\ 1 & 2 \end{bmatrix}, \quad A_3 = \begin{bmatrix} 0 & 2 & 2 \\ 2 & 0 & 2 \\ 2 & 2 & 0 \end{bmatrix}.$$

Solution   We determine the eigenvalues of each of the matrices. It follows that (verify)

$$\det(\lambda I_2 - A_1) = \lambda(\lambda - 4) = 0, \qquad \text{so } \lambda_1 = 0,\ \lambda_2 = 4,\ \text{and}$$
$$\text{In}(A_1) = (1, 0, 1).$$

$$\det(\lambda I_2 - A_2) = (\lambda - 1)(\lambda - 3) = 0, \qquad \text{so } \lambda_1 = 1,\ \lambda_2 = 3,\ \text{and}$$
$$\text{In}(A_2) = (2, 0, 0).$$

$$\det(\lambda I_3 - A_3) = (\lambda + 2)^2(\lambda - 4) = 0, \qquad \text{so } \lambda_1 = \lambda_2 = -2,\ \lambda_3 = 4,\ \text{and}$$
$$\text{In}(A_3) = (1, 2, 0).$$
■

From Section 8.8 the signature of a quadratic form $\mathbf{x}^T A \mathbf{x}$ is the difference between the number of positive eigenvalues and the number of negative eigenvalues of A. In terms of inertia, the signature of $\mathbf{x}^T A \mathbf{x}$ is $s = \text{pos} - \text{neg}$.

In order to use inertia for classification of quadric surfaces (or conic sections), we assume that the eigenvalues of an $n \times n$ symmetric matrix A of a quadratic form in n variables are denoted by

$$\lambda_1 \geq \cdots \geq \lambda_{\text{pos}} > 0$$

$$\lambda_{\text{pos}+1} \leq \cdots \leq \lambda_{\text{pos}+\text{neg}} < 0$$

$$\lambda_{\text{pos}+\text{neg}+1} = \cdots = \lambda_n = 0.$$

The largest positive eigenvalue is denoted by λ_1 and the smallest one by λ_{pos}. We also assume that $\lambda_1 > 0$ and $j \geq 0$ in (2), which eliminates redundant and impossible cases. For example, if

$$A = \begin{bmatrix} -1 & 0 & 0 \\ 0 & -2 & 0 \\ 0 & 0 & -3 \end{bmatrix}, \quad B = \begin{bmatrix} 0 & 0 & 0 \end{bmatrix}, \quad \text{and} \quad j = 5,$$

then the second-degree polynomial is $-x^2 - 2y^2 - 3z^2 = 5$, which has an empty solution set. That is, the surface represented has no points. However, if $j = -5$, then the second-degree polynomial is $-x^2 - 2y^2 - 3z^2 = -5$, which is identical to $x^2 + 2y^2 + 3z^2 = 5$. The assumptions $\lambda_1 > 0$ and $j \geq 0$ avoid such a redundant representation.

EXAMPLE 2 ■ Consider a quadratic form in two variables with matrix A and assume that $\lambda_1 > 0$ and $f \geq 0$ in Equation (1) of Section 8.9. Then there are only three possible cases for the inertia of A, which we summarize as follows.

1. $\text{In}(A) = (2, 0, 0)$; then the quadratic form represents an ellipse.
2. $\text{In}(A) = (1, 1, 0)$; then the quadratic form represents a hyperbola.
3. $\text{In}(A) = (1, 0, 1)$; then the quadratic form represents a parabola.

This classification is identical to that given in Table 8.2, taking the assumptions into account. ■

Note that the classification of the conic sections in Example 2 does not distinguish between special cases within a particular geometric class. For example, both $y = x^2$ and $x = y^2$ have inertia $(1, 0, 1)$.

Before classifying quadric surfaces using inertia, we present the quadric surfaces in the standard forms met in analytic geometry and calculus. (In the following, a, b, and c are positive unless otherwise stated.)

Ellipsoid

(See Figure 8.32.)

$$\frac{x^2}{a^2} + \frac{y^2}{b^2} + \frac{z^2}{c^2} = 1.$$

The special case $a = b = c$ is a sphere.

FIGURE 8.32
Ellipsoid

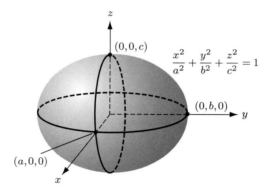

Elliptic Paraboloid

(See Figure 8.33.)

$$z = \frac{x^2}{a^2} + \frac{y^2}{b^2}, \qquad y = \frac{x^2}{a^2} + \frac{z^2}{c^2}, \qquad x = \frac{y^2}{b^2} + \frac{z^2}{c^2}.$$

FIGURE 8.33
Elliptic paraboloid

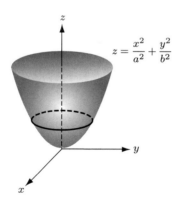

A degenerate case of a parabola is a line, so a degenerate case of an elliptic paraboloid is an **elliptic cylinder** (see Figure 8.34), which is given by

$$\frac{x^2}{a^2} + \frac{y^2}{b^2} = 1, \qquad \frac{x^2}{a^2} + \frac{z^2}{c^2} = 1, \qquad \frac{y^2}{b^2} + \frac{z^2}{c^2} = 1.$$

FIGURE 8.34
Elliptic cylinder

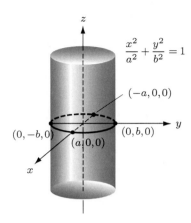

$$\frac{x^2}{a^2} + \frac{y^2}{b^2} = 1$$

$(-a,0,0)$

$(0,-b,0)$ $(0,b,0)$

$(a,0,0)$

Hyperboloid of One Sheet

(See Figure 8.35.)

$$\frac{x^2}{a^2} + \frac{y^2}{b^2} - \frac{z^2}{c^2} = 1, \qquad \frac{x^2}{a^2} - \frac{y^2}{b^2} + \frac{z^2}{c^2} = 1, \qquad -\frac{x^2}{a^2} + \frac{y^2}{b^2} + \frac{z^2}{c^2} = 1.$$

FIGURE 8.35
Hyperboloid of one sheet

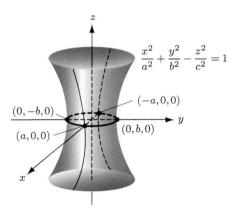

$$\frac{x^2}{a^2} + \frac{y^2}{b^2} - \frac{z^2}{c^2} = 1$$

$(-a,0,0)$

$(0,-b,0)$

$(a,0,0)$ $(0,b,0)$

A degenerate case of a hyperboloid is a pair of lines through the origin; hence a degenerate case of a hyperboloid of one sheet is a **cone** (Figure 8.36), which is given by

$$\frac{x^2}{a^2} + \frac{y^2}{b^2} - \frac{z^2}{c^2} = 0, \qquad \frac{x^2}{a^2} - \frac{y^2}{b^2} + \frac{z^2}{c^2} = 0, \qquad -\frac{x^2}{a^2} + \frac{y^2}{b^2} + \frac{z^2}{c^2} = 0.$$

FIGURE 8.36
Cone

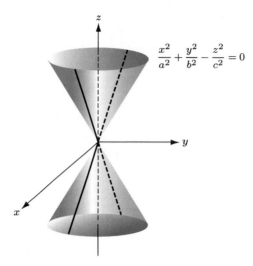

$$\frac{x^2}{a^2} + \frac{y^2}{b^2} - \frac{z^2}{c^2} = 0$$

Hyperboloid of Two Sheets

(See Figure 8.37.)

$$\frac{x^2}{a^2} - \frac{y^2}{b^2} - \frac{z^2}{c^2} = 1, \qquad -\frac{x^2}{a^2} - \frac{y^2}{b^2} + \frac{z^2}{c^2} = 1, \qquad -\frac{x^2}{a^2} + \frac{y^2}{b^2} - \frac{z^2}{c^2} = 1.$$

FIGURE 8.37
Hyperboloid of two sheets

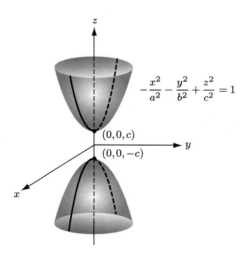

$$-\frac{x^2}{a^2} - \frac{y^2}{b^2} + \frac{z^2}{c^2} = 1$$

$(0, 0, c)$

$(0, 0, -c)$

Hyperbolic Paraboloid

(See Figure 8.38.)

$$\pm z = \frac{x^2}{a^2} - \frac{y^2}{b^2}, \qquad \pm y = \frac{x^2}{a^2} - \frac{z^2}{b^2}, \qquad \pm x = \frac{y^2}{a^2} - \frac{z^2}{b^2}.$$

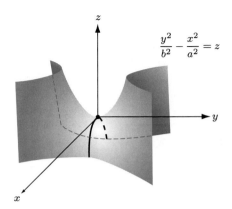

$$\frac{y^2}{b^2} - \frac{x^2}{a^2} = z$$

A degenerate case of a parabola is a line, so a degenerate case of a hyperbolic paraboloid is a hyperbolic cylinder (see Figure 8.39), which is given by

$$\frac{x^2}{a^2} - \frac{y^2}{b^2} = \pm 1, \qquad \frac{x^2}{a^2} - \frac{z^2}{b^2} = \pm 1, \qquad \frac{y^2}{a^2} - \frac{z^2}{b^2} = \pm 1.$$

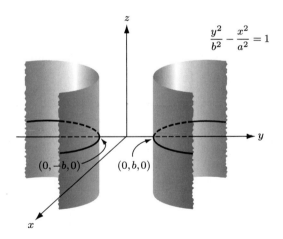

$$\frac{y^2}{b^2} - \frac{x^2}{a^2} = 1$$

$(0, -b, 0)$ $(0, b, 0)$

Parabolic Cylinder

(See Figure 8.40.) One of a or b is not zero.

$$x^2 = ay + bz, \qquad y^2 = ax + bz, \qquad z^2 = ax + by.$$

FIGURE 8.40
Parabolic cylinder

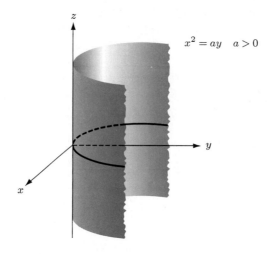

$$x^2 = ay \quad a > 0$$

For a quadratic form in three variables with matrix A, under the assumptions $\lambda_1 > 0$ and $j \geq 0$ in (2), there are exactly six possibilities for the inertia of A. We present these in Table 8.3. As with the conic section classification in Example 2, the classification of quadric surfaces in Table 8.3 does not distinguish between special cases within a particular geometric class.

Table 8.3: Identification of the Quadric Surfaces

$\text{In}(A) = (3, 0, 0)$	Ellipsoid
$\text{In}(A) = (2, 0, 1)$	Elliptic paraboloid
$\text{In}(A) = (2, 1, 0)$	Hyperboloid of one sheet
$\text{In}(A) = (1, 2, 0)$	Hyperboloid of two sheets
$\text{In}(A) = (1, 1, 1)$	Hyperbolic paraboloid
$\text{In}(A) = (1, 0, 2)$	Parabolic cylinder

EXAMPLE 3 ■ Classify the quadric surface represented by the quadratic form $\mathbf{x}^T A \mathbf{x} = 3$, where

$$A = \begin{bmatrix} 0 & 2 & 2 \\ 2 & 0 & 2 \\ 2 & 2 & 0 \end{bmatrix} \quad \text{and} \quad \mathbf{x} = \begin{bmatrix} x \\ y \\ z \end{bmatrix}.$$

Solution   From Example 1 we have that $\text{In}(A) = (1, 2, 0)$ and hence the quadric surface is a hyperboloid of two sheets. ■

EXAMPLE 4 ■ Classify the quadric surface given by

$$2x^2 + 4y^2 - 4z^2 + 6yz - 5x + 3y = 2.$$

Solution   Rewrite the second-degree polynomial as a quadratic form in three variables to identify the matrix A of the quadratic form. We have

$$A = \begin{bmatrix} 2 & 0 & 0 \\ 0 & 4 & 3 \\ 0 & 3 & -4 \end{bmatrix}.$$

Its eigenvalues are $\lambda_1 = 5$, $\lambda_2 = 2$, and $\lambda_3 = -5$ (verify). Thus $\text{In}(A) = (2, 1, 0)$ and hence the quadric surface is a hyperboloid of one sheet. ∎

The classification of a quadric surface is much easier than the problem of transforming it to the standard forms that are used in analytic geometry and calculus. The algebraic steps to obtain an equation in standard form from a second-degree polynomial equation (1) require, in general, a rotation and translation of axes, as mentioned earlier. The rotation requires both the eigenvalues and eigenvectors of the matrix A of the quadratic form. The eigenvectors of A are used to form an orthogonal matrix P so that $\det(P) = 1$, and hence the change of variables $\mathbf{x} = P\mathbf{y}$ represents a rotation. The resulting associated form is that obtained in the principal axes theorem, Theorem 8.9; that is, all cross-product terms are eliminated. We illustrate this with the next example.

EXAMPLE 5 ■ For the quadric surface in Example 4,

$$\mathbf{x}^T A \mathbf{x} + \begin{bmatrix} -5 & 3 & 0 \end{bmatrix} \mathbf{x} = 2,$$

determine the rotation so that all cross-product terms are eliminated.

Solution   The eigenvalues and eigenvectors of

$$A = \begin{bmatrix} 2 & 3 & 0 \\ 0 & 4 & 3 \\ 0 & 3 & -4 \end{bmatrix}$$

are (verify), respectively,

$$\lambda_1 = 5, \qquad \lambda_2 = 2, \qquad \lambda_3 = -5$$

and

$$\mathbf{v}_1 = \begin{bmatrix} 0 \\ 3 \\ 1 \end{bmatrix}, \qquad \mathbf{v}_2 = \begin{bmatrix} 1 \\ 0 \\ 0 \end{bmatrix}, \qquad \mathbf{v}_3 = \begin{bmatrix} 0 \\ 1 \\ -3 \end{bmatrix}.$$

The eigenvectors $\mathbf{v}_i$ are mutually orthogonal, since they correspond to distinct eigenvalues of a symmetric matrix (see Theorem 5.6 in Section 5.2). We normalize the eigenvectors as

$$\mathbf{u}_1 = \frac{1}{\sqrt{10}} \begin{bmatrix} 0 \\ 3 \\ 1 \end{bmatrix}, \qquad \mathbf{u}_2 = \mathbf{v}_2, \qquad \mathbf{u}_3 = \frac{1}{\sqrt{10}} \begin{bmatrix} 0 \\ 1 \\ -3 \end{bmatrix}$$

and define $P = \begin{bmatrix} \mathbf{u}_1 & \mathbf{u}_2 & \mathbf{u}_3 \end{bmatrix}$. Then $|P| = 1$ (verify), so we let $\mathbf{x} = P\mathbf{y}$ and obtain the representation

$$(P\mathbf{y})^T A (P\mathbf{y}) + \begin{bmatrix} -5 & 3 & 0 \end{bmatrix} P\mathbf{y} = 2$$

$$\mathbf{y}^T (P^T A P)\mathbf{y} + \begin{bmatrix} -5 & 3 & 0 \end{bmatrix} P\mathbf{y} = 2.$$

Since $P^T A P = D$, and letting $\mathbf{y} = \begin{bmatrix} x' \\ y' \\ z' \end{bmatrix}$, we have

$$\mathbf{y}^T D \mathbf{y} + \begin{bmatrix} -5 & 3 & 0 \end{bmatrix} P \mathbf{y} = 2,$$

$$\mathbf{y}^T \begin{bmatrix} 5 & 0 & 0 \\ 0 & 2 & 0 \\ 0 & 0 & -5 \end{bmatrix} \mathbf{y} + \begin{bmatrix} \dfrac{9}{\sqrt{10}} & -5 & \dfrac{3}{\sqrt{10}} \end{bmatrix} \mathbf{y} = 2$$

(if $|P| \neq 1$, we redefine P by reordering its columns until we get its determinant to be 1), or

$$5x'^2 + 2y'^2 - 5z'^2 + \frac{9}{\sqrt{10}} x' - 5y' + \frac{3}{\sqrt{10}} z' = 2. \qquad \blacksquare$$

To complete the transformation to standard form, we introduce a change of variable to perform a translation that eliminates any first-degree terms. Algebraically, we complete the square in each of the three variables.

EXAMPLE 6 ■ Continue with Example 5 to eliminate the first-degree terms.

Solution   The last expression for the quadric surface in Example 5 can be written as

$$5x'^2 + \frac{9}{\sqrt{10}} x' + 2y'^2 - 5y' - 5z'^2 + \frac{3}{\sqrt{10}} z' = 2.$$

Completing the square in each variable, we have

$$5\left(x'^2 + \frac{9}{5\sqrt{10}} x' + \frac{81}{1000} \right) + 2\left(y'^2 - \frac{5}{2} y' + \frac{25}{16} \right) - 5\left(z'^2 - \frac{3}{5\sqrt{10}} z' + \frac{9}{1000} \right)$$

$$= 5\left(x' + \frac{9}{10\sqrt{10}} \right)^2 + 2\left(y' - \frac{5}{4} \right)^2 - 5\left(z' - \frac{3}{10\sqrt{10}} \right)^2$$

$$= 2 + \frac{405}{1000} + \frac{50}{16} - \frac{45}{1000}.$$

Letting

$$x'' = x' + \frac{9}{10\sqrt{10}}, \qquad y'' = y' - \frac{5}{4}, \qquad z'' = z' - \frac{3}{10\sqrt{10}},$$

we can write the equation of the quadric surface as

$$5x''^2 + 2y''^2 - 5z''^2 = \frac{5485}{1000} = 5.485.$$

This can be written in standard form as

$$\frac{x''^2}{\frac{5.485}{5}} + \frac{y''^2}{\frac{5.485}{2}} - \frac{z''^2}{\frac{5.485}{5}} = 1. \qquad \blacksquare$$

8.10   EXERCISES

In Exercises 1 through 14, use inertia to classify the quadric surface given by each equation.

1. $x^2 + y^2 + 2z^2 - 2xy - 4xz - 4yz + 4x = 8$.

2. $x^2 + 3y^2 + 2z^2 - 6x - 6y + 4z - 2 = 0$.

3. $z = 4xy$.

4. $x^2 + y^2 + z^2 + 2xy = 4$.

5. $x^2 - y = 0$.

6. $2xy + z = 0$.

7. $5y^2 + 20y + z - 23 = 0$.

8. $x^2 + y^2 + 2z^2 - 2xy + 4xz + 4yz = 16$.

9. $4x^2 + 9y^2 + z^2 + 8x - 18y - 4z - 19 = 0$.

10. $y^2 - z^2 - 9x - 4y + 8z - 12 = 0$.

11. $x^2 + 4y^2 + 4x + 16y - 16z - 4 = 0$.

12. $4x^2 - y^2 + z^2 - 16x + 8y - 6z + 5 = 0$.

13. $x^2 - 4z^2 - 4x + 8z = 0$.

14. $2x^2 + 2y^2 + 4z^2 + 2xy - 2xz - 2yz + 3x - 5y + z = 7$.

In Exercises 15 through 28, classify the quadric surface given by each equation and determine its standard form.

15. $x^2 + 2y^2 + 2z^2 + 2yz = 1$.

16. $x^2 + y^2 + 2z^2 - 2xy + 4xz + 4yz = 16$.

17. $2xz - 2z - 4y - 4z + 8 = 0$.

18. $x^2 + 3y^2 + 3z^2 - 4yz = 9$.

19. $x^2 + y^2 + z^2 + 2xy = 8$.

20. $-x^2 - y^2 - z^2 + 4xy + 4xz + 4yz = 3$.

21. $2x^2 + 2y^2 + 4z^2 - 4xy - 8xz - 8yz + 8x = 15$.

22. $4x^2 + 4y^2 + 8z^2 + 4xy - 4xz - 4yz + 6x - 10y + 2z = \frac{9}{4}$.

23. $2y^2 + 2z^2 + 4yz + \frac{16}{\sqrt{2}}\, x + 4 = 0$.

24. $x^2 + y^2 - 2z^2 + 2xy + 8xz + 8yz + 3x + z = 0$.

25. $-x^2 - y^2 - z^2 + 4xy + 4xz + 4yz + \frac{3}{\sqrt{2}}\, x - \frac{3}{\sqrt{2}}\, y = 6$.

26. $2x^2 + 3y^2 + 3z^2 - 2yz + 2x + \frac{1}{\sqrt{2}}\, y + \frac{1}{\sqrt{2}}\, z = \frac{3}{8}$.

27. $x^2 + y^2 - z^2 - 2x - 4y - 4z + 1 = 0$.

28. $-8x^2 - 8y^2 + 10z^2 + 32xy - 4xz - 4yz = 24$.

8.11 ▾ The Theory of Games

Prerequisite. Chapter 7, Linear Programming.

There are many problems in economics, politics, warfare, business, and so on, that require decisions to be made in conflicting or competitive situations. The theory of games is a rather new area of applied mathematics that attempts to analyze conflict situations and provides a basis for rational decision making.

The theory of games was developed in the 1920s by John von Neumann[*] and E. Borel,[†] but the subject did not come to fruition until the publication, in

---

[*]John von Neumann (1903–1957) was born in Hungary and came to the United States in 1930. His talents in mathematics were recognized early, and he is considered one of the greatest mathematicians of the twentieth century. He made fundamental contributions to the foundations of mathematics, quantum mechanics, operator calculus, and the theory of computing machines and automata. He also participated in many defense projects during World War II and helped to develop the atomic bomb. Beginning in 1926 he developed the theory of games, culminating in his joint work with Oskar Morgenstern in 1944, *Theory of Games and Economic Behavior*.

[†]Emile Borel (1871–1956) was born in Saint Affrique, France. A child prodigy and prolific author of mathematical papers, he married an author, was part of French literary circles, and was active in politics, serving in the National Assembly for a time and joining the Resistance during World War II. In mathematics he contributed to the theory of functions, measure theory,

1944, of the decisive book *Theory of Games and Economic Behavior* by John von Neumann and Oskar Morgenstern.[‡]

A **game** is a competitive situation in which each of a number of players is pursuing his objective in direct conflict with the other players. Each player is doing everything he can to gain as much as possible for himself. Essentially, games are of two types. First, there are **games of chance**, such as roulette, which require no skill on the part of the players; the outcomes and winnings are determined solely by the laws of probability and can in no way be affected by any actions of the players. Second, there are **games of strategy**, such as chess, checkers, bridge, and poker, which require skill on the part of the players; the outcomes and winnings are determined by the skills of the players. By **game**, we mean a game of strategy. In addition to such parlor games there are many games in economic competition, warfare, geological exploration, farming, administration of justice, and so on, in which each player (competitor) can choose one of a set of possible moves and the outcomes depend on the player's (competitor's) skill. The theory of games attempts to determine the best course of action for each player. The subject is still being developed, and many new theoretical and applied results are needed for handling the complex games that occur in everyday situations.

We shall limit our treatment to games played by two players, usually denoted by R and C. We shall also assume that R has m possible moves (or courses of action) and that C has n moves. We now form an $m \times n$ matrix by labeling its rows, from top to bottom, with the moves of R, and labeling its columns, from left to right, with the moves of C. Entry a_{ij}, in row i and column j, indicates the amount (money or some other valuable item) received by R if R makes his ith move and C makes his jth move. The entry a_{ij} is called a **payoff** and the matrix $A = \begin{bmatrix} a_{ij} \end{bmatrix}$ is called the **payoff matrix** (for R). Such games are called **two-person games**. We also refer to them as **matrix games**.

We could also, if we wished, construct a second matrix that represents payoffs for C. However, in a class of games called **constant-sum games**, the sum of the payoff to R and the payoff to C is constant for all mn pairs of RC moves. The more R gains, the less C gains (or the more C loses), and vice versa. A special kind of constant-sum game is the **zero-sum game**, in which the amount won by one player is exactly the amount lost by the other player. Because of this strict interrelation between the payoff matrix for C and that for R, in the case of a constant-sum game it is sufficient to study only the payoff matrix for R. In the following discussion we study only constant-sum games, and the matrix involved is the payoff matrix for R.

In the study of matrix games, it is always assumed that both players are equally capable, that each is playing as well as he can possibly play, and that each player makes his move without knowing what his opponent's move will be.

---

and probability theory. In the 1920s he wrote the first papers in game theory, defining the basic terms, stating the minimax theorem, and discussing applications to war and economics.

[‡]Oskar Morgenstern (1902–1977) was born in Germany and came to the United States in 1938. He was a professor of economics at Princeton University until 1970 and at New York University until his death. In 1944 he collaborated with von Neumann on the influential book *Theory of Games and Economic Behavior.*

EXAMPLE 1   ■   Consider the game of matching pennies, consisting of two players, R and C, each of whom has a penny in his hand. Each player shows one side of the coin without knowing his opponent's choice. If both players are showing the same side of the coin, then R wins \$1 from C; otherwise, C wins \$1 from R.

In this two-person, zero-sum game each player has two possible moves: he can show a tail or he can show a head. The payoff matrix is thus

$$
\begin{array}{cc}
 & C \\
 & \begin{array}{cc} H & T \end{array} \\
R \begin{array}{c} H \\ T \end{array} & \begin{bmatrix} 1 & -1 \\ -1 & 1 \end{bmatrix}.
\end{array}
$$

   ■

EXAMPLE 2   ■   There are two suppliers, firms R and C, of a new specialized type of tire that has 100,000 customers. Each company can advertise its product on TV or in the newspapers. A marketing firm determines that if both firms advertise on TV, then firm R gets 40,000 customers (and firm C gets 60,000 customers). If they both use newspapers, then each gets 50,000 customers. If R uses newspapers and C uses TV, then R gets 60,000 customers (and C gets 40,000 customers). If R uses TV and C uses newspapers, they each get 50,000 customers.

We can consider this situation as a game between firms R and C, with the payoff matrix as shown in Figure 8.41. *The entries in the matrix indicate the number of customers secured by firm R.* This is a constant-sum game, for the sum of the R customers and the C customers is always the total population of 100,000 tire customers.   ■

FIGURE 8.41

Firm C

$$
\text{Firm } R \quad
\begin{array}{c} \\ \text{TV} \\ \text{Newspapers} \end{array}
\begin{array}{cc} \text{TV} & \text{Newspapers} \\ \begin{bmatrix} 40{,}000 & 50{,}000 \\ 60{,}000 & 50{,}000 \end{bmatrix} \end{array}
$$

Consider now a two-person, constant-sum game with the $m \times n$ payoff matrix $A = [a_{ij}]$, so that player R has m moves and player C has n moves. If player R plays his ith move, he is assured of winning at least the smallest entry in the ith row of A, no matter what C does. Thus R's best course of action is to choose that move which will maximize his assured winnings in spite of C's best countermove. Player R will get his largest payoff by maximizing his smallest gain. Player C's goals are in direct conflict with those of player R: he is trying to keep R's winnings to a minimum. If C plays his jth move, he is assured of losing no more than the largest entry in the jth column of A, no matter what R does. Thus C's best course of action is to choose that move which will minimize his assured losses in spite of R's best countermove. Player C will do his best by minimizing his largest loss.

DEFINITION | If the payoff matrix of a matrix game contains an entry a_{rs}, which is at the same time the minimum of row r and the maximum of column s, then a_{rs} is called a **saddle point**. Also, a_{rs} is called the **value** of the game. If the value of a zero-sum game is zero, the game is said to be **fair**.

DEFINITION | A matrix game is said to be **strictly determined** if its payoff matrix has a saddle point.

If a_{rs} is a saddle point for a matrix game, then player R will be assured of winning at least a_{rs} by playing his rth move and player C will be guaranteed that he will lose no more than a_{rs} by playing his sth move. This is the best that each player can do.

EXAMPLE 3 ■ Consider a game with payoff matrix

$$R \begin{array}{c} C \\ \begin{bmatrix} 0 & -3 & -1 & 3 \\ 3 & 2 & 2 & 4 \\ 1 & 4 & 0 & 6 \end{bmatrix} \end{array}.$$

To determine whether this game has a saddle point, we write the minimum of each row to the right of the row and the maximum of each column at the bottom of each column. Thus we have

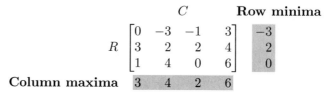

Entry $a_{23} = 2$ is both the least entry in the second row and the largest entry in the third column. Hence it is a saddle point for the game, which is then a strictly determined game. The value of the game is 2 and player R has an advantage. The best course of action for R is to play his second move; he will win at least 2 units from C, no matter what C does. The best course of action for C is to play his third move; he will limit his loss to not more than 2 units, no matter what R does. ■

EXAMPLE 4 ■ Consider the advertising game of Example 2. The payoff matrix is shown in Figure 8.42. Thus entry $a_{22} = 50,000$ is a saddle point. The best course of action for both firms is to advertise in newspapers. The game is strictly determined with value 50,000. ■

There are many games that are not strictly determined.

FIGURE 8.42

Firm C

		TV	Newspapers	**Row minima**
Firm R	TV	$\begin{bmatrix} 40,000$	$50,000 \end{bmatrix}$	40,000
	Newspapers	$60,000$	$50,000$	50,000
Column maxima		60,000	50,000	

EXAMPLE 5 ■ Consider the game with payoff matrix

Row minima

$$\begin{bmatrix} 1 & 6 & -1 \\ 3 & -2 & 4 \\ 4 & 5 & -3 \end{bmatrix} \quad \begin{matrix} -1 \\ -2 \\ -3 \end{matrix}$$

Column maxima 4 6 4

It is clear that there is no saddle point. ■

On the other hand, a game may have more than one saddle point. However, it can be proved that all saddle points must have the same value.

EXAMPLE 6 ■ Consider the game with payoff matrix

Row minima

$$\begin{bmatrix} 5 & 4 & 5 & 4 \\ 6 & -1 & 3 & 2 \\ 6 & 4 & 6 & 4 \end{bmatrix} \quad \begin{matrix} 4 \\ -1 \\ 4 \end{matrix}$$

Column maxima 6 4 6 4

Entries a_{12}, a_{14}, a_{32}, and a_{34} are all saddle points and have the same value, 4. They appear shaded in the payoff matrix. The value of the game is also 4. ■

Consider now the penny-matching game of Example 1 with payoff matrix

$$C$$

		H	T	**Row minima**
R	H	$\begin{bmatrix} 1$	$-1 \end{bmatrix}$	-1
	T	-1	1	-1
Column maxima		1	1	

It is clear that this game is not strictly determined; that is, it has no saddle point.

To analyze this type of situation, we assume that a game is played repeatedly and that each player is trying to determine his best course of action. Thus player R tries to maximize his winnings while player C tries to minimize his losses. A **strategy** for a player is a decision for choosing his moves.

Consider now the above penny-matching game. Suppose that in the repeated play of the game, player R always chooses the first row (he chooses to show heads), in the hope that player C will always choose the first column (play heads), thereby ensuring a win of $1 for himself. However, as player C begins to notice that R always chooses his first row, then player C will choose his second column, resulting in a loss of $1 for R. Similarly, if R always chooses the second row, then C will choose the first column, resulting in a loss of $1 for R. We can thus conclude that each player must somehow keep the other player from anticipating his choice of moves. This situation is in marked contrast with the case in strictly determined games. In a strictly determined game each player will make the same move whether or not he has advanced knowledge of his opponent's move. Thus, in a nonstrictly determined game, each player will make each move with a certain relative frequency.

DEFINITION   Suppose that we have a matrix game with an $m \times n$ payoff matrix A. Let p_i, $1 \leq i \leq m$, be the probability that R chooses the ith row of A (that is, chooses his ith move). Let q_j, $1 \leq j \leq n$, be the probability that C chooses the jth column of A. The vector $\mathbf{p} = \begin{bmatrix} p_1 & p_2 & \cdots & p_m \end{bmatrix}$ is called a **strategy** for player R; the vector

$$\mathbf{q} = \begin{bmatrix} q_1 \\ q_2 \\ \vdots \\ q_n \end{bmatrix}$$

is called a **strategy** for player C.

Of course, the probabilities p_i and q_j in the definition satisfy

$$p_1 + p_2 + \cdots + p_m = 1$$
$$q_1 + q_2 + \cdots + q_n = 1.$$

If a matrix game is strictly determined, then optimal strategies for R and C are strategies having 1 as one component and zero for all other components. Such strategies are called **pure strategies**. A strategy that is not pure is called a **mixed strategy**. Thus, in Example 3, the pure strategy for R is

$$\mathbf{p} = \begin{bmatrix} 0 & 1 & 0 \end{bmatrix},$$

and the pure strategy for C is

$$\mathbf{q} = \begin{bmatrix} 0 \\ 0 \\ 1 \\ 0 \end{bmatrix}.$$

Consider now a matrix game with payoff matrix

$$A = \begin{bmatrix} a_{11} & a_{12} \\ a_{21} & a_{22} \end{bmatrix}. \tag{1}$$

Suppose that

$$\mathbf{p} = \begin{bmatrix} p_1 & p_2 \end{bmatrix} \quad \text{and} \quad \mathbf{q} = \begin{bmatrix} q_1 \\ q_2 \end{bmatrix}$$

are strategies for R and C, respectively. Then if R plays his first row with probability p_1 and if C plays his first column with probability q_1, then R's expected payoff is $p_1 q_1 a_{11}$. Similarly, we can examine the remaining three possibilities, obtaining Table 8.4. The expected payoff $E(\mathbf{p}, \mathbf{q})$ of the game to R is then the sum of the four quantities in the rightmost column. We obtain

$$E(\mathbf{p}, \mathbf{q}) = p_1 q_1 a_{11} + p_1 q_2 a_{12} + p_2 q_1 a_{21} + p_2 q_2 a_{22},$$

which can be written in matrix form (verify) as

$$E(\mathbf{p}, \mathbf{q}) = \mathbf{p} A \mathbf{q}. \tag{2}$$

Table 8.4

| Moves | | | Payoff to | Expected Payoff |
Player R	*Player C*	*Probability*	*Player R*	*to Player R*
Row 1	Column 1	$p_1 q_1$	a_{11}	$p_1 q_1 a_{11}$
Row 1	Column 2	$p_1 q_2$	a_{12}	$p_1 q_2 a_{12}$
Row 2	Column 1	$p_2 q_1$	a_{21}	$p_2 q_1 a_{21}$
Row 2	Column 2	$p_2 q_2$	a_{22}	$p_2 q_2 a_{22}$

The same analysis applies to a matrix game with an $m \times n$ payoff matrix A. Thus if

$$\mathbf{p} = \begin{bmatrix} p_1 & p_2 & \cdots & p_m \end{bmatrix} \quad \text{and} \quad \mathbf{q} = \begin{bmatrix} q_1 \\ q_2 \\ \vdots \\ q_n \end{bmatrix}$$

are strategies for R and C, respectively, then the payoff to player R is given by (2).

EXAMPLE 7 ■ Consider a matrix game with payoff matrix

$$A = \begin{bmatrix} 2 & -2 & 3 \\ 4 & 0 & -3 \end{bmatrix}.$$

If

$$\mathbf{p} = \begin{bmatrix} \frac{1}{4} & \frac{3}{4} \end{bmatrix} \quad \text{and} \quad \mathbf{q} = \begin{bmatrix} \frac{1}{3} \\ \frac{1}{3} \\ \frac{1}{3} \end{bmatrix}$$

are strategies for R and C, respectively, then the expected payoff to R is

$$E(\mathbf{p}, \mathbf{q}) = \mathbf{p}A\mathbf{q} = \begin{bmatrix} \frac{1}{4} & \frac{3}{4} \end{bmatrix} \begin{bmatrix} 2 & -2 & 3 \\ 4 & 0 & -3 \end{bmatrix} \begin{bmatrix} \frac{1}{3} \\ \frac{1}{3} \\ \frac{1}{3} \end{bmatrix} = \frac{1}{2}.$$

If

$$\mathbf{p} = \begin{bmatrix} \frac{3}{4} & \frac{1}{4} \end{bmatrix} \quad \text{and} \quad \mathbf{q} = \begin{bmatrix} \frac{1}{3} \\ \frac{2}{3} \\ 0 \end{bmatrix}$$

are strategies for R and C, respectively, then the expected payoff to R is $-\frac{1}{6}$. Thus, in the first case, R gains $\frac{1}{2}$ from C, whereas in the second case R loses $\frac{1}{6}$ to C. ■

A strategy for player R is said to be **optimal** if it guarantees R the largest possible payoff no matter what his opponent may do. Similarly, a strategy for player C is said to be **optimal** if it guarantees the smallest possible payoff to R no matter what R may do.

If $\mathbf{p}$ and $\mathbf{q}$ are optimal strategies for R and C, respectively, then the expected payoff to R, $v = E(\mathbf{p}, \mathbf{q})$, is called the **value** of the game. Although $E(\mathbf{p}, \mathbf{q})$ is a 1×1 matrix, we think of it merely as a number v. If the value of a zero-sum game is zero, the game is said to be **fair**. The principal task of the theory of games is the determination of optimal strategies for each player.

Consider again a matrix game with the 2×2 payoff matrix (1) and suppose that the game is not strictly determined. It can then be shown that

$$a_{11} + a_{22} - a_{12} - a_{21} \neq 0.$$

To determine an optimal strategy for R, we proceed as follows. Suppose that R's strategy is $\begin{bmatrix} p_1 & p_2 \end{bmatrix}$. Then if C plays his first column, the expected payoff to R is

$$a_{11}p_1 + a_{21}p_2. \tag{3}$$

If C plays his second column, the expected payoff to R is

$$a_{12}p_1 + a_{22}p_2. \tag{4}$$

If v is the minimum of the expected payoffs (3) and (4), then R expects to gain at least v units from C no matter what C does. Thus we have

$$a_{11}p_1 + a_{21}p_2 \geq v \tag{5}$$
$$a_{12}p_1 + a_{22}p_2 \geq v. \tag{6}$$

Moreover, player R seeks to make v as large as possible. Thus player R seeks to find p_1, p_2, and v such that

$$v \text{ is a maximum}$$

and

$$a_{11}p_1 + a_{21}p_2 - v \geq 0$$
$$a_{12}p_1 + a_{22}p_2 - v \geq 0$$
$$p_1 + p_2 = 1 \tag{7}$$
$$p_1 \geq 0, \quad p_2 \geq 0, \quad v \geq 0.$$

We shall see below (in a more general situation) that problem (7) is a linear programming problem. It can be shown that a solution to (7), giving an optimal strategy for R, is

$$p_1 = \frac{a_{22} - a_{21}}{a_{11} + a_{22} - a_{12} - a_{21}}, \qquad p_2 = \frac{a_{11} - a_{12}}{a_{11} + a_{22} - a_{12} - a_{21}} \tag{8}$$

and

$$v = \frac{a_{11}a_{22} - a_{12}a_{21}}{a_{11} + a_{22} - a_{12} - a_{21}}. \tag{9}$$

We now find an optimal strategy for C. Suppose that C's strategy is

$$\begin{bmatrix} q_1 \\ q_2 \end{bmatrix}.$$

If R plays the first row, then the expected payoff to R is

$$a_{11}q_1 + a_{12}q_2, \tag{10}$$

while if R plays the second row, the expected payoff to R is

$$a_{21}q_1 + a_{22}q_2. \tag{11}$$

If v' is the maximum of the expected payoffs (10) and (11), then

$$a_{11}q_1 + a_{12}q_2 \leq v'$$
$$a_{21}q_1 + a_{22}q_2 \leq v'.$$

Since player C wishes to lose as little as possible, he seeks to make v' as small as possible. Thus C wants to find q_1, q_2, and v such that

$$v' \text{ is a minimum}$$

and

$$a_{11}q_1 + a_{12}q_2 - v' \leq 0$$
$$a_{21}q_1 + a_{22}q_2 - v' \leq 0$$
$$q_1 + q_2 = 1 \tag{12}$$
$$q_1 \geq 0, \quad q_2 \geq 0, \quad v' \geq 0.$$

Problem (12) is also a linear programming problem. It can be shown that a solution to (12), giving an optimal strategy for C, is

$$q_1 = \frac{a_{22} - a_{12}}{a_{11} + a_{22} - a_{12} - a_{21}}, \qquad q_2 = \frac{a_{11} - a_{21}}{a_{11} + a_{22} - a_{12} - a_{21}} \tag{13}$$

and

$$v' = \frac{a_{11}a_{22} - a_{12}a_{21}}{a_{11} + a_{22} - a_{12} - a_{21}}. \tag{14}$$

Thus $v = v'$ when both players use their optimal strategies.

EXAMPLE 8 ■ For the zero-sum penny-matching game of Example 1, we have upon substituting in (8), (9), and (13),

$$p_1 = p_2 = \tfrac{1}{2} \quad \text{and} \quad q_1 = q_2 = \tfrac{1}{2}, \quad v = 0,$$

so that optimal strategies for R and C are

$$\begin{bmatrix} \frac{1}{2} & \frac{1}{2} \end{bmatrix} \quad \text{and} \quad \begin{bmatrix} \frac{1}{2} \\ \frac{1}{2} \end{bmatrix},$$

respectively. This means that half the time R should show heads and half the time he should show tails; likewise for player C. The value of the game is zero, so the game is fair. ■

EXAMPLE 9 ■ Consider a matrix game with payoff matrix

$$\begin{bmatrix} 2 & -5 \\ 1 & 3 \end{bmatrix}.$$

Again substituting in (8), (9), and (13), we obtain

$$p_1 = \frac{3-1}{2+3-1+5} = \frac{2}{9}, \qquad p_2 = \frac{2+5}{2+3-1+5} = \frac{7}{9},$$

$$q_1 = \frac{3+5}{2+3-1+5} = \frac{8}{9}, \qquad q_2 = \frac{2-1}{2+3-1+5} = \frac{1}{9},$$

$$v = \frac{6+5}{2+3-1+5} = \frac{11}{9}.$$

Thus optimal strategies for R and C are

$$\begin{bmatrix} \frac{2}{9} & \frac{7}{9} \end{bmatrix} \quad \text{and} \quad \begin{bmatrix} \frac{8}{9} \\ \frac{1}{9} \end{bmatrix},$$

respectively; when both players use their optimal strategies, the value of the game (the expected payoff to R) is $\frac{11}{9}$. If this matrix represents a zero-sum game, the game is not fair and in the long run favors player R. ■

We can now generalize our discussion to a game with an $m \times n$ payoff matrix $A = [a_{ij}]$. First, let us observe that if we add a constant r to every entry of A, then the optimal strategies for R and C do not change, and the value of the new game is r plus the value of the old game (Exercise T.2). Thus we can assume that, by adding a suitable constant to every entry of the payoff matrix, every entry of A is positive.

Player R seeks to find $p_1, p_2, \ldots, p_m$, and v such that

$$v \text{ is a maximum}$$

subject to

$$
\begin{aligned}
a_{11}p_1 + a_{21}p_2 + \cdots + a_{m1}p_m - v &\geq 0 \\
a_{12}p_1 + a_{22}p_2 + \cdots + a_{m2}p_m - v &\geq 0 \\
\vdots \qquad \vdots \qquad\qquad \vdots \quad \vdots \; \vdots & \\
a_{1n}p_1 + a_{2n}p_2 + \cdots + a_{mn}p_m - v &\geq 0 \\
p_1 + p_2 + \cdots + p_m &= 1 \\
p_1 \geq 0, p_2 \geq 0, \ldots, p_m \geq 0, \quad v &\geq 0.
\end{aligned}
\tag{15}
$$

Since every entry of A is positive, we may assume that $v > 0$. Now divide each of the constraints in (15) by v, and let

$$y_i = \frac{p_i}{v}.$$

Observe that

$$y_1 + y_2 + \cdots + y_m = \frac{p_1}{v} + \frac{p_2}{v} + \cdots + \frac{p_m}{v} = \frac{1}{v}(p_1 + p_2 + \cdots + p_m) = \frac{1}{v}.$$

Thus v is a maximum if and only if $y_1 + y_2 + \cdots + y_m$ is a minimum. We can now restate problem (15), R's problem, as follows:

$$\text{Minimize} \quad y_1 + y_2 + \cdots + y_m$$

subject to

$$
\begin{aligned}
a_{11}y_1 + a_{21}y_2 + \cdots + a_{m1}y_m &\geq 1 \\
a_{12}y_1 + a_{22}y_2 + \cdots + a_{m2}y_m &\geq 1 \\
\vdots \qquad \vdots \qquad\qquad \vdots \quad \vdots & \\
a_{1n}y_1 + a_{2n}y_2 + \cdots + a_{mn}y_m &\geq 1 \\
y_1 \geq 0, y_2 \geq 0, \ldots, y_m &\geq 0.
\end{aligned}
\tag{16}
$$

Observe that (16) is a linear programming problem and that it has one fewer constraint and one fewer variable than (15).

Turning next to C's problem, we note that he seeks to find $q_1, q_2, \ldots, q_n$ and v' such that

$$v' \text{ is a minimum}$$

subject to

$$
\begin{aligned}
a_{11}q_1 + a_{12}q_2 + \cdots + a_{1n}q_n - v' &\leq 0 \\
a_{21}q_1 + a_{22}q_2 + \cdots + a_{2n}q_n - v' &\leq 0 \\
\vdots \qquad \vdots \qquad\qquad \vdots \quad \vdots \; \vdots & \\
a_{m1}q_1 + a_{m2}q_2 + \cdots + a_{mn}q_n - v' &\leq 0 \\
q_1 + q_2 + \cdots + q_n &= 1 \\
q_1 \geq 0, q_2 \geq 0, \ldots, q_n \geq 0, \quad v' &\geq 0.
\end{aligned}
\tag{17}
$$

The fundamental theorem of matrix games, which we now state, says that every matrix game has a solution.

THEOREM 8.12 ■
(*Fundamental Theorem of Matrix Games*)

Every matrix game has a solution. That is, there are optimal strategies for R and C. Moreover, $v = v'$. ■

Since $v = v'$, we can divide each of the constraints in (17) by $v = v'$ and let

$$x_i = \frac{q_i}{v}.$$

Now,

$$x_1 + x_2 + \cdots + x_n = \frac{1}{v},$$

so v is a minimum if and only if $x_1 + x_2 + \cdots + x_n$ is a maximum. We can now restate problem (17), C's problem, as follows:

$$\text{Maximize} \quad x_1 + x_2 + \cdots + x_n$$

subject to

$$\begin{aligned}
a_{11}x_1 + a_{12}x_2 + \cdots + a_{1n}x_n &\leq 1 \\
a_{21}x_1 + a_{22}x_2 + \cdots + a_{2n}x_n &\leq 1 \\
\vdots \qquad \vdots \qquad\qquad \vdots \qquad \vdots \\
a_{m1}x_1 + a_{m2}x_2 + \cdots + a_{mn}x_n &\leq 1 \\
x_1 \geq 0, x_2 \geq 0, \ldots, x_n &\geq 0.
\end{aligned} \qquad (18)$$

Observe that (18) is a linear programming problem in standard form, which is the dual of (16). From the results of Section 7.3, it follows that when (18) is solved by the simplex method, the final tableau will contain the optimal strategies for R in the objective row under the columns of the slack variables. That is, y_1 is found in the objective row under the first slack variable, y_2 is found in the objective row under the second slack variable, and so on.

EXAMPLE 10 ■   Consider a game with payoff matrix

$$\begin{bmatrix} 2 & -3 & 0 \\ 3 & 1 & -2 \end{bmatrix}.$$

Adding 4 to each element of the matrix, we obtain a matrix A with positive entries.

$$A = \begin{bmatrix} 6 & 1 & 4 \\ 7 & 5 & 2 \end{bmatrix}.$$

We now find optimal strategies for the game with payoff matrix A. Problem (18), C's problem, becomes:

$$\text{Maximize} \quad x_1 + x_2 + x_3$$

subject to

$$\begin{aligned}
6x_1 + x_2 + 4x_3 &\leq 1 \\
7x_1 + 5x_2 + 2x_3 &\leq 1 \\
x_1 \geq 0, \quad x_2 \geq 0, \quad x_3 &\geq 0.
\end{aligned}$$

If we introduce the slack variables x_4 and x_5, our problem becomes:

$$\text{Maximize} \quad x_1 + x_2 + x_3$$

subject to

$$6x_1 + x_2 + 4x_3 + x_4 \qquad = 1$$
$$7x_1 + 5x_2 + 2x_3 \qquad + x_5 = 1$$
$$x_1 \geq 0, \ x_2 \geq 0, \ x_3 \geq 0, \ x_4 \geq 0, \ x_5 \geq 0.$$

Using the simplex method, we obtain

	x_1	x_2	x_3	x_4	x_5	z	
$\leftarrow x_4$	6	1	④	1	0	0	1
x_5	7	5	2	0	1	0	1
	-1	-1	-1	0	0	1	0

	x_1	x_2	x_3	x_4	x_5	z	
x_3	$\frac{3}{2}$	$\frac{1}{4}$	1	$\frac{1}{4}$	0	0	$\frac{1}{4}$
$\leftarrow x_5$	4	$\textcircled{\frac{9}{2}}$	0	$-\frac{1}{2}$	1	0	$\frac{1}{2}$
	$\frac{1}{2}$	$-\frac{3}{4}$	0	$\frac{1}{4}$	0	1	$\frac{1}{4}$

	x_1	x_2	x_3	x_4	x_5	z	
x_3	$\frac{23}{18}$	0	1	$\frac{5}{18}$	$-\frac{1}{18}$	0	$\frac{2}{9}$
x_2	$\frac{8}{9}$	1	0	$-\frac{1}{9}$	$\frac{2}{9}$	0	$\frac{1}{9}$
	$\frac{7}{6}$	0	0	$\frac{1}{6}$	$\frac{1}{6}$	1	$\frac{1}{3}$

Thus we have

$$x_1 = 0, \quad x_2 = \tfrac{1}{9}, \quad \text{and} \quad x_3 = \tfrac{2}{9}.$$

The maximum value of $x_1 + x_2 + x_3$ is $\frac{1}{3}$, so the minimum value of v is 3. Hence

$$q_1 = x_1 v = 0, \qquad q_2 = x_2 v = \left(\tfrac{1}{9}\right)(3) = \tfrac{1}{3},$$

and

$$q_3 = x_3 v = \left(\tfrac{2}{9}\right)(3) = \tfrac{2}{3}.$$

Thus an optimal strategy for C is

$$\mathbf{q} = \begin{bmatrix} 0 \\ \frac{1}{3} \\ \frac{2}{3} \end{bmatrix}.$$

An optimal solution to (16), R's problem, is found in the objective row under the columns of slack variables. Thus, under the slack variables x_4 and x_5, we find

$$y_1 = \tfrac{1}{6} \quad \text{and} \quad y_2 = \tfrac{1}{6},$$

respectively. Since $v = 3$,

$$p_1 = y_1 v = \left(\tfrac{1}{6}\right)(3) = \tfrac{1}{2} \quad \text{and} \quad p_2 = y_2 v = \left(\tfrac{1}{6}\right)(3) = \tfrac{1}{2}.$$

Thus an optimal strategy for R is

$$\mathbf{p} = \begin{bmatrix} \tfrac{1}{2} & \tfrac{1}{2} \end{bmatrix}.$$

The zero-sum game with payoff matrix A is not fair, since the value is 3, with player R having the advantage. Since matrix A was obtained from the game as initially proposed by adding $r = 4$ to all matrix entries, the value of the *initial* game is $3 - 4 = -1$. That initial game is to C's advantage. ■

Sometimes it is possible to solve a matrix game by reducing the size of the payoff matrix A. If each element of the rth *row* of A is *less than or equal* to the corresponding element of the sth row of A, then the rth row is called **recessive** and the sth row is said to **dominate** the rth row. If each element of the rth *column* of A is *greater than or equal* to the corresponding element in the sth column of A, then the rth column is called **recessive** and the sth column is said to **dominate** the rth column.

EXAMPLE 11 ■ In the payoff matrix

$$\begin{bmatrix} 2 & -1 & 3 \\ 0 & 3 & 4 \\ 3 & 2 & 4 \end{bmatrix},$$

the first row is recessive; the third row dominates the first row. In the payoff matrix

$$\begin{bmatrix} -2 & 4 & 3 \\ 3 & -3 & -3 \\ 5 & 2 & 1 \end{bmatrix},$$

the second column is recessive; the third column dominates the second column. ■

Consider a matrix game in which the rth row is recessive and the sth row dominates the rth row. It is then obvious that player R will always tend to choose the sth row rather than the rth row, since he will be guaranteed a gain equal to or greater than the gain realized by choosing the rth row. Thus, since the rth row will never be chosen, it can be dropped from further consideration. Suppose now that the rth column is recessive and that the sth column dominates the rth column. Since player C wishes to keep his losses to a minimum, by choosing the sth column he will be guaranteed a loss equal to or smaller than the loss incurred by choosing the rth column. Since the rth column will never be chosen, it can be dropped from further consideration. These techniques, when applicable, result in a smaller payoff matrix.

EXAMPLE 12 ■ Consider the matrix game with payoff matrix

$$A = \begin{bmatrix} 2 & -1 & 3 \\ -2 & 2 & 4 \\ 3 & 0 & 4 \end{bmatrix}.$$

Since the third row of A dominates its first row, the latter can be dropped, obtaining

$$A_1 = \begin{bmatrix} -2 & 2 & 4 \\ 3 & 0 & 4 \end{bmatrix}.$$

Since the second column of A_1 dominates its third column, the latter can be dropped, obtaining

$$A_2 = \begin{bmatrix} -2 & 2 \\ 3 & 0 \end{bmatrix},$$

which has no saddle point. The solution to the matrix game with payoff matrix A_2 can be obtained from Equations (8), (9), and (13). We have

$$p_1 = \frac{0-3}{-2+0-2-3} = \frac{-3}{-7} = \frac{3}{7},$$

$$p_2 = \frac{-2-2}{-2+0-2-3} = \frac{-4}{-7} = \frac{4}{7},$$

$$q_1 = \frac{0-2}{-2+0-2-3} = \frac{-2}{-7} = \frac{2}{7},$$

$$q_2 = \frac{-2-3}{-2+0-2-3} = \frac{-5}{-7} = \frac{5}{7},$$

and

$$v = \frac{0-6}{-2+0-2-3} = \frac{-6}{-7} = \frac{6}{7}.$$

Since in A, the original payoff matrix, the first row and third column were dropped, we obtain

$$\mathbf{p} = \begin{bmatrix} 0 & \frac{3}{7} & \frac{4}{7} \end{bmatrix}$$

as an optimal strategy for player R. Similarly,

$$\mathbf{q} = \begin{bmatrix} \frac{2}{7} \\ \frac{5}{7} \\ 0 \end{bmatrix}$$

is an optimal strategy for player C. ■

Further Readings ▶

OWEN, G. *Game Theory*, 3rd ed. Orlando, Fla.: Academic Press, 1995.

STRAFFIN, PHILIP D. *Game Theory and Strategy*. Washington, D.C.: New Mathematical Library, No. 36, 1993.

THIE, PAUL R. *An Introduction to Linear Programming and Game Theory*, 2nd ed. New York: Wiley, 1988.

8.11   EXERCISES

In Exercises 1 through 4, write the payoff matrix for the given game.

1. Each of two players shows two or three fingers. If the sum of the fingers shown is even, then R pays C an amount equal to the sum of the numbers shown; if the sum is odd, then C pays R an amount equal to the sum of the numbers shown.

2. (**Stone, Scissors, Paper**) Each of two players selects one of the words *stone, scissors, paper*. Stone beats scissors, scissors beats paper, and paper beats stone. In case of a tie, there is no payoff. In case of a win, the winner collects $1.

3. Firms A and B, both handling specialized sporting equipment, are planning to locate in either Abington or Wyncote. If they both locate in the same town, each will capture 50 percent of the trade. If A locates in Abington and B locates in Wyncote, then A will capture 60 percent of the business (and B will keep 40 percent); if A locates in Wyncote and B locates in Abington, then A will hold on to 25 percent of the business (and B to 75 percent).

4. Player R has a nickel and a dime with him. He chooses one of the coins and player C must guess R's choice. If C guesses correctly, he keeps the coin; if he guesses incorrectly, he must give R an amount equal to the coin shown.

5. Find all saddle points for the following matrix games.

   (a) $\begin{bmatrix} 5 & 4 \\ 3 & -2 \end{bmatrix}$.   (b) $\begin{bmatrix} 2 & 1 & 0 \\ 3 & 1 & -2 \\ 4 & 2 & -4 \end{bmatrix}$.

   (c) $\begin{bmatrix} 3 & 4 & 5 \\ -2 & 5 & 1 \\ -1 & 0 & 1 \end{bmatrix}$.

   (d) $\begin{bmatrix} 5 & 2 & 4 & 2 \\ 0 & -1 & 2 & 0 \\ 3 & 2 & 3 & 2 \\ 1 & 0 & -1 & -1 \end{bmatrix}$.

6. Find optimal strategies for the following strictly determined games. Give the payoff for R.

   (a) $\begin{bmatrix} -3 & 4 \\ 3 & 5 \end{bmatrix}$.   (b) $\begin{bmatrix} -1 & -3 & -2 \\ 3 & -1 & 4 \\ -1 & -2 & 5 \end{bmatrix}$.

   (c) $\begin{bmatrix} -2 & 3 & -2 & 4 \\ -1 & 2 & -2 & 4 \\ -2 & 3 & -3 & 5 \\ -1 & 2 & -3 & 1 \end{bmatrix}$.

7. Find optimal strategies for the following strictly determined games. Give the payoff for R.

   (a) $\begin{bmatrix} 2 & 1 & 3 \\ -2 & 0 & 2 \end{bmatrix}$.

   (b) $\begin{bmatrix} -2 & -2 & 4 & 5 \\ -2 & -2 & 1 & 0 \\ 0 & 1 & 1 & 2 \end{bmatrix}$.

   (c) $\begin{bmatrix} 6 & 4 \\ 7 & 4 \end{bmatrix}$.

8. Consider a matrix game with payoff matrix

   $$\begin{bmatrix} 2 & -3 & -2 \\ -4 & 5 & 6 \end{bmatrix}.$$

   Find $E(\mathbf{p}, \mathbf{q})$, the expected payoff to R, if

   (a) $\mathbf{p} = \begin{bmatrix} \frac{1}{4} & \frac{3}{4} \end{bmatrix}$ and $\mathbf{q} = \begin{bmatrix} \frac{1}{3} \\ \frac{1}{6} \\ \frac{1}{2} \end{bmatrix}$.

   (b) $p_1 = \frac{2}{3}, p_2 = \frac{1}{3}$; $q_1 = \frac{1}{2}, q_2 = \frac{1}{4}$, and $q_3 = \frac{1}{4}$.

9. Consider a matrix game with payoff matrix

$$\begin{bmatrix} 3 & -3 \\ 2 & 5 \\ 1 & 0 \end{bmatrix}.$$

Find $E(\mathbf{p}, \mathbf{q})$, the expected payoff to R, if

(a) $\mathbf{p} = \begin{bmatrix} \frac{1}{2} & \frac{1}{3} & \frac{1}{6} \end{bmatrix}$ and $\mathbf{q} = \begin{bmatrix} \frac{1}{6} \\ 0 \\ \frac{5}{6} \end{bmatrix}$.

(b) $p_1 = 0$, $p_2 = 0$, $p_3 = 1$; $q_1 = \frac{1}{7}$, $q_2 = \frac{6}{7}$.

In Exercises 10 and 11, solve the given matrix game using (8) and (13). Find the value of the game using (9).

10. $\begin{bmatrix} 4 & 8 \\ 6 & -2 \end{bmatrix}.$ 11. $\begin{bmatrix} -3 & 2 \\ 4 & -5 \end{bmatrix}.$

In Exercises 12 and 13, solve the given matrix game by linear programming.

12. $\begin{bmatrix} -2 & 3 \\ 4 & 5 \\ 5 & 2 \end{bmatrix}.$ 13. $\begin{bmatrix} 2 & -3 & 4 \\ 4 & 0 & 1 \\ 3 & 2 & -2 \end{bmatrix}.$

In Exercises 14 and 15, solve the given matrix game using the method of Example 11.

14. $\begin{bmatrix} -3 & 1 & 3 \\ 1 & -2 & 2 \\ 2 & -1 & 3 \end{bmatrix}.$

15. $\begin{bmatrix} 0 & -4 & 3 & 0 \\ 2 & -3 & 4 & 1 \\ -1 & 2 & 2 & 2 \\ 1 & -4 & 3 & 0 \end{bmatrix}.$

16. Solve Exercise 1.

17. Solve Exercise 2.

18. Solve Exercise 3.

19. Solve Exercise 4.

20. In a labor–management dispute, labor can make one of three different moves, L_1, L_2, and L_3, while management can make one of two moves, M_1 and M_2. Suppose that the following payoff matrix is obtained (the entries represent millions of dollars). Determine the best courses of action for both labor and management.

$$L \begin{array}{c} \\ L_1 \\ L_2 \\ L_3 \end{array} \overset{\begin{array}{cc} & M \\ M_1 & M_2 \end{array}}{\begin{bmatrix} 2 & 4 \\ 3 & 2 \\ 2 & 5 \end{bmatrix}}.$$

THEORETICAL EXERCISES

T.1. Consider a matrix game with $m \times n$ payoff matrix A. Verify that if player R uses strategy $\mathbf{p}$ and player C uses strategy $\mathbf{q}$, then the expected payoff to R is $\mathbf{p}A\mathbf{q}$.

T.2. Consider a matrix game with payoff matrix A. Show that if a constant r is added to each entry

of A, then we get a new game whose optimal strategies are the same as those for the original game, and the value of the new game is r plus the value of the old game.

KEY IDEAS FOR REVIEW

☐ **Theorem 8.1.** Let $A(G)$ be the adjacency matrix of a digraph G and let the rth power of $A(G)$ be B_r:
$$[A(G)]^r = B_r = [b_{ij}^{(r)}].$$
Then the i, jth element in B_r, $b_{ij}^{(r)}$, is the number of ways in which P_i has access to P_j in r stages.

☐ **Theorem 8.2.** Let $A(G)$ be the adjacency matrix of a digraph and let $S = [s_{ij}]$ be the symmetric matrix defined by $s_{ij} = 1$ if $a_{ij} = 1$ and 0 otherwise, with $S^3 = [s_{ij}^{(3)}]$, where $s_{ij}^{(3)}$ is the i, jth element in

S^3. Then P_i belongs to a clique if and only if the diagonal entry $s_{ii}^{(3)}$ is positive.

☐ **Theorem 8.3.** A digraph with n vertices is strongly connected if and only if its adjacency matrix $A(G)$ has the property that
$$A(G) + [A(G)]^2 + \cdots + [A(G)]^{n-1} = E$$
has no zero entries.

☐ **Theorem 8.4.** If T is the transition matrix of a Markov process, then the state vector $\mathbf{x}^{(k+1)}$, at the

$(k+1)$th observation period, can be determined by the state vector $\mathbf{x}^{(k)}$, at the kth observation period, as

$$\mathbf{x}^{(k+1)} = T\mathbf{x}^{(k)}.$$

☐ **Theorem 8.5.** If T is the transition matrix of a regular Markov process, then:

(a) As $n \to \infty$, T^n approaches a matrix

$$A = \begin{bmatrix} u_1 & u_1 & \cdots & u_1 \\ u_2 & u_2 & \cdots & u_2 \\ \vdots & \vdots & & \vdots \\ u_n & u_n & \cdots & u_n \end{bmatrix},$$

all of whose columns are identical.

(b) Every column

$$\mathbf{u} = \begin{bmatrix} u_1 \\ u_2 \\ \vdots \\ u_n \end{bmatrix}$$

of A is a probability vector all of whose components are positive. That is, $u_i > 0$ ($1 \le i \le n$) and

$$u_1 + u_2 + \cdots + u_n = 1.$$

☐ **Theorem 8.6.** If T is a regular transition matrix and A and U are as in Theorem 8.5, then:

(a) For any probability vector $\mathbf{x}$, $T^n\mathbf{x} \to \mathbf{u}$ as $n \to \infty$, so that $\mathbf{u}$ is a steady-state vector.

(b) The steady-state vector $\mathbf{u}$ is the unique probability vector satisfying the matrix equation $T\mathbf{u} = \mathbf{u}$.

☐ **Method of least squares** for finding the straight line $y = b_1 x + b_0$ that best fits the data $(x_1, y_1), (x_2, y_2), \ldots, (x_n, y_n)$. Let

$$\mathbf{b} = \begin{bmatrix} y_1 \\ y_2 \\ \vdots \\ y_n \end{bmatrix}, \quad A = \begin{bmatrix} x_1 & 1 \\ x_2 & 1 \\ \vdots & \vdots \\ x_n & 1 \end{bmatrix}, \quad \text{and} \quad \mathbf{x} = \begin{bmatrix} b_1 \\ b_0 \end{bmatrix}.$$

Then b_1 and b_0 can be found by solving the normal system $A^T A \mathbf{x} = A^T \mathbf{b}$ for $\mathbf{x}$ by Gauss–Jordan reduction.

☐ **Method of least squares** for finding the least squares polynomial

$$y = a_m x^m + a_{m-1} x^{m-1} + \cdots + a_1 x + a_0$$

that best fits the data $(x_1, y_1), (x_2, y_2), \ldots, (x_n, y_n)$. Let

$$\mathbf{b} = \begin{bmatrix} y_1 \\ y_2 \\ \vdots \\ y_n \end{bmatrix},$$

$$A = \begin{bmatrix} x_1^m & x_1^{m-1} & \cdots & x_1^2 & x_1 & 1 \\ x_2^m & x_2^{m-1} & \cdots & x_2^2 & x_2 & 1 \\ \vdots & \vdots & & \vdots & \vdots & \vdots \\ x_n^m & x_n^{m-1} & \cdots & x_n^2 & x_n & 1 \end{bmatrix},$$

and

$$\mathbf{x} = \begin{bmatrix} a_m \\ a_{m-1} \\ \vdots \\ a_1 \\ a_0 \end{bmatrix}.$$

Solve the normal system $A^T A \mathbf{x} = A^T \mathbf{b}$ for $\mathbf{x}$ by Gauss–Jordan reduction.

☐ **Leontief closed model:** Given an exchange matrix A, find a vector $\mathbf{p} \ge \mathbf{0}$ with at least one positive component satisfying $(I_n - A)\mathbf{p} = \mathbf{0}$.

☐ **Leontief open model:** Given a consumption matrix C, and a demand vector $\mathbf{d}$, find a production vector $\mathbf{x} \ge \mathbf{0}$ satisfying the equation $(I_n - C)\mathbf{x} = \mathbf{d}$.

☐ **Theorem 8.8.** If the $n \times n$ matrix A has n linearly independent eigenvectors $\mathbf{p}_1, \mathbf{p}_2, \ldots, \mathbf{p}_n$ associated with the eigenvalues $\lambda_1, \lambda_2, \ldots, \lambda_n$, respectively, then the general solution to the system of differential equations

$$\mathbf{x}' = A\mathbf{x}$$

is given by

$$\mathbf{x}(t) = b_1 \mathbf{p}_1 e^{\lambda_1 t} + b_2 \mathbf{p}_2 e^{\lambda_2 t} + \cdots + b_n \mathbf{p}_n e^{\lambda_n t}.$$

☐ **Fibonacci sequence.**

$$u_n = \frac{1}{\sqrt{5}} \left[\left(\frac{1 + \sqrt{5}}{2} \right)^{n+1} - \left(\frac{1 - \sqrt{5}}{2} \right)^{n+1} \right].$$

☐ **Theorem 8.9 (Principal Axes Theorem).** Any quadratic form in n variables $g(\mathbf{x}) = \mathbf{x}^T A \mathbf{x}$ is equivalent to a quadratic form,

$$h(\mathbf{y}) = \lambda_1 y_1^2 + \lambda_2 y_2^2 + \cdots + \lambda_n y_n^2,$$

where

$$\mathbf{y} = \begin{bmatrix} y_1 \\ y_2 \\ \vdots \\ y_n \end{bmatrix}.$$

and $\lambda_1, \lambda_2, \ldots, \lambda_n$ are the eigenvalues of the matrix A of Q.

☐ **Theorem 8.10.** A quadratic form $g(\mathbf{x}) = \mathbf{x}^T A \mathbf{x}$ in n variables is equivalent to a quadratic form

$$h(\mathbf{y}) = y_1^2 + y_2^2 + \cdots + y_p^2 - y_{p+1}^2 - y_{p+2}^2 - \cdots - y_r^2.$$

☐ If a_{rs} is a saddle point for a matrix game, then the optimal strategy for player R is his rth move, the optimal strategy for player C is his sth move. The value of the game is a_{rs}.

☐ The optimal strategies

$$\mathbf{p} = \begin{bmatrix} p_1 & p_2 \end{bmatrix} \quad \text{and} \quad \mathbf{q} = \begin{bmatrix} q_1 \\ q_2 \end{bmatrix}$$

for players R and C, respectively, in a 2×2 matrix game are given by

$$p_1 = \frac{a_{22} - a_{21}}{a_{11} + a_{22} - a_{12} - a_{21}}$$

$$p_2 = \frac{a_{11} - a_{12}}{a_{11} + a_{22} - a_{12} - a_{21}}$$

$$q_1 = \frac{a_{22} - a_{12}}{a_{11} + a_{22} - a_{12} - a_{21}}$$

$$q_2 = \frac{a_{11} - a_{21}}{a_{11} + a_{22} - a_{12} - a_{21}}.$$

The value of the game is

$$v = \frac{a_{11}a_{22} - a_{12}a_{21}}{a_{11} + a_{22} - a_{12} - a_{21}} = v'.$$

☐ **Theorem 8.12 (Fundamental Theorem of Matrix Games).** Every matrix game has a solution. That is, there are optimal strategies for R and C. Moreover, $v = v'$.

SUPPLEMENTARY EXERCISES ▓

1. Consider a communications network among six individuals with adjacency matrix

$$\begin{array}{c} & \begin{array}{cccccc} P_1 & P_2 & P_3 & P_4 & P_5 & P_6 \end{array} \\ \begin{array}{c} P_1 \\ P_2 \\ P_3 \\ P_4 \\ P_5 \\ P_6 \end{array} & \begin{bmatrix} 0 & 1 & 0 & 0 & 0 & 1 \\ 0 & 0 & 1 & 1 & 0 & 0 \\ 0 & 0 & 0 & 1 & 0 & 0 \\ 1 & 0 & 1 & 0 & 1 & 0 \\ 0 & 0 & 0 & 0 & 0 & 1 \\ 0 & 0 & 0 & 1 & 0 & 0 \end{bmatrix} \end{array}.$$

In how many ways does P_1 have access to P_3 through two individuals?

2. Determine the unknown currents in the following circuit.

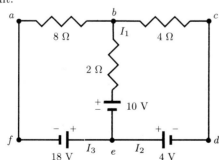

3. A marketing research organization has determined the following behavior of the average student at a certain college. If the student plays a video game

on a given day, there is a probability of 0.2 that he or she will play a game on the following day, whereas if the student has not played on a given day, there is a probability of 0.6 that he or she will play on the following day.

   (a) Write the transition matrix for the Markov process.

   (b) If the average student plays a video game on Monday, what is the probability that he or she will play on Friday of the same week?

   (c) In the long run, what is the probability that the average student will play a video game?

4. Find the least squares line for the following data points:

$$(0, 1), \quad (3, 2), \quad (5, 4), \quad (8, 10).$$

5. Find the least squares quadratic polynomial for the following data points:

$$(-1.5, 1.3), \quad (-1, 1), \quad (0, 2.8), \quad (0.5, 3.2),$$
$$(1, 3), \quad (1.5, 3.3), \quad (2, 3.6), \quad (3, 2.8).$$

6. Consider an isolated village in a remote part of Australia consisting of three households: a rancher (R) who raises all the cattle needed and nothing else; a dairy farmer (D) who produces all the dairy products needed and nothing else; and a vegetable farmer (V) who produces all the necessary vegetables and nothing else. Suppose that the units

are selected so that each person produces one unit of each commodity. Suppose that during the year the portion of each commodity that is consumed by each individual is given in Table 8.5. Let p_1, p_2, and p_3 be the prices per unit of cattle, dairy produce and vegetable produce, respectively. Suppose that everyone pays the same price for a commodity. What prices p_1, p_2, and p_3 should be assigned to the commodities so that we have a state of equilibrium?

Table 8.5

Goods Consumed by:	Goods Produced by:		
	R	D	V
R	$\frac{1}{2}$	$\frac{3}{8}$	$\frac{1}{3}$
D	$\frac{1}{4}$	$\frac{1}{4}$	$\frac{1}{3}$
V	$\frac{1}{4}$	$\frac{3}{8}$	$\frac{1}{3}$

7. Consider the homogeneous linear system of differential equations
$$\begin{bmatrix} x_1' \\ x_2' \end{bmatrix} = \begin{bmatrix} 1 & 1 \\ 3 & -1 \end{bmatrix} \begin{bmatrix} x_1 \\ x_2 \end{bmatrix}.$$
   (a) Find the general solution.
   (b) Find the solution to the initial value problem determined by the initial value conditions $x_1(0) = 4$, $x_2(0) = 6$.

8. Using a hand calculator or MATLAB, compute the Fibonacci number u_{25}.

9. Find a quadratic form of the type in Theorem 8.10 that is equivalent to the quadratic form
$$x^2 + 2y^2 + z^2 - 2xy - 2yz.$$

10. Solve the following matrix game:
$$\begin{bmatrix} 6 & 2 & 3 \\ 3 & 4 & 2 \\ 4 & 1 & 2 \end{bmatrix}.$$

CHAPTER TEST ◾

1. Determine a clique, if there is one, for the digraph with the following adjacency matrix:
$$\begin{bmatrix} 0 & 1 & 1 & 0 & 1 & 1 \\ 0 & 0 & 0 & 1 & 0 & 1 \\ 1 & 0 & 0 & 0 & 1 & 1 \\ 0 & 1 & 0 & 0 & 1 & 0 \\ 1 & 0 & 1 & 1 & 0 & 0 \\ 1 & 1 & 1 & 0 & 0 & 0 \end{bmatrix}.$$

2. Determine the unknowns in the following circuit.

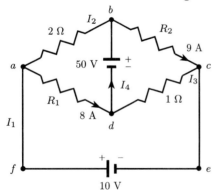

3. A political analyst who has been studying the voting patterns of nonregistered voters in a certain city has come to the following conclusions. If on a given year, a voter votes Republican, then the probability that on the following year he or she will again vote Republican is 0.4. If on a given year, a voter votes Democratic, then the probability that on the following year he or she will again vote Democratic is 0.5.
   (a) Write the transition matrix for the Markov process.
   (b) If a voter votes Republican in 1996, what is the probability that he or she will vote Republican in 2000?
   (c) In the long run, what is the probability of a voter voting Democratic?

4. A physical equipment center determines the relationship between the length of time spent on a certain piece of equipment and the number of calories lost. The following data are obtained:

Time on Equipment (minutes)	5	8	10	12	15
Lost Calories (hundreds)	3.2	5.5	6.8	7.8	9.2

Let x denote the number of minutes on the equipment, and let y denote the number of calories lost (in hundreds).

(a) Find the least squares line relating x and y.

(b) Use the equation obtained in part (a) to estimate the number of calories lost after 20 minutes on the equipment.

5. Consider a city that has three basic industries: a steel plant, a coal mine, and a railroad. To produce $1 of steel, the steel plant uses $0.50 of steel, $0.30 of coal, and $0.10 of transportation. To mine $1 of coal, the coal mine uses $0.10 of steel, $0.20 of coal, and $0.30 of transportation. To provide $1 of transportation, the railroad uses $0.10 of steel, $0.40 of coal, and $0.05 of transportation. Suppose that during the month of December there is an outside demand of 2 million dollars for steel, 1.5 million dollars for coal, and 0.5 million dollars for transportation. How much should each industry produce to satisfy the demands?

6. Find the general solution to the linear system of differential equations:
$$\begin{bmatrix} x_1' \\ x_2' \end{bmatrix} = \begin{bmatrix} 3 & 2 \\ 6 & -1 \end{bmatrix} \begin{bmatrix} x_1 \\ x_2 \end{bmatrix}.$$

7. Using a hand calculator or MATLAB, compute the Fibonacci number u_{30}.

8. Let $g(\mathbf{x}) = 2x^2 + 6xy + 2y^2 = 1$ be the equation of a central conic. Identify the conic by finding a quadratic form of the type in Theorem 8.10 that is equivalent to g.

9. Solve the following matrix game by linear programming:
$$\begin{bmatrix} -3 & 2 & 4 \\ 4 & 1 & 5 \end{bmatrix}.$$

10. Show that the following matrix game is strictly determined regardless of the value of a:
$$\begin{bmatrix} 2 & 3 \\ 1 & a \end{bmatrix}.$$

PART III

NUMERICAL LINEAR ALGEBRA

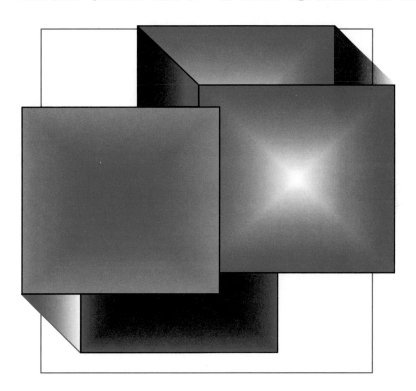

9

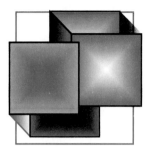

Numerical Linear Algebra

In this chapter we give a brief sketch of some widely used numerical methods for solving linear systems and for finding eigenvalues and eigenvectors. Almost all computational work in applied linear algebra using large matrices is done on a computer. These efforts have given added impetus to the relatively new area of **numerical linear algebra**, which seeks to evaluate various methods, improve existing methods, and find new methods. Computer programs that implement these methods are widely available.

9.1 ▾ Error Analysis

Prerequisite. None.

Before discussing the evaluation of different numerical methods, we look at the way in which numbers are handled in a computer. Most of the computational work done by the computer is carried out in **floating-point arithmetic**. In this system each number is written as

$$\pm 0.d_1 d_2 \cdots d_n \times 10^e, \tag{1}$$

where e is an integer that satisfies

$$-M_1 \le e \le M_2,$$

and the digits d_i satisfy

$$1 \le d_1 \le 9 \quad \text{and} \quad 0 \le d_i \le 9 \qquad (2 \le i \le n).$$

The numbers M_1 and M_2 depend on the computer and software being used. Common values for single-precision floating-point arithmetic are $M_1 = 39$ and $M_2 = 47$. In MATLAB, $M_1 = M_2 = 308$. The number n of digits in (1) is called the **number of significant digits**. In some computers, seven significant digits are used; in others, eight digits are used. The fraction $.d_1 d_2 \cdots d_n$ in (1) is called the **mantissa** and e is called the **exponent**. A number of the form in (1) is said to be a **floating-point number**.

Examples of floating-point numbers for $n = 4$ are

$$+0.2310 \times 10^2, \qquad -0.6844 \times 10^4, \qquad -0.7510 \times 10^{-2},$$
$$+0.4000 \times 10^2, \qquad +0.6300 \times 10^0, \qquad +0.3142 \times 10^1.$$

Suppose that we now want to represent a number N in the computer. If N has a nonterminating decimal representation (for example, $N = \frac{7}{3} = 2.33\ldots$), then we must approximate the number as it is entered into the computer, for we only have n significant digits at our disposal. First, we write the number as

$$\pm 0.d_1 d_2 \cdots d_n d_{n+1} \cdots \times 10^e, \tag{2}$$

$1 \le d_1 \le 9$, and $0 \le d_i \le 9$ $(i \ge 2)$. Now we proceed in one of two ways.

Truncation or Chopping

Keep the first n digits in (2). Thus $\frac{7}{3} = 2.333 \cdots = 0.233 \cdots \times 10^1$, which when truncated to four significant figures becomes 0.2333×10^1; $\frac{8}{3} = 2.666 \cdots = 0.2666 \cdots \times 10^1$, which when truncated to four significant figures becomes 0.2666×10^1.

Rounding

If $d_{n+1} \ge 5$, add 1 to d_n and truncate. Otherwise, merely truncate. Another way of stating this procedure is: Add 5 to d_{n+1} and then truncate. Thus $\frac{7}{3}$ rounded to four significant figures is 0.2333×10^1 and $\frac{8}{3}$ rounded to four significant figures is 0.2667×10^1. Some computers deal with the problem of representing numbers by rounding, others by truncating. Also, many computers make available **double-precision arithmetic** (through either hardware or software), thereby providing twice as many digits as in single-precision floating-point arithmetic.

Numerical errors that occur due to either rounding or truncating are called **roundoff errors**. There are two types of errors that can be studied. Let N be a real number and let $\widehat{N}$ be an approximation to it. The **absolute error** ε_a is defined as

$$\varepsilon_a = \widehat{N} - N,$$

and the **relative error** ε_r is

$$\varepsilon_r = \frac{\widehat{N} - N}{N} \quad \text{if } N \ne 0.$$

The relative error is not defined if $N = 0$. The relative error is the more important quantity, since a small error in a large number may not be too harmful, while the same error in a small number can be disastrous. Thus an error of 0.1234 in the number 20,642.3217 is rather trivial, but the same error in 0.8266 poses some serious problems. The three major causes of roundoff error are (1) adding a large number to a relatively small number, (2) subtracting two numbers that are almost equal, and (3) dividing a number by a very small number. We might also note that today's personal computers and handheld calculators have, except for storage and speed, many of the capabilities of a large computer. They also have the same problems with roundoff.

Much effort in numerical linear algebra goes into the development of methods that reduce roundoff error and methods that estimate the roundoff error incurred in a numerical procedure.

9.1   EXERCISES

In Exercises 1 through 4, write the given number as a floating-point number with four significant digits.

1. 34.7213

2. -0.0002135

3. -284

4. 24.00

In Exercises 5 through 8, write the given number as a floating-point number with four significant digits, first by rounding and then by truncating.

5. 1.230

6. -4.25678

7. $\frac{17}{3}$

8. $\frac{13}{6}$

In Exercises 9 through 12, find the absolute error and the relative error.

9. $N = 12.341$, $\widehat{N} = 12.362$.

10. $N = -0.4821$, $\widehat{N} = -0.4215$.

11. $N = 6482.0$, $\widehat{N} = 6483.1$.

12. $N = 0.00724$, $\widehat{N} = 0.00742$.

MATLAB EXERCISES ■

MATLAB uses a floating-point number system with approximately 16 significant digits. The exponent range has $M = 308$, hence very large and very small numbers can be represented.

ML.1. MATLAB has a built-in constant **pi** that approximates the irrational number π. To illustrate MATLAB's floating-point number system and its display formats (see Section 10.1), use the following MATLAB commands:

> **format short e**
>
> **pi**
>
> **format long e**
>
> **pi**
>
> **format short**

ML.2. Roundoff error is a problem in every floating-point number system. Enter the following commands for a demonstration of adding 0.1 to itself 15 times.

> **format long e**
>
> **v(1)=0.1; for i=2:15, v(i)=v(i−1)+.1; end, v′**
>
> **format short**

Is the display what you expected? Change the 15 to 35 and inspect the output.

9.2 ▾ Linear Systems

Prerequisite. Section 1.5, Solutions of Linear Systems of Equations.

In Section 1.5 we discussed the Gauss–Jordan reduction method for solving a linear system of m equations in n unknowns by transforming the augmented matrix to reduced row echelon form. We first present here a similar method, which is more efficient from a computational point of view.

DEFINITION   An $m \times n$ matrix A is said to be in **row echelon form** when it satisfies the following properties:

(a) All rows consisting entirely of zeros, if any, are at the bottom of the matrix.

(b) The first nonzero entry in each row that does not consist entirely of zeros is a 1, called the **leading entry** of its row.

(c) If rows i and $i+1$ are two successive rows that do not consist entirely of zeros, then the leading entry of row $i+1$ must be to the right of the leading entry of row i.

Row echelon form differs from the earlier concept of reduced row echelon form in that in a column that contains the leading entry of some row, the entries *above* the leading entry need not be zero.

EXAMPLE 1 ■ The matrices

$$
\begin{bmatrix} 1 & -3 & 2 & 4 \\ 0 & 1 & 4 & 5 \\ 0 & 0 & 1 & 2 \end{bmatrix},
\qquad
\begin{bmatrix} 1 & 2 & 3 & 1 & 2 \\ 0 & 1 & -2 & 4 & 3 \\ 0 & 0 & 1 & -2 & 6 \\ 0 & 0 & 0 & 0 & 0 \\ 0 & 0 & 0 & 0 & 0 \end{bmatrix}
$$

are in row echelon form.   ■

It is not difficult to see that by using the first six steps of Theorem 1.5 of Section 1.5, we can show that every $m \times n$ matrix is row equivalent to a matrix in row echelon form, which need not be unique. We transform a matrix to row echelon form by using the first six steps of Theorem 1.5 in the following example.

EXAMPLE 2 ■ Let A be the matrix of Example 5 in Section 1.5.

$$
A = \begin{bmatrix} 0 & 2 & 3 & -4 & 1 \\ 0 & 0 & 2 & 3 & 4 \\ 2 & 2 & -5 & 2 & 4 \\ 2 & 0 & -6 & 9 & 7 \end{bmatrix}.
$$

Proceeding as far as Step 6, we have

$$
B_3 = \begin{bmatrix} 1 & 1 & -\frac{5}{2} & 1 & 2 \\ 0 & 1 & \frac{3}{2} & -2 & \frac{1}{2} \\ 0 & 0 & 2 & 3 & 4 \\ 0 & 0 & 2 & 3 & 4 \end{bmatrix}.
$$

We now identify C by deleting but not erasing the first row of B_3.

$$C = \begin{bmatrix} 1 & 1 & -\frac{5}{2} & 1 & 2 \\ 0 & 1 & \frac{3}{2} & -2 & \frac{1}{2} \\ 0 & 0 & ② & 3 & 4 \\ 0 & 0 & 2 & 3 & 4 \end{bmatrix}$$

$$C_1 = C_2 = \begin{bmatrix} 1 & 1 & -\frac{5}{2} & 1 & 2 \\ 0 & 1 & \frac{3}{2} & -2 & \frac{1}{2} \\ 0 & 0 & 1 & \frac{3}{2} & 2 \\ 0 & 0 & 2 & 3 & 4 \end{bmatrix}$$

No row of C had to be interchanged. The first row of C was multiplied by $\frac{1}{2}$.

$$C_3 = \begin{bmatrix} 1 & 1 & -\frac{5}{2} & 1 & 2 \\ 0 & 1 & \frac{3}{2} & -2 & \frac{1}{2} \\ 0 & 0 & 1 & \frac{3}{2} & 2 \\ 0 & 0 & 0 & 0 & 0 \end{bmatrix}$$

(-2) times the first row of C_2 was added to its second row.

The final matrix

$$\begin{bmatrix} 1 & 1 & -\frac{5}{2} & 1 & 2 \\ 0 & 1 & \frac{3}{2} & -2 & \frac{1}{2} \\ 0 & 0 & 1 & \frac{3}{2} & 2 \\ 0 & 0 & 0 & 0 & 0 \end{bmatrix}$$

is in row echelon form.   ∎

If we transform the augmented matrix $\begin{bmatrix} A \vdots \mathbf{b} \end{bmatrix}$ of the linear system $A\mathbf{x} = \mathbf{b}$ to row echelon form, we get another method for solving linear systems, which is called **Gaussian elimination**.

The Gaussian elimination method for solving the linear system $A\mathbf{x} = \mathbf{b}$ is as follows.

Step 1. Form the augmented matrix $\begin{bmatrix} A \vdots \mathbf{b} \end{bmatrix}$.

Step 2. Transform the augmented matrix to row echelon form by using elementary row operations.

Step 3. Solve the linear system that corresponds to the matrix in row echelon form that has been obtained in Step 2 by using back substitution.

EXAMPLE 3 ■ Consider Example 6 of Section 1.5:

$$x + 2y + 3z = 9$$
$$2x - y + z = 8$$
$$3x \quad - z = 3.$$

The augmented matrix is row equivalent to the matrix (verify),

$$\begin{bmatrix} 1 & 2 & 3 & \vdots & 9 \\ 0 & 1 & 1 & \vdots & 2 \\ 0 & 0 & 1 & \vdots & 3 \end{bmatrix}, \tag{1}$$

which is in row echelon form. The solution can be obtained by using **back substitution** as follows.

From the last row of (1) we have

$$0x + 0y + 1z = 3$$

so that $z = 3$. From the second row of (1),

$$0x + 1y + 1z = 2,$$

so

$$y = 2 - z = 2 - 3 = -1.$$

Now from the first row of (1),

$$x + 2y + 3z = 9,$$

so

$$x = 9 - 2y - 3z = 9 - 2(-1) - 3(3) = 2.$$

Thus the solution is $x = 2$, $y = -1$, $z = 3$. ■

Throughout the rest of this section we limit our discussion to linear systems of n equations with n unknowns.

One of the troubles with Gaussian elimination is that we have to divide by the pivot; if the latter is a very small number, then the roundoff error can cast considerable doubt upon the final answer. A partial remedy to this problem consists of using the technique called **partial pivoting**. This method calls for choosing the largest (in absolute value) nonzero entry in the pivotal column as the pivot element.

EXAMPLE 4 ■ To solve the linear system

$$14x_1 + 2x_2 + 4x_3 = -10$$
$$16x_1 + 40x_2 - 4x_3 = 55 \tag{2}$$
$$-2x_1 + 4x_2 - 16x_3 = -38$$

by Gaussian elimination with partial pivoting, we proceed as follows. Form the augmented matrix

$$\begin{bmatrix} 14 & 2 & 4 & \vdots & -10 \\ 16 & 40 & -4 & \vdots & 55 \\ -2 & 4 & -16 & \vdots & -38 \end{bmatrix}.$$

Our computations will be carried out to three decimal places and we round after each multiplication and division.

$$\begin{bmatrix} 16 & 40 & -4 & \vdots & 55 \\ 14 & 2 & 4 & \vdots & -10 \\ -2 & 4 & -16 & \vdots & -38 \end{bmatrix}$$

The first and second rows were interchanged.

$$\begin{bmatrix} 1.000 & 2.500 & -0.250 & \vdots & 3.438 \\ 14.000 & 2.000 & 4.000 & \vdots & -10.000 \\ -2.000 & 4.000 & -16.000 & \vdots & -38.000 \end{bmatrix}$$

The first row was multiplied by $\frac{1}{16}$.

$$\begin{bmatrix} 1.000 & 2.500 & -0.250 & \vdots & 3.438 \\ 0.000 & -33.000 & 7.500 & \vdots & -58.132 \\ 0.000 & 9.000 & -16.500 & \vdots & -31.124 \end{bmatrix}$$

(-14) times the first row was added to the second row; 2 times the first row was added to its third row.

$$\begin{bmatrix} 1.000 & 2.500 & -0.250 & \vdots & 3.438 \\ 0.000 & 1.000 & -0.227 & \vdots & 1.762 \\ 0.000 & 9.000 & -16.500 & \vdots & -31.124 \end{bmatrix}$$

The second row was multiplied by $\left(\frac{1}{-33}\right)$; that is, -33 is the pivot for the 2×4 submatrix obtained by removing the first row.

$$\begin{bmatrix} 1.000 & 2.500 & -0.250 & \vdots & 3.438 \\ 0.000 & 1.000 & -0.227 & \vdots & 1.762 \\ 0.000 & 0.000 & -14.457 & \vdots & -46.982 \end{bmatrix}$$

(-9) times the second row was added to the third row.

$$\begin{bmatrix} 1.000 & 2.500 & -0.250 & \vdots & 3.438 \\ 0.000 & 1.000 & -0.227 & \vdots & 1.762 \\ 0.000 & 0.000 & 1.000 & \vdots & 3.250 \end{bmatrix}$$

The third row was multiplied by $\left(\frac{1}{-14.457}\right)$; that is, -14.457 is the pivot for the 1×4 submatrix obtained by removing the first two rows.

Then

$$x_3 = 3.250$$
$$x_2 = 1.762 + (0.227)(3.250) = 2.500$$
$$x_1 = 3.438 + (0.250)(3.250) - (2.500)(2.500) = -1.999.$$

Thus the solution obtained is

$$x_1 = -1.999, \qquad x_2 = 2.500, \qquad x_3 = 3.250,$$

which agrees rather well with the exact solution:

$$x_1 = -2, \qquad x_2 = 2.5, \qquad x_3 = 3.25. \qquad \blacksquare$$

Full pivoting consists of choosing the largest nonzero entry in the entire coefficient matrix as the pivot. Although this variant sometimes gives better results than partial pivoting, it requires a relabeling of the variables. The additional programming and bookkeeping required make Gaussian elimination with partial pivoting a more popular method than Gaussian elimination with full pivoting.

Ill-Conditioned Systems

There are linear systems, $A\mathbf{x} = \mathbf{b}$, where A is $n \times n$, in which relatively small changes in some of the elements of the augmented matrix $\begin{bmatrix} A \mid \mathbf{b} \end{bmatrix}$ may lead to relatively large changes in the solution. Such systems are called **ill-conditioned systems**. Gaussian elimination with partial pivoting or full pivoting may reduce the roundoff error for some of these systems. Sometimes **double-precision arithmetic** (requiring twice as much storage as single-precision and longer running times) may provide some help. However, for some ill-conditioned systems there may be no possible remedy.

EXAMPLE 5 ■ Consider the linear system

$$\begin{aligned} x - \quad y &= 1 \\ x - 1.01y &= 0 \end{aligned} \tag{3}$$

whose exact solution is

$$x = 101, \qquad y = 100.$$

Suppose that the coefficient of y in the second equation of (3) now changes from 1.01 to 0.99. The resulting system

$$\begin{aligned} x - \quad y &= 1 \\ x - 0.99y &= 0 \end{aligned} \tag{4}$$

has the exact solution

$$x = -99, \qquad y = -100.$$

Thus the systems (3) and (4), which are nearly identical, have vastly different solutions. System (3) is an ill-conditioned system. If we sketch the lines for (3) (Figure 9.1), we see that they are nearly parallel. Thus a minor shift of one of the lines results in a major shift of the point of intersection. ■

Iterative Methods

Gaussian elimination with partial pivoting is a **direct** method for solving $A\mathbf{x} = \mathbf{b}$, where A is $n \times n$. That is, the solution will be obtained after a finite number of steps. Moreover, the number of steps required to obtain the solution can be estimated in advance.

We shall now outline two additional methods, one due to Jacobi, the other to Gauss and Seidel, for solving such linear systems. These methods are **iterative** in nature; that is, we start out with an initial approximation to the solution, which we successively try to improve. If the successive approximations tend to

FIGURE 9.1

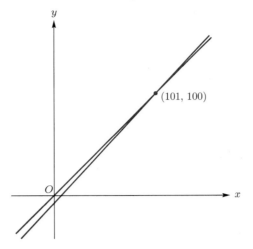

approach the solution, we say that the method **converges**. Otherwise, we say that the method **diverges**.

The Jacobi and Gauss–Seidel methods do not always converge, and we give below a sufficient condition for their convergence. These methods are more efficient, when applicable, than Gaussian elimination with partial pivoting especially for very large matrices that are **sparse** matrices (very large matrices having a large number of zero elements).

To discuss these methods, we start with the linear system $A\mathbf{x} = \mathbf{b}$, where A is $n \times n$. Assume that the determinant of A is nonzero and that the diagonal entries of A are all nonzero.

We shall illustrate the two methods (see pages 544 and 545) with the linear system (2). Throughout the rest of this section we work with three decimal places and *round* after each multiplication and division.

In Step 3 of the Jacobi iteration method, we substitute the values of the variables calculated in the last iteration into the right side of (5) to obtain a new approximation. However, in the next iteration, once we have a new value for x_1, we can use it to calculate x_2 in Equation (5); once we have new values for x_1 and x_2, we can use them to calculate x_3; and so on. Thus we can replace Step 3 of the Jacobi iteration method by the following procedure.

Step 3. The kth iteration consists of the following. For the ith equation of (5), substitute the most recently calculated values of the variables into the right side to obtain the next approximation $x_i^{(k)}$. Do this for $i = 1, 2, \ldots, n$. The resulting method is called the **Gauss–Seidel iteration method**.

Jacobi* Iteration Method

The procedure for carrying out the Jacobi iteration method for solving n equations in n unknowns is as follows.

Step 1. Rewrite (2) by expressing x_i in the ith equation in terms of the remaining variables. Thus

$$x_1 = -\tfrac{10}{14} - \tfrac{2}{14}x_2 - \tfrac{4}{14}x_3$$
$$x_2 = \tfrac{55}{40} - \tfrac{16}{40}x_1 + \tfrac{4}{40}x_3$$
$$x_3 = \tfrac{38}{16} - \tfrac{2}{16}x_1 + \tfrac{4}{16}x_2$$

or

$$\begin{aligned} x_1 &= -0.714 - 0.143x_2 - 0.286x_3 \\ x_2 &= 1.375 - 0.400x_1 + 0.100x_3 \\ x_3 &= 2.375 - 0.125x_1 + 0.250x_2. \end{aligned} \tag{5}$$

Step 2. Choose an initial approximation $x_1^{(0)}, x_2^{(0)}, \ldots, x_n^{(0)}$ to the solution. In the absence of other information, let

$$x_1^{(0)} = x_2^{(0)} = \cdots = x_n^{(0)} = 0.$$

In our example, $x_1^{(0)} = x_2^{(0)} = x_3^{(0)} = 0$.

Step 3. Substitute the values of the variables calculated in the previous iteration [$(k-1)$st iteration] into the right side of (5) to obtain a new approximation, $x_1^{(k)}, x_2^{(k)}, \ldots, x_n^{(k)}$. For our example,

$$x_1^{(1)} = -0.714, \qquad x_2^{(1)} = 1.375, \qquad x_3^{(1)} = 2.375.$$

Using Step 3 again, we now obtain our second approximation,

$$x_1^{(2)} = -1.590, \qquad x_2^{(2)} = 1.898, \qquad x_3^{(2)} = 2.808.$$

The first nine approximations ($k = 8$) to the solution are shown in Table 9.1 on page 546. Thus our approximation to the solution is

$$x_1 \approx -1.998, \qquad x_2 \approx 2.497, \qquad x_3 \approx 3.247$$

and it was obtained in nine iterations. The exact solution is

$$x_1 = -2, \qquad x_2 = 2.5, \qquad x_3 = 3.25.$$

The symbol $\approx$ means "approximately."

---

*Karl Gustav Jacobi (1804–1851) was born in Potsdam, Germany, and was educated at the University of Berlin. He was a professor at the University of Königsberg and made many important contributions to the theory of elliptic functions, determinants, and differential equations.

Gauss–Seidel** Iteration Method

The procedure for carrying out the Gauss–Seidel iteration method for solving n equations in n unknowns is as follows.

Step 1. Rewrite the given system by expressing x_i in the ith equation in terms of the remaining variables. Thus

$$
\begin{aligned}
x_1 &= -\tfrac{10}{14} - \tfrac{2}{14}x_2 - \tfrac{4}{14}x_3 \\
x_2 &= \tfrac{55}{40} - \tfrac{16}{40}x_1 + \tfrac{4}{40}x_3 \\
x_3 &= \tfrac{38}{16} - \tfrac{2}{16}x_1 + \tfrac{4}{16}x_2.
\end{aligned}
\tag{6}
$$

Step 2. Choose an initial approximation $x_1^{(0)} = x_2^{(0)} = \cdots = x_n^{(0)}$ to the solution. In the absence of other information, let

$$x_1^{(0)} = x_2^{(0)} = \cdots = x_n^{(0)} = 0.$$

In our example, $x_1^{(0)} = x_2^{(0)} = x_3^{(0)} = 0$.

Step 3. Substitute the most recently calculated values of the variables into the right side of the ith equation in the equations in Step 1 to obtain the new approximation $x_i^{(k)}$. For our example, we substitute these values into the right side of the first equation of (6), and obtain

$$x_1^{(1)} = -0.714.$$

We next substitute

$$x_1^{(1)} = -0.714 \quad \text{and} \quad x_3^{(0)} = 0$$

into the right side of the second equation of (5) to obtain $x_2^{(1)} = 1.661$. Now substitute

$$x_1^{(1)} = -0.714 \quad \text{and} \quad x_2^{(1)} = 1.661$$

into the right side of the third equation of (5) to obtain $x_3^{(1)} = 2.880$. We next use the values of $x_2^{(1)} = 1.661$ and $x_3^{(1)} = 2.880$ in the first equation of (5) to obtain $x_1^{(2)} = -1.775$. Now use $x_1^{(2)} = -1.775$ and $x_3^{(1)} = 2.880$ in the second equation of (5) to obtain $x_2^{(2)} = 2.373$. The first six approximations $(k = 5)$ to the solution are shown in Table 9.2 on page 546. Thus our approximation to the solution is

$$x_1 \approx -2.000, \qquad x_2 \approx 2.500, \qquad x_3 \approx 3.250,$$

and it was obtained in six iterations.

***

**Philipp Ludwig von Seidel (1821–1896) was born in Bavaria, Germany. He studied at Berlin, Königsberg, and Munich, and taught at Munich before he was forced to retire early by failing eyesight. He participated in the general development of analysis during the nineteenth century and also made contributions to astronomy.

Table 9.1

kth *Iteration*	$x_1^{(k)}$	$x_2^{(k)}$	$x_3^{(k)}$
0	0	0	0
1	−0.714	1.375	2.375
2	−1.590	1.898	2.808
3	−1.789	2.292	3.048
4	−1.913	2.395	3.172
5	−1.964	2.457	3.213
6	−1.984	2.482	3.235
7	−1.994	2.492	3.244
8	−1.998	2.497	3.247

Table 9.2

kth *Iteration*	$x_1^{(k)}$	$x_2^{(k)}$	$x_3^{(k)}$
0	0	0	0
1	−0.714	1.661	2.880
2	−1.775	2.373	3.190
3	−1.966	2.480	3.241
4	−1.996	2.498	3.249
5	−2.000	2.500	3.250

Although Gauss–Seidel iteration took fewer iterations than the Jacobi iteration method for our illustrative problem, there are problems where the opposite is true. Moreover, one cannot predict, in advance, which method is better for a given problem unless the coefficient matrix has a special structure. (Such topics are studied in numerical linear algebra.)

Convergence

Consider the linear system $A\mathbf{x} = \mathbf{b}$, where A is $n \times n$. A sufficient condition for the convergence of the Jacobi and Gauss–Seidel iteration methods is

$$|a_{ii}| > \sum_{\substack{j=1 \\ j \neq i}}^{n} |a_{ij}| \qquad (1 \leq i \leq n). \tag{7}$$

Equation (7) means that every diagonal element is in absolute value larger than the sum of the absolute values of the other entries in its row. A matrix A satisfying (6) is called **diagonally dominant**.

If A is diagonally dominant, then the Jacobi and Gauss–Seidel methods converge. Of course, the methods may converge even if this condition fails to hold (see Exercise 15).

EXAMPLE 6 ■ The matrix

$$\begin{bmatrix} 3 & -2 & 0 \\ 2 & -6 & 3 \\ 5 & 2 & 8 \end{bmatrix}$$

is diagonally dominant, whereas

$$\begin{bmatrix} 4 & 2 & -2 \\ 2 & 4 & 1 \\ 5 & 2 & 8 \end{bmatrix}$$

is not, since $|4|$ is not greater than $|2| + |-2|$.   ■

Sometimes it is necessary to interchange the equations so that the coefficient matrix will be diagonally dominant.

EXAMPLE 7 ■ Consider the linear system

$$\begin{aligned} 12x_1 - 14x_2 + x_3 &= 4 \\ -6x_1 + 4x_2 + 11x_3 &= -6 \\ 6x_1 + 3x_2 + 2x_3 &= 8. \end{aligned}$$

The coefficient matrix is obviously not diagonally dominant. However, if we interchange the first and second rows and then the first and third rows of the resulting system (verify), we obtain the coefficient matrix

$$\begin{bmatrix} 6 & 3 & 2 \\ 12 & -14 & 1 \\ -6 & 4 & 11 \end{bmatrix},$$

which is diagonally dominant:

$$|6| > |3| + |2|; \qquad |-14| > |12| + |1|; \qquad |11| > |-6| + |4|. \qquad ■$$

9.2   EXERCISES

In Exercises 1 and 2, transform the given matrix to row echelon form.

1. $\begin{bmatrix} 1 & -2 & 0 \\ 2 & -3 & -1 \\ 1 & 3 & 2 \end{bmatrix}$.   **2.** $\begin{bmatrix} 0 & -1 & 2 & 3 \\ 2 & 3 & 4 & 5 \\ 1 & 3 & -1 & 2 \\ 3 & 2 & 4 & 1 \end{bmatrix}$.

In Exercises 3 and 4, solve the given linear system by Gaussian elimination.

3. $\begin{aligned} x_1 + 2x_2 + x_3 &= 0 \\ -3x_1 + 3x_2 + 2x_3 &= -7 \\ 4x_1 - 2x_2 - 3x_3 &= 2. \end{aligned}$

4. $3x_1 - 3x_2 + 2x_3 = 12$
$4x_1 + 6x_2 - 4x_3 = -2$
$5x_1 - 9x_2 - 6x_3 = 6.$

In Exercises 5 through 8, solve the given linear system by Gaussian elimination with partial pivoting. Carry out your computations to three decimal places and round after each multiplication and division.

5. $3x_1 - 2x_2 + 3x_3 = -8$
$6x_1 - 4x_2 + 5x_3 = -14$
$-12x_1 + 6x_2 + 7x_3 = -8.$

6. $-2x_1 + 3x_2 + 4x_3 = 8$
$3x_1 + 6x_2 - x_3 = -8$
$x_1 + 2x_2 + x_3 = 4.$

7. $2.5x_1 + 3.5x_2 - 4.25x_3 = 37.3$
$3.4x_1 + 2.5x_2 - 2.01x_3 = 26.8$
$5.3x_1 - 2.4x_2 + 6.21x_3 = -20.68.$

8. $2.2x_1 - 3.5x_2 + 1.2x_3 = -18.3$
$4.4x_1 + 4.6x_2 - 3.8x_3 = -25.6$
$-6.3x_1 + 3.8x_2 + 2.4x_3 = 57.1.$

9. Verify that the linear system

$$2.121x + 3.421y = 13.205$$
$$2.12x + 3.42y = 13.200$$

is ill-conditioned by solving the linear system

$$2.121x + 3.421y = 13.205$$
$$2.12x + 3.42y = 13.203.$$

Carry out your computations to *four* decimal places and round after each multiplication and division.

10. Which of the following matrices are diagonally dominant?

(a) $\begin{bmatrix} 2 & 1 \\ 4 & -5 \end{bmatrix}.$ (b) $\begin{bmatrix} 3 & 2 \\ -4 & 3 \end{bmatrix}.$

(c) $\begin{bmatrix} 3 & 1 & 2 \\ 0 & 4 & -2 \\ 1 & -2 & 4 \end{bmatrix}.$ (d) $\begin{bmatrix} -4 & 3 & 0 \\ 4 & 6 & 1 \\ 2 & -2 & 8 \end{bmatrix}.$

(e) $\begin{bmatrix} 3 & -2 & -1 \\ 4 & 5 & 1 \\ -2 & -4 & 6 \end{bmatrix}.$

In Exercises 11 through 14, solve the given linear system by (1) Jacobi's iteration method and (2) the Gauss–Seidel iteration method. Carry out the computations to three decimal places for six iterations ($k = 5$) and round after each multiplication and division.

11. $16x + 5y = -7$
$4x + 15y = 67.$

12. $6x_1 - x_2 - 2x_3 = 12$
$3x_1 + 5x_2 + x_3 = 3$
$2x_1 - x_2 + 7x_3 = 36.$

13. $9x_1 - 2x_2 + 6x_3 = 9$
$3x_1 + 6x_2 - 2x_3 = 15$
$6x_1 - x_2 + 8x_3 = 4.5.$

14. $15x_1 - 4x_2 - 2x_3 = -2$
$5x_1 - 8x_2 + x_3 = -9$
$10x_1 - 2x_2 + 15x_3 = -50.$

15. (a) Consider the linear system

$$2x + 4y = 8$$
$$x - y = -5.$$

The coefficient matrix is not diagonally dominant. Show that the Gauss–Seidel iteration method diverges. Use seven iterations ($k = 6$).

(b) Consider the linear system

$$2x - y = -8$$
$$x + y = -1.$$

The coefficient matrix is not diagonally dominant. Show that the Gauss–Seidel iteration method converges. Use seven iterations ($k = 6$).

MATLAB EXERCISES ▇

Routine **reduce** can be used to obtain the row echelon form of a matrix $\begin{bmatrix} A & \vdots & \mathbf{b} \end{bmatrix}$. Then back substitution can be performed by hand or by using routine **bksub** as described below.

ML.1. In Example 3, denote the coefficient matrix as A and the right-hand side as $\mathbf{b}$. Enter them

into MATLAB and find the row echelon form using **reduce**. Use command

$$\mathbf{Q} = \mathbf{reduce}([\mathbf{A} \quad \mathbf{b}])$$

Your final matrix should be that given in Equation (1). To solve the system using

bksub, use the command

$$x = \text{bksub}(\mathbf{Q}(:,1:3),\mathbf{Q}(:,4))$$

You should get the solution $x = 2$, $y = -1$, $z = 3$ as in Example 3. (If A were $n \times n$, then we would use the command

$$x = \text{bksub}(\mathbf{Q}(:,1:n),\mathbf{Q}(:,n+1))$$

Use the appropriate number in place of n.)

ML.2. Follow the procedure in Exercise ML.1 to solve each of the following linear systems.

(a) $\begin{aligned} 3x_1 + 4x_2 - x_3 &= -2 \\ 2x_1 - 2x_2 \phantom{{}- 2x_3} &= 8 \\ x_1 - 3x_2 + 4x_3 &= 8. \end{aligned}$

(b) $\begin{aligned} 2x_1 + 3x_2 - 2x_3 + x_4 &= 3 \\ 4x_2 - 2x_3 + 3x_4 &= 1 \\ x_1 + x_2 - 2x_3 + x_4 &= 6 \\ x_1 + 2x_2 + 3x_3 \phantom{{}+ 3x_4} &= -12. \end{aligned}$

ML.3. Routine **reduce** can be used to perform the partial pivoting method illustrated in Example 4. Once the row echelon form is obtained, routine **bksub** can be used as described in Exercise ML.1. Solve Exercise 5 using MATLAB's arithmetic and the routine **reduce** followed by **bksub**.

ML.4. The \ command for solving linear systems, which is described in Exercise ML.12 in Section 1.5, uses partial pivoting. Check your answer to Exercise ML.3 using the \ command.

9.3 ▾ LU-Factorization (Optional)

Prerequisites. Section 1.5, Solutions of Linear Systems of Equations. Section 9.2, Linear Systems.

In this section we discuss a variant of Gaussian elimination (presented in Section 9.2) that decomposes a matrix as a product of a lower triangular matrix and an upper triangular matrix. This decomposition leads to an algorithm for solving a linear system $A\mathbf{x} = \mathbf{b}$ that is the most widely used method on computers for solving a linear system. A main reason for the popularity of this method is that it provides the cheapest way of solving a linear system for which we repeatedly have to change the right side. This type of situation occurs often in applied problems. For example, an electric utility company must determine the inputs (the unknowns) needed to produce some required outputs (the right sides). The inputs and outputs might be related by a linear system, whose coefficient matrix is fixed, while the right side changes from day to day, or even hour to hour. The decomposition discussed in this section is also useful in solving other problems in linear algebra.

When U is an upper triangular matrix all of whose diagonal entries are different from zero, then the linear system $U\mathbf{x} = \mathbf{b}$ can be solved without transforming the augmented matrix $\begin{bmatrix} U \mid \mathbf{b} \end{bmatrix}$ to reduced row echelon form or to row echelon form. The augmented matrix of such a system is given by

$$\begin{bmatrix} u_{11} & u_{12} & u_{13} & \cdots & u_{1n} & b_1 \\ 0 & u_{22} & u_{23} & \cdots & u_{2n} & b_2 \\ 0 & 0 & u_{33} & \cdots & u_{3n} & b_3 \\ \vdots & \vdots & \vdots & \cdots & \vdots & \vdots \\ 0 & 0 & 0 & \cdots & u_{nn} & b_n \end{bmatrix}.$$

The solution is obtained by the following algorithm:

$$x_n = \frac{b_n}{u_{nn}}$$

$$x_{n-1} = \frac{b_{n-1} - u_{n-1\,n}x_n}{u_{n-1\,n-1}}$$

$$\vdots$$

$$x_j = \frac{b_j - \displaystyle\sum_{k=n}^{j-1} u_{jk}x_k}{u_{jj}}, \qquad j = n, n-1, \ldots, 2, 1.$$

This procedure is merely **back substitution**, which we used in conjunction with Gaussian elimination in Section 9.2, where it was additionally required that the diagonal entries be 1.

In a similar manner, if L is a lower triangular matrix all of whose diagonal entries are different from zero, then the linear system $L\mathbf{x} = \mathbf{b}$ can be solved by **forward substitution**, which consists of the following procedure: The augmented matrix has the form

$$\begin{bmatrix}
\ell_{11} & 0 & 0 & \cdots & 0 & \vdots & b_1 \\
\ell_{21} & \ell_{22} & 0 & \cdots & 0 & \vdots & b_2 \\
\ell_{31} & \ell_{32} & \ell_{33} & \cdots & 0 & \vdots & b_3 \\
\vdots & \vdots & \vdots & \cdots & \vdots & \vdots & \vdots \\
\ell_{n1} & \ell_{n2} & \ell_{n3} & \cdots & \ell_{nn} & \vdots & b_n
\end{bmatrix}$$

and the solution is given by

$$x_1 = \frac{b_1}{\ell_{11}}$$

$$x_2 = \frac{b_2 - \ell_{21}x_1}{\ell_{22}}$$

$$\vdots$$

$$x_j = \frac{b_j - \displaystyle\sum_{k=1}^{j-1} \ell_{jk}x_k}{\ell_{jj}}, \qquad j = 2, \ldots, n.$$

That is, we proceed from the first equation downward, solving for one variable from each equation.

We illustrate forward substitution in the following example.

EXAMPLE 1 ■ To solve the linear system

$$\begin{aligned}
5x_1 &&&= 10 \\
4x_1 - 2x_2 &&&= 28 \\
2x_1 + 3x_2 + 4x_3 &&= 26
\end{aligned}$$

we use forward substitution. Hence we obtain from the previous algorithm

$$x_1 = \frac{10}{5} = 2$$

$$x_2 = \frac{28 - 4x_1}{-2} = -10$$

$$x_3 = \frac{26 - 2x_1 - 3x_2}{4} = 13,$$

which implies that the solution to the given lower triangular system of equations is

$$\mathbf{x} = \begin{bmatrix} 2 \\ -10 \\ 13 \end{bmatrix}.$$

■

As illustrated above, the ease with which systems of equations with upper or lower triangular coefficient matrices can be solved is quite attractive. The forward substitution and back substitution algorithms are fast and simple to use. These are used in another important numerical procedure for solving linear systems of equations, which we develop next.

Suppose that an $n \times n$ matrix A can be written as a product of a matrix L in lower triangular form and a matrix U in upper triangular form; that is,

$$A = LU.$$

In this case we say that A has an **LU-factorization** or an **LU-decomposition**. The LU-factorization of a matrix A can be used to efficiently solve a linear system $A\mathbf{x} = \mathbf{b}$. Substituting LU for A, we have

$$(LU)\mathbf{x} = \mathbf{b}$$

or by (a) of Theorem 1.2 in Section 1.4,

$$L(U\mathbf{x}) = \mathbf{b}.$$

Letting $U\mathbf{x} = \mathbf{z}$, this matrix equation becomes

$$L\mathbf{z} = \mathbf{b}.$$

Since L is in lower triangular form, we solve directly for $\mathbf{z}$ by forward substitution. Once we determine $\mathbf{z}$, since U is in upper triangular form, we solve $U\mathbf{x} = \mathbf{z}$ by back substitution. In summary, if an $n \times n$ matrix A has an LU-factorization, then the solution of $A\mathbf{x} = \mathbf{b}$ can be determined by a forward substitution followed by a back substitution. We illustrate this procedure in the next example.

EXAMPLE 2 ■ Consider the linear system

$$\begin{aligned}
6x_1 - 2x_2 - 4x_3 + 4x_4 &= 2 \\
3x_1 - 3x_2 - 6x_3 + x_4 &= -4 \\
-12x_1 + 8x_2 + 21x_3 - 8x_4 &= 8 \\
-6x_1 \qquad - 10x_3 + 7x_4 &= -43
\end{aligned}$$

whose coefficient matrix

$$A = \begin{bmatrix} 6 & -2 & -4 & 4 \\ 3 & -3 & -6 & 1 \\ -12 & 8 & 21 & -8 \\ -6 & 0 & -10 & 7 \end{bmatrix}$$

has an LU-factorization where

$$L = \begin{bmatrix} 1 & 0 & 0 & 0 \\ \frac{1}{2} & 1 & 0 & 0 \\ -2 & -2 & 1 & 0 \\ -1 & 1 & -2 & 1 \end{bmatrix} \quad \text{and} \quad U = \begin{bmatrix} 6 & -2 & -4 & 4 \\ 0 & -2 & -4 & -1 \\ 0 & 0 & 5 & -2 \\ 0 & 0 & 0 & 8 \end{bmatrix}$$

(verify). To solve the given system using this LU-factorization, we proceed as follows. Let

$$\mathbf{b} = \begin{bmatrix} 2 \\ -4 \\ 8 \\ -43 \end{bmatrix}.$$

Then we solve $A\mathbf{x} = \mathbf{b}$ by writing it as $LU\mathbf{x} = \mathbf{b}$. First, let $U\mathbf{x} = \mathbf{z}$ and now solve $L\mathbf{z} = \mathbf{b}$:

$$\begin{bmatrix} 1 & 0 & 0 & 0 \\ \frac{1}{2} & 1 & 0 & 0 \\ -2 & -2 & 1 & 0 \\ -1 & 1 & -2 & 1 \end{bmatrix} \begin{bmatrix} z_1 \\ z_2 \\ z_3 \\ z_4 \end{bmatrix} = \begin{bmatrix} 2 \\ -4 \\ 8 \\ -43 \end{bmatrix}$$

by forward substitution. We obtain

$$z_1 = 2$$
$$z_2 = -4 - \tfrac{1}{2}z_1 = -5$$
$$z_3 = 8 + 2z_1 + 2z_2 = 2$$
$$z_4 = -43 + z_1 - z_2 + 2z_3 = -32.$$

Next we solve $U\mathbf{x} = \mathbf{z}$,

$$\begin{bmatrix} 6 & -2 & -4 & 4 \\ 0 & -2 & -4 & -1 \\ 0 & 0 & 5 & -2 \\ 0 & 0 & 0 & 8 \end{bmatrix} \begin{bmatrix} x_1 \\ x_2 \\ x_3 \\ x_4 \end{bmatrix} = \begin{bmatrix} 2 \\ -5 \\ 2 \\ -32 \end{bmatrix},$$

by back substitution. We obtain

$$x_4 = \frac{-32}{8} = -4$$
$$x_3 = \frac{2 + 2x_4}{5} = -1.2$$
$$x_2 = \frac{-5 + 4x_3 + x_4}{-2} = 6.9$$
$$x_1 = \frac{2 + 2x_2 + 4x_3 - 4x_4}{6} = 4.5.$$

Thus the solution to the given linear system is

$$\mathbf{x} = \begin{bmatrix} 4.5 \\ 6.9 \\ -1.2 \\ -4 \end{bmatrix}.$$

■

Next we show how to obtain an LU-decomposition of a matrix by modifying the Gaussian elimination procedure from Section 9.2. No row interchanges will be permitted and we do not require that the diagonal entries have value 1. The following LU-factorization procedure may fail when row interchanges are not permitted. At the end of this section we discuss how to enhance the LU-factorization scheme presented to prevent such failures. We observe that the only elementary row operation permitted is the one that adds a multiple of one row to a different row.

To describe the LU-factorization, we present a step-by-step procedure in the next example.

EXAMPLE 3 ■ Let A be the coefficient matrix of the linear system of Example 2.

$$A = \begin{bmatrix} 6 & -2 & -4 & 4 \\ 3 & -3 & -6 & 1 \\ -12 & 8 & 21 & -8 \\ -6 & 0 & -10 & 7 \end{bmatrix}.$$

We proceed to "zero out" entries below the diagonal entries using only the row operation that adds a multiple of one row to a different row.

Procedure	Matrices Used

Step 1. "Zero out" below the first diagonal entry of A. Add $\left(-\frac{1}{2}\right)$ times the first row of A to the second row of A. Add 2 times the first row of A to the third row of A. Add 1 times the first row of A to the fourth row of A. Call the new resulting matrix U_1.

$$U_1 = \begin{bmatrix} 6 & -2 & -4 & 4 \\ 0 & -2 & -4 & -1 \\ 0 & 4 & 13 & 0 \\ 0 & -2 & -14 & 11 \end{bmatrix}$$

We begin building a lower triangular matrix, L_1, with 1's on the main diagonal, to record the row operations. Enter the *negatives of the multipliers* used in the row operations in the first column of L_1, below the first diagonal entry of L_1.

$$L_1 = \begin{bmatrix} 1 & 0 & 0 & 0 \\ \frac{1}{2} & 1 & 0 & 0 \\ -2 & * & 1 & 0 \\ -1 & * & * & 1 \end{bmatrix}$$

Step 2. "Zero out" below the second diagonal entry of U_1. Add 2 times the second row of U_1 to the third row of U_1. Add (-1) times the second row of U_1 to the fourth row of U_1. Call the new resulting matrix U_2.

$$U_2 = \begin{bmatrix} 6 & -2 & -4 & 4 \\ 0 & -2 & -4 & -1 \\ 0 & 0 & 5 & -2 \\ 0 & 0 & -10 & 12 \end{bmatrix}$$

Enter the negatives of the multipliers from the row operations below the second diagonal entry of L_1. Call the new matrix L_2.

$$L_2 = \begin{bmatrix} 1 & 0 & 0 & 0 \\ \frac{1}{2} & 1 & 0 & 0 \\ -2 & -2 & 1 & 0 \\ -1 & 1 & * & 1 \end{bmatrix}$$

Step 3. "Zero out" below the third diagonal entry of U_2. Add 2 times the third row of U_2 to the fourth row of U_2. Call the new resulting matrix U_3.

$$U_3 = \begin{bmatrix} 6 & -2 & -4 & 4 \\ 0 & -2 & -4 & -1 \\ 0 & 0 & 5 & -2 \\ 0 & 0 & 0 & 8 \end{bmatrix}$$

Enter the negative of the multiplier below the third diagonal entry of L_2. Call the new matrix L_3.

$$L_3 = \begin{bmatrix} 1 & 0 & 0 & 0 \\ \frac{1}{2} & 1 & 0 & 0 \\ -2 & -2 & 1 & 0 \\ -1 & 1 & -2 & 1 \end{bmatrix}$$

Let $L = L_3$ and $U = U_3$. Then the product LU gives the original matrix A (verify). The linear system of equations in Example 2 was solved there using the LU-factorization just obtained. ∎

REMARK   In general, a given matrix may have more than one LU-factorization. For example, if A is the coefficient matrix considered in Example 2, then another LU-factorization is LU, where

$$L = \begin{bmatrix} 2 & 0 & 0 & 0 \\ 1 & -1 & 0 & 0 \\ -4 & 2 & 1 & 0 \\ -2 & -1 & -2 & 2 \end{bmatrix} \quad \text{and} \quad U = \begin{bmatrix} 3 & -1 & -2 & 2 \\ 0 & 2 & 4 & 1 \\ 0 & 0 & 5 & -2 \\ 0 & 0 & 0 & 4 \end{bmatrix}.$$

There are many methods for obtaining an LU-factorization of a matrix, besides the scheme for **storage of multipliers** described in Example 3. It is important to note that if $a_{11} = 0$, then the procedure used in Example 3 fails. Moreover, if the second diagonal entry of U_1 is zero or if the third diagonal entry of U_2 is zero, then the procedure also fails. In such cases we can try rearranging the equations of the system and beginning again or using one of the other methods for LU-factorization. Most computer programs for LU-factorization incorporate row interchanges into the storage of multipliers scheme and use additional strategies to help control roundoff error. If row interchanges are required, then the product of L and U is not necessarily A—it is a matrix that is a permutation of the rows of A. For example, if row interchanges occur when using the **lu** command in

MATLAB in the form $[\mathbf{L,U}] = \mathbf{lu(A)}$, then MATLAB responds as follows: the matrix that it yields as L is not lower triangular, U is upper triangular, and LU is A. The book *Experiments in Computational Matrix Algebra*, by David R. Hill (New York: Random House, 1988, distributed by McGraw-Hill) explores such a modification of the procedure for LU-factorization.

9.3 EXERCISES

In Exercises 1 through 4, solve the linear system $A\mathbf{x} = \mathbf{b}$ with the given LU-factorization of the coefficient matrix A. Solve the linear system using a forward substitution followed by a back substitution.

1. $A = \begin{bmatrix} 2 & 8 & 0 \\ 2 & 2 & -3 \\ 1 & 2 & 7 \end{bmatrix}$, $B = \begin{bmatrix} 18 \\ 3 \\ 12 \end{bmatrix}$,

$L = \begin{bmatrix} 2 & 0 & 0 \\ 2 & -3 & 0 \\ 1 & -1 & 4 \end{bmatrix}$, $U = \begin{bmatrix} 1 & 4 & 0 \\ 0 & 2 & 1 \\ 0 & 0 & 2 \end{bmatrix}$.

2. $A = \begin{bmatrix} 8 & 12 & -4 \\ 6 & 5 & 7 \\ 2 & 1 & 6 \end{bmatrix}$, $B = \begin{bmatrix} -36 \\ 11 \\ 16 \end{bmatrix}$,

$L = \begin{bmatrix} 4 & 0 & 0 \\ 3 & 2 & 0 \\ 1 & 1 & 1 \end{bmatrix}$, $U = \begin{bmatrix} 2 & 3 & -1 \\ 0 & -2 & 5 \\ 0 & 0 & 2 \end{bmatrix}$.

3. $A = \begin{bmatrix} 2 & 3 & 0 & 1 \\ 4 & 5 & 3 & 3 \\ -2 & -6 & 7 & 7 \\ 8 & 9 & 5 & 21 \end{bmatrix}$, $B = \begin{bmatrix} -2 \\ -2 \\ -16 \\ -66 \end{bmatrix}$,

$L = \begin{bmatrix} 1 & 0 & 0 & 0 \\ 2 & 1 & 0 & 0 \\ -1 & 3 & 1 & 0 \\ 4 & 3 & 2 & 1 \end{bmatrix}$,

$U = \begin{bmatrix} 2 & 3 & 0 & 1 \\ 0 & -1 & 3 & 1 \\ 0 & 0 & -2 & 5 \\ 0 & 0 & 0 & 4 \end{bmatrix}$.

4. $A = \begin{bmatrix} 4 & 2 & 1 & 0 \\ -4 & -6 & 1 & 3 \\ 8 & 16 & -3 & -4 \\ 20 & 10 & 4 & -3 \end{bmatrix}$, $B = \begin{bmatrix} 6 \\ 13 \\ -20 \\ 15 \end{bmatrix}$,

$L = \begin{bmatrix} 1 & 0 & 0 & 0 \\ -1 & 1 & 0 & 0 \\ 2 & -3 & 1 & 0 \\ 5 & 0 & -1 & 1 \end{bmatrix}$,

$U = \begin{bmatrix} 4 & 2 & 1 & 0 \\ 0 & -4 & 2 & 3 \\ 0 & 0 & 1 & 5 \\ 0 & 0 & 0 & 2 \end{bmatrix}$.

In Exercises 5 through 10, find an LU-factorization of the coefficient matrix of the given linear system $A\mathbf{x} = \mathbf{b}$. Solve the linear system using a forward substitution followed by a back substitution.

5. $A = \begin{bmatrix} 2 & 3 & 4 \\ 4 & 5 & 10 \\ 4 & 8 & 2 \end{bmatrix}$, $B = \begin{bmatrix} 6 \\ 16 \\ 2 \end{bmatrix}$.

6. $A = \begin{bmatrix} -3 & 1 & -2 \\ -12 & 10 & -6 \\ 15 & 13 & 12 \end{bmatrix}$, $B = \begin{bmatrix} 15 \\ 82 \\ -5 \end{bmatrix}$.

7. $A = \begin{bmatrix} 4 & 2 & 3 \\ 2 & 0 & 5 \\ 1 & 2 & 1 \end{bmatrix}$, $B = \begin{bmatrix} 1 \\ -1 \\ -3 \end{bmatrix}$.

8. $A = \begin{bmatrix} -5 & 4 & 0 & 1 \\ -30 & 27 & 2 & 7 \\ 5 & 2 & 0 & 2 \\ 10 & 1 & -2 & 1 \end{bmatrix}$, $B = \begin{bmatrix} -17 \\ -102 \\ -7 \\ -6 \end{bmatrix}$.

9. $A = \begin{bmatrix} 2 & 1 & 0 & -4 \\ 1 & 0 & 0.25 & -1 \\ -2 & -1.1 & 0.25 & 6.2 \\ 4 & 2.2 & 0.3 & -2.4 \end{bmatrix}$, $B = \begin{bmatrix} -3 \\ -1.5 \\ 5.6 \\ 2.2 \end{bmatrix}$.

10. $A = \begin{bmatrix} 4 & 1 & 0.25 & -0.5 \\ 0.8 & 0.6 & 1.25 & -2.6 \\ -1.6 & -0.08 & 0.01 & 0.2 \\ 8 & 1.52 & -0.6 & -1.3 \end{bmatrix}$,

$B = \begin{bmatrix} -0.15 \\ 9.77 \\ 1.69 \\ -4.576 \end{bmatrix}$.

$U = \begin{bmatrix} 4 & 2 & 1 & 0 \\ 0 & -4 & 2 & 3 \\ 0 & 0 & 1 & 5 \\ 0 & 0 & 0 & 2 \end{bmatrix}$.

MATLAB EXERCISES ■

Routine **lupr** provides a step-by-step procedure in MATLAB for obtaining the LU-factorization discussed in this section. Once we have the LU-factorization, routines **forsub** and **bksub** can be used to perform the forward and back substitution, respectively. Use **help** for further information on these routines.

ML.1. Use **lupr** in MATLAB to find an LU-factorization of

$$A = \begin{bmatrix} 2 & 8 & 0 \\ 2 & 2 & -3 \\ 1 & 2 & 7 \end{bmatrix}.$$

ML.2. Use **lupr** in MATLAB to find an LU-factorization of

$$A = \begin{bmatrix} 8 & -1 & 2 \\ 3 & 7 & 2 \\ 1 & 1 & 5 \end{bmatrix}.$$

ML.3. Solve the linear system in Example 2 using **lupr**, **forsub**, and **bksub** in MATLAB. Check your LU-factorization using Example 3.

ML.4. Solve Exercises 7 and 8 using **lupr**, **forsub**, and **bksub** in MATLAB.

9.4 ▼ QR-Factorization (Optional)

Prerequisite. Section 4.8, Orthonormal Bases in R^n.

In Section 9.3 we discussed the LU-factorization of a matrix and showed how it leads to a very efficient method for solving a linear system. We now discuss another factorization of a matrix A, called the **QR-factorization** of A. This type of factorization is widely used in computer codes to find the eigenvalues of a matrix, to solve linear systems, and to find least squares approximations (see Section 8.4 for a discussion of least squares).

THEOREM 9.1 ■ *If A is an $m \times n$ matrix with linearly independent columns, then A can be factored as $A = QR$, where Q is an $m \times n$ matrix whose columns form an orthonormal basis for the column space of A and R is an $n \times n$ nonsingular upper triangular matrix.*

Proof   Let $\mathbf{u}_1, \mathbf{u}_2, \ldots, \mathbf{u}_n$ denote the linearly independent columns of A, which form a basis for the column space of A. By using the Gram–Schmidt process (see Theorem 4.18 in Section 4.8), we can obtain an orthonormal basis $\mathbf{w}_1, \mathbf{w}_2, \ldots, \mathbf{w}_n$ for the column space of A. Recall how this orthonormal basis was obtained. We first constructed an orthogonal basis $\mathbf{v}_1, \mathbf{v}_2, \ldots, \mathbf{v}_n$ as follows: $\mathbf{v}_1 = \mathbf{u}_1$ and then for $i = 2, 3, \ldots, n$ we have

$$\mathbf{v}_i = \mathbf{u}_i - \frac{\mathbf{u}_i \cdot \mathbf{v}_1}{\mathbf{v}_1 \cdot \mathbf{v}_1} \mathbf{v}_1 - \frac{\mathbf{u}_i \cdot \mathbf{v}_2}{\mathbf{v}_2 \cdot \mathbf{v}_2} \mathbf{v}_2 - \cdots - \frac{\mathbf{u}_i \cdot \mathbf{v}_{i-1}}{\mathbf{v}_{i-1} \cdot \mathbf{v}_{i-1}} \mathbf{v}_{i-1}. \tag{1}$$

Finally, $\mathbf{w}_i = \dfrac{1}{\|\mathbf{v}_i\|} \mathbf{v}_i$ for $i = 1, 2, 3, \ldots, n$. Now each of the vectors $\mathbf{u}_i$ can be written as a linear combination of the **w**-vectors:

$$\mathbf{u}_1 = r_{11}\mathbf{w}_1 + r_{21}\mathbf{w}_2 + \cdots + r_{n1}\mathbf{w}_n$$
$$\mathbf{u}_2 = r_{12}\mathbf{w}_1 + r_{22}\mathbf{w}_2 + \cdots + r_{n2}\mathbf{w}_n$$
$$\vdots \tag{2}$$
$$\mathbf{u}_n = r_{1n}\mathbf{w}_1 + r_{2n}\mathbf{w}_2 + \cdots + r_{nn}\mathbf{w}_n.$$

From Theorem 4.17 we have

$$r_{ji} = \mathbf{u}_i \cdot \mathbf{w}_j.$$

Moreover, from Equation (1), we see that $\mathbf{u}_i$ lies in

$$\text{span } \{\mathbf{v}_1, \mathbf{v}_2, \ldots, \mathbf{v}_i\} = \text{span } \{\mathbf{w}_1, \mathbf{w}_2, \ldots, \mathbf{w}_i\}.$$

Since $\mathbf{w}_j$ is orthogonal to span $\{\mathbf{w}_1, \mathbf{w}_2, \ldots, \mathbf{w}_i\}$ for $j > i$, it is orthogonal to $\mathbf{u}_i$. Hence $r_{ji} = 0$ for $j > i$. Let Q be the matrix whose columns are $\mathbf{w}_1, \mathbf{w}_2, \ldots, \mathbf{w}_j$. Let

$$\mathbf{r}_j = \begin{bmatrix} r_{1j} \\ r_{2j} \\ \vdots \\ r_{nj} \end{bmatrix}.$$

Then the equations in (2) can be written in matrix form (see Exercise T.9 in Section 1.3) as

$$A = \begin{bmatrix} \mathbf{u}_1 & \mathbf{u}_2 & \cdots & \mathbf{u}_n \end{bmatrix} = \begin{bmatrix} Q\mathbf{r}_1 & Q\mathbf{r}_2 & \cdots & Q\mathbf{r}_n \end{bmatrix} = QR,$$

where R is the matrix whose columns are $\mathbf{r}_1, \mathbf{r}_2, \ldots, \mathbf{r}_n$. Thus

$$R = \begin{bmatrix} r_{11} & r_{12} & \cdots & r_{1n} \\ 0 & r_{22} & \cdots & r_{2n} \\ 0 & 0 & \cdots & \\ \vdots & \vdots & & \vdots \\ 0 & 0 & \cdots & r_{nn} \end{bmatrix}.$$

We now show that R is nonsingular. Let $\mathbf{x}$ be a solution to the linear system $R\mathbf{x} = \mathbf{0}$. Multiplying this equation by Q on the left, we have

$$Q(R\mathbf{x}) = (QR)\mathbf{x} = A\mathbf{x} = Q\mathbf{0} = \mathbf{0}.$$

As we know from Equation (3) in Section 1.3, the homogeneous system $A\mathbf{x} = \mathbf{0}$ can be written as

$$x_1\mathbf{u}_1 + x_2\mathbf{u}_2 + \cdots + x_n\mathbf{u}_n = \mathbf{0},$$

where $x_1, x_2, \ldots, x_n$ are the components of the vector $\mathbf{x}$. Since the columns of A are linearly independent,

$$x_1 = x_2 = \cdots = x_n = 0,$$

so $\mathbf{x}$ must be the zero vector. Then Theorem 1.12 implies that R is nonsingular. In Exercise T.1 we ask you to show that the diagonal entries r_{ii} of R are nonzero by first expressing $\mathbf{u}_i$ as a linear combination of $\mathbf{v}_1, \mathbf{v}_2, \ldots, \mathbf{v}_i$ and then computing $r_{ii} = \mathbf{u}_i \cdot \mathbf{w}_i$. This provides another proof of the nonsingularity of R. ■

The procedure for finding the QR-factorization of an $m \times n$ matrix A with linearly independent columns is as follows.

Step 1. Let the columns of A be denoted by $\mathbf{u}_1, \mathbf{u}_2, \ldots, \mathbf{u}_n$ and let W be the subspace of R^n with these vectors as basis.

Step 2. Transform the basis $\{\mathbf{u}_1, \mathbf{u}_2, \ldots, \mathbf{u}_n\}$ for W by using the Gram–Schmidt process to an orthonormal basis $\{\mathbf{w}_1, \mathbf{w}_2, \ldots, \mathbf{w}_n\}$. Let

$$Q = \begin{bmatrix} \mathbf{w}_1 & \mathbf{w}_2 & \cdots & \mathbf{w}_n \end{bmatrix}$$

be the matrix whose columns are $\mathbf{w}_1, \mathbf{w}_2, \ldots, \mathbf{w}_n$.

Step 3. Compute $R = \begin{bmatrix} r_{ij} \end{bmatrix}$, where

$$r_{ji} = \mathbf{u}_i \cdot \mathbf{w}_j.$$

EXAMPLE 1 ■ Find the QR-factorization of

$$A = \begin{bmatrix} 1 & -1 & -1 \\ 1 & 0 & 0 \\ 1 & -1 & 0 \\ 0 & 1 & -1 \end{bmatrix}.$$

Solution   Letting the columns of A be denoted by $\mathbf{u}_1$, $\mathbf{u}_2$, and $\mathbf{u}_3$, we let W be the subspace of R^4 with $\mathbf{u}_1$, $\mathbf{u}_2$, $\mathbf{u}_3$ as a basis. Applying the Gram–Schmidt process to this basis, starting with $\mathbf{u}_1$, we find (verify) the following orthonormal basis for the column space of A:

$$\mathbf{w}_1 = \begin{bmatrix} \frac{1}{\sqrt{3}} \\ \frac{1}{\sqrt{3}} \\ \frac{1}{\sqrt{3}} \\ 0 \end{bmatrix}, \quad \mathbf{w}_2 = \begin{bmatrix} -\frac{1}{\sqrt{15}} \\ \frac{2}{\sqrt{15}} \\ -\frac{1}{\sqrt{15}} \\ \frac{3}{\sqrt{15}} \end{bmatrix}, \quad \mathbf{w}_3 = \begin{bmatrix} -\frac{4}{\sqrt{35}} \\ \frac{3}{\sqrt{35}} \\ \frac{1}{\sqrt{35}} \\ -\frac{3}{\sqrt{35}} \end{bmatrix}.$$

Then

$$Q = \begin{bmatrix} \frac{1}{\sqrt{3}} & -\frac{1}{\sqrt{15}} & -\frac{4}{\sqrt{35}} \\ \frac{1}{\sqrt{3}} & \frac{2}{\sqrt{15}} & \frac{3}{\sqrt{35}} \\ \frac{1}{\sqrt{3}} & -\frac{1}{\sqrt{15}} & \frac{1}{\sqrt{35}} \\ 0 & \frac{3}{\sqrt{15}} & -\frac{3}{\sqrt{35}} \end{bmatrix} \approx \begin{bmatrix} 0.5774 & -0.2582 & -0.6761 \\ 0.5774 & 0.5164 & 0.5071 \\ 0.5774 & -0.2582 & 0.1690 \\ 0 & 0.7746 & -0.5071 \end{bmatrix}$$

and

$$R = \begin{bmatrix} r_{11} & r_{12} & r_{13} \\ 0 & r_{22} & r_{23} \\ 0 & 0 & r_{33} \end{bmatrix},$$

where $r_{ji} = \mathbf{u}_i \cdot \mathbf{w}_j$. Thus

$$
R = \begin{bmatrix} \frac{3}{\sqrt{3}} & -\frac{2}{\sqrt{3}} & -\frac{1}{\sqrt{3}} \\ 0 & \frac{5}{\sqrt{15}} & -\frac{2}{\sqrt{15}} \\ 0 & 0 & \frac{7}{\sqrt{35}} \end{bmatrix} \approx \begin{bmatrix} 1.7321 & -1.1547 & -0.5774 \\ 0 & 1.2910 & -0.5164 \\ 0 & 0 & 1.1832 \end{bmatrix}.
$$

As you can verify, $A = QR$. ∎

REMARK   State-of-the-art computer implementations (such as in MATLAB) yield an alternate QR-factorization of an $m \times n$ matrix A as the product of an $m \times m$ matrix Q and an $m \times n$ matrix $R = \begin{bmatrix} r_{ij} \end{bmatrix}$, where $r_{ij} = 0$ if $i > j$. Thus, if A is 5×3, then

$$
R = \begin{bmatrix} * & * & * \\ 0 & * & * \\ 0 & 0 & * \\ 0 & 0 & 0 \\ 0 & 0 & 0 \end{bmatrix}.
$$

Least Squares Using QR-Factorization

Recall from Section 8.4 that a least squares solution $\widehat{\mathbf{x}}$ to the linear system $A\mathbf{x} = \mathbf{b}$ is a solution to

$$
A^T A\mathbf{x} = A^T \mathbf{b}. \tag{3}
$$

Recall also that (3) is called the normal system associated with $A\mathbf{x} = \mathbf{b}$.

Let $A = QR$ be a QR-factorization of A. Substituting this expression for A into Equation (3), we obtain

$$
(QR)^T (QR)\mathbf{x} = (QR)^T \mathbf{b}
$$

or

$$
R^T (Q^T Q) R\mathbf{x} = R^T Q^T \mathbf{b}.
$$

Since the columns of Q form an orthonormal set, we have $Q^T Q = I_m$, so

$$
R^T R\mathbf{x} = R^T Q^T \mathbf{b}.
$$

Since R^T is a nonsingular matrix, we obtain

$$
R\mathbf{x} = Q^T \mathbf{b}.
$$

Using the fact that R is upper triangular, we easily solve this linear system by back substitution to obtain $\widehat{\mathbf{x}}$.

EXAMPLE 2 ∎ Solve Example 1 in Section 8.4 using the QR-factorization of A.

Solution We use the Gram–Schmidt process, carrying out all computations in MATLAB. We find that Q is given by (verify)

$$Q = \begin{bmatrix} -0.1961 & -0.3851 & 0.5099 & 0.3409 \\ -0.3922 & -0.1311 & -0.1768 & 0.4244 \\ 0.3922 & -0.7210 & -0.4733 & -0.2177 \\ -0.7845 & -0.2622 & -0.1041 & -0.5076 \\ 0 & -0.4260 & 0.0492 & 0.4839 \\ -0.1961 & 0.2540 & -0.6867 & 0.4055 \end{bmatrix}$$

and R is given by (verify)

$$R = \begin{bmatrix} -5.0990 & -0.9806 & 0.1961 & -0.9806 \\ 0 & -4.6945 & -2.8102 & -3.4164 \\ 0 & 0 & -4.0081 & 0.8504 \\ 0 & 0 & 0 & 3.1054 \end{bmatrix}.$$

Then

$$Q^T \mathbf{b} = \begin{bmatrix} 4.7068 \\ -0.1311 \\ 2.8172 \\ 5.8699 \end{bmatrix}.$$

Finally, solving

$$R\mathbf{x} = Q^T \mathbf{b}$$

we find (verify)

$$\mathbf{x} = \begin{bmatrix} 0.9990 \\ -2.0643 \\ 1.1039 \\ 1.8902 \end{bmatrix},$$

exactly the same $\widehat{\mathbf{x}}$ as in the solution to Example 1 in Section 8.4. ∎

9.4 EXERCISES

In Exercises 1 through 6, compute the QR-factorization of A.

1. $A = \begin{bmatrix} 1 & 2 \\ -1 & 3 \end{bmatrix}$.

2. $A = \begin{bmatrix} 1 & 2 \\ -1 & -2 \\ 1 & 1 \end{bmatrix}$.

3. $A = \begin{bmatrix} 1 & 0 & -1 \\ 2 & -3 & 3 \\ -1 & 2 & 4 \end{bmatrix}$.

4. $A = \begin{bmatrix} 2 & -1 \\ -1 & 3 \\ 0 & 1 \end{bmatrix}$.

5. $A = \begin{bmatrix} 1 & 0 & 2 \\ -1 & 2 & 0 \\ -1 & -2 & 2 \end{bmatrix}$.

6. $A = \begin{bmatrix} 2 & -1 & 1 \\ 1 & 2 & -2 \\ 0 & 1 & -2 \end{bmatrix}$.

7. Solve Exercise 1 in Section 8.4 using the QR-factorization of A.

8. Solve Exercise 3 in Section 8.4 using the QR-factorization of A.

THEORETICAL EXERCISES ■

T.1. In the proof of Theorem 9.1, show that r_{ii} is nonzero by first expressing $\mathbf{u}_i$ as a linear combination of $\mathbf{v}_1, \mathbf{v}_2, \ldots, \mathbf{v}_i$ and then computing $r_{ii} = \mathbf{u}_i \cdot \mathbf{w}_i$.

T.2. Show that every nonsingular matrix has a QR-factorization.

9.5 ▼ Eigenvalues and Eigenvectors

Prerequisite. Chapter 5, Eigenvalues and Eigenvectors.

In this section we present two numerical methods for finding eigenvalues and eigenvectors of matrices. The first method finds the eigenvalue of largest absolute value and an associated eigenvector. The second method finds all the eigenvalues and associated eigenvectors of a symmetric matrix.

The Power Method

Suppose that A is an $n \times n$ matrix that is diagonalizable and the eigenvalues λ_i have been indexed in such a way that

$$|\lambda_1| \geq |\lambda_2| \geq |\lambda_3| \geq \cdots \geq |\lambda_n| .$$

Further, assume that $|\lambda_1| > |\lambda_2|$: the magnitude of λ_1 is strictly greater than that of λ_2. The power method, an easily implemented technique, computes λ_1 and an associated eigenvector.

The key idea in the power method is the following result, whose proof we omit. If $\mathbf{u}_0$ is an arbitrary nonzero vector, then the vectors $A^k \mathbf{u}_0$ (k a nonnegative integer) will usually form a sequence of vectors whose directions converge to the direction of $\mathbf{x}_1$, an eigenvector of A associated with the dominant eigenvalue λ_1.

If $\mathbf{x}_1$ were exactly known, then using the dot product in R^n, we form the ratio

$$\frac{(A\mathbf{x}_1) \cdot \mathbf{x}_1}{\mathbf{x}_1 \cdot \mathbf{x}_1} = \frac{(\lambda_1 \mathbf{x}_1) \cdot \mathbf{x}_1}{\mathbf{x}_1 \cdot \mathbf{x}_1} = \frac{\lambda_1 (\mathbf{x}_1 \cdot \mathbf{x}_1)}{\mathbf{x}_1 \cdot \mathbf{x}_1} = \lambda_1$$

and thus obtain λ_1.

Suppose now that for k sufficiently large, we have a good approximation $\widehat{\mathbf{x}}_1$ to $\mathbf{x}_1$. Then

$$\frac{(A\widehat{\mathbf{x}}_1) \cdot \widehat{\mathbf{x}}_1}{\widehat{\mathbf{x}}_1 \cdot \widehat{\mathbf{x}}_1}$$

is an approximation to λ_1.

The components of $A^k \mathbf{u}_0$ may become increasingly large numbers as k increases, which leads to a large roundoff error. This problem can be avoided by multiplying $A^k \mathbf{u}_0$ by a suitable scalar for each k.

The procedure for carrying out the power method is as follows.

Step 1. Choose an arbitrary nonzero vector $\mathbf{u}_0$ as an initial approximation to an eigenvector of A associated with λ_1. For example, $\mathbf{u}_0$ can be taken as the vector all of whose components are 1.

Step 2. Compute $A\mathbf{u}_0 = \mathbf{u}_1$. Let $k = 1$.

Step 3. Let u_{kr} be the component of $\mathbf{u}_k$ with largest absolute value. Define

$$\mathbf{v}_k = \frac{1}{|u_{kr}|}\, \mathbf{u}_k \qquad (k \geq 1).$$

Thus $\mathbf{v}_k$ is parallel to and in the same direction as $\mathbf{u}_k$, and is an approximation to an eigenvector of A associated with λ_1.

Step 4. Form an approximation to λ_1:

$$\frac{(A\mathbf{v}_k) \cdot \mathbf{v}_k}{\mathbf{v}_k \cdot \mathbf{v}_k} \approx \lambda_1 \qquad (k \geq 1).$$

Step 5. Increase k by 1, let $\mathbf{u}_{k+1} = A\mathbf{v}_k$, and return to Step 3.

EXAMPLE 1 ■ Let

$$A = \begin{bmatrix} 4 & 1 \\ 2 & 5 \end{bmatrix}.$$

If we carry three decimal places and round after each multiplication and division, our results for nine iterations ($k = 8$) are shown in Table 9.3.

Table 9.3

kth *Iteration*	$\mathbf{v}_k$	*Approximate Value of* λ_1
0	$\begin{bmatrix} 1 & 1 \end{bmatrix}^T$	—
1	$\begin{bmatrix} 0.714 & 1 \end{bmatrix}^T$	6.080
2	$\begin{bmatrix} 0.600 & 1 \end{bmatrix}^T$	6.059
3	$\begin{bmatrix} 0.548 & 1 \end{bmatrix}^T$	6.035
4	$\begin{bmatrix} 0.524 & 1 \end{bmatrix}^T$	6.016
5	$\begin{bmatrix} 0.512 & 1 \end{bmatrix}^T$	6.010
6	$\begin{bmatrix} 0.506 & 1 \end{bmatrix}^T$	6.005
7	$\begin{bmatrix} 0.503 & 1 \end{bmatrix}^T$	6.002
8	$\begin{bmatrix} 0.501 & 1 \end{bmatrix}^T$	6.001

We conclude that the dominant eigenvalue is approximately 6.001 and that an associated eigenvector is approximately

$$\widehat{\mathbf{x}}_1 = \begin{bmatrix} 0.501 \\ 1 \end{bmatrix}.$$

The exact answers are

$$\lambda_1 = 6 \quad \text{and} \quad \mathbf{x}_1 = \begin{bmatrix} \frac{1}{2} \\ 1 \end{bmatrix}.$$

∎

The rate of convergence of the power method depends on the ratio $|\lambda_2/\lambda_1|$ and on the choice of the vector $\mathbf{u}_0$. If the ratio $|\lambda_2/\lambda_1|$ is close to 1, then convergence is very slow. There is no way of knowing how to choose a good $\mathbf{u}_0$. There are many ways of extending the power method to handle cases excluded in our discussion.

If A is an $n \times n$ symmetric matrix with eigenvalues $\lambda_1, \lambda_2, \ldots, \lambda_n$, where

$$|\lambda_1| > |\lambda_2| > |\lambda_3| \geq \cdots \geq |\lambda_n|,$$

it is possible to find λ_2 and an associated eigenvector $\mathbf{x}_2$ by a method called **deflation**, which applies the power method to the matrix $A_1 = A - \lambda_1 \mathbf{x}_1 \mathbf{x}_1^T$. This procedure works very well if we have a *very good* approximation to λ_1 and $\mathbf{x}_1$. Otherwise, the results can be rather poor. A much better procedure for finding all the eigenvalues and associated eigenvectors of a *symmetric* matrix is Jacobi's method, which is widely used.

Jacobi's Method

Let A be a symmetric $n \times n$ matrix and let E_{pq} $(p \neq q)$ be the $n \times n$ matrix obtained from the identity matrix I_n by replacing the following elements of I_n as indicated: p, pth element replaced by $\cos\theta$; p, qth element replaced by $\sin\theta$; q, pth element replaced by $-\sin\theta$; q, qth element replaced by $\cos\theta$. Thus, if we start with I_4, then

$$E_{24} = \begin{bmatrix} 1 & 0 & 0 & 0 \\ 0 & \cos\theta & 0 & \sin\theta \\ 0 & 0 & 1 & 0 \\ 0 & -\sin\theta & 0 & \cos\theta \end{bmatrix}.$$

The E_{pq} matrices are called **rotation matrices** because for $n = 2$, they represent a rotation of axes through the angle θ.

The matrices E_{pq} are orthogonal; that is, $E_{pq}^T E_{pq} = I_n$ (Exercise T.1). If we now form the matrix

$$B_1 = E_{pq}^T A E_{pq}, \tag{1}$$

then the entries of B_1 differ from those of A only in the pth and qth rows and in the pth and qth columns. More specifically, the p, qth element of B_1 is

$$(a_{pp} - a_{qq}) \cos\theta \sin\theta + a_{pq}(\cos^2\theta - \sin^2\theta). \tag{2}$$

Since
$$\cos\theta \sin\theta = \tfrac{1}{2}\sin 2\theta \quad \text{and} \quad \cos^2\theta - \sin^2\theta = \cos 2\theta,$$
we can write (2) as
$$\tfrac{1}{2}(a_{pp} - a_{qq})\sin 2\theta + a_{pq}\cos 2\theta. \tag{3}$$

Let a_{pq} be in absolute value the largest nonzero element of A that is not on the main diagonal, and choose θ so that the p, qth element of B_1 is zero. If there are several candidates for a_{pq}, choose any one. From (3) we see that θ is chosen so that
$$\tan 2\theta = -\frac{2a_{pq}}{a_{pp} - a_{qq}} \quad \text{for } a_{pp} - a_{qq} \neq 0. \tag{4}$$

If $a_{pp} - a_{qq} = 0$, then from (3) the p, qth element of B_1 is
$$a_{pq}(\cos^2\theta - \sin^2\theta) = a_{pq}\cos 2\theta,$$
which can be made zero by taking $\theta = \pi/4$. We now repeat the same procedure with B_1 taking the place of A. That is, if b_{rs} is in absolute value the largest off-diagonal element of B_1, then we form the matrix
$$B_2 = E_{rs}^T B_1 E_{rs},$$
where E_{rs} is the rotation matrix chosen so that the r, sth element of B_2 will be zero; that is, θ is chosen so that
$$\tan 2\theta = -\frac{2b_{rs}}{b_{rr} - b_{ss}} \quad \text{for } b_{rr} - b_{ss} \neq 0.$$

If $b_{rr} - b_{ss} = 0$, let $\theta = \pi/4$.

It can then be proved that the sequence of matrices $B_1, B_2, \ldots, B_t, \ldots,$ converges to a diagonal matrix D, whose diagonal elements are the eigenvalues of A.

Thus, when we stop our computations, we have
$$E_{k+1}^T \cdots E_2^T E_1^T A E_1 E_2 \cdots E_{k+1} = B_{k+1}, \tag{5}$$
where $E_1, E_2, \ldots, E_{k+1}$ are rotation matrices. If we let
$$E_1 E_2 \cdots E_{k+1} = P, \tag{6}$$
then we can write (5) as
$$P^T A P = B_{k+1}. \tag{7}$$

Now P is an orthogonal matrix and we may recall from Section 5.2 that the columns of P are eigenvectors of A. Thus, to find approximations to the eigenvectors of A associated with $\lambda_1, \lambda_2, \ldots, \lambda_n$, we have to form the products of the rotation matrices as in (6).

From (7) it follows that
$$A = P B_{k+1} P^T.$$

As an additional check on the accuracy of our results, we may compute $P B_{k+1} P^T$ and see how close it is to the given matrix A.

In using the Jacobi method, it is not necessary to evaluate trigonometric functions to compute $\sin\theta$ and $\cos\theta$ from (4); instead we proceed as follows. Let

$$f = -a_{pq} \quad \text{and} \quad g = \tfrac{1}{2}(a_{pp} - a_{qq}).$$

Compute

$$h = \text{sign}(g)\,\frac{f}{\sqrt{f^2 + g^2}}. \tag{8}$$

Then

$$\sin\theta = \frac{h}{\sqrt{2(1 + \sqrt{1 - h^2})}}, \qquad \cos\theta = \sqrt{1 - \sin^2\theta}. \tag{9}$$

Observe that if $a_{pp} - a_{qq} = 0$, then taking $\text{sign}(g)$ as $+$, we obtain $h = 1$ in (8), and from (9), $\sin\theta = \frac{1}{\sqrt{2}}$, so that $\theta = \frac{\pi}{4}$, as has already been mentioned earlier. Thus we can use (8) and (9) in all cases, without having to check whether $a_{pp} - a_{qq} \neq 0$.

For a detailed discussion of a computer implementation of Jacobi's method, the reader is referred to the article by Greenstadt listed in Further Readings at the end of the chapter.

The procedure for carrying out Jacobi's method for diagonalizing a symmetric matrix A is as follows.

Step 1. Identify a_{pq} as the largest (in absolute value) nonzero element of A that is not on the main diagonal.

Step 2. Compute

$$f = -a_{pq}, \qquad g = \tfrac{1}{2}(a_{pp} - a_{qq}), \qquad h = \text{sign}(g)\,\frac{f}{\sqrt{f^2 + g^2}}.$$

Step 3. Compute

$$\sin\theta = \frac{h}{\sqrt{2(1 + \sqrt{1 - h^2})}}, \qquad \cos\theta = \sqrt{1 - \sin^2\theta}.$$

Step 4. Compute the $n \times n$ matrix E_{pq} from I_n by replacing the following elements of I_n as indicated:

p, pth element by $\cos\theta$ $\qquad$ p, qth element by $\sin\theta$

q, pth element by $-\sin\theta$ $\qquad$ q, qth element by $\cos\theta$.

Step 5. Compute $B_1 = E_{pq}^T A E_{pq}$.

Step 6. Repeat Steps 1–5 with B_1 taking the place of A, obtaining matrices $B_1, B_2, \ldots, B_{k+1}$. Then

$$P = E_1 E_2 \cdots E_{k+1}$$

is orthogonal and

$$B = P^T A P$$

is approximately diagonal.

EXAMPLE 2 ■ Consider the symmetric matrix

$$A = \begin{bmatrix} -1 & 1 & -2 \\ 1 & -1 & -2 \\ -2 & -2 & -2 \end{bmatrix}.$$

We now find the eigenvalues and associated eigenvectors of A by Jacobi's method. We carry three decimal places and round after each multiplication and division. Moreover, the matrix product ABC will be calculated by first computing BC and then $A(BC)$. Proceed as follows.

Using Jacobi's method we could begin with the 1, 3; 2, 3; 3, 1; or 3, 2 element of A. We choose to begin with the 1, 3 element of A, obtaining

$$E_{13} = \begin{bmatrix} 0.789 & 0.000 & 0.615 \\ 0.000 & 1.000 & 0.000 \\ -0.615 & 0.000 & 0.789 \end{bmatrix} = E_1 \qquad \begin{array}{l} f = 2.000 \\ g = 0.500 \\ h = 0.970; \end{array}$$

$$B_1 = \begin{bmatrix} 0.562 & 2.019 & -0.003 \\ 2.019 & -1.000 & -0.963 \\ -0.004 & -0.963 & -3.565 \end{bmatrix}.$$

We now reduce the 2, 1 element of B_1 to zero, obtaining

$$E_{21} = \begin{bmatrix} 0.825 & -0.565 & 0.000 \\ 0.565 & 0.825 & 0.000 \\ 0.000 & 0.000 & 1.000 \end{bmatrix} = E_2 \qquad \begin{array}{l} f = -2.019 \\ g = -0.781 \\ h = 0.933; \end{array}$$

$$B_2 = \begin{bmatrix} 1.946 & 0.001 & -0.546 \\ 0.001 & -2.384 & -0.792 \\ -0.547 & -0.792 & -3.565 \end{bmatrix}.$$

We next reduce the 2, 3 element of B_2 to zero, obtaining

$$E_{23} = \begin{bmatrix} 1.000 & 0.000 & 0.000 \\ 0.000 & 0.893 & 0.449 \\ 0.000 & -0.449 & 0.893 \end{bmatrix} = E_3 \qquad \begin{array}{l} f = 0.792 \\ g = 0.591 \\ h = 0.802; \end{array}$$

$$B_3 = \begin{bmatrix} 1.946 & 0.246 & -0.488 \\ 0.247 & -1.984 & 0.002 \\ -0.488 & 0.002 & -3.959 \end{bmatrix}.$$

We now reduce the 1, 3 element of B_3 to zero, obtaining

$$E_{13} = \begin{bmatrix} 0.996 & 0.000 & 0.082 \\ 0.000 & 1.000 & 0.000 \\ -0.082 & 0.000 & 0.996 \end{bmatrix} = E_4 \qquad \begin{array}{l} f = 0.488 \\ g = 2.953 \\ h = 0.163; \end{array}$$

$$B_4 = \begin{bmatrix} 1.983 & 0.245 & 0.002 \\ 0.246 & -1.984 & 0.022 \\ 0.002 & 0.022 & -3.994 \end{bmatrix}.$$

We next reduce the 2, 1 element of B_4 to zero, obtaining

$$E_{21} = \begin{bmatrix} 0.998 & -0.062 & 0.000 \\ 0.062 & 0.998 & 0.000 \\ 0.000 & 0.000 & 1.000 \end{bmatrix} = E_5 \qquad \begin{aligned} f &= -0.246 \\ g &= -1.984 \\ h &= 0.123; \end{aligned}$$

$$B_5 = \begin{bmatrix} 1.998 & -0.002 & 0.003 \\ -0.001 & -1.999 & 0.022 \\ 0.003 & 0.022 & -3.994 \end{bmatrix}.$$

We now stop; the diagonal elements of B_5 are approximations to the eigenvalues of A. Thus we have $\lambda_1 \approx 1.998$, $\lambda_2 \approx -1.999$, and $\lambda_3 \approx -3.994$, which agree rather well with the exact values,

$$\lambda_1 = 2, \qquad \lambda_2 = -2, \qquad \lambda_3 = -4.$$

To find associated eigenvectors, we form $E_1 E_2 E_3 E_4 E_5 = P$. We have

$$E_1 E_2 E_3 E_4 E_5 = P = \begin{bmatrix} 0.576 & -0.707 & 0.409 \\ 0.578 & 0.708 & 0.407 \\ -0.578 & -0.001 & 0.816 \end{bmatrix}.$$

The columns of P give the approximate eigenvectors

$$\mathbf{x}_1 \approx \begin{bmatrix} 0.576 \\ 0.578 \\ -0.578 \end{bmatrix}, \qquad \mathbf{x}_2 \approx \begin{bmatrix} -0.707 \\ 0.708 \\ -0.001 \end{bmatrix}, \qquad \mathbf{x}_3 \approx \begin{bmatrix} 0.409 \\ 0.407 \\ 0.816 \end{bmatrix},$$

which are associated with λ_1, λ_2, and λ_3, respectively.

As a final check of our results, we compute $P B_5 P^T$, obtaining

$$P B_5 P^T = \begin{bmatrix} -1.015 & 1.002 & -2.010 \\ 1.001 & -0.982 & -1.981 \\ -2.010 & -1.981 & -1.995 \end{bmatrix},$$

which is in very close agreement with the given matrix A. ■

Further Readings ▶

COLEMAN, T. F., and C. VAN LOAN. *Handbook for Matrix Computations.* Philadelphia: SIAM, 1989.

FOX, L. *An Introduction to Numerical Linear Algebra.* New York: Oxford University Press, Inc., 1965.

GOLUB, G. H., and C. VAN LOAN. *Matrix Computations*, 2nd. ed. Baltimore: Johns Hopkins University Press, 1989.

GREENSTADT, JOHN. "The Determination of the Characteristic Roots of a Matrix by the Jacobi Method," in *Mathematical Methods for Digital Computers*, ed. by A. Ralston and H. S. Wilf. New York: John Wiley & Sons, Inc., 1960.

KAHANER, D., C. B. MOLER, and S. NASH. *Numerical Methods and Software.* Upper Saddle River, N.J.: Prentice Hall, Inc., 1989.

NOBEL, BEN. *Applied Linear Algebra*, 3rd ed. Upper Saddle River, N.J.: Prentice Hall, Inc., 1988.

STEWART, G. W. *Introduction to Matrix Computations.* New York: Academic Press, Inc., 1973.

WILKINSON, J. H. *The Algebraic Eigenvalue Problem.* New York: Oxford University Press, Inc., 1965.

9.5   EXERCISES

In Exercises 1 through 5, compute an approximation to the eigenvalue of largest magnitude and an associated eigenvector by the power method. Carry out the computations to three decimal places and round after each multiplication and division for six iterations $(k = 5)$.

1. $\begin{bmatrix} -4 & 2 \\ 3 & 1 \end{bmatrix}$.   **2.** $\begin{bmatrix} 0 & -1 \\ 2 & 3 \end{bmatrix}$.

3. $\begin{bmatrix} 4 & 3 \\ 2 & 3 \end{bmatrix}$.   **4.** $\begin{bmatrix} 6 & 2 \\ -4 & -3 \end{bmatrix}$.

5. $\begin{bmatrix} 8 & 8 \\ 3 & -2 \end{bmatrix}$.

In Exercises 6 through 9, find the eigenvalues and associated eigenvectors by Jacobi's method. Carry out the computations to three decimal places and round after each multiplication and division. In each case calculate ABC by first calculating BC.

6. $\begin{bmatrix} 2 & 4 \\ 4 & 2 \end{bmatrix}$. Find B_3.

7. $\begin{bmatrix} 2 & 2 \\ 2 & 2 \end{bmatrix}$. Find B_3.

8. $\begin{bmatrix} -1 & -4 & -8 \\ -4 & -7 & 4 \\ -8 & 4 & -1 \end{bmatrix}$. Find B_5.

9. $\begin{bmatrix} 8 & -2 & 0 \\ -2 & 9 & -2 \\ 0 & -2 & 10 \end{bmatrix}$. Find B_5.

THEORETICAL EXERCISE ▪

T.1. Show that the matrices E_{pq} are orthogonal.

KEY IDEAS FOR REVIEW ▪

☐ **Gaussian elimination method** (for solving the linear system $A\mathbf{x} = \mathbf{b}$). Transform the augmented matrix $\begin{bmatrix} A & \vdots & \mathbf{b} \end{bmatrix}$ to row echelon form and use back substitution.

☐ **Iterative methods** (for solving $A\mathbf{x} = \mathbf{b}$): **Jacobi's iteration method**. See page 544. **Gauss–Seidel iteration method**. See page 545.

☐ **LU-factorization** (for writing an $n \times n$ matrix A as LU, where L is in lower triangular form and U is in upper triangular form). See page 553.

☐ **QR-factorization** (for writing an $m \times n$ matrix A with linearly independent columns as QR, where Q is an $m \times n$ matrix whose columns form an orthonormal basis for the column space of A and R is an $n \times n$ nonsingular upper triangular matrix). See page 558.

☐ **Power method** (for finding the eigenvalue of largest absolute value and an associated eigenvector). See page 562.

☐ **Jacobi's method** (for finding the eigenvalues and eigenvectors of a symmetric matrix). See page 565.

SUPPLEMENTARY EXERCISES ▪

1. Solve the following linear system by (a) Jacobi's iteration method and (b) the Gauss–Seidel iteration method. Carry out the computations to three decimal places for six iterations ($k = 5$) and round after each multiplication and division.

$$10x + \ 3y = -24$$
$$2x + 11y = \ \ \ 16.$$

In Exercises 2 and 3, find an LU-factorization of the coefficient matrix of the given linear system $A\mathbf{x} = \mathbf{b}$. Solve the linear system using a forward substitution followed by a back substitution.

2. $A = \begin{bmatrix} 2 & 2 & 3 \\ 6 & 5 & 7 \\ -6 & -8 & -10 \end{bmatrix}, \quad \mathbf{b} = \begin{bmatrix} -6 \\ -13 \\ 22 \end{bmatrix}.$

3. $A = \begin{bmatrix} -2 & 1 & -2 \\ 6 & 1 & 9 \\ -4 & 18 & 5 \end{bmatrix}, \quad \mathbf{b} = \begin{bmatrix} -6 \\ 19 \\ -17 \end{bmatrix}.$

4. Find the QR-factorization of

$$A = \begin{bmatrix} 1 & 0 & 1 \\ -1 & 1 & 2 \\ 2 & 2 & -1 \end{bmatrix}.$$

5. Find the least squares line for the following data:

$$(1, -3), \quad (2, 1), \quad (3, -2), \quad (5, 0), \quad (6, 2).$$

6. For the following matrix, compute an approximation to the eigenvalue of largest magnitude and an associated eigenvector by the power method. Carry out the computations to three decimal places and round after each multiplication and division for six iterations ($k = 5$).

$$\begin{bmatrix} 7 & 6 \\ -5 & -4 \end{bmatrix}.$$

7. For the following matrix, find the eigenvalues and associated eigenvectors by Jacobi's method. Carry out the computations to three decimal places and round after each multiplication and division. In each case calculate ABC by first calculating BC.

$$\begin{bmatrix} 1 & -1 & -1 \\ -1 & 1 & -1 \\ -1 & -1 & 1 \end{bmatrix}.$$

Find B_2.

CHAPTER TEST ▪

1. Solve the following linear system by (a) Jacobi's iteration method and (b) the Gauss–Seidel iteration method. Carry out the computations to three decimal places for six iterations ($k = 5$) and round after each multiplication and division.

$$16x + \ 3y = \ \ \ 55$$
$$x + 10y = -26.$$

2. Find an LU-factorization of the coefficient matrix of the linear system $A\mathbf{x} = \mathbf{b}$. Solve the linear system by using a forward substitution followed by a back substitution.

$$A = \begin{bmatrix} 2 & 2 & -1 \\ -8 & -11 & 5 \\ 4 & 13 & -7 \end{bmatrix}, \qquad \mathbf{b} = \begin{bmatrix} 3 \\ -14 \\ -5 \end{bmatrix}.$$

3. Find a QR-factorization of

$$A = \begin{bmatrix} 2 & 1 \\ -1 & -1 \\ -2 & 3 \end{bmatrix}.$$

4. For the following matrix, compute an approximation to the eigenvalue of largest magnitude and an associated eigenvector by the power method. Carry out the computations to three decimal places and round after each multiplication and division for six iterations ($k = 5$).

$$\begin{bmatrix} 4 & 3 \\ 5 & 6 \end{bmatrix}.$$

5. For the following matrix, find the eigenvalues and associated eigenvectors by Jacobi's method. Carry out the computations to three decimal places and round after each multiplication and division. In each case calculate ABC by first calculating BC.

$$\begin{bmatrix} -1 & 2 & 2 \\ 2 & -1 & 2 \\ 2 & 2 & -1 \end{bmatrix}.$$

Find B_2.

PART IV

MATLAB FOR
LINEAR ALGEBRA

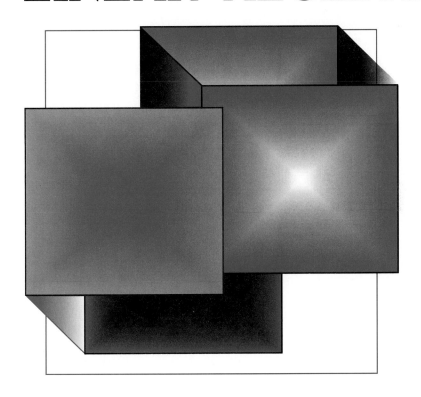

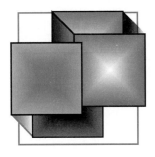

10

MATLAB for Linear Algebra

Introduction*

MATLAB is a versatile piece of computer software with linear algebra capabilities as its core. MATLAB stands for MATrix LABoratory. It incorporates portions of professionally developed projects of quality computer routines for linear algebra computation. The code employed by MATLAB is written in the C language and is upgraded as new versions of MATLAB are released.

MATLAB has a wide range of capabilities. In this book we will use only a small portion of its features. We will find that MATLAB's command structure is very close to the way we write algebraic expressions and linear algebra operations. The names of many MATLAB commands closely parallel those of the operations and concepts of linear algebra. We give descriptions of commands and features of MATLAB that relate directly to this course. A more detailed discussion of MATLAB commands can be found in *MATLAB User's Guide* that accompanies the software and in the books, *Experiments in Computational Matrix Algebra*, by David R. Hill (New York: Random House, 1988) and *Linear Algebra LABS with MATLAB*, 2nd ed., by David R. Hill and David E. Zitarelli (Upper Saddle River, N.J.: Prentice Hall, Inc., 1996). Alternatively, the MATLAB software provides immediate on-screen descriptions using the **help** command. Typing

help

displays a list of MATLAB subdirectories and alternate directories containing files corresponding to commands and data sets. Typing **help name**, where **name** is the name of a command, accesses information on the specific command named. In some cases the description displayed goes much further than

---

*This material on MATLAB refers to the Windows version. There are minor variations in the Macintosh version.

we need for this course. Hence you may not fully understand all of the description displayed by **help**. We provide a list of the majority of MATLAB commands we use in this book in Section 10.9.

Once you initiate the MATLAB software, you will see the MATLAB logo appear and the MATLAB prompt ≫. The prompt ≫ indicates that MATLAB is awaiting a command. In Section 10.1 we describe how to enter matrices into MATLAB and give explanations of several commands. However, there are certain MATLAB features you should be aware of before you begin the material in Section 10.1.

☐ *Starting execution of a command.*
After you have typed a command name and any arguments or data required, you must press ENTER before it will begin to execute.

☐ *The command stack.*
As you enter commands, MATLAB saves a number of the most recent commands in a stack. Previous commands saved on the stack can be recalled using the **up arrow** key. The number of commands saved on the stack varies depending on the length of the commands and other factors.

☐ *Editing commands.*
If you make an error or mistype something in a command, you can use the **left arrow** and **right arrow** keys to position the cursor for corrections. The **home** key moves the cursor to the beginning of a command, and the **end** key moves the cursor to the end. The **backspace** and **delete** keys can be used to remove characters from a command line. The **insert** key is used to initiate the insertion of characters. Pressing the insert key a second time exits the insert mode. If MATLAB recognizes an error after you have pressed ENTER, then MATLAB responds with a beep and a message that helps define the error. You can recall the command line using the up arrow key in order to edit the line.

☐ *Continuing commands.*
MATLAB commands that do not fit on a single line can be continued to the next line using an ellipsis, which is three consecutive periods, followed by ENTER.

☐ *Stopping a command.*
To stop execution of a MATLAB command, press **Ctrl** and **C** simultaneously, then press ENTER. Sometimes this sequence must be repeated.

☐ *Quitting.*
To quit MATLAB, type **exit** or **quit** followed by ENTER.

10.1 ▼ Input and Output in MATLAB

Matrix Input

To enter a matrix into MATLAB just type the entries enclosed in square brackets
[...], with entries separated by a space and rows terminated with a semicolon.
Thus, matrix

$$\begin{bmatrix} 9 & -8 & 7 \\ -6 & 5 & -4 \\ 11 & -12 & 0 \end{bmatrix}$$

is entered by typing

$$[9 \quad -8 \quad 7; -6 \quad 5 \quad -4; 11 \quad -12 \quad 0]$$

and the accompanying display is

ans =

$$\begin{matrix} 9 & -8 & 7 \\ -6 & 5 & -4 \\ 11 & -12 & 0 \end{matrix}$$

Notice that no brackets are displayed and that **MATLAB** has assigned this
matrix the name **ans**. Every matrix in MATLAB must have a name. If you
do not assign a matrix a name, then MATLAB assigns it **ans**, which is called
the **default variable name**. To assign a matrix name we use the assignment
operator =. For example,

$$\mathbf{A} = [4 \quad 5 \quad 8; 0 \quad -1 \quad 6]$$

is displayed as

A =

$$\begin{matrix} 4 & 5 & 8 \\ 0 & -1 & 6 \end{matrix}$$

WARNING 1. All rows must have the same number of entries.

2. MATLAB distinguishes between uppercase and lowercase letters. So matrix
B is not the same as matrix **b**.

3. A matrix name can be reused. In such a case the "old" contents are lost.

To assign a matrix but *suppress the display of its entries*, follow the closing
square bracket,], with a semicolon.

$$\mathbf{A} = [4 \quad 5 \quad 8; 0 \quad -1 \quad 6];$$

assigns the same matrix to name A as above, but no display appears. To assign a
currently defined matrix a new name, use the assignment operator =. Command
$\mathbf{Z} = \mathbf{A}$ assigns the contents of A to Z. Matrix **A** is still defined.

To determine the matrix names that are in use, use the **who** command. To delete a matrix, use the **clear** command followed by a space and then the matrix name. For example, the command

$$\textbf{clear A}$$

deletes name A and its contents from MATLAB. The command **clear** by itself deletes all currently defined matrices.

To determine the number of rows and columns in a matrix, use the **size** command, as in

$$\textbf{size(A)}$$

which, assuming that A has not been cleared, displays

```
ans =

    2    3
```

meaning that there are two rows and three columns in matrix A.

Seeing a Matrix

To see all of the components of a matrix, type its name. If the matrix is large, the display may be broken into subsets of columns that are shown successively. For example, use the command

$$\textbf{hilb(9)}$$

which displays the first seven columns followed by columns 8 and 9. (For information on command **hilb**, use **help hilb**.) If the matrix is quite large, the screen display will scroll too fast for you to see the matrix. To see a portion of a matrix, type command **more on** followed by ENTER, then type the matrix name or a command to generate it. Press the Space Bar to reveal more of the matrix. Continue pressing the Space Bar until the "--more--" no longer appears near the bottom of the screen. Try this with **hilb(20)**. To disable this paging feature, type command **more off**. If a scroll bar is available, you can use your mouse to move the scroll bar to reveal previous portions of displays.

We have the following conventions to see a portion of a matrix in MATLAB. For purposes of illustration, suppose matrix A has been entered into MATLAB as a 5×5 matrix.

☐ To see the $(2, 3)$ entry of A, type

$$\textbf{A(2,3)}$$

☐ To see the fourth row of A, type

$$\textbf{A(4,:)}$$

☐ To see the first column of A, type

$$\textbf{A(:,1)}$$

In the above situations the : is interpreted to mean "all." The colon can also be used to represent a range of rows or columns. For example, typing

$$2:8$$

displays

```
ans =

    2    3    4    5    6    7    8
```

We can use this feature to display a subset of rows or columns of a matrix. As an illustration, to display rows 3 through 5 of matrix **A**, type

$$\mathbf{A(3:5,:)}$$

Similarly, columns 1 through 3 are displayed by typing

$$\mathbf{A(:,1:3)}$$

For more information on the use of the colon operator, type **help colon**. The colon operator in MATLAB is very versatile, but we will not need to use all of its features.

Display Formats

MATLAB stores matrices in decimal form and does its arithmetic computations using a decimal-type arithmetic. This decimal form retains about 16 digits, but not all digits must be shown. Between what goes on in the machine and what is shown on the screen are routines that convert or format the numbers into displays. Here we give an overview of the display formats that we will use. (For more information, see the *MATLAB User's Guide* or type **help format**.)

☐ If the matrix contains *all* integers, then the entire matrix is displayed as integer values; that is, no decimal points appear.

☐ If any entry in the matrix is not exactly represented as an integer, then the entire matrix is displayed in what is known as **format short**. Such a display shows four places behind the decimal point and the last place may have been rounded. The exception to this is zero. If an entry is exactly zero, then it is displayed as an integer zero. Enter the matrix

$$\mathbf{Q = [5 \quad 0 \quad 1/3 \quad 2/3 \quad 7.123456]}$$

into MATLAB. The display is

```
Q =

   5.0000        0    0.3333    0.6667    7.1235
```

WARNING   If a value is displayed as 0.0000, then it is not identically zero. You should change to **format long**, discussed below, and display the matrix again.

☐ To see more than four places, change the display format. One way to proceed is to use the command

format long

which shows 15 places. The matrix Q above in format long is

```
Q =
    Columns 1 through 4
    5.00000000000000    0    0.33333333333333    0.66666666666667
    Column 5
    7.12345600000000
```

Other display formats use an exponent of 10. They are **format short e** and **format long e**. The "e-formats" are often used in numerical analysis. Try these formats with matrix Q.

☐ MATLAB can display values in rational form. The command **format rat**, short for rational display, is used. Inspect the output from the following sequence of MATLAB commands.

format short
$$\mathbf{V} = [1 \quad 1/2 \quad 1/6 \quad 1/12]$$

displays

```
V =
    1.0000    0.5000    0.1667    0.0833
```

and

format rat
$$\mathbf{V}$$

displays

```
V =
    1    1/2    1/6    1/12
```

Finally type **format short** to return to a decimal display form.

WARNING Rational output is displayed in what is called "string" form. Strings are not numeric data and hence cannot be used with arithmetic operators. Thus rational output is for "looks" only.

When MATLAB starts, the format in effect is **format short**. If you change the format, it remains in effect until another format command is executed. Some MATLAB routines change the format within the routine.

10.2 ▾ Matrix Operations in MATLAB

The operations of addition, subtraction, and multiplication of matrices in MATLAB follow the same definitions as in Sections 1.2 and 1.3. If A and B are $m \times n$ matrices that have been entered into MATLAB, then their sum in MATLAB is computed using command

$$\mathbf{A+B}$$

and their difference by command

$$\mathbf{A-B}$$

(spaces can be used on either side of $+$ or $-$). If A is $m \times n$ and C is $n \times k$, then the product of A and C in MATLAB must be written as

$$\mathbf{A*C}$$

In MATLAB, $*$ must be specifically placed between the names of matrices to be multiplied. In MATLAB, writing AC does not perform an implied multiplication. In fact, MATLAB considers AC as a new matrix name and, if it has not been previously defined, an error will result. If the matrices involved are not compatible for the operation specified, then an error message will be displayed. Compatibility for addition and subtraction means that the matrices are the same size. Matrices are compatible for multiplication if the number of columns in the first matrix equals the number of rows in the second.

EXAMPLE 1 ■ Enter the matrices

$$A = \begin{bmatrix} 1 & 2 \\ 2 & 4 \end{bmatrix}, \quad \mathbf{b} = \begin{bmatrix} -3 \\ 1 \end{bmatrix}, \quad \text{and} \quad C = \begin{bmatrix} 3 & -5 \\ 5 & 2 \end{bmatrix}$$

into MATLAB and compute the following expressions. We display the results from MATLAB.

Solution   (a) **A+C** displays

```
ans =

     4   -3
     7    6
```

(b) **A*C** displays

```
ans =

    13   -1
    26   -2
```

(c) **b*A** displays   ??? Error using ==> *

                Inner matrix dimensions must agree.   ■

Scalar multiplication in MATLAB requires the use of the multiplication symbol ∗. For the matrix A in Example 1, $5A$ designates scalar multiplication in the book, while **5∗A** is required in MATLAB.

In MATLAB the transpose operator (or symbol) is the single quotation mark or prime, ′. Using the matrices in Example 1, in MATLAB

$$\mathbf{Q} = \mathbf{C}'$$ displays   Q =

$$\begin{array}{rr} 3 & 5 \\ -5 & 2 \end{array}$$

and

$$\mathbf{p} = \mathbf{b}'$$ displays   p =

$$\begin{array}{rr} -3 & 1 \end{array}$$

As a convenience, we can enter column matrices into MATLAB using ′. To enter the matrix

$$\mathbf{x} = \begin{bmatrix} 1 \\ 3 \\ -5 \end{bmatrix},$$

we can use either the command

$$\mathbf{x} = [\mathbf{1};\mathbf{3};-\mathbf{5}]$$

or the command

$$\mathbf{x} = [\mathbf{1} \quad \mathbf{3} \quad -\mathbf{5}]'$$

Suppose that we are given the linear system $A\mathbf{x} = \mathbf{b}$, where coefficient matrix A and right-hand side $\mathbf{b}$ have been entered into MATLAB. The augmented matrix $\begin{bmatrix} A & \mathbf{b} \end{bmatrix}$ is formed in MATLAB by typing

$$[\mathbf{A} \quad \mathbf{b}]$$

or, if we want to name it **aug**, by typing

$$\mathbf{aug} = [\mathbf{A} \quad \mathbf{b}]$$

No bar will be displayed separating the right-hand side from the coefficient matrix. Using matrices A and $\mathbf{b}$ from Example 1, form the augmented matrix in MATLAB for the system $A\mathbf{x} = \mathbf{b}$.

Forming augmented matrices is a special case of building matrices in MATLAB. Essentially we can "paste together" matrices as long as sizes are appropriate. Using the matrices A, **b**, and C in Example 1, we give some examples:

[A   C]   displays   ans =

```
1   2   3  -5
2   4   5   2
```

[A;C]   displays   ans =

```
1   2
2   4
3  -5
5   2
```

[A   b   C]   displays   ans =

```
1   2  -3   3  -5
2   4   1   5   2
```

[C   A;A   C]   displays   ans =

```
3  -5   1   2
5   2   2   4
1   2   3  -5
2   4   5   2
```

MATLAB has a command to build diagonal matrices by entering only the diagonal entries. The command is **diag**, and

$$D = \mathbf{diag}([1 \quad 2 \quad 3])$$ displays   D =

```
1   0   0
0   2   0
0   0   3
```

Command **diag** also works to "extract" a set of diagonal entries. If

$$R = \begin{bmatrix} 5 & 2 & 1 \\ -3 & 7 & 0 \\ 6 & 4 & -8 \end{bmatrix}$$

has been entered into MATLAB, then

diag(R)   displays   ans =

```
 5
 7
-8
```

Note that

diag(diag(R))   displays   ans =

```
5   0   0
0   7   0
0   0  -8
```

For more information on **diag**, use **help**. Commands related to **diag** are **tril** and **triu**.

10.3 ▼ Matrix Powers and Some Special Matrices

In MATLAB, to raise a matrix to a power we must use the exponentiation operator $\wedge$. If A is square and k is a positive integer, then A^k is denoted in MATLAB by

$$\mathbf{A}\wedge\mathbf{k}$$

which corresponds to a matrix product of A with itself k times. The rules for exponents given in Section 1.4 apply in MATLAB. In particular,

$$\mathbf{A}\wedge\mathbf{0}$$

displays an identity matrix having the same size as A.

EXAMPLE 1 ■ Enter matrices

$$A = \begin{bmatrix} 1 & -1 \\ 1 & 1 \end{bmatrix} \quad \text{and} \quad B = \begin{bmatrix} 1 & -2 \\ 2 & 1 \end{bmatrix}$$

into MATLAB and compute the following expressions. We display the MATLAB results.

(a) **A^2**   displays   ans =

$$\begin{array}{rr} 0 & -2 \\ 2 & 0 \end{array}$$

(b) **(A∗B)^2**   displays   ans =

$$\begin{array}{rr} -8 & 6 \\ -6 & -8 \end{array}$$

(c) **(B−A)^3**   displays   ans =

$$\begin{array}{rr} 0 & 1 \\ -1 & 0 \end{array}$$

■

The $n \times n$ identity matrix is denoted by I_n throughout this book. MATLAB has a command to generate I_n when it is needed. The command is **eye**, and it behaves as follows:

eye(2)	displays a 2×2 identity matrix.
eye(5)	displays a 5×5 identity matrix.
t = 10;eye(t)	displays a 10×10 identity matrix.
eye(size(A))	displays an identity matrix the same size as A.

Two other MATLAB commands, **zeros** and **ones**, behave in a similar manner. The command **zeros** produces a matrix of all zeros, while the command **ones** generates a matrix of all ones. Rectangular matrices of size $m \times n$ can be generated using

$$\mathbf{zeros(m,n)}, \quad \mathbf{ones(m,n)}$$

where m and n have been previously defined with positive integer values in MATLAB. Using this convention we can generate a column with four zeros using the command

$$\textbf{zeros(4,1)}$$

From algebra you are familiar with polynomials in x such as

$$4x^3 - 5x^2 + x - 3 \quad \text{and} \quad x^4 - x - 6.$$

The evaluation of such polynomials at a value of x is easily handled in MAT-LAB using the command **polyval**. Define the coefficients of the polynomial as a vector (a row or column matrix) with the coefficient of the largest power first, the coefficient of the next largest power second, and so on down to the constant term. If any power is explicitly missing, its coefficient must be set to zero in the corresponding position in the coefficient vector. In MATLAB, for the polynomials above we have coefficient vectors

$$\mathbf{v} = \begin{bmatrix} 4 & -5 & 1 & -3 \end{bmatrix} \quad \text{and} \quad \mathbf{w} = \begin{bmatrix} 1 & 0 & 0 & -1 & -6 \end{bmatrix}$$

respectively. The command

$$\textbf{polyval(v,2)}$$

evaluates the first polynomial at $x = 2$ and displays the computed value of 11. Similarly, the command

$$\textbf{t} = -\textbf{1;polyval(w,t)}$$

evaluates the second polynomial at $x = -1$ and displays the value -4.

Polynomials in a square matrix A have the form

$$5A^3 - A^2 + 4A - 7I.$$

Note that the constant term in a matrix polynomial is an identity matrix of the same size as A. This convention is a natural one if we recall that the constant term in an ordinary polynomial is the coefficient of x^0 and that $A^0 = I$. We often meet matrix polynomials when evaluating a standard polynomial such as $p(x) = x^4 - x - 6$ at an $n \times n$ matrix A. The resulting matrix polynomial is

$$p(A) = A^4 - A - 6I_n.$$

Matrix polynomials can be evaluated in MATLAB using the command **polyvalm**. Define the square matrix A and the coefficient vector

$$\mathbf{w} = \begin{bmatrix} 1 & 0 & 0 & -1 & -6 \end{bmatrix}$$

in MATLAB. Then the command

$$\textbf{polyvalm(w,A)}$$

produces the value of $p(A)$, which will be a matrix the same size as A.

EXAMPLE 2 ■ Let

$$A = \begin{bmatrix} 1 & -1 & 2 \\ -1 & 0 & 1 \\ 0 & 3 & 1 \end{bmatrix} \quad \text{and} \quad p(x) = 2x^3 - 6x^2 + 2x + 3.$$

To compute $p(A)$ in MATLAB, use the following commands. We show the MATLAB display below the commands.

$$\mathbf{A} = [1 \quad -1 \quad 2; -1 \quad 0 \quad 1; 0 \quad 3 \quad 1];$$
$$\mathbf{v} = [2 \quad -6 \quad 2 \quad 3];$$
$$\mathbf{Q} = \mathbf{polyvalm(v,A)}$$

```
Q =

   -13   -18    10
    -6   -25    10
     6    18   -17
```
■

At times you may want a matrix with integer entries to use in testing some matrix relationship. MATLAB commands can generate such matrices quite easily. Type

$$\mathbf{C} = \mathbf{fix(10*rand(4))}$$

and you will see displayed a 4×4 matrix C with integer entries. To investigate what this command does, use **help** with the commands **fix** and **rand**.

EXAMPLE 3 ■ In MATLAB, generate several $k \times k$ matrices A for $k = 3$, 4, 5 and display $B = A + A^T$. Look over the matrices displayed and try to determine a property that these matrices share. We show several such matrices below. Your results may not be the same because of the random number generator **rand**.

$$\mathbf{k} = 3;$$
$$\mathbf{A} = \mathbf{fix(10*rand(k))};$$
$$\mathbf{B} = \mathbf{A+A'}$$

The display is

```
B =

     4     6    11
     6    18    11
    11    11     0
```

Using the **up arrow** key recall the previous commands one at a time, pressing ENTER after each command. This time the matrix displayed is

```
B =

     0     5    10
     5     6     6
    10     6    10
```

See Exercise T.27 at the end of Section 1.4. ■

10.4 ▼ Elementary Row Operations in MATLAB

The solution of linear systems of equations as discussed in Section 1.5 uses elementary row operations to obtain a sequence of linear systems whose augmented matrices are row equivalent. Row equivalent linear systems have the same solutions, hence we choose elementary row operations to produce row equivalent systems that are easy to solve. It is shown that linear systems in **reduced row echelon form** are easily solved using the Gauss–Jordan procedure and systems in **row echelon form** are easily solved using Gaussian elimination with back substitution. Using either of these procedures requires that we perform row operations that introduce zeros into the augmented matrix of the linear system. We show how to perform such row operations using MATLAB. The arithmetic involved is done by the MATLAB software and we are able to concentrate on the strategy to produce the reduced row echelon form or row echelon form.

Given a linear system $A\mathbf{x} = \mathbf{b}$, we enter the coefficient matrix A and the right-hand side $\mathbf{b}$ into MATLAB. We form the augmented matrix (see Section 10.2) as

$$\mathbf{C} = [\mathbf{A} \quad \mathbf{b}]$$

Now we are ready to begin applying row operations to the augmented matrix C. Each row operation replaces an existing row by a new row. Our strategy is to construct the row operation so that the resulting new row moves us closer to the goal of reduced row echelon form or row echelon form. There are many different choices that can be made for the sequence of row operations to transform $\begin{bmatrix} A & \vdots & \mathbf{b} \end{bmatrix}$ to one of these forms. Naturally we try to use the fewest number of row operations, but many times it is convenient to avoid introducing fractions (if possible), especially when doing calculations by hand. Since MATLAB will be doing the arithmetic for us, we need not be concerned about fractions, but it is visually pleasing to avoid them anyway.

As described in Section 1.5, there are three row operations. They are

☐ Interchange two rows.

☐ Multiply a row by a nonzero number.

☐ Add a multiple of one row to another row.

To perform these operations on an augmented matrix $C = \begin{bmatrix} A & \vdots & \mathbf{b} \end{bmatrix}$ in MATLAB, we employ the colon operator, which was discussed in Section 10.1. We illustrate the technique on the linear system in Example 6 of Section 1.5. When the augmented matrix is entered into MATLAB, we have

```
C =
    1    2    3    9
    2   -1    1    8
    3    0   -1    3
```

To produce the reduced row echelon form given in Equation (2) of Section 1.5, we proceed as follows.

	Description	*MATLAB Commands and Display*

add (−2) times row 1 to row 2 $C(2,:) = -2 * C(1,:) + C(2,:)$

[Explanation of MATLAB
command: Row 2 is replaced by
(or set equal to) the sum of −2
times row 1 and row 2.]

```
C   =
      1      2      3      9
      0     -5     -5    -10
      3      0     -1      3
```

add (−3) times row 1 to row 3 $C(3,:) = -3 * C(1,:) + C(3,:)$

```
C   =
      1      2      3      9
      0     -5     -5    -10
      0     -6    -10    -24
```

multiply row 2 by (−1/5) $C(2,:) = (-1/5) * C(2,:)$

[Explanation of MATLAB
command: Row 2 is replaced by
(or set equal to) $\left(-\frac{1}{5}\right)$ times row
2.]

```
C   =
      1      2      3      9
      0      1      1      2
      0     -6    -10    -24
```

add (−2) times row 2 to row 1 $C(1,:) = -2 * C(2,:) + C(1,:)$

```
C   =
      1      0      1      5
      0      1      1      2
      0     -6    -10    -24
```

add 6 times row 2 to row 3 $C(3,:) = 6 * C(3,:) + C(3,:)$

```
C   =
      1      0      1      5
      0      1      1      2
      0      0     -4    -12
```

multiply row 3 by (−1/4) $C(3,:) = (-1/4) * C(3,:)$

```
C   =
      1      0      1      5
      0      1      1      2
      0      0      1      3
```

add (-1) times row 3 to row 2 $C(2,:) = -1 * C(3,:) + C(2,:)$

```
C   =
        1   0   1    5
        0   1   0   -1
        0   0   1    3
```

add (-1) times row 3 to row 1 $C(1,:) = -1 * C(3,:) + C(1,:)$

```
C   =
        1   0   0    2
        0   1   0   -1
        0   0   1    3
```

This last augmented matrix implies that the solution of the linear system is $x = 2$, $y = -1$, $z = 3$.

In the preceding reduction of the augmented matrix to reduced row echelon form, no row interchanges were required. Suppose at some stage we had to interchange rows 2 and 3 of the augmented matrix C. To accomplish this we use a temporary storage area. (We choose to name this area **temp** here.) In MATLAB we proceed as follows.

Description	*MATLAB Commands*
Assign row 2 to temporary storage.	**temp = C(2,:);**
Assign the contents of row 3 to row 2.	**C(2,:) = C(3,:);**
Assign the contents of row 2 contained in temporary storage to row 3.	**C(3,:) = temp;**

(The semicolons at the end of each command just suppress the display of the contents.)

Using the colon operator and the assignment operator, $=$, as above, we can instruct MATLAB to perform row operations to obtain the reduced row echelon form or row echelon form of a matrix. MATLAB does the arithmetic and we concentrate on choosing the row operations to perform the reduction. We also must enter the appropriate MATLAB command. If we mistype a multiplier or row number, the error can be corrected, but the correction process requires a number of steps. To permit us to concentrate completely on choosing row operations for the reduction process, there is a routine called **reduce** on the diskette of auxiliary MATLAB routines available to users of this book.* Once you have incorporated these routines into MATLAB, you can type **help reduce** and see the following display.

---

*Some MATLAB commands that follow require the incorporation of the instructional routines that are on the diskette.

> REDUCE   Perform row reduction on matrix A by explicitly
> choosing row operations to use. A row operation can
> be ''undone,'' but this feature cannot be used in
> succession.
>
> Use the form ===> **reduce(A)** <===.

Routine **reduce** alleviates all the command typing and instructs MAT-LAB to perform the associated arithmetic. To use **reduce**, enter the augmented matrix C of your system as discussed previously and type

reduce(C)

We display the first three steps of **reduce** for Example 6 in Section 1.5. The matrices involved will be the same as those in the first three steps of the reduction process above, where we made direct use of the colon operator to perform the row operations in MATLAB. Screen displays are shown between rows of plus signs below, and all input appears in boxes.

```
+++++++++++++++++++++++++++++++++++++++++++++++++

          ***** "REDUCE" a Matrix by Row Reduction *****

The current matrix is:

A =
    1    2    3    9
    2   -1    1    8
    3    0   -1    3


                  OPTIONS
   <1>  Row(i) <===> Row(j)
   <2>  k * Row(i)     (k not zero)
   <3>  k * Row(i) + Row(j) ===> Row(j)
   <4>  Turn on rational display.
   <5>  Turn off rational display.
  <-1>  "Undo" previous row operation.
   <0>  Quit reduce!
        ENTER your choice ===>  3

Enter multiplier.  -2

Enter first row number.  1

Enter number of row that changes.  2
```

$$\left\{\begin{array}{c} \texttt{Comment: Option 3 in the above menu means the same as} \\ \textbf{add a multiple of one row to another row} \\ \texttt{The input above performs the operation in the form} \\ \texttt{multiplier * (first row) + (second row)} \end{array}\right\}$$

```
+++++++++++++++++++++++++++++++++++++++++++++++++++++
     ***** Replacement by Linear Combination Complete *****

The current matrix is:

A =
    1    2    3    9
    0   -5   -5  -10
    3    0   -1    3


               OPTIONS
  <1>   Row(i) <===> Row(j)
  <2>   k * Row(i)      (k not zero)
  <3>   k * Row(i) + Row(j) ===> Row(j)
  <4>   Turn on rational display.
  <5>   Turn off rational display.
  <-1>  "Undo" previous row operation.
  <0>   Quit reduce!
        ENTER your choice ===>  3
```

Enter multiplier. -3

Enter first row number. 1

Enter number of row that changes. 3

```
+++++++++++++++++++++++++++++++++++++++++++++++++++++
     ***** Replacement by Linear Combination Complete *****

The current matrix is:

A =
    1    2    3    9
    0   -5   -5  -10
    0   -6  -10  -24
```

```
                        OPTIONS
        <1>  Row(i) <===> Row(j)
        <2>  k * Row(i)     (k not zero)
        <3>  k * Row(i) + Row(j) ===> Row(j)
        <4>  Turn on rational display.
        <5>  Turn off rational display.
        <-1> "Undo" previous row operation.
        <0>  Quit reduce!
              ENTER your choice ===> │ 2 │
```

Enter multiplier. │ -1/5 │

Enter row number. │ 2 │

++

              ***** Multiplication Complete *****

The current matrix is:

A =
```
    1    2    3    9
    0    1    1    2
    0   -6  -10  -24
```

```
                        OPTIONS
        <1>  Row(i) <===> Row(j)
        <2>  k * Row(i)     (k not zero)
        <3>  k * Row(i) + Row(j) ===> Row(j)
        <4>  Turn on rational display.
        <5>  Turn off rational display.
        <-1> "Undo" previous row operation.
        <0>  Quit reduce!
              ENTER your choice ===>
```

++

At this point you should complete the reduction of this matrix to reduced row echelon form using **reduce**.

COMMENTS

1. Although options 1–3 in **reduce** appear in symbols, they have the same meaning as the phrases used to describe the row operations near the beginning of this section. Option <3> forms a *linear combination* of rows to replace a row. This terminology will be used later in this course and appear in certain displays of **reduce**. (See Sections 10.7 and 1.5.)

2. Within routine **reduce**, the matrix on which the row operations are performed is called *A*, regardless of the name of your input matrix.

EXAMPLE 1 ■ Solve the following linear system using **reduce**.

$$\tfrac{1}{3}x + \tfrac{1}{4}y = \tfrac{13}{6}$$

$$\tfrac{1}{7}x + \tfrac{1}{9}y = \tfrac{59}{63}$$

Solution   Enter the augmented matrix into MATLAB and name it C.

$$\mathbf{C} = [1/3 \quad 1/4 \quad 13/6; 1/7 \quad 1/9 \quad 59/63]$$

```
C =
     0.3333  0.2500  2.1667
     0.1429  0.1111  0.9365
```

Then type

$$\mathbf{reduce(C)}$$

The steps from **reduce** are displayed below. The steps are shown with decimal displays unless you choose the rational display option <4>. The corresponding rational displays are shown in braces in the following examples for illustration purposes. Ordinarily the decimal and rational displays are not shown simultaneously.

```
++++++++++++++++++++++++++++++++++++++++++++++++++
       ***** "REDUCE" a Matrix by Row Reduction *****
The current matrix is:

A =
     0.3333    0.2500    2.1667          {1/3    1/4    13/6 }
     0.1429    0.1111    0.9365          {1/7    1/9    59/63}

                OPTIONS
  <1>   Row(i) <===> Row(j)
  <2>   k * Row(i)     (k not zero)
  <3>   k * Row(i) + Row(j) ===> Row(j)
  <4>   Turn on rational display.
  <5>   Turn off rational display.
 <-1>   "Undo" previous row operation.
  <0>   Quit reduce!
        ENTER your choice ===>  2

Enter multiplier.  1/A(1,1)

Enter row number.  1
```

```
+++++++++++++++++++++++++++++++++++++++++++++++++
        ***** Row Multiplication Complete *****

The current matrix is:

A =
    1.0000    0.7500    6.5000            {1        3/4      13/2 }
    0.1429    0.1111    0.9365            {1/7      1/9      59/63}

                    OPTIONS
  <1>   Row(i) <===> Row(j)
  <2>   k * Row(i)      (k not zero)
  <3>   k * Row(i) + Row(j) ===> Row(j)
  <4>   Turn on rational display.
  <5>   Turn off rational display.
 <-1>   "Undo" previous row operation.
  <0>   Quit reduce!
            ENTER your choice ===>  3

Enter multiplier.  -A(2,1)

Enter first row number.  1

Enter number of row that changes.  2

+++++++++++++++++++++++++++++++++++++++++++++++++
    ***** Replacement by Linear Combination Complete *****

The current matrix is:

A =
    1.0000    0.7500    6.5000            {1        3/4      13/2 }
         0    0.0040    0.0079            {0        1/252    1/126}

                    OPTIONS
  <1>   Row(i) <===> Row(j)
  <2>   k * Row(i)      (k not zero)
  <3>   k * Row(i) + Row(j) ===> Row(j)
  <4>   Turn on rational display.
  <5>   Turn off rational display.
 <-1>   "Undo" previous row operation.
  <0>   Quit reduce!
            ENTER your choice ===>  2

Enter multiplier.  1/A(2,2)

Enter row number.  2
```

++

***** Row Multiplication Complete *****

The current matrix is:

A =

```
    1.0000    0.7500    6.5000              {1      3/4     13/2}
         0    1.0000    2.0000              {0      1       2    }
```

                        OPTIONS
 <1>   Row(i) <===> Row(j)
 <2>   k * Row(i)   (k not zero)
 <3>   k * Row(i) + Row(j) ===> Row(j)
 <4>   Turn on rational display.
 <5>   Turn off rational display.
<-1>   "Undo" previous row operation.
 <0>   Quit reduce!
       ENTER your choice ===> 3

Enter multiplier. -A(1,2)

Enter first row number. 2

Enter number of row that changes. 1

++

***** Replacement by Linear Combination Complete *****

The current matrix is:

A =

```
    1.0000         0    5.0000              {1      0      5}
         0    1.0000    2.0000              {0      1      2}
```

                        OPTIONS
 <1>   Row(i) <===> Row(j)
 <2>   k * Row(i)   (k not zero)
 <3>   k * Row(i) + Row(j) ===> Row(j)
 <4>   Turn on rational display.
 <5>   Turn off rational display.
<-1>   "Undo" previous row operation.
 <0>   Quit reduce!
       ENTER your choice ===> 0

         * * * * ===> REDUCE is over. Your final matrix is:

```
A  =

        1.0000         0    5.0000
             0    1.0000    2.0000
```

++

It follows that the solution of the system is $x = 5$, $y = 2$.   ■

The **reduce** routine forces you to concentrate on the strategy of the row reduction process. Once you have used **reduce** on a number of linear systems, the reduction process becomes a fairly systematic computation. The reduced row echelon form of a matrix is used in many places in linear algebra to provide information related to concepts (we study a number of these later). As such, the reduced row echelon form of a matrix becomes one step of more involved computational processes. Hence MATLAB provides an automatic way to obtain the reduced row echelon form. The command is **rref**. Once you have entered the matrix A under consideration, where A could represent an augmented matrix, just type

$$\textbf{rref(A)}$$

and MATLAB responds by displaying the reduced row echelon form of A.

EXAMPLE 2   ■   In Section 1.5, Example 11 asks for the solution of the homogeneous system

$$x +\ y + z + w = 0$$
$$x \qquad\qquad + w = 0$$
$$x + 2y + z \qquad = 0.$$

Form the augmented matrix C in MATLAB to obtain

```
C  =

        1    1    1    1    0
        1    0    0    1    0
        1    2    1    0    0
```

Next type

$$\textbf{rref(C)}$$

and MATLAB displays

```
ans  =

        1    0    0    1    0
        0    1    0   -1    0
        0    0    1    1    0
```

It follows that unknown w can be chosen arbitrarily—say, $w = r$, where r is any real number. Hence the solution is

$$x = -r, \qquad y = r, \qquad z = -r, \qquad w = r.$$ ■

10.5 ▼ Matrix Inverses in MATLAB

As discussed in Section 1.6, for a square matrix A to be nonsingular, the reduced row echelon form of A must be the identity matrix. Hence in MATLAB we can determine if A is singular or nonsingular by computing the reduced row echelon form of A using either **reduce** or **rref**. If the result is the identity matrix, then A is nonsingular. Such a computation determines whether or not an inverse exists, but does not explicitly compute the inverse when it exists. To compute the inverse of A we can proceed as in Section 1.6 and find the reduced row echelon form of $\begin{bmatrix} A \vdots I \end{bmatrix}$. If the resulting matrix is $\begin{bmatrix} I \vdots Q \end{bmatrix}$, then $Q = A^{-1}$. In MATLAB, once a nonsingular matrix A has been entered, the inverse can be found step by step using

$$\textbf{reduce}([\textbf{A}\quad \textbf{eye(size(A))}])$$

or computed immediately using

$$\textbf{rref}([\textbf{A}\quad \textbf{eye(size(A))}])$$

For example, if we use the matrix A in Example 5 of Section 1.6, then

$$A = \begin{bmatrix} 1 & 1 & 1 \\ 0 & 2 & 3 \\ 5 & 5 & 1 \end{bmatrix}.$$

Entering matrix A into MATLAB and typing the command

$$\textbf{rref}([\textbf{A}\quad \textbf{eye(size(A))}])$$

displays

```
ans =
```

1.0000	0	0	1.6250	−0.5000	−0.1250
0	1.0000	0	−1.8750	0.5000	0.3750
0	0	1.0000	1.2500	0	−0.2500

To extract the inverse matrix we use

$$\textbf{Ainv} = \textbf{ans}(:,\textbf{4:6})$$

and obtain

```
Ainv =
```

1.6250	−0.5000	−0.1250
−1.8750	0.5000	0.3750
1.2500	0	−0.2500

To see the result in rational display, use

$$\textbf{format rat}$$
$$\textbf{Ainv}$$

which gives

$$\text{Ainv} =$$

$$\begin{array}{ccc} 13/8 & -1/2 & -1/8 \\ -15/8 & 1/2 & 3/8 \\ 5/4 & 0 & -1/4 \end{array}$$

Type command

format short

Thus our previous MATLAB commands can be used in a manner identical to the way the hand computations are described in Section 1.6.

For convenience, there is a routine that computes inverses directly. The command is **invert**. For the preceding matrix A we would type

invert(A)

and the result would be identical to that obtained in **Ainv** by using **rref**. If the matrix is not square or is singular, an error message will appear.

10.6 ▾ Vectors in MATLAB

An n-vector $\mathbf{x}$ (see Section 3.2) in MATLAB can be represented either as a column matrix with n elements,

$$\mathbf{x} = \begin{bmatrix} x_1 \\ x_2 \\ \vdots \\ x_n \end{bmatrix},$$

or as a row matrix with n elements,

$$\mathbf{x} = \begin{bmatrix} x_1 & x_2 & \cdots & x_n \end{bmatrix}.$$

In a particular problem or exercise, choose one way of representing the n-vectors and stay with that form.

The vector operations of Section 3.2 correspond to operations on $n \times 1$ matrices or columns. If the n-vector is represented by row matrices in MATLAB, then the vector operations correspond to operations on $1 \times n$ matrices. These are just special cases of addition, subtraction, and scalar multiplication of matrices, which were discussed in Section 10.2.

The norm or length of vector $\mathbf{x}$ in MATLAB is obtained by using the command

norm(x)

This command computes the square root of the sum of the squares of the components of $\mathbf{x}$, which is equal to $\|\mathbf{x}\|$, as discussed in Section 3.2.

The distance between vectors $\mathbf{x}$ and $\mathbf{y}$ in R^n in MATLAB is given by

norm(x − y)

EXAMPLE 1 ■ Let

$$\mathbf{u} = \begin{bmatrix} 2 \\ 1 \\ 1 \\ -1 \end{bmatrix} \quad \text{and} \quad \mathbf{v} = \begin{bmatrix} 3 \\ 1 \\ 2 \\ 0 \end{bmatrix}.$$

Enter these vectors in R^4 into MATLAB as columns. Then

norm(u)

displays

        ans =

            2.6458

while

norm(v)

gives

        ans =

            3.7417

and

norm(u − v)

gives

        ans =

            1.7321   ■

The dot product of a pair of vectors $\mathbf{u}$ and $\mathbf{v}$ in R^n in MATLAB is computed by the command

dot(u,v)

For the vectors in Example 1, MATLAB gives the dot product as

        ans =
            9

As discussed in Section 3.2, the notion of a dot product is useful to define the angle between n-vectors. Equation (4) in Section 3.2 tells us that the cosine of the angle θ between $\mathbf{u}$ and $\mathbf{v}$ is given by

$$\cos\theta = \frac{\mathbf{u} \cdot \mathbf{v}}{\|\mathbf{u}\| \, \|\mathbf{v}\|}.$$

In MATLAB the cosine of the angle between $\mathbf{u}$ and $\mathbf{v}$ is computed by the command

dot(u,v)/(norm(u) ∗ norm(v))

The angle θ can be computed by taking the arccosine of the value of the previous expression. In MATLAB the arccosine function is denoted by **acos**. The result will be an angle in radians.

EXAMPLE 2 ■ For the vectors **u** and **v** in Example 1, the angle between the vectors is computed as

$$c = \mathbf{dot(u,v)/(norm(u) * norm(v));}$$
$$\mathbf{angle = acos(c)}$$

which displays

```
angle =
      0.4296
```

and is approximately $24.61°$. ■

10.7 ▾ Applications of Linear Combinations in MATLAB

The notion of a linear combination as discussed in Section 4.2 is fundamental to a wide variety of topics in linear algebra. The ideas of span, linear independence, linear dependence, and basis are based on forming linear combinations of vectors. In addition, the elementary row operations discussed in Sections 1.5 and 10.4 are essentially of the form, "Replace an existing row by a linear combination of rows." This is clearly the case when we add a multiple of one row to another row. (See the menu for the routine **reduce** in Section 10.4.) From this point of view, it follows that the reduced row echelon form and the row echelon form are processes for implementing linear combinations of rows of a matrix. Hence the MATLAB routines **reduce** and **rref** should be useful in solving problems that involve linear combinations.

Here we discuss how to use MATLAB to solve problems dealing with linear combinations, span, linear independence, linear dependence, and basis. The basic strategy is to set up a linear system related to the problem and ask questions such as "Is there a solution?" or "Is the only solution the trivial solution?"

The Linear Combination Problem

Given a vector space V and a set of vectors $S = \{\mathbf{v}_1, \mathbf{v}_2, \ldots, \mathbf{v}_k\}$ in V, determine if $\mathbf{v}$, belonging to V, can be expressed as a linear combination of the members of S. That is, can we find some set of scalars $c_1, c_2, \ldots, c_k$ so that

$$c_1\mathbf{v}_1 + c_2\mathbf{v}_2 + \cdots + c_k\mathbf{v}_k = \mathbf{v}?$$

There are several common situations.

Case 1. If the vectors in S are row matrices, then we construct (as shown in Example 10 of Section 4.2) a linear system whose coefficient matrix A is

$$A = \begin{bmatrix} \mathbf{v}_1 \\ \mathbf{v}_2 \\ \vdots \\ \mathbf{v}_k \end{bmatrix}^T$$

and whose right-hand side is $\mathbf{v}^T$. That is, the columns of A are the row matrices of set S converted to columns. Let $\mathbf{c} = \begin{bmatrix} c_1 & c_2 & \cdots & c_k \end{bmatrix}$ and $\mathbf{b} = \mathbf{v}^T$, then transform the linear system $A\mathbf{c} = \mathbf{b}$ using **reduce** or **rref** in MATLAB. If the system is shown to be consistent, so that no rows of the form $\begin{bmatrix} 0 & 0 & \cdots & 0 \mid q \end{bmatrix}$, $q \neq 0$, occur, then the vector $\mathbf{v}$ can be written as a linear combination of the vectors in S. In that case the solution of the system gives the values of the coefficients. *Caution*: Many times we need only determine if the system is consistent to decide whether $\mathbf{v}$ is a linear combination of the members of S. Read the question carefully.

EXAMPLE 1 ■ To apply MATLAB to Example 10 of Section 4.2, proceed as follows. Define

$$A = \begin{bmatrix} 1 & 2 & 1; 1 & 0 & 2; 1 & 1 & 0 \end{bmatrix}'$$
$$b = \begin{bmatrix} 2 & 1 & 5 \end{bmatrix}'$$

Then use the command

$$\mathbf{rref}([A \quad b])$$

to give

```
ans =

        1    0    0    1
        0    1    0    2
        0    0    1   -1
```

Recall that this display represents the reduced row echelon form of an augmented matrix. It follows that the system is consistent, with solution

$$c_1 = 1, \qquad c_2 = 2, \qquad c_3 = -1.$$

Hence $\mathbf{v}$ is a linear combination of $\mathbf{v}_1$, $\mathbf{v}_2$, and $\mathbf{v}_3$. ■

Case 2. If the vectors in S are column matrices, then just lay the columns side by side to form the coefficient matrix

$$A = \begin{bmatrix} \mathbf{v}_1 & \mathbf{v}_2 & \cdots & \mathbf{v}_k \end{bmatrix}$$

and set $\mathbf{b} = \mathbf{v}$. Proceed as described in Case 1.

Case 3. If the vectors in S are polynomials, then associate with each polynomial a column of coefficients. Make sure any missing terms in the polynomial are associated with a zero coefficient. One way to proceed is to use the coefficient of the highest-power term as the first entry of the column, the coefficient of the next-highest-power term as the second entry, and so on. For example,

$$t^2 + 2t + 1 \longrightarrow \begin{bmatrix} 1 \\ 2 \\ 1 \end{bmatrix}, \qquad t^2 + 2 \longrightarrow \begin{bmatrix} 1 \\ 0 \\ 2 \end{bmatrix}, \qquad 3t - 2 \longrightarrow \begin{bmatrix} 0 \\ 3 \\ -2 \end{bmatrix}.$$

The linear combination problem is now solved as in Case 2.

Case 4. If the vectors in S are $m \times n$ matrices, then associate with each such matrix A_j a column $\mathbf{v}_j$ formed by stringing together its columns one after the other. In MATLAB this transformation is done using the **reshape** command. Then we proceed as in Case 2.

EXAMPLE 2 ■ Given matrix

$$P = \begin{bmatrix} 1 & 2 & 3 \\ 4 & 5 & 6 \end{bmatrix}.$$

To associate a column matrix as described above within MATLAB, first enter P into MATLAB, then type the command

$$\mathbf{v} = \mathbf{reshape(P,6,1)}$$

which gives

```
v =

     1
     4
     2
     5
     3
     6
```

For more information, type **help reshape**.   ■

The Span Problem

There are two common types of problems related to span. The first is:

Given the set of vectors $S = \{\mathbf{v}_1, \mathbf{v}_2, \ldots, \mathbf{v}_k\}$ and the vector $\mathbf{v}$ in a vector space V, is $\mathbf{v}$ in span S?

This is identical to the linear combination problem addressed above because we want to know if $\mathbf{v}$ is a linear combination of the members of S. As shown above, we can use MATLAB in many cases to solve this problem.

The second type of problem related to span is:

Given vectors $S = \{\mathbf{v}_1, \mathbf{v}_2, \ldots, \mathbf{v}_k\}$ in a vector space V, does span $S = V$?

Here we are asked if every vector in V can be written as a linear combination of the vectors in S. In this case the linear system constructed has a right-hand side that contains arbitrary values that correspond to an arbitrary vector in V. (See Example 1 in Section 4.3.) Since MATLAB manipulates only numerical values in routines such as **reduce** and **rref**, we cannot use MATLAB here to (fully) answer this question.

For the second type of spanning question there is a special case that arises frequently and can be handled in MATLAB. In Section 4.4 the concept of the dimension of a vector space is discussed. The dimension of a vector space V is the number of vectors in a basis (see Section 4.4), which is the smallest number of vectors that can span V. If we know that V has dimension k and the set S has k vectors, then we can proceed as follows to see if span $S = V$. Develop a linear system $A\mathbf{c} = \mathbf{b}$ associated with the span question. If the reduced row echelon form of the coefficient matrix A has the form

$$\begin{bmatrix} I_k \\ \mathbf{0} \end{bmatrix},$$

where $\mathbf{0}$ is a submatrix of all zeros, then any vector in V is expressible in terms of the members of S. In fact, S is a basis for V. In MATLAB we can use routine **reduce** or **rref** on matrix A. If A is square, we can also use **det**. Try this strategy on Example 1 in Section 4.3.

Another spanning question involves finding a set that spans the set of solutions of a homogeneous system of equations $A\mathbf{x} = \mathbf{0}$. The strategy in MATLAB is to find the reduced row echelon form of $\begin{bmatrix} A \mathbin{\vdots} \mathbf{0} \end{bmatrix}$ using the command

$$\mathbf{rref(A)}$$

(There is no need to include the augmented column since it is all zeros.) Then form the general solution of the system and express it as a linear combination of columns. The columns form a spanning set for the solution set of the system. See Example 6 in Section 4.3.

The Linear Independence/Dependence Problem

The linear independence or dependence of a set of vectors $S = \{\mathbf{v}_1, \mathbf{v}_2, \dots, \mathbf{v}_k\}$ is a linear combination question. Set S is linearly independent if the *only* time the linear combination $c_1\mathbf{v}_1 + c_2\mathbf{v}_2 + \cdots + c_k\mathbf{v}_k$ gives the zero vector is when $c_1 = c_2 = \cdots = c_k = 0$. If we can produce the zero vector with any one of the coefficients $c_j \neq 0$, then S is linearly dependent. Following the discussion on linear combination problems, we produce the associated linear system

$$A\mathbf{c} = \mathbf{0}.$$

Note that the linear system is homogeneous. We have the following result:

> S is linearly independent if and only if $A\mathbf{c} = \mathbf{0}$
> has only the trivial solution.

Otherwise, S is linearly dependent. See Examples 8 and 9 in Section 4.3. Once we have the homogeneous system $A\mathbf{c} = \mathbf{0}$, we can use MATLAB routine **reduce** or **rref** to analyze whether or not the system has a nontrivial solution.

A special case arises if we have k vectors in a set S in a vector space V whose dimension is k (see Section 4.4). Let the linear system associated with the linear combination problem be $A\mathbf{c} = \mathbf{0}$. It can be shown that

> S is linearly independent if and only if
> the reduced row echelon form of A is $\begin{bmatrix} I_k \\ \mathbf{0} \end{bmatrix}$,

where $\mathbf{0}$ is a submatrix of all zeros. In fact we can extend this further to say S is a basis for V. (See Theorem 4.9.) In MATLAB we can use **reduce** or **rref** on A to aid in the analysis of such a situation.

10.8 ▼ Linear Transformations in MATLAB

We consider the special case of linear transformations $L: R^n \to R^m$. Every such linear transformation can be represented by an $m \times n$ matrix A. (See Section 3.3.) Then, for $\mathbf{x}$ in R^n,

$$L(\mathbf{x}) = A\mathbf{x},$$

which is in R^m. For example, suppose that $L: R^4 \to R^3$ is given by $L(\mathbf{x}) = A\mathbf{x}$, where matrix

$$A = \begin{bmatrix} 1 & -1 & -2 & -2 \\ 2 & -3 & -5 & -6 \\ 1 & -2 & -3 & -4 \end{bmatrix}.$$

The image of

$$\mathbf{x} = \begin{bmatrix} 1 \\ 2 \\ -1 \\ 0 \end{bmatrix}$$

under L is

$$L(\mathbf{x}) = A\mathbf{x} = \begin{bmatrix} 1 & -1 & -2 & -2 \\ 2 & -3 & -5 & -6 \\ 1 & -2 & -3 & -4 \end{bmatrix} \begin{bmatrix} 1 \\ 2 \\ -1 \\ 0 \end{bmatrix} = \begin{bmatrix} 1 \\ 1 \\ 0 \end{bmatrix}.$$

The **range of a linear transformation** L is the subspace of R^m consisting of the set of all images of vectors from R^n. It is easily shown that

$$\text{range } L = \text{column space of } A.$$

(See Example 11 in Section 6.2.) It follows that we "know the range of L" when we have a basis for the column space of A. There are two simple ways to find a basis for the column space of A:

1. The transposes of the nonzero rows of $\mathbf{rref(A')}$ form a basis for the column space. (See Example 4 in Section 4.6.)
2. If the columns containing the leading 1's of $\mathbf{rref(A)}$ are $k_1 < k_2 < \cdots < k_r$, then columns $k_1, k_2, \ldots, k_r$ of A are a basis for the column space of A. (See Example 4 in Section 4.6.)

For the matrix A given above, we have

$$\mathbf{rref(A')} = \begin{bmatrix} 1 & 0 & -1 \\ 0 & 1 & 1 \\ 0 & 0 & 0 \\ 0 & 0 & 0 \end{bmatrix},$$

and hence $\left\{ \begin{bmatrix} 1 \\ 0 \\ -1 \end{bmatrix}, \begin{bmatrix} 0 \\ 1 \\ 1 \end{bmatrix} \right\}$ is a basis for the range of L. Using method 2,

$$\mathbf{rref(A)} = \begin{bmatrix} 1 & 0 & -1 & 0 \\ 0 & 1 & 1 & 2 \\ 0 & 0 & 0 & 0 \end{bmatrix}.$$

Thus it follows that columns 1 and 2 of A are a basis for the column space of A and hence a basis for the range of L. In addition, routine **lisub** can be used. Use **help** for directions.

The **kernel of a linear transformation** is the subspace of all vectors in R^n whose image is the zero vector in R^m. This corresponds to the set of all vectors $\mathbf{x}$ satisfying

$$L(\mathbf{x}) = A\mathbf{x} = \mathbf{0}.$$

Hence it follows that the kernel of L, denoted $\ker L$, is the set of all solutions of the homogeneous system

$$A\mathbf{x} = \mathbf{0},$$

which is the null space of A. Thus we "know the kernel of L" when we have a basis for the null space of A. To find a basis for the null space of A, we form the general solution of $A\mathbf{x} = \mathbf{0}$ and "separate it into a linear combination of columns using the arbitrary constants that are present." The columns employed form a basis for the null space of A. This procedure uses **rref(A)**. For the matrix A given above, we have

$$\mathbf{rref(A)} = \begin{bmatrix} 1 & 0 & -1 & 0 \\ 0 & 1 & 1 & 2 \\ 0 & 0 & 0 & 0 \end{bmatrix}.$$

If we choose the variables corresponding to columns without leading 1's to be arbitrary, we have

$$x_3 = r \quad \text{and} \quad x_4 = t.$$

It follows that the general solution to $A\mathbf{x} = \mathbf{0}$ is given by

$$\mathbf{x} = \begin{bmatrix} x_1 \\ x_2 \\ x_3 \\ x_4 \end{bmatrix} = \begin{bmatrix} r \\ -r - 2t \\ r \\ t \end{bmatrix} = r \begin{bmatrix} 1 \\ -1 \\ 1 \\ 0 \end{bmatrix} + t \begin{bmatrix} 0 \\ -2 \\ 0 \\ 1 \end{bmatrix}.$$

Thus columns

$$\begin{bmatrix} 1 \\ -1 \\ 1 \\ 0 \end{bmatrix} \quad \text{and} \quad \begin{bmatrix} 0 \\ -2 \\ 0 \\ 1 \end{bmatrix}$$

form a basis for $\ker L$. See also routine **homsoln**, which will display the general solution of a homogeneous linear system. In addition, the command **null** will produce an orthonormal basis for the null space of a matrix. Use **help** for further information on these commands.

In summary, appropriate use of the **rref** command in MATLAB will give bases for both the kernel and range of the linear transformation $L(\mathbf{x}) = A\mathbf{x}$.

10.9 ▼ MATLAB Command Summary

In this section we list the principal MATLAB commands and operators used in this book. The list is divided into two parts: commands that come with the MATLAB software, and special instructional routines on the diskette available to users of this book. Both parts are cross-indexed with sections of Chapter 10 where they are discussed and/or initial sections that have MATLAB exercises specifically referring to them. For ease of reference we have included a brief description of each instructional routine that is on the diskette available to users of this book. These descriptions are also available using MATLAB's **help** command once the installation procedures are complete. A description of any MATLAB command can be obtained by using **help**. (See the introduction to this chapter.)

Built-in MATLAB Commands

ans ⟨10.1⟩	**inv** ⟨10.5⟩	**roots** ⟨5.1⟩
clear ⟨10.1, 1.2⟩	**norm** ⟨10.6⟩	**rref** ⟨10.4, 1.6⟩
conj ⟨A.1⟩	**null** ⟨10.8⟩	**size** ⟨10.1⟩
det ⟨10.7, 2.1⟩	**ones** ⟨10.3, 1.4⟩	**sqrt** ⟨A.1⟩
diag ⟨10.2, 1.2⟩	**pi** ⟨9.1⟩	**sum** ⟨8.3⟩
dot ⟨10.6, 1.3, 3.2⟩	**poly** ⟨5.1⟩	**tril** ⟨10.2, 1.4⟩
eig ⟨5.2⟩	**polyval** ⟨10.3⟩	**triu** ⟨10.2, 1.4⟩
exit ⟨10⟩	**polyvalm** ⟨10.3, 1.4⟩	**zeros** ⟨10.3⟩
eye ⟨10.3, 2.1⟩	**quit** ⟨10⟩	\ ⟨1.5⟩
fix ⟨10.3, 1.4⟩	**rand** ⟨10.3, 1.4⟩	; ⟨10.1⟩
format ⟨10.1, 1.2⟩	**rank** ⟨4.6⟩	: ⟨10.1⟩
help ⟨10⟩	**rat** ⟨10.1⟩	′ (prime) ⟨10.2, 1.2⟩
hilb ⟨10.1, 1.2⟩	**real** ⟨A.1⟩	+, −, *, /, ^ ⟨10.2, 10.3⟩
image ⟨A.1⟩	**reshape** ⟨10.7⟩	

Supplemental Instructional Commands

adjoint ⟨2.2⟩	**forsub** ⟨9.3⟩	**lsqline** ⟨8.4⟩
bksub ⟨9.2, 9.3⟩	**gschmidt** ⟨4.8⟩	**lupr** ⟨9.3⟩
cofactor ⟨2.2⟩	**homsoln** ⟨10.8, 4.5⟩	**planelt** ⟨3.4⟩
crossprd ⟨3.5⟩	**invert** ⟨10.5⟩	**reduce** ⟨10.4, 1.5⟩
crossdemo ⟨3.5⟩	**linprog** ⟨7.2⟩	**vec2demo** ⟨3.1⟩
dotprod ⟨10.6, 3.2⟩	**lpstep** ⟨7.2⟩	**vec3demo** ⟨3.2⟩

Notes: ⟨10⟩ refers to the introduction to Chapter 10. Both **rref** and **reduce** are used in many sections. Several utilities required by the instructional commands also appear on the diskette: **arrowh**, **mat2strh**, and **blkmat**. The description given below is that displayed in response to the **help** command. In the description of several commands, the notation differs slightly from that in the text.

Description of Instructional Commands

ADJOINT

Compute the classical adjoint of a square matrix A. If A is not square an empty matrix is returned.
*** This routine should only be used by students to check adjoint computations and should not be used as part of a routine to compute inverses. See invert or inv.

Use in the form ==> adjoint(A) <==

BKSUB

Perform back substitution on upper triangular system Ax = b. If A is not square, upper triangular, and nonsingular, an error message is displayed. In case of an error the solution returned is all zeros.

Use in the form ==> bksub(A,b) <==

COFACTOR

Computes the (i,j)-cofactor of matrix A. If A is not square, an error message is displayed.
*** This routine should only be used by students to check cofactor computations.

Use in the form ==> cofactor(i,j,A) <==

CROSSDEMO

Display a pair of three-dimensional vectors and their cross product.

The input vectors X and Y are displayed in a three-dimensional perspective along with their cross product. For visualization purposes a set of coordinate 3-D axes are shown.

Use in the form ==> crossdemo(X,Y) <==

CROSSPRD

Compute the cross product of vectors x and y in 3-space. The output is a vector orthogonal to both of the original vectors x and y. The output is returned as a row matrix with 3 components [v1 v2 v3] which is interpreted as v1*i + v2*j + v3*k where i, j, k are the unit vectors in the x, y, and z directions respectively.

Use in the form ==> v = crossprd(x,y) <==

DOTPROD

The dot product of two n-vectors x and y is computed. The vectors can be either rows, columns, or matrices of the same size. For complex vectors the dot product of x and y is computed as the conjugate transpose of the first times the second.

Use in the form ==> dotprod(x,y) <==

FORSUB

Perform forward substitution on a lower triangular system Ax = b. If A is not square, lower triangular, and nonsingular, an error

	message is displayed. In case of an error the solution returned is all zeros.

Use in the form ==> forsub(A,b) <==

GSCHMIDT   The Gram-Schmidt process on the columns in matrix x. The orthonormal basis appears in the columns of y unless there is a second argument, in which case y contains only an orthogonal basis. The second argument can have any value.

Use in the form ==> y = gschmidt(x) <==
or ==> y = gschmidt(x,v) <==

HOMSOLN   Find the general solution of a homogeneous system of equations. The routine returns a set of basis vectors for the null space of $Ax = 0$.

Use in the form ==> ns = homsoln(A) <==

If there is a second argument, the general solution is displayed.

Use in the form ==> homsoln(A,1) <==

This option assumes that the general solution has at most 10 arbitrary constants.

INVERT   Compute the inverse of a matrix A by using the reduced row echelon form applied to [A I]. If A is singular, a warning is given.

Use in the form ==> B = invert(A) <==

LINPROG   Directly solves the standard linear programming problem using slack variables as formulated in Introductory Linear Algebra with Applications by B. Kolman. This routine is only designed for small problems.

To form the initial tableau A the coefficients of the constraints are entered into the rows where the equations are of the form:

$$a_1X_1 + a_2X_2 + a_3X_3 + \cdots + a_nX_n = C_m$$

and the bottom row consists of the objective function written in the form:

$$z_1X_1 + z_2X_2 + z_3X_3 + \cdots + z_nX_n = 0.$$

Then, as long as there is at least one negative entry in the last row, linprog will find the optimal solution. If no argument containing the initial tableau is present, the routine prompts for the tableau.

Use in the form ==> linprog(A) or linprog <==

LPSTEP   A step-by-step solver for small standard linear programming problems. At each stage you are asked to determine the pivot.

Incorrect responses initiate a set of questions to aid in pivot selection. The screens reflect the problem form developed in Introductory Linear Algebra with Applications by B. Kolman.

This routine solves the standard linear programming problem using the Simplex Method with slack variables. To form the initial tableau A the coefficients of the constraints are entered into the rows where the equations are of the form

$$a_1 X_1 + a_2 X_2 + a_3 X_3 + \cdots + a_n X_n = C_m$$

and the bottom row consists of the objective function written in the form

$$z_1 X_1 + z_2 X_2 + z_3 X_3 + \cdots + z_n X_n = 0.$$

Then, as long as there is at least one negative entry in the last row, LPSTEP will find an optimal solution. If the tableau A is not supplied as an argument, the user will be prompted to enter it. <<requires utility mat2strh.m>>

Use in the form ==> lpstep(A) or lpstep <==

LSQLINE

This routine will construct the equation of the least square line to a data set of ordered pairs and then graph the line and the data set. A short menu of options is available, including evaluating the equation of the line at points.

Use in the form ==> c = lsqline(x,y) or lsqline(x,y) <==

Here x is a vector containing the x-coordinates and y is a vector containing the corresponding y-coordinates. On output, c contains the coefficients of the least squares line:

$$y = c(1) * x + c(2)$$

LUPR

Perform LU-factorization on matrix A by explicitly choosing row operations to use. No row interchanges are permitted, hence it is possible that the factorization cannot be found. It is recommended that the multipliers be constructed in terms of the elements of matrix U, like $-U(3,2)/U(2,2)$, since the displays of matrices L and U do not show all the decimal places available. A row operation can be "undone," but this feature cannot be used in succession.

This routine uses the utilities mat2strh and blkmat.

Use in the form ==> [L,U] = lupr(A) <==

PLANELT

Demonstration of plane linear transformations:

Rotations, Reflections, Expansions/Compressions, Shears

Or you may specify your own transformation.

Graphical results of successive plane linear transformations can be seen using a multiple window display. Standard figures can be chosen or you may choose to use your own figure.

Use in the form ==> planelt <==

REDUCE

Perform row reduction on matrix A by explicitly choosing row operations to use. A row operation can be "undone," but this feature cannot be used in succession.

Use in the form ==> reduce(A) <==

VEC2DEMO

A graphical demonstration of vector operations for two-dimensional vectors.

Select vectors X = [x1 x2] and Y = [y1 y2]. They will be displayed graphically along with their sum, difference, and a scalar multiple.

Use in the form ==> vec2demo(X,Y) <==

or ==> vec2demo <==

In the latter case you will be prompted for input.

VEC3DEMO

Display a pair of three-dimensional vectors, their sum, difference and scalar multiples.

The input vectors X and Y are displayed in a 3-dimensional perspective along with their sum, difference and selected scalar multiples. For visualization purposes a set of coordinate 3-D axes are shown.

Use in the form ==> vec3demo(X,Y) <==

Appendix A

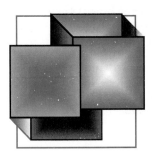

Complex Numbers

A.1 ▾ Complex Numbers

Complex numbers are usually introduced in an algebra course to "complete" the solution to the quadratic equation

$$ax^2 + bx + c = 0, \qquad a \neq 0.$$

In using the quadratic formula

$$x = \frac{-b \pm \sqrt{b^2 - 4ac}}{2a},$$

the case in which $b^2 - 4ac < 0$ is not resolved unless we can cope with the square roots of negative numbers. In the sixteenth century mathematicians and scientists justified this "completion" of the solution of quadratic equations by intuition. Naturally, a controversy arose, with some mathematicians denying the existence of these numbers and others using them along with real numbers. The use of complex numbers did not lead to any contradictions, and the idea proved to be an important milestone in the development of mathematics.

A **complex number** c is of the form $c = a + bi$, where a and b are real numbers and where $i = \sqrt{-1}$; a is called the **real part** of c and b is called the **imaginary part** of c. The term "imaginary part" arose from the mysticism surrounding the beginnings of complex numbers; however, these numbers are as "real" as the real numbers.

EXAMPLE 1 ■ (a) $5 - 3i$ has real part 5 and imaginary part -3; (b) $-6 + \sqrt{2}\,i$ has real part -6 and imaginary part $\sqrt{2}$. ■

The symbol $i = \sqrt{-1}$ has the property that $i^2 = -1$ and we can deduce the following relationships:

$$i^3 = -i, \quad i^4 = 1, \quad i^5 = i, \quad i^6 = -1, \quad i^7 = -i, \dots .$$

These results will be handy for simplifying operations involving complex numbers.

We say that two complex numbers $c_1 = a_1 + b_1 i$ and $c_2 = a_2 + b_2 i$ are **equal** if their real and imaginary parts are equal, that is, if $a_1 = a_2$ and $b_1 = b_2$. Of course, every real number a is a complex number with its imaginary part zero: $a = a + 0i$.

Operations on Complex Numbers

If $c_1 = a_1 + b_1 i$ and $c_2 = a_2 + b_2 i$ are complex numbers, then their **sum** is

$$c_1 + c_2 = (a_1 + a_2) + (b_1 + b_2)i,$$

and their **difference** is

$$c_1 - c_2 = (a_1 - a_2) + (b_1 - b_2)i.$$

In words, to form the sum of two complex numbers, add the real parts and add the imaginary parts. The **product** of c_1 and c_2 is

$$c_1 c_2 = (a_1 + b_1 i) \cdot (a_2 + b_2 i) = a_1 a_2 + (a_1 b_2 + b_1 a_2)i + b_1 b_2 i^2$$
$$= (a_1 a_2 - b_1 b_2) + (a_1 b_2 + b_1 a_2)i.$$

A special case of multiplication of complex numbers occurs when c_1 is real. In this case we obtain the simple result

$$c_1 c_2 = c_1 \cdot (a_2 + b_2 i) = c_1 a_2 + c_1 b_2 i.$$

If $c = a + bi$ is a complex number, then the **conjugate** of c is the complex number $\bar{c} = a - bi$. It is easy to show that if c and d are complex numbers, then the following basic properties of complex arithmetic hold:

1. $\bar{\bar{c}} = c$.
2. $\overline{c + d} = \bar{c} + \bar{d}$.
3. $\overline{cd} = \bar{c}\,\bar{d}$.
4. c is a real number if and only if $c = \bar{c}$.
5. $c\bar{c}$ is a nonnegative real number and $c\bar{c} = 0$ if and only if $c = 0$.

We prove property 4 here and leave the others as exercises. Let $c = a + bi$ so that $\bar{c} = a - bi$. If $c = \bar{c}$, then $a + bi = a - bi$, so $b = 0$ and c is real. On the other hand, if c is real, then $c = a$ and $\bar{c} = a$, so $c = \bar{c}$.

EXAMPLE 2 ■ Let $c_1 = 5 - 3i$, $c_2 = 4 + 2i$, and $c_3 = -3 + i$.

(a) $c_1 + c_2 = (5 - 3i) + (4 + 2i) = 9 - i$.

(b) $c_2 - c_3 = (4 + 2i) - (-3 + i) = (4 - (-3)) + (2 - 1)i = 7 + i$.

(c) $c_1 c_2 = (5 - 3i) \cdot (4 + 2i) = 20 + 10i - 12i - 6i^2 = 26 - 2i$.

(d) $c_1 \bar{c_3} = (5 - 3i) \cdot \overline{(-3 + i)} = (5 - 3i) \cdot (-3 - i) = -15 - 5i + 9i + 3i^2$
$\quad = -18 + 4i$.

(e) $3c_1 + 2\bar{c}_2 = 3(5 - 3i) + 2\overline{(4 + 2i)} = (15 - 9i) + 2(4 - 2i)$
$$= (15 - 9i) + (8 - 4i) = 23 - 13i.$$

(f) $c_1 \bar{c}_1 = (5 - 3i)\overline{(5 - 3i)} = (5 - 3i)(5 + 3i) = 34.$ ■

When we consider systems of linear equations with complex coefficients, we will need to divide complex numbers to complete the solution process and obtain a reasonable form for the solution. Let $c_1 = a_1 + b_1 i$ and $c_2 = a_2 + b_2 i$. If $c_2 \neq 0$, that is, if $a_2 \neq 0$ or $b_2 \neq 0$, then we can **divide** c_1 by c_2:

$$\frac{c_1}{c_2} = \frac{a_1 + b_1 i}{a_2 + b_2 i}.$$

To conform to our practice of expressing a complex number in the form real part + imaginary part $\cdot$ i, we must simplify the foregoing expression for c_1/c_2. To simplify this complex fraction, we multiply the numerator and the denominator by the conjugate of the denominator. Thus, dividing c_1 by c_2 gives the complex number

$$\frac{c_1}{c_2} = \frac{a_1 + b_1 i}{a_2 + b_2 i} = \frac{(a_1 + b_1 i)(a_2 - b_2 i)}{(a_2 + b_2 i)(a_2 - b_2 i)} = \frac{a_1 a_2 + b_1 b_2}{a_2^2 + b_2^2} - \frac{a_1 b_2 + a_2 b_1}{a_2^2 + b_2^2} i.$$

EXAMPLE 3 ■ Let $c_1 = 2 - 5i$ and $c_2 = -3 + 4i$. Then

$$\frac{c_1}{c_2} = \frac{2 - 5i}{-3 + 4i} = \frac{(2 - 5i)(-3 - 4i)}{(-3 + 4i)(-3 - 4i)} = \frac{-26 + 7i}{(-3)^2 + (4)^2} = -\frac{26}{25} + \frac{7}{25} i.$$ ■

Finding the reciprocal of a complex number is a special case of division of complex numbers. If $c = a + bi$, $c \neq 0$, then

$$\frac{1}{c} = \frac{1}{a + bi} = \frac{a - bi}{(a + bi)(a - bi)} = \frac{a - bi}{a^2 + b^2}$$
$$= \frac{a}{a^2 + b^2} - \frac{b}{a^2 + b^2} i.$$

EXAMPLE 4 ■ (a) $\dfrac{1}{2 + 3i} = \dfrac{2 - 3i}{(2 + 3i)(2 - 3i)} = \dfrac{2 - 3i}{2^2 + 3^2} = \dfrac{2}{13} - \dfrac{3}{13} i.$

(b) $\dfrac{1}{i} = \dfrac{-i}{i(-i)} = \dfrac{-i}{-i^2} = \dfrac{-i}{-(-1)} = -i.$ ■

Summarizing, we can say that complex numbers are mathematical objects for which addition, subtraction, multiplication, and division are defined in such a way that these operations on real numbers can be derived as special cases. In fact, it is easy to show that complex numbers form a mathematical system that is called a field.

Geometric Representation of Complex Numbers

A complex number $c = a + bi$ may be regarded as an ordered pair (a, b) of real numbers. This ordered pair of real numbers corresponds to a point in the plane. Such a correspondence naturally suggests that we represent $a + bi$ as a point in the **complex plane**, where the horizontal axis is used to represent the real part of c and the vertical axis is used to represent the imaginary part of c. To simplify matters, we call these the **real axis** and **imaginary axis**, respectively (see Figure A.1).

FIGURE A.1
Complex plane

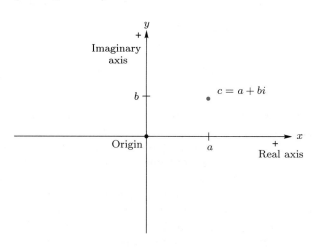

EXAMPLE 5 ■ Plot the complex numbers $c = 2 - 3i$, $d = 1 + 4i$, $e = -3$, and $f = 2i$ in the complex plane.

Solution   See Figure A.2.   ■

FIGURE A.2

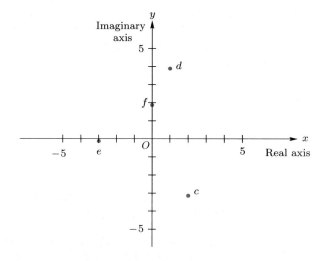

The rules concerning inequality of real numbers, such as less than and greater than, *do not apply to complex numbers*. There is no way to arrange the complex numbers according to size. However, using the geometric representation from the complex plane, we can attach a notion of size to a complex number by measuring its distance from the origin. The distance from the origin to $c = a + bi$ is called the **absolute value** or **modulus** of the complex number, and is denoted by $|c| = |a + bi|$. Using the formula for the distance between ordered pairs of real numbers, we obtain

$$|c| = |a + bi| = \sqrt{a^2 + b^2}.$$

It follows that $c\bar{c} = |c|^2$ (verify).

EXAMPLE 6 ■ Referring to Example 5: $|c| = \sqrt{13}$; $|d| = \sqrt{17}$; $|e| = 3$; $|f| = 2$. ■

A different, but related, interpretation of a complex number is obtained if we associate with $c = a + bi$ the vector OP, where O is the origin $(0, 0)$ and P is the point (a, b). There is an obvious correspondence between this representation and vectors in the plane discussed in calculus, which we reviewed in Section 3.1. Using a vector representation, addition and subtraction of complex numbers can be viewed as the corresponding vector operations. These are represented in Figures 3.14, 3.15, and 3.19. We will not pursue the manipulation of complex numbers by vector operations here, but such a point of view is important for the development and study of complex variables.

Matrices with Complex Entries

If the entries of a matrix are complex numbers, we can perform the matrix operations of addition, subtraction, multiplication, and scalar multiplication in a manner completely analogous to that for real matrices. The validity of these operations can be verified using properties of complex arithmetic, just imitating the proofs for real matrices presented in the text. We illustrate these concepts in the following example.

EXAMPLE 7 ■ Let

$$A = \begin{bmatrix} 4 + i & -2 + 3i \\ 6 + 4i & -3i \end{bmatrix}, \quad B = \begin{bmatrix} 2 - i & 3 - 4i \\ 5 + 2i & -7 + 5i \end{bmatrix}, \quad C = \begin{bmatrix} 1 + 2i & i \\ 3 - i & 8 \\ 4 + 2i & 1 - i \end{bmatrix}.$$

(a) $A + B = \begin{bmatrix} (4 + i) + (2 - i) & (-2 + 3i) + (3 - 4i) \\ (6 + 4i) + (5 + 2i) & (-3i) + (-7 + 5i) \end{bmatrix} = \begin{bmatrix} 6 & 1 - i \\ 11 + 6i & -7 + 2i \end{bmatrix}.$

(b) $B - A = \begin{bmatrix} (2 - i) - (4 + i) & (3 - 4i) - (-2 + 3i) \\ (5 + 2i) - (6 + 4i) & (-7 + 5i) - (-3i) \end{bmatrix} = \begin{bmatrix} -2 - 2i & 5 - 7i \\ -1 - 2i & -7 + 8i \end{bmatrix}.$

(c) $CA = \begin{bmatrix} 1+2i & i \\ 3-i & 8 \\ 4+2i & 1-i \end{bmatrix} \begin{bmatrix} 4+i & -2+3i \\ 6+4i & -3i \end{bmatrix}$

$$= \begin{bmatrix} (1+2i)(4+i)+(i)(6+4i) & (1+2i)(-2+3i)+(i)(-3i) \\ (3-i)(4+i)+(8)(6+4i) & (3-i)(-2+3i)+(8)(-3i) \\ (4+2i)(4+i)+(1-i)(6+4i) & (4+2i)(-2+3i)+(1-i)(-3i) \end{bmatrix}$$

$$= \begin{bmatrix} -2+15i & -5-i \\ 61+31i & -3-13i \\ 24+10i & -17+5i \end{bmatrix}.$$

(d) $(2+i)B = \begin{bmatrix} (2+i)(2-i) & (2+i)(3-4i) \\ (2+i)(5+2i) & (2+i)(-7+5i) \end{bmatrix} = \begin{bmatrix} 5 & 10-5i \\ 8+9i & -19+3i \end{bmatrix}.$ ∎

Just as we can compute the conjugate of a complex number, we can compute the **conjugate of a matrix** by computing the conjugate of each entry of the matrix. We denote the conjugate of a matrix A by $\overline{A}$, and write

$$\overline{A} = \left[\overline{a_{ij}}\right].$$

EXAMPLE 8 ∎ Referring to Example 7, we find that

$$\overline{A} = \begin{bmatrix} 4-i & -2-3i \\ 6-4i & 3i \end{bmatrix} \quad \text{and} \quad \overline{B} = \begin{bmatrix} 2+i & 3+4i \\ 5-2i & -7-5i \end{bmatrix}.$$ ∎

The following properties of the conjugate of a matrix hold:

1. $\overline{\overline{A}} = A$.

2. $\overline{A+B} = \overline{A} + \overline{B}$.

3. $\overline{AB} = \overline{A}\,\overline{B}$.

4. For any real number k, $\overline{kA} = k\,\overline{A}$.

5. For any complex number c, $\overline{cA} = \overline{c}\,\overline{A}$.

6. $(\overline{A})^T = \overline{A^T}$.

7. If A is nonsingular, then $(\overline{A})^{-1} = \overline{A^{-1}}$.

We prove properties 5 and 6 here and leave the others as exercises. First property 5: If c is complex, the (k, j) entry of $\overline{cA}$ is

$$\overline{ca_{kj}} = \overline{c}\,\overline{a_{kj}},$$

which is the (k, j) entry of $\overline{c}\,\overline{A}$. Next, property 6: The (k, j) entry of $(\overline{A})^T$ is $\overline{a_{jk}}$, which is the (k, j) entry of $\overline{A^T}$.

Special Types of Complex Matrices

As we have already seen, certain types of real matrices satisfy some important properties. The same situation applies to complex matrices and we now discuss several of these types of matrices.

An $n \times n$ complex matrix A is called **Hermitian** if

$$\overline{A^T} = A.$$

This is equivalent to saying that $\overline{a_{jk}} = a_{kj}$ for all k and j. Every real symmetric matrix is Hermitian [Exercise T.3(c)], so we may consider Hermitian matrices as the analogs of real symmetric matrices.

EXAMPLE 9   ■   The matrix

$$A = \begin{bmatrix} 2 & 3+i \\ 3-i & 5 \end{bmatrix}$$

is Hermitian, since

$$\overline{A^T} = \overline{\begin{bmatrix} 2 & 3-i \\ 3+i & 5 \end{bmatrix}} = \begin{bmatrix} 2 & 3+i \\ 3-i & 5 \end{bmatrix} = A.$$ ■

An $n \times n$ complex matrix A is called **unitary** if

$$(\overline{A^T})A = A(\overline{A^T}) = I_n.$$

This is equivalent to saying that $\overline{A^T} = A^{-1}$. Every real orthogonal matrix is unitary [Exercise T.4(a)], so we may consider unitary matrices as the analogs of real orthogonal matrices.

EXAMPLE 10   ■   The matrix

$$A = \begin{bmatrix} \dfrac{1}{\sqrt{3}} & \dfrac{1+i}{\sqrt{3}} \\ \dfrac{1-i}{\sqrt{3}} & -\dfrac{1}{\sqrt{3}} \end{bmatrix}$$

is unitary, since (verify)

$$(\overline{A^T})A = \begin{bmatrix} \dfrac{1}{\sqrt{3}} & \dfrac{1+i}{\sqrt{3}} \\ \dfrac{1-i}{\sqrt{3}} & -\dfrac{1}{\sqrt{3}} \end{bmatrix} \begin{bmatrix} \dfrac{1}{\sqrt{3}} & \dfrac{1+i}{\sqrt{3}} \\ \dfrac{1-i}{\sqrt{3}} & -\dfrac{1}{\sqrt{3}} \end{bmatrix} = I_2$$

and similarly, $A(\overline{A^T}) = I_2$.   ■

There is one more type of complex matrix that is important. An $n \times n$ complex matrix is called **normal** if

$$(\overline{A^T})A = A(\overline{A^T}).$$

EXAMPLE 11 ■ The matrix

$$A = \begin{bmatrix} 5 - i & -1 + i \\ -1 - i & 3 - i \end{bmatrix}$$

is normal, since (verify)

$$(\overline{A^T}) A = A (\overline{A^T}) = \begin{bmatrix} 28 & -8 + 8i \\ -8 - 8i & 12 \end{bmatrix}.$$

Moreover, A is not Hermitian, since $\overline{A^T} \neq A$ (verify). ■

Complex Numbers and Roots of Polynomials

A polynomial of degree n with real coefficients has n complex roots, some, all, or none of which may be real numbers. Thus the polynomial $f_1(x) = x^4 - 1$ has the roots i, $-i$, 1, and -1; the polynomial $f_2(x) = x^2 - 1$ has the roots 1 and -1; and the polynomial $f_3(x) = x^2 + 1$ has the roots i and $-i$. We make use of this result in Section 5.1 where we extend the definition of an eigenvalue to allow it to be a complex number.

A.1   EXERCISES

1. Let $c_1 = 3 + 4i$, $c_2 = 1 - 2i$, and $c_3 = -1 + i$. Compute each of the following and simplify as much as possible.

(a) $c_1 + c_2$. (b) $c_3 - c_1$.

(c) $c_1 c_2$. (d) $c_2 \overline{c_3}$.

(e) $4c_3 + \overline{c_2}$. (f) $(-i) \cdot c_2$.

(g) $\overline{3c_1 - ic_2}$. (h) $c_1 c_2 c_3$.

2. Write in the form $a + bi$.

(a) $\dfrac{1 + 2i}{3 - 4i}$. (b) $\dfrac{2 - 3i}{3 - i}$.

(c) $\dfrac{(2 + i)^2}{i}$. (d) $\dfrac{1}{(3 + 2i)(1 + i)}$.

3. Represent each complex number as a point and as a vector in the complex plane.

(a) $4 + 2i$. (b) $-3 + i$.

(c) $3 - 2i$. (d) $i(4 + i)$.

4. Find the modulus of each complex number in Exercise 3.

5. In the complex plane sketch the vectors corresponding to c and $\overline{c}$ for $c = 2 + 3i$ and $c = -1 + 4i$. Geometrically, we can say that $\overline{c}$ is the reflection of c with respect to the real axis. (See also Example 4 in Section 3.3.)

6. Let

$$A = \begin{bmatrix} 2 + 2i & -1 + 3i \\ -2 & 1 - i \end{bmatrix},$$

$$B = \begin{bmatrix} 2i & 1 + 2i \\ 0 & 3 - i \end{bmatrix}, \quad C = \begin{bmatrix} 2 + i \\ -i \end{bmatrix}.$$

Compute each of the following and simplify each entry as $a + bi$.

(a) $A + B$. (b) $(1 - 2i)C$. (c) AB.

(d) BC. (e) $A - 2I_2$. (f) $\overline{B}$.

(g) $A\overline{C}$. (h) $(\overline{A + B})C$.

7. If

$$A = \begin{bmatrix} 0 & i \\ i & 0 \end{bmatrix},$$

compute A^2, A^3, and A^4. Give a general rule for A^n, n a positive integer.

8. Which of the following matrices are Hermitian, unitary, or normal?

(a) $\begin{bmatrix} 3 & 2 + i \\ 2 - i & 4 \end{bmatrix}$. (b) $\begin{bmatrix} 2 & 1 - i \\ 3 + i & -2 \end{bmatrix}$.

(c) $\begin{bmatrix} \dfrac{1-i}{2} & \dfrac{1+i}{2} \\ \dfrac{1+i}{2} & \dfrac{1-i}{2} \end{bmatrix}$. (d) $\begin{bmatrix} 1 & -1 \\ 1 & 1 \end{bmatrix}$.

(e) $\begin{bmatrix} 1 & 3-i & 4-i \\ 3+i & -2 & 2+i \\ 4+i & 2-i & 3 \end{bmatrix}$.

(f) $\begin{bmatrix} 3 & \dfrac{3-i}{2} & \dfrac{4-i}{2} \\ \dfrac{3-i}{2} & -2 & 2+i \\ \dfrac{4-i}{2} & 2-i & 5 \end{bmatrix}$.

(g) $\begin{bmatrix} 3+2i & -1 \\ -i & 2+i \end{bmatrix}$. (h) $\begin{bmatrix} i & i \\ -i & 1 \end{bmatrix}$.

(i) $\begin{bmatrix} 1 & 0 & 0 \\ 0 & \dfrac{1+i}{\sqrt{3}} & \dfrac{1}{\sqrt{3}} \\ 0 & -\dfrac{1}{\sqrt{3}} & \dfrac{1-i}{\sqrt{3}} \end{bmatrix}$.

(j) $\begin{bmatrix} 4+7i & -2-i \\ 1-2i & 3+4i \end{bmatrix}$.

9. Find all the roots.
   (a) $x^2 + x + 1 = 0$.   (b) $x^3 + 2x^2 + x + 2 = 0$.
   (c) $x^5 + x^4 - x - 1 = 0$.

10. Let $p(x)$ denote a polynomial and let A be a square matrix. Then $p(A)$ is called a **matrix**

polynomial or a **polynomial in the matrix** A. For $p(x) = 2x^2 + 5x - 3$, compute $p(A) = 2A^2 + 5A - 3I_n$ for each of the following.

   (a) $A = \begin{bmatrix} -3 & 0 \\ 0 & -3 \end{bmatrix}$.   (b) $A = \begin{bmatrix} 1 & 2 \\ 0 & 1 \end{bmatrix}$.

   (c) $A = \begin{bmatrix} 0 & i \\ i & 0 \end{bmatrix}$.   (d) $A = \begin{bmatrix} 1 & i \\ 0 & 0 \end{bmatrix}$.

11. Let $p(x) = x^2 + 1$.
   (a) Determine two different 2×2 matrices A of the form kI_2 that satisfy $p(A) = O_2$.
   (b) Verify that $p(A) = O_2$, for $A = \begin{bmatrix} 1 & 2 \\ -1 & -1 \end{bmatrix}$.

12. Find all the 2×2 matrices A of the form kI_2 that satisfy $p(A) = O_2$ for $p(x) = x^2 - x - 2$.

13. In Supplementary Exercise 27 in Chapter 1, we introduced the concept of a square root of a matrix with real entries. We can generalize the notion of a square root of a matrix if we permit complex entries.
   (a) Compute a complex square root of

$$A = \begin{bmatrix} -1 & 0 \\ 0 & 0 \end{bmatrix}.$$

   (b) Compute a complex square root of

$$A = \begin{bmatrix} -2 & 2 \\ 2 & -2 \end{bmatrix}.$$

THEORETICAL EXERCISES

T.1. If $c = a + bi$, then we can denote the real part of c by $\text{Re}\,(c)$ and the imaginary part of c by $\text{Im}\,(c)$.

   (a) For any complex numbers $c_1 = a_1 + b_1 i$ and $c_2 = a_2 + b_2 i$, show that
   $\text{Re}\,(c_1 + c_2) = \text{Re}\,(c_1) + \text{Re}\,(c_2)$ and
   $\text{Im}\,(c_1 + c_2) = \text{Im}\,(c_1) + \text{Im}\,(c_2)$.

   (b) For any real number k, show that
   $\text{Re}\,(kc) = k\,\text{Re}\,(c)$ and $\text{Im}\,(kc) = k\,\text{Im}\,(c)$.

   (c) Is part (b) true if k is a complex number?

   (d) Prove or disprove:

$$\text{Re}\,(c_1 c_2) = \text{Re}\,(c_1) \cdot \text{Re}\,(c_2).$$

T.2. Let A and B be $m \times n$ complex matrices, and let C be an $n \times n$ nonsingular matrix.
   (a) Show that $\overline{A + B} = \overline{A} + \overline{B}$.

   (b) Show that for any real number k, $\overline{kA} = k\overline{A}$.
   (c) Show that $(\overline{C})^{-1} = \overline{C^{-1}}$.

T.3. (a) Show that the diagonal entries of a Hermitian matrix must be real.

   (b) Show that every Hermitian matrix A can be written as $A = B + iC$, where B is real and symmetric and C is real and skew symmetric (see Exercise T.24 in Section 1.4). [*Hint:* Consider $B = (A + \overline{A})/2$ and $C = (A - \overline{A})/2i$.]

   (c) Show that every real symmetric matrix is Hermitian.

T.4. (a) Show that every real orthogonal matrix is unitary.

   (b) Show that if A is a unitary matrix, then A^T is unitary.

(c) Show that if A is a unitary matrix, then A^{-1} is unitary.

T.5. Let A be an $n \times n$ complex matrix.

(a) Show that A can be written as $B + iC$, where B and C are Hermitian.

(b) Show that A is normal if and only if

$$BC = CB.$$

[*Hint*: Consider $B = (A + \overline{A^T})/2$ and $C = (A - \overline{A^T})/2i$.]

T.6. (a) Show that any Hermitian matrix is normal.

(b) Show that any unitary matrix is normal.

(c) Find a 2×2 normal matrix that is neither Hermitian nor unitary.

T.7. An $n \times n$ complex matrix A is called **skew Hermitian** if

$$\overline{A^T} = -A.$$

Show that a matrix $A = B + iC$, where B and C are real matrices, is skew Hermitian if and only if B is skew symmetric and C is symmetric.

MATLAB EXERCISES ▪

MATLAB does complex arithmetic automatically. To enter a complex number into MATLAB, first assign the complex unit $\sqrt{-1}$ to a variable name i with command

$$\mathbf{i = sqrt(-1)}$$

Then to store $3 - 5i$ into variable v, type

$$\mathbf{v = 3 - 5i}$$

If we enter a second complex number $-2 + 7i$ into w with command

$$\mathbf{w = -2 + 7i}$$

we can add, subtract, multiply and divide v and w using the standard arithmetic symbols. Command

$$\mathbf{conj(v)}$$

displays the conjugate of v, while **real(v)** and **imag(v)** display the real and imaginary parts of v, respectively. We define a complex matrix just by entering its elements as complex numbers, for example,

$$\mathbf{A = [2 - i \quad 3 + 5i; 6 \quad -2i]}$$

Do not leave any extra spaces in complex numbers, otherwise MATLAB takes it as two separate numbers. The commands **conj**, **real**, and **imag** apply to matrices as well. $\mathbf{A'}$ gives the conjugate transpose of matrix A.

ML.1. For Exercise 1, enter the three values into MATLAB as $c1$, $c2$, and $c3$, no subscripts, and compute the values in parts (a)–(h).

ML.2. Do Exercise 6 in MATLAB.

A.2 ▼ Complex Numbers in Linear Algebra

Almost everything in the first six chapters of this book remains true if we cross out the word *real* and replace it by the word *complex*. In Section A.1 we introduced matrices with complex entries. Here we consider linear systems of equations with complex entries, determinants of complex matrices, complex vector spaces, complex inner products, and eigenvalues and eigenvectors of complex matrices. The theory we have developed for these topics using real numbers is immediately applicable to the case with complex numbers. We could have developed Chapters 1 through 5 with complex numbers, but we limited ourselves to the real numbers for the following reasons:

1. The results would not look markedly different.

2. For a good many applications the real number situation is adequate; the transition to complex numbers is, as we have just remarked, not too difficult.

3. The computational effort of complex arithmetic is, in most cases, more than double that resulting with only real numbers.

As a matter of fact, we can do most of the first six chapters not only over the real numbers and the complex numbers, but also over any *field* (the real numbers and the complex numbers are familiar examples of a field).

The primary goal of this appendix is to provide an easy transition to complex numbers in linear algebra. This is of particular importance in Chapter 5, where complex eigenvalues and eigenvectors arise naturally for matrices with real entries. Hence we only restate the main theorems in the complex case and provide a discussion and examples of the major ideas needed to accomplish this transition. It will soon be evident that the increased computational effort of complex arithmetic becomes quite tedious if done by hand. However, such calculations can be easily done by computers.

Solving Linear Systems with Complex Entries

The results and techniques dealing with the solution of linear systems that we developed in Chapter 1 carry over directly to linear systems with complex coefficients. We shall illustrate row operations and echelon forms for such systems with Gauss–Jordan reduction using complex arithmetic.

EXAMPLE 1 ■ Solve the following linear system by Gauss–Jordan reduction:

$$(1 + i)x_1 + (2 + i)x_2 = 5$$
$$(2 - 2i)x_1 + \qquad ix_2 = 1 + 2i.$$

Solution We form the augmented matrix and use elementary row operations to transform it to reduced row echelon form. For the augmented matrix $\begin{bmatrix} A \vdots B \end{bmatrix}$,

$$\begin{bmatrix} 1+i & 2+i & \vdots & 5 \\ 2-2i & i & \vdots & 1+2i \end{bmatrix},$$

multiply the first row by $\dfrac{1}{1+i}$ to obtain

$$\begin{bmatrix} 1 & \frac{3}{2} - \frac{1}{2}i & \vdots & \frac{5}{2} - \frac{5}{2}i \\ 2-2i & i & \vdots & 1+2i \end{bmatrix}.$$

We now add $[-(2 - 2i)]$ times the first row to the second row to get

$$\begin{bmatrix} 1 & \frac{3}{2} - \frac{1}{2}i & \vdots & \frac{5}{2} - \frac{5}{2}i \\ 0 & -2+5i & \vdots & 1+12i \end{bmatrix}.$$

Multiply the second row by $\dfrac{1}{-2+5i}$ to obtain

$$\begin{bmatrix} 1 & \frac{3}{2} - \frac{1}{2}i & \vdots & \frac{5}{2} - \frac{5}{2}i \\ 0 & 1 & \vdots & 2-i \end{bmatrix},$$

which is in row echelon form. To get to reduced row echelon form, we add $-\left(\frac{3}{2} - \frac{1}{2}i\right)$ times the second row to the first row to obtain

$$\left[\begin{array}{cc:c} 1 & 0 & 0 \\ 0 & 1 & 2-i \end{array}\right].$$

Hence the solution is $x_1 = 0$ and $x_2 = 2 - i$.   ■

If you carry out the arithmetic for the row operations in the preceding example, you will feel the burden of the complex arithmetic even though there were just two equations in two unknowns.

Gaussian elimination with back substitution can also be used on linear systems with complex coefficients.

EXAMPLE 2 ■ Suppose that the augmented matrix of a linear system has been transformed to the following matrix in row echelon form:

$$\left[\begin{array}{ccc:c} 1 & 0 & 1+i & -1 \\ 0 & 1 & 3i & 2+i \\ 0 & 0 & 1 & 2i \end{array}\right].$$

The back-substitution procedure gives us

$$\begin{aligned} x_3 &= 2i \\ x_2 &= 2 + i - 3i(2i) = 2 + i + 6 = 8 + i \\ x_1 &= -1 - (1+i)(2i) = -1 - 2i + 2 = 3 - 2i. \end{aligned}$$
   ■

We can alleviate the tediousness of complex arithmetic by using computers to solve linear systems with complex entries. However, we must still pay a high price because the execution time will be approximately twice as long as that for the same size linear system with all real entries. We can illustrate this by showing how to transform an $n \times n$ linear system with complex coefficients to a $2n \times 2n$ linear system with only real coefficients.

EXAMPLE 3 ■ Consider the linear system

$$\begin{aligned} (2+i)x_1 + (1+i)x_2 &= 3 + 6i \\ (3-i)x_1 + (2-2i)x_2 &= 7 - i. \end{aligned}$$

If we let $x_1 = a_1 + b_1 i$ and $x_2 = a_2 + b_2 i$, with a_1, b_1, a_2, and b_2 real numbers, then we can write this system in matrix form as

$$\begin{bmatrix} 2+i & 1+i \\ 3-i & 2-2i \end{bmatrix} \begin{bmatrix} a_1 + b_1 i \\ a_2 + b_2 i \end{bmatrix} = \begin{bmatrix} 3 + 6i \\ 7 - i \end{bmatrix}.$$

We first rewrite the given linear system as

$$\left(\begin{bmatrix} 2 & 1 \\ 3 & 2 \end{bmatrix} + i \begin{bmatrix} 1 & 1 \\ -1 & -2 \end{bmatrix}\right) \left(\begin{bmatrix} a_1 \\ a_2 \end{bmatrix} + i \begin{bmatrix} b_1 \\ b_2 \end{bmatrix}\right) = \begin{bmatrix} 3 \\ 7 \end{bmatrix} + i \begin{bmatrix} 6 \\ -1 \end{bmatrix}.$$

Multiplying, we have

$$
\left(\begin{bmatrix} 2 & 1 \\ 3 & 2 \end{bmatrix} \begin{bmatrix} a_1 \\ a_2 \end{bmatrix} - \begin{bmatrix} 1 & 1 \\ -1 & -2 \end{bmatrix} \begin{bmatrix} b_1 \\ b_2 \end{bmatrix} \right)
$$

$$
+ i \left(\begin{bmatrix} 2 & 1 \\ 3 & 2 \end{bmatrix} \begin{bmatrix} b_1 \\ b_2 \end{bmatrix} + \begin{bmatrix} 1 & 1 \\ -1 & -2 \end{bmatrix} \begin{bmatrix} a_1 \\ a_2 \end{bmatrix} \right) = \begin{bmatrix} 3 \\ 7 \end{bmatrix} + i \begin{bmatrix} 6 \\ -1 \end{bmatrix}.
$$

The real and imaginary parts on both sides of the equation must agree, respectively , and so we have

$$
\begin{bmatrix} 2 & 1 \\ 3 & 2 \end{bmatrix} \begin{bmatrix} a_1 \\ a_2 \end{bmatrix} - \begin{bmatrix} 1 & 1 \\ -1 & -2 \end{bmatrix} \begin{bmatrix} b_1 \\ b_2 \end{bmatrix} = \begin{bmatrix} 3 \\ 7 \end{bmatrix}
$$

and

$$
\begin{bmatrix} 2 & 1 \\ 3 & 2 \end{bmatrix} \begin{bmatrix} b_1 \\ b_2 \end{bmatrix} + \begin{bmatrix} 1 & 1 \\ -1 & -2 \end{bmatrix} \begin{bmatrix} a_1 \\ a_2 \end{bmatrix} = \begin{bmatrix} 6 \\ -1 \end{bmatrix}.
$$

This leads to the linear system

$$
\begin{aligned}
2a_1 + a_2 - b_1 - b_2 &= 3 \\
3a_1 + 2a_2 + b_1 + 2b_2 &= 7 \\
a_1 + a_2 + 2b_1 + b_2 &= 6 \\
-a_1 - 2a_2 + 3b_1 + 2b_2 &= -1,
\end{aligned}
$$

which can be written as

$$
\begin{bmatrix} 2 & 1 & -1 & -1 \\ 3 & 2 & 1 & 2 \\ 1 & 1 & 2 & 1 \\ -1 & -2 & 3 & 2 \end{bmatrix} \begin{bmatrix} a_1 \\ a_2 \\ b_1 \\ b_2 \end{bmatrix} = \begin{bmatrix} 3 \\ 7 \\ 6 \\ -1 \end{bmatrix}.
$$

This linear system of four equations in four unknowns is now solved as in Chapter 1. The solution is (verify) $a_1 = 1$, $a_2 = 2$, $b_1 = 2$, and $b_2 = -1$. Thus $x_1 = 1 + 2i$ and $x_2 = 2 - i$ is the solution to the given linear system.   ■

Determinants of Complex Matrices

The definition of a determinant and all the properties derived in Chapter 2 apply to matrices with complex entries. The following example is an illustration.

EXAMPLE 4 ■ Compute $|A|$ for the coefficient matrix A in Example 3.

Solution

$$
\begin{vmatrix} 2+i & 1+i \\ 3-i & 2-2i \end{vmatrix} = (2+i)(2-2i) - (3-i)(1+i)
$$

$$
= (6 - 2i) - (4 + 2i)
$$

$$
= 2 - 4i.
$$

■

Complex Vector Spaces

A **complex vector space** is defined exactly as was a real vector space in Definition 1 in Section 4.1, except that the scalars in properties (e) through (h) are permitted to be complex numbers. The terms *complex* vector space and *real* vector space emphasize the set from which the scalars are chosen. It happens that in order to satisfy the closure property of scalar multiplication [Definition $1(\beta)$ in Section 4.1] in a complex vector space, we must, in most examples, consider vectors that involve complex numbers.

Most of the real vector spaces of Chapter 4 have complex vector space analogs.

EXAMPLE 5 ■ (a) Consider C^n, the set of all $n \times 1$ matrices

$$\begin{bmatrix} a_1 \\ a_2 \\ \vdots \\ a_n \end{bmatrix}$$

with complex entries. Let the operation $\oplus$ be matrix addition and let the operation $\odot$ be multiplication of a matrix by a complex number. We can verify that C^n is a complex vector space by using the properties of matrices established in Section 1.4 and the properties of complex arithmetic established in Section A.1. (Note that if the operation $\odot$ is taken as multiplication of a matrix by a real number, then C^n is a real vector space whose vectors have complex components.)

(b) The set of all $m \times n$ matrices with complex entries with matrix addition as $\oplus$ and multiplication of a matrix by a complex number as $\odot$ is a complex vector space (verify). We denote this vector space by C_{mn}.

(c) The set of polynomials with complex coefficients with polynomial addition as $\oplus$ and multiplication of a polynomial by a complex constant as $\odot$ forms a complex vector space. Verification follows the pattern of Example 8 in Section 4.1.

(d) The set of complex-valued functions that are continuous on the interval $[a, b]$ (i.e., all functions of the form $f(t) = f_1(t) + if_2(t)$, where f_1 and f_2 are real-valued functions that are continuous on $[a, b]$), with $\oplus$ defined by $(f \oplus g)(t) = f(t) + g(t)$ and $\odot$ defined by $(c \odot f)(t) = c \cdot f(t)$ for a complex scalar c, forms a complex vector space. The corresponding real vector space is given in Example 7 in Section 4.2 for the interval $(-\infty, \infty)$. ■

A **complex vector subspace** W of a complex vector space V is defined as in Section 4.1, but with real scalars replaced by complex ones. The analog of Theorem 4.2 can be proved to show that a nonempty subset W of a complex vector space V is a complex vector subspace if and only if the following conditions hold:

(a) If $\mathbf{u}$ and $\mathbf{v}$ are any vectors in W, then $\mathbf{u} \oplus \mathbf{v}$ is in W.

(b) If c is any complex number and $\mathbf{u}$ is any vector in W, then $c \odot \mathbf{u}$ is in W.

EXAMPLE 6 ■ (a) Let W be the set of all vectors in C^3 of the form $(a, 0, b)$, where a and b are complex numbers. It easily follows that

$$(a, 0, b) \oplus (d, 0, e) = (a + d, 0, b + e)$$

belongs to W and for any complex scalar c,

$$c \odot (a, 0, b) = (ca, 0, cb)$$

belongs to W. Hence W is a complex vector subspace of C^3.

(b) Let W be the set of all vectors in C_{mn} having only real entries. If $A = \begin{bmatrix} a_{ij} \end{bmatrix}$ and $B = \begin{bmatrix} b_{kj} \end{bmatrix}$ belong to W, then so will $A \oplus B$, because if a_{kj} and b_{kj} are real, then so is their sum. However, if c is any complex scalar and A belongs to W, then $c \odot A = cA$ can have entries ca_{kj} that need not be real numbers. It follows that $c \odot A$ need not belong to W, so W is not a complex vector subspace. ■

Linear Independence and Basis in Complex Vector Spaces

The notions of linear combinations, spanning sets, linear dependence, linear independence, and basis are unchanged for complex vector spaces, except that we use complex scalars (see Sections 4.2, 4.3, and 4.4).

EXAMPLE 7 ■ Let V be the complex vector space C^3. Let

$$\mathbf{v}_1 = (1, i, 0), \quad \mathbf{v}_2 = (i, 0, 1 + i), \quad \text{and} \quad \mathbf{v}_3 = (1, 1, 1).$$

(a) Determine whether $\mathbf{v} = (-1, -3 + 3i, -4 + i)$ is a linear combination of $\{\mathbf{v}_1, \mathbf{v}_2, \mathbf{v}_3\}$.

(b) Determine whether $\{\mathbf{v}_1, \mathbf{v}_2, \mathbf{v}_3\}$ spans C^3.

(c) Determine whether $\{\mathbf{v}_1, \mathbf{v}_2, \mathbf{v}_3\}$ is a linearly independent subset of C^3.

(d) Is $\{\mathbf{v}_1, \mathbf{v}_2, \mathbf{v}_3\}$ a basis for C^3?

Solution (a) We proceed as in Example 10 of Section 4.2. We form a linear combination of $\mathbf{v}_1$, $\mathbf{v}_2$, and $\mathbf{v}_3$ with unknown coefficients c_1, c_2, and c_3, respectively, and set it equal to $\mathbf{v}$:

$$c_1 \mathbf{v}_1 + c_2 \mathbf{v}_2 + c_3 \mathbf{v}_3 = \mathbf{v}.$$

If we substitute the vectors $\mathbf{v}_1$, $\mathbf{v}_2$, $\mathbf{v}_3$, and $\mathbf{v}$ into this expression, we obtain (verify) the linear system

$$\begin{aligned} c_1 + \quad\quad ic_2 + c_3 &= -1 \\ ic_1 \quad\quad\quad\quad + c_3 &= -3 + 3i \\ (1 + i)c_2 + c_3 &= -4 + i. \end{aligned}$$

We next investigate the consistency of this linear system by using elementary row operations to transform its augmented matrix to either row echelon or reduced row echelon form. The row echelon form is (verify)

$$\left[\begin{array}{ccc:c} 1 & i & 1 & -1 \\ 0 & 1 & 1-i & -3+4i \\ 0 & 0 & 1 & -3 \end{array}\right],$$

which implies that the system is consistent, hence $\mathbf{v}$ is a linear combination of $\mathbf{v}_1$, $\mathbf{v}_2$, $\mathbf{v}_3$. In fact, back substitution gives (verify) $c_1 = 3$, $c_2 = i$, and $c_3 = -3$.

(b) Let $\mathbf{v} = (a, b, c)$ be an arbitrary vector of C^3. We form the linear combination

$$c_1\mathbf{v}_1 + c_2\mathbf{v}_2 + c_3\mathbf{v}_3 = \mathbf{v}$$

and solve for c_1, c_2, and c_3. The resulting linear system is

$$\begin{aligned} c_1 + \qquad ic_2 + c_3 &= a \\ ic_1 \qquad\qquad + c_3 &= b \\ (1+i)c_2 + c_3 &= c. \end{aligned}$$

Transforming the augmented matrix to row echelon form, we obtain (verify)

$$\left[\begin{array}{ccc:c} 1 & i & 1 & a \\ 0 & 1 & 1-i & b-ia \\ 0 & 0 & 1 & -c+(1+i)(b-ia) \end{array}\right].$$

Hence we can solve for c_1, c_2, c_3 for any choice of complex numbers a, b, c, which implies that $\{\mathbf{v}_1, \mathbf{v}_2, \mathbf{v}_3\}$ spans C^3.

(c) Proceeding as in Example 8 of Section 4.3, we form the equation

$$c_1\mathbf{v}_1 + c_2\mathbf{v}_2 + c_3\mathbf{v}_3 = \mathbf{0}$$

and solve for c_1, c_2, and c_3. The resulting homogeneous system is

$$\begin{aligned} c_1 + \qquad ic_2 + c_3 &= 0 \\ ic_1 \qquad\qquad + c_3 &= 0 \\ (1+i)c_2 + c_3 &= 0. \end{aligned}$$

Transforming the augmented matrix to row echelon form, we obtain (verify)

$$\left[\begin{array}{ccc:c} 1 & i & 1 & 0 \\ 0 & 1 & 1-i & 0 \\ 0 & 0 & 1 & 0 \end{array}\right],$$

and hence the only solution is $c_1 = c_2 = c_3 = 0$, showing that $\{\mathbf{v}_1, \mathbf{v}_2, \mathbf{v}_3\}$ is linearly independent.

(d) Yes, because $\mathbf{v}_1$, $\mathbf{v}_2$, and $\mathbf{v}_3$ span C^3 [part (b)] and they are linearly independent [part (c)].   ∎

Just as for real vector spaces, the questions of spanning sets, linearly independent or linearly dependent sets, and basis in a complex vector space are resolved by using an appropriate linear system. The definition of the dimension of a complex vector space is the same as that given in Section 4.4. In discussing the dimension of a complex vector space like C^n, we must adjust our intuitive picture. For example, C^1 consists of all complex multiples of a single nonzero vector. This collection can be put into one–one correspondence with the complex numbers themselves, that is, with all the points in the complex plane (see Figure A.1). Since the elements of a two-dimensional real vector space can be put into a one-to-one correspondence with the points of R^2 (see Section 3.1), we see that a complex vector space of dimension 1 has a geometric model that is in one-to-one correspondence with a geometric model of a two-dimensional real vector space. Similarly, a complex vector space of dimension 2 is the same, geometrically, as a four-dimensional real vector space.

If

$$\mathbf{u} = \begin{bmatrix} u_1 \\ u_2 \\ \vdots \\ u_n \end{bmatrix} \quad \text{and} \quad \mathbf{v} = \begin{bmatrix} v_1 \\ v_2 \\ \vdots \\ v_n \end{bmatrix}$$

are vectors in C^n, we define the dot product $\mathbf{u} \cdot \mathbf{v}$ as

$$\mathbf{u} \cdot \mathbf{v} = \overline{u_1}v_1 + \overline{u_2}v_2 + \cdots + \overline{u_n}v_n,$$

which can also be expressed as

$$\mathbf{u} \cdot \mathbf{v} = \overline{\mathbf{u}}^T \mathbf{v}.$$

EXAMPLE 8 ■ Let

$$\mathbf{u} = \begin{bmatrix} 1 - i \\ 2 \\ -3 + 2i \end{bmatrix} \quad \text{and} \quad \mathbf{v} = \begin{bmatrix} 3 + 2i \\ 3 - 4i \\ -3i \end{bmatrix}.$$

Compute $\mathbf{u} \cdot \mathbf{v}$.

Solution   We have

$$\begin{aligned} \mathbf{u} \cdot \mathbf{v} &= \overline{(1 - i)}(3 + 2i) + 2(3 - 4i) + \overline{(-3 + 2i)}(-3i) \\ &= (1 + 5i) + (6 - 8i) + (-6 + 9i) \\ &= 1 + 6i. \end{aligned}$$

■

We can also define the length of a vector $\mathbf{u}$ in C^n exactly as in the real case:

$$\|\mathbf{u}\| = \sqrt{\mathbf{u} \cdot \mathbf{u}}.$$

Moreover, the vectors $\mathbf{u}$ and $\mathbf{v}$ in C^n are said to be **orthogonal** if $\mathbf{u} \cdot \mathbf{v} = 0$.

EXAMPLE 9 ■ Let

$$\mathbf{u} = \begin{bmatrix} 1+i \\ 2-i \\ 3+i \end{bmatrix} \quad \text{and} \quad \mathbf{v} = \begin{bmatrix} 6-2i \\ 2i \\ -1+i \end{bmatrix}.$$

Then

$$\mathbf{u} \cdot \mathbf{v} = \overline{\mathbf{u}}^T \mathbf{v} = (1-i)(6-2i) + (2+i)(2i) + (3-i)(-1+i) = 0,$$

so $\mathbf{u}$ and $\mathbf{v}$ are orthogonal. Also,

$$\|\mathbf{u}\| = \sqrt{\mathbf{u} \cdot \mathbf{u}} = \sqrt{\overline{\mathbf{u}}^T \mathbf{u}}$$
$$= \sqrt{(1-i)(1+i) + (2+i)(2-i) + (3-i)(3+i)}$$
$$= \sqrt{17}.$$

■

We can prove the following properties for the dot product on C^n (Exercise T.6):

(a) $\mathbf{u} \cdot \mathbf{u} > 0$ for $\mathbf{u} \neq \mathbf{0}$ in C^n; $\mathbf{u} \cdot \mathbf{u} = 0$ if and only if $\mathbf{u} = \mathbf{0}$ in C^n.

(b) $\mathbf{u} \cdot \mathbf{v} = \overline{\mathbf{v} \cdot \mathbf{u}}$ for any $\mathbf{u}$, $\mathbf{v}$ in C^n.

(c) $(\mathbf{u} + \mathbf{v}) \cdot \mathbf{w} = \mathbf{u} \cdot \mathbf{w} + \mathbf{v} \cdot \mathbf{w}$ for any $\mathbf{u}$, $\mathbf{v}$, $\mathbf{w}$ in C^n.

(d) $(c\mathbf{u}) \cdot \mathbf{v} = \overline{c}(\mathbf{u} \cdot \mathbf{v})$ for any $\mathbf{u}$, $\mathbf{v}$ in C^n and c a complex scalar.

REMARK   Observe that these properties are somewhat different from those satisfied by the dot product on R^n. (See Theorem 3.3 in Section 3.2.)

Complex Eigenvalues and Eigenvectors

Let A be an $n \times n$ matrix. The complex number λ is an **eigenvalue** of A if there exists a nonzero vector $\mathbf{x}$ in C^n such that

$$A\mathbf{x} = \lambda\mathbf{x}. \tag{1}$$

Every nonzero vector $\mathbf{x}$ satisfying (1) is called an **eigenvector of A associated with the eigenvalue** λ. Equation (1) can be rewritten as the homogeneous system

$$(\lambda I_n - A)\mathbf{x} = \mathbf{0}.$$

This homogeneous system has a nonzero solution $\mathbf{x}$ if and only if

$$\det(\lambda I_n - A) = 0$$

has a solution. Following the development in Section 5.1, we shall call $\det(\lambda I_n - A)$ the **characteristic polynomial** of the matrix A, which is a complex polynomial of degree n in λ. As in Section 5.1, the eigenvalues of the complex matrix A are the complex roots of the characteristic polynomial. According to the Fundamental Theorem of Algebra, an nth-degree polynomial has exactly n roots, if multiple roots are counted. Even if $\det(\lambda I_n - A)$ has only one root of multiplicity n, we know that the characteristic polynomial of any complex matrix A always has at least one root (possibly complex). This is quite different from the case of real vector spaces discussed in Section 5.1, because there are real matrices that do not have real eigenvalues (see Example 7 in Section 5.1).

EXAMPLE 10 ■ Let
$$A = \begin{bmatrix} 0 & 1 \\ -1 & 0 \end{bmatrix}.$$

Then the characteristic polynomial of A is
$$\det(\lambda I_2 - A) = \lambda^2 + 1.$$

The corresponding (complex) eigenvectors are obtained by finding a nontrivial solution of the homogeneous systems
$$(iI_2 - A)\mathbf{x} = \mathbf{0} \quad \text{and} \quad (-iI_2 - A)\mathbf{x} = \mathbf{0},$$

respectively. An eigenvector associated with eigenvalue i is
$$\begin{bmatrix} 1 \\ i \end{bmatrix} \quad \text{(verify)}$$

and an eigenvector associated with eigenvalue $-i$ is
$$\begin{bmatrix} -1 \\ i \end{bmatrix} \quad \text{(verify)}.$$

■

In the case of complex matrices, we have the following analogs of the theorem presented in Section 5.2, which show the role played by the special matrices discussed in Section A.1.

THEOREM A.1 ■ *If A is a Hermitian matrix, then the eigenvalues of A are all real. Moreover, eigenvectors belonging to distinct eigenvalues are orthogonal (complex analog of Theorems 5.5 and 5.6).* ■

THEOREM A.2 ■ *If A is a Hermitian matrix, then there exists a unitary matrix U such that $U^{-1}AU = D$, a diagonal matrix. The eigenvalues of A lie on the main diagonal of D (complex analog of Theorem 5.8).* ■

In Section 5.2 we proved that if A is a real symmetric matrix, then there exists an orthogonal matrix P such that $P^{-1}AP = D$, a diagonal matrix, and conversely, if there is an orthogonal matrix P such that $P^{-1}AP$ is a diagonal matrix, then A is a symmetric matrix. For complex matrices, the situation is more complicated. The converse of Theorem A.2 is not true. That is, if A is a matrix for which there exists a unitary matrix U such that $U^{-1}AU = D$, a diagonal matrix, then A need not be a Hermitian matrix. The correct statement involves normal matrices. The following result can be established.

THEOREM A.3 ■ *If A is a normal matrix, then there exists a unitary matrix U such that $U^{-1}AU = D$, a diagonal matrix. Conversely, if A is a matrix for which there exists a unitary matrix U such that $U^{-1}AU = D$, a diagonal matrix, then A is a normal matrix.* ■

Every real $n \times n$ matrix can be considered as an $n \times n$ complex matrix. Hence the eigenvalue and eigenvector computations for real matrices carried out in Chapter 5 can "be fixed up" by permitting complex roots of the characteristic polynomial to be called eigenvalues. Of course, this means that we would need to use complex arithmetic to solve for the associated eigenvector. So if, as in Chapter 5, we restrict ourselves to real vector spaces, we must permit only real numbers to be eigenvalues. The eigen-problem for symmetric matrices, Section 5.2, needs no "fix" because the eigenvalues of any real symmetric matrix are all real numbers (see Theorem 5.5).

A.2 EXERCISES

1. Solve using Gauss–Jordan reduction.

   (a) $(1+2i)x_1 + (-2+i)x_2 = 1 - 3i$
   $(2+i)x_1 + (-1+2i)x_2 = -1 - i.$

   (b) $2ix_1 - (1-i)x_2 = 1 + i$
   $(1-i)x_1 + x_2 = 1 - i.$

   (c) $(1+i)x_1 - x_2 = -2 + i$
   $2ix_1 + (1-i)x_2 = i.$

2. Transform the given augmented matrix of a linear system to row echelon form and solve by back substitution.

   (a) $\begin{bmatrix} 2 & i & 0 & \vdots & 1-i \\ 0 & 3i & -2+i & \vdots & 4 \\ 0 & 0 & 2+i & \vdots & 2-i \end{bmatrix}.$

   (b) $\begin{bmatrix} i & 2 & 1+i & \vdots & 3i \\ 0 & 1-i & 0 & \vdots & 2+i \\ 0 & 0 & 3 & \vdots & 6-3i \end{bmatrix}.$

3. Solve by Gaussian elimination with back substitution.

   (a) $ix_1 + (1+i)x_2 = i$
   $(1-i)x_1 + x_2 - ix_3 = 1$
   $ix_2 + x_3 = 1.$

   (b) $x_1 + ix_2 + (1-i)x_3 = 2 + i$
   $ix_1 + (1+i)x_3 = -1 + i$
   $2ix_2 - x_3 = 2 - i.$

4. Compute the determinant and simplify as much as possible.

   (a) $\begin{vmatrix} 1+i & -1 \\ 2i & 1+i \end{vmatrix}.$

   (b) $\begin{vmatrix} 2-i & 1+i \\ 1+2i & -(1-i) \end{vmatrix}.$

   (c) $\begin{vmatrix} 1+i & 2 & 2-i \\ i & 0 & 3+i \\ -2 & 1 & 1+2i \end{vmatrix}.$

   (d) $\begin{vmatrix} 2 & 1-i & 0 \\ 1+i & -1 & i \\ 0 & -i & 2 \end{vmatrix}.$

5. Find the inverse of each of the following matrices if possible.

   (a) $\begin{bmatrix} i & 2 \\ 1+i & -i \end{bmatrix}.$
   (b) $\begin{bmatrix} 2 & i & 3 \\ 1+i & 0 & 1-i \\ 2 & 1 & 2+i \end{bmatrix}.$

6. Determine whether or not the following subsets W of C_{22} are complex vector subspaces.

   (a) W is the set of all 2×2 complex matrices with zeros on the main diagonal.

   (b) W is the set of all 2×2 complex matrices that have diagonal entries with real part equal to zero.

   (c) W is the set of all symmetric 2×2 complex matrices.

7. Let $W = \text{span}\{\mathbf{v}_1, \mathbf{v}_2, \mathbf{v}_3\}$, where

   $$\mathbf{v}_1 = (-1+i, 2, 1), \qquad \mathbf{v}_2 = (1, 1+i, i),$$
   $$\mathbf{v}_3 = (-5+2i, -1-3i, 2-3i).$$

   (a) Does $\mathbf{v} = (i, 0, 0)$ belong to W?

   (b) Is the set $\{\mathbf{v}_1, \mathbf{v}_2, \mathbf{v}_3\}$ linearly independent or linearly dependent?

8. Let $\{\mathbf{v}_1, \mathbf{v}_2, \mathbf{v}_3\}$ be a basis for a complex vector space V. Determine whether or not $\mathbf{w}$ is in span $\{\mathbf{w}_1, \mathbf{w}_2\}$.

   (a) $\mathbf{w}_1 = i\mathbf{v}_1 + (1-i)\mathbf{v}_2 + 2\mathbf{v}_3$
   $\mathbf{w}_2 = (2+i)\mathbf{v}_1 + 2i\mathbf{v}_2 + (3-i)\mathbf{v}_3$
   $\mathbf{w} = (-2-3i)\mathbf{v}_1 + (3-i)\mathbf{v}_2 + (-2-2i)\mathbf{v}_3.$

(b) $\mathbf{w}_1 = 2i\mathbf{v}_1 + \mathbf{v}_2 + (1-i)\mathbf{v}_3$
   $\mathbf{w}_2 = 3i\mathbf{v}_1 + (1+i)\mathbf{v}_2 + 3\mathbf{v}_3$
   $\mathbf{w} = (2+3i)\mathbf{v}_1 + (2+i)\mathbf{v}_2 + (4-2i)\mathbf{v}_3$.

9. Determine whether $\mathbf{u}$ and $\mathbf{v}$ are orthogonal.
   (a) $\mathbf{u} = (3+i, 2-i)$, $\mathbf{v} = (1-i, 2+i)$.
   (b) $\mathbf{u} = (i, 1+i, 1-i)$, $\mathbf{v} = (3-i, 1+2i, -1+3i)$.
   (c) $\mathbf{u} = (1+i, 2-i, 3)$, $\mathbf{v} = (i, -2i, 1+i)$.
   (d) $\mathbf{u} = (1+2i, 2i, 2-i)$,
       $\mathbf{v} = (-3+5i, 5-5i, 1-i)$.

10. Compute $\|\mathbf{u}\|$.
    (a) $\mathbf{u} = (1-i, 2, 3+i)$.
    (b) $\mathbf{u} = (i, -2-3i, 1+i)$.
    (c) $\mathbf{u} = (i, -i, 1, 0)$.

(d) $\mathbf{u} = (1+i, 1-i, 2+i, 3-i)$.

11. Find the eigenvalues and associated eigenvectors of the following complex matrices.

   (a) $A = \begin{bmatrix} 1 & 1 \\ -1 & 1 \end{bmatrix}$.   (b) $A = \begin{bmatrix} 1 & i \\ -i & 1 \end{bmatrix}$.

   (c) $A = \begin{bmatrix} 2 & 0 & 0 \\ 0 & 2 & i \\ 0 & -i & 2 \end{bmatrix}$.

12. For each of the parts in Exercise 11, find a matrix P such that $P^{-1}AP = D$, a diagonal matrix. For part (c), find three different matrices P that diagonalize A.

THEORETICAL EXERCISES ■

T.1. (a) Prove or disprove: The set W of all $n \times n$ Hermitian matrices is a complex vector subspace of C_{nn}.

   (b) Prove or disprove: The set W of all $n \times n$ Hermitian matrices is a real vector subspace of the real vector space of all $n \times n$ complex matrices.

T.2. Prove or disprove: The set W of all $n \times n$ unitary matrices is a complex vector subspace of C_{nn}.

T.3. (a) Show that if A is Hermitian, then the eigenvalues of A are real.

   (b) Verify that A in Exercise 11(c) is Hermitian.

   (c) Are the eigenvectors associated with an eigenvalue of a Hermitian matrix guaranteed to be real vectors? Explain.

T.4. Show that an $n \times n$ complex matrix A is unitary

if and only if the columns (and rows) of A form an orthonormal set with respect to the dot product on C^n. (*Hint*: See Theorem 5.7.)

T.5. Show that if A is a skew Hermitian matrix (see Exercise T.7 in Section A.1) and λ is an eigenvalue of A, then the real part of λ is zero.

T.6. Show that the dot product on C^n satisfies the following properties.

   (a) $\mathbf{u} \cdot \mathbf{u} > 0$ for $\mathbf{u} \neq \mathbf{0}$ in C^n; $\mathbf{u} \cdot \mathbf{u} = 0$ if and only if $\mathbf{u} = \mathbf{0}$.

   (b) $\mathbf{u} \cdot \mathbf{v} = \overline{\mathbf{v} \cdot \mathbf{u}}$ for any $\mathbf{u}$, $\mathbf{v}$ in C^n.

   (c) $(\mathbf{u} + \mathbf{v}) \cdot \mathbf{w} = \mathbf{u} \cdot \mathbf{w} + \mathbf{v} \cdot \mathbf{w}$ for any $\mathbf{u}$, $\mathbf{v}$, $\mathbf{w}$ in C^n.

   (d) $(c\mathbf{u}) \cdot \mathbf{v} = \overline{c}(\mathbf{u} \cdot \mathbf{v})$ for any $\mathbf{u}$, $\mathbf{v}$ in C^n and c a complex scalar.

MATLAB EXERCISES ■

All the routines for solving linear systems, such as **reduce**, **rref**, and ****, and the commands **det**, **invert**, **eig**, **roots**, **poly**, and so on, apply to complex matrices.

ML.1. Solve the linear system in Exercise 1 using ****.

ML.2. Solve Exercise 4 using **det**.

ML.3. Solve Exercise 5 using **invert**.

ML.4. Solve Exercise 11 using **eig**.

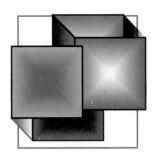

Appendix B

Further Directions

B.1 ▼ Inner Product Spaces (Calculus Required)

In this section we use the properties of the standard inner product or dot product on R^3 listed in Theorem 3.1 as our foundation for generalizing the notion of the inner product to any real vector space. Here V is an arbitrary real vector space, not necessarily finite dimensional.

DEFINITION — Let V be any real vector space. An **inner product** on V is a function that assigns to each ordered pair of vectors $\mathbf{u}$, $\mathbf{v}$ in V a real number denoted by $(\mathbf{u}, \mathbf{v})$ satisfying:

(a) $(\mathbf{u}, \mathbf{u}) > 0$ for $\mathbf{u} \neq \mathbf{0}_V$; $(\mathbf{u}, \mathbf{u}) = 0$ if and only if $\mathbf{u} = \mathbf{0}_V$, where $\mathbf{0}_V$ is the zero vector in V.

(b) $(\mathbf{v}, \mathbf{u}) = (\mathbf{u}, \mathbf{v})$ for any $\mathbf{u}$, $\mathbf{v}$ in V.

(c) $(\mathbf{u} + \mathbf{v}, \mathbf{w}) = (\mathbf{u}, \mathbf{w}) + (\mathbf{v}, \mathbf{w})$ for any $\mathbf{u}$, $\mathbf{v}$, $\mathbf{w}$ in V.

(d) $(c\mathbf{u}, \mathbf{v}) = c(\mathbf{u}, \mathbf{v})$ for $\mathbf{u}$, $\mathbf{v}$ in V and c a real scalar.

From these properties it follows that $(\mathbf{u}, c\mathbf{v}) = c(\mathbf{u}, \mathbf{v})$ because $(\mathbf{u}, c\mathbf{v}) = (c\mathbf{v}, \mathbf{u}) = c(\mathbf{v}, \mathbf{u}) = c(\mathbf{u}, \mathbf{v})$.

EXAMPLE 1 ■ The dot product on R^n, as defined in Section 1.3, is

$$(\mathbf{u}, \mathbf{v}) = \mathbf{u} \cdot \mathbf{v} = u_1 v_1 + u_2 v_2 + \cdots + u_n v_n,$$

where $\mathbf{u} = (u_1, u_2, \ldots, u_n)$ and $\mathbf{v} = (v_1, v_2, \ldots, v_n)$ is an inner product. This inner product will be called the **standard inner product** on R^n. ■

EXAMPLE 2 ■ Let V be any finite-dimensional vector space and let $S = \{\mathbf{u}_1, \mathbf{u}_2, \ldots, \mathbf{u}_n\}$ be a basis for V. If

$$\mathbf{v} = a_1 \mathbf{u}_1 + a_2 \mathbf{u}_2 + \cdots + a_n \mathbf{u}_n$$

A23

and

$$\mathbf{w} = b_1\mathbf{u}_1 + b_2\mathbf{u}_2 + \cdots + b_n\mathbf{u}_n,$$

we define

$$(\mathbf{v}, \mathbf{w}) = \left([\mathbf{v}]_S, [\mathbf{w}]_S\right) = a_1b_1 + a_2b_2 + \cdots + a_nb_n.$$

It is not difficult to verify that this defines an inner product on V. This definition of $(\mathbf{v}, \mathbf{w})$ as an inner product on V uses the standard inner product on R^n. ∎

Example 2 shows that we can define an inner product on any finite-dimensional vector space. Of course, if we change the basis for V in Example 2, we obtain a different inner product.

EXAMPLE 3 ∎ Let $\mathbf{u} = (u_1, u_2)$ and $\mathbf{v} = (v_1, v_2)$ be vectors in R^2. We define

$$(\mathbf{u}, \mathbf{v}) = u_1v_1 - u_2v_1 - u_1v_2 + 3u_2v_2.$$

Show that this gives an inner product on R^2.

Solution   We have

$$(\mathbf{u}, \mathbf{u}) = u_1^2 - 2u_1u_2 + 3u_2^2 = u_1^2 - 2u_1u_2 + u_2^2 + 2u_2^2$$
$$= (u_1 - u_2)^2 + 2u_2^2 > 0 \quad \text{if } \mathbf{u} \neq \mathbf{0}.$$

Moreover, if $(\mathbf{u}, \mathbf{u}) = 0$, then $u_1 = u_2$ and $u_2 = 0$, so $\mathbf{u} = \mathbf{0}$. Conversely, if $\mathbf{u} = \mathbf{0}$, then $(\mathbf{u}, \mathbf{u}) = 0$. We can also verify (see Exercise 1) the remaining three properties of the definition above. This inner product is, of course, not the standard inner product on R^2. ∎

Example 3 shows that on one vector space we may have many different inner products, since we also have the standard inner product on R^2.

EXAMPLE 4 ∎ Let V be the vector space $C[0, 1]$ consisting of all real-valued continuous func-
(*Calculus Required*)   tions that are defined on the unit interval $[0, 1]$. For f and g in V, we let

$$(f, g) = \int_0^1 f(t)g(t)\, dt.$$

We now verify that this is an inner product on V, that is, that the properties of the definition above are satisfied.

Using results from calculus, we have for $f \neq 0$, the zero function,

$$(f, f) = \int_0^1 (f(t))^2\, dt > 0.$$

Moreover, if $(f, f) = 0$, then $f = 0$. Conversely, if $f = 0$, then $(f, f) = 0$. Also,

$$(f, g) = \int_0^1 f(t)g(t)\, dt = \int_0^1 g(t)f(t)\, dt = (g, f).$$

Next,

$$(f+g, h) = \int_0^1 (f(t) + g(t))h(t)\, dt = \int_0^1 f(t)h(t)\, dt + \int_0^1 g(t)h(t)\, dt$$
$$= (f, h) + (g, h).$$

Finally,

$$(cf, g) = \int_0^1 (cf(t))g(t)\, dt = c \int_0^1 f(t)g(t)\, dt = c(f, g).$$

Thus, if f and g are the functions defined by $f(t) = t + 1$, $g(t) = 2t + 3$, then

$$(f, g) = \int_0^1 (t+1)(2t+3)\, dt = \int_0^1 (2t^2 + 5t + 3)\, dt = \tfrac{37}{6}. \qquad \blacksquare$$

DEFINITION   A real vector space that has an inner product defined on it is called an **inner product space**.

EXAMPLE 5 ■   Let $V = P$, the set of all polynomials. Since P is a subspace of $C[0, 1]$, if we use the inner product defined in Example 4, we see that P is an inner product space. ■

If V is an inner product space, then by the **dimension** of V we mean the dimension of V as a real vector space, and a set S is a **basis** for V if S is a basis for the real vector space V. In an inner product space we define the **length** of a vector $\mathbf{u}$ by

$$\|\mathbf{u}\| = \sqrt{(\mathbf{u}, \mathbf{u})}.$$

This definition of length seems reasonable because at least we have $\|\mathbf{u}\| > 0$ if $\mathbf{u} \neq \mathbf{0}$. We can show [see Exercise T.1(a)] that $\|\mathbf{0}\| = 0$.

Any result in Sections 3.2, 4.8, and 4.9 dealing with R^n is also valid for any *inner product space* if a basis is not involved in the statement; it is true for any finite-dimensional inner product *space* if a basis is involved in the statement. Thus, let V be an inner product space. We define the **distance** between two vectors $\mathbf{u}$ and $\mathbf{v}$ in V as

$$d(\mathbf{u}, \mathbf{v}) = \|\mathbf{u} - \mathbf{v}\|.$$

The vectors $\mathbf{u}$ and $\mathbf{v}$ are defined to be **orthogonal** if $(\mathbf{u}, \mathbf{v}) = 0$. An orthogonal set of vectors $\mathbf{u}_1$, $\mathbf{u}_2$, ..., $\mathbf{u}_k$ in V is called **orthonormal** if the vectors are all of unit length.

Moreover, the Cauchy–Schwarz inequality (Theorem 3.4), the Triangle inequality (Theorem 3.5), Theorems 4.16, 4.18 (the Gram–Schmidt process), 4.20, 4.21, and 4.23 are all valid in any inner product space. Theorem 4.17 holds in a finite-dimensional inner product space. We illustrate these ideas in the following examples.

EXAMPLE 6 ■ Let V be the inner product space P_2 with inner product defined as in Example
(*Calculus Required*) 4. If $p(t) = t + 2$, then the length of $p(t)$ is

$$\|p(t)\| = \sqrt{(p(t), p(t))} = \sqrt{\int_0^1 (t+2)^2 \, dt} = \sqrt{\frac{19}{3}}.$$

If $q(t) = 2t - 3$, then to find the cosine of the angle θ between $p(t)$ and $q(t)$, we
proceed as follows. First,

$$\|q(t)\| = \sqrt{\int_0^1 (2t-3)^2 \, dt} = \sqrt{\frac{13}{3}}.$$

Next,

$$(p(t), q(t)) = \int_0^1 (t+2)(2t-3) \, dt = \int_0^1 (2t^2 + t - 6) \, dt = -\frac{29}{6}.$$

Then

$$\cos \theta = \frac{(p(t), q(t))}{\|p(t)\| \, \|q(t)\|} = \frac{-\frac{29}{6}}{\sqrt{\frac{19}{3}} \sqrt{\frac{13}{3}}} = \frac{-29}{2\sqrt{(19)(13)}}.$$

■

EXAMPLE 7 ■ Let V be the inner product space P_2 considered in Example 6. The vectors t
and $t - \frac{2}{3}$ are orthogonal, since

$$\left(t, t - \frac{2}{3}\right) = \int_0^1 t\left(t - \frac{2}{3}\right) \, dt = \int_0^1 \left(t^2 - \frac{2t}{3}\right) \, dt = 0.$$

■

EXAMPLE 8 ■ Let V be the vector space $C[-\pi, \pi]$ of all real-valued continuous functions that
(*Calculus Required*) are defined on $[-\pi, \pi]$. For f and g in V, we let $(f, g) = \int_{-\pi}^{\pi} f(t)g(t) \, dt$, which
is easily shown to be an inner product on V (see Example 4). Consider the
functions

$$1, \quad \cos t, \quad \sin t, \quad \cos 2t, \quad \sin 2t, \ldots, \quad \cos nt, \quad \sin nt, \ldots, \qquad (1)$$

which are clearly in V. The relationships

$$\int_{-\pi}^{\pi} \cos nt \, dt = \int_{-\pi}^{\pi} \sin nt \, dt = \int_{-\pi}^{\pi} \sin nt \cos nt \, dt = 0,$$

$$\int_{-\pi}^{\pi} \cos mt \cos nt \, dt = \int_{-\pi}^{\pi} \sin mt \sin nt \, dt = 0 \quad \text{if } m \neq n$$

demonstrate that $(f, g) = 0$ whenever f and g are distinct functions from (1).
Hence every finite subset of functions from (1) is an orthogonal set. Theorem
4.16 generalized to inner product spaces then implies that any finite subset of
functions from (1) is linearly independent. ■

EXAMPLE 9 ■ Let V be the inner product space P_3 with the inner product defined in Example
4. Let W be the subspace of P_3 with basis $\{1, t^2\}$. Find a basis for $W^{\perp}$.

Solution   Let $p(t) = at^3 + bt^2 + ct + d$ be an element of $W^\perp$. Since $p(t)$ must be orthogonal to each of the vectors in the given basis for W, we have

$$(p(t), 1) = \int_0^1 (at^3 + bt^2 + ct + d)\, dt = \frac{a}{4} + \frac{b}{3} + \frac{c}{2} + d = 0$$

$$(p(t), t^2) = \int_0^1 (at^5 + bt^4 + ct^3 + dt^2)\, dt = \frac{a}{6} + \frac{b}{5} + \frac{c}{4} + \frac{d}{3} = 0.$$

Solving the homogeneous system

$$\frac{a}{4} + \frac{b}{3} + \frac{c}{2} + d = 0$$

$$\frac{a}{6} + \frac{b}{5} + \frac{c}{4} + \frac{d}{3} = 0,$$

we obtain (verify)

$$a = 3r + 16s, \qquad b = -\tfrac{15}{4} r - 15s, \qquad c = r, \qquad d = s.$$

Then

$$p(t) = (3r + 16s)t^3 + \left(-\tfrac{15}{4} r - 15s\right)t^2 + rt + s$$
$$= r\left(3t^3 - \tfrac{15}{4}t^2 + t\right) + s(16t^3 - 15t^2 + 1).$$

Hence the vectors $3t^3 - \tfrac{15}{4}t^2 + t$ and $16t^3 - 15t^2 + 1$ span $W^\perp$. Since they are not multiples of each other, they are linearly independent and thus form a basis for $W^\perp$. ∎

EXAMPLE 10 ■   Let V be the inner product space P_3 with the inner product defined in Example 4. Let W be the subspace of P_3 having $S = \{t^2, t\}$ as a basis. Find an orthonormal basis for W.

Solution   First, let $\mathbf{u}_1 = t^2$ and $\mathbf{u}_2 = t$. Now let $\mathbf{v}_1 = \mathbf{u}_1 = t^2$. Then

$$\mathbf{v}_2 = \mathbf{u}_2 - \frac{(\mathbf{u}_2, \mathbf{v}_1)}{(\mathbf{v}_1, \mathbf{v}_1)}\, \mathbf{v}_1 = t - \frac{\frac{1}{4}}{\frac{1}{5}}\, t^2 = t - \tfrac{5}{4}t^2,$$

where

$$(\mathbf{v}_1, \mathbf{v}_1) = \int_0^1 t^2 t^2\, dt = \int_0^1 t^4\, dt = \frac{1}{5}$$

and

$$(\mathbf{u}_2, \mathbf{v}_1) = \int_0^1 tt^2\, dt = \int_0^1 t^3\, dt = \frac{1}{4}.$$

Then

$$T^* = \left\{t^2, t - \tfrac{5}{4}t^2\right\}$$

is an orthogonal basis for W. We now have to normalize the vectors in T^* to obtain an orthonormal basis T for W. We have already computed

$$(\mathbf{v}_1, \mathbf{v}_1) = \tfrac{1}{5}, \quad \text{so } \|\mathbf{v}_1\| = \sqrt{\tfrac{1}{5}}.$$

We also have

$$(\mathbf{v}_2, \mathbf{v}_2) = \int_0^1 \left(t - \tfrac{5}{4}t^2\right)^2 dt = \tfrac{1}{48}, \quad \text{so } \|\mathbf{v}_2\| = \sqrt{\tfrac{1}{48}}.$$

We now let

$$\mathbf{w}_1 = \frac{1}{\|\mathbf{v}_1\|}\,\mathbf{v}_1 = \sqrt{5}\,t^2, \qquad \mathbf{w}_2 = \frac{1}{\|\mathbf{v}_2\|}\,\mathbf{v}_2 = \sqrt{48}\left(t - \tfrac{5}{4}t^2\right).$$

Then $T = \left\{\sqrt{5}\,t^2, \sqrt{48}\left(t - \tfrac{5}{4}t^2\right)\right\}$ is an orthonormal basis for W. If we choose $\mathbf{u}_1 = t$ and $\mathbf{u}_2 = t^2$, then we obtain (verify) the orthonormal basis

$$\left\{\sqrt{3}\,t, \sqrt{30}\left(t^2 - \tfrac{1}{2}t\right)\right\}$$

for W. ∎

Approximation of Functions (Calculus Required)

In the study of power series in calculus, the functions

$$f_0(t) = 1, \quad f_1(t) = t, \quad f_2(t) = t^2, \dots, \quad f_n(t) = t^n, \dots$$

play a central role. Most of the functions encountered in calculus can be expanded in a power series. The coefficients in Taylor and Maclaurin series expansions involve successive derivatives of the given function evaluated at a point, the center of the expansion.

The function $f(t) = |t|$ does not have a Taylor series expansion with center of expansion $t_0 = 0$ (a Maclaurin series), because f does not have a derivative at $t = 0$. Thus there is no way to compute the coefficients in such an expansion. However, it may be possible to obtain series expansions for such functions as $f(t) = |t|$ by using a different type of expansion. One such important expansion involves the set of functions

$$1, \quad \cos t, \quad \sin t, \quad \cos 2t, \quad \sin 2t, \dots, \quad \cos nt, \quad \sin nt, \dots, \tag{2}$$

which we discussed briefly in Example 8. Obtaining a series expansion of $f(t) = |t|$ in terms of the functions in (2) would take us too far afield at this time, but we will show how to obtain an approximation from a finite subset of these functions in the next example.

EXAMPLE 11 ■ Let V be the vector space $C[-\pi, \pi]$ of all real-valued continuous functions that are defined on $[-\pi, \pi]$. Then $W = \text{span}\,\{1, \cos t, \sin t\}$ is a finite-dimensional subspace of V. If f and g belong to V, then $(f, g) = \int_{-\pi}^{\pi} f(t)g(t)\,dt$ defines an inner product on V as in Example 8. It also follows that $S = \{1, \cos t, \sin t\}$ is an orthogonal set, but as the proof of Theorem 4.20 shows, to compute $\text{proj}_W \mathbf{v}$ using Equation (1) in Section 4.9, we need an orthonormal basis. It is easy to show (verify) that

$$\int_{-\pi}^{\pi} 1\,dt = 2\pi, \qquad \int_{-\pi}^{\pi} \cos^2 t\,dt = \pi, \qquad \int_{-\pi}^{\pi} \sin^2 t\,dt = \pi.$$

If we let

$$\mathbf{w}_1 = \frac{1}{\sqrt{2\pi}}, \qquad \mathbf{w}_2 = \frac{1}{\sqrt{\pi}}\cos t, \quad \text{and} \quad \mathbf{w}_3 = \frac{1}{\sqrt{\pi}}\sin t,$$

then $T = \{\mathbf{w}_1, \mathbf{w}_2, \mathbf{w}_3\}$ is an orthonormal set in V and $W = \text{span } T$ (verify). Using Equation (1) in Section 4.9 we can compute $\text{proj}_W \mathbf{v}$, for $\mathbf{v} = |t|$, as

$$\text{proj}_W \mathbf{v} = (\mathbf{v}, \mathbf{w}_1)\mathbf{w}_1 + (\mathbf{v}, \mathbf{w}_2)\mathbf{w}_2 + (\mathbf{v}, \mathbf{w}_3)\mathbf{w}_3$$

or

$$\text{proj}_W |t| = \left(|t|, \frac{1}{\sqrt{2\pi}}\right) \frac{1}{\sqrt{2\pi}} + \left(|t|, \frac{1}{\sqrt{\pi}} \cos t\right) \frac{1}{\sqrt{\pi}} \cos t$$
$$+ \left(|t|, \frac{1}{\sqrt{\pi}} \sin t\right) \frac{1}{\sqrt{\pi}} \sin t.$$

We have

$$\left(|t|, \frac{1}{\sqrt{2\pi}}\right) = \int_{-\pi}^{\pi} |t| \frac{1}{\sqrt{2\pi}} \, dt$$
$$= \frac{1}{\sqrt{2\pi}} \int_{-\pi}^{0} -t \, dt + \frac{1}{\sqrt{2\pi}} \int_{0}^{\pi} t \, dt = \frac{\pi^2}{\sqrt{2\pi}},$$

$$\left(|t|, \frac{1}{\sqrt{\pi}} \cos t\right) = \int_{-\pi}^{\pi} |t| \frac{1}{\sqrt{\pi}} \cos t \, dt$$
$$= \frac{1}{\sqrt{\pi}} \int_{-\pi}^{0} -t \cos t \, dt + \frac{1}{\sqrt{\pi}} \int_{0}^{\pi} t \cos t \, dt$$
$$= -\frac{2}{\sqrt{\pi}} - \frac{2}{\sqrt{\pi}} = -\frac{4}{\sqrt{\pi}},$$

and

$$\left(|t|, \frac{1}{\sqrt{\pi}} \sin t\right) = \int_{-\pi}^{\pi} |t| \frac{1}{\sqrt{\pi}} \sin t \, dt$$
$$= \frac{1}{\sqrt{\pi}} \int_{-\pi}^{0} -t \sin t \, dt + \frac{1}{\sqrt{\pi}} \int_{0}^{\pi} t \sin t \, dt$$
$$= -\sqrt{\pi} + \sqrt{\pi} = 0.$$

Then

$$\text{proj}_W |t| = \frac{\pi^2}{\sqrt{2\pi}} \frac{1}{\sqrt{2\pi}} - \frac{4}{\sqrt{\pi}} \frac{1}{\sqrt{\pi}} \cos t = \frac{\pi}{2} - \frac{4}{\pi} \cos t.$$

Figure B.1 shows a graph of both $|t|$ and $\text{proj}_W |t|$. ■

REMARK   Theorem 4.23 generalized to inner product spaces shows that in Example 11, the function $\frac{\pi}{2} - \frac{4}{\pi} \cos t$ is the vector in span $\{1, \cos t, \sin t\}$ that is closest to the function $|t|$. Theorem 4.23 is important in the theory of approximation of functions and can be generalized to more abstract settings. Example 11 is a preview of ideas from the study of Fourier series that play an important role in the study of heat distribution and the analysis of sound waves.

FIGURE B.1

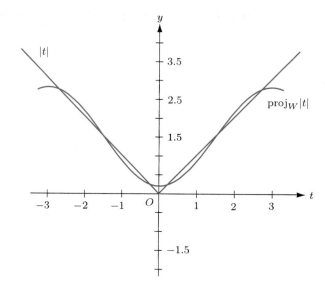

B.1   EXERCISES

1. Verify that the function in Example 3 satisfies the remaining three properties of an inner product.

2. (**Calculus Required**). Verify that the function defined on P, the vector space of all polynomials, in Example 5 is an inner product.

3. Let $V = R^2$. If

$$\mathbf{u} = (u_1, u_2) \quad \text{and} \quad \mathbf{v} = (v_1, v_2),$$

define

$$(\mathbf{u}, \mathbf{v}) = u_1 v_1 + 5u_2 v_2.$$

Show that this function is an inner product on R^2.

4. Let $V = M_{22}$. If

$$A = \begin{bmatrix} a_{11} & a_{12} \\ a_{21} & a_{22} \end{bmatrix} \quad \text{and} \quad B = \begin{bmatrix} b_{11} & b_{12} \\ b_{21} & b_{22} \end{bmatrix},$$

define

$$(A, B) = a_{11}b_{11} + a_{12}b_{12} + a_{21}b_{21} + a_{22}b_{22}.$$

Show that this function is an inner product on V.

5. Let $V = M_{nn}$ be the real vector space of all $n \times n$ matrices. If A and B are in V, we define $(A, B) = \text{Tr}(B^T A)$, where Tr is the trace function defined in Supplementary Exercise T.1 of Chapter 1. Show that this function is an inner product on V.

6. Let V be the vector space $C[a, b]$ consisting of all real-valued continuous functions that are defined on $[a, b]$. If f and g are in V, let $(f, g) = \int_a^b f(x)g(x)\,dx$. Show that this function is an inner product on V.

In Exercises 7 and 8, use the inner product in Example 3 and compute $(\mathbf{u}, \mathbf{v})$.

7. $\mathbf{u} = (1, 2), \mathbf{v} = (3, -1)$.

8. $\mathbf{u} = (0, 1), \mathbf{v} = (-2, 5)$.

In Exercises 9 and 10, use the inner product defined in Example 4 and compute (f, g).

9. $f(t) = 1, g(t) = 3 + 2t$.

10. $f(t) = \sin t, g(t) = \cos t$.

In Exercises 11 and 12, use the inner product space defined in Exercise 4 and compute (A, B).

11. $A = \begin{bmatrix} 1 & 2 \\ -1 & 3 \end{bmatrix}, B = \begin{bmatrix} 1 & 0 \\ 2 & -1 \end{bmatrix}$.

12. $A = \begin{bmatrix} 0 & 1 \\ 2 & 3 \end{bmatrix}, B = \begin{bmatrix} 1 & 1 \\ 2 & -1 \end{bmatrix}$.

13. Let V be the inner product space in Example 3. Compute the length of the given vector.
    (a) $(1, 3)$.   (b) $(-2, -4)$.   (c) $(3, -1)$.

14. Let V be the inner product space in Example 4. Compute the length of the given vector.

(a) t^2.   (b) e^t.

15. Let V be the inner product space in Example 4. Find the distance between $\mathbf{u}$ and $\mathbf{v}$.

(a) $\mathbf{u} = t$, $\mathbf{v} = t^2$.   (b) $\mathbf{u} = e^t$, $\mathbf{v} = e^{-t}$.

16. Let V be the inner product space in Exercise 3. Find the distance between $\mathbf{u}$ and $\mathbf{v}$.

(a) $\mathbf{u} = (0, 1)$, $\mathbf{v} = (1, -1)$.

(b) $\mathbf{u} = (-2, -1)$, $\mathbf{v} = (2, 3)$.

17. Let V be the inner product space in Example 4. Find the cosine of the angle between each pair of given vectors in V.

(a) $p(t) = t$, $q(t) = t - 1$.

(b) $p(t) = \sin t$, $q(t) = \cos t$.

18. Let V be the inner product space in Exercise 3. Find the cosine of the angle between each pair of given vectors in V.

(a) $\mathbf{u} = (2, 1)$, $\mathbf{v} = (3, 2)$.

(b) $\mathbf{u} = (1, 1)$, $\mathbf{v} = (-2, -3)$.

In Exercises 19 and 20, let V be the inner product space of Example 4.

19. Let $p(t) = 3t + 1$ and $q(t) = at$. For what values of a are $p(t)$ and $q(t)$ orthogonal?

20. Let $p(t) = 3t + 1$ and $q(t) = at + b$. For what values of a and b are $p(t)$ and $q(t)$ orthogonal?

21. Let

$$A = \begin{bmatrix} 1 & 2 \\ 3 & 4 \end{bmatrix}.$$

Find a matrix $B \neq O_2$ such that A and B are orthogonal in the inner product space defined in Exercise 5. Can there be more than one matrix B that is orthogonal to A?

22. Let V be the inner product space in Example 4.

(a) If $p(t) = \sqrt{t}$, find $q(t) = a + bt \neq 0$ such that $p(t)$ and $q(t)$ are orthogonal.

(b) If $p(t) = \sin t$, find $q(t) = a + be^t \neq 0$ such that $p(t)$ and $q(t)$ are orthogonal.

23. Consider the inner product space R^4 with the standard inner product and let

$$\mathbf{u}_1 = (1, 0, 0, 1) \quad \text{and} \quad \mathbf{u}_2 = (0, 1, 0, 1).$$

(a) Show that the set W consisting of all vectors in R^4 that are orthogonal to both $\mathbf{u}_1$ and $\mathbf{u}_2$ is a subspace of R^4.

(b) Find a basis for W.

24. State the Cauchy–Schwarz inequality and the Triangle inequality for the inner product space in Example 4.

In Exercises 25 through 28, the inner product on the given vector space is as defined in Example 4.

25. (a) Let $S = \{t, 1\}$ be a basis for a subspace W of the inner product space P_2. Use the Gram–Schmidt process to find an orthonormal basis for W.

(b) Use Theorem 4.17 to write $2t - 1$ as a linear combination of the orthonormal basis obtained in part (a).

26. (a) Repeat Exercise 25 with $S = \{t + 1, t - 1\}$.

(b) Use Theorem 4.17 to find the coordinate vector of $3t + 2$ with respect to the orthonormal basis found in part (a).

27. Let $S = \{t, \sin 2\pi t\}$ be a basis for a subspace W of the inner product space in Example 9. Use the Gram–Schmidt process to find an orthonormal basis for W.

28. Let $S = \{t, e^t\}$ be a basis for a subspace W of the inner product space in Example 4. Use the Gram–Schmidt process to find an orthonormal basis for W.

29. Let V be the inner product space P_3 with the inner product defined in Example 4. Let W be the subspace of P_3 spanned by $\{t - 1, t^2\}$. Find a basis for $W^\perp$.

30. Let V be the inner product space P_4 with the inner product defined in Example 4. Let W be the subspace of P_4 spanned by $\{1, t\}$. Find a basis for $W^\perp$.

In Exercises 31 and 32, let W be the subspace of continuous functions on $[-\pi, \pi]$ defined in Example 11. Find $\text{proj}_W \mathbf{v}$ for the given vector $\mathbf{v}$.

31. $\mathbf{v} = t$.   **32.** $\mathbf{v} = e^t$.

In Exercises 33 and 34, let W be the subspace of continuous functions on $[-\pi, \pi]$ defined in Example 11. Write the vector $\mathbf{v}$ as $\mathbf{w} + \mathbf{u}$, with $\mathbf{w}$ in W and $\mathbf{u}$ in $W^\perp$.

33. $\mathbf{v} = t - 1$.   **34.** $\mathbf{v} = t^2$.

In Exercises 35 and 36, let W be the subspace of continuous functions on $[-\pi, \pi]$ defined in Example 11. Find the distance from $\mathbf{v}$ to W.

35. $\mathbf{v} = t$.   **36.** $\mathbf{v} = 1 - \cos t$.

THEORETICAL EXERCISES ■

T.1. Let V be an inner product space. Show the following.

(a) $\|\mathbf{0}\| = 0$.

(b) $(\mathbf{u}, \mathbf{0}) = (\mathbf{0}, \mathbf{u}) = 0$ for any $\mathbf{u}$ in V.

(c) If $(\mathbf{u}, \mathbf{v}) = 0$ for all $\mathbf{v}$ in V, then $\mathbf{u} = \mathbf{0}$.

(d) If $(\mathbf{u}, \mathbf{w}) = (\mathbf{v}, \mathbf{w})$ for all $\mathbf{w}$ in V, then $\mathbf{u} = \mathbf{v}$.

(e) If $(\mathbf{w}, \mathbf{u}) = (\mathbf{w}, \mathbf{v})$ for all $\mathbf{w}$ in V, then $\mathbf{u} = \mathbf{v}$.

T.2. Let V be an inner product space. If $\mathbf{u}$ and $\mathbf{v}$ are vectors in V, we define the distance between $\mathbf{u}$ and $\mathbf{v}$ as

$$d(\mathbf{u}, \mathbf{v}) = \|\mathbf{u} - \mathbf{v}\|.$$

Let $\mathbf{u}$, $\mathbf{v}$, and $\mathbf{w}$ be in V. Show that

(a) $d(\mathbf{u}, \mathbf{v}) \geq 0$.

(b) $d(\mathbf{u}, \mathbf{v}) = 0$ if and only if $\mathbf{u} = \mathbf{v}$.

(c) $d(\mathbf{u}, \mathbf{v}) = d(\mathbf{v}, \mathbf{u})$.

(d) $d(\mathbf{u}, \mathbf{v}) \leq d(\mathbf{u}, \mathbf{w}) + d(\mathbf{w}, \mathbf{v})$.

T.3. Show that if T is an orthonormal basis for a finite-dimensional inner product space and

$$[\mathbf{v}]_T = \begin{bmatrix} a_1 \\ a_2 \\ \vdots \\ a_n \end{bmatrix},$$

then $\|\mathbf{v}\| = \sqrt{a_1^2 + a_2^2 + \cdots + a_n^2}$.

T.4. Let $S = \{\mathbf{v}_1, \mathbf{v}_2, \ldots, \mathbf{v}_n\}$ be an orthonormal basis for a finite-dimensional inner product space V

and let $\mathbf{v}$ and $\mathbf{w}$ be vectors in V with

$$[\mathbf{v}]_S = \begin{bmatrix} a_1 \\ a_2 \\ \vdots \\ a_n \end{bmatrix} \quad \text{and} \quad [\mathbf{w}]_S = \begin{bmatrix} b_1 \\ b_2 \\ \vdots \\ b_n \end{bmatrix}.$$

Show that

$$d(\mathbf{v}, \mathbf{w}) = \sqrt{(a_1 - b_1)^2 + (a_2 - b_2)^2 + \cdots + (a_n - b_n)^2}.$$

T.5. Prove the **parallelogram law** for any two vectors $\mathbf{u}$ and $\mathbf{v}$ in an inner product space:

$$\|\mathbf{u} + \mathbf{v}\|^2 + \|\mathbf{u} - \mathbf{v}\|^2 = 2\|\mathbf{u}\|^2 + 2\|\mathbf{v}\|^2.$$

T.6. Let V be an inner product space. Show that $\|c\mathbf{u}\| = |c|\|\mathbf{u}\|$ for any vector $\mathbf{u}$ and any scalar c.

T.7. Let V be an inner product space. Show that if $\mathbf{u}$ and $\mathbf{v}$ are any vectors in V, then

$$\|\mathbf{u} + \mathbf{v}\|^2 = \|\mathbf{u}\|^2 + \|\mathbf{v}\|^2$$

if and only if $(\mathbf{u}, \mathbf{v}) = 0$, that is, if and only if $\mathbf{u}$ and $\mathbf{v}$ are orthogonal. This result is known as the **Pythagorean theorem**.

T.8. Let $\{\mathbf{u}, \mathbf{v}, \mathbf{w}\}$ be an orthonormal set of vectors in an inner product space V. Compute $\|\mathbf{u} + \mathbf{v} + \mathbf{w}\|^2$.

T.9. Let V be an inner product space. Show that if $\mathbf{v}$ is orthogonal to $\mathbf{w}_1, \mathbf{w}_2, \ldots, \mathbf{w}_k$, then $\mathbf{v}$ is orthogonal to every vector in span $\{\mathbf{w}_1, \mathbf{w}_2, \ldots, \mathbf{w}_k\}$.

B.2 ▼ Composite and Invertible Linear Transformations

We have already seen that nonsingular matrices are important and lead to many useful results. In this section we examine the analogous notion for linear transformations.

Composite Linear Transformations

DEFINITION   Let V_1 be an n-dimensional vector space, V_2 an m-dimensional vector space, and V_3 a p-dimensional vector space. Let $L_1 : V_1 \to V_2$ and $L_2 : V_2 \to V_3$ be linear transformations. The function $L_2 \circ L_1 : V_1 \to V_3$ defined by

$$(L_2 \circ L_1)(\mathbf{u}) = L_2(L_1(\mathbf{u}))$$

for $\mathbf{u}$ in V_1 is called the **composite of L_2 with L_1**. See Figure B.2.

FIGURE B.2

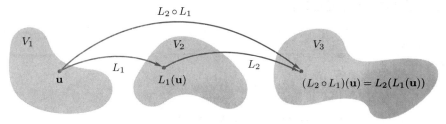

$L_2 \circ L_1$: The composite of L_2 with L_1.

If $V_1 = V_2 = V_3$, and $L_1 = L_2$, we write $L \circ L$ as L^2.

THEOREM B.1 ■ *Let $L_1: V_1 \to V_2$ and $L_2: V_2 \to V_3$ be linear transformations. Then*
$$L_2 \circ L_1: V_1 \to V_3$$
is a linear transformation.

Proof   Exercise T.1.   ■

EXAMPLE 1 ■ Let $L_1: R^2 \to R^3$ and $L_2: R^3 \to R^4$ be defined by
$$L_1\left(\begin{bmatrix} a_1 \\ a_2 \end{bmatrix}\right) = \begin{bmatrix} a_1 + a_2 \\ a_1 - a_2 \\ a_1 + 2a_2 \end{bmatrix}; \quad L_2\left(\begin{bmatrix} b_1 \\ b_2 \\ b_3 \end{bmatrix}\right) = \begin{bmatrix} b_1 + b_2 \\ b_1 - b_2 \\ b_2 + b_3 \\ 2b_1 + 3b_3 \end{bmatrix}.$$

Then $L_2 \circ L_1: R^2 \to R^4$ is given by
$$(L_2 \circ L_1)\left(\begin{bmatrix} a_1 \\ a_2 \end{bmatrix}\right) = L_2\left(L_1\left(\begin{bmatrix} a_1 \\ a_2 \end{bmatrix}\right)\right) = L_2\left(\begin{bmatrix} a_1 + a_2 \\ a_1 - a_2 \\ a_1 + 2a_2 \end{bmatrix}\right)$$
$$= \begin{bmatrix} (a_1 + a_2) + (a_1 - a_2) \\ (a_1 + a_2) - (a_1 - a_2) \\ (a_1 - a_2) + (a_1 + 2a_2) \\ 2(a_1 + a_2) + 3(a_1 + 2a_2) \end{bmatrix} = \begin{bmatrix} 2a_1 \\ 2a_2 \\ 2a_1 + a_2 \\ 5a_1 + 8a_2 \end{bmatrix}.$$
                                                                          ■

EXAMPLE 2 ■ Let $L_1: P_2 \to P_2$ and $L_2: P_2 \to P_2$ be defined by
$$L_1(at^2 + bt + c) = 2at + b$$
$$L_2(at^2 + bt + c) = 2at^2 + bt.$$

Compute

(a) $L_2 \circ L_1$.   (b) $L_1 \circ L_2$.

Solution   (a) We have
$$(L_2 \circ L_1)(at^2 + bt + c) = L_2(L_1(at^2 + bt + c))$$
$$= L_2(2at + b) = 7at.$$

(b) We have

$$(L_1 \circ L_2)(at^2 + bt + c) = L_1(L_2(at^2 + bt + c))$$
$$= L_1(7at^2 + bt) = 14at + b. \qquad \blacksquare$$

REMARK   Example 2 shows that in general, $L_2 \circ L_1 \neq L_1 \circ L_2$.

THEOREM B.2 ■ *Let V_1 be an n-dimensional vector space with basis P, V_2 an m-dimensional vector space with basis S, and V_3 a p-dimensional vector space with basis T. Let $L_1: V_1 \to V_2$ and $L_2: V_2 \to V_3$ be linear transformations. If A_1 represents L_1 with respect to P and S and A_2 represents L_2 with respect to S and T, then $A_2 A_1$ represents $L_2 \circ L_1$ with respect to P and T.*

Proof   From Theorem 6.8, it follows that if $\mathbf{x}$ is any vector in V_1 and $\mathbf{y}$ is any vector in V_2, then

$$\left[L_1(\mathbf{x})\right]_S = A_1 \left[\mathbf{x}\right]_P$$
$$\left[L_2(\mathbf{y})\right]_T = A_2 \left[\mathbf{y}\right]_S.$$

Then

$$\left[(L_2 \circ L_1)(\mathbf{x})\right]_T = \left[L_2(L_1(\mathbf{x}))\right]_T$$
$$= A_2 \left[L_1(\mathbf{x})\right]_S = A_2 \left(A_1 \left[\mathbf{x}\right]_P\right) = A_2 A_1 \left[\mathbf{x}\right]_P.$$

Since the matrix representing a given linear transformation with respect to two given bases is unique, we conclude that $A_2 A_1$ is the matrix representing $L_2 \circ L_1$ with respect to P and T. ■

REMARK   Since AB need not equal BA for matrices A and B, it is not surprising that $L_1 \circ L_2$ need not be the same linear transformation as $L_2 \circ L_1$, as we have just seen in Example 2.

EXAMPLE 3 ■ Let $L_1: R^2 \to R^2$ and $L_2: R^2 \to R^3$ be defined by

$$L_1\left(\begin{bmatrix} a_1 \\ a_2 \end{bmatrix}\right) = \begin{bmatrix} a_2 \\ a_1 \end{bmatrix}; \quad L_2\left(\begin{bmatrix} a_1 \\ a_2 \end{bmatrix}\right) = \begin{bmatrix} a_1 + a_2 \\ a_1 - a_2 \\ a_2 \end{bmatrix}.$$

The matrix representing L_1 with respect to the natural basis for R^2 is (verify)

$$A_1 = \begin{bmatrix} 0 & 1 \\ 1 & 0 \end{bmatrix}.$$

The matrix representing L_2 with respect to the natural bases for R^2 and R^3 is (verify)

$$A_2 = \begin{bmatrix} 1 & 1 \\ 1 & -1 \\ 0 & 1 \end{bmatrix}.$$

Then by Theorem 6.12, the matrix representing $L_2 \circ L_1 \colon R^2 \to R^3$ with respect to the natural bases for R^2 and R^3 is

$$A_2 A_1 = \begin{bmatrix} 1 & 1 \\ 1 & -1 \\ 0 & 1 \end{bmatrix} \begin{bmatrix} 0 & 1 \\ 1 & 0 \end{bmatrix} = \begin{bmatrix} 1 & 1 \\ -1 & 1 \\ 1 & 0 \end{bmatrix}.$$

Computing $L_2 \circ L_1$, we find that

$$(L_2 \circ L_1) \left(\begin{bmatrix} a_1 \\ a_2 \end{bmatrix} \right) = L_2 \left(L_1 \left(\begin{bmatrix} a_1 \\ a_2 \end{bmatrix} \right) \right)$$

$$= L_2 \left(\begin{bmatrix} a_2 \\ a_1 \end{bmatrix} \right) = \begin{bmatrix} a_2 + a_1 \\ a_2 - a_1 \\ a_1 \end{bmatrix} = \begin{bmatrix} a_1 + a_2 \\ -a_1 + a_2 \\ a_1 \end{bmatrix}.$$

We can then compute the matrix of $L_2 \circ L_1$ directly, obtaining (verify) the same answer as we got from $A_2 A_1$ above. ■

Invertible Linear Transformations

DEFINITION

A linear transformation $L \colon V \to W$ is called **invertible** if there exists a unique function $L^{-1} \colon W \to V$ such that $L^{-1} \circ L = I_V$, the identity linear operator on V, defined by $I_V(\mathbf{v}) = \mathbf{v}$, and $L \circ L^{-1} = I_W$, the identity linear operator on W, defined by $I_W(\mathbf{w}) = \mathbf{w}$. The function L^{-1} is called the **inverse** of L.

THEOREM B.3 ■ *A linear transformation $L \colon V \to W$ is invertible if and only if L is one-to-one and onto. Moreover, L^{-1} is a linear transformation and $(L^{-1})^{-1} = L$.*

Proof

Let L be one-to-one and onto. We define a function $H \colon W \to V$ as follows. If $\mathbf{w}$ is in W, then since L is onto, $\mathbf{w} = L(\mathbf{v})$ for some $\mathbf{v}$ in V, and since L is one-to-one, $\mathbf{v}$ is unique. Let $H(\mathbf{w}) = \mathbf{v}$; H is a function and $L(H(\mathbf{w})) = L(\mathbf{v}) = \mathbf{w}$, so that $L \circ H = I_W$. Also, $H(L(\mathbf{v})) = H(\mathbf{w}) = \mathbf{v}$, so $H \circ L = I_V$. Thus H is an inverse of L. Now H is unique, for if $H_1 \colon W \to V$ is a function such that $L \circ H_1 = I_W$ and $H_1 \circ L = I_V$, then $L(H(\mathbf{w})) = \mathbf{w} = L(H_1(\mathbf{w}))$ for any $\mathbf{w}$ in W. Since L is one-to-one, we conclude that $H(\mathbf{w}) = H_1(\mathbf{w})$. Hence $H = H_1$. Thus $H = L^{-1}$ and L is invertible.

Conversely, let L be invertible; that is, $L \circ L^{-1} = I_W$ and $L^{-1} \circ L = I_V$. We show that L is one-to-one and onto. Suppose that $L(\mathbf{v}_1) = L(\mathbf{v}_2)$ for $\mathbf{v}_1$, $\mathbf{v}_2$ in V. Then $L^{-1}(L(\mathbf{v}_1)) = L^{-1}(L(\mathbf{v}_2))$, so $\mathbf{v}_1 = \mathbf{v}_2$, which means that L is one-to-one. Also, if $\mathbf{w}$ is a vector in W, then $L(L^{-1}(\mathbf{w})) = \mathbf{w}$, so if we let $L^{-1}(\mathbf{w}) = \mathbf{v}$, then $L(\mathbf{v}) = \mathbf{w}$. Thus L is onto.

We now show that L^{-1} is a linear transformation. Let $\mathbf{w}_1$, $\mathbf{w}_2$ be in W, where $L(\mathbf{v}_1) = \mathbf{w}_1$ and $L(\mathbf{v}_2) = \mathbf{w}_2$ for $\mathbf{v}_1$, $\mathbf{v}_2$ in V. Then since

$$L(a\mathbf{v}_1 + b\mathbf{v}_2) = aL(\mathbf{v}_1) + bL(\mathbf{v}_2) = a\mathbf{w}_1 + b\mathbf{w}_2 \quad \text{for } a, b \text{ real numbers,}$$

we have

$$L^{-1}(a\mathbf{w}_1 + b\mathbf{w}_2) = a\mathbf{v}_1 + b\mathbf{v}_2 = aL^{-1}(\mathbf{w}_1) + bL^{-1}(\mathbf{w}_2),$$

which implies (by Exercise T.3 in Section 6.1) that L^{-1} is a linear transformation.

Finally, since $L \circ L^{-1} = I_W$, $L^{-1} \circ L = I_V$, and inverses are unique, we conclude that $(L^{-1})^{-1} = L$. ∎

EXAMPLE 4 ∎ Let $L: R^4 \to R^2$ be the linear transformation defined in Example 5 of Section 6.2:

$$L\left(\begin{bmatrix} x \\ y \\ z \\ w \end{bmatrix}\right) = \begin{bmatrix} x + y \\ z + w \end{bmatrix}.$$

As we have seen in Example 5, $\ker L$ has dimension 2, so L is not one-to-one and thus is not invertible. ∎

EXAMPLE 5 ∎ Consider the linear operator $L: R^3 \to R^3$ defined by

$$L\left(\begin{bmatrix} a_1 \\ a_2 \\ a_3 \end{bmatrix}\right) = \begin{bmatrix} 1 & 1 & 1 \\ 2 & 2 & 1 \\ 0 & 1 & 1 \end{bmatrix} \begin{bmatrix} a_1 \\ a_2 \\ a_3 \end{bmatrix}.$$

Since $\ker L = \{\mathbf{0}\}$ (verify), L is one-to-one and by Corollary 6.2 it is also onto, so it is invertible. To obtain L^{-1}, we proceed as follows. Since $L^{-1}(\mathbf{w}) = \mathbf{v}$, we must solve $L(\mathbf{v}) = \mathbf{w}$ for $\mathbf{v}$. We have

$$L(\mathbf{v}) = L\left(\begin{bmatrix} a_1 \\ a_2 \\ a_3 \end{bmatrix}\right) = \begin{bmatrix} a_1 + a_2 + a_3 \\ 2a_1 + 2a_2 + a_3 \\ a_2 + a_3 \end{bmatrix} = \mathbf{w} = \begin{bmatrix} b_1 \\ b_2 \\ b_3 \end{bmatrix}.$$

We are then solving the linear system

$$
\begin{aligned}
a_1 + a_2 + a_3 &= b_1 \\
2a_1 + 2a_2 + a_3 &= b_2 \\
a_2 + a_3 &= b_3
\end{aligned}
$$

for a_1, a_2, and a_3. We find that (verify)

$$\begin{bmatrix} a_1 \\ a_2 \\ a_3 \end{bmatrix} = \mathbf{v} = L^{-1}(\mathbf{w}) = L^{-1}\left(\begin{bmatrix} b_1 \\ b_2 \\ b_3 \end{bmatrix}\right) = \begin{bmatrix} b_1 - b_3 \\ -2b_1 + b_2 + b_3 \\ 2b_1 - b_2 \end{bmatrix}.$$ ∎

REMARK In Example 5 it was rather straightforward to find $L^{-1}(\mathbf{w})$. In general, if $L: V \to W$ is an invertible linear transformation, it is not always so easy to find an expression for $L^{-1}(\mathbf{w})$ for $\mathbf{w}$ in W. In Example 6 we shall solve this problem quite readily by using the matrix representing L^{-1} with respect to a basis S for V.

THEOREM B.4 ∎ *Let $L: V \to V$ be an invertible linear operator and let A be a matrix representing L with respect to a basis S for V. Then A^{-1} is the matrix representing L^{-1} with respect to S.*

Proof   Let B be the matrix representing L^{-1} with respect to S. Since $L \circ L^{-1} = I_W$, the identity linear operator on W, the matrix representing $L \circ L^{-1}$ with respect to S is I_n (see Exercise T.2 in Section 6.3). From Theorem B.2, it follows that the matrix representing $L \circ L^{-1}$ with respect to S is AB. Thus

$$AB = I_n,$$

which implies (by Theorem 1.10 in Section 1.6) that $B = A^{-1}$. ∎

We can now complete our List of Nonsingular Equivalences.

List of Nonsingular Equivalences

The following statements are equivalent for an $n \times n$ matrix A.

1. A is nonsingular.
2. $A\mathbf{x} = \mathbf{0}$ has only the trivial solution.
3. A is row equivalent to I_n.
4. The linear system $A\mathbf{x} = \mathbf{b}$ has a unique solution for every $n \times 1$ matrix $\mathbf{b}$.
5. $\det(A) \neq 0$.
6. A has rank n.
7. A has nullity 0.
8. The rows of A form a linearly independent set of n vectors in R^n.
9. The columns of A form a linearly independent set of n vectors in R^n.
10. Zero is *not* an eigenvalue of A.
11. The linear operator $L: R^n \to R^n$ defined by $L(\mathbf{x}) = A\mathbf{x}$, for $\mathbf{x}$ in R^n, is one-to-one and onto.
12. The linear operator $L: R^n \to R^n$ defined by $L(\mathbf{x}) = A\mathbf{x}$, for $\mathbf{x}$ in R^n, is invertible.

EXAMPLE 6 ■ Let $L: P_2 \to P_2$ be the linear operator defined by

$$L(at^2 + bt + c) = 2at^2 + bt + c.$$

The matrix representing L with respect to the basis $\{t^2 + 1, t - 1, t\}$ for P_2 is (verify)

$$A = \begin{bmatrix} 2 & 0 & 0 \\ 1 & 1 & 0 \\ -1 & 0 & 1 \end{bmatrix}.$$

We then find (verify) that

$$A^{-1} = \begin{bmatrix} \frac{1}{2} & 0 & 0 \\ -\frac{1}{2} & 1 & 0 \\ \frac{1}{2} & 0 & 1 \end{bmatrix}$$

is the matrix of L^{-1} with respect to S.

We now use A^{-1} to obtain a formula for $L^{-1}(at^2 + bt + c)$ as follows. Since A^{-1} is the matrix of L^{-1} with respect to S, we have

$$\left[L^{-1}(at^2 + bt + c)\right]_S = A^{-1}\left[at^2 + bt + c\right]_S. \tag{1}$$

To compute $\left[at^2 + bt + c\right]_S$ we form

$$at^2 + bt + c = k_1(t^2 + 1) + k_2(t - 1) + k_3 t$$

and solve the resulting linear system for k_1, k_2, and k_3, obtaining (verify)

$$k_1 = a, \quad k_2 = a - c, \quad k_3 = b + c - a.$$

Thus

$$\left[at^2 + bt + c\right]_S = \begin{bmatrix} a \\ a - c \\ b + c - a \end{bmatrix}.$$

Substituting this coordinate vector into Equation (1) yields

$$\left[L^{-1}(at^2 + bt + c)\right]_S = \begin{bmatrix} \frac{1}{2} & 0 & 0 \\ -\frac{1}{2} & 1 & 0 \\ \frac{1}{2} & 0 & 1 \end{bmatrix} \begin{bmatrix} a \\ a - c \\ b + c - a \end{bmatrix} = \begin{bmatrix} \frac{1}{2}a \\ \frac{1}{2}a - c \\ b + c - \frac{1}{2}a \end{bmatrix}.$$

Then

$$\begin{aligned} L^{-1}(at^2 + bt + c) &= \tfrac{1}{2}a(t^2 + 1) + \left(\tfrac{1}{2}a - c\right)(t - 1) + \left(b + c - \tfrac{1}{2}a\right)t \\ &= \tfrac{1}{2}at^2 + bt + c. \end{aligned}$$

■

B.2   EXERCISES

1. Let $L_1 \colon R^2 \to R^3$ and $L_2 \colon R^3 \to R^3$ be defined by

$$L_1(x, y) = (x + y, x - y, 2x + y),$$
$$L_2(x, y, z) = (x + y + z, y + z, x + z).$$

Compute

(a) $(L_2 \circ L_1)(-1, 1)$.   (b) $(L_2 \circ L_1)(x, y)$.

2. Let $L_1 \colon R^2 \to R^2$ and $L_2 \colon R^2 \to R^3$ be defined by

$$L_1\left(\begin{bmatrix} x \\ y \end{bmatrix}\right) = \begin{bmatrix} x \\ 2y - x \end{bmatrix},$$

$$L_2\left(\begin{bmatrix} x \\ y \end{bmatrix}\right) = \begin{bmatrix} 3x - 2y \\ x + y \\ x - y \end{bmatrix}.$$

Compute

(a) $(L_2 \circ L_1)\left(\begin{bmatrix} 2 \\ -1 \end{bmatrix}\right)$.   (b) $(L_2 \circ L_1)\left(\begin{bmatrix} x \\ y \end{bmatrix}\right)$.

3. Let $L_1 \colon P_1 \to P_1$ and $L_2 \colon P_1 \to P_2$ be defined by

$$L_1(at + b) = 2at - b$$

and

$$L_2(at + b) = t(at + b).$$

Compute

(a) $(L_2 \circ L_1)(3t + 2)$.   (b) $(L_2 \circ L_1)(at + b)$.

4. Let $L_1 \colon P_2 \to P_2$ and $L_2 \colon P_2 \to P_2$ be defined by

$$L_1(at^2 + bt + c) = 2at + b$$

and

$$L_2(at^2 + bt + c) = at + c.$$

Compute

(a) $(L_2 \circ L_1)(2t^2 - 3t + 1)$.

(b) $(L_1 \circ L_2)(2t^2 - 3t + 1)$.

(c) $(L_2 \circ L_1)(at^2 + bt + c)$.

(d) $(L_1 \circ L_2)(at^2 + bt + c)$.

5. Let $L_1 \colon R^2 \to R^2$ and $L_2 \colon R^2 \to R^2$ be defined by

$$L_1(x, y) = (x + y, x - 2y)$$

and

$$L_2(x, y) = (y, x - y).$$

Compute

(a) $(L_2 \circ L_1)(1, 2)$.   (b) $(L_1 \circ L_2)(1, 2)$.

(c) $(L_2 \circ L_1)(x, y)$.   (d) $(L_1 \circ L_2)(x, y)$.

6. Let L_1 and L_2 be as defined in Exercise 2. Let

$$S = \{(1, 1), (0, 1)\}$$

and

$$T = \{(1, 0, 0), (0, 1, -1), (1, 1, 0)\}$$

be bases for R^2 and R^3, respectively.

(a) Compute the matrix B of $L_2 \circ L_1$ with respect to S and T.

(b) Compute the matrix A_1 of L_1 with respect to S and the matrix A_2 of L_2 with respect to S and T. Verify that $B = A_2 A_1$.

7. Repeat Exercise 6 with L_1 and L_2 as defined in Exercise 3, and let

$$S = \{t + 1, t - 1\} \quad \text{and} \quad T = \{t^2 + 1, t, t - 1\}$$

be bases for P_1 and P_2, respectively.

8. Let L_1 and L_2 be as defined in Exercise 5, and let $S = \{(1, -1), (0, 1)\}$ and $T = \{(1, 0), (2, 1)\}$ be bases for R^2. Compute the matrix of

(a) $L_2 \circ L_1$ with respect to S.

(b) $L_1 \circ L_2$ with respect to S.

(c) $L_2 \circ L_1$ with respect to S and T.

(d) $L_1 \circ L_2$ with respect to S and T.

9. Let $L_1 \colon R^2 \to R^2$ and $L_2 \colon R^2 \to R^2$ be linear transformations whose matrices with respect to bases S and T for R^2 are

$$A_1 = \begin{bmatrix} 1 & 2 \\ -1 & 3 \end{bmatrix} \quad \text{and} \quad A_2 = \begin{bmatrix} 0 & 1 \\ -2 & 3 \end{bmatrix}.$$

(a) Compute the matrix of $L_2 \circ L_1$ with respect to S and T.

(b) Compute the matrix of $L_1 \circ L_2$ with respect to S and T.

10. Let $L \colon R^3 \to R^3$ be defined by

$$L\left(\begin{bmatrix} 1 \\ 0 \\ 0 \end{bmatrix}\right) = \begin{bmatrix} 1 \\ 2 \\ 3 \end{bmatrix}, \quad L\left(\begin{bmatrix} 0 \\ 1 \\ 0 \end{bmatrix}\right) = \begin{bmatrix} 0 \\ 1 \\ 1 \end{bmatrix},$$

$$L\left(\begin{bmatrix} 0 \\ 0 \\ 1 \end{bmatrix}\right) = \begin{bmatrix} 1 \\ 1 \\ 0 \end{bmatrix}.$$

(a) Show that L is invertible.

(b) Find $L^{-1}\left(\begin{bmatrix} 2 \\ 3 \\ 4 \end{bmatrix}\right)$.

In Exercises 11 through 18, determine whether the given linear transformation is invertible. If it is, find its inverse.

11. $L \colon R^2 \to R^3$ defined by
$L(x, y) = (x + y, x - y, x + 2y)$.

12. $L \colon R^2 \to R^2$ defined by $L(x, y) = (x - y, x + 3y)$.

13. $L \colon R^3 \to R^3$ defined by

$$L\left(\begin{bmatrix} x \\ y \\ z \end{bmatrix}\right) = \begin{bmatrix} 1 & 0 & 1 \\ 0 & 1 & 1 \\ 1 & 0 & 2 \end{bmatrix} \begin{bmatrix} x \\ y \\ z \end{bmatrix}.$$

14. $L \colon R^2 \to R^2$ defined by $L(x, y) = (x - y, x - y)$.

15. $L \colon R^3 \to R^3$ defined by

$$L\left(\begin{bmatrix} x \\ y \\ z \end{bmatrix}\right) = \begin{bmatrix} 1 & 1 & 1 \\ 0 & 1 & 2 \\ -2 & -1 & 0 \end{bmatrix} \begin{bmatrix} x \\ y \\ z \end{bmatrix}.$$

16. $L \colon P_1 \to P_1$ defined by $L(at + b) = -bt + a$.

17. $L \colon P_2 \to P_2$ defined by
$L(at^2 + bt + c) = -at^2 + bt - c$.

18. $L \colon P_2 \to P_2$ defined by $L(at^2 + bt + c) = 2at^2 + bt$.

In Exercises 19 through 22, determine whether L is invertible from the given information. [Recall that nullity $L = \dim(\ker L)$ and rank $L = \dim(\text{range } L)$.]

19. $L \colon R^4 \to R^4$, rank $L = 4$.

20. $L \colon R^4 \to R^4$, nullity $L = 2$.

21. $L \colon P_2 \to P_2$, nullity $L = 1$.

22. $L \colon P_3 \to P_3$, rank $L = 4$.

23. Let $L \colon R^3 \to R^3$ be the linear transformation defined in Exercise 10. Find the matrix representing L^{-1} with respect to the natural basis for R^3.

24. Let $L: R^3 \to R^3$ be the linear transformation defined by $L(\mathbf{x}) = A\mathbf{x}$, where
$$A = \begin{bmatrix} 1 & 1 & 1 \\ 0 & 1 & 2 \\ 1 & 2 & 2 \end{bmatrix}.$$

(a) Show that L is invertible.

(b) Find the matrix representing L^{-1} with respect to the natural basis for R^3.

25. Let $L: R^3 \to R^3$ be the invertible linear transformation represented by
$$A = \begin{bmatrix} 2 & 0 & 4 \\ -1 & 1 & -2 \\ 2 & 3 & 3 \end{bmatrix}$$

with respect to a basis S for R^3. Find the matrix of L^{-1} with respect to S.

26. Let $L: P_1 \to P_1$ be the invertible linear transformation represented by

$$A = \begin{bmatrix} 2 & 3 \\ 1 & 2 \end{bmatrix}$$

with respect to a basis S for P_1. Find the matrix of L^{-1} with respect to S.

THEORETICAL EXERCISES

T.1. Prove Theorem B.1.

T.2. Let $L: V \to W$ be a linear transformation and let I_V and I_W be the identity linear transformations on V and W, respectively. Show that
$$L \circ I_V = L$$
$$I_W \circ L = L.$$

T.3. Let $L: V \to V$ be a linear operator and let O_V be the zero linear transformation on V. Show that
$$L \circ O_V = O_V$$
$$O_V \circ L = O_V$$

T.4. Let $L: V \to V$ be a linear operator whose matrix with respect to a basis S for V is A. Show that A^2 is the matrix of $L^2 = L \circ L$ with respect to S. Moreover, show that if k is a positive integer, then A^k is the matrix of $L^k = L \circ L \circ \cdots \circ L$ (k times) with respect to S.

T.5. Let $L_1: V \to V$ and $L_2: V \to V$ be invertible linear operators. Show that $L_2 \circ L_1$ is also invertible and that $(L_2 \circ L_1)^{-1} = L_1^{-1} \circ L_2^{-1}$.

T.6. Let $L: V \to V$ be an invertible linear operator and let c be a nonzero scalar. Show that cL is an invertible linear operator and that
$$(cL)^{-1} = \frac{1}{c} L^{-1}.$$

T.7. Let $L: M_{22} \to M_{22}$ be defined by $L(A) = A^T$. Is L invertible? If it is, find L^{-1}.

T.8. Let $L: M_{22} \to M_{22}$ be defined by $L(A) = BA$, where
$$B = \begin{bmatrix} 1 & 2 \\ -2 & -3 \end{bmatrix}.$$
Is L invertible? If it is, find L^{-1}.

T.9. Let $L: V \to V$ be a linear operator, where V is an n-dimensional vector space. Show that the following are equivalent:

(a) L is invertible.

(b) Rank $L = n$.
    [Recall that rank $L = \dim(\text{range } L)$.]

(c) Nullity $L = 0$.
    [Recall that nullity $L = \dim(\ker L)$.]

T.10. Let $L_1: V \to V$ and $L_2: V \to V$ be linear transformations on a vector space V. Show that
$$(L_1 + L_2)^2 = L_1^2 + 2L_1 \circ L_2 + L_2^2$$
if and only if $L_1 \circ L_2 = L_2 \circ L_1$.

T.11. Let V be an inner product space, and let $\mathbf{w}$ be a fixed vector in V. Let $L: V \to V$ be defined by $L(\mathbf{v}) = (\mathbf{v}, \mathbf{w})$ for $\mathbf{v}$ in V. Show that L is a linear transformation.

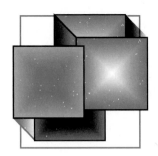

Answers to Odd-Numbered Exercises and Chapter Tests

Chapter 1

Section 1.1, page 9

1. $x = 4, y = 2$.

3. $x = -4, y = 2, z = 10$.

5. $x = 2, y = -1, z = -2$.

7. $x = -20, y = \frac{1}{4}z + 8, z = r$, where r is any real number.

9. No solution.   **11.** $x = 5, y = 1$.

13. No solution.

15. (a) $t = 10$.   (b) One value is $t = 3$.
   (c) The choice $t = 3$ in part (b) was arbitrary. Any choice for t, other than $t = 10$ makes the system inconsistent. Hence there are infinitely many ways to choose a value of t in part (b).

17. $x = 1, y = 1, z = 4$.   **19.** $r = -3$.

21. 20 tons of each type of fuel.

23. 3.2 ounces of A, 4.2 ounces of B, 2.0 ounces of C.

Section 1.2, page 16

1. (a) $-3, -5, 4$.   (b) $4, 5$.
   (c) $2, 6, -1$.

3. $a = 0, b = 2, c = 1, d = 2$.

5. (a) $\begin{bmatrix} 1 & 4 \\ 10 & 18 \end{bmatrix}$.

   (b) $3(2A) = 6A = \begin{bmatrix} 6 & 12 & 18 \\ 12 & 6 & 24 \end{bmatrix}$.

(c) $3A + 2A = 5A = \begin{bmatrix} 5 & 10 & 15 \\ 10 & 5 & 20 \end{bmatrix}$.

(d) $2(D + F) = 2D + 2F = \begin{bmatrix} -2 & 6 \\ 8 & 14 \end{bmatrix}$.

(e) $(2 + 3)D = 2D + 3D = \begin{bmatrix} 15 & -10 \\ 10 & 20 \end{bmatrix}$.

(f) Impossible.

7. (a) $\begin{bmatrix} 2 & 4 \\ 4 & 2 \\ 6 & 8 \end{bmatrix}$.   (b) Impossible.

(c) $\begin{bmatrix} 1 & -4 \\ 2 & 1 \\ 3 & -2 \end{bmatrix}$.

(d) Impossible.

(e) $(-A)^T = -(A^T) = \begin{bmatrix} -1 & -2 \\ -2 & -1 \\ -3 & -4 \end{bmatrix}$.

(f) Impossible.

ML.1. (a) Commands: $\mathbf{A(2,3)}$, $\mathbf{B(3,2)}$, $\mathbf{B(1,2)}$.
   (b) For $\mathrm{row}_1(\mathbf{A})$, use command $\mathbf{A(1,:)}$.
   For $\mathrm{col}_3(\mathbf{A})$, use command $\mathbf{A(:,3)}$.
   For $\mathrm{row}_2(\mathbf{B})$, use command $\mathbf{B(2,:)}$.
   (In this context the colon means "all.")
   (c) Matrix B in **format long** is

$$\begin{bmatrix} 8.00000000000000 & 0.666666666666667 \\ 0.00497512437811 & -3.200000000000000 \\ 0.00001000000000 & 4.333333333333333 \end{bmatrix}.$$

Section 1.3, page 30

1. (a) 2. (b) 1.

(c) 4. (d) 1.

3. ± 2.

5. (a) $\begin{bmatrix} 10 & -6 \\ 14 & -6 \end{bmatrix}$. (b) $\begin{bmatrix} 7 & 6 & -11 \\ 18 & 4 & -14 \\ 19 & -2 & -7 \end{bmatrix}$.

(c) Impossible. (d) $\begin{bmatrix} 26 & -9 \\ 4 & -5 \end{bmatrix}$.

(e) Impossible.

7. (a) 4. (b) 13. (c) 3. (d) 12.

9. $AB = \begin{bmatrix} -4 & 7 \\ 0 & 5 \end{bmatrix}$; $BA = \begin{bmatrix} -1 & 2 \\ 9 & 2 \end{bmatrix}$.

11. (a) $\begin{bmatrix} 6 \\ 25 \\ 10 \\ 25 \end{bmatrix}$. (b) $\begin{bmatrix} 12 \\ 11 \\ 17 \\ 20 \end{bmatrix}$.

13. $2\begin{bmatrix} 2 \\ -1 \\ 5 \end{bmatrix} + 1\begin{bmatrix} -3 \\ 2 \\ -1 \end{bmatrix} + 4\begin{bmatrix} 4 \\ 3 \\ -2 \end{bmatrix}$.

15. (a) $\begin{bmatrix} 2 & 0 & 0 & 1 \\ 3 & 2 & 3 & 0 \\ 2 & 3 & -4 & 0 \\ 1 & 0 & 3 & 0 \end{bmatrix}$.

(b) $\begin{bmatrix} 2 & 0 & 0 & 1 \\ 3 & 2 & 3 & 0 \\ 2 & 3 & -4 & 0 \\ 1 & 0 & 3 & 0 \end{bmatrix} \begin{bmatrix} x \\ y \\ z \\ w \end{bmatrix} = \begin{bmatrix} 7 \\ -2 \\ 3 \\ 5 \end{bmatrix}$.

(c) $\begin{bmatrix} 2 & 0 & 0 & 1 & \vdots & 7 \\ 3 & 2 & 3 & 0 & \vdots & -2 \\ 2 & 3 & -4 & 0 & \vdots & 3 \\ 1 & 0 & 3 & 0 & \vdots & 5 \end{bmatrix}$.

17. $2x \quad\ - 4z = \ \ 3$
$\qquad y + 2z = \ \ 5$
$\ x + 3y + 4z = -1$.

19. They are equivalent.

21. (a) $x\begin{bmatrix} 1 \\ 2 \end{bmatrix} + y\begin{bmatrix} 2 \\ -1 \end{bmatrix} = \begin{bmatrix} 3 \\ 5 \end{bmatrix}$.

(b) $x\begin{bmatrix} 2 \\ 1 \end{bmatrix} + y\begin{bmatrix} -3 \\ 4 \end{bmatrix} + z\begin{bmatrix} 5 \\ -1 \end{bmatrix} = \begin{bmatrix} -2 \\ 3 \end{bmatrix}$.

23. (a) $r = -5$. (b) BA^T.

25. $A + B = \begin{bmatrix} 4 & 5 & \vdots & 0 \\ 0 & 4 & \vdots & 1 \\ 6 & -2 & \vdots & 6 \end{bmatrix}$ is one possible answer.

27. AB tells the total cost of producing each kind of product in each city:

$$\begin{array}{c} \text{Salt Lake} \\ \text{City} \quad \text{Chicago} \\ \begin{bmatrix} 38 & 44 \\ 67 & 78 \end{bmatrix} \begin{array}{l} \text{Chair} \\ \text{Table} \end{array} \end{array}$$

29. (a) 2800 g. (b) 6000 g.

ML.1. (a) $\begin{bmatrix} 4.5000 & 2.2500 & 3.7500 \\ 1.5833 & 0.9167 & 1.5000 \\ 0.9667 & 0.5833 & 0.9500 \end{bmatrix}$.

(b) ??? Error using ==> *
Inner matrix dimensions must agree.

(c) $\begin{bmatrix} 5.0000 & 1.5000 \\ 1.5833 & 2.2500 \\ 2.4500 & 3.1667 \end{bmatrix}$.

(d) ??? Error using ==> *
Inner matrix dimensions must agree.

(e) ??? Error using ==> *
Inner matrix dimensions must agree.

(f) ??? Error using ==> —
Inner matrix dimensions must agree.

(g) $\begin{bmatrix} 18.2500 & 7.4583 & 12.2833 \\ 7.4583 & 5.7361 & 8.9208 \\ 12.2833 & 8.9208 & 14.1303 \end{bmatrix}$.

ML.3. $\begin{bmatrix} 4 & -3 & 2 & -1 & -5 \\ 2 & 1 & -3 & 0 & 7 \\ -1 & 4 & 1 & 2 & 8 \end{bmatrix}$.

ML.5. (a) $\begin{bmatrix} 1 & 0 & 0 & 0 \\ 0 & 2 & 0 & 0 \\ 0 & 0 & 3 & 0 \\ 0 & 0 & 0 & 4 \end{bmatrix}$.

(b) $\begin{bmatrix} 0 & 0 & 0 & 0 & 0 \\ 0 & 1.0000 & 0 & 0 & 0 \\ 0 & 0 & 0.5000 & 0 & 0 \\ 0 & 0 & 0 & 0.3333 & 0 \\ 0 & 0 & 0 & 0 & 0.2500 \end{bmatrix}$.

(c) $\begin{bmatrix} 5 & 0 & 0 & 0 & 0 & 0 \\ 0 & 5 & 0 & 0 & 0 & 0 \\ 0 & 0 & 5 & 0 & 0 & 0 \\ 0 & 0 & 0 & 5 & 0 & 0 \\ 0 & 0 & 0 & 0 & 5 & 0 \\ 0 & 0 & 0 & 0 & 0 & 5 \end{bmatrix}$.

Section 1.4, page 44

1. $A + B = \begin{bmatrix} 3 & 2 & -1 \\ 6 & 2 & 10 \end{bmatrix}$.

$A + B + C = \begin{bmatrix} -1 & -4 & 0 \\ 8 & 5 & 10 \end{bmatrix}$.

3. $A(B + C) = \begin{bmatrix} -10 & -8 & 16 \\ 10 & 14 & -28 \end{bmatrix}$.

5. $A(rB) = \begin{bmatrix} -6 & 18 & -42 \\ 9 & -27 & 0 \end{bmatrix}$.

7. $(AB)^T = \begin{bmatrix} 11 & 5 \\ 15 & -4 \end{bmatrix}$.

9. (a) $\begin{bmatrix} 2 & -62 \\ 25 & 33 \\ 30 & 15 \end{bmatrix}$.

(b) $\begin{bmatrix} 3 & -5 \\ 1 & -3 \\ -11 & -3 \end{bmatrix}$.

(c) $\begin{bmatrix} 6 & 10 & 16 \\ -9 & 7 & 18 \end{bmatrix}$.

(d) $\begin{bmatrix} -2 & 30 \\ -6 & 38 \\ -4 & -20 \end{bmatrix}$.

(e) $\begin{bmatrix} 1 & 11 & 28 \\ 7 & 17 & 30 \end{bmatrix}$.

11. $AB = AC = \begin{bmatrix} 8 & -6 \\ -8 & 6 \end{bmatrix}$.

13. (a) $\begin{bmatrix} 30 & 20 \\ 10 & 20 \end{bmatrix}$.   (b) $\begin{bmatrix} 247 & 206 \\ 103 & 144 \end{bmatrix}$.

15. $r = 3$.

17. $A^T A = \begin{bmatrix} \mathbf{a}_1^T \\ \mathbf{a}_2^T \\ \mathbf{a}_3^T \end{bmatrix} \begin{bmatrix} \mathbf{a}_1 & \mathbf{a}_2 & \mathbf{a}_3 \end{bmatrix}$

$= \begin{bmatrix} 25 & 14 & -3 \\ 14 & 29 & 2 \\ -3 & 2 & 1 \end{bmatrix}$.

19. (a) $\begin{bmatrix} \frac{4}{9} \\ \frac{5}{9} \end{bmatrix}$.   (b) $\begin{bmatrix} \frac{3}{7} \\ \frac{4}{7} \end{bmatrix}$.

ML.1. (a) $k = 3$.   (b) $k = 5$.

ML.3. (a) $\begin{bmatrix} 0 & -2 & 4 \\ 4 & 0 & -2 \\ -2 & 4 & 0 \end{bmatrix}$.   (b) $\begin{bmatrix} 0 & 0 & 0 \\ 0 & 0 & 0 \\ 0 & 0 & 0 \end{bmatrix}$.

ML.5. The sequence seems to be converging to

$$\begin{bmatrix} 1.0000 & 0.7500 \\ 0 & 0 \end{bmatrix}.$$

ML.7. (a) $A^T A = \begin{bmatrix} 2 & -3 & -1 \\ -3 & 9 & 2 \\ -1 & 2 & 6 \end{bmatrix}$,

$AA^T = \begin{bmatrix} 6 & -1 & -3 \\ -1 & 6 & 4 \\ -3 & 4 & 5 \end{bmatrix}$.

(b) $B = \begin{bmatrix} 2 & -3 & 1 \\ -3 & 2 & 4 \\ 1 & 4 & 2 \end{bmatrix}$,

$C = \begin{bmatrix} 0 & -1 & 1 \\ 1 & 0 & 0 \\ -1 & 0 & 0 \end{bmatrix}$.

(c) $B + C = \begin{bmatrix} 2 & -4 & 2 \\ -2 & 2 & 4 \\ 0 & 4 & 2 \end{bmatrix}$,

$B + C = 2A$.

Section 1.5, page 65

1. A, E, G.

3. (a) $\begin{bmatrix} 2 & 0 & 4 & 2 \\ -1 & 3 & 1 & 1 \\ 3 & -2 & 5 & 6 \end{bmatrix}$.

(b) $\begin{bmatrix} 2 & 0 & 4 & 2 \\ -12 & 8 & -20 & -24 \\ -1 & 3 & 1 & 1 \end{bmatrix}$.

(c) $\begin{bmatrix} 0 & 6 & 6 & 4 \\ 3 & -2 & 5 & 6 \\ -1 & 3 & 1 & 1 \end{bmatrix}$.

5. Possible answers:

(a) $\begin{bmatrix} 4 & 3 & 7 & 5 \\ 2 & 0 & 1 & 4 \\ -2 & 4 & -2 & 6 \end{bmatrix}$.

(b) $\begin{bmatrix} 3 & 5 & 6 & 8 \\ -4 & 8 & -4 & 12 \\ 2 & 0 & 1 & 4 \end{bmatrix}$.

(c) $\begin{bmatrix} 4 & 3 & 7 & 5 \\ -1 & 2 & -1 & 3 \\ 0 & 4 & -1 & 10 \end{bmatrix}$.

7. $\begin{bmatrix} 1 & 0 & 0 & 0 \\ 0 & 1 & 0 & 0 \\ 0 & 0 & 1 & 0 \\ 0 & 0 & 0 & 1 \end{bmatrix}.$

9. (a) $x = -2 + r$, $y = -1$, $z = 8 - 2r$, $w = r$,
   $r = $ any real number.
   (b) $x = 1$, $y = \frac{2}{3}$, $z = -\frac{2}{3}$.
   (c) No solution.

11. (a) $a = -2$.   (b) $a \neq \pm 2$.   (c) $a = 2$.

13. (a) $a = \pm\sqrt{6}$.   (b) $a \neq \pm\sqrt{6}$.   (c) None.

15. (a) $x = -1$, $y = 4$, $z = -3$.
   (b) $x = 0$, $y = 0$, $z = 0$.

17. (a) $x = 1 - r$, $y = 2$, $z = 1$, $w = r$,
   $r = $ any real number.
   (b) No solution.

19. $x = -r$, $y = 0$, $z = r$, $r = $ any real number.

21. $-3a - b + c = 0$.

23. $\mathbf{x} = \begin{bmatrix} r \\ 0 \end{bmatrix}$, where $r \neq 0$.

25. $\mathbf{x} = \begin{bmatrix} -\frac{1}{4}r \\ \frac{1}{4}r \\ r \end{bmatrix}$, where $r \neq 0$.

27. $y = \frac{1}{2}x^2 - \frac{3}{2}x + 3$.

29. $y = \frac{11}{6}x^3 - 2x^2 + \frac{7}{6}x - 1$.

31. 30 chairs, 30 coffee tables, and 20 dining-room tables.

ML.1. (a) $\begin{bmatrix} 1.0000 & 0.5000 & 0.5000 \\ -3.0000 & 1.0000 & 4.0000 \\ 1.0000 & 0 & 3.0000 \\ 5.0000 & -1.0000 & 5.0000 \end{bmatrix}.$

   (b) $\begin{bmatrix} 1.0000 & 0.5000 & 0.5000 \\ 0 & 2.5000 & 5.5000 \\ 1.0000 & 0 & 3.0000 \\ 5.0000 & -1.0000 & 5.0000 \end{bmatrix}.$

   (c) $\begin{bmatrix} 1.0000 & 0.5000 & 0.5000 \\ 0 & 2.5000 & 5.5000 \\ 0 & -0.5000 & 2.5000 \\ 5.0000 & -1.0000 & 5.0000 \end{bmatrix}.$

   (d) $\begin{bmatrix} 1.0000 & 0.5000 & 0.5000 \\ 0 & 2.5000 & 5.5000 \\ 0 & -0.5000 & 2.5000 \\ 0 & -3.5000 & 2.5000 \end{bmatrix}.$

   (e) $\begin{bmatrix} 1.0000 & 0.5000 & 0.5000 \\ 0 & -3.5000 & 2.5000 \\ 0 & -0.5000 & 2.5000 \\ 0 & 2.5000 & 5.5000 \end{bmatrix}.$

ML.3. $\begin{bmatrix} 1 & 0 & 0 \\ 0 & 1 & 0 \\ 0 & 0 & 1 \\ 0 & 0 & 0 \end{bmatrix}.$

ML.5. $x = -2 + r$, $y = -1$, $z = 8 - 2r$, $w = r$,
   $r = $ any real number.

ML.7. Only the trivial solution.

ML.9. $\mathbf{x} = \begin{bmatrix} 0.5r \\ r \end{bmatrix}.$

ML.11. Exercise 15:
   (a) Unique solution: $x = -1$, $y = 4$, $z = -3$.
   (b) The only solution is the trivial one.

   Exercise 16:
   (a) $x = r$, $y = -2r$, $z = r$, where r is any real number.
   (b) Unique solution: $x = 1$, $y = 2$, $z = 2$.

ML.13. The \ command yields a matrix showing that the system is inconsistent. The **rref** command leads to the display of a warning that the result may contain large roundoff errors.

Section 1.6, page 83

1. $A^{-1} = \begin{bmatrix} \frac{3}{8} & -\frac{1}{8} \\ \frac{1}{4} & \frac{1}{4} \end{bmatrix}.$

3. Nonsingular: $A^{-1} = \begin{bmatrix} 4 & -1 \\ -3 & 1 \end{bmatrix}.$

5. (a) $\begin{bmatrix} \frac{1}{2} & -\frac{1}{4} \\ \frac{1}{6} & \frac{1}{12} \end{bmatrix}.$ (b) $\begin{bmatrix} 0 & 1 & -1 \\ 2 & -2 & -1 \\ -1 & 1 & 1 \end{bmatrix}.$

   (c) $\begin{bmatrix} \frac{7}{3} & -\frac{1}{3} & -\frac{1}{3} & -\frac{2}{3} \\ \frac{4}{9} & -\frac{1}{9} & -\frac{4}{9} & \frac{1}{9} \\ -\frac{1}{9} & -\frac{2}{9} & \frac{1}{9} & \frac{2}{9} \\ -\frac{5}{3} & \frac{2}{3} & \frac{2}{3} & \frac{1}{3} \end{bmatrix}.$

7. (a) $\begin{bmatrix} -2 & \frac{3}{2} \\ 1 & -\frac{1}{2} \end{bmatrix}.$ (b) Singular.

(c) $\begin{bmatrix} \frac{3}{2} & -1 & \frac{1}{2} \\ \frac{1}{2} & 0 & -\frac{1}{2} \\ -\frac{3}{2} & 1 & \frac{1}{2} \end{bmatrix}$.

9. (a) Singular. (b) $\begin{bmatrix} 1 & -1 & 0 \\ 1 & -2 & 1 \\ -\frac{3}{2} & \frac{5}{2} & -\frac{1}{2} \end{bmatrix}$.

(c) $\begin{bmatrix} -1 & \frac{3}{2} & \frac{1}{2} \\ 1 & -\frac{3}{2} & \frac{1}{2} \\ 0 & \frac{1}{2} & -\frac{1}{2} \end{bmatrix}$.

11. (a) and (b). **13.** $\begin{bmatrix} \frac{4}{5} & -\frac{3}{5} \\ -\frac{1}{5} & \frac{2}{5} \end{bmatrix}$.

17. (a) $\begin{bmatrix} -30 \\ 60 \\ 10 \end{bmatrix}$. (b) $\begin{bmatrix} 23 \\ -31 \\ -1 \end{bmatrix}$. **19.** Yes.

21. $\lambda = -1$, $\lambda = 3$.

23. $\begin{bmatrix} \frac{1}{4} & 0 & 0 \\ 0 & -\frac{1}{2} & 0 \\ 0 & 0 & \frac{1}{3} \end{bmatrix}$. **25.** $\mathbf{x} = \begin{bmatrix} 19 \\ 23 \end{bmatrix}$.

ML.1. (a) and (c).

ML.3. (a) $\begin{bmatrix} -2 & 3 \\ 1 & -1 \end{bmatrix}$.

(b) $\begin{bmatrix} -\frac{1}{4} & \frac{3}{4} & -\frac{1}{4} \\ -\frac{1}{4} & -\frac{1}{4} & \frac{3}{4} \\ \frac{3}{4} & -\frac{1}{4} & -\frac{1}{4} \end{bmatrix}$.

ML.5. (a) $t = 4$. (b) $t = 3$.

Supplementary Exercises, page 86

1. $\begin{bmatrix} -1 & -3 \\ 26 & 6 \end{bmatrix}$. **3.** $\begin{bmatrix} 19 & 10 \\ -6 & 1 \end{bmatrix}$.

5. (a) $\begin{bmatrix} 1 & 2 & -1 & 1 & \vdots & 7 \\ 2 & -1 & 0 & 2 & \vdots & -8 \end{bmatrix}$.

(b) $3x + 2y = -4$
$5x + y = 2$
$3x + 2y = 6$.

7. $x = 1$, $y = 2$, $z = -2$. **9.** (a) $a = -3$.
(b) $a \neq \pm 3$.
(c) $a = 3$.

11. $x = -3r$, $y = r$, $z = 0$, $r = $ any real number.

13. $\begin{bmatrix} -40 & 16 & 9 \\ 13 & -5 & -3 \\ 5 & -2 & -1 \end{bmatrix}$.

15. Yes. **17.** $\mathbf{x} = \begin{bmatrix} 4 \\ 1 \\ 4 \end{bmatrix}$.

21. (a) $a \neq 15$. (b) None. (c) $a = 15$.

23. $a = 1, -1$.

25. (a) $k = 1$; $B = \begin{bmatrix} b_1 \\ 0 \end{bmatrix}$.

$k = 2$; $B = \begin{bmatrix} b_{11} & b_{12} \\ 0 & 0 \end{bmatrix}$.

$k = 3$; $B = \begin{bmatrix} b_{11} & b_{12} & b_{13} \\ 0 & 0 & 0 \end{bmatrix}$.

$k = 4$; $B = \begin{bmatrix} b_{11} & b_{12} & b_{13} & b_{14} \\ 0 & 0 & 0 & 0 \end{bmatrix}$.

(b) The answers are not unique. The only requirement is that row 2 of B have all zero entries.

27. (a) $\begin{bmatrix} 1 & \frac{1}{2} \\ 0 & 1 \end{bmatrix}$. (b) $\begin{bmatrix} 1 & 0 & 0 \\ 0 & 0 & 0 \\ 0 & 0 & 0 \end{bmatrix} = B$.

(c) I_4.

29. $A^2 = \begin{bmatrix} 1 & \frac{3}{4} \\ 0 & \frac{1}{4} \end{bmatrix}$, $A^3 = \begin{bmatrix} 1 & \frac{7}{8} \\ 0 & \frac{1}{8} \end{bmatrix}$,

$A^4 = \begin{bmatrix} 1 & \frac{15}{16} \\ 0 & \frac{1}{16} \end{bmatrix}$, $A^5 = \begin{bmatrix} 1 & \frac{31}{32} \\ 0 & \frac{1}{32} \end{bmatrix}$.

It appears that $A^n = \begin{bmatrix} 1 & (2^n - 1)/2^n \\ 0 & 1/2^n \end{bmatrix}$.

Chapter Test, page 89

1. No solution.

2. (a) $a = 2, 3$. (b) $a \neq 2, 3$. (c) None.

3. $\begin{bmatrix} -\frac{1}{2} & 1 & \frac{3}{2} \\ \frac{1}{2} & 0 & -\frac{1}{2} \\ -\frac{1}{2} & 1 & \frac{1}{2} \end{bmatrix}$. **4.** $-2, 3$.

5. (a) $\begin{bmatrix} 3 & 6 & 5 \\ -2 & 2 & -8 \\ 0 & 5 & -3 \end{bmatrix}$.   (b) $\mathbf{x} = \begin{bmatrix} -4 \\ 14 \\ 25 \end{bmatrix}$.

6. (a) F. (b) T. (c) F. (d) T. (e) F.

Chapter 2

Section 2.1, page 100

1. (a) 5. (b) 7. (c) 4. (d) 4. (e) 7.   (f) 0.

3. (a) $-$. (b) $+$. (c) $-$. (d) $-$. (e) $+$.   (f) $+$.

5. (a) 7. (b) 30. (c) -24. (d) 4.

7. There are 24 terms.

9. $|B| = 3$, $|C| = 9$, $|D| = -3$.

11. (a) $\lambda^2 - 3\lambda - 4$.   (b) $\lambda^2 - 5\lambda + 6$.

13. (a) $-1, 4$.   (b) 2, 3.

15. (a) 72. (b) 0. (c) -24.

17. (a) -30. (b) 0. (c) 6.   **21.** $-\frac{3}{2}$.

ML.1. (a) -18. (b) 5.

ML.3. (a) 4. (b) 0.

ML.5. $t = 3$, $t = 4$.

Section 2.2, page 115

1. $A_{11} = -11$, $A_{12} = 29$, $A_{13} = 1$,
$A_{21} = -4$, $A_{22} = 7$, $A_{23} = -2$,
$A_{31} = 2$, $A_{32} = -10$, $A_{33} = 1$.

3. (a) -43. (b) 75. (c) 0.

5. (a) 0. (b) -6. (c) -36.

9. (a) $\begin{bmatrix} 24 & -42 & -30 \\ 19 & -2 & -30 \\ -4 & 32 & 30 \end{bmatrix}$. (b) 150.

11. (a) Singular. (b) $\begin{bmatrix} \frac{2}{7} & -\frac{3}{7} \\ \frac{1}{7} & \frac{2}{7} \end{bmatrix}$.

(c) $\begin{bmatrix} \frac{1}{4} & -\frac{1}{20} & \frac{3}{20} \\ 0 & \frac{1}{5} & \frac{2}{5} \\ 0 & \frac{1}{10} & -\frac{3}{10} \end{bmatrix}$.

13. (a) $\begin{bmatrix} 0 & \frac{1}{2} \\ 1 & \frac{3}{2} \end{bmatrix}$. (b) $\begin{bmatrix} \frac{1}{4} & 0 & 0 \\ 0 & -\frac{1}{3} & 0 \\ 0 & 0 & \frac{1}{2} \end{bmatrix}$.

(c) $\begin{bmatrix} \frac{15}{14} & \frac{5}{28} & -\frac{9}{28} & -\frac{23}{14} \\ \frac{8}{7} & -\frac{1}{7} & -\frac{1}{7} & -\frac{9}{7} \\ \frac{3}{7} & \frac{1}{14} & \frac{1}{14} & -\frac{6}{7} \\ -\frac{4}{7} & \frac{1}{14} & \frac{1}{14} & \frac{8}{7} \end{bmatrix}$.

15. (d) is nonsingular.

17. (a) 1, 4.   (b) $-5, 0, 3$.

19. (a) Has only the trivial solution.
(b) Has nontrivial solutions.

21. $x = 1$, $y = -1$, $z = 0$, $w = 2$.

23. No solution.

ML.1. $A_{11} = -11$, $A_{23} = -2$, $A_{31} = 2$.

ML.3. 0.

ML.5. (a) The matrix is singular.

(b) $\begin{bmatrix} \frac{2}{7} & -\frac{3}{7} \\ \frac{1}{7} & \frac{2}{7} \end{bmatrix}$.

(c) $\begin{bmatrix} \frac{1}{4} & -\frac{1}{20} & \frac{3}{20} \\ 0 & \frac{1}{5} & \frac{2}{5} \\ 0 & \frac{1}{10} & -\frac{3}{10} \end{bmatrix}$.

Supplementary Exercises, page 119

1. (a) -24. (b) 24.

3. (a) $\frac{1}{5}$. (b) 80. (c) $\frac{16}{5}$. (d) $\frac{1}{80}$.

5. $0, -1, -4$.   **7.** 172.

9. -218.   **11.** $\frac{1}{5}\begin{bmatrix} 0 & 5 & -5 \\ -2 & 7 & -4 \\ 1 & -1 & 2 \end{bmatrix}$.

13. $\lambda \neq -1, 0, 1$.   **17.** $a \neq 0$, $a \neq 2$.

Chapter Test, page 121

1. 17.   **2.** (a) 54. (b) $\frac{27}{2}$. (c) $\frac{1}{54}$.

3. $\frac{14}{3}$.   **4.** $-3, 0, 3$.

5. $x = 1$, $y = 0$, $z = -2$.

6. (a) T. (b) F.   (c) F.   (d) F.   (e) T.
(f) T.   (g) T.   (h) T.   (i) F.   (j) T.

Chapter 3

Section 3.1, page 139

1.

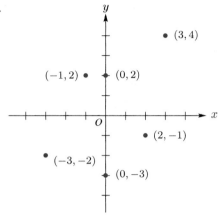

3. (a) $(1, 7)$.

5. (a) $\mathbf{u} + \mathbf{v} = (0, 8)$, $\mathbf{u} - \mathbf{v} = (4, -2)$,
$2\mathbf{u} = (4, 6)$, $3\mathbf{u} - 2\mathbf{v} = (10, -1)$.

(b) $\mathbf{u} + \mathbf{v} = (3, 5)$, $\mathbf{u} - \mathbf{v} = (-3, 1)$,
$2\mathbf{u} = (0, 6)$, $3\mathbf{u} - 2\mathbf{v} = (-6, 5)$.

(c) $\mathbf{u} + \mathbf{v} = (5, 8)$, $\mathbf{u} - \mathbf{v} = (-1, 4)$,
$2\mathbf{u} = (4, 12)$, $3\mathbf{u} - 2\mathbf{v} = (0, 14)$.

7. (a) $\mathbf{w}_1 = 2$.   (b) $\mathbf{x}_2 = \frac{8}{3}$.

(c) $\mathbf{w}_1 = 3$, $\mathbf{x}_2 = -2$.

9. (a) $\sqrt{5}$.   (b) 5.   (c) 2.   (d) 5.

11. (a) $\sqrt{2}$.   (b) 5.   (c) $\sqrt{10}$.   (d) $\sqrt{13}$.

13. $(-5, 6) = 19(1, 2) - 8(3, 4)$.

15. 6.   **17.** 6.

19. (a) $\left(\frac{3}{5}, \frac{4}{5}\right)$.   (b) $\left(-\frac{2}{\sqrt{13}}, -\frac{3}{\sqrt{13}}\right)$.

(c) $(1, 0)$.

21. (a) $\dfrac{-4}{\sqrt{5} \cdot \sqrt{13}}$.   (b) 0.

(c) 0.   (d) -1.

25. $a = \frac{8}{5}$.

27. (a) $\mathbf{i} + 3\mathbf{j}$.   (b) $-2\mathbf{i} - 3\mathbf{j}$.

(c) $-2\mathbf{i}$.   (d) $3\mathbf{j}$.

29.

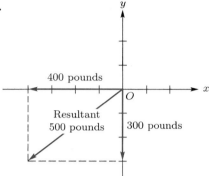

Section 3.2, page 155

1. (a) $\mathbf{u} + \mathbf{v} = (1, 3, -5)$, $\mathbf{u} - \mathbf{v} = (1, 1, -1)$,
$2\mathbf{u} = (2, 4, -6)$, $3\mathbf{u} - 2\mathbf{v} = (3, 4, -5)$.

(b) $\mathbf{u} + \mathbf{v} = (3, 0, 6, -1)$, $\mathbf{u} - \mathbf{v} = (5, -4, -4, 7)$,
$2\mathbf{u} = (8, -4, 2, 6)$, $3\mathbf{u} - 2\mathbf{v} = (14, -10, -7, 17)$.

3. (a) $a = \frac{1}{2}$, $b = \frac{3}{2}$.   (b) $a = 4$, $b = 0$.

(c) $a = -6$, $b = 1$, $c = 0$.

7.

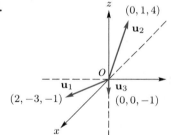

9. $(4, 2, 2)$.

11. (a) $\sqrt{29}$.   (b) $\sqrt{14}$.   (c) $\sqrt{5}$.   (d) $\sqrt{30}$.

13. (a) $\sqrt{18}$.   (b) $\sqrt{6}$.   (c) $\sqrt{50}$.   (d) $\sqrt{10}$.

15. Possible answer: $c_1 = -2$, $c_2 = -1$, $c_3 = 1$.

17. $a = 4$ or $a = -1$.

21. (a) 0.   (b) $-\dfrac{1}{\sqrt{3}}$.   (c) $\dfrac{1}{\sqrt{5}}$.   (d) 0.

23. (a) $\mathbf{u}_1$ and $\mathbf{u}_2$, $\mathbf{u}_1$ and $\mathbf{u}_6$, $\mathbf{u}_2$ and $\mathbf{u}_3$,
$\mathbf{u}_3$ and $\mathbf{u}_6$, $\mathbf{u}_4$ and $\mathbf{u}_6$.

(b) $\mathbf{u}_1$ and $\mathbf{u}_3$.   (c) None.

25. Possible answer: $a = 1$, $b = 0$, $c = -1$.

27. (a) $\left(\frac{2}{\sqrt{14}}, -\frac{1}{\sqrt{14}}, \frac{3}{\sqrt{14}}\right)$.

(b) $\left(\frac{1}{\sqrt{30}}, \frac{2}{\sqrt{30}}, \frac{3}{\sqrt{30}}, \frac{4}{\sqrt{30}}\right)$.

(c) $\left(0, \frac{1}{\sqrt{2}}, -\frac{1}{\sqrt{2}}\right)$.

(d) $\left(0, -\frac{1}{\sqrt{6}}, \frac{2}{\sqrt{6}}, -\frac{1}{\sqrt{6}}\right)$.

29. (a) $\mathbf{i} + 2\mathbf{j} - 3\mathbf{k}$. (b) $2\mathbf{i} + 3\mathbf{j} - \mathbf{k}$.

(c) $\mathbf{j} + 2\mathbf{k}$. (d) $-2\mathbf{k}$.

33. $1.08\mathbf{u}$.   **35.** $\frac{1}{2}(\mathbf{t} + \mathbf{b})$.

ML.3. (a) 2.2361. (b) 5.4772. (c) 3.1623.

ML.5. (a) 19. (b) -11. (c) -55.

ML.9. (a) $\begin{bmatrix} 0.6667 \\ 0.6667 \\ -0.3333 \end{bmatrix}$ or in rational form $\begin{bmatrix} \frac{2}{3} \\ \frac{2}{3} \\ -\frac{1}{3} \end{bmatrix}$.

(b) $\begin{bmatrix} 0 \\ 0.8000 \\ -0.6000 \\ 0 \end{bmatrix}$ or in rational form $\begin{bmatrix} 0 \\ \frac{4}{5} \\ -\frac{3}{5} \\ 0 \end{bmatrix}$.

(c) $\begin{bmatrix} 0.3015 \\ 0 \\ 0.3015 \\ 0 \end{bmatrix}$.

Section 3.3, page 169

1. (b).

3. (a).

5.

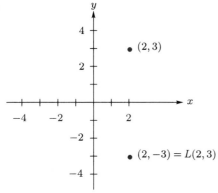

7.

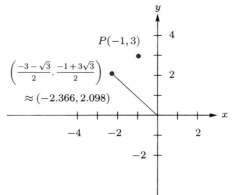

9.

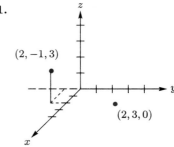

11.

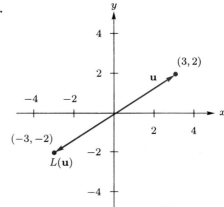

13. (a) Yes. (b) Yes.

15. $c - a + b = 0$.   **17.** $\begin{bmatrix} 11 \\ 6 \end{bmatrix}$.

19. $\mathbf{x} = \begin{bmatrix} 0 \\ 0 \\ r \end{bmatrix}$, where r is any real number.

21. (a) Reflection about the y-axis;

(b) Reflection about the origin;

(c) Rotate counterclockwise through $\frac{\pi}{2}$.

23. $\begin{bmatrix} -1 & 0 \\ 0 & 1 \end{bmatrix}$.　　**25.** $\begin{bmatrix} \frac{\sqrt{2}}{2} & -\frac{\sqrt{2}}{2} \\ \frac{\sqrt{2}}{2} & \frac{\sqrt{2}}{2} \end{bmatrix}$.

27. $\begin{bmatrix} 1 & -1 & 0 \\ 1 & 0 & 1 \\ 0 & 1 & -1 \end{bmatrix}$.

29. (a) 71 52 33 47 30 26 84 56 43 99 69 55.

　　(b) Message: CERTAINLY NOT.

Section 3.4, page 177

1. (a) $\begin{bmatrix} -1 & 0 \\ 0 & 1 \end{bmatrix}$. (b) $\begin{bmatrix} 0 & 1 \\ 1 & 0 \end{bmatrix}$.

　　(c) $\begin{bmatrix} -\frac{\sqrt{2}}{2} & -\frac{\sqrt{2}}{2} \\ \frac{\sqrt{2}}{2} & -\frac{\sqrt{2}}{2} \end{bmatrix}$. (d) $\begin{bmatrix} \frac{\sqrt{3}}{2} & \frac{1}{2} \\ -\frac{1}{2} & \frac{\sqrt{3}}{2} \end{bmatrix}$.

3. (a) $\begin{bmatrix} 1 & 0 \\ k & 1 \end{bmatrix}$.

　　(b)

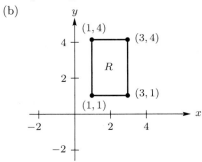

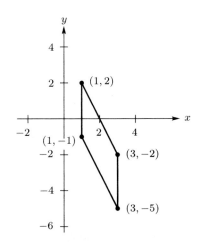

5. (a) $\begin{bmatrix} k & 0 \\ 0 & 1 \end{bmatrix}$.

　　(b)

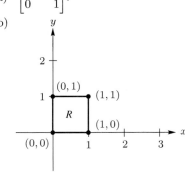

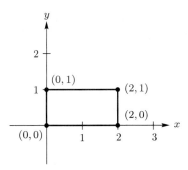

7. (a) 8.　(b) $L\left(\begin{bmatrix} 5 \\ 0 \end{bmatrix}\right) = \begin{bmatrix} -10 \\ 15 \end{bmatrix}$,

　　$L\left(\begin{bmatrix} 0 \\ 3 \end{bmatrix}\right) = \begin{bmatrix} 3 \\ 12 \end{bmatrix}$, $L\left(\begin{bmatrix} -2 \\ 1 \end{bmatrix}\right) = \begin{bmatrix} 5 \\ -2 \end{bmatrix}$.

　　(d) 88.

9.

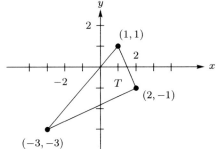

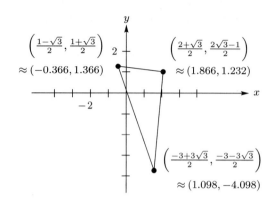

$\left(\frac{1-\sqrt{3}}{2}, \frac{1+\sqrt{3}}{2}\right)$
$\approx (-0.366, 1.366)$

$\left(\frac{2+\sqrt{3}}{2}, \frac{2\sqrt{3}-1}{2}\right)$
$\approx (1.866, 1.232)$

$\left(\frac{-3+3\sqrt{3}}{2}, \frac{-3-3\sqrt{3}}{2}\right)$
$\approx (1.098, -4.098)$

Section 3.5, page 185

1. (a) $-15\mathbf{i} - 2\mathbf{j} + 9\mathbf{k}$. (b) $-3\mathbf{i} + 3\mathbf{j} + 3\mathbf{k}$.
   (c) $7\mathbf{i} + 5\mathbf{j} - \mathbf{k}$. (d) $0\mathbf{i} + 0\mathbf{j} + 0\mathbf{k}$.

9. $\frac{1}{2}\sqrt{478}$. 11. $\sqrt{150}$. 13. 39.

ML.1. (a) $\begin{bmatrix} -11 & 2 & 5 \end{bmatrix}$. (b) $\begin{bmatrix} 3 & 1 & -1 \end{bmatrix}$.
   (c) $\begin{bmatrix} 1 & -8 & -5 \end{bmatrix}$.

ML.5. 8.

Section 3.6, page 192

1. (a) $-7x + 5y + 1 = 0$. (b) $9x + 5y + 7 = 0$.
   (c) $-5x - 3y = 0$. (d) $-7x + 3y - 6 = 0$.

3. (d).

5. (a) $x = 3 + 4t, y = 4 - 5t, z = -2 + 2t, \infty < t < \infty$.
   (b) $x = 3 - 2t, y = 2 + 5t, z = 4 + t, -\infty < t < \infty$.
   (c) $x = t, y = t, z = t, -\infty < t < \infty$.
   (d) $x = -2 + 2t, y = -3 + 3t, z = 1 + 4t,$
   $-\infty < t < \infty$.

7. (a) $\dfrac{x-2}{2} = \dfrac{y+3}{5} = \dfrac{z-1}{4}$.
   (b) $\dfrac{x+3}{8} = \dfrac{y+2}{7} = \dfrac{z+2}{6}$.
   (c) $\dfrac{x+2}{4} = \dfrac{y-3}{-6} = \dfrac{z-4}{1}$.
   (d) $\dfrac{x}{4} = \dfrac{y}{5} = \dfrac{z}{2}$.

9. (a) $3x - 2y + 4z + 16 = 0$.
   (b) $y - 3z + 3 = 0$.
   (c) $-z + 4 = 0$.
   (d) $-x - 2y + 4z - 3 = 0$.

11. (a) $x = \frac{8}{13} + 23t, y = -\frac{27}{13} + 2t, z = 13t,$
    $-\infty < t < \infty$.

(b) $x = -\frac{28}{13} + 7t, y = -\frac{16}{13} - 22t, z = 13t,$
   $-\infty < t < \infty$.

(c) $x = -\frac{16}{3} + 5t, y = -\frac{8}{3} + 4t, z = -3t,$
   $-\infty < t < \infty$.

13. Yes. 15. $(5, 1, 2)$.

19. $4x - 4y + z + 16 = 0$.

21. $7x + 2y - 2z - 19 = 0$.

23. $x = -2 + 2t, y = 5 - 3t, z = -3 + 4t$.

Supplementary Exercises, page 194

1.

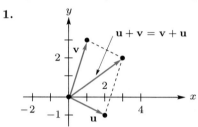

3.

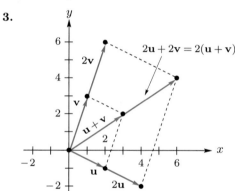

5. $\mathbf{x} = \left(-\frac{1}{2}, -\frac{3}{8}, \frac{7}{8}\right)$.

7. (a) $\sqrt{15}$. (b) $3\sqrt{2}$. (c) $\sqrt{43}$.
   (d) -5. (e) $-\frac{1}{3}\sqrt{\frac{5}{6}}$.

9. $c = \pm 3$. 11. No.

13. $\frac{k}{\sqrt{14}}(-1, 2, 3)$, where $k = \pm 1$.

15. $a = -2, b = 2$. 17. 26.

19. $\begin{bmatrix} \frac{\sqrt{3}}{2} & \frac{1}{2} \\ -\frac{1}{2} & \frac{\sqrt{3}}{2} \end{bmatrix}$.

21. (a) $(1, 3, -2) = 1(1, 1, 0) + 2(0, 1, 1) - 4(0, 0, 1)$.
    (b) $(4, 7)$.

23. $n = 3, m = 5$. 25. $c + 3a - b = 0$.

27. $x = -\frac{5}{3} - \frac{5}{3}t, y = \frac{2}{3} - \frac{1}{3}t, \quad -\infty < t < \infty$.

Chapter Test, page 195

1. $-\frac{1}{\sqrt{22}\sqrt{30}}$.   **2.** $\frac{1}{\sqrt{15}}(2,-1,1,3)$.

3. Yes.   **4.** Yes.   **5.** $\begin{bmatrix} 2 & 3 \\ -2 & 3 \\ 1 & 1 \end{bmatrix}$.

6. $x = 5 + 3t$, $y = -2 - 2t$, $z = 1 + 5t$, $-\infty < t < \infty$.

7. $x - y + 1 = 0$.

8. (a) F.   (b) F.   (c) T.   (d) F.   (e) F.
  (f) T.   (g) F.   (h) T.   (i) T.   (j) F.

Chapter 4

Section 4.1, page 202

1. Closed under $\oplus$; not closed under $\odot$.

3. Not closed under $\oplus$; not closed under $\odot$.

5. Not a vector space; (a), (c), (d), and (f) do not hold.

13. Vector space.

15. Not a vector space; (β) and (d) do not hold.

17. Vector space.

Section 4.2, page 210

1. (b).   **3.** (b) and (c).

5. (a) and (c).   **13.** (c).

15. (b).   **17.** (a), (b), (c), and (d).

19. (b).

21. (a) No.   (b) No.   (c) No.   (d) No.

23. (a) No.   (b) No.   (c) No.   (d) Yes.

ML.3. (a) No.   (b) Yes.

ML.5. (a) $0\mathbf{v}_1 + \mathbf{v}_2 - \mathbf{v}_3 - \mathbf{v}_4 = \mathbf{v}$.
  (b) $p_1(t) + 2p_2(t) + 2p_3(t) = p(t)$.

ML.7. (a) Yes.   (b) Yes.   (c) Yes.

Section 4.3, page 222

1. (a), (c), and (d).   **3.** (a) and (d).

5. No.   **7.** $\left\{ \begin{bmatrix} 1 \\ 0 \\ 0 \\ 1 \end{bmatrix}, \begin{bmatrix} 0 \\ 1 \\ -2 \\ 0 \end{bmatrix} \right\}$.

9. Yes.

11. (a) $(4,6,8,6) = 3(1,1,2,1) + (1,0,0,2) + (0,3,2,1)$.
  (b) $(-2,4,-6,2) = -2(1,-2,3,-1)$.
  (d) $(6,5,-5,1) = 2(4,2,-1,3) - (2,-1,3,5)$.

13. (b) and (c) are linearly independent, (a) is linearly dependent.

$$\begin{bmatrix} 2 & 6 \\ 4 & 6 \end{bmatrix} = 3\begin{bmatrix} 1 & 1 \\ 1 & 2 \end{bmatrix} - \begin{bmatrix} 1 & 0 \\ 0 & 2 \end{bmatrix} + \begin{bmatrix} 0 & 3 \\ 1 & 2 \end{bmatrix}.$$

15. $c = 1$.

ML.1. (a) Linearly dependent.
  (b) Linearly independent.
  (c) Linearly independent.

Section 4.4, page 234

1. (a) and (d).   **3.** (a) and (d).   **5.** (c).

7. (a) $(2,1,3) = \frac{3}{2}(1,1,1) + \frac{1}{2}(1,2,3) - \frac{3}{2}(0,1,0)$.

9. (a) Forms a basis, $5t^2 - 3t + 8 = 5(t^2 + t) - 8(t - 1)$.

11. Possible answer: $\{\mathbf{v}_1, \mathbf{v}_2\}$; $\dim W = 2$.

13. Possible answer: $\{t^3 + t^2 - 2t + 1, t^2 + 1\}$;
  $\dim W = 2$.

15. $\left\{ \begin{bmatrix} 1 & 0 & 0 \\ 0 & 0 & 0 \end{bmatrix}, \begin{bmatrix} 0 & 1 & 0 \\ 0 & 0 & 0 \end{bmatrix}, \begin{bmatrix} 0 & 0 & 1 \\ 0 & 0 & 0 \end{bmatrix}, \right.$
$\left. \begin{bmatrix} 0 & 0 & 0 \\ 1 & 0 & 0 \end{bmatrix}, \begin{bmatrix} 0 & 0 & 0 \\ 0 & 1 & 0 \end{bmatrix}, \begin{bmatrix} 0 & 0 & 0 \\ 0 & 0 & 1 \end{bmatrix} \right\}$;
  $\dim M_{23} = 6$; $\dim M_{mn} = mn$.

17. (a) Possible answer: $\{(1,1,0),(0,1,1)\}$.
  (b) Possible answer: $\{(1,1,0),(0,0,1)\}$.
  (c) Possible answer: $\{(1,0,2),(0,1,1)\}$.

19. (a) 3.   (b) 2.

21. $\{t^2 - 1, t - 1\}$.

23. (a) 2.   (b) 1.   (c) 2.   (d) 2.

25. (a) 4.   (b) 3.   (c) 3.   (d) 4.

27. Possible answer:
  $\{(1,0,1,0),(0,1,-1,0),(1,0,0,0),(0,0,0,1)\}$.

29. $\left\{ \begin{bmatrix} 1 & 0 & 0 \\ 0 & 0 & 0 \\ 0 & 0 & 0 \end{bmatrix}, \begin{bmatrix} 0 & 1 & 0 \\ 1 & 0 & 0 \\ 0 & 0 & 0 \end{bmatrix}, \begin{bmatrix} 0 & 0 & 1 \\ 0 & 0 & 0 \\ 1 & 0 & 0 \end{bmatrix}, \right.$
$\left. \begin{bmatrix} 0 & 0 & 0 \\ 0 & 1 & 0 \\ 0 & 0 & 0 \end{bmatrix}, \begin{bmatrix} 0 & 0 & 0 \\ 0 & 0 & 1 \\ 0 & 1 & 0 \end{bmatrix}, \begin{bmatrix} 0 & 0 & 0 \\ 0 & 0 & 0 \\ 0 & 0 & 1 \end{bmatrix} \right\}$.

31. The set of all vectors of the form
  $(a, a + 2b, -2a + b, a - 2b)$, where a, b, c, and d are
  real numbers.

ML.1. Basis.

ML.3. Basis.

ML.5. Basis.

ML.7. dim span $S = 3$, span $S \neq R^4$.

ML.9. dim span $S = 3$, span $S = P_2$.

ML.11. $\{t^3 - t + 1, t^3 + 2, t, 1\}$.

Section 4.5, page 243

1. $\left\{ \begin{bmatrix} 0 \\ -1 \\ 0 \\ 1 \end{bmatrix}, \begin{bmatrix} 2 \\ -3 \\ 1 \\ 0 \end{bmatrix} \right\}$; dimension $= 2$.

3. $\left\{ \begin{bmatrix} 1 \\ -\frac{8}{3} \\ -\frac{4}{3} \\ 1 \end{bmatrix} \right\}$; dimension $= 1$.

5. $\left\{ \begin{bmatrix} -2 \\ 1 \\ 0 \\ 0 \\ 0 \end{bmatrix}, \begin{bmatrix} -3 \\ 0 \\ 1 \\ 1 \\ 0 \end{bmatrix} \right\}$; dimension $= 2$.

7. $\left\{ \begin{bmatrix} -2 \\ \frac{1}{2} \\ 0 \\ 0 \\ 1 \end{bmatrix}, \begin{bmatrix} 1 \\ 1 \\ 0 \\ 1 \\ 0 \end{bmatrix} \right\}$; dimension $= 2$.

9. $\left\{ \begin{bmatrix} -3 \\ 2 \\ 0 \\ 1 \end{bmatrix}, \begin{bmatrix} 5 \\ -4 \\ 1 \\ 0 \end{bmatrix} \right\}$.   **11.** $\left\{ \begin{bmatrix} -1 \\ 1 \end{bmatrix} \right\}$.

13. $\left\{ \begin{bmatrix} 1 \\ -2 \\ 1 \end{bmatrix} \right\}$.   **15.** $\lambda = 3$ or -4.

17. $\lambda = 0$ or 1.

ML.1. $\left\{ \begin{bmatrix} -2 \\ 0 \\ 1 \\ 0 \\ 0 \end{bmatrix}, \begin{bmatrix} -1 \\ -1 \\ 0 \\ 1 \\ 0 \end{bmatrix}, \begin{bmatrix} -2 \\ 1 \\ 0 \\ 0 \\ 1 \end{bmatrix} \right\}$.

ML.3. $\left\{ \begin{bmatrix} 1 \\ -2 \\ 1 \\ 0 \end{bmatrix}, \begin{bmatrix} \frac{4}{3} \\ -\frac{1}{3} \\ 0 \\ 1 \end{bmatrix} \right\}$.

ML.5. $\mathbf{x} = \begin{bmatrix} t \\ t \\ t \end{bmatrix}$, where t is any nonzero real number.

Section 4.6, page 253

1. Possible answer: $\{(1,0,0), (0,1,0), (0,0,1)\}$.

3. Possible answer: $\left\{ \begin{bmatrix} 1 \\ 0 \\ 1 \\ 0 \end{bmatrix}, \begin{bmatrix} 0 \\ 1 \\ 0 \\ 1 \end{bmatrix} \right\}$.

5. (a) $\{(1,0,-1), (0,1,0)\}$.

     (b) $\{(1,2,-1), (1,9,-1)\}$.

7. (a) $\left\{ \begin{bmatrix} 1 \\ 0 \\ 0 \\ 0 \end{bmatrix}, \begin{bmatrix} 0 \\ 1 \\ 0 \\ \frac{1}{5} \end{bmatrix}, \begin{bmatrix} 0 \\ 0 \\ 1 \\ \frac{3}{5} \end{bmatrix} \right\}$.

     (b) $\left\{ \begin{bmatrix} 1 \\ 1 \\ 3 \\ 2 \end{bmatrix}, \begin{bmatrix} -2 \\ -1 \\ 2 \\ 1 \end{bmatrix}, \begin{bmatrix} 0 \\ 0 \\ 5 \\ 3 \end{bmatrix} \right\}$.

9. Row rank $=$ column rank $= 3$.

11. Rank $A = 2$, nullity $A = 2$.

13. Rank $A = 3$, nullity $A = 0$.

15. Rank $A = 2$, nullity $A = 1$.

21. Singular.   **23.** Nonsingular.

25. Has a unique solution.

27. Linearly dependent.

29. Nontrivial solution.   **31.** Has a solution.

33. Has no solution.

ML.3. (a) $\left\{ \begin{bmatrix} 1 \\ 2 \\ 4 \\ 6 \end{bmatrix}, \begin{bmatrix} 3 \\ 5 \\ 11 \\ 9 \end{bmatrix}, \begin{bmatrix} 1 \\ 0 \\ 2 \\ 1 \end{bmatrix} \right\}$.

     (b) $\left\{ \begin{bmatrix} 2 \\ 0 \\ 1 \\ 4 \\ 3 \end{bmatrix}, \begin{bmatrix} 1 \\ 0 \\ 2 \\ 5 \\ 3 \end{bmatrix} \right\}$.

ML.5. (a) Consistent.   (b) Inconsistent.

      (c) Inconsistent.

Section 4.7, page 265

1. $\begin{bmatrix} 3 \\ -2 \end{bmatrix}$. **3.** $\begin{bmatrix} 2 \\ -1 \end{bmatrix}$. **5.** $\begin{bmatrix} 1 \\ -1 \\ 0 \\ 2 \end{bmatrix}$.

7. $\begin{bmatrix} 0 \\ 3 \end{bmatrix}$. **9.** $4t - 3$. **11.** $\begin{bmatrix} -1 & 0 \\ 9 & 7 \end{bmatrix}$.

13. (a) $[\mathbf{v}]_T = \begin{bmatrix} -7 \\ 4 \end{bmatrix}$; $[\mathbf{w}]_T = \begin{bmatrix} 7 \\ -1 \end{bmatrix}$.

(b) $\begin{bmatrix} 1 & 2 \\ -1 & -1 \end{bmatrix}$.

(c) $[\mathbf{v}]_S = \begin{bmatrix} 1 \\ 3 \end{bmatrix}$; $[\mathbf{w}]_S = \begin{bmatrix} 5 \\ -6 \end{bmatrix}$.

(d) Same as (c). (e) $\begin{bmatrix} -1 & -2 \\ 1 & 1 \end{bmatrix}$.

(f) Same as (a).

15. (a) $[\mathbf{v}]_T = \begin{bmatrix} 3 \\ 2 \\ -7 \end{bmatrix}$; $[\mathbf{w}]_T = \begin{bmatrix} 2 \\ 3 \\ -3 \end{bmatrix}$.

(b) $\begin{bmatrix} 2 & 1 & 0 \\ 1 & -\frac{2}{5} & \frac{3}{5} \\ 0 & \frac{2}{5} & \frac{2}{5} \end{bmatrix}$.

(c) $[\mathbf{v}]_S = \begin{bmatrix} 8 \\ -2 \\ -2 \end{bmatrix}$; $[\mathbf{w}]_S = \begin{bmatrix} 7 \\ -1 \\ 0 \end{bmatrix}$.

(d) Same as (c).

(e) $\begin{bmatrix} \frac{1}{3} & \frac{1}{3} & -\frac{1}{2} \\ \frac{1}{3} & -\frac{2}{3} & 1 \\ -\frac{1}{3} & \frac{2}{3} & \frac{3}{2} \end{bmatrix}$.

(f) Same as (a).

17. (a) $[\mathbf{v}]_T = \begin{bmatrix} 1 \\ 1 \\ 1 \\ 0 \end{bmatrix}$; $[\mathbf{w}]_T = \begin{bmatrix} 2 \\ -2 \\ 1 \\ -1 \end{bmatrix}$.

(b) $\begin{bmatrix} 1 & 0 & 0 & 1 \\ \frac{1}{3} & \frac{2}{3} & -\frac{2}{3} & 0 \\ \frac{1}{3} & -\frac{1}{3} & \frac{1}{3} & 0 \\ -\frac{1}{3} & \frac{1}{3} & \frac{2}{3} & 0 \end{bmatrix}$.

(c) $[\mathbf{v}]_S = \begin{bmatrix} 1 \\ \frac{1}{3} \\ \frac{1}{3} \\ \frac{2}{3} \end{bmatrix}$; $[\mathbf{w}]_S = \begin{bmatrix} 1 \\ -\frac{4}{3} \\ \frac{5}{3} \\ -\frac{2}{3} \end{bmatrix}$.

(d) Same as (c).

(e) $\begin{bmatrix} 0 & 1 & 2 & 0 \\ 0 & 1 & 0 & 1 \\ 0 & 0 & 1 & 1 \\ 1 & -1 & -2 & 0 \end{bmatrix}$.

(f) Same as (a).

19. $\begin{bmatrix} 5 \\ 3 \end{bmatrix}$. **21.** $\begin{bmatrix} 4 \\ -1 \\ 3 \end{bmatrix}$.

23. $\left\{ \begin{bmatrix} 3 \\ 2 \\ 0 \end{bmatrix}, \begin{bmatrix} 2 \\ 1 \\ 0 \end{bmatrix}, \begin{bmatrix} 3 \\ 1 \\ 3 \end{bmatrix} \right\}$. **25.** $\left\{ \begin{bmatrix} 2 \\ 5 \end{bmatrix}, \begin{bmatrix} 1 \\ 3 \end{bmatrix} \right\}$.

ML.1. $\left\{ \begin{bmatrix} 1 \\ 2 \\ 3 \end{bmatrix}, \begin{bmatrix} -1 \\ 2 \\ -1 \end{bmatrix}, \begin{bmatrix} 1 \\ 1 \\ 1 \end{bmatrix} \right\}$.

ML.3. (a) $\begin{bmatrix} 0.5000 \\ -0.5000 \\ 0 \\ -0.5000 \end{bmatrix}$. (b) $\begin{bmatrix} 1.0000 \\ 0.5000 \\ 0.3333 \\ 0 \end{bmatrix}$.

(c) $\begin{bmatrix} 0.5000 \\ 0.1667 \\ -0.3333 \\ -1.5000 \end{bmatrix}$.

ML.5. $\begin{bmatrix} -0.5000 & -1.0000 & -0.5000 & 0 \\ -0.5000 & 0 & 1.5000 & 0 \\ 1.0000 & 0 & -1.0000 & 1.0000 \\ 0 & 0 & 0 & 1.0000 \end{bmatrix}$.

ML.7. (a) $\begin{bmatrix} 1.0000 & -1.6667 & 2.3333 \\ 1.0000 & 0.6667 & -1.3333 \\ 0 & 1.3333 & -0.6667 \end{bmatrix}$.

(b) $\begin{bmatrix} 2 & 0 & 1 \\ -1 & 1 & -1 \\ 0 & -1 & 2 \end{bmatrix}$.

(c) $\begin{bmatrix} 2 & -2 & 4 \\ 0 & 1 & -3 \\ -1 & 2 & 0 \end{bmatrix}$. (d) QP.

Section 4.8, page 275

1. (b). **3.** $a = 5$.

5. $\left\{ \left(\frac{1}{\sqrt{2}}, -\frac{1}{\sqrt{2}}, 0 \right), \left(\frac{1}{\sqrt{3}}, \frac{1}{\sqrt{3}}, \frac{1}{\sqrt{3}} \right) \right\}$.

7. $\left\{\left(\frac{1}{\sqrt{3}}, -\frac{1}{\sqrt{3}}, 0, \frac{1}{\sqrt{3}}\right), \left(\frac{5}{\sqrt{42}}, \frac{1}{\sqrt{42}}, 0, -\frac{4}{\sqrt{42}}\right),\right.$

$\left.(0, 0, 1, 0)\right\}.$

9. (a) $\{(1, 2), (-4, 2)\}.$

(b) $\left\{\left(\frac{1}{\sqrt{5}}, \frac{2}{\sqrt{5}}\right), \left(-\frac{2}{\sqrt{5}}, \frac{1}{\sqrt{5}}\right)\right\}.$

11. $\left\{\left(\frac{2}{3}, -\frac{2}{3}, \frac{1}{3}\right), \left(\frac{2}{3}, \frac{1}{3}, -\frac{2}{3}\right), \left(\frac{1}{3}, \frac{2}{3}, \frac{2}{3}\right)\right\}.$

13. Possible answer: $\left\{\left(\frac{1}{\sqrt{2}}, \frac{1}{\sqrt{2}}, 0, 0\right),\right.$

$\left(\frac{3}{\sqrt{22}}, -\frac{3}{\sqrt{22}}, 0, \frac{2}{\sqrt{22}}\right),$

$\left.\left(\frac{1}{\sqrt{11}}, -\frac{1}{\sqrt{11}}, 0, -\frac{3}{\sqrt{11}}\right)\right\}.$

15. $\left\{\left(\frac{1}{\sqrt{2}}, \frac{1}{\sqrt{2}}, 0, 0\right),\right.$

$\left(-\frac{1}{\sqrt{6}}, \frac{1}{\sqrt{6}}, 0, \frac{2}{\sqrt{6}}\right),$

$\left.\left(\frac{1}{\sqrt{12}}, -\frac{1}{\sqrt{12}}, \frac{3}{\sqrt{12}}, \frac{1}{\sqrt{12}}\right)\right\}.$

17. $\left\{\left(\frac{1}{\sqrt{2}}, \frac{1}{\sqrt{2}}, 0, 0\right),\right.$

$\left(\frac{1}{\sqrt{3}}, -\frac{1}{\sqrt{3}}, \frac{1}{\sqrt{3}}, 0\right),$

$\left.\left(-\frac{1}{\sqrt{42}}, \frac{1}{\sqrt{42}}, \frac{2}{\sqrt{42}}, \frac{6}{\sqrt{42}}\right)\right\}.$

19. $\left\{\frac{1}{\sqrt{42}} \begin{bmatrix} -4 \\ 5 \\ 1 \end{bmatrix}\right\}.$

21. $\frac{4}{\sqrt{5}}\left(\frac{1}{\sqrt{5}}, 0, \frac{2}{\sqrt{5}}\right) - \frac{3}{\sqrt{5}}\left(-\frac{2}{\sqrt{5}}, 0, \frac{1}{\sqrt{5}}\right) - 3(0, 1, 0) =$
$(2, -3, 1).$

ML.1. $\left\{\begin{bmatrix} 0.7071 \\ 0.7071 \\ 0 \end{bmatrix}, \begin{bmatrix} 0.7071 \\ -0.7071 \\ 0 \end{bmatrix}, \begin{bmatrix} 0 \\ 0 \\ 1.0000 \end{bmatrix}\right\}$

$= \left\{\begin{bmatrix} \frac{\sqrt{2}}{2} \\ \frac{\sqrt{2}}{2} \\ 0 \end{bmatrix}, \begin{bmatrix} \frac{\sqrt{2}}{2} \\ -\frac{\sqrt{2}}{2} \\ 0 \end{bmatrix}, \begin{bmatrix} 0 \\ 0 \\ 1 \end{bmatrix}\right\}.$

ML.3. (a) $\begin{bmatrix} -1.4142 \\ 1.4142 \\ 1.0000 \end{bmatrix}.$ (b) $\begin{bmatrix} 0 \\ 1.4142 \\ 1.0000 \end{bmatrix}.$

(c) $\begin{bmatrix} 0.7071 \\ 0.7071 \\ -1.0000 \end{bmatrix}.$

Section 4.9, page 284

1. (a) $\left\{\left(\frac{3}{2}, 1, 0\right), \left(-\frac{1}{2}, 0, 1\right)\right\}.$

(b) The set of all points $P(x, y, z)$ such that $2x - 3y + z = 0$. $W^{\perp}$ is the plane whose normal is **w**.

3. $\left\{\begin{bmatrix} -\frac{17}{5} \\ \frac{6}{5} \\ 5 \\ 1 \\ 0 \end{bmatrix}, \begin{bmatrix} \frac{8}{5} \\ \frac{1}{5} \\ -3 \\ 0 \\ 1 \end{bmatrix}\right\}.$

5. Basis for null space of A: $\left\{\begin{bmatrix} -\frac{1}{3} \\ \frac{7}{3} \\ 1 \\ 0 \end{bmatrix}, \begin{bmatrix} -\frac{7}{3} \\ -\frac{2}{3} \\ 0 \\ 1 \end{bmatrix}\right\}.$

Basis for row space of A: $\left\{\left(1, 0, \frac{1}{3}, \frac{7}{3}\right),\right.$
$\left.\left(0, 1, -\frac{7}{3}, \frac{2}{3}\right)\right\}.$

Basis for null space of A^T: $\left\{\begin{bmatrix} -\frac{1}{2} \\ \frac{1}{2} \\ 1 \\ 0 \end{bmatrix}, \begin{bmatrix} -\frac{1}{2} \\ \frac{3}{2} \\ 0 \\ 1 \end{bmatrix}\right\}.$

Basis for column space of A: $\left\{\begin{bmatrix} 1 \\ 0 \\ \frac{1}{2} \\ \frac{1}{2} \end{bmatrix}, \begin{bmatrix} 0 \\ 1 \\ -\frac{1}{2} \\ -\frac{3}{2} \end{bmatrix}\right\}.$

7. (a) $\left(\frac{7}{5}, \frac{11}{5}, \frac{9}{5}, -\frac{3}{5}\right).$

(b) $\left(\frac{2}{5}, -\frac{1}{5}, \frac{1}{5}, -\frac{2}{5}\right).$

(c) $\left(\frac{1}{10}, \frac{9}{5}, \frac{1}{5}, \frac{31}{10}\right).$

9. $\mathbf{w} = (1, 0, 2, 3)$, $\mathbf{u} = (0, 0, 0, 0).$ **11.** 2.

ML.1. (a) $\begin{bmatrix} 0 \\ \frac{5}{6} \\ \frac{5}{3} \\ \frac{5}{6} \end{bmatrix}.$ (b) $\begin{bmatrix} \frac{3}{5} \\ \frac{3}{5} \\ \frac{3}{5} \\ \frac{3}{5} \\ \frac{3}{5} \end{bmatrix}.$

ML.3. (a) $\begin{bmatrix} 2.4286 \\ 3.9341 \\ 7.9011 \end{bmatrix}.$

(b) $\sqrt{\dfrac{(2.4286-2)^2}{+(3.9341-4)^2+(7.9011-8)^2}}$

$\approx 0.4448.$

ML.5. $\mathbf{p} = \begin{bmatrix} 0.8571 \\ 0.5714 \\ 1.4286 \\ 0.8571 \\ 0.8571 \end{bmatrix}.$

Supplementary Exercises, page 287

1. No.   **3.** No.

5. Linearly dependent; possible answer:
$-t-3 = (2t^2+3t+1)-2(t^2+2t+2).$

7. Possible answer: $\{(1,0,1,0),(1,1,-1,1)\}$;
dimension is 2.

9. Possible answer: $\left\{ \begin{bmatrix} 0 \\ -1 \\ 0 \\ 0 \\ 1 \end{bmatrix}, \begin{bmatrix} 7 \\ -4 \\ 0 \\ 1 \\ 0 \end{bmatrix}, \begin{bmatrix} -5 \\ 3 \\ 1 \\ 0 \\ 0 \end{bmatrix} \right\}$;

dimension is 3.

11. $\lambda \neq \pm 2.$ **13.** $a = 1.$

17. (a) m arbitrary and $b = 0.$ (b) $r = 0.$

21. (b) $k = 0.$ **23.** 3.

27. (a) $\begin{bmatrix} \mathbf{v} \end{bmatrix}_T = \begin{bmatrix} -6 \\ 11 \\ 8 \end{bmatrix}.$ (b) $\begin{bmatrix} \mathbf{v} \end{bmatrix}_S = \begin{bmatrix} 2 \\ 3 \\ 4 \end{bmatrix}.$

(c) $P_{S \leftarrow T} = \begin{bmatrix} 1 & 0 & 1 \\ 0 & 1 & -1 \\ 1 & -2 & 4 \end{bmatrix}.$ (d) Same as (b).

(e) $Q_{T \leftarrow S} = \begin{bmatrix} 2 & -2 & -1 \\ -1 & 3 & 1 \\ -1 & 2 & 1 \end{bmatrix}.$

(f) Same as (a).

29. $a = b = 0.$

31. (a) One such basis is
$\left\{ \dfrac{1}{\sqrt{30}}\begin{bmatrix} -5 \\ 2 \\ 1 \\ 0 \end{bmatrix}, \dfrac{1}{\sqrt{30}}\begin{bmatrix} 2 \\ 5 \\ 0 \\ 1 \end{bmatrix} \right\}.$

(b) One such basis is
$\left\{ \dfrac{1}{\sqrt{30}}\begin{bmatrix} -5 \\ 2 \\ 1 \\ 0 \end{bmatrix}, \dfrac{1}{\sqrt{255}}\begin{bmatrix} -5 \\ -14 \\ 3 \\ 5 \end{bmatrix} \right\}.$

33. Possible answer:

$\left\{ \left(\dfrac{1}{\sqrt{2}},0,0,-\dfrac{1}{\sqrt{2}} \right), \left(\dfrac{1}{\sqrt{6}},-\dfrac{2}{\sqrt{6}},0,\dfrac{1}{\sqrt{6}} \right), \right.$
$\left. \left(\dfrac{1}{\sqrt{3}},\dfrac{1}{\sqrt{3}},0,\dfrac{1}{\sqrt{3}} \right) \right\}.$

35. (a) Possible answer: $\{(-1,0,1)\}.$

(c) (i) $\mathbf{w} = \left(\tfrac{1}{2},0,\tfrac{1}{2} \right), \mathbf{u} = \left(\tfrac{1}{2},0,-\tfrac{1}{2} \right).$

(ii) $\mathbf{w} = (2,2,2), \mathbf{u} = (-1,0,1).$

37. Basis for null space of $A = \left\{ \begin{bmatrix} -\frac{37}{11} \\ \frac{20}{11} \\ \frac{8}{11} \\ 1 \end{bmatrix} \right\}.$

Basis for row space of
$A = \left\{ \left(1,0,0,\frac{37}{11}\right), \left(0,1,0,-\frac{20}{11}\right), \left(0,0,1,-\frac{8}{11}\right) \right\}.$
There is no basis for the null space of A^T, since the
null space of $A^T = \{\mathbf{0}\}.$

Basis for the column space of
$A = \left\{ \begin{bmatrix} 1 \\ 0 \\ 0 \end{bmatrix}, \begin{bmatrix} 0 \\ 1 \\ 0 \end{bmatrix}, \begin{bmatrix} 0 \\ 0 \\ 1 \end{bmatrix} \right\}.$

Chapter Test, page 290

1. Yes

2. Possible answer: $\left\{ \begin{bmatrix} 0 \\ -1 \\ 1 \\ 0 \\ 0 \end{bmatrix}, \begin{bmatrix} 4 \\ -1 \\ 0 \\ 1 \\ 0 \end{bmatrix}, \begin{bmatrix} -2 \\ 0 \\ 0 \\ 0 \\ 1 \end{bmatrix} \right\}.$

3. Yes.   **4.** $\lambda = \pm 3.$

5. Possible answer:

$\left\{ \left(\dfrac{1}{\sqrt{2}},0,-\dfrac{1}{\sqrt{2}},0 \right), \left(\dfrac{1}{\sqrt{6}},-\dfrac{2}{\sqrt{6}},\dfrac{1}{\sqrt{6}},0 \right), \right.$
$\left. \left(\dfrac{1}{\sqrt{3}},\dfrac{1}{\sqrt{3}},\dfrac{1}{\sqrt{3}},0 \right) \right\}.$

6. (a) T.   (b) F.   (c) F.   (d) F.   (e) T.
(f) F.   (g) T.   (h) F.   (i) F.   (j) T.

Chapter 5

Section 5.1, page 310

1. $\lambda^3 - 4\lambda^2 + 7$.

3. $(\lambda - 4)(\lambda - 2)(\lambda - 3) = \lambda^3 - 9\lambda^2 + 26\lambda - 24$.

5. $f(\lambda) = (\lambda - 1)(\lambda - 3)(\lambda + 2)$;
$\lambda_1 = 1, \lambda_2 = 3, \lambda_3 = -2$;
$$\mathbf{x}_1 = \begin{bmatrix} 6 \\ 3 \\ 8 \end{bmatrix}, \mathbf{x}_2 = \begin{bmatrix} 0 \\ 5 \\ 2 \end{bmatrix}, \mathbf{x}_3 = \begin{bmatrix} 0 \\ 0 \\ 1 \end{bmatrix}.$$

7. $f(\lambda) = \lambda^2 - 5\lambda + 6; \lambda_1 = 2, \lambda_2 = 3$;
$$\mathbf{x}_1 = \begin{bmatrix} 1 \\ -1 \end{bmatrix}, \mathbf{x}_2 = \begin{bmatrix} 1 \\ -2 \end{bmatrix}.$$

9. $f(\lambda) = \lambda^3 - 5\lambda^2 + 2\lambda + 8$;
$\lambda_1 = -1, \lambda_2 = 2, \lambda_3 = 4$;
$$\mathbf{x}_1 = \begin{bmatrix} 1 \\ 0 \\ -1 \end{bmatrix}, \mathbf{x}_2 = \begin{bmatrix} -2 \\ -3 \\ 2 \end{bmatrix}, \mathbf{x}_3 = \begin{bmatrix} 8 \\ 5 \\ 2 \end{bmatrix}.$$

11. $f(\lambda) = (\lambda - 1)(\lambda + 1)(\lambda - 3)(\lambda - 2)$;
$\lambda_1 = 1, \lambda_2 = -1, \lambda_3 = 3, \lambda_4 = 2$;
$$\mathbf{x}_1 = \begin{bmatrix} 1 \\ 0 \\ 0 \\ 0 \end{bmatrix}, \mathbf{x}_2 = \begin{bmatrix} 1 \\ -1 \\ 0 \\ 0 \end{bmatrix}, \mathbf{x}_3 = \begin{bmatrix} 9 \\ 3 \\ 4 \\ 0 \end{bmatrix}, \mathbf{x}_4 = \begin{bmatrix} 29 \\ 7 \\ 9 \\ -3 \end{bmatrix}.$$

13. Not diagonalizable; $\lambda_1 = \lambda_2 = 1$.

15. Diagonalizable; $\lambda_1 = 1, \lambda_2 = -1, \lambda_3 = 2$.

17. Not diagonalizable; $\lambda_1 = 1, \lambda_2 = 1, \lambda_3 = 3$.

19. $P = \begin{bmatrix} 1 & -3 & 1 \\ 0 & 0 & -6 \\ 1 & 2 & 4 \end{bmatrix}$;
$\lambda_1 = 4, \lambda_2 = -1, \lambda_3 = 1$.

21. $P = \begin{bmatrix} 1 & 2 & 1 \\ 0 & 1 & 0 \\ 0 & 0 & -3 \end{bmatrix}$;
$\lambda_1 = 3, \lambda_2 = 2, \lambda_3 = 0$.

23. Basis for eigenspace associated with $\lambda_1 = \lambda_2 = 2$ is
$$\left\{ \begin{bmatrix} 1 \\ 0 \\ 0 \\ 0 \end{bmatrix} \right\}.$$
Basis for eigenspace associated with $\lambda_3 = \lambda_4 = 1$ is
$$\left\{ \begin{bmatrix} 3 \\ -3 \\ 1 \\ 0 \end{bmatrix} \right\}.$$

25. $256 \begin{bmatrix} 3 & -5 \\ 1 & -3 \end{bmatrix}$.

ML.1. (a) $\lambda^2 - 5$.   (b) $\lambda^3 - 6\lambda^2 + 4\lambda + 8$.
(c) $\lambda^4 - 3\lambda^3 - 3\lambda^2 + 11\lambda - 6$.

ML.3. (a) $\begin{bmatrix} 1 \\ 1 \end{bmatrix}$.   (b) $\begin{bmatrix} 0 \\ 0 \\ 1 \end{bmatrix}$.   (c) $\begin{bmatrix} 1 \\ -2 \\ 1 \end{bmatrix}$.

ML.5. $\begin{bmatrix} 1 & -1 & 1 \\ 0 & 0 & 1 \\ 0 & 0 & 1 \end{bmatrix}$.

ML.7. The sequence $A, A^3, A^5, \ldots$ converges to
$$\begin{bmatrix} -1 & 1 & -1 \\ -2 & 2 & -1 \\ -2 & 2 & -1 \end{bmatrix}.$$
The sequence $A^2, A^4, A^6, \ldots$ converges to
$$\begin{bmatrix} 1 & -1 & 1 \\ 0 & 0 & 1 \\ 0 & 0 & 1 \end{bmatrix}.$$

Section 5.2, page 322

5. $\begin{bmatrix} 0 & 0 \\ 0 & 4 \end{bmatrix}$; $P = \begin{bmatrix} \frac{1}{\sqrt{2}} & \frac{1}{\sqrt{2}} \\ -\frac{1}{\sqrt{2}} & \frac{1}{\sqrt{2}} \end{bmatrix}$.

7. $\begin{bmatrix} 0 & 0 & 0 \\ 0 & 0 & 0 \\ 0 & 0 & 4 \end{bmatrix}$; $P = \begin{bmatrix} 1 & 0 & 0 \\ 0 & -\frac{1}{\sqrt{2}} & \frac{1}{\sqrt{2}} \\ 0 & \frac{1}{\sqrt{2}} & \frac{1}{\sqrt{2}} \end{bmatrix}$.

9. $\begin{bmatrix} -2 & 0 & 0 \\ 0 & 1 & 0 \\ 0 & 0 & 1 \end{bmatrix}$; $P = \begin{bmatrix} \frac{1}{\sqrt{3}} & -\frac{1}{\sqrt{2}} & -\frac{1}{\sqrt{6}} \\ \frac{1}{\sqrt{3}} & \frac{1}{\sqrt{2}} & -\frac{1}{\sqrt{6}} \\ \frac{1}{\sqrt{3}} & 0 & \frac{2}{\sqrt{6}} \end{bmatrix}$.

11. $\begin{bmatrix} 3 & 0 \\ 0 & 1 \end{bmatrix}$.

13. $\begin{bmatrix} 1 & 0 & 0 \\ 0 & 2 & 0 \\ 0 & 0 & 0 \end{bmatrix}$.   **15.** $\begin{bmatrix} 1 & 0 & 0 \\ 0 & 0 & 0 \\ 0 & 0 & 2 \end{bmatrix}$.

17. $\begin{bmatrix} 1 & 0 & 0 \\ 0 & 1 & 0 \\ 0 & 0 & 4 \end{bmatrix}$.

ML.1. (a) $\lambda_1 = 0, \lambda_2 = 12$; $P = \begin{bmatrix} 0.7071 & 0.7071 \\ -0.7071 & 0.7071 \end{bmatrix}$.

(b) $\lambda_1 = -1$, $\lambda_2 = -1$, $\lambda_3 = 5$;

$$P = \begin{bmatrix} 0.7743 & -0.2590 & 0.5774 \\ -0.6115 & -0.5411 & 0.5774 \\ -0.1629 & 0.8001 & 0.5774 \end{bmatrix}.$$

(c) $\lambda_1 = 5.4142$, $\lambda_2 = 4.0000$, $\lambda_3 = 2.5858$.

$$P = \begin{bmatrix} 0.5000 & -0.7071 & -0.5000 \\ 0.7071 & -0.0000 & 0.7071 \\ 0.5000 & 0.7071 & -0.5000 \end{bmatrix}.$$

Supplementary Exercises, page 324

1. $f(\lambda) = (\lambda + 2)(\lambda^2 - 8\lambda + 15)$; $\lambda_1 = -2$, $\lambda_2 = 3$, $\lambda_3 = 5$;

$$\mathbf{x}_1 = \begin{bmatrix} -35 \\ 12 \\ 19 \end{bmatrix}, \mathbf{x}_2 = \begin{bmatrix} 0 \\ 3 \\ 1 \end{bmatrix}, \mathbf{x}_3 = \begin{bmatrix} 0 \\ 1 \\ 1 \end{bmatrix}.$$

3. Yes.

5. Not diagonalizable; the roots of the characteristic polynomial are not all real.

7. For $\lambda = 0$, possible answer: $\left\{ \begin{bmatrix} 1 \\ 0 \\ 0 \end{bmatrix} \right\}$;

for $\lambda = 2$, possible answer: $\left\{ \begin{bmatrix} 1 \\ 0 \\ 2 \end{bmatrix}, \begin{bmatrix} 0 \\ 1 \\ 0 \end{bmatrix} \right\}$.

9. $P = \begin{bmatrix} -\frac{1}{\sqrt{2}} & \frac{1}{\sqrt{6}} & \frac{1}{\sqrt{3}} \\ \frac{1}{\sqrt{2}} & -\frac{1}{\sqrt{6}} & \frac{1}{\sqrt{3}} \\ 0 & \frac{2}{\sqrt{6}} & \frac{1}{\sqrt{3}} \end{bmatrix}$,

$D = \begin{bmatrix} 0 & 0 & 0 \\ 0 & 0 & 0 \\ 0 & 0 & 3 \end{bmatrix}.$

$\lambda = 0$, $\lambda = 1$.

11.

Chapter Test, page 325

1. Not diagonalizable; $\lambda_1 = 1$, $\lambda_2 = \lambda_3 = 2$.

2. Check that $AA^T = I_3$.

3. $P = \begin{bmatrix} -2 & 1 & 1 \\ 1 & 2 & 0 \\ 2 & 0 & 1 \end{bmatrix}$, $D = \begin{bmatrix} 9 & 0 & 0 \\ 0 & -9 & 0 \\ 0 & 0 & -9 \end{bmatrix}.$

4. (a) F. (b) F. (c) T. (d) T. (e) F.

Chapter 6

Section 6.1, page 332

1. (a) and (c).

3. (a) Yes. (b) No. (c) Yes.

5. (a) No. (b) Yes. (c) Yes.

7. Yes.   **9.** Yes.

11. (a) $\begin{bmatrix} 15 & 5 & 4 & 8 \\ -5 & -1 & 10 & 2 \end{bmatrix}.$

17. (a) $\begin{bmatrix} 8 & 5 \end{bmatrix}.$ (b) $\begin{bmatrix} \dfrac{-a_1 + 3a_2}{2} & \dfrac{-5a_1 + a_2}{2} \end{bmatrix}.$

19. (a) $17t - 7$. (b) $\left(\dfrac{5a - b}{2} \right) t + \dfrac{a + 5b}{2}.$

Section 6.2, page 343

1. (a) Yes. (b) No. (c) Yes. (d) No.
   (e) $\{(0, r)\}$, $r =$ any real number.
   (f) $\{(r, 0)\}$, $r =$ any real number.

3. (a) $\{(0, 0)\}$. (b) Yes. (c) No.

5. (a) Possible answer: $\left\{ \begin{bmatrix} -2 \\ 0 \\ 1 \\ 1 \\ 0 \end{bmatrix}, \begin{bmatrix} 0 \\ 1 \\ 0 \\ 0 \\ 0 \end{bmatrix} \right\}.$

(b) Possible answers:

$\left\{ \begin{bmatrix} 1 \\ 0 \\ 0 \\ 1 \end{bmatrix}, \begin{bmatrix} 0 \\ 1 \\ 0 \\ -1 \end{bmatrix}, \begin{bmatrix} 0 \\ 0 \\ 1 \\ 0 \end{bmatrix} \right\}, \left\{ \begin{bmatrix} 1 \\ -1 \\ 3 \\ -1 \end{bmatrix}, \begin{bmatrix} 1 \\ 0 \\ 2 \\ -1 \end{bmatrix}, \begin{bmatrix} 0 \\ -1 \\ 1 \\ 0 \end{bmatrix} \right\}.$

7. (a) Yes. (b) 1.

11. (a) No. (b) Yes. (c) Yes. (d) No.
   (e) Possible answer: $\{-t^2 - t + 1\}$.
   (f) Possible answer: $\{t^2, t\}$.

13. (a) $\ker L = \left\{ \begin{bmatrix} 0 & 0 \\ 0 & 0 \end{bmatrix} \right\}$, so $\ker L$ has no basis.

(b) $\left\{ \begin{bmatrix} 1 & 0 \\ 1 & 0 \end{bmatrix}, \begin{bmatrix} 1 & 1 \\ 0 & 1 \end{bmatrix}, \begin{bmatrix} 0 & 1 \\ 0 & 0 \end{bmatrix}, \begin{bmatrix} 0 & 0 \\ 1 & 1 \end{bmatrix} \right\}.$

15. (a) Possible answer: $\left\{ \begin{bmatrix} 1 & 0 \\ 0 & 1 \end{bmatrix}, \begin{bmatrix} 0 & 1 \\ \frac{1}{2} & 0 \end{bmatrix} \right\}.$

(b) Possible answer: $\left\{ \begin{bmatrix} 0 & -2 \\ 1 & 0 \end{bmatrix}, \begin{bmatrix} -1 & 0 \\ 0 & 1 \end{bmatrix} \right\}.$

17. (a) Possible answer: $\{1\}$.
   (b) Possible answer: $\{t, 1\}$.

19. (a) 2. (b) 1.

ML.1. Basis for ker L: $\left\{\begin{bmatrix} -1 \\ -2 \\ 1 \\ 0 \end{bmatrix}, \begin{bmatrix} 1 \\ -3 \\ 0 \\ 1 \end{bmatrix}\right\}.$

Basis for range L: $\left\{\begin{bmatrix} 1 \\ 0 \end{bmatrix}, \begin{bmatrix} 0 \\ 1 \end{bmatrix}\right\}.$

ML.3. Basis for ker L: $\left\{\begin{bmatrix} -2 \\ 0 \\ 1 \\ -2 \\ 1 \end{bmatrix}, \begin{bmatrix} -1 \\ 1 \\ 0 \\ 0 \\ 0 \end{bmatrix}\right\}.$

Basis for range L: $\left\{\begin{bmatrix} 1 \\ 0 \\ 0 \end{bmatrix}, \begin{bmatrix} 0 \\ 1 \\ 0 \end{bmatrix}, \begin{bmatrix} 0 \\ 0 \\ 1 \end{bmatrix}\right\}.$

Section 6.3, page 358

1. (a) $\begin{bmatrix} 3 & -2 \\ 2 & 0 \end{bmatrix}.$ (b) $\begin{bmatrix} 3 & -2 \\ -1 & 2 \end{bmatrix}.$

(c) $\begin{bmatrix} 1 & -2 \\ 2 & 0 \end{bmatrix}.$ (d) $\begin{bmatrix} 1 & -2 \\ 1 & 2 \end{bmatrix}.$ (e) $(4,0).$

3. (a) $\begin{bmatrix} 1 & -2 \\ 2 & 1 \\ 1 & 1 \end{bmatrix}.$ (b) $\begin{bmatrix} \frac{7}{3} & -\frac{4}{3} \\ -\frac{2}{3} & \frac{5}{3} \\ \frac{2}{3} & -\frac{2}{3} \end{bmatrix}.$ (c) $\begin{bmatrix} -3 \\ 4 \\ 3 \end{bmatrix}.$

5. (a) $\begin{bmatrix} 1 & 1 & 0 \\ 0 & 1 & -1 \end{bmatrix}.$ (b) $\begin{bmatrix} -1 & -\frac{1}{3} & 0 \\ 1 & \frac{2}{3} & 0 \end{bmatrix}.$

(c) $\begin{bmatrix} 3 \\ -1 \end{bmatrix}.$

7. (a) $\begin{bmatrix} 1 & 0 \\ 0 & 1 \\ 0 & 0 \\ 0 & 0 \end{bmatrix}.$ (b) $\begin{bmatrix} 1 & 1 \\ 0 & 1 \\ 0 & -1 \\ 0 & 1 \end{bmatrix}.$

9. (a) $\begin{bmatrix} 1 & 0 & 0 & 0 \\ 0 & 0 & 1 & 0 \\ 0 & 1 & 0 & 0 \\ 0 & 0 & 0 & 1 \end{bmatrix}.$

(b) $\begin{bmatrix} 1 & 1 & 0 & -1 \\ -1 & -1 & 1 & 1 \\ 0 & 1 & 0 & 0 \\ 0 & -1 & 0 & 1 \end{bmatrix}.$

(c) $\begin{bmatrix} 1 & 0 & 0 & 1 \\ 0 & 0 & 1 & 0 \\ 1 & 1 & 0 & 0 \\ 0 & 0 & 1 & 1 \end{bmatrix}.$

(d) $\begin{bmatrix} 1 & 0 & -1 & 0 \\ -1 & 0 & 2 & 0 \\ 1 & 1 & 0 & 0 \\ 0 & 0 & 1 & 1 \end{bmatrix}.$

11. (a) $\begin{bmatrix} 10 \\ 5 \\ 5 \end{bmatrix}.$ (b) $\begin{bmatrix} 4 \\ 2 \\ 2 \end{bmatrix}.$

13. (a) $\left[L(\mathbf{v_1})\right]_S = \begin{bmatrix} 2 \\ -1 \end{bmatrix}, \left[L(\mathbf{v_2})\right]_S = \begin{bmatrix} -3 \\ 4 \end{bmatrix}.$

(b) $L(\mathbf{v_1}) = \begin{bmatrix} 1 \\ 5 \end{bmatrix}, L(\mathbf{v_2}) = \begin{bmatrix} 1 \\ -10 \end{bmatrix}.$

(c) $\begin{bmatrix} -2 \\ 25 \end{bmatrix}.$

15. (a) $\left[L(\mathbf{v_1})\right]_T = \begin{bmatrix} 1 \\ 2 \\ -1 \end{bmatrix}, \left[L(\mathbf{v_2})\right]_T = \begin{bmatrix} 0 \\ 1 \\ -2 \end{bmatrix}.$

(b) $L(\mathbf{v_1}) = t^2 + t + 2, L(\mathbf{v_2}) = -t + 2.$
(c) $L(2t + 1) = \frac{3}{2}t^2 + t + 4.$

(d) $L(at + b) = \left(\dfrac{a + b}{2}\right) t^2 + bt + 2a.$

17. (a) $\begin{bmatrix} 0 & \frac{3}{2} \\ 1 & \frac{1}{2} \end{bmatrix}.$ (b) $\frac{3}{2}t - 3.$

(c) $\left(\dfrac{3a - b}{2}\right) t - b.$

19. $\begin{bmatrix} 1 & 0 \\ 4 & -1 \end{bmatrix}.$

21. (a) $\begin{bmatrix} 1 & 0 & 2 & 0 \\ 0 & 0 & 0 & 6 \\ 0 & 0 & 0 & 0 \\ 0 & 0 & 0 & 0 \end{bmatrix}.$

(b) $\begin{bmatrix} 0 & 0 & 0 & 0 \\ 0 & 0 & 0 & 0 \\ 6 & 0 & 0 & 0 \\ 0 & 1 & 0 & 1 \end{bmatrix}.$ (c) Same as (b).

23. $\begin{bmatrix} 1 & 0 \\ 0 & -1 \end{bmatrix}.$

ML.1. $A = \begin{bmatrix} -1 & 0 & 3 \\ 1 & 0 & -2 \end{bmatrix}.$

ML.3. (a) $A = \begin{bmatrix} 1.3333 & -0.3333 \\ -1.6667 & -3.3333 \end{bmatrix}.$

(b) $B = \begin{bmatrix} -3.6667 & 0.3333 \\ -3.3333 & 1.6667 \end{bmatrix}.$

(c) $P = \begin{bmatrix} -0.3333 & 0.6667 \\ 1.6667 & -0.3333 \end{bmatrix}.$

Supplementary Exercises, page 364

1. Yes.   **3.** $8t + 7$.

5. (a) Possible answer: $\{(1, 1, 1), (1, -1, 2)\}$.
   (b) No.

7. (a) $\begin{bmatrix} 2 \\ 2 \\ -4 \end{bmatrix}$.   (b) $2t^2 + 2t - 7$.

9. (a) $\begin{bmatrix} 2 \\ 1 \end{bmatrix}, \begin{bmatrix} -3 \\ 2 \end{bmatrix}$.   (b) $3t - 3, -t + 8$.
   (c) $-\frac{7}{3}t + \frac{35}{3}$.

11. $\begin{bmatrix} 1 & 0 & 0 \\ 0 & 1 & 0 \\ 0 & 0 & 1 \end{bmatrix}$.

13. Yes.

15. (b) The kernel of L consists of any continuous function f such that $L(f) = f(0) = 0$. That is, f is in ker L provided the value of f at $x = 0$ is zero.
   (c) Yes.

Chapter Test, page 365

1. $\begin{bmatrix} 2 \\ 7 \\ 4 \end{bmatrix}$.

2. (a) ker $L = \{(0, 0, 0)\}$, so there is no basis.
   (b) Yes.

3. (a) Possible answer: $\{(1, 1, 2), (1, -1, 1)\}$.
   (b) No.

4. 2.

5. $\begin{bmatrix} 0 & \frac{3}{2} \\ 1 & -\frac{5}{2} \end{bmatrix}$.

6. (a) F.   (b) T.   (c) T.   (d) F.   (e) F.

Cumulative Review of Part I, page 366

1. F.	**2.** T.	**3.** F.	**4.** F.	**5.** F.
6. T.	**7.** T.	**8.** T.	**9.** F.	**10.** T.
11. T.	**12.** F.	**13.** F.	**14.** F.	**15.** T.
16. T.	**17.** F.	**18.** F.	**19.** F.	**20.** T.
21. T.	**22.** F.	**23.** T.	**24.** T.	**25.** F.
26. T.	**27.** F.	**28.** F.	**29.** F.	**30.** T.

31. T.	**32.** T.	**33.** T.	**34.** T.	**35.** T.
36. F.	**37.** F.	**38.** F.	**39.** F.	**40.** T.
41. F.	**42.** F.	**43.** F.	**44.** T.	**45.** T.
46. F.	**47.** T.	**48.** T.	**49.** T.	**50.** T.
51. T.	**52.** F.	**53.** T.	**54.** T.	**55.** T.
56. F.	**57.** T.	**58.** T.	**59.** T.	**60.** F.
61. F.	**62.** T.	**63.** F.	**64.** F.	**65.** F.
66. F.	**67.** F.	**68.** T.	**69.** T.	**70.** T.
71. T.	**72.** F.	**73.** T.	**74.** T.	**75.** F.

Chapter 7

Section 7.1, page 387

1. Maximize $z = 120x + 100y$
subject to

$$2x + 2y \leq 8$$
$$5x + 3y \leq 15$$
$$x \geq 0, \quad y \geq 0.$$

3. Maximize $z = 0.08x + 0.10y$
subject to

$$x + y \leq 6000$$
$$x \geq 1500$$
$$y \leq 4000$$
$$y \leq \tfrac{1}{2}x$$
$$x \geq 0, \quad y \geq 0.$$

5. Maximize $z = 40{,}000 + 45{,}000y$
subject to

$$x + y \leq 30$$
$$y \geq 24$$
$$x \geq 2$$
$$x \leq 4$$
$$x \geq 0, \quad y \geq 0.$$

7. Maximize $z = 4x + 6y$
subject to

$$x + 2y \leq 10$$
$$x + y \leq 7$$
$$x \geq 0, \quad y \geq 0.$$

9. Minimize $z = 10x + 12y$
subject to

$$2x + 3y \geq 18$$
$$x + 3y \geq 12$$
$$80x + 60y \geq 480$$
$$x \geq 0, \quad y \geq 0.$$

11.

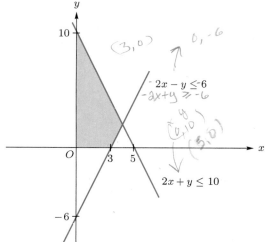

13.

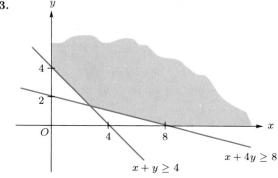

15. $x = \frac{8}{11}$, $y = \frac{45}{11}$, optimal value of $z = -\frac{21}{11}$.

17. Invest \$2000 in bond A and \$4000 in bond B;
maximum return is \$560.

19. Carry no containers from the Smith corporation
and 1500 containers from the Johnson Corporation,
or 120 containers from the Smith Corporation and
1440 containers from the Johnson Corporation. In
either case the maximum revenue is \$900.

21. Use $\frac{5}{2}$ gallons of L and $\frac{3}{2}$ gallons of H; minimum
cost is \$2.25.

23. No solution.   **25.** (a).

27. Maximize $z = 3x_1 - x_2 + 6x_3$
subject to

$$2x_1 + 4x_2 + x_3 \leq 4$$
$$3x_1 - 2x_2 + 3x_3 \leq 4$$
$$2x_1 + x_2 - x_3 \leq 8$$
$$x_1 \geq 0, \quad x_2 \geq 0, \quad x_3 \geq 0.$$

29. Maximize $z = 2x_1 + 3x_2 + 7x_3$
subject to

$$3x_1 + x_2 - 4x_3 + x_4 = 3$$
$$x_1 - 2x_2 + 6x_3 + x_5 = 21$$
$$x_1 - x_2 - x_3 + x_6 = 9$$
$$x_1 \geq 0, x_2 \geq 0, x_3 \geq 0, x_4 \geq 0, x_5 \geq 0, x_6 \geq 0.$$

Section 7.2, page 406

1.

	x	y	u	v	w	z	
u	3	-2	1	0	0	0	7
v	2	5	0	1	0	0	6
w	2	3	0	0	1	0	8
	-3	-7	0	0	0	1	0

3.

	x_1	x_2	x_3	x_4	x_5	x_6	x_7	z	
x_5	3	-2	1	1	1	0	0	0	6
x_6	1	1	1	1	0	1	0	0	8
x_7	2	-3	-1	2	0	0	1	0	10
	-2	-2	-3	-1	0	0	0	1	0

5. $x = 2$, $y = 0$, optimal $z = 4$.

7. No finite optimal solution.

9. $x_1 = 0$, $x_2 = \frac{33}{20}$, $x_3 = \frac{27}{10}$, optimal $z = \frac{69}{10}$.

11. $x_1 = 0$, $x_2 = 0$, $x_3 = 49$, $x_4 = 41$,
optimal $z = 156$.

13. Carry no containers from the Smith corporation
and 1500 containers from the Johnson Corporation,
or 120 containers from the Smith Corporation and
1440 containers from the Johnson Corporation. In
either case the maximum revenue is \$900.

15. Use 4 tons of gas, no coal and no oil; maximum
energy generated is 2000 kilowatt hours.

ML.3. $x = 0$, $y = 0.8571$, optimal $z = 4.286$.

ML.5. $x_1 = 1$, $x_2 = 0.3333$, $x_3 = 0$, optimal $z = 11$.

ML.7. Exercise 10: $x_1 = 0$, $x_2 = 2.5$, $x_3 = 0$, optimal
$z = 10$. Exercise 12: 1.5 tons of regular steel and
2.5 tons of special steel; maximum profit is \$430.

Section 7.3, page 414

1. Minimize $z' = 7w_1 + 6w_2 + 9w_3$
subject to

$$4w_1 + 5w_2 + 6w_3 \geq 3$$
$$3w_1 - 2w_2 + 8w_3 \geq 2$$
$$w_1 \geq 0, \quad w_2 \geq 0, \quad w_3 \geq 0.$$

3. Maximize $z' = 7w_1 + 12w_2 + 18w_3$
subject to

$$2w_1 + 8w_2 + 10w_3 \leq 3$$
$$3w_1 - 9w_2 + 15w_3 \leq 5$$
$$w_1 \geq 0, \quad w_2 \geq 0, \quad w_3 \geq 0.$$

7. $w_1 = \frac{5}{7}$, $w_2 = 0$, $w_3 = 0$, optimal $z' = \frac{30}{7}$.

9. $w_1 = 2$, $w_2 = 0$, $w_3 = 0$, optimal $z' = 10$.

Supplementary Exercises, page 415

1. $x = \frac{6}{5}$, $y = \frac{12}{5}$, optimal $z = \frac{48}{5}$.

3. $x = 0$, $y = 8$, or $x = 2$, $y = 7$, optimal $z = 800$.

5. $x_1 = \frac{18}{11}$, $x_2 = \frac{10}{11}$, optimal $z = \frac{158}{11}$.

Chapter Test, page 416

1. Plant no corn, plant 20 acres of wheat.

2. $x_1 = 1$, $x_2 = \frac{1}{3}$, $x_3 = 0$, optimal $z = 11$.

3. Maximize $z' = 8y_1 + 12y_2 + 6y_3$
subject to

$$y_1 + 2y_2 + 2y_3 \leq 3$$
$$4y_1 + 3y_2 + y_3 \leq 4$$
$$y_1 \geq 0, \quad y_2 \geq 0, \quad y_3 \geq 0.$$

Chapter 8

Section 8.1, page 432

1.

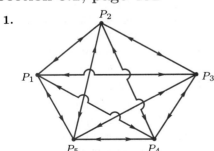

3. (a) and (c).

5. (a)

	P_1	P_2	P_3	P_4	P_5
P_1	0	1	0	0	0
P_2	1	0	1	0	1
P_3	1	0	0	1	0
P_4	0	1	0	0	0
P_5	0	0	0	1	0

(b)

	P_1	P_2	P_3	P_4	P_5	P_6
P_1	0	1	1	0	0	0
P_2	1	0	0	1	0	0
P_3	0	1	0	0	0	0
P_4	0	0	1	0	1	1
P_5	0	0	1	1	0	1
P_6	1	0	0	0	0	0

7. (b).

9. (a) One way: $P_2 \rightarrow P_5 \rightarrow P_1$.

(b) Two ways: $P_2 \rightarrow P_5 \rightarrow P_1 \rightarrow P_2$.
$P_2 \rightarrow P_5 \rightarrow P_3 \rightarrow P_2$.

11. P_2, P_3, and P_4.

13. There is no clique.

15. (a) Strongly connected.

(b) Not strongly connected.

17. P_1, P_2, or P_3.

ML.1. P_2, P_3, and P_4 form a clique.

ML.3. (a) Strongly connected.

(b) Not strongly connected.

Section 8.2, page 438

1. $I_1 = 15$ A from e to a, $I_2 = 8$ A from a to b,
$I_3 = 7$ A from a to c, $I_4 = 1$ A from d to c,
$I_5 = 16$ A from c to e.

3. $I_1 = 25$ A from f to c, $I_2 = 10$ A from c to b,
$I_3 = 15$ A from c to d, $I_4 = 5$ A from f to d,
$I_5 = 20$ A from e to a.

5. $I_1 = 5$ A from b to a, $I_2 = 8$ A from c to d,
$E = 40$ V.

7. $I_1 = 4$ A from f to a, $I_2 = 14$ A from c to b,
$I_3 = 18$ A from b to e, $I_4 = 24$ A from d to e,
$R = 1\,\Omega$, $E = 100$ V.

Section 8.3, page 447

1. (b) and (c).

3. (a) $\mathbf{x}^{(1)} = \begin{bmatrix} 0.7 \\ 0.3 \end{bmatrix}$, $\mathbf{x}^{(2)} = \begin{bmatrix} 0.61 \\ 0.39 \end{bmatrix}$, $\mathbf{x}^{(3)} = \begin{bmatrix} 0.583 \\ 0.417 \end{bmatrix}$.

   (b) $\mathbf{T} > \mathbf{0}$, hence it is regular; $\mathbf{u} = \begin{bmatrix} 0.571 \\ 0.429 \end{bmatrix}$.

5. (a) and (d).

9. (a)

$$T = \begin{matrix} & A & B \\ & \begin{bmatrix} 0.3 & 0.4 \\ 0.7 & 0.6 \end{bmatrix} & \begin{matrix} A \\ B \end{matrix} \end{matrix}$$

   (b) 0.364.

   (c) $\mathbf{u} = \begin{bmatrix} \frac{4}{11} \\ \frac{7}{11} \end{bmatrix} = \begin{bmatrix} 0.364 \\ 0.636 \end{bmatrix}$.

11. (a) 0.69.

   (b) 20.7 percent of the population will be farmers.

13. (a) 35 percent, 37.5 percent.

   (b) 40 percent.

ML.3. (a).

Section 8.4, page 459

1. $\widehat{\mathbf{x}} = \begin{bmatrix} \frac{24}{17} \\ -\frac{8}{17} \end{bmatrix} \approx \begin{bmatrix} 1.4118 \\ -0.4706 \end{bmatrix}$.

3. $\widehat{\mathbf{x}} \approx \begin{bmatrix} -1.5333 \\ -1.8667 \\ 4.2667 \end{bmatrix}$.

5. $y = 0.4x + 0.6$.

7. $y = 0.086x + 3.114$.

9. $y = 0.5718x^2 - 3.1314x + 3.4627$.

11. (a) $y = 0.426x + 0.827$.

   (b) 5.087.

13. (a) $y = 0.974x - 2.657$.

   (b) 10.979 millions of dollars.

15. $\widehat{\mathbf{x}} = \begin{bmatrix} -\frac{5}{11} \\ \frac{4}{11} \\ 0 \end{bmatrix}$.

ML.1. $y = 0.08571x + 3.114$.

ML.3. (a) $T = -8.278t + 188.1$, where $t = $ time.

   (b) $T(1) = 179.7778°$ F.
   $T(6) = 138.3889°$ F.
   $T(8) = 121.8333°$ F.

   (c) 3.3893 minutes.

ML.5. $y = 1.0204x^2 + 3.1238x + 1.0507$,
   when $x = 7$, $y = 72.9169$.

Section 8.5, page 468

1. (b) and (d).

3. $\begin{bmatrix} 4 \\ 0 \\ 3 \end{bmatrix}$.

5.

	Farmer	Carpenter	Tailor
Farmer	$\frac{2}{5}$	$\frac{1}{3}$	$\frac{1}{2}$
Carpenter	$\frac{2}{5}$	$\frac{1}{3}$	$\frac{1}{2}$
Tailor	$\frac{1}{5}$	$\frac{1}{3}$	0

$$\mathbf{p} = \begin{bmatrix} 75 \\ 75 \\ 40 \end{bmatrix}.$$

7. Not productive.   **9.** Productive.

11. (a) $\begin{bmatrix} 18 \\ 16 \end{bmatrix}$.   (b) $\begin{bmatrix} 12 \\ 8 \end{bmatrix}$.

Section 8.6, page 479

1. (a) $\mathbf{x}(t) = \begin{bmatrix} x_1(t) \\ x_2(t) \\ x_3(t) \end{bmatrix} = \begin{bmatrix} b_1 e^{-3t} \\ b_2 e^{4t} \\ b_3 e^{2t} \end{bmatrix}$

$$= b_1 \begin{bmatrix} 1 \\ 0 \\ 0 \end{bmatrix} e^{-3t} + b_2 \begin{bmatrix} 0 \\ 1 \\ 0 \end{bmatrix} e^{4t} + b_3 \begin{bmatrix} 0 \\ 0 \\ 1 \end{bmatrix} e^{2t}.$$

   (b) $\begin{bmatrix} 3e^{-3t} \\ 4e^{4t} \\ 5e^{2t} \end{bmatrix} = 3 \begin{bmatrix} 1 \\ 0 \\ 0 \end{bmatrix} e^{-3t} + 4 \begin{bmatrix} 0 \\ 1 \\ 0 \end{bmatrix} e^{4t} + 5 \begin{bmatrix} 0 \\ 0 \\ 1 \end{bmatrix} e^{2t}.$

3. $\mathbf{x}(t) = b_1 \begin{bmatrix} 6 \\ 2 \\ 7 \end{bmatrix} e^{4t} + b_2 \begin{bmatrix} 0 \\ 7 \\ -1 \end{bmatrix} e^{-5t} + b_3 \begin{bmatrix} 0 \\ 0 \\ 1 \end{bmatrix} e^{2t}.$

5. $\mathbf{x}(t) = b_1 \begin{bmatrix} 1 \\ 0 \\ 0 \end{bmatrix} e^{5t} + b_2 \begin{bmatrix} 0 \\ 1 \\ 3 \end{bmatrix} e^{5t} + b_3 \begin{bmatrix} 0 \\ -3 \\ 1 \end{bmatrix} e^{-5t}.$

7. $\mathbf{x}(t) = b_1 \begin{bmatrix} 1 \\ 0 \\ 1 \end{bmatrix} e^{4t} + b_2 \begin{bmatrix} -3 \\ 0 \\ 2 \end{bmatrix} e^{-t} + b_3 \begin{bmatrix} 1 \\ -6 \\ 4 \end{bmatrix} e^{t}.$

9. $\mathbf{x}(t) = 220 \begin{bmatrix} 2 \\ 1 \end{bmatrix} + 20 \begin{bmatrix} 3 \\ -1 \end{bmatrix} e^{-5t} = \begin{bmatrix} 440 + 60e^{-5t} \\ 220 - 20e^{-5t} \end{bmatrix}.$

ML.1. $\mathbf{x}(t) = b_1 \begin{bmatrix} -0.5774 \\ -0.5774 \\ -0.5774 \end{bmatrix} e^t +$

$b_2 \begin{bmatrix} 0.2182 \\ 0.4364 \\ 0.8729 \end{bmatrix} e^{2t} + b_3 \begin{bmatrix} 0.0605 \\ 0.2421 \\ 0.9684 \end{bmatrix} e^{4t}.$

ML.3. $\mathbf{x}(t) = b_1 \begin{bmatrix} -0.8321 \\ 0 \\ 0.5547 \end{bmatrix} e^{-t} +$

$b_2 \begin{bmatrix} -0.7071 \\ 0 \\ -0.7071 \end{bmatrix} e^{4t} + b_3 \begin{bmatrix} -0.1374 \\ 0.8242 \\ -0.5494 \end{bmatrix} e^t.$

Section 8.7, page 483

3. (a) $u_8 = 34.$ (b) $u_{12} = 233.$
(c) $u_{20} = 10{,}946.$

Section 8.8, page 492

1. (a) $\begin{bmatrix} x & y \end{bmatrix} \begin{bmatrix} -3 & \frac{5}{2} \\ \frac{5}{2} & -2 \end{bmatrix} \begin{bmatrix} x \\ y \end{bmatrix}.$

(b) $\begin{bmatrix} x_1 & x_2 & x_3 \end{bmatrix} \begin{bmatrix} 2 & \frac{3}{2} & -\frac{5}{2} \\ \frac{3}{2} & 0 & \frac{7}{2} \\ -\frac{5}{2} & \frac{7}{2} & 0 \end{bmatrix} \begin{bmatrix} x_1 \\ x_2 \\ x_3 \end{bmatrix}.$

(c) $\begin{bmatrix} x_1 & x_2 & x_3 \end{bmatrix} \begin{bmatrix} 3 & \frac{1}{2} & -1 \\ \frac{1}{2} & 1 & -2 \\ -1 & -2 & -2 \end{bmatrix} \begin{bmatrix} x_1 \\ x_2 \\ x_3 \end{bmatrix}.$

3. (a) $\begin{bmatrix} -1 & 0 & 0 \\ 0 & 2 & 0 \\ 0 & 0 & 0 \end{bmatrix}.$ (b) $\begin{bmatrix} 3 & 0 & 0 \\ 0 & 0 & 0 \\ 0 & 0 & 0 \end{bmatrix}.$

5. $2x'^2 - 3y'^2.$ **7.** $y_2^2 - y_3^2.$

9. $-2y_1^2 + 5y_2^2 - 5y_3^2.$ **11.** $y_1'^2.$

13. $y_1^2 + y_2^2 - y_3^2.$ **15.** $y_1^2 - y_2^2.$

17. $h(\mathbf{y}) = y_1^2 - y_2^2,$ rank of g is 2 and signature of g is 0.

19. $y_1^2 + y_2^2 = 1$ is a circle.

$-y_1^2 - y_2^2 = 1$ is empty; it represents no conic.

$y_1^2 - y_2^2 = 1$ is a hyperbola.

$y_1^2 = 1$ is a pair of lines; $y_1 = 1,\ y_1 = -1.$

$-y_1^2 = 1$ is empty; it represents no conic.

21. $g_1,\ g_2,$ and $g_4.$

23. (a), (b), and (c).

ML.1. (a) rank $= 2,$ signature $= 0.$
(b) rank $= 1,$ signature $= 1.$
(c) rank $= 4,$ signature $= 2.$
(d) rank $= 4,$ signature $= 4.$

Section 8.9, page 501

1. Ellipse. **3.** Hyperbola.

5. Two intersecting lines. **7.** Circle.

9. Point.

11. Ellipse; $\dfrac{x'^2}{2} + y'^2 = 1.$

13. Circle; $\dfrac{x'^2}{5^2} + \dfrac{y'^2}{5^2} = 1.$

15. Pair of parallel lines; $y' = 2,\ y' = -2;\ y'^2 = 4.$

17. Point $(1, 3);\ x'^2 + y'^2 = 0.$

19. Possible answer: ellipse; $\dfrac{x'^2}{12} + \dfrac{y'^2}{4} = 1.$

21. Possible answer: pair of parallel lines $y' = \frac{2}{5}$ and $y' = -\frac{2}{5};\ y'^2 = \frac{4}{10}.$

23. Possible answer: two intersecting lines $y' = 3x'$ and $y' = -3x';\ 9x'^2 - y'^2 = 0.$

25. Possible answer: parabola; $y''^2 = -4x''.$

27. Possible answer: hyperbola; $\dfrac{x''^2}{4} - \dfrac{y''^2}{9} = 1.$

29. Possible answer: hyperbola; $\dfrac{x''^2}{\frac{9}{8}} - \dfrac{y''^2}{\frac{9}{8}} = 1.$

Section 8.10, page 511

1. Hyperboloid of one sheet.

3. Hyperbolic paraboloid.

5. Parabolic cylinder.

7. Parabolic cylinder.

9. Ellipsoid.

11. Elliptic paraboloid.

13. Hyperbolic paraboloid.

15. Ellipsoid; $x'^2 + y'^2 + \dfrac{z'^2}{\frac{1}{3}} = 1.$

17. Hyperbolic paraboloid; $\dfrac{x''^2}{4} - \dfrac{y''^2}{4} = z''$.

19. Elliptic paraboloid; $\dfrac{x'^2}{4} + \dfrac{y'^2}{8} = 1$.

21. Hyperboloid of one sheet; $\dfrac{x''^2}{2} + \dfrac{y''^2}{4} - \dfrac{z''}{4} = 1$.

23. Parabolic cylinder; $x''^2 = \dfrac{4}{\sqrt{2}}\, y''$.

25. Hyperboloid of two sheets;

$$\frac{x''^2}{\frac{7}{4}} - \frac{y''^2}{\frac{7}{4}} - \frac{z''^2}{\frac{7}{4}} = 1.$$

27. Cone; $x''^2 + y''^2 - z''^2 = 0$.

Section 8.11, page 526

1.

$$
\begin{array}{c}
 & C \\
 & \begin{array}{cc} \text{2 fingers shown} & \text{3 fingers shown} \end{array} \\
R\begin{array}{c} \text{2 fingers shown} \\ \text{3 fingers shown} \end{array} & \left[\begin{array}{cc} -4 & 5 \\ 5 & -6 \end{array}\right].
\end{array}
$$

3.

$$
\begin{array}{c}
 & \text{Firm } B \\
 & \begin{array}{cc} \text{Abington} & \text{Wyncote} \end{array} \\
\text{Firm } A\begin{array}{c} \text{Abington} \\ \text{Wyncote} \end{array} & \left[\begin{array}{cc} 50 & 60 \\ 25 & 50 \end{array}\right].
\end{array}
$$

5. (a) $\begin{bmatrix} 5 & ④ \\ 3 & -2 \end{bmatrix}$. (b) $\begin{bmatrix} 2 & 1 & ⓪ \\ 3 & 1 & -2 \\ 4 & 2 & -4 \end{bmatrix}$.

(c) $\begin{bmatrix} ③ & 4 & 5 \\ -2 & 5 & 1 \\ -1 & 0 & 1 \end{bmatrix}$.

(d) $\begin{bmatrix} 5 & ② & 4 & ② \\ 0 & -1 & 2 & 0 \\ 3 & ② & 3 & ② \\ 1 & 0 & -1 & -1 \end{bmatrix}$.

7. (a) $\mathbf{p} = \begin{bmatrix} 1 & 0 \end{bmatrix}$, $\mathbf{q} = \begin{bmatrix} 0 \\ 1 \\ 0 \end{bmatrix}$, $v = 1$.

(b) $\mathbf{p} = \begin{bmatrix} 0 & 0 & 1 \end{bmatrix}$, $\mathbf{q} = \begin{bmatrix} 1 \\ 0 \\ 0 \\ 0 \end{bmatrix}$, $v = 0$.

(c) $\mathbf{p} = \begin{bmatrix} 1 & 0 \end{bmatrix}$ or $\begin{bmatrix} 0 & 1 \end{bmatrix}$, $\mathbf{q} = \begin{bmatrix} 0 \\ 1 \end{bmatrix}$, $v = 4$.

9. (a) $\frac{19}{36}$. (b) $\frac{1}{7}$.

11. $p_1 = \frac{9}{14}$, $p_2 = \frac{5}{14}$, $q_1 = \frac{1}{2}$, $q_2 = \frac{1}{2}$, $v = -\frac{1}{2}$.

13. $p_1 = 0$, $p_2 = \frac{4}{5}$, $p_3 = \frac{1}{5}$, $q_1 = 0$, $q_2 = \frac{3}{5}$, $q_3 = \frac{2}{5}$, $v = \frac{22}{5}$.

15. $\mathbf{p} = \begin{bmatrix} 0 & \frac{3}{8} & -\frac{5}{8} & 0 \end{bmatrix}$, $\mathbf{q} = \begin{bmatrix} \frac{5}{8} \\ \frac{3}{8} \\ 0 \\ 0 \end{bmatrix}$, $v = \frac{1}{8}$.

17. $\mathbf{p} = \begin{bmatrix} \frac{1}{3} & \frac{1}{3} & \frac{1}{3} \end{bmatrix}$, $\mathbf{q} = \begin{bmatrix} \frac{1}{3} \\ \frac{1}{3} \\ \frac{1}{3} \end{bmatrix}$, $v = 0$.

19. $\mathbf{p} = \begin{bmatrix} \frac{2}{3} & \frac{1}{3} \end{bmatrix}$, $\mathbf{q} = \begin{bmatrix} \frac{1}{2} \\ \frac{1}{2} \end{bmatrix}$, $v = 0$.

Supplementary Exercises, page 529

1. 2.

3. (a) $\begin{bmatrix} 0.2 & 0.6 \\ 0.8 & 0.4 \end{bmatrix}$. (b) 0.4432. (c) $\frac{3}{7}$.

5. $y = -0.2413x^2 + 0.8118x + 2.6627$.

7. (a) $\mathbf{x} = b_1 \begin{bmatrix} 1 \\ 1 \end{bmatrix} e^{2t} + b_2 \begin{bmatrix} -1 \\ 3 \end{bmatrix} e^{-2t}$.

(b) $\mathbf{x} = \frac{9}{2} \begin{bmatrix} 1 \\ 1 \end{bmatrix} e^{2t} + \frac{1}{2} \begin{bmatrix} -1 \\ 3 \end{bmatrix} e^{-2t}$.

9. $y_1^2 + y_2^2$.

Chapter Test, page 530

1. There are two cliques: P_1, P_3, P_5 and P_1, P_3, P_6.

2. $I_1 = 5$ A from a to f, $I_2 = 13$ A from b to a, $R_1 = 3\,\Omega$, $R_2 = 4\,\Omega$, $I_3 = 14$ A from c to d, $I_4 = 22$ A from d to b.

3. (a) $\begin{bmatrix} 0.4 & 0.5 \\ 0.6 & 0.5 \end{bmatrix}$. (b) 0.4546. (c) $\frac{6}{11}$.

4. (a) $y = \frac{173}{290}x + \frac{31}{58}$.

(b) Approximately 1247 calories.

5. \$5.65 million of steel, \$5.41 million of coal, and \$2.83 million of transportation.

6. $\mathbf{x} = b_1 \begin{bmatrix} 1 \\ 1 \end{bmatrix} e^{5t} + b_2 \begin{bmatrix} -1 \\ 3 \end{bmatrix} e^{-3t}$.

7. 1,346,269.

8. $y_1^2 - y_2^2$, a hyperbola.

9. $\mathbf{p} = \begin{bmatrix} \frac{3}{8} & \frac{5}{8} \end{bmatrix}$, $\mathbf{q} = \begin{bmatrix} \frac{1}{8} \\ \frac{7}{8} \\ 0 \end{bmatrix}$, $v = \frac{11}{8}$.

10. $a_{11} = 2$ is a saddle point for any value of a.

Chapter 9

Section 9.1, page 537

1. 0.3472×10^2.

3. -0.2840×10^3.

5. 0.1230×10^1, 0.1230×10^1.

7. 0.5666×10^1, 0.5667×10^1.

9. $\varepsilon_a = 0.21 \times 10^{-1}$, $\varepsilon_r = 0.17 \times 10^{-2}$.

11. $\varepsilon_a = 0.11 \times 10^1$, $\varepsilon_r = 0.17 \times 10^{-3}$.

ML.1.
```
format short e
pi
ans =
        3.1416e+000
format long e
pi
ans =
        3.141592653589793e+000
format short
```

Section 9.2, page 547

1. $\begin{bmatrix} 1 & -2 & 0 \\ 0 & 1 & -1 \\ 0 & 0 & 1 \end{bmatrix}$.

3. $x_1 = 2$, $x_2 = -3$, $x_3 = 4$.

5. $x_1 \approx 0.004$, $x_2 \approx 1.006$, $x_3 \approx -2.000$.

7. $x_1 \approx 2.393$, $x_2 \approx 4.565$, $x_3 \approx -3.607$.

9. The solution to the first linear system is $x \approx 3.2305$, $y \approx 1.8571$. The solution to the second linear system is $x \approx -3.6821$, $y \approx 6.1429$.

11.

	Jacobi	Gauss–Seidel	Exact
x	-1.993	-2.004	-2
y	4.998	5.002	5

13.

	Jacobi	Gauss–Seidel	Exact
x_1	1.559	1.658	$\frac{5}{3}$
x_2	1.685	1.513	$\frac{3}{2}$
x_3	-0.313	-0.492	$-\frac{1}{2}$

15. (a) Exact solution is $x = -2$, $y = 3$.
    (b) Exact solution is $x = -3$, $y = 2$.

ML.3. $x_1 = 0.0000$, $x_2 = 1.0000$, $x_3 = -2.0000$.

Section 9.3, page 555

1. $\mathbf{x} = \begin{bmatrix} 1 \\ 2 \\ 1 \end{bmatrix}$.   **3.** $\mathbf{x} = \begin{bmatrix} 1 \\ 0 \\ 2 \\ -4 \end{bmatrix}$.

5. $L = \begin{bmatrix} 1 & 0 & 0 \\ 2 & 1 & 0 \\ 2 & -2 & 1 \end{bmatrix}$, $U = \begin{bmatrix} 2 & 3 & 4 \\ 0 & -1 & 2 \\ 0 & 0 & -2 \end{bmatrix}$,

$\mathbf{x} = \begin{bmatrix} 4 \\ -2 \\ 1 \end{bmatrix}$.

7. $L = \begin{bmatrix} 1 & 0 & 0 \\ 0.5 & 1 & 0 \\ 0.25 & -1.5 & 1 \end{bmatrix}$, $U = \begin{bmatrix} 4 & 2 & 3 \\ 0 & -1 & 3.5 \\ 0 & 0 & 5.5 \end{bmatrix}$,

$\mathbf{x} = \begin{bmatrix} 2 \\ -2 \\ -1 \end{bmatrix}$.

9. $L = \begin{bmatrix} 1 & 0 & 0 & 0 \\ 0.5 & 1 & 0 & 0 \\ -1 & 0.2 & 1 & 0 \\ 2 & 0.4 & 2 & 1 \end{bmatrix}$,

$U = \begin{bmatrix} 2 & 1 & 0 & -4 \\ 0 & -0.5 & 0.25 & 1 \\ 0 & 0 & 0.2 & 2 \\ 0 & 0 & 0 & 2 \end{bmatrix}$,

$\mathbf{x} = \begin{bmatrix} 0.5 \\ 2 \\ -2 \\ 1.5 \end{bmatrix}$.

ML.1. $L = \begin{bmatrix} 1 & 0 & 0 \\ 1 & 1 & 0 \\ 0.5 & 0.3333 & 1 \end{bmatrix}$,

$U = \begin{bmatrix} 2 & 8 & 0 \\ 0 & -6 & -3 \\ 0 & 0 & 8 \end{bmatrix}$.

ML.3. $L = \begin{bmatrix} 1.0000 & 0 & 0 & 0 \\ 0.5000 & 1.0000 & 0 & 0 \\ -2.0000 & -2.0000 & 1.0000 & 0 \\ -1.0000 & 1.0000 & -2.0000 & 1.0000 \end{bmatrix}$,

$U = \begin{bmatrix} 6 & -2 & -4 & 4 \\ 0 & -2 & -4 & -1 \\ 0 & 0 & 5 & -2 \\ 0 & 0 & 0 & 8 \end{bmatrix}$,

$\mathbf{z} = \begin{bmatrix} 2 \\ -5 \\ 2 \\ -32 \end{bmatrix}$, $\mathbf{x} = \begin{bmatrix} 4.5000 \\ 6.9000 \\ -1.2000 \\ -4.0000 \end{bmatrix}$.

Section 9.4, page 560

1. $Q = \begin{bmatrix} \frac{1}{\sqrt{2}} & \frac{1}{\sqrt{2}} \\ -\frac{1}{\sqrt{2}} & \frac{1}{\sqrt{2}} \end{bmatrix} \approx \begin{bmatrix} 0.7071 & 0.7071 \\ -0.7071 & 0.7071 \end{bmatrix}$,

$R = \begin{bmatrix} \sqrt{2} & -\frac{1}{\sqrt{2}} \\ 0 & \frac{5}{\sqrt{2}} \end{bmatrix} \approx \begin{bmatrix} 1.4142 & 0.7071 \\ 0 & 3.5355 \end{bmatrix}$.

3. $Q = \begin{bmatrix} \frac{1}{\sqrt{6}} & \frac{4}{\sqrt{21}} & -\frac{1}{\sqrt{14}} \\ \frac{2}{\sqrt{6}} & -\frac{1}{\sqrt{21}} & \frac{2}{\sqrt{14}} \\ -\frac{1}{\sqrt{6}} & \frac{2}{\sqrt{21}} & \frac{3}{\sqrt{14}} \end{bmatrix}$

$\approx \begin{bmatrix} 0.4082 & 0.8729 & -0.2673 \\ 0.8165 & -0.2182 & 0.5345 \\ -0.4082 & 0.4364 & 0.8018 \end{bmatrix}$,

$R = \begin{bmatrix} \frac{6}{\sqrt{6}} & -\frac{8}{\sqrt{6}} & \frac{1}{\sqrt{6}} \\ 0 & \frac{7}{\sqrt{21}} & \frac{1}{\sqrt{21}} \\ 0 & 0 & \frac{19}{\sqrt{14}} \end{bmatrix}$

$\approx \begin{bmatrix} 2.4495 & -3.2660 & 0.4082 \\ 0 & 1.5275 & 0.2182 \\ 0 & 0 & 5.0780 \end{bmatrix}$.

5. $Q = \begin{bmatrix} -\frac{1}{\sqrt{3}} & 0 & \frac{2}{\sqrt{6}} \\ \frac{1}{\sqrt{3}} & -\frac{1}{\sqrt{2}} & \frac{1}{\sqrt{6}} \\ \frac{1}{\sqrt{3}} & \frac{1}{\sqrt{2}} & \frac{1}{\sqrt{6}} \end{bmatrix}$

$\approx \begin{bmatrix} -0.5774 & 0 & 0.8165 \\ 0.5774 & -0.7071 & 0.4082 \\ 0.5774 & 0.7071 & 0.4082 \end{bmatrix}$,

$R = \begin{bmatrix} -\sqrt{3} & 0 & 0 \\ 0 & -\sqrt{8} & \sqrt{2} \\ 0 & 0 & \sqrt{6} \end{bmatrix}$

$\approx \begin{bmatrix} -1.7321 & 0 & 0 \\ 0 & -2.8284 & 1.4142 \\ 0 & 0 & -2.4495 \end{bmatrix}$.

Section 9.5, page 568

		Approximate			Exact	
	Eigenvalue	Eigenvector		Eigenvalue	Eigenvector	
1.	-4.974	$\begin{bmatrix} -1.000 & 0.527 \end{bmatrix}^T$		-5	$\begin{bmatrix} -1 & \frac{1}{2} \end{bmatrix}^T$	
3.	6.000	$\begin{bmatrix} 1.000 & 0.667 \end{bmatrix}^T$		6	$\begin{bmatrix} 1 & \frac{2}{3} \end{bmatrix}^T$	
5.	9.974	$\begin{bmatrix} 1.000 & 0.244 \end{bmatrix}^T$		10	$\begin{bmatrix} 1 & \frac{1}{4} \end{bmatrix}^T$	
7.	3.998	$\begin{bmatrix} 0.707 & 0.707 \end{bmatrix}^T$		4	$\begin{bmatrix} \frac{1}{\sqrt{2}} & \frac{1}{\sqrt{2}} \end{bmatrix}^T$	
	0.000	$\begin{bmatrix} -0.707 & 0.707 \end{bmatrix}^T$		0	$\begin{bmatrix} -\frac{1}{\sqrt{2}} & \frac{1}{\sqrt{2}} \end{bmatrix}^T$	
9.	6.001	$\begin{bmatrix} 0.666 & 0.666 & 0.334 \end{bmatrix}^T$		6	$\begin{bmatrix} \frac{2}{3} & \frac{2}{3} & \frac{1}{3} \end{bmatrix}^T$	
	11.989	$\begin{bmatrix} -0.333 & 0.666 & -0.666 \end{bmatrix}^T$		12	$\begin{bmatrix} -\frac{1}{3} & \frac{2}{3} & -\frac{2}{3} \end{bmatrix}^T$	
	8.998	$\begin{bmatrix} -0.666 & 0.331 & 0.667 \end{bmatrix}^T$		9	$\begin{bmatrix} -\frac{2}{3} & \frac{1}{3} & \frac{2}{3} \end{bmatrix}^T$	

Supplementary Exercises, page 569

1.

	Jacobi	Gauss–Seidel	Exact
x	−2.9999	−3.000	−3
y	1.9999	2.001	2

$$U = \begin{bmatrix} -2 & 1 & -2 \\ 0 & 4 & 3 \\ 0 & 0 & -3 \end{bmatrix}, \ \mathbf{x} = \begin{bmatrix} -1 \\ -2 \\ 3 \end{bmatrix}.$$

5. $y = 0.6860x - 2.7326.$

3. $L = \begin{bmatrix} 1 & 0 & 0 \\ -3 & 1 & 0 \\ 2 & 4 & 1 \end{bmatrix},$

7.

	Approximate		Exact	
Eigenvalue	Eigenvector	Eigenvalue	Eigenvector	
2.000	$\begin{bmatrix} 0.707 & -0.707 & 0.000 \end{bmatrix}^T$	2	$\begin{bmatrix} 1 & 1 & 1 \end{bmatrix}^T$	
−1.001	$\begin{bmatrix} 0.578 & 0.578 & -0.577 \end{bmatrix}^T$	−1	$\begin{bmatrix} 1 & -1 & 0 \end{bmatrix}^T$	
2.000	$\begin{bmatrix} -0.408 & -0.408 & 0.817 \end{bmatrix}^T$	−2	$\begin{bmatrix} -1 & -1 & 2 \end{bmatrix}^T$	

Chapter Test, page 569

1.

	Jacobi	Gauss–Seidel	Exact
x	4.002	4.002	4
y	−3.000	−3.000	−3

3. $Q = \begin{bmatrix} \frac{2}{3} & \frac{5}{\sqrt{90}} \\ -\frac{1}{3} & -\frac{4}{\sqrt{90}} \\ -\frac{2}{3} & \frac{7}{\sqrt{90}} \end{bmatrix} \approx \begin{bmatrix} 0.6667 & 0.5270 \\ -0.3333 & -0.4216 \\ -0.6667 & 0.7379 \end{bmatrix},$

2. $L = \begin{bmatrix} 1 & 0 & 0 \\ -4 & 1 & 0 \\ 2 & -3 & 1 \end{bmatrix},$

$U = \begin{bmatrix} 2 & 2 & -1 \\ 0 & -3 & 1 \\ 0 & 0 & -2 \end{bmatrix}, \ \mathbf{x} = \begin{bmatrix} 2.25 \\ 3.50 \\ 8.50 \end{bmatrix}.$

$R = \begin{bmatrix} 3 & -1 \\ 0 & \frac{10}{\sqrt{10}} \end{bmatrix} \approx \begin{bmatrix} 3.0000 & -1.0000 \\ 0 & 3.1623 \end{bmatrix}.$

		Approximate		Exact	
	Eigenvalue	Eigenvector	Eigenvalue	Eigenvector	
4.	9.001	$\begin{bmatrix} 0.600 & 1.000 \end{bmatrix}^T$	9	$\begin{bmatrix} 3 & 5 \end{bmatrix}^T$	
5.	3.000	$\begin{bmatrix} 0.577 & 0.577 & 0.577 \end{bmatrix}^T$	3	$\begin{bmatrix} 1 & 1 & 1 \end{bmatrix}^T$	
	−3.000	$\begin{bmatrix} -0.707 & 0.707 & 0.000 \end{bmatrix}^T$	−3	$\begin{bmatrix} -1 & 1 & 0 \end{bmatrix}^T$	
	−3.000	$\begin{bmatrix} -0.408 & -0.408 & -0.816 \end{bmatrix}^T$	−3	$\begin{bmatrix} -1 & -1 & 2 \end{bmatrix}^T$	

Appendix A

Section A.1, page A8

1. (a) $4 + 2i$.   (b) $-4 - 3i$.   (c) $11 - 2i$.
   (d) $-3 + i$.   (e) $-3 + 6i$.   (f) $-2 - i$.
   (g) $7 - 11i$.   (h) $-9 + 13i$.

3. (a)

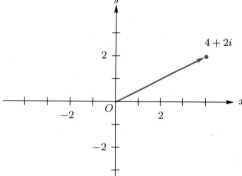

(b)

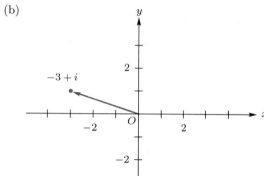

(c)

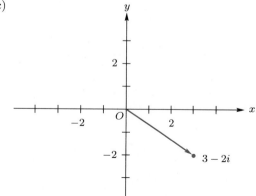

(d)

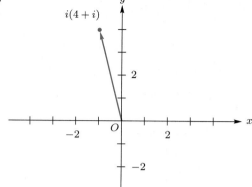

5.

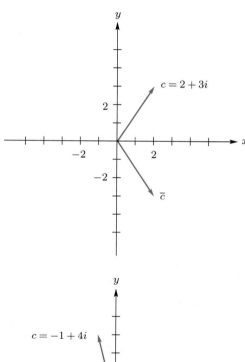

7. $A^2 = \begin{bmatrix} -1 & 0 \\ 0 & -1 \end{bmatrix}$, $A^3 = \begin{bmatrix} 0 & -i \\ -i & 0 \end{bmatrix}$,

$A^4 = \begin{bmatrix} 1 & 0 \\ 0 & 1 \end{bmatrix}$, $A^{4n} = I_2$, $A^{4n+1} = A$,

$A^{4n+2} = A^2 = -I_2$, $A^{4n+3} = A^3 = -A$.

9. (a) $\dfrac{-1 \pm i\sqrt{3}}{2}$. (b) $-2, \pm i$. (c) $\pm 1, \pm i$.

11. (a) Possible answers: $A_1 = \begin{bmatrix} i & 0 \\ 0 & i \end{bmatrix}$,

$A_2 = \begin{bmatrix} -i & 0 \\ 0 & -i \end{bmatrix}$.

13. (a) Possible answers: $\begin{bmatrix} i & 0 \\ 0 & 0 \end{bmatrix}$, $\begin{bmatrix} -i & 0 \\ 0 & 0 \end{bmatrix}$.

(b) Possible answers: $\begin{bmatrix} i & -i \\ -i & i \end{bmatrix}$, $\begin{bmatrix} -i & i \\ i & -i \end{bmatrix}$.

Section A.2, page A20

1. (a) No solution. (b) No solution.

(c) $x_1 = \frac{3}{4} + \frac{5}{4}i$, $x_2 = \frac{3}{2} + i$.

3. (a) $x_1 = i$, $x_2 = 1$, $x_3 = 1 - i$.

(b) $x_1 = 0$, $x_2 = -i$, $x_3 = i$.

5. (a) $\dfrac{1}{5} \begin{bmatrix} 2+i & 2-4i \\ 3-i & -2-i \end{bmatrix}$.

(b) $\dfrac{1}{6} \begin{bmatrix} i & 1-3i & 1 \\ -2-3i & 2i & 3+2i \\ 1 & 2i & -i \end{bmatrix}$.

7. (a) Yes. (b) Linearly independent.

9. (a) No. (b) No. (c) No. (d) Yes.

11. (a) The eigenvalues are $\lambda_1 = 1 + i$, $\lambda_2 = 1 - i$. Associated eigenvectors are

$$\mathbf{x}_1 = \begin{bmatrix} -i \\ 1 \end{bmatrix} \quad \text{and} \quad \mathbf{x}_2 = \begin{bmatrix} i \\ 1 \end{bmatrix}.$$

(b) The eigenvalues are $\lambda_1 = 0$, $\lambda_2 = 2$. Associated eigenvectors are

$$\mathbf{x}_1 = \begin{bmatrix} -i \\ 1 \end{bmatrix} \quad \text{and} \quad \mathbf{x}_2 = \begin{bmatrix} i \\ 1 \end{bmatrix}.$$

(c) The eigenvalues are $\lambda_1 = 1$, $\lambda_2 = 2$, $\lambda_3 = 3$. Associated eigenvectors are

$$\mathbf{x}_1 = \begin{bmatrix} 0 \\ -i \\ 1 \end{bmatrix}, \quad \mathbf{x}_2 = \begin{bmatrix} 1 \\ 0 \\ 0 \end{bmatrix}, \quad \text{and} \quad \mathbf{x}_3 = \begin{bmatrix} 0 \\ i \\ 1 \end{bmatrix}.$$

Appendix B

Section B.1, page A30

7. -8. **9.** 4. **11.** -4.

13. (a) $\sqrt{22}$. (b) 6. (c) $\sqrt{18}$.

15. (a) $\sqrt{\frac{1}{30}}$. (b) $\sqrt{\frac{1}{2}(e^2 - e^{-2}) - 2}$.

17. (a) $-\frac{1}{2}$. (b) $\dfrac{2\sin^2 1}{\sqrt{4 - \sin^2 2}}$. **19.** $a = 0$.

21. $B = \begin{bmatrix} b_{11} & b_{12} \\ b_{21} & b_{22} \end{bmatrix}$ with $b_{11} + 3b_{21} + 2b_{12} + 4b_{22} = 0$.

23. (b) $\{(0, 0, 1, 0), (-1, 1, 0, 1)\}$.

25. (a) $\{\sqrt{3}\,t, 2 - 3t\}$.

(b) $2t - 1 = \frac{\sqrt{3}}{6}(\sqrt{3}\,t) - \frac{1}{2}(2 - 3t)$.

27. $\left\{ \sqrt{3}\,t, \dfrac{\sin 2\pi t + \left(\frac{3}{2\pi}\right)t}{\sqrt{\frac{1}{2} - \frac{3}{4\pi^2}}} \right\}$.

29. $\left\{ \frac{45}{14}t^3 - \frac{55}{14}t^2 + t, \ \frac{130}{7}t^3 - \frac{120}{7}t^2 + 1 \right\}$.

31. $2\sin t$.

33. $\mathbf{w} = 2\sin t - 1$, $\mathbf{u} = t - 1 - (2\sin t - 1) = t - 2\sin t$.

35. $\sqrt{\frac{2}{3}\pi^3 - 4\pi}$.

Section B.2, page A38

1. (a) $(-3, -3, -1)$. (b) $(4x + y, 3x, 3x + 2y)$.

3. (a) $6t^2 - 2t$. (b) $2at^2 - bt$.

5. (a) $(-3, 6)$. (b) $(1, 4)$.

(c) $(x - 2y, 3y)$. (d) $(x, -2x + 3y)$.

7. (a) $B = \begin{bmatrix} 2 & 2 \\ -3 & -1 \\ 2 & 2 \end{bmatrix}$.

(b) $A_1 = \begin{bmatrix} \frac{1}{2} & \frac{3}{2} \\ \frac{3}{2} & \frac{1}{2} \end{bmatrix}$, $A_2 = \begin{bmatrix} 2 & 2 \\ -3 & -1 \\ 2 & 2 \end{bmatrix}$.

9. (a) $\begin{bmatrix} -1 & 3 \\ -5 & 5 \end{bmatrix}$. (b) $\begin{bmatrix} -4 & 7 \\ -6 & 8 \end{bmatrix}$.

11. Not invertible.

13. Invertible. $L^{-1}\left(\begin{bmatrix} b_1 \\ b_2 \\ b_3 \end{bmatrix} \right) = \begin{bmatrix} 2b_1 - b_3 \\ b_1 + b_2 - b_3 \\ -b_1 + b_3 \end{bmatrix}$.

15. Not invertible.

17. Invertible. $L^{-1}(dt^2 + et + f) = -dt^2 + et - f$.

19. Invertible.   **21.** Not invertible.

23. $\begin{bmatrix} \frac{1}{2} & -\frac{1}{2} & \frac{1}{2} \\ -\frac{3}{2} & \frac{3}{2} & -\frac{1}{2} \\ \frac{1}{2} & \frac{1}{2} & -\frac{1}{2} \end{bmatrix}$.

25. $\begin{bmatrix} -\frac{9}{2} & -6 & 2 \\ \frac{1}{2} & 1 & 0 \\ \frac{5}{2} & 3 & -1 \end{bmatrix}$.

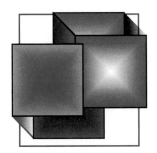

Index

LIST OF FREQUENTLY USED SYMBOLS

$A = \begin{bmatrix} a_{ij} \end{bmatrix}$	An $m \times n$ matrix, p. 11
A^T	The transpose of the matrix A, p. 16
$\begin{bmatrix} A \vdots B \end{bmatrix}$	An augmented matrix, p. 25
$\sum\limits_{i=1}^{n} a_i$	Summation notation, p. 28
O	The zero $m \times n$ matrix, p. 35
I_n	The $n \times n$ identity matrix, p. 38
A^{-1}	The inverse of the matrix A, p. 69
$j_1 j_2 \cdots j_n$	A permutation of $S = \{1, 2, \ldots, n\}$, p. 91
$\det(A)$	The determinant of the matrix A, p. 92
$|A|$	The determinant of the matrix A, p. 92
$\det(M_{ij})$	The minor of a_{ij}, p. 103
A_{ij}	The cofactor of a_{ij}, p. 103
$\text{adj } A$	The adjoint of the matrix A, p. 108
$\mathbf{u}, \mathbf{v}, \mathbf{w}, \mathbf{x}, \mathbf{y}, \mathbf{z}$	Vectors in a vector space, pp. 123, 197
$\overrightarrow{OP}$	A directed line segment, p. 126
$\|\mathbf{u}\|$	Length of the vector $\mathbf{u}$, pp. 129, 148
$\mathbf{0}$	The zero vector, pp. 134, 198
$-\mathbf{u}$	The negative of the vector $\mathbf{u}$, pp. 134, 198
$\mathbf{x} \cdot \mathbf{y}$	Dot product, standard inner product, pp. 136, 148
θ	Angle between two nonzero vectors, pp. 136, 151
R^n	n-space; the vector space of all n-vectors, p. 141
$\|\mathbf{u} - \mathbf{v}\|$	Distance between the vectors $\mathbf{u}$ and $\mathbf{v}$ in R^n, p. 148
L	A linear transformation, p. 158, 327
$L(\mathbf{u})$	The image of $\mathbf{u}$, p. 158
range L	The range of the linear transformation L, p. 158, 337
$\times$	The cross product operator, p. 180
V, W	Vector spaces, p. 197
M_{mn}	The vector space of all $m \times n$ matrices, p. 199
$p(t)$	A polynomial in t, p. 199
P_n	The vector space of all polynomials of degree $\leq n$ and the zero polynomial, p. 200
P	The vector space of all polynomials, p. 201
$C(-\infty, \infty)$	The vector space of all real-valued continuous functions that are defined on $(-\infty, \infty)$, p. 206
$C[a, b]$	The vector space of all real-valued continuous functions that are defined on the interval $[a, b]$, p. 206
S, T, S', T'	Sets of vectors in a vector space, pp. 208, 231, 331
span $\{\mathbf{v}_1, \mathbf{v}_2, \ldots, \mathbf{v}_k\}$	The set of all vectors of the form $a_1\mathbf{v}_1 + a_2\mathbf{v}_2 + \cdots + a_k\mathbf{v}_k$, p. 208
$\mathbf{e}_i$	An n-vector whose ith component is 1 and all other components are zero, p. 215